V International Conference "Modeling of Nonlinear Processes and Systems" (MNPS-2020)

EPJ Web of Conferences Volume 248 (2020)

Moscow, Russia
16 - 20 November 2020

Editors:

A. Nadykto
N. Aleksic
P. Lima
P. Pivkin
L. Uvarova
X. Jiang
A. Zelensky

ISBN: 978-1-7138-2926-3

Printed from e-media with permission by:

Curran Associates, Inc.
57 Morehouse Lane
Red Hook, NY 12571

Some format issues inherent in the e-media version may also appear in this print version.

For additional information, please contact EDP Sciences – Web of Conferences
at the address below.

EDP Sciences – Web of Conferences
17, Avenue du Hoggar
Parc d'Activité de Courtabœuf
BP 112
F-91944 Les Ulis Cedex A
France

Phone: +33 (0) 1 69 18 75 75
Fax: +33 (0) 1 69 28 84 91

contact-edps@webofconferences.org

Additional copies of this publication are available from:

Curran Associates, Inc.
57 Morehouse Lane
Red Hook, NY 12571 USA
Phone: 845-758-0400
Fax: 845-758-2633
Email: curran@proceedings.com
Web: www.proceedings.com

TABLE OF CONTENTS

MODELLING OF SELECTIVE LASER MELTING PROCESS OF QUARTZ GLASS AT ELEVATED TEMPERATURES 1
Gridnev M. A., Khmyrov R. S., Gusarov A. V.

THE OSCILLATION NUMBERS AND THE ABRAMOV METHOD OF SPECTRAL COUNTING FOR LINEAR HAMILTONIAN SYSTEMS 7
Elyseeva Julia

PHASE TRANSITIONS IN ION-EXCHANGE MATERIALS DURING THE WATER VAPOR SORPTION 13
Ivanov Igor, Ivanova Oksana

S-SYNCHRONIZATION AND EXCITATION CONSTANCY IN STRUCTURAL IDENTIFIABILITY PROBLEM OF NONLINEAR SYSTEMS 18
Karabutov Nikolay

ANALYSIS OF TEAR FLUID FOR EVALUATION ON THE CONCENTRATION OF THE GENERIC ANTI-GLAUCOMA DRUG BETOPTIK USING MULTISENSORY STRIPPING VOLTAMMETRY AND MULTIVARIATE STATISTICS 25
Kolesnichenko I. I., Balashova L. M.

SIMULATION OF THE PROCESSING PROCESS IN A GLOW DISCHARGE PLASMA GENERATOR 31
Logvin Vladimir A., Yvarova Ludmila A.

RHYTHM DRIVERS, PHYSICAL PROPERTIES, MECHANISMS OF FORMATION 36
Mazurov Mikhail

THE NATURE OF CVC NONLINEARITY IN LOW-VOLTAGE SCANNING TUNNELING SPECTROSCOPY OF SEMICONDUCTORS 42
Loskutov A. I., Mandel A. M., Karpova E. E., Oshurko V. B., Veselko S. G., Sharts A. A., Loskutov S. A.

IDENTIFICATION OF THE OCULO-MOTOR SYSTEM IN THE FORM VOLTERRA MODEL BASED ON EYE-TRACKING DATA 50
Pavlenko Vitaliy D., Shamanina Tetiana V., Chori Vladislav V.

NON-EUCLIDEAN BIOSYMMETRIES AND ALGEBRAIC HARMONY IN GENOMES OF HIGHER AND LOWER ORGANISMS 56
Petoukhov Sergey, Petukhova Elena, Svirin Vitaly

HARMONIC PROGRESSION IN BIOINFORMATICS AND RECURRENT SERIES IN INHERITED BIOSTRUCTURES 61
Verevkin Vladimir, Petoukhov Sergey

GRADIENT-BASED ALGORITHM FOR TRACKING THE ACTIVITY OF NEURAL NETWORK WEIGHTS CHANGING 67
Starodub Anton, Eliseeva Natalia, Georgiev Milen

COMPUTER VISUALIZATION OF JULIA SETS FOR MAPS BEYOND COMPLEX ANALYTICITY 73
Toporensky Alexey, Stepanyan Ivan

"FUTURE EARTH": NIGMATULIN HYPOTHESIS AND DYNAMIC MODEL OF RADIATION FIELD OF OCEAN-ATMOSPHERE SYSTEM 76
Sushkevich Tamara, Strelkov Sergey, Maksakova Svetlana

MODELLING BIOMOLECULAR STRUCTURES IN CATEGORICAL SYSTEMS THEORY 88
Tolokonnikov Georgy

DIFFERENT-SCALE SIMULATION OF FLOWS IN POROUS MEDIA 94
Trapeznikova Marina, Churbanova Natalia, Chechina Antonina

IMPROVING THE LEARNING POWER OF ARTIFICIAL INTELLIGENCE USING MULTIMODAL DEEP LEARNING 99
Shchetinin Eugene Yu., Sevastianov Leonid

PLASTIC FLOW IN A THIN LAYER: THE THEORY, FORMULATIONS OF THE BOUNDARY-VALUE PROBLEMS AND THE APPLICATIONS 103
Kadymov Vagid, Yanovskaya Elena

TRANSMUTATION OPERATOR METHOD FOR SOLVING HEAT CONDUCTION PROBLEM 109
Yaremko Oleg, Yaremko Natalia

EXCITATION OF SOME NITROGEN-LIKE IONS: CALCULATIONS AND COMPARISON WITH MEASUREMENTS 117
Nadykto O. B., Nadykto B. A., Nadykto A. B.

MATHEMATICAL MODELING OF WORKING MEMORY IN THE PRESENCE OF RANDOM DISTURBANCE USING NEURAL FIELD EQUATIONS 124
Lima Pedro M., Erlhagen Wolfram, Kulikov Gennady Yu., Kulikova Maria V.

DISTRIBUTIONS OF TWO ATOMS COLLISIONS OVER THE SURFACE OF THE CONDENSED PHASE 136
Zheltov Sergey, Pletnev Leonid

PROBABILITIES OF ATOMIC DEPARTURES FROM THE SLIT SYSTEM 141
Pletnev Leonid

MODELING COGNITIVE ACTIVITY OF THE HUMAN BRAIN BY THE MATHEMATICAL APPARATUS OF QUANTUM MECHANICS 148
Petukhov Alexandr

APPLICATION OF ALGORITHMS FOR PLACEMENT OF ORTHOGONAL POLYHEDRONS FOR SOLVING THE PROBLEMS OF PACKING OBJECTS OF COMPLEX GEOMETRIC SHAPE 153
Chekanin Vladislav, Chekanin Alexander

DRIVER BEHAVIOUR ALGORITHMS FOR THE CELLULAR AUTOMATA-BASED MATHEMATICAL MODEL OF TRAFFIC FLOWS 159
Chechina Antonina, Churbanova Natalia, Trapeznikova Marina

MATHEMATICAL MODELING OF TECHNOLOGICAL MODES OF OPERATION OF LIQUID STAMPING OF ALUMINUM ALLOY AK7 164
Kadymov Vagid, Sosenushkin Evgeny

THE INTEGRATED STATISTICAL CONTROL METHOD OF TECHNICAL CONDITION OF RADIO-ELECTRONIC EQUIPMENT FOR PROTOTYPING OF ROBOTIC SYSTEMS 171
Budko Pavel, Goydenko Vladimir, Rozhnov Aleksei, Vinogradenko Aleksei

INFORMATION-ACTIVE SYSTEMS DESIGN TOOL: "IS-2" INTEGRATED ENVIRONMENT 175
Gavrilov Andrey, Volkova Galina, Novoselova Olga

DESIGN OF WATER FIRE EXTINGUISHING SYSTEM FOR PROTECTION OF METRO ROLLING STOCK STAGNATION ZONES 180
Filippov Ilya, Yusupdzhanov Vladimir, Zybina Olga, Klochihin Ilya, Chalovskaya Ekaterina

APPLICATION OF DIGITAL TWIN TECHNOLOGY FOR MODELLING OF INFORMATION SECURITY LEVEL OF INDUSTRIAL PLANT 187
Kuznetsova Natalia Mikhailovna, Karlova Tatyana Vladimirovna

DEVELOPMENT OF EXPERIMENTAL PROTOTYPES FOR THE TEACHING OF SCIENCE 191
Terán Guillermo, Ávalos Fausto, Lomas Washington, Quintana Iliana

MATHEMATICAL MODELING OF VIBRATION DAMPERS OF VIBRATION-INSULATED STRUCTURES 195
Sosenushkin Evgeny, Ivanova Oksana, Yanovskaya Elena, Vinogradova Yuliya

MULTIROTOR HEAVY LIFT DRONE 200
Carrera Fausto Freire, Tamayo Pablo, Chadrina Olga, Kharlamov Sergei

A SYSTEM FOR MEASURING WATER LEVELS IN OPEN-AIR IRRIGATION CANALS 206
Carrera Fausto Freire, Sanchez Harry Saltos, Garcia-Orellana Yelitza, Chadrina Olga

ACOUSTIC SAFETY MANAGEMENT AT THE LOCKSMITH'S WORKPLACE MECHANIC-REPAIRMAN TO REPLACE THE FILTER MATERIAL IN THE TECHNOLOGICAL PROCESS OF OBTAINING OF APATITE CONCENTRATE 212
Chalovskaya Ekaterina, Kaverzneva Tatiana, Skripnik Igor, Senchenko Vladimir, Klochihin Ilya

MATHEMATICAL ESTIMATION METHODS AND MODELS FOR INDUSTRIAL COMPANIES 217
Stikhova Olga

DESIGNING AN EXPERT SYSTEM FOR RECOGNIZING THE EMOTIONAL STATE OF AN ENTERPRISE EMPLOYEE 222
Sekin A. A., Bychkova N. A.

NEOPATRIMONIAL MODEL OF PUBLIC ADMINISTRATION AND ITS TRANSFORMATION IN THE CONTEXT OF GLOBAL UNCERTAINTY 226
Maracha Viacheslav, Bespalov Sergey

REGIONAL INVESTMENT ATTRACTIVENESS'S MODEL ANALYSIS AT THE HEART OF RUSSIAN INVESTMENT POLICY 236
Mitrofanov Evgenii, Kulagina Alevtina, Panferova Liya

THE INTERDEPENDENCE OF HUMANISTIC VALUES AND THE LEVEL OF EXISTENTIAL FULFILLMENT OF AN INTELLECTUAL LEADER 240
Mikhaylova Marianna V., Dmitrieva Viktoria A.

STUDY OF THE DISTRIBUTION AND MODIFICATION OF WRITING IN PROTO-CHINESE LANGUAGE COMMUNITIES: CORRECT IDENTIFICATION OF HIEROGLYPHIC SIGNS IN THE JIAGUWEN INSCRIPTIONS 247
Egorova Maia, Egorov Aleksandr
ANALYTICAL PROCESSING OF INFOLOGICAL MODELS OF APPLIED TASKS AT DESIGN OF INFORMATION-ACTIVE SYSTEMS 255
Novoselova Olga, Ruzheynikov Anton
REPRESENTATION OF SOCIAL SYSTEMS AS NONLINEAR, COMPLEX, SELF-ORGANIZING AND CLOSELY RELATED TO BIOSYSTEMS AND MATHEMATICAL MODEL OF THEIR EVOLUTION 260
Romanova Ekaterina, Uvarova Ludmila
MATHEMATICAL MODEL FOR COLLECTING AND EVALUATING COMPLEX SECURITY FACTORS AND ENSURING THE PREVENTION OF THREATS TO INDIVIDUALS AND SOCIETY IN CYBERSPACE 265
Poltoranov Denis V., Karlova Tatiana V.
GENERATION OF ELECTRON AND FAST ATOM BEAMS BY A GRID IMMERSED IN PLASMA 270
Metel Alexander, Mustafaev Enver, Melnik Yury, Hamdy Khaled
COLLISIONAL RELAXATION OF FAST ELECTRONS ENERGY IN GAS 275
Metel Alexander, Mustafaev Enver, Melnik Yury, Hamdy Khaled
MATHEMATICAL MODELS OF THE GEOMETRY OF MICRO MILLING CUTTERS 281
Isaev Alexander, Khamzin Ramil, Ershov Artem, Leonesio Marco
SOLUTION OF THE RODRIGUEZ EQUATION IN MODELING VOLUMETRIC GEOMETRIC ACCURACY OF MULTI-COORDINATE SYSTEMS 288
Stebulyanin M. M., Pimushkin Ya. I.
INVESTIGATION OF THE STRESS-STRAIN STATE OF THE CAVITY PROFILES FOR OSTEOSYNTHESIS SCREW HEADS 293
Balova Daria, Rechenko Denis
SYNERGISTIC EFFECTS UNDER ION-BEAM MODIFICATION OF METALS 297
Borisov Anatoly, Krit Boris, Suminov Igor, Ovchinnikov Mikhail, Tikhonov Sergey
INTEGRATED PROBABILISTIC SURFACE ROUGHNESS ASSESSMENT FOR VARIOUS PROCESSING METHODS 301
Bobrovskij Igor, Zibrov Petr, Luk'Yanov Aleksey, Levitskih Olesja
DEVELOPMENT AND POWER CALCULATION OF A GRINDING WHEEL DESIGN FOR ULTRA-HIGH-SPEED GRINDING 306
Rechenko Denis, Kamenov Renat
MODEL OF THE ULTIMATE STRESS STATE OF THE COATED TOOL CUTTER MATERIAL 313
Osinnii Aleksei, Bublikov Jury, Chigarev Anatoli, Okun'Kova Anna, Kapustina Nataliya
RELIABILITY CALCULATION FOR A MULTILAYERED COATING BASED ON A MODEL WITH INDEPENDENTLY (SEQUENTIALLY) DEGRADING LAYERS 319
Osinnii Aleksei, Chigarev Anatoli, Bublikov Jury, Mosyanov Mikhail, Kekalo Anton
DETERMINATION OF RATIONAL DESIGN AND GEOMETRIC PARAMETERS OF A HIGH-PERFORMANCE DRILL BASED ON A MATHEMATICAL MODEL OF THE CUTTING PART 324
Pivkin Petr, Grechishnikov Vladimir, Ershov Artem, Kuptsov Vladimir, Jiang Xiaohui
EFFECT OF GEOMETRICAL PARAMETER CUTTING EDGE ON THE TOROIDAL FORMING SURFACE OF A SOLID END MILL 329
Pivkin Petr, Minin Ilya
ANALYSIS OF POSITIONING ADJUSTMENT APPROACHES FOR CUTTING INSERTS OF THREAD CUTTERS 334
Grechishnikov Vladimir, Isaev Alexander, Kozochkin Mikhail, Aung Pyae Phyo
MATHEMATICAL MODELING OF PRODUCTION PROCESSES OF DISCRETE MACHINE-BUILDING ENTERPRISES BASED ON THE INTERACTION OF SIMULATION SYSTEMS AND OPERATIONAL PLANNING SYSTEMS 338
Dolgov Vitalii A., Nikishechkin Petr A., Leonov Aleksandr A., Ivashin Sergey S., Dolgov Nikita V.
MODELS FOR MANAGING PRODUCTION SYSTEMS OF MACHINE-BUILDING ENTERPRISES BASED ON THE DEVELOPMENT AND USING OF THEIR DIGITAL TWINS 342
Dolgov Vitalii A., Nikishechkin Petr. A., Arkhangelskii Vladimir E., Umnov Pavel I., Podkidyshev Alexey A.
APPLICATION OF CIRCUIT MODELING IN THE STUDY OF SPARK FORMATION DURING ELECTROEROSION TREATMENT 347
Porvatov Artur, Kozochkin Mikhail, Minin Ilya, Maslov Andrey, Kuptsov Vladimir, Morozkin Marian, Kuznetsov Vladimir, Leonesio Marko, Zelensky Alexander, Pivkin Petr

THERMODYNAMIC MODEL OF WEAR INTENSITY OF CUTTING TOOLS WITH COATINGS 352
Grigoriev Sergey, Migranov Mars, Migranov Artur, Seitkulov Abdumalik
MODELING OF NON-STATIONARY PROCESSES WHEN CUTTING HARD-TO-PROCESS MATERIALS 357
Grigoriev Sergey, Migranov Mars, Seitkulov Abdumalik
MODELING OF THERMOPHYSICAL PHENOMENA WHEN CUTTING WITH COATED TOOLS 363
Volosova Marina, Migranov Artur, Rai Maksim
NON-STATIONARY TEMPERATURE BEHAVIOR OF SPINDLE ASSEMBLIES AND BALL SCREW GEARS FOR HIGH PRECISION METAL CUTTING MACHINES 368
Kuznetsov A. P.
USING THE MATHEMATICAL CRITERION OF GAS PIPELINE FAILURE TO DETERMINE THE REMAINING OPERATION LIFE 376
Bayburova Minsariya, Ismagilova Zulfiya
METHODS OF MATHEMATICAL MODELING OF AUTOMATED MACHINE SYSTEMS OF MULTI-NOMENCLATURE PRODUCTION 381
Dumler Elena, Vakhitova Roza, Saracheva Diana, Makarova Tatyana
SIMULATION OF AUTOMATED MACHINE TOOL SYSTEMS 387
Dumler Elena, Vakhitova Roza, Saracheva Diana, Makarova Tatyana
DEVELOPMENT OF A NONLINEAR DEPENDENCE FOR DETERMINING THE STRESS-STRAIN STATE OF THE PIPELINE-COMPOSITE BANDAGE SYSTEM 391
Ramazanov Robert À., Kantemirov Igor F., Gulin Denis A., Makhmudova Indira F., Ismagilova Zulfiya F.
STRUCTURAL FEATURES OF METAL POWDERS OBTAINED BY PLASMA SPRAYING 396
Gulikhandanov Evgeny, Ermakov Sergey
Author Index

MNPS-2020: V International Conference "Modeling of Nonlinear Processes and Systems"

V International Conference "Modeling of Nonlinear Processes and Systems" was held at the Moscow State Technological University "STANKIN" on November 17 -21 2020. This was the eleventh issue of the Conference that was held since 1995 in Moscow, Dubna, Tver and Nizhny Novgorod. Since 2008, its home is MSTU STANKIN.

The Conference was dedicated to a wide range of contemporary problems of mathematical modeling of nonlinear systems and processes in physics, chemistry, biology, ecology, materials sciences, robotics, information and intelligent environments, including artificial intelligence, numerical methods and high-performance computing. Plenary talks at the Conference were given by leading Russian and international scholars.

In addition to the main Conference program, there was a Special Event, Workshop on Advanced Materials Processing and Smart Manufacturing, held on November 20-21, 2020.

Conference Organizers

- Moscow State University of Technology "STANKIN"
- Centro de Matematica Computacional e Estocastica Instituto Superior Tecnico, Lisbon (Portugal)
- National Research Nuclear University MEPHI
- Institute of Cell Biophysics of Russian Academy of Sciences
- Keldysh Institute of Applied Mathematics of Russian Academy of Sciences
- Institute of Mathematical Problems of Biology of Russian Academy of Sciences
- Arctic-SG Foundation (Russia-Belarus Union)
- Murmansk Arctic State University (MASU)

Organizing Committee

International Scientific Committee

- E.R. Schukin (Russia)
- E. Weinmuller (Austria)
- T.V. Rijkova(Russia)
- Q.Zhang (China)
- A.E.Arinshtein (Israel)
- L.V. Yakushevich (Russia)
- L.Dimitrov (Bulgaria)
- I.V. Krivenko (Russia)
- G.G.Gogoberidze (Russia)
- F.Xu (China)
- K.Egizarian (Finland)

edp sciences

Web of Conferences

Statement of Peer review

In submitting conference proceedings to *Web of Conferences*, the editors of the proceedings certify to the Publisher that

1. They adhere to its Policy on Publishing Integrity in order to safeguard good scientific practice in publishing.
2. All articles have been subjected to peer review administered by the proceedings editors.
3. Reviews have been conducted by expert referees, who have been requested to provide unbiased and constructive comments aimed, whenever possible, at improving the work.
4. Proceedings editors have taken all reasonable steps to ensure the quality of the materials they publish and their decision to accept or reject a paper for publication has been based only on the merits of the work and the relevance to the journal.

Title, date and place of the conference

V International Conference „Modeling of Nonlinear Processes and Systems" Moscow State University of Technology „STANKIN", Moscow, Russian Federation

November 16-20, 2020

Proceedings editor(s): Alexey Nadykto

edp sciences

Web of Conferences

Statement of Peer review

In submitting conference proceedings to *Web of Conferences*, the editors of the proceedings certify to the Publisher that

1. They adhere to its Policy on Publishing Integrity in order to safeguard good scientific practice in publishing.
2. All articles have been subjected to peer review administered by the proceedings editors.
3. Reviews have been conducted by expert referees, who have been requested to provide unbiased and constructive comments aimed, whenever possible, at improving the work.
4. Proceedings editors have taken all reasonable steps to ensure the quality of the materials they publish and their decision to accept or reject a paper for publication has been based only on the merits of the work and the relevance to the journal.

Title, date and place of the conference

V International Conference „Modeling of Nonlinear Processes and Systems" (MNPS-2020) Moscow State University of Technology „Stankin", Moscow, Russian Federation
November 16-20, 2020

Proceedings editor(s):

Najdan Aleksic

Date and editor's signature

edp sciences

Web of Conferences

Statement of Peer review

In submitting conference proceedings to *Web of Conferences*, the editors of the proceedings certify to the Publisher that

1. They adhere to its Policy on Publishing Integrity in order to safeguard good scientific practice in publishing.
2. All articles have been subjected to peer review administered by the proceedings editors.
3. Reviews have been conducted by expert referees, who have been requested to provide unbiased and constructive comments aimed, whenever possible, at improving the work.
4. Proceedings editors have taken all reasonable steps to ensure the quality of the materials they publish and their decision to accept or reject a paper for publication has been based only on the merits of the work and the relevance to the journal.

Title, date and place of the conference

V International Conference „Modeling of Nonlinear Processes and Systems" Moscow State of Technology „Stankin", Moscow, Russian Federation

November 16-20, 2020

Proceedings editor(s):

Pedro Lima

Date and editor's signature

20.03.2021

17, avenue du Hoggar - PA de Courtabœuf – BP 112 - 91944 Les Ulis Cedex A (France)
Tél. : 33 (0)1 69 18 75 75 - Fax : 33(0)1 69 07 45 17 – www.edpsciences.org

edp sciences

Web of Conferences

Statement of Peer review

In submitting conference proceedings to *Web of Conferences*, the editors of the proceedings certify to the Publisher that

1. They adhere to its **Policy on Publishing Integrity** in order to safeguard good scientific practice in publishing.
2. All articles have been subjected to peer review administered by the proceedings editors.
3. Reviews have been conducted by expert referees, who have been requested to provide unbiased and constructive comments aimed, whenever possible, at improving the work.
4. Proceedings editors have taken all reasonable steps to ensure the quality of the materials they publish and their decision to accept or reject a paper for publication has been based only on the merits of the work and the relevance to the journal.

Title, date and place of the conference

V International Conference "Modeling of Nonlinear Processes and Systems" (MNPS-2020)
16-20 November
Moscow State University of Technology "STANKIN" (MSUT "STANKIN")

Proceedings editor(s): Petr Pivkin

Date and editor's signature 10.04.2021

17, avenue du Hoggar - PA de Courtabœuf – BP 112 - 91944 Les Ulis Cedex A (France)
Tél. : 33 (0)1 69 18 75 75 - Fax : 33(0)1 69 07 45 17 – www.edpsciences.org

edp sciences

Web of Conferences

Statement of Peer review

In submitting conference proceedings to *Web of Conferences*, the editors of the proceedings certify to the Publisher that

1. They adhere to its **Policy on Publishing Integrity** in order to safeguard good scientific practice in publishing.
2. All articles have been subjected to peer review administered by the proceedings editors.
3. Reviews have been conducted by expert referees, who have been requested to provide unbiased and constructive comments aimed, whenever possible, at improving the work.
4. Proceedings editors have taken all reasonable steps to ensure the quality of the materials they publish and their decision to accept or reject a paper for publication has been based only on the merits of the work and the relevance to the journal.

Title, date and place of the conference

V International Conference „Modeling of Nonlinear process and Systems" 16-20 november
Moscow State University of Technology „Stankin", Moscow, Russia

Proceedings editor(s): Lydmila Uvarova

Date and editor's signature

14.04.2021

17, avenue du Hoggar - PA de Courtabœuf – BP 112 - 91944 Les Ulis Cedex A (France)
Tél. : 33 (0)1 69 18 75 75 - Fax : 33(0)1 69 07 45 17 – www.edpsciences.org

Statement of Peer review

In submitting conference proceedings to *Web of Conferences*, the editors of the proceedings certify to the Publisher that

1. They adhere to its **Policy on Publishing Integrity** in order to safeguard good scientific practice in publishing.
2. All articles have been subjected to peer review administered by the proceedings editors.
3. Reviews have been conducted by expert referees, who have been requested to provide unbiased and constructive comments aimed, whenever possible, at improving the work.
4. Proceedings editors have taken all reasonable steps to ensure the quality of the materials they publish and their decision to accept or reject a paper for publication has been based only on the merits of the work and the relevance to the journal.

Title, date and place of the conference 16 - 20 November 2020

V International Conference "Modeling of Nonlinear Processes and Systems" (MNPS-2020), Moscow, Russia

Proceedings editor(s): Xiaohui Jiang

Date and editor's signature

Statement of Peer review

In submitting conference proceedings to *Web of Conferences*, the editors of the proceedings certify to the Publisher that

1. They adhere to its **Policy on Publishing Integrity** in order to safeguard good scientific practice in publishing.
2. All articles have been subjected to peer review administered by the proceedings editors.
3. Reviews have been conducted by expert referees, who have been requested to provide unbiased and constructive comments aimed, whenever possible, at improving the work.
4. Proceedings editors have taken all reasonable steps to ensure the quality of the materials they publish and their decision to accept or reject a paper for publication has been based only on the merits of the work and the relevance to the journal.

Title, date and place of the conference

V International Conference „Modeling of Nonlinear Processes and Systems" (MNPS-2020), November 16-20, 2020, Moscow State University of Technology „STANKIN", Moscow, Russian Federation

Proceedings editor(s): Alexander Zelensky

Date and editor's signature

17, avenue du Hoggar - PA de Courtabœuf – BP 112 - 91944 Les Ulis Cedex A (France)
Tél. : 33 (0)1 69 18 75 75 - Fax : 33(0)1 69 07 45 17 – www.edpsciences.org

https://doi.org/10.1051/epjconf/202124801001

Modelling of Selective Laser Melting Process of Quartz Glass at Elevated Temperatures

M.A. Gridnev[1,*], *R.S.* Khmyrov[1], and *A.V.* Gusarov[1]

[1]Moscow State University of Technology "STANKIN", RU-127055, Moscow, Russia

Abstract. Selective laser melting (SLM) to date is the method of additive manufacturing allowing fabricating products from powder layer-by-layer according to a 3D model. However, when applying this method to fragile materials, parts crack while fabricating due to high temperatures. Quartz glass is a promising material for fabricating products by SLM without cracks due to a low thermal expansion. However, quality of fabricated material differs from the fused cast ones. This article aims to test the method of SLM with preheating to improve the material quality. Experiments on single track formation in SLM are analysed by modelling the coupled processes of heat transfer and powder consolidation in the laser-interaction zone. The mathematical model is validated by the experiments. It is shown that the preheating can improve the material quality and increase the process productivity but overheating may result in undesirable crystallization.

1 Introduction

The emerging 3D printing technologies can be advantageous for manufacturing components of glass. They do not require fabrication of exact molds necessary for die pressing of optical elements and can be applied to manufacture parts of complicated shape, for example hollow elements that are difficult to fabricate by conventional processes like die pressing [1]. Selective laser melting (SLM) is an additive manufacturing (3D printing) process consisting in layer-by-layer consolidation of a powder bed by a scanning laser beam that fuses powder particles together to form a solid part. Successful application of this material is demonstrated in specialized literature. Material does not contain cracks, however, pores remain inside the fused material [2]. In addition, the authors of the article [3] predict that consolidation can be improved provided that sample heating occurs while laser processing. The SLM was successively applied to borosilicate [4], soda-lime [5], and quartz [6] glass. The advantages of the process are flexibility and high precision. The disadvantages are residual porosity and stresses. In SLM, a portion of a powder bed is generally exposed to the laser beam for a fraction of a millisecond [3]. It can be insufficient for complete consolidation of powder and result in excessive residual porosity. The residual stresses arise because of a highly non-uniform temperature distribution in the laser-interaction zone, which launches thermo-mechanical processes [7].

Quartz glass is a promising material for SLM because of extremely low thermal expansion. That is why it does not crack in the conditions of a strong thermal shock at laser processing [6]. Consolidation of quartz glass powder is controlled by viscous flow at coalescence of molten powder particles [3]. Unfortunately, the viscosity of quartz glass is rather high even at elevated temperatures [8]. That is why the residual porosity can be unsatisfactory [3]. To reduce the porosity, one can reduce the scan speed at SLM [3]. This extends the thermal cycle, and thus increases the extent of consolidation. The disadvantage of such an approach is decreasing the productivity that is proportional to the scan speed. Another drawback is a strong non-uniformity of pore distribution in the laser-interaction zone. Even if the central domain is free of pores, a high porosity can be observed in the periphery of the heat affected zone where the temperature is lower and the viscosity is greater [3]. To attain a low and more uniform porosity and increase the productivity of SLM, it was proposed to apply a preheat [3]. The preheat is maintaining the working zone of the SLM machine at an elevated temperature during the whole process. It was successfully applied to other materials, for example to reduce cracking at SLM [9].

The present work aims to test experimentally the influence of preheating on SLM of quartz glass and to model the coupled physical processes of heat transfer and powder consolidation in the laser-interaction zone at SLM. The comparison of the experiments with the modelling is useful to analyze the influence of SLM process parameters on the quality of the resulting material.

2 Model and experiment

2.1. Mathematical model

This work uses a mathematical model of laser-influence zone on a layer of quartz glass powder at different process parameters of selective laser melting (such as

* Corresponding author: Gridnev.M@stankin.ru

laser power, scanning speed, layer thickness and preheat temperature). Heat transfer is calculated in this model by the following equation:

$$\partial H/\partial t=\nabla\cdot(\lambda_e\cdot\nabla T)+u\cdot\partial H/\partial x \quad (1)$$

where H is the volumetric enthalpy, λe the effective thermal conductivity, ∇ the nabla operator, t the time, T the temperature, u the scanning speed, x the longitudinal coordinate. Figure 1 shows scanning process.

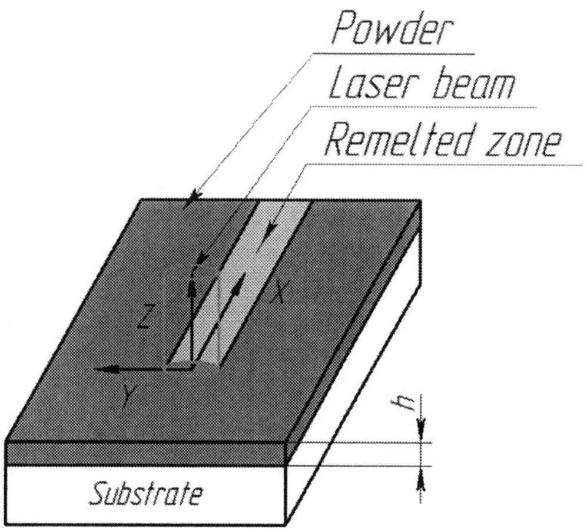

Fig. 1. Scanning process

Also enthalpy is related with temperature by the following equation [2+]:

$$H=(1-\varepsilon)C_pT \quad (2)$$

where ε is the porosity equal to the volume fraction of the gas phase, and C_p the volumetric specific heat.

Effective thermal conductivity is calculated by the following equation [3]:

$$\lambda_e=\lambda_s\xi(1-\varepsilon)N/\pi \quad (3)$$

where λ_s is the thermal conductivity of solid phase, N the average coordination number, $\xi=a/D$ the consolidation degree meaning that the softened particles of fused silica behave like the droplets of viscous liquid. They tend to coalesce as shown in Fig. 2. The extent of coalescence is characterised by neck diameter a. The kinetics of neck growth between two spherical particles of diameter D is estimated as [3] calculated by the following equation [3]:

$$(\partial\xi^2)/\partial t=3\delta/(\eta D) \quad (4)$$

where D is the particle diameter, δ the surface tension coefficient, η the dynamic viscosity. The efficiency of this model has been demonstrated in Ref. [3] and has been proven with our experiments.

The mathematical model considers laser radiation influence on a single layer and the substrate because every single layer becomes substrate for previous ones in the technology of SLM. Scanning of the surface by a laser beam occurs in a straight line at a constant speed u. The thermal conditions of laser scanning can be different for initial stages of consolidation, however, eventually all processes become steady.

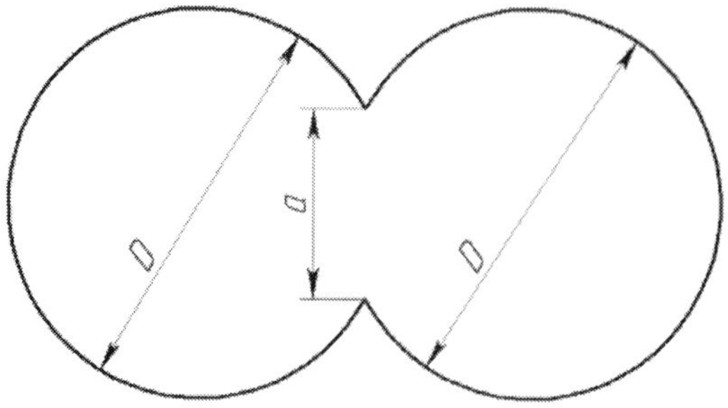

Fig. 2. Coalescence of two spherical particles of diameter D with formation of a neck of diameter a.

According to predictions made in Ref. 1, the laser radiation is absorbed on the surface of the target. This gives a heat flux through plane $z = 0$ with the density distribution equal to the energy density flux in the cross-section of the laser beam [3]:

$$q_0=P_0\cdot\exp(-r^2/r_0^2)/(\pi\cdot r_0^2) \quad (5)$$

where P_0 is the absorbed power of the laser beam, r_0 its nominal radius, and $r^2 = x^2 + y^2$.

Also this model considers heat losses from sample's surface due to evaporation. The Model proposes evaporation speed as function of surface temperature, boiling point T_b, latent heat of vaporation Q_b and vapour molecular mass. Heat losses on evaporation can be evaluated the mass loss multiplied with the latent heat of evaporation.

All necessary parameters are shown in Table 1 [2], [3].

Table 1. Calculation parameters.

Name	Letter	Value
Boiling point	T_b	3223 K
Latent heat of evaporation	Q_b	9.6 MJ/kg
Porosity	ε	0.4
Mean coordination number	N	7.1
Particle diameter	D	20 μm
Solid phase thermal conductivity	λ_s	2.2 (W/m*K)
Layer thickness	h	0.2 mm
Volumetric specific heat	C_p	2.68 MJ/(m^{3}K)

2.2. Experiment

In this work, we made experiments with silica glass powder with fraction 80-100 microns. The substrates are plates of quartz glass TU-21RSFSR-644-83 with non-polished surface of thickness 5 mm.

Laser CO_2 source provides 10.6 micron wavelength in continuous mode which quartz absorbs. The source is integrated in commercial engraving machine Qualitech 203 mini. Also, a heater is integrated by our team in this machine. The scheme of the experimental setup is shown in Fig.3 laser beam 2 from sealed CO_2 tube 1 is deflected by fixed mirror 3 and moving mirror 4 that is standing on the support moving at X axis. Laser head 5 moves along this rail in Y direction. The laser head contains a mirror at the top and a ZnSe lens of 2 inches focal length on the bottom. Target 6 is mounted in the heater 7 for making samples preheat. Metallographic microscope Olympus BX51M is used for getting photos of samples.

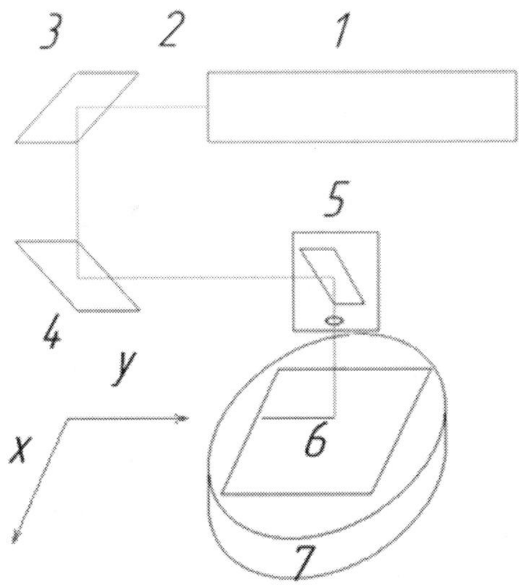

Fig. 3. Principal scheme of laser engraving machine Qualitech 203 mini.

3 Results and discussion

Figure 4 presents the results of calculation. As we can see, the heat affected zone in SLM often become bigger than the powder layer thickness. Thereby, powder sinters with the previous layer. However, using only laser processing is sometimes not enough for good sintering with the substrate. Preliminary heating of the sample is perspective for attaining high consolidation. Figure 5 shows comparative results of the consolidation degree. As one can see, the influence of preheat is not only elevated temperatures, it is also a better sintering due to the decreasing of viscosity with temperature. However, we can also try to predict more effective parameters of scanning. In our case, other parameters such as the layer thickness, the scanning speed and the beam power can be changed

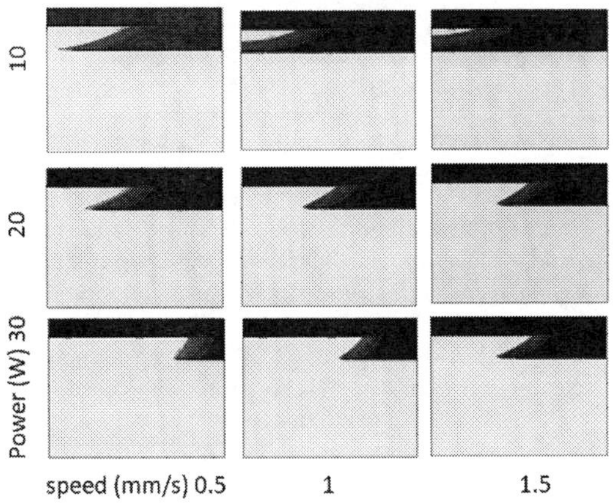

Fig. 6. Comparative results of modelling consolidation degree (white means ξ >0.9) of nine different situations with have different changing parameters: from top to bottom power changes from 10 to 30 W, from left to right scanning speed changes from 0.5 to 1.5 mm/s. Ambient temperature is 1100 K.

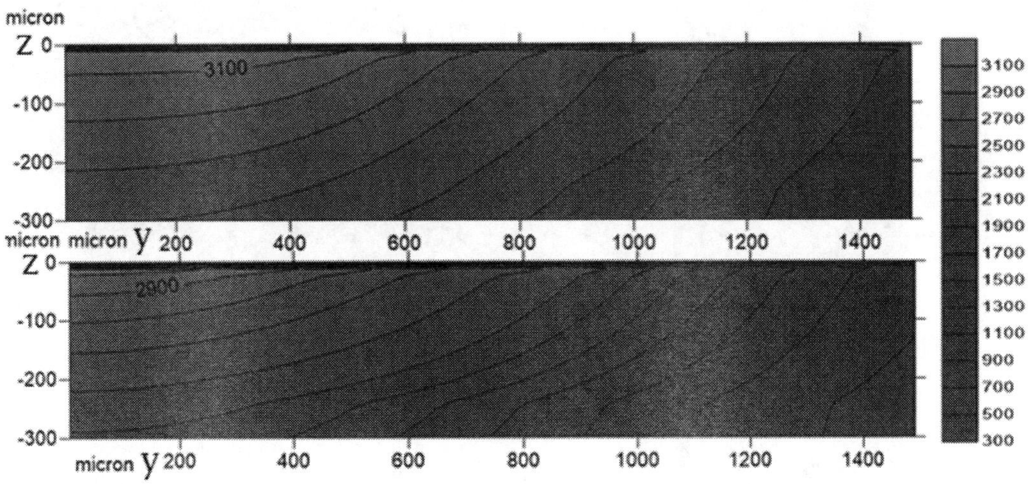

Fig. 4. Distribution of the maximum temperature (K) attained in the transversal cross section (YZ) for the modeled processes of laser melting of glass. On the top, processes modeled with ambient temperature = 1100 K. Below, ambient temperature = 300

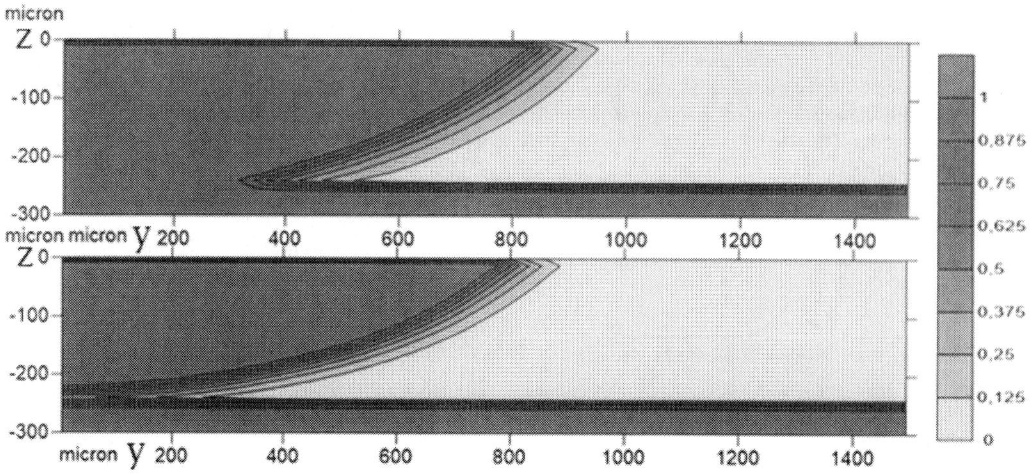

Fig.5. Sintering degree (χ2) for the modeled processes of laser melting of glass. On the top,ambient temperature = 1100K. Below, ambient temperature = 300 K.

Figure 6 shows the influence of changing the scanning speed and beam power. The 10 W beam power cannot be used in our case because it gives bad sintering results, but we can use 20 or 30 W and the difference there is not as big as between 10 and 20 W. We can also see that scanning speed influence is bigger for 30 W laser power. We can make conclusion that 30 W and 0.5 mm/s is the most effective mode.

After getting modelling results, we made some experiments with fused silica powder according to them. Figure 7 presents sintering tracks. Figures 8 and 9 present cross section of these tracks.

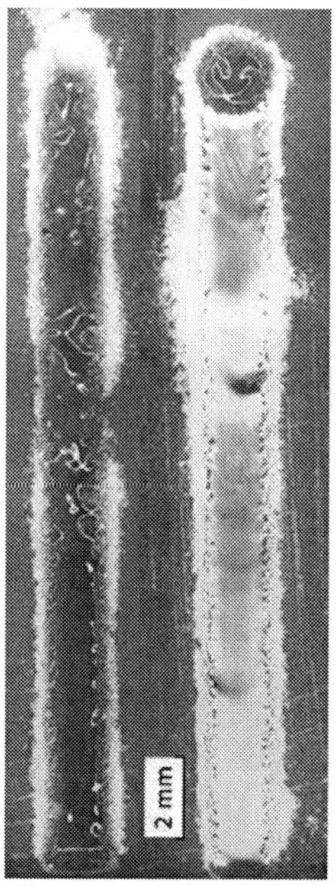

Fig. 7. Top view of the tracks made at scanning speed = 1 mm/s and power = 20 W (left) and 30 W (right), preheat temperature 1100K, layer thickness 200 microns.

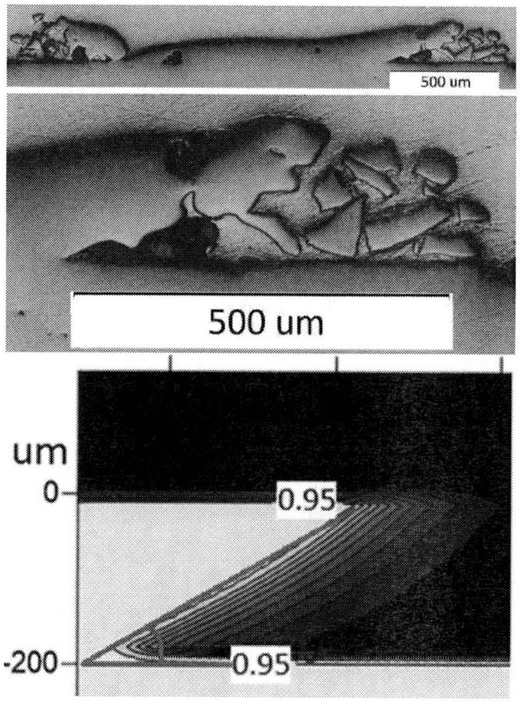

Fig. 8. Cross section of tracks made at 1 mm/s 20 W(on the top) same zoomed track (in the middle) and the modelling (on the bottom). The red lines show the angle between the substrate and the consolidation border.

As one can see in Figs. 8 and 9, this calculation method works even with elevated temperatures. At 20 W beam power, we can see similar angle between substrate and sintered material in calculations and in experiments. At 30 W beam power, we also can see this angle. The angle between the substrate and consolidation border increases when power increases.

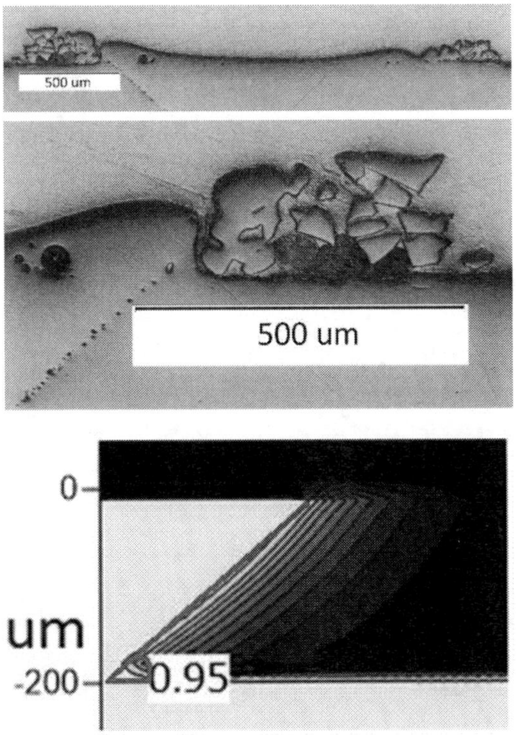

Fig. 9. Cross section of tracks made at 1 mm/s 30 W(on the top) same zoomed track (in the middle) and his modelling (on the bottom). The red lines show the angle between the substrate and the consolidation border.

Also at 30 W we predict that we got visible cristobalite (turbid track in Fig.7. right) because in [10] authors show results of quartz crystallization with transformation into crystobalite experiments and describe it [10]. It cannot be tridymite because the used silica glass powder is clean [11]. Getting crystobalite means that we reached temperature above the crystallization one for a time interval enough to crystallization [10]. The preheating temperature must be lower than the glass transition temperature, otherwise glass will not form after cooling the melt obtained by laser treatment. Laser heating of glass in the crystallization interval must be for a short term, otherwise the glass will devitrify. The numerical values of the characteristic temperatures are shown in Table 2. The preheat temperature 1100 K used in this work is less than both the glass transition temperature and the minimum crystallization temperature. However, the laser treatment results in formation of a heat affected zone in the periphery of the laser-interaction zone where the temperature is within the crystallization range (see Fig. 4). This may initiate crystallization observed as turbid domains in Fig. 7.

4 Conclusions

A mathematical model is proposed to describe interdependent processes of heat transfer and powder consolidation at selective laser melting (SLM) of quartz glass. The model is validated by comparison with experiments on formation of single tracks. Numerical calculations indicate that the preheating accelerates powder consolidation. This result is in line with experiments. Thus, the preheating is favourable to increase the productivity of SLM and to improve the quality of the material. However, quartz glass can crystallize at a too high preheat temperature.

Table 2. Temperatures of glass transition and crystallization.

Name	Value	Ref.
Glass transition temperature	1475 K	3
Crystallization interval	1273-1923 K	10

This work was supported by the Russian Science Foundation, grant #20-79-00274.

References

1. D. Zhang, X. Liu, J. Qiu, Frontiers of Optoelectronics, 1-15 (2020) https://doi.org/10.1007/s12200-020-1009-z
2. R.S. Khmyrov, S.N. Grigoriev, A. A. Okunkova, A.V. Gusarov, Physics Procedia, **56**, 345-356 (2014) https://doi.org/10.1016/j.phpro.2014.08.117
3. C.E. Protasov, R. S. Khmyrov, S.N. Grigoriev, A.V. Gusarov, International Journal of Heat and Mass Transfer, **104**, 665-674 (2017) https://doi.org/10.1016/j.ijheatmasstransfer.2016.08.107
4. F. Klocke, A. McClung, C. Ader, In 2004 International Solid Freeform Fabrication Symposium (2004) http://dx.doi.org/10.26153/tsw/6986
5. J. Luo, H. Pan, E.C. Kinzel, Journal of Manufacturing Science and Engineering, **136**(6) (2014) https://doi.org/10.1115/1.4028531
6. R.S. Khmyrov, C.E. Protasov, S.N. Grigoriev, A.V. Gusarov, The International Journal of Advanced Manufacturing Technology, **85**, 1461-1469 (2016) https://doi.org/10.1007/s00170-015-8051-9
7. A.V. Gusarov, M. Pavlov, I. Smurov, Physics Procedia, **12**, 248-254 (2011) https://doi.org/10.1016/j.phpro.2011.03.032
8. R.H. Doremus, Journal of Applied Physics, **92**(12), 7619-7629 (2002) https://doi.org/10.1063/1.1515132
9. Q. Liu, Y. Danlos, B. Song, B. Zhang, S. Yin, H. Liao, Journal of Materials Processing Technology, **222**, 61-74 (2015) https://doi.org/10.1016/j.jmatprotec.2015.02.036
10. A.I. Nepomnyschih, A.A. Shalaev, T.Y. Sizova, A.S. Palkin, A.N. Sapozhnikov, L.A. Pavlova Starting temperature and kinetics of crystallisation quartz glass, Crystallography, **63**, 314-319 (2018)

https://doi.org/10.1051/epjconf/202124801001

11. P.J. Heaney, D. R. Veblen, J .E. Post, American Mineralogist, **79**(5-6), 452-460 (1994)

The Oscillation Numbers and the Abramov Method of Spectral Counting for Linear Hamiltonian Systems

Julia Elyseeva[1,*]

[1]Department of Applied Mathematics, Moscow State Technological University "STANKIN", RU-127055, Moscow, Russia

Abstract. In this paper we consider linear Hamiltonian differential systems which depend in general nonlinearly on the spectral parameter and with Dirichlet boundary conditions. For the Hamiltonian problems we do not assume any controllability and strict normality assumptions which guarantee that the classical eigenvalues of the problems are isolated. We also omit the Legendre condition for their Hamiltonians. We show that the Abramov method of spectral counting can be modified for the more general case of finite eigenvalues of the Hamiltonian problems and then the constructive ideas of the Abramov method can be used for stable calculations of the oscillation numbers and finite eigenvalues of the Hamiltonian problems.

1 Introduction

In this paper we consider the spectral and oscillation theory for the linear Hamiltonian systems

$$y' = JH(t,\lambda)y,\ t\in[a,b],\ y=\begin{pmatrix} x(t,\lambda)\\ u(t,\lambda)\end{pmatrix},\ \lambda\in R \quad (1)$$

with the Dirichlet boundary conditions

$$x(a,\lambda)=x(b,\lambda), \quad (2)$$

where

$$H(t,\lambda)=H^T(t,\lambda),\ J=\begin{bmatrix} 0 & I\\ -I & 0\end{bmatrix},\ \frac{d}{d\lambda}H(t,\lambda)\ge 0. \quad (3)$$

Here $A\ge 0$ means that the symmetric matrix A is nonnegative definite, and $I,0$ denote the identity and zero matrices of appropriate dimensions. The Hamiltonian $H(t,\lambda)\in R^{2n\times 2n}$ and it's derivative in (3) are piece-wise continuous matrix functions with respect to $t\in[a,b]$ and $\lambda\in R$.

The linear Hamiltonian system (1) covers as a special case scalar and matrix second order Sturm–Liouville differential equations, as well as Sturm–Liouville differential equations of arbitrary even order, other self-conjugate differential equations and systems [3-4].

Based on the consideration in [5] we do not suppose any conditions connected with strict monotonicity of some matrix functions associated with (1). Such conditions will guarantee that the classical eigenvalues of (1), (2) are isolated (see [1-3]). For example, in [3] problem (1), (2) is considered under the strict normality assumptions with respect to $t\in[a,b]$ and $\lambda\in R$. In particular, the strict normality supposes that solutions of (1) are not "degenerate" with respect to change in $\lambda\in R$, i.e., if $y(t,\lambda)$ solves system (1) for different $\lambda_1,\lambda_2\in R$ on some non-degenerate interval of $[a,b]$, then necessary $y(t,\lambda)=0,\ t\in[a,b]$.

For a special case of the second order Sturm–Liouville differential equation

$$(r(t,\lambda)x')'+q(t,\lambda)x=0,\ t\in[a,b]$$

condition (3) means that $r(t,\lambda)\ne 0$ is nonincreasing while $q(t,\lambda)$ is nondecreasing in λ, but strict monotonicity of $q(t,\lambda)$ is not required in this paper. This corresponds to removing the strict normality assumption compared to [3, Assumption (8.3.7), p. 245].

The spectral and oscillation theory for the Hamiltonian systems without the strict normality and the controllability assumptions are developed in [5-7] (see also the references given therein). In [5] the notion of a finite eigenvalue of (1), (2) was introduced. Let $Y(t,\lambda)=(X^T(t,\lambda),U^T(t,\lambda))^T$ be $2n\times n$ matrix solutions of (1) with the conditions

$$rank\,Y(t,\lambda)=n,\ Y^T(t,\lambda)JY(t,\lambda)=0 \quad (4)$$

(the so-called conjoined bases). Consider a conjoined basis $Y(t,\lambda)$ with the initial condition $Y(a,\lambda)=(0\ I)^T$ (the principal solution of (1) at $t=a$). Then, under assumption (3) $rank\,X(t,\lambda)$ is piecewise constant in $\lambda\in R$ (see [3]) and $\lambda_0\in R$ is called a left (right) finite eigenvalue of (1),(2) with the multiplicity $\theta^-(\lambda_0)$ (resp. $\theta^+(\lambda_0)$) provided

$$\theta^{\pm}(\lambda_0)=rank\,X(b,\lambda_0^{\pm})-rank\,X(b,\lambda_0)\ge 1 \quad (5)$$

* Corresponding author: elyseeva@gmail.com

Here $rank\, X(b,\lambda_0^{\pm})$ denote the left-hand (the right-hand) limits of $rank\, X(b,\lambda)$ at λ_0. Under the strict normality assumption the matrix $X(t,\lambda)$ is invertible except at isolated values of $\lambda \in R$ and then left (right) finite eigenvalues reduce to the classical ones which are determined by the condition $\det X(b,\lambda_0)=0$ with the multiplicities

$$\theta^-(\lambda_0)=\theta^+(\lambda_0)=n-rank\, X(b,\lambda_0). \quad (6)$$

In general the multiplicities $\theta^-(\lambda_0)$ and $\theta^+(\lambda_0)$ are connected by the equality

$$\theta^+(\lambda_0)-\theta^-(\lambda_0)=rank\, X(b,\lambda_0^+)-rank\, X(b,\lambda_0^-).$$

The global oscillation theorem (see Theorem 3.5 in [5]) relates the number of left finite eigenvalues of (1),(2) in the interval $(-\infty,\beta]$ with the number of the so-called proper focal points of the principal solution of system (1) in $(a,b]$ evaluated for $\lambda=\beta$. This result is derived under the additional monotonicity assumption

$$B(t,\lambda)=(0I)H(t,\lambda)(0I)^T \ge 0 \quad (7)$$

for the Hamiltonian $H(t,\lambda)$ (the Legendre condition). In [6-7], following the main ideas in [1-2] we generalized Theorem 3.5 in [3] introducing the so-called oscillation numbers which calculate the number of left finite eigenvalues in the interval $(\alpha,\beta], \alpha<\beta$ without assumption (7). In the recent paper (J. Elyseeva, *Relative oscillation theory for linear Hamiltonian systems with nonlinear dependence on the spectral parameter*, submitted) we introduced the dual oscillation numbers which make possible to evaluate right finite eigenvalues of (1),(2) in $[\alpha,\beta), \alpha<\beta$.

The main results of this paper (see Theorems 2,4) relate the spectral count introduced by A. Abramov in [1] with the oscillation numbers in [6-7]. The Abramov method of spectral counting [1-2] is developed for the case when classical eigenvalues are isolated and their multiplicities are defined by (6). In this paper we show how the approach in [1] can be modified for the more general case of finite eigenvalues with the multiplicities (5). The practical value of these results is in possible implementations of the outstanding ideas of [1-2] for stable calculations of the oscillation numbers and finite eigenvalues of (1), (2).

2 Oscillation numbers and the Abramov spectral count

In this section we recall the main notions used in the paper such that the comparative index (see [8] and Chapter 3 in [9]), the oscillation numbers [6-7], as well as the Abramov method of spectral counting in [1].

2.1. The comparative index

According to the definition of the comparative index in [8-9] (see also the references given therein) we consider $2n\times n$ matrices $Y=(X^TU^T)^T, \hat{Y}=(\hat{X}^T\hat{U}^T)^T$ with condition (4) using the notation $M=(I-XX^{\dagger})\hat{X}$, $T=I-M^{\dagger}M$, where $A^{\dagger}$ denotes Moore-Penrose pseudoinverse of A. We also define the symmetric matrix

$$P=T(\hat{X}^T(\hat{Q}-Q)\hat{X})T, \quad (8)$$

where the symmetric matrices $Q,\hat{Q}$ solve the matrix equations

$$X^TQX=X^TU,\ \hat{X}^T\hat{Q}\hat{X}=\hat{X}^T\hat{U}. \quad (9)$$

The comparative index is defined by

$$\mu(Y,\hat{Y})=\mu_1(Y,\hat{Y})+\mu_2(Y,\hat{Y}),$$

where

$$\mu_1(Y,\hat{Y})=rank\, M,\ \mu_2(Y,\hat{Y})=ind\, P.$$

Here $ind\, P$ denotes the number of negative eigenvalues of the symmetric matrix P. For the case $\det X\neq 0$ we have $\mu_1(Y,\hat{Y})=0$ and $Q=UX^{-1}$. If, additionally, $\det\hat{X}\neq 0$ we have

$$\mu(Y,\hat{Y})=ind(\hat{Q}-Q).$$

We also define the dual comparative index

$$\mu^*(Y,\hat{Y})=\mu_1(Y,\hat{Y})+\mu_2^*(Y,\hat{Y}),$$

where

$\mu_2^*(Y,\hat{Y})=ind(-P)$. For $\mu(Y,\hat{Y})$ and $\mu^*(Y,\hat{Y})$ we have the following connections

$$\mu^*(Y,\hat{Y})-\mu(Y,\hat{Y})=\mu_2^*(Y,\hat{Y})-\mu_2(Y,\hat{Y})=i(P), \quad (10)$$

where $i(P)=ind(-P)-ind(P)$ denotes the inertia of $P=P^T$. We also have (combining Properties 5,6 of the comparative index in [8])

$$\mu^*(Y,\hat{Y})+\mu(Y,\hat{Y})=rank\,\hat{X}-rank\, X + rank\, w(Y,\hat{Y}), \quad (11)$$

where we use the Wronskian according to

$$w(Y,\hat{Y})=Y^TJ\hat{Y}.$$

2.2. The oscillation numbers

Consider the definition of the oscillation numbers for conjoined bases $Y(t,\lambda)$ of (1) according to [6-7], where $\lambda \in R$ is a fixed number. Introduce a partition of the interval $[a,b]$:

$$\Pi : a = t_0 < t_1 < \ldots < t_{p-1} < t_p = b\,, \qquad (12)$$

that for any $[t_i,t_{i+1}] \subset [a,b], i = 0,1,\ldots,p-1$ there exists a constant symplectic matrix $R_i(\lambda) = \begin{pmatrix} K_i(\lambda) & M_i(\lambda) \\ L_i(\lambda) & S_i(\lambda) \end{pmatrix}$, $R_i^T(\lambda) J R_i(\lambda) = J$ with the condition

$$\det \tilde{X}_i(t,\lambda) \neq 0, t \in [t_i,t_{i+1}], i = 0,1,\ldots,p-1, \quad (13)$$

where

$$\tilde{Y}_i(t,\lambda) = R_i^{-1}(\lambda) Y(t,\lambda) = (\tilde{X}_i^T(t,\lambda)\ \tilde{U}_i^T(t,\lambda)\,)^T.$$

Then we define the oscillation number

$$N(Y(\lambda),[a,b]) = \sum_{i=0}^{p-1} \mu(Y(t,\lambda), R_i(\lambda)(0\,I)^T)\Big|_{t_i}^{t_{i+1}}$$
$$= -\sum_{i=0}^{p-1} \mu(\tilde{Y}(t,\lambda), R_i^{-1}(\lambda)(0\,I)^T)\Big|_{t_i}^{t_{i+1}}, \qquad (14)$$

where we use the notation for the substitution $A(t)\big|_x^y = A(y) - A(x)$. According to Section 2.1 and (13) the comparative index $\mu(\tilde{Y}(t,\lambda), R_i^{-1}(\lambda)(0\,I)^T)$ in (14) takes the form

$$\begin{aligned} &\mu(\tilde{Y}(t,\lambda), R_i^{-1}(\lambda)(0\,I)^T) = ind\, P(t,\lambda), \\ &P(t,\lambda) = M_i(\lambda)(Q_i(\lambda) - \tilde{Q}_i(t,\lambda)) M_i^T(\lambda), \end{aligned} \qquad (15)$$

where

$$M_i(\lambda) Q_i(\lambda) M_i^T(\lambda) = -M_i(\lambda) K_i^T(\lambda),$$
$$\tilde{Q}_i(t,\lambda) = \tilde{U}_i(t,\lambda) \tilde{X}_i^{-1}(t,\lambda).$$

According to the main results in [6-7] oscillation numbers (14) are invariant with respect to the choice of partitions and transformation matrices $R_i(\lambda)$.

Moreover, under the Legendre condition (7) the oscillation numbers are nonnegative and coincide with the total number of left proper focal points of $Y(t,\lambda)$ in $(a,b]$

$$N(Y(\lambda),[a,b]) = \sum_{t_0 \in (a,b]} m^-(t_0,\lambda) \geq 0,$$
$$m^-(t_0,\lambda) = rank\, X(t_0^-,\lambda) - rank\, X(t_0,\lambda).$$

Under the additional assumptions that (1) is controllable for $t \in [a,b]$ we have by analogy with (6) that the points $t_0 \in [a,b]$ such that $\det X(t_0,\lambda) = 0$ are isolated and $m^-(t_0,\lambda) = n - rank\, X(t_0,\lambda)$.

Assume (3), then for the principal solution $Y(t,\lambda)$ of (1) we have (see Theorem 3.1 in [6])

$$\begin{aligned} &N(Y(\lambda_0^+), [a,b]) = N((Y(\lambda_0), [a,b]), \\ &N((Y(\lambda_0), [a,b]) - N(Y(\lambda_0^-), [a,b]) = \theta^-(\lambda_0), \end{aligned} \qquad (16)$$

where the multiplicity of a left eigenvalue of (1),(2) $\theta^-(\lambda_0)$ is defined by (5). Then for arbitrary

$$\alpha < \beta, \alpha, \beta \in R$$

$$N(Y(\beta),[a,b]) - N(Y(\alpha),[a,b])$$
$$= \sum_{\lambda_0 \in (\alpha,\beta]} \theta^-(\lambda_0) \qquad (17)$$

(see Theorem 3.2 in [6]). Remark that the sum in the right-hand side of (17) is finite because $rank\, X(t,\lambda)$ is piecewise constant in $\lambda \in R$ (see [5]).

2.3 The Abramov spectral count

Here we recall the construction of the spectral count in [1] for problem (1),(2) when the left boundary condition $x(a,\lambda) = 0$ is transferred to the right point $t = b$. The general case described in [1] can be considered analogously.

For a conjoined basis $Y(t,\lambda)$ of (1) with a fixed λ consider a partition

$$\Pi : a = s_0 < s_1 < \ldots < s_{l-1} < t_l = b$$

and the symplectic orthogonal transformation matrices

$$R_{\beta_k} = \begin{pmatrix} \cos(\beta_k)I & -\sin(\beta_k)I \\ \sin(\beta_k)I & \cos(\beta_k)I \end{pmatrix}, \beta_k(\lambda) \neq \frac{\pi m}{2}, m \in Z$$

such that

$$\det \breve{U}_k(t,\lambda) \neq 0, t \in [s_k,s_{k+1}], k = 0,1,\ldots,l-1,$$

where

$$\breve{Y}_k(t,\lambda) = R_{\beta_k}^{-1}(\lambda) Y(t,\lambda) = (\breve{X}_k^T(t,\lambda)\ \breve{U}_k^T(t,\lambda)\,)^T.$$

Then there exist the symmetric matrices

$$\breve{Q}_k(t,\lambda) = \breve{X}_k(t,\lambda) \breve{U}_k^{-1}(t,\lambda), t \in [s_k,s_{k+1}], \qquad (18)$$

and

$$\sigma_k(t,\lambda) = \cos\beta_k(\cos\beta_k \breve{Q}_k(t,\lambda) - \sin\beta_k I), t \in [s_k,s_{k+1}], \quad (19)$$

for all $k = 0,1,\ldots,l-1$.

Let $Y(t,\lambda)$ be the principal solution of (1). Consider problem (1), (2) under assumptions which guarantee that all classical eigenvalues which are zeros of $\det X(b,\lambda)$

are isolated and their multiplicities in (5) are defined by (6). For the case $\det X(b,\lambda)\neq 0$ define the number

$$N_A(Y(\lambda),[a,b]) = \sum_{k=0}^{l-1} i(\sigma_k(t,\lambda))\Big|_{s_k}^{s_{k+1}}, \qquad (20)$$

where $i(\sigma_k(t,\lambda_0))$ is the inertia of $\sigma_k(t,\lambda_0)$ given by (19).

Then, by Theorem 5 in [1]

$$N_A(Y(\lambda_0^+),[a,b]) - N_A(Y(\lambda_0^-),[a,b]) = 2\theta(\lambda_0)$$

and by Theorem 6 in [1] for any $\alpha<\beta,\ \alpha,\beta\in R$ such that $\det X(b,\alpha)\neq 0,\ \det X(b,\beta)\neq 0$

$$N(Y(\beta),[a,b]) - N(Y(\alpha),[a,b]) = 2\sum_{\lambda_0\in(\alpha,\beta)}\theta(\lambda_0).$$

Remark that the symmetric matrices $\tilde{Q}_i(t,\lambda)$ and $\breve{Q}_k(t,\lambda)$ in (15) and (18) obey transformed matrix Riccati equations for all $t\in[t_i,t_{i+1}]$ and $t\in[s_k,s_{k+1}]$ with coefficients associated with the transformed Hamiltonians

$$R_i^T(\lambda)H(t,\lambda)R_i(\lambda) \quad\text{and}\quad R_{\beta_k}^T(\lambda)H(t,\lambda)R_{\beta_k}(\lambda),$$

respectively. Moreover, the angles $\beta_k(\lambda)\neq \frac{\pi m}{2},\ m\in Z$ can be chosen in such a way that for $\breve{Q}_k(t,\lambda)$ defined by (18) we have

$$\left\|\breve{Q}_k(s_k,\lambda)\right\|\le C(n),$$

where $C(n)$ depends only on the dimension of the problem. A similar estimate can be derived for the matrices $\tilde{Q}_i(t,\lambda)$ in (15) for the special case of symplectic orthogonal transformation matrices in form R_{β_k}.

3 Main results

In this section we prove main connections between the oscillation numbers (14) and the Abramov spectral function (20).

Lemma 1. For arbitrary conjoined basis $Y(t,\lambda)$ of (1) we have the following representation for oscillation number (14)

$$\begin{aligned}N(Y(\lambda),[a,b]) &= -\frac{1}{2}\left(\sum_{i=0}^{p-1} i(P_i(t,\lambda))\Big|_{t_i}^{t_{i+1}} + rank\,X(t,\lambda)\Big|_a^b\right) \qquad (21)\\ &= \frac{1}{2}\left(\sum_{i=0}^{p-1} i(\tilde{P}_i(t,\lambda))\Big|_{t_i}^{t_{i+1}} - rank\,X(t,\lambda)\Big|_a^b\right),\end{aligned}$$

where $P_i(t,\lambda)$ and $\tilde{P}_i(t,\lambda)$ are the symmetric matrices defined by (8) for the comparative indices $\mu(Y(t,\lambda),R_i(\lambda)(0I)^T)$ and $\mu(\tilde{Y}(t,\lambda),R_i^{-1}(\lambda)(0I)^T)$ in (14), respectively.

Proof. The proof is based on the identity

$$\mu(Y,\hat{Y}) = \frac{1}{2}(rank\,\hat{X} - rank\,X + rank\,w(Y,\hat{Y}) - i(P)), \qquad (22)$$

which follows from (10), (11). Indeed, by the definition of $\mu(Y,\hat{Y})$ and $\mu^*(Y,\hat{Y})$ we have in the left-hand side of (11)

$$\mu(Y,\hat{Y}) + \mu^*(Y,\hat{Y}) = 2\mu_1(Y,\hat{Y}) + 2\mu_2(Y,\hat{Y}) + \mu_2^*(Y,\hat{Y}) - \mu_2(Y,\hat{Y}).$$

Then, by (10) it follows (22).

Applying (22) for the case $Y:=Y(t,\lambda),\ \hat{Y}:=R_i(\lambda)(0I)^T$ (see the first representation of the oscillation number in (14)) we have

$$\mu(Y(t,\lambda),R_i(0\,I)^T) = \frac{1}{2}(rank\,M_i - rank\,X(t,\lambda) + rank\,\tilde{X}_i(t,\lambda) - i(P_i(t,\lambda))),$$

then by (13)

$$\mu(Y(t,\lambda),R_i(0\,I)^T)\Big|_{t_i}^{t_{i+1}} = -\frac{1}{2}(rank\,X(t,\lambda) + i(P_i(t,\lambda)))\Big|_{t_i}^{t_{i+1}}.$$

Substituting the last representation into (14) we derive the first representation in (21). The second one follows from (22) for the case $Y:=\tilde{Y}_i(t,\lambda),\ \hat{Y}:=R_i^{-1}(\lambda)(0I)^T$. For this case we have instead of (22)

$$\mu(\tilde{Y}_i(t,\lambda),R_i^{-1}(0\,I)^T) = \frac{1}{2}(rank(-M_i^T) - rank\,\tilde{X}_i(t,\lambda) + rank\,(-X^T(t,\lambda)) - i(\tilde{P}_i(t,\lambda))),$$

then, by (13)

$$\mu(\tilde{Y}_i(t,\lambda),R_i^{-1}(0\,I)^T)\Big|_{t_i}^{t_{i+1}} = \frac{1}{2}(rank\,(X^T(t,\lambda)) - i(\tilde{P}_i(t,\lambda)))\Big|_{t_i}^{t_{i+1}}.$$

Sustituting the last representation into (14) we derive the second representation in (21). The proof is completed.

From Lemma 1 we derive the main result of the paper.

Theorem 2. Consider problem (1),(2) under condition (3) coupled with the assumption that all zeros of $\det X(b,\lambda)$ are isolated for $\lambda\in R$, where $X(b,\lambda)$ is the upper block of the principal solution $Y(t,\lambda)$ of (1) at $t=a$. Then, for any $\tilde{\lambda}\in R$ such that $\det X(b,\tilde{\lambda})\neq 0$ we have

$$N_A(Y(\tilde{\lambda}),[a,b])=2N(Y(\tilde{\lambda}),[a,b])+n, \quad (23)$$

where $N_A(Y(\lambda),[a,b])$ is the Abramov spectral count given by (20) and $N(Y(\lambda),[a,b])$ is the oscillation number (14).

Proof. According to the definition of $N_A(Y(\lambda),[a,b])$ in Section 2.3, we have the following representation

$$\breve{Y}_k(t,\lambda)=\begin{pmatrix}\breve{Q}_k(t,\lambda)\\ I\end{pmatrix}\breve{U}_k(t,\lambda)=R_{\beta_k}^{-1}(\lambda)Y(t,\lambda),$$

$$\det\breve{U}_k^T(t,\lambda)\neq 0,\quad t\in[s_k,s_{k+1}].$$

Then the matrix

$$R_{\beta_k}(\lambda)J^T=\begin{pmatrix}\sin(\beta_k)I & \cos(\beta_k)I\\ -\cos(\beta_k)I & \sin(\beta_k)I\end{pmatrix},$$

$$\beta_k(\lambda)\neq\frac{\pi m}{2},\ m\in Z$$

can be used in the definition of the oscillation numbers (14) for the same partition as for $N_A(Y(\lambda),[a,b])$. Rewriting (14) for the partition

$$\Pi: a=s_0<s_1<\ldots<s_{l-1}<t_l=b$$

we see that the comparative index in the second representation takes the form

$$\mu(J\breve{Y}_k(t,\lambda),JR_{\beta_k}^{-1}(0\,I)^T)=ind\,\sigma_k(t,\lambda),$$

then the second formula in (21) can be rewritten as follows

$$N(Y(\lambda),[a,b])=\frac{1}{2}\left(\sum_{k=0}^{l-1}i(\sigma_k(t,\lambda))\Big|_{t_i}^{t_{i+1}}-rank\,X(t,\lambda)\Big|_a^b\right).$$

For the special case when $Y(t,\lambda)$ is the principal solution of (1): $X(a,\lambda)=0$ and for $\lambda:=\tilde{\lambda}$ with $\det X(b,\tilde{\lambda})\neq 0$ the last formula tranfers into

$$N(Y(\tilde{\lambda}),[a,b])=\frac{1}{2}\left(\sum_{k=0}^{l-1}i(\sigma_k(t,\tilde{\lambda}))\Big|_{t_i}^{t_{i+1}}-n\right)$$
$$=\frac{1}{2}\left(N_A(Y(\tilde{\lambda}),[a,b])-n\right).$$

The proof of (23) is completed.

3.1 Modifications of the oscillation numbers

In this section we present different forms of representation of the oscillation numbers (14). The theoretical ground of these representations is the following separation theorem for the oscillation numbers (see Theorem 4.1 in [5]).

Theorem 2. Let $Y(t,\lambda)$ and $\hat{Y}(t,\lambda)$ be conjoined bases of (1), then for arbitrary $[\tau_0,\tau_1]\subset[a,b]$

$$\begin{aligned}&N(Y(\lambda),[\tau_0,\tau_1])-N(\hat{Y}(\lambda),[\tau_0,\tau_1])\\&=\mu(Y(\tau_1,\lambda),\hat{Y}(\tau_1,\lambda))-\mu(Y(\tau_0,\lambda),\hat{Y}(\tau_0,\lambda)).\end{aligned} \quad (24)$$

Remark that in (24) $Y(t,\lambda)$ and $\hat{Y}(t,\lambda)$ are considered for the same fixed value of λ. We present the following modification of the oscillation number which can be used as a spectral count for (1), (2).

Theorem 3. Let $Y(t,\lambda)$ and $\hat{Y}(t,\lambda)$ be the principal solutions of (1) at $t=a$ and $t=b$, i.e., $Y(a,\lambda)=\hat{Y}(b,\lambda)=(0\,I)^T$. Then, under assumption (3) the oscillation number $N(Y(\lambda),[a,b])$ in formulas (16), (17) can be replaced according to

$$\begin{aligned}&N(Y(\lambda),[a,b])=N(Y(\lambda),[a,\tau_0])+N(\hat{Y}(\lambda),[\tau_0,b])\\&-\mu(Y(\tau_0,\lambda),\hat{Y}(\tau_0,\lambda)),\ \tau_0\in[a,b].\end{aligned} \quad (25)$$

Proof. For the case $\tau_1:=b$ and $\hat{Y}(b,\lambda)=(0\,I)^T$ (in this case $\hat{Y}(t,\lambda)$ is called the principal solution of (1) at $t=b$) formula (24) takes the form

$$N(Y(\lambda),[\tau_0,b])=N(\hat{Y}(\lambda),[\tau_0,b])-\mu(Y(\tau_0,\lambda),\hat{Y}(\tau_0,\lambda)).$$

Finally, using the additivity property of the oscillation numbers [6] for $\tau_0\in[a,b]$

$$N(Y(\lambda),[a,b])=N(Y(\lambda),[a,\tau_0])+N(Y(\lambda),[\tau_0,b])$$

we complete the proof of (25).

Based on (25) one can present a complete analogue of the results in [1] related to the transferring of the left and right boundary conditions of (1),(2) to arbitrary point $\tau_0\in[a,b]$. Introduce partitions of the intervals $[a,\tau_0]$ and $[\tau_0,b]$ according to (12), (13) for $Y(t,\lambda)$ and $\hat{Y}(t,\lambda)$, respectively. Then introduce unified numbering of the points of the partitions and transformation matrices R_i in such a way, that condition (13) holds for $Y(t,\lambda)$ when $i=0,1,\ldots,r-1$ (so we have $\tau_0:=t_r$) and similarly, (13) holds for $\hat{Y}(t,\lambda)$ when $i=r,r+1,\ldots,p-1$. Under this agreement we prove the following result.

Theorem 4. Under the assumptions of Theorem 3 formula (25) can we rewritten in form

$$\begin{aligned}N(Y(\lambda),[a,b])=\frac{1}{2}\Bigg(&\sum_{i=0}^{r-1}i(\tilde{P}_i(t,\lambda))\Big|_{t_i}^{t_{i+1}}\\&-\sum_{i=r}^{p-1}i(\hat{P}_i(t,\lambda))\Big|_{t_{i+1}}^{t_i}+i(P(\tau_0,\lambda))-rank\,X(b,\lambda)\Bigg),\end{aligned} \quad (26)$$

where the symmetric matrices $\tilde{P}_i(t,\lambda)$, $\hat{P}_i(t,\lambda)$, and $P(\tau_0,\lambda)$ are defined by (8) for the comparative indices

$$\mu(R_i^{-1}(\lambda)Y(t,\lambda),R_i^{-1}(\lambda)(0I)^T),\mu(R_i^{-1}(\lambda)\hat{Y}(t,\lambda),R_i^{-1}(\lambda)(0I)^T)$$

in (14) and for $\mu(Y(\tau_0,\lambda),\hat{Y}(\tau_0,\lambda))$ in (25).

Proof. According to the second representation in (21) we have

$$N(Y(\lambda),[a,\tau_0]) = \frac{1}{2}\left(\sum_{i=0}^{r-1} i(\tilde{P}_i(t,\lambda))\Big|_{t_i}^{t_{i+1}} - rank\, X(\tau_0,\lambda)\right)$$

$$N(\hat{Y}(\lambda),[\tau_0,b]) = \frac{1}{2}\left(-\sum_{i=r}^{p-1} i(\hat{P}_i(t,\lambda))\Big|_{t_{i+1}}^{t_i} + rank\, \hat{X}(\tau_0,\lambda)\right)$$

and by (22)

$$\mu(Y(\tau_0,\lambda),\hat{Y}(\tau_0,\lambda)) = \frac{1}{2}(rank\,\hat{X}(\tau_0,\lambda) - rank\, X(\tau_0,\lambda)$$
$$+rank\;w(Y(\tau_0,\lambda),\hat{Y}(\tau_0,\lambda)) - i(P(\tau_0,\lambda))).$$

Remark that the Wronskian $w(Y(\tau_0,\lambda),\hat{Y}(\tau_0,\lambda))$ is constant

$$w(Y(\tau_0,\lambda),\hat{Y}(\tau_0,\lambda)) = w(Y(b,\lambda),\hat{Y}(b,\lambda)) = X^T(b,\lambda).$$

Substituting the derived above representations into (25) and cancelling the same terms we prove (26). The proof of Theorem 4 is completed.

Remark. Observe that the matrix $P(\tau_0,\lambda)$ in (26) corresponds to the symmetric matrix ω defined by (3.3), (3.4) in [1] for $\tilde{\lambda}$ such that $\det X(b,\tilde{\lambda}) \neq 0$. One can proof using the properties of the comparative index (see Properties 1,3,4 p. 448 in [8]) that $P(\tau_0,\tilde{\lambda}) = \sigma_{r-1}(\tau_0,\tilde{\lambda})(\sigma_{r-1}(\tau_0,\tilde{\lambda}) - \hat{\sigma}_r(\tau_0,\tilde{\lambda}))^{-1}\hat{\sigma}_r(\tau_0,\tilde{\lambda})$, where $\sigma_{r-1}(\tau_0,\tilde{\lambda})$ is defined by (19) for $Y(t,\lambda)$ and $\hat{\sigma}_r(\tau_0,\tilde{\lambda})$ is defined similarly for $\hat{Y}(t,\lambda)$. Moreover, by analogy with [1] in the proof we assume that the angles in the definitions of these matrices obey the condition $\beta_{r-1} = \beta_r$.

4 Conclusions

In this paper we show that the results in [1] can be extended in several directions. First of all, in this paper we consider a generalization of classical eigenvalues to the case of left finite eigenvalues with the multiplicities (5). Comparing with [1], in the main construction of the oscillation numbers (14) we do not use terms which are not defined at the eigenvalues of the Hamiltonian problems, moreover, oscillation numbers are defined for arbitrary constant symplectic transformations with condition (13).

The results in [1] concern general separated boundary conditions and boundary conditions with joint end points which include the Dirichlet boundary conditions considered in this paper as a special case. However, the results in this paper can be extended to such separated (and even jointly varying) endpoints by a standard method, which is based on adding two isolated points to the interval $[a,b]$, see [3-4].

This work was supported by the Ministry of Science and Higher Education of the Russian Federation under project 0707-2020-0034. This research is supported by Federal Programme of Ministry of Education and Science of the Russian Federation in the framework of the state order (under grant NoFSFS-2020-0031). This work was carried out using equipment provided by the Center of Collective Use of MSUT "STANKIN".

References

1. A. Abramov, Comput. Math. Math. Phys., **41** (2001)
2. A. Abramov, Comput. Math. Math. Phys., **51**(1) (2011)
3. W. Kratz, *Quadratic Functionals in Variational Analysis and Control Theory* (Akademie Verlag, Berlin, 1995)
4. R. Johnson, R.Obaya, S. Novo, C. Núñez, R. Fabbri, *Nonautonomous Linear Hamiltonian Systems:Oscillation, Spectral Theory and Control.* (Springer, Berlin 2016)
5. M. Bohner, W. Kratz, R. Šimon Hilscher, Math. Nachr., **285**(11-12) (2012)
6. J. Elyseeva, Appl. Math. Lett., **90** (2019)
7. J. Elyseeva, Monatsh. Math., **193** (2020)
8. Yu. Eliseeva, Differ. Equ., **45** (2009)
9. O. Došlý, J. Elyseeva, R. Šimon Hilscher, *Symplectic Difference Systems: Oscillation and Spectral Theory* (Birkhäuser Basel, 2019)

Phase Transitions in Ion-Exchange Materials during the Water Vapor Sorption

Igor Ivanov[1,*] and *Oksana* Ivanova[2]

[1]Russian State Agrarian University, Kaluga branch, RU-248007, Kaluga, Russia
[2]Moscow State Technological University "STANKIN", RU-127055, Moscow, Russia

Abstract. Water vapor sorption process on ion-exchange materials was investigated using a model taking into account the mutual influence of variations in hydration energy and osmotic effect. At realistic parameter values, phase transitions of the first kind of the thin adsorbed liquid film and thick adsorbed liquid film type can occur. Possible reasons and criteria of the phase transitions that led to stepwise change of ionite resistance were discussed. The phase transitions obtained in this work manifest themselves as hysteresis loops on water vapor adsorption-desorption experimental isotherms. Hysteresis was observed experimentally both at high and at low values of relative air humidity. It was explained using different models: for high and low humidity separately. The present work made it possible to explain these critical phenomena using single mathematical model.

1 Introduction

An adequate description of the water vapor sorption by ion-exchange materials (ionites or ion exchangers (IEs)) is possible only on the basis of models that take into account the specific feature of IEs: the relationship between the amount of absorbed water and the dissociation of ion-exchange groups (ionogenic groups (IGs)) [1-8]. On the one hand, an increase in water content enhances hydration and, consequently, dissociation of IGs. On the other hand, an osmotic effect takes place [3], i.e. the dependence of the moisture content on the degree of dissociation. In the presence of such a mutual dependence, a situation is possible when a fluctuating increase in moisture content enhances the dissociation of IGs, which, due to the osmotic effect, leads to the condensation of additional water molecules into the IE phase. As a result, under certain conditions, fluctuations greater than a certain value will not decay, but grow until the system enters a new state. This state is characterized by a moisture content and degree of dissociation significantly different from the initial ones, so this transition can be considered as a phase transition of the first kind, similar to the phase transitions of a thin film – a thick film of an adsorbed substance [9] or a strong electrolyte – a weak electrolyte [10]. Since a stepwise change in the dissociation degree leads to a corresponding change in the conductivity of IE, the phase transitions can be observed experimentally in form of a stepwise change in IE sample resistance.

In work [5], based on a model that takes into account the mutual influence of variations in hydration energy and osmotic effect, the possibility of phase transitions of the thin adsorbed liquid film and thick adsorbed liquid film type for high humidity ($0{,}6 < \theta < 0{,}8$) was shown.

Based on thermodynamic approach consistently developed in [5, 11, 12], the possibility of phase transitions during the water vapor sorption in IEs over the entire range of relative air humidity has been investigated in the present work. Given model made it possible to refine the phase transition criterion obtained in [5] and to find the phase transition at low humidity ($\theta < 0{,}3$) at physically reasonable values of system parameters. The adequacy of these parameters is substantiated by comparing the theoretical and experimental dependences of the sulfocationite membranes samples resistance on the relative air humidity. This approach allows one to obtain basic (noncritical) system parameters at which a phase transition is impossible. And the sequential change of these parameters makes it possible to simulate a phase transition and find out the criterion for its occurrence. The parameters obtained in this way will be critical, and the relationship between them will be phase transition criterion.

As a result of mathematical modeling, the critical dependences of the resistance of IEs samples on the relative air humidity were obtained. These dependences have an *S*-shaped character, typical for systems with a phase transition of the first kind. We note once again that the study of the phase transitions on the above dependences is equivalent to the study of this phenomenon on the adsorption isotherms (the dependence of the moisture content on the relative air humidity): due to the mutual dependence of the degree of IGs dissociation and the moisture content, the phase transition also manifest itself in a stepwise change in the

* mcxa_iv@inbox.ru

concentration of free ions in solution, and, therefore, in the corresponding change in the IE sample resistance.

The phase transitions obtained in this work manifest themselves as hysteresis loops on water vapor adsorption-desorption experimental isotherms. Hysteresis was observed experimentally both at high and at low values of relative air humidity [13]. It was explained using different models: for high and low humidity separately. The present work made it possible to explain these critical phenomena using single mathematical model.

2 Theoretical Analysis

Mathematical model of the water vapor sorption in IEs was built by minimizing the total thermodynamic potential of the system. This potential consists of the thermodynamic potential of the ionite in the standard state, which was defined as the state in which the ionite contains only the water entering into the hydration shells; the change in the thermodynamic potential caused by the subsequent condensation of water; the ionic subsystem potential, more specifically, this associated with the dissociation of IGs, the formation of their hydration shells, and the formation of a solution composed of a hydrated ions and free water, as well as the entropy contribution associated with the different configurations in the positioning of dissociated and nondissociated IGs in the polymer matrix; and finally the potential associated with elastic deformation of the polymer chains during moisture sorption. In comparison with previous models, this model calculates the IGs dissociation energy more precisely. In addition to the effects of the surface charge and the Debye screening, the ion-dipole interaction during the sorption of polarized molecules is taken into account. The contribution of these factors is most significant at low humidity.

Without going into details of calculation described in work [12], we present the system of equations describing water vapor sorption on the surface of IE:

$$\begin{cases} P/P_S = \dfrac{\beta-\beta_0-\alpha n}{1+\beta-\beta_0-\alpha(n-1)}\exp\left(\dfrac{\alpha}{T}\dfrac{\partial E_D}{\partial\beta}+U\beta\right), & (1)\\ \dfrac{\alpha^2}{1-\alpha}\dfrac{(1+\beta-\beta_0-\alpha(n-1))^{n-1}}{(\beta-\beta_0-\alpha n)^n} = \exp\left(-\dfrac{1}{T}\dfrac{\partial(\alpha E_D)}{\partial\alpha}\right). & (2)\end{cases}$$

The system includes the following parameters: $P/P_s = \theta$ is the relative air humidity, P_s is the saturation vapor pressure; $\beta = N_W/N$ is the relative moisture content, $\beta_0 = N_{W0}/N$ is the conventional number of hydration of IGs, were N is the number of IGs bonded to the polymer matrix, N_{W0} is the number of water molecules entering into the hydration shells of the IGs, and N_W is the total number of water molecules in the system; α is the degree of dissociation of IGs; n is the conventional number of hydration of ions in a single-component solution; T is the absolute temperature expressed in energy units; $E_D = E_0 + E_{EL}$ is the dissociation energy, which consists of the energy of abstraction of the ion from the polymer chain (E_0) and the change in its electrostatic energy (E_{EL}); $U\beta = 1/2\partial(U\beta^2)/\partial\beta$, were $UT\beta^2/2$ is the energy of elastic deformation of the polymer chains per IG, $U = 6(V_W/V_P)^2$, V_W is the molecular volume of water, and V_P is the polymer chain volume per IG.

Passage to the limits transforms this system into standard expressions: at large values of the relative moisture content $\partial(\alpha E_D)/\partial\alpha \to E_0$, $\partial E_D/\partial\beta \to 0$, and, therefore, formula (1) transforms into the standard expression for the concentration dependence of the vapor pressure over solution [14], while formula (2) takes the form of the law of mass action [15]. Formulas (1) and (2) make it possible not only to determine the values of α and θ for the bulk phase, but also to examine how they change in the case of sorbed thin water films.

As shown in work [12], the dependence of the energy E_D on the degree of dissociation α and the water content β takes the form

$$E_D = E_0 - \frac{CT}{\beta}\left(\frac{\xi\sqrt{\alpha\beta}}{2} + \ln\left(1 - f\exp(-\xi\sqrt{\alpha\beta})\right)\right) - \frac{\xi T\sqrt{\alpha\beta}\exp(-\xi\sqrt{\alpha\beta}/2)}{1-\exp(-\xi\sqrt{\alpha\beta})} - \frac{n\delta}{12^{1/3}}(C\xi)^{2/3}T, \quad (3)$$

where $C = 2q^2s/(\varepsilon T V_W)$ and $\xi = [4\pi q^2 V_W/(\varepsilon T s^2)]^{1/2}$ are dimensionless quantities that include the following parameters: q is the ion charge, s is the surface area per IG, ε is the dielectric permittivity of the solution; $f = (\varepsilon - \varepsilon_1)/(\varepsilon + \varepsilon_1)$, ε_1 is the dielectric permittivity of the polymer matrix; δq is effective charge of dipole edges (polarized water molecule $\delta \sim 0{,}1$).

The terms in (3), starting from the second, correspond to the contributions to the electrostatic energy from the bulk of the screening cloud and the charges induced in the solid dielectric layers due to the finite thickness of the liquid layer, the charge of the surface, and the ion-dipole interaction respectively.

The expression (3) is valid at all physically reasonable values of system parameters (water content and degree of dissociation), even at lowest humidity, when hydration shells of ions are not formed completely (the dependence $n(\beta)$ is needed). For this reason, this expression can be used over the full relative humidity range.

Solving the system of equations (1), (2) with consideration given to (3) makes it possible to determine the dependence of the relative moisture content β on the relative humidity θ, i.e. adsorption isotherm $\beta(\theta)$, and the dependence of the degree of dissociation α on the relative humidity θ, i.e. $\alpha(\theta)$. In terms of searching for phase transitions, these dependences are equivalent: hysteresis is shown on each of them. Since we have experimental data of dependences of the resistance of IE sample R on the relative humidity θ at various temperatures [12], we consider it appropriate to investigate the theoretical dependences $R(\theta)$.

To describe the conduction in IEs, which are nonuniform media, we used the concepts underlying the theory of percolation [16]. According to [16], the current flowing through a nonuniform medium is determined by the lowest conductivity areas on the route between the electrodes of the measuring unit. Conductivity areas look like channels connecting the electrodes. Water molecules are adsorbed in these channels that leads to channel water-filling, IGs dissociation, and appearance of a current between the electrodes. Thus, it is sufficient to calculate the conductivity of these areas.

To calculate the resistance of the system from the degree of dissociation α, we assumed that the mobility of the ions μ is humidity-independent, i.e. the conductance λ depends only on concentration c of free carriers. Then the resistance of IE sample

$$R = \frac{1}{\lambda}\frac{l}{S} = \frac{1}{\mu q c}\frac{l}{S},$$

where l is the characteristic length of conducting channel between electrodes and S is the summary cross section area of all conducting channels.

Considered that the concentration of free ions in solution $c = \alpha N/Sl$ we obtained

$$R = \frac{l^2}{\mu q N}\frac{1}{\alpha} = \frac{R_0}{\alpha}, \qquad (4)$$

where parameter $R_0 = l^2/(\mu q N)$ depends on the geometry of the sample and the concentration of active IGs but does not depend on the humidity.

The empirical temperature dependence of the ion mobility μ can be presented as [17]

$$\mu = \mu_0(1 + \gamma(t - t_0)), \qquad (5)$$

where $\gamma \approx 0{,}1$ and $t_0 = 15$ °C.

Taking into account formulas (3), (4), and (5), let us proceed to the analysis of equations system (1) and (2) for the possibility of phase transitions. As follows from these equations, the effect of changing the hydration energy E_{EL} with moisture content β increasing is determined by the value of the coefficient C. At low values of C, the change in hydration energy is insignificant. With the growth of parameter C, the mutual influence of hydration and osmotic effect increases, and unstable states and phase transitions can occur. Obviously, the criterion of a phase transition is the existence of such θ values at which $dR/d\theta \to \infty$. Let's try to get the "extreme" values of the system parameters, i.e. such when phase transition are still possible.

3 Results and Discussion

The analytical study of $R(\theta)$ dependence is rather cumbersome. Therefore, in general, for arbitrary dissociation degrees, the answer to the question of the possibility of phase transitions can be obtained only on the basis of a numerical solution of the system (1) and (2). At first, the "basic" (noncritical) dependences were plotted at different temperatures (Fig. 1, (*1*)). These dependences were calculated using the following values of the model parameters: $q = 4{,}8\cdot 10^{-10}$ esu, $\varepsilon = 79$, $\varepsilon_1 = 4$, $V_W = 3\cdot 10^{-23}$ cm^3, $s = 1{,}2\cdot 10^{-15}$ cm^2, $l = 1$ cm, $n = 4$, and $U = 1$; at $t = t_0 = 15$ °C, $\mu_0 = 3{,}2\cdot 10^{-4}$ cm^2/(V·s), $K_0 = 1{,}2$, and $R_0 = 2$ kΩ. The elastic deformation energy parameter U was assumed to be temperature-independent. The circles represent the experiment data obtained on the sulfocationite membranes samples [12]. It should be noted that a good agreement between theoretical and experimental values at quite realistic system parameters confirms the adequacy of the proposed model. By successively increasing the value of parameter C, it is possible to obtain critical dependences (Fig. 1, (*2*), (*3*)) with pronounced areas where $dR/d\theta \to \infty$, corresponding to the phase transitions beginning. A further increase in the parameter C leads to the appearance of characteristic *S*-shaped curve pieces (Fig. 2, (*2*), (*3*)), which are shown as stepwise sections or hysteresis loops on an experimental curves. Indeed, like other systems in which phase transitions of the first kind are possible, the transition to a new phase occurs when fluctuations exceed a certain critical value [15]. The probability of such fluctuations is usually quite small that can practically mean the existence of long-lived metastable states. As a result, in experiment the transition from the thin film (weak electrolyte) phase to the thick film (strong electrolyte) phase can occur at humidity different from that at which the reverse transition is observed (i.e. hysteresis on water vapor adsorption-desorption isotherms can be an experimental manifestation of the phase transitions obtained in the present work).

Thus, the numerical analysis of equations (1) and (2) gives rise to the following criterion of phase transition:

$$C \geq 4{,}7\beta_0, \qquad (6)$$

where numerical coefficient before β_0 is slightly decreased with n reduction (if $n = 1$ this coefficient equals 4,1), so that it can be set constant.

In addition the present model explains phase transitions appearance both at high and at low humidity values. For example, such situation can appear in ionites with a small number of hydration of surface IGs ($\beta_0 = 1$, Fig. 2, (*2*)) or with rarely placed IGs (large surface area s per IG, Fig. 2, (*3*)). Therefore, it is necessary to take into account these facts in measurements of IEs conductance. During the manufacturing of hygrometers with IEs sensors, phase transitions should be avoided by selecting proper materials with noncritical parameters (Fig. 2, (*1*)).

According to [13], the water vapor adsorption hysteresis on sulfocationite membranes was discovered by some authors at high humidity ($0{,}6 < \theta < 0{,}8$). Other authors discovered the hysteresis at low humidity ($\theta \leq 0{,}3$). Note, these experimental results are in close agreement with the results of the present work (as can be seen from Fig. 2, dependence *3* has hysteresis at high humidity and dependence *2* – at low). These authors explain the appearance of hysteresis as a result of capillary condensation. The present model explains the hysteresis by correlation between the water content and

https://doi.org/10.1051/epjconf/202124801003

the dissociation of IGs: the fluctuating increase of water content leads to increase of IGs dissociation and, because of osmotic effect, the additional water molecules come into ionite. Moreover, this model explains the hysteresis at low humidity, when the capillary condensation is impossible in principle.

Thus, the results of the present work show that using a single model, we can theoretically explain the experimentally observed phase transitions both at high and at low humidity [13].

4 Conclusion

Obviously, in order to obtain more accurate quantitative results for real systems, the proposed calculation method can be somewhat refined or modified. For this purpose it is possible to consider details of adsorption of water molecules specifically interacting with active centers [3], dependence of dissociation energy on geometry of growing clusters [18], effect of self-dissociation of electrolyte [10, 15]. However, these effects will slightly

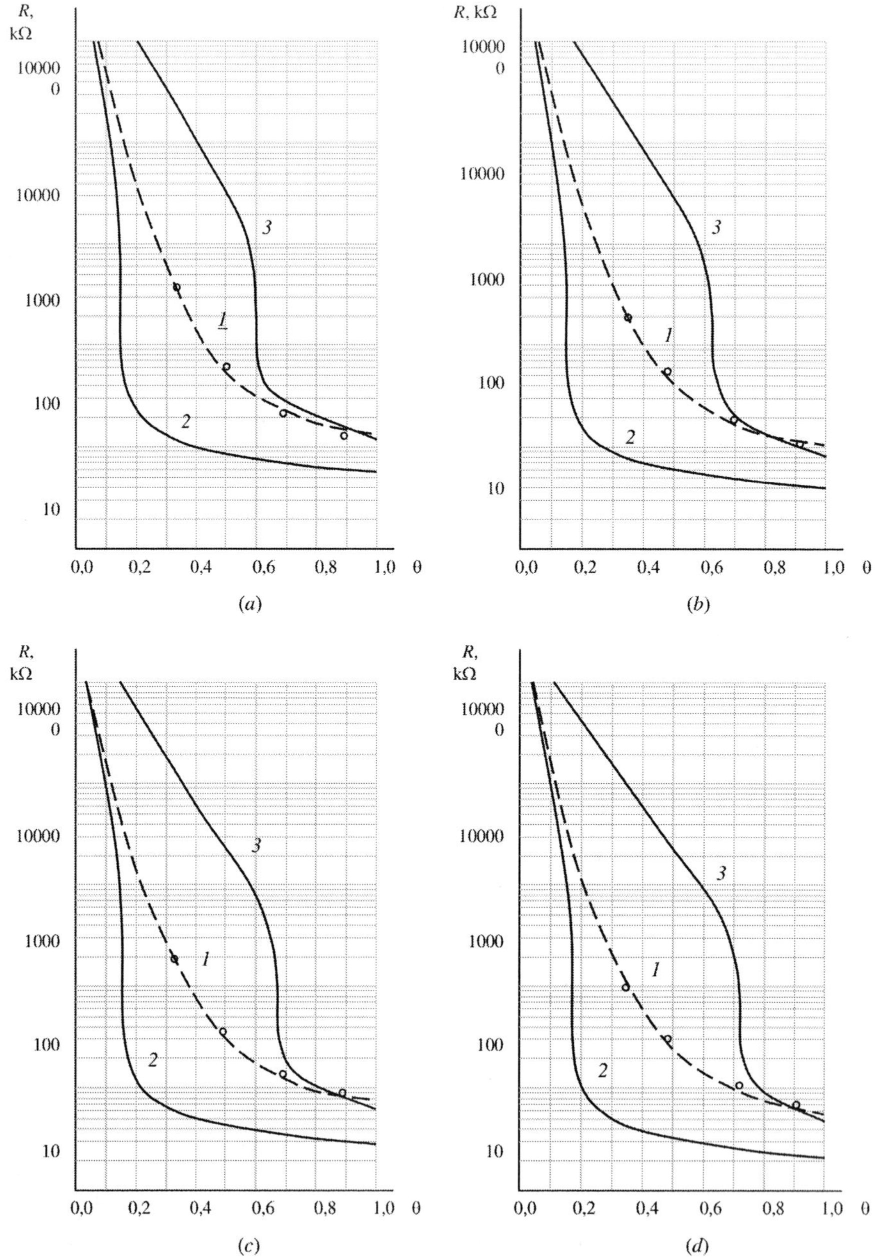

Fig. 1. The $R(\theta)$ dependence at various temperatures:
(*a*) $t = 15$ °C, $\beta_0 = 2$, $C = 5{,}87$ (*1*), $\beta_0 = 1$, $C = 4{,}65$ (2), $\beta_0 = 2$, $C = 9{,}54$ (*3*);
(*b*) $t = 20$ °C, $\beta_0 = 2$, $C = 5{,}77$ (*1*), $\beta_0 = 1$, $C = 4{,}67$ (2), $\beta_0 = 2$, $C = 9{,}57$ (*3*);
(*c*) $t = 30$ °C, $\beta_0 = 2$, $C = 5{,}58$ (*1*), $\beta_0 = 1$, $C = 4{,}69$ (2), $\beta_0 = 2$, $C = 9{,}67$ (*3*);
(*d*) $t = 40$ °C, $\beta_0 = 2$, $C = 5{,}40$ (*1*), $\beta_0 = 1$, $C = 4{,}70$ (2), $\beta_0 = 2$, $C = 9{,}77$ (*3*).

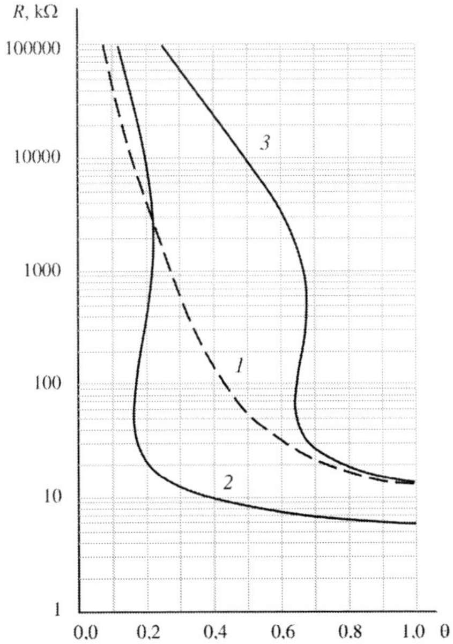

Fig. 2. The $R(\theta)$ dependences at $\beta_0 = 2$, $C = 5{,}87$ (*1*); $\beta_0 = 1$, $C = 5{,}87$ (*2*); and $\beta_0 = 2$, $C = 10{,}27$ (*3*).

change the quantitative parameters of adsorption and will not affect the overall physical picture of the occurrence of phase transitions proposed in the present work.

The question of the possibility of experimental observation of the phenomenon described in the present work at high humidity, when capillary condensation is possible, remains open. The fact is that the effects of capillary condensation also manifest themselves in steps and hysteresis loops on adsorption isotherms and the dependences of the resistance of IEs samples on the relative air humidity. The contribution of these two phenomena can be unambiguously estimated by making measurements according to the following scheme. At first, the adsorption isotherms are measured on some porous matrix. After that, the surface of the pores is covered with IGs by precipitation from a condensed solution, and after drying, the adsorption isotherms are measured. The new loops and steps arising in this case can be attributed to the phenomenon described in this work. Another possible method is the study of weakly crosslinked polymer matrices where the appearance of pores with rigid walls, and, consequently, capillary condensation, seems unlikely. Therefore, the hysteresis loops observed at $\theta \sim 0{,}6$-$0{,}8$ in such materials can also be attributed to this phenomenon.

In conclusion, it should be noted that the obtained results force us to be cautious about the possibility of using capillary condensation methods in IEs porosimetry [19]. For example, when condition (6) is satisfied, a hysteresis loop, as shown in the present work, can be observed even during adsorption on a smooth surface. At the same time, if we assume that its appearance is due to capillary condensation, we can get the wrong conclusion about the existence of pores of a sufficiently small radius in the system. We emphasize, however, that if an IE contains closely placed IGs ($C < 4{,}7\beta_0$), condition (6) is violated, and the anomalies observed in this case can be unambiguously attributed to the effect of capillary condensation.

References

1. Y. Tanaka, *Ion Exchange Membranes: Fundamentals and Applications* (Elsevier, Tokyo, 2015)
2. Yu.A. Kokotov, *Ion Exchangers and Ion Exchange* (Khimiya, Leningrad, 1980)
3. A.I. Gantman, Zh. Fiz. Khim., **66**, 1713 (1992) [in Russia]
4. B. Conway, *Ionic Hydration in Chemistry and Biophysics* (Elsevier, Amsterdam, 1981)
5. V.M. Agafonov, I.V. Ivanov, Russ. J. Phys. Chem., **70**, 825 (1996)
6. V.I. Vasil'eva, E.M. Akberova, V.A. Shaposhnik, M.D. Malykin, Russ. J. Electrochem., **50**, 789 (2014)
7. V. Soldatov, V. Zelenkovskii, E. Kosandrovich, React. Funct. Polym., **102**, 147 (2016)
8. N. Kononenko, V. Nikonenko, M. Fomenko, D. Grande, C. Larchet, L. Dammak, Y. Volfkovich, Adv. Colloid Interfac., **246**, 196 (2017)
9. J. Cahn, J. Chem. Phys., **66**, 3667 (1977)
10. V.A. Kozlov, S.V. Sokolova, N.A. Trufanov, Zh. Eksp. Teor. Fiz., **98**, 2176 (1990) [in Russia]
11. I.V. Ivanov, Russ. J. Phys. Chem., **79**, 448 (2005)
12. I. Ivanov, O. Ivanova, EPJ Web Conf., **224**, 03001 (2019)
13. Yu.A. Kokotov, V.A. Pasechnik, *Equilibrium and Kinetic of Ion Exchange* (Khimiya, Leningrad, 1970)
14. L.D. Landau, E.M. Lifshits, *Course of Theoretical Physics*, Vol. 5: *Statistical Physics*, 2nd ed. (Nauka, Moscow, 1964; Pergamon, Oxford, 1980)
15. Ya.I. Frenkel', *Kinetic Theory of Liquids* (Nauka, Leningrad, 1975; Clarendon, Oxford, 1946)
16. S.F. Timashev, S.N. Cladkikh, Dokl. Akad. Nauk SSSR, **262**, 656 (1982)
17. L.I. Antropov, *Theoretical Electrochemistry* (Vysshaya Shkola, Moscow, 1975)
18. R.R. Dogonadze, F.F. Kornyshev, A.M. Kuznetsov, Teor. Mat. Fiz., **15**, 127 (1973)
19. S. Gregg, K. Sing, *Adsorption, Surface Area and Porosity* (Academic Press, London, 1982)

EPJ Web of Conferences **248**, 01004 (2021)
MNPS-2020
https://doi.org/10.1051/epjconf/202124801004

S-synchronization and Excitation Constancy in Structural Identifiability Problem of Nonlinear Systems

Nikolay Karabutov[1,*]

[1]Mirea-Russian Technological University, Department of Problems Control, RU-119454, Moscow, Russia

Abstract. An approach to analysis the structural identifiability (SI) of nonlinear dynamical systems under uncertainty was proposed. S-synchronizability condition of an input is the basis for the structural identifiability estimation of the nonlinear system. A method for obtaining a set containing information about the nonlinear part of the system wasproposed. The decision on SI of the system was based on the analysis of geometric frameworks reflected the state of the system nonlinear part. Geometric frameworks were defined on the specified set. Conditions for structural indistinguishability of geometric frameworks and local identifiability of the nonlinear part were obtained. It shown that a non-S-synchronizing input gives an insignificant geometric framework. This input is a sign of structural non-identifiability of the nonlinear system. The method for estimating the structural identifiability of the nonlinear system was proposed. We show that the structural identifiability is the basis for structural identification of the system. The structural identifiability degree was introduced, and the method of its estimation was proposed.

1 Introduction

Analysis of recent publications shows that the system identifiability is performed in a parametric space. Many publications are devoted to the study of this problem. Identifiability results are typically presented in the form accepted in the parametric estimation problem. Various methods were proposed for estimating the structural identifiability [1, 2].

Many authors study the parametric identifiability of nonlinear systems (see, e.g. [2-5]). In [2], an approach based on the sensitivity analysis of system output was applied to study identifiability and the analysis of experimental data was used for obtaining parametric identifiability conditions. The critical analysis of approaches applied to the estimate the identifiability of biological models is given in [3]. Models for estimating the identifiability of nonlinear systems are typically based on Taylor series expansion, identifiability tables, and Lie algebra study. Research on practical identifiability was discussed in [4].

The analysis shows that the identifiability of the model is normally understood as the possibility of estimating its parameters and the proposed methods use the nondegeneracy estimation of the information matrix. Similar results were obtained in the parametric estimation theory as the condition of nondegeneracy (completeness of rank) of the matrix input. As a rule, the model structure is set a priori, and, thus, the use of local structural identifiability is not obvious. The concept of structure is widely used in problems of identifiability. The identifiability of a nonlinear system reduces to the parametric identifiability problem and is based on the application of various linearization methods. This extensive field of research does not cover the structural identifiability problem of nonlinear dynamical systems in regarding to deciding on the structure (form, dependence) of the nonlinear system under uncertainty. This statement corresponds to the analysis of structural aspects of the identifiability (identification) system. Also, the question not considered which an input having the excitement constancy property guarantees the structural identifiability of the system. This problem proposed in [6] first.

Also, an important question: "Which input having the excitement constancy property can provide guarantees the structural identifiability of the system" was not considered [6].

In this paper, we study the structural identifiability problem of a nonlinear system. It is important to note that this problem is very complex problem and that methods of formalizing the system structure have not been developed yet. The concept of the structural identifiability (h-identifiability) was introduced in [6]. We propose the approach for solving the structure estimating problem of a nonlinear dynamical system. The approach is based on the analysis of a specific class of frameworks that adequately describe the state of the nonlinear system. The main results obtained in this study can be considered as a generalization of those obtained in [6, 7].

2 Problem statement

Consider the system

* Corresponding author: kn22@yandex.ru

$$\dot{X} = AX + B_{\varphi}\varphi(y) + B_u u,\\ y = C^T X, \tag{1}$$

where $u \in R$, $y \in R$ are the input and output, $\varphi(y)$ is scalar nonlinear function, $B_u \in R^q$, $B_{\varphi} \in R^q$; $B_{\varphi} = B_u = I = [0,0,\ldots,0,1]^T$, $A \in R^{q\times q}$ is Hurwitz matrix, $C = [1,0,\ldots,0]^T$.

$\varphi(y)$ satisfies the condition

$$\chi \in \mathcal{F}_{\varphi} = \{\gamma_1\xi^2 \le \varphi(\xi)\xi \le \gamma_2\xi^2, \xi \ne 0,\\ \varphi(0) = 0,\ \gamma_1 \ge 0, \gamma_2 < \infty\}. \tag{2}$$

The information set

$$\mathrm{I}_o = \{u(t), y(t),\ t \in J = [t_0, t_k]\} \tag{3}$$

is known for the system (1).

Problem: use the analysis I_o and estimate the structural identifiability (SI) of the system (1).

3 Method of design S_{ey}-framework

The proposed approach based on the analysis of geometric framework S_{ey}. The S_{ey}-framework design based on the formation of a set $\mathrm{I}_{N,g}$ contained information about the function $\varphi(y)$ [9]. Apply the differentiation operation to $y(t)$ and denote the obtained variable as x_1. Accounting x_1 extends the information the set I_o: $\mathrm{I}_{ent} = \{\mathrm{I}_o, x_1\}$. Accounting x_1 extends the information the set I_o: $\mathrm{I}_{ent} = \{\mathrm{I}_o, x_1\}$.

Select the subset $\mathrm{I}_g \subset \mathrm{I}_{ent}$ corresponding to the particular solution of the system (5) (steady state). The set $\mathrm{I}_g = \mathrm{I}_{ent} \setminus \mathrm{I}_{tr}$ does not contain data about the transition process in the system Apply the mathematical model

$$\hat{x}_1^l(t) = H^T [1\ u(t)\ y(t)]^T \tag{4}$$

for obtaining the linear component x_1. The variable x_1 is defined on the interval $J_g = J \setminus J_{tr}$. Here $H \in R^3$ is the vector of model parameters.

Determine the vector H as the solution to the problem

$$\min_H Q(e)\Big|_{e=\hat{x}_1^l - x_1} \to H_{opt},$$

where $Q(e) = 0.5e^2$.

Find the forecast for variable x_1 by applying model (4) $\forall t \in \mathrm{I}_g$, and generate an error $e(t) = \hat{x}_1^l(t) - x_1(t)$. $e(t)$ depends on the nonlinearity $\varphi(y)$ in the system (1).

So, set $\mathrm{I}_{N,g} = \{y(t), e(t)\ t \in J_g\}$ has obtained. We apply the designation $y(t)$ supposing $y(t) \in \mathrm{I}_{N,g}$.

Remark 1. Choosing the model structure (4) is the stage for structural identification of the system (1). Simulation results show that the model (4) is applicable in identification systems of objects with static nonlinearity. The model (4) structure choice proposed for another nonlinearity in [9].

The phase portrait S described by function $\Gamma: \{y\} \to \{y'\}$ does not always guarantee decision-making about the nonlinear properties of the system under uncertainty. Therefore, analyze the set $\mathrm{I}_{N,g}$ and go to the space $\mathcal{P}_{ye} = (y, e)$ that we call structural.

Consider the function $\Gamma_{ey}: \{y\} \to \{e\}$ $\forall t \in J_g$ described on the plane (y, e) the change in the framework S_{ey}.

Apply the models (4) and present the system (1) as

$$S_y: \begin{cases} \dot{\tilde{X}} = A\tilde{X} + I\zeta, \\ \tilde{y} = C^T \tilde{X}, \end{cases}\\ S_{\varphi}: e = f(y, x_1), \tag{5}$$

where $\tilde{X} \in R^q$ is the variable describing the general solution of the system (1); $\zeta \in R$ is a limited disturbance, which is the result of a procedure applied for the variable definition e.

4 Structural identifiability of system (1)

Consider the system S_{φ} and properties $\mathrm{I}_{N,g}$ that allow solving the problem of structural identification (h-identifiability).

Let the following conditions be satisfied.

B1. The input $u(t)$ is constantly excited at the interval J.

B2. The input $u(t)$ ensures an informative framework S_{ey}.

Definition 1. If $u(t)$ satisfies B1 and B2 conditions, then the input $u(t)$ is representative.

Let the framework S_{ey} closed, and the area S_{ey} is not zero. Denote height S_{ey} as $h(S_{ey})$, where $h(S_{ey})$ is the distance between two points opposite sides of the framework S_{ey}.

Theorem 1. Let (i) the linear part of the system (1) is stable; (ii) the nonlinearity $\varphi(\cdot)$ satisfies the condition (2); (iii) the input $u(t)$ is bounded, and constantly excited; (iv) $h(S_{ey}) \ge \delta_S$, where $\delta_S > 0$. Then the framework S_{ey} is h-identifiable on the set $\mathrm{I}_{N,g}$.

Definition 2. The framework S_{ey} is called h-identifiable if theorem 1 holds for S_{ey}.

Let S_{ey} be h-identifiable. h-identifiability features consider in [6].

A "bad" input that is constantly excited exists. This input gives a so-called "insignificant" S_{ey}-framework ($\mathcal{NS}_{ey}$-framework). But the $\mathcal{NS}_{ey}$-framework can be h-identifiable. Identification the system (1) with $\mathcal{NS}_{ey}$-framework gives results not typical in this case.

Give the conditions for the existence of the framework. Consider a class of nonlinear functions to which the homotopy operation is applies [10]. Let $S_{ey} = \mathcal{F}^l_{S_{ey}} \cup \mathcal{F}^r_{S_{ey}}$, where $\mathcal{F}^l_{S_{ey}}, \mathcal{F}^r_{S_{ey}}$ are left and right fragments of S_{ey}. Define secants for $\mathcal{F}^l_{S_{ey}}, \mathcal{F}^r_{S_{ey}}$

$$\gamma^r_S = a^r y, \gamma^l_S = a^l y, \qquad (6)$$

where a^l, a^r are estimations obtained least-squares method (LSM).

Theorem 2. Let (i) the framework S_{ey} is h-identifiable and has the form $S_{ey} = F^l_{S_{ey}} \cup F^r_{S_{ey}}$, where $\mathcal{F}^l_{S_{ey}}, \mathcal{F}^r_{S_{ey}}$ are the left and right fragments of S_{ey}; (ii) secants for $\mathcal{F}^l_{S_{ey}}, \mathcal{F}^r_{S_{ey}}$ have forms (6), where a^l, a^r are LSM-estimations. Then S_{ey} is $\mathcal{NS}_{ey}$-framework if

$$\left\| a^l \right| - \left| a^r \right\| > \delta_h,\ \delta_h > 0. \qquad (7)$$

Remark 2. $\mathcal{NS}_{ey}$-framework is typical for systems with multiple-valued nonlinearity. Often the framework is the result of inadequate application of an input.

Consider the framework S_{ey}. Introduced the designations: $\mathcal{D}_y = \mathrm{dom}(S_{ey})$ is the definitional domain S_{ey}, $D_y = D_y(\mathcal{D}_y) = \max_t y(t) - \min_t y(t)$ is the diameter $\mathcal{D}_y$. The set U contains representative inputs.

Definition 3. If $\mathcal{D}_y$ of the framework S_{ey} has the maximum diameter D_y, then the input S-synchronizes the system (1).

Consider a reference framework S^{ref}_{ey}. S^{ref}_{ey} is the framework S_{ey} reflecting all properties of the function $\varphi(y)$. Designate by the diameter $D_y(S^{ref}_{ey})$ as D^{ref}_y. D^{ref}_y exists if the input the system (1) is S-synchronizing.

Definitions 2, 3 show if $S_{ey} \cong S^{ref}_{ey}$, then $\left| D_y - D^{ref}_y \right| \le \varepsilon_y$, where $\varepsilon_y \ge 0$, $\cong$ is the proximity sign. Elements of the subset $\mathrm{U_S}$ have property

$$\left| D_y\left(S_{ey}\left(u(t)\big|_{u \in \mathrm{U_S}}\right)\right) - D^{ref}_y \right| \le \varepsilon_y. \qquad (8)$$

We interpret the choice $u_h(t) \in \mathrm{U}$ as ensuring synchronization between structures of a model and the system. Therefore, the fulfillment of the condition $d_{h,y} = \max_{u_h} D_y$ gives h-identifiability of the system.

The conditions for h-identifiability and for the framework $\mathcal{NS}_{ey}$ have the form

$$\left| D_y\left(S_{ey}\left(u(t)\big|_{u \in \mathrm{U_S}}\right)\right) - d_{h,y} \right| \le \varepsilon_y, \qquad (9)$$

$$\left| D_y\left(S_{ey}\left(u(t)\big|_{u \in \mathrm{U \backslash U_S}}\right)\right) - d_{h,y} \right| > \varepsilon_y. \qquad (10)$$

Condition (13) can interpreted as the proximity domain

$$Q_D = \left| S_{ey}\left(u(t)\big|_{u \in \mathrm{U_S}}\right) - S^{ref}_{ey} \right|,$$

which is understood as $\left| \dot{y}(t) - \dot{y}^{ref}(t) \right| \le \varepsilon_y$ for almost $\forall t \ge t_0$.

We will write $\delta Q_D \le \varepsilon_y$ if frameworks are close. If the condition $\delta Q_D \le \varepsilon_y$ is satisfied for almost $\forall t \ge t^*$, then the region Q_D call the S-synchronizability region on the set $\{u_h(t)\}$ or the structural identifiability region on the set $\{S_h(u_h(t))\}$, where S_h is the phase portrait of the system (1).

So, we present two criteria (7) and (10) for the existence of the insignificant framework S_{ey}. The system S_φ structure and hence the system (1) is structurally unidentifiable in this case.

Definition 4. If S_{ey} is h-identifiable and conditions $\left\| a^r \right| - \left| a^l \right\| \le \delta_h$, (8) are satisfied, then the framework S_{ey} or the system (1) is structurally identifiable or h_{δ_h}-identifiable.

Definition 4 shows if the system (1) is h_{δ_h}-identified, then the framework S_{ey} has the maximum diameter of domain $\mathcal{D}_y$.

Let the framework S contain m singularities. We understand features of the function $\varphi(y)$ as loss of continuity on the interval I^j_y, inflection points or extremum. These features are the sign of the function nonlinearity.

Definition 5. The model (4) is SM-identifying if the framework S_{ey} is h_{δ_h}-identifiable.

Theorem 3 [8]. Let (i) the input $u(t) \in \mathrm{S}$ is constantly excited; (ii) the system (1) phase portrait have m features; (iii) S_{ey}-framework is h_{δ_h}-identifiable and contains fragments corresponding to features of the system (1). Then the model (4) is SM-identifying.

Theorem 3 shows if the model (4) is not SM-identifying then model (4) structure or the informational set (3) need to change.

Let c_S is the center of the framework S_{ey} on the set $J_y = \{y(t)\}$, c_{D_y} is the center of the domain $\mathcal{D}_y$.

Theorem 4. Consider the S-synchronizing set U_S of the system (1) and let (i) exists $\varepsilon \geq 0$ such that $|c_S - c_{D_y}| \leq \varepsilon$; (ii) $\| |a^l| - |a^r| \| \leq \delta_h$, where a^l, a^r are coefficients of secants for $(\mathcal{F}^l_{S_{ey}}, \mathcal{F}^r_{S_{ey}}) \subset S_{ey}$. Then the system (1) is h_{δ_h}-identifiable, and the input $u_h(t) \in S$.

Some subset $\{u_{h,i}(t)\} \subset U_S \subseteq U$ $(i \geq 1)$ which elements have property of S-synchronizability exists. The framework $S_{ey,i}(u_{h,i})$ with the diameter $D_{y,i}$ of the domain $\mathcal{D}_{y,i}$ corresponds to every $u_{h,i}(t)$. As $u_{h,i}(t) \in S$, the diameter $D_{y,i}$ has the property $d_{h,\Sigma}$-optimality.

Definition 6. The framework $S_{ey,i}$ has $d_{h,\Sigma}$-optimality property on the set U_h if $\varepsilon_\Sigma > 0$ such that $|d_{h,\Sigma} - D_{y,i}| \leq \varepsilon_\Sigma$ $\forall i = \overline{1, \# U_h}$.

Definition 7. If $(\{u_{h,i}(t)\} = U_h \subset U) \& (u_{h,i}(t) \in S)$, $i \geq 1$ and frameworks $S_{ey,i}(u_{h,i})$ have $d_{h,\Sigma}$-optimality property, then frameworks $S_{ey,i}(u_{h,i})$ are structurally indiscernible on sets $\{u_{h,i}(t)\}$, $J_y(u(t) = u_{h,i}(t))$.

Definition 8. If frameworks $S_{ey,i}(u_{h,i})$ have $d_{h,\Sigma}$-optimality property, then $S_{ey,i}(u_{h,i})$ is locally structurally identifiable on the set U_h.

Let the framework $S_{ey,i}(u_{h,i})$ having $d_{h,\Sigma}$-optimality property is $S^{\Sigma}_{ey,i}$, and the locally structurally identifiable framework $S_{ey,i}(u_{h,i})$ is $S^{LSI}_{ey,i}$.

Therefore, the framework S_{ey} is locally structurally identifiable on the set $U_h \subseteq U_S$ if

$$(\exists u_h \in S), \text{что} \left(S_{ey} \cong S^{\Sigma}_{ey}\right) \rightarrow S_{ey} \cong S^{LSI}_{ey}. \quad (11)$$

Remark 3. We have considered the case of symmetric nonlinearity. Therefore, remark and conditions for the existence of the $\mathcal{NS}_{ey}$-framework are valid. If a nonlinear function does not have the symmetry property, then the further study of this problem is required. Any nonlinearity has features and their use based on the analysis of a prior data.

Definition 9. The framework $S_{ey,i}(u_i \notin U_S)$ that does not have the $d_{h,\Sigma}$-optimality property is locally structurally unidentifiable.

Remark 4. The described approach applies to the nonlinear system with a dynamic law of the nonlinearity changing, where multi-level structure identifiability analysis used.

The identifiability of the system S_y considered in [11].

5 On constant excitation influence on system identifiability

In [11], the influence of the excitation constancy (EC) condition on the system identifiability estimation was shown. It is noted that not every input had the EC-property guarantees the structural identifiability of the system. We present results allowing us to evaluate the effect of this condition.

The property of excitation constancy for input $u(t)$

$$\mathcal{PE}_\alpha : u^2(t) \geq \alpha$$

for $(\exists \alpha > 0) \& (\forall t \geq t_0), t \in [t_0, T]$. We will write $u \in \mathcal{PE}_\alpha$.

Let the input $u(t) \in \mathcal{PEF}_{\alpha,\omega_k}$, $u_k \in U_k$, $U_k = U \backslash U_S$, where

$$\begin{gathered} u_k(t): \left(u \in \mathcal{PE}_\alpha\right) \& \left(u \in \mathcal{PF}_{\omega_k}\right) \& \left(\overline{u_h \in S}\right) \\ \mathcal{PF}_{\omega_k}: u_k(t) = \mathcal{RF}_k(\Omega_k) \end{gathered}, \quad (12)$$

$\mathcal{RF}(\Omega_k)$ is model based on a Fourier series, $\Omega_k = \{\omega_1, \omega_2, \ldots, \omega_k\}$ is a finite set of frequencies. Therefore, $u_k \notin S$.

We have for $u_h \in S$

$$\begin{gathered} u_h(t): \left(u_h \in \mathcal{PE}_\alpha\right) \& \left(u_h \in \mathcal{PF}_{\omega_h}\right) \& \left(u_h \in S\right), \\ \mathcal{PF}_{\omega_h}: u_h(t) = \mathcal{RF}_h(\Omega_h), \end{gathered} \quad (13)$$

where $\Omega_h \neq \Omega_k$.

Obtain from (12), (13)

$$\left(\mathcal{RF}_h(\Omega_h) \neq \mathcal{RF}_k(\Omega_k)\right) \Rightarrow S^h_{ey} \neq S^k_{ey} \Rightarrow S^k_{ey} = \mathcal{NS}_{ey}. \quad (14)$$

and from (14)

$$\left(\mathcal{D}_y\left(S^h_{ey}\right) \neq \mathcal{D}_y\left(S^k_{ey}\right)\right) \Rightarrow \left[D_y\left(S^h_{ey}\right) \geq D_y\left(S^k_{ey}\right)\right]. \quad (15)$$

The domains of the frameworks S^h_{ey}, S^k_{ey} do not coincide, and S^h_{ey} is $d_{h,\Sigma}$-optimal on the set U_h, then the fulfilment of condition (10) follows from inequality (15). The condition (15) shows that the system (1) nonlinear part structure with the input u_k does not coincide with structurally identifiable parameters with u_h.

Theorem 5. Let 1) the input u_k satisfy the condition (12); 2) the framework S^k_{ey} corresponds to the input u_k; 3) condition (13) held for $u_h \in S$; 4) conditions (14) and (15) held. Then the S_φ-system is structurally unidentifiable by input u_k, and structural parameters of the S_φ-system do not correspond to the system (1) with the identifiable framework S^h_{ey}.

Estimate the degree of non-identifiability of the S_φ-system. Let the phase portrait $\mathcal{S}$ constructed for the system (1). $\mathcal{S}$ and S_{ey}-frameworks have the same definitional domains. Therefore, the diameter $D(S_{ey})$ of the definitional domain S_{ey} is known. Consider the input set $\{u_i(t)\}$ had the property $\mathcal{PE}_\alpha$. Construct framework $S_{ey,i}$ and determine $D_{y,i}(S_{ey,i})$ for each $u_i(t)$. Let

$$d_{h,y} = \max_{u_i} \left| D_y\left(\mathcal{D}\left(S_{ey,i}\right)\right) \right|$$

and $d_{h,y}$ correspond to the input u_h. Determine diameters

$$d_{y,j} = \left| D_{y,j}\left(\mathcal{D}\left[S_{ey,j}\left(u_j \in \mathcal{U}\right)\right]\right) \right|$$

for all inputs $\mathcal{U} = \{u_i(t)\} \setminus \{u_h\}$. As $u_h \in \mathrm{S}$ then $d_{h,y} > D_{y,j}$ $\forall j \geq 1$. Then the degree of unidentifiable is

$$SI_j = SI\left(S_{ey,j}\right) = \frac{d_{h,y} - d_{y,j}}{d_{h,y}}. \quad (16)$$

We see that system (1) is structurally identifiable if $SI_j \to 0$. The structural identifiability domain Q_D determined by the condition (9).

Remark 5. Select fragments $\mathcal{F}_S^l, \mathcal{F}_S^r$ in the phase portrait $\mathcal{S}$. Obtain the non-identifiability degree estimate of the system as

$$SI = SI(\mathcal{S}) = \frac{d_y^l\left(\mathcal{F}_S^l\right)}{d_y^r\left(\mathcal{F}_S^r\right)},$$

where $d_y^l\left(\mathcal{F}_S^l\right)$, $d_y^r\left(\mathcal{F}_S^r\right)$ are diameters of the fragments $\mathcal{F}_S^l, \mathcal{F}_S^r$.

The SI of nonlinear systems may depend on the input amplitude. Modify the conditions (12), (13)

$$\begin{gathered} u_k(t): \left(u \in \mathcal{PE}_\alpha\right) \& \left(u \in \mathcal{PF}_{\omega_k}\right) \& \left(\overline{u_h \in \mathrm{S}}\right), \\ \mathcal{PF}_{\omega_k}: u_k(t) = \mathcal{RF}_k\left(G_k, \Omega_k\right), \end{gathered} \quad (12a)$$

$$\begin{gathered} u_h(t): \left(u_h \in \mathcal{PE}_\alpha\right) \& \left(u_h \in \mathcal{PF}_{\omega_h}\right) \& \left(u_h \in \mathrm{S}\right), \\ \mathcal{PF}_{\omega_h}: u_h(t) = \mathcal{RF}_h\left(G_h, \Omega_h\right), \end{gathered} \quad (13a)$$

where G_k, G_h are parameter vectors of models $\mathcal{RF}_k$, $\mathcal{RF}_h$.

Present the model $\mathcal{RF}_k$ and $\mathcal{RF}_h$ in the form

$$\mathcal{RF}_h\left(G_h, \Omega_h\right) = g_h \tilde{\mathcal{RF}}_h\left(\tilde{G}_h, \Omega_h\right),$$

$$\mathcal{RF}_k\left(G_k, \Omega_k\right) = g_k \tilde{\mathcal{RF}}_k\left(\tilde{G}_k, \Omega_k\right),$$

where $\tilde{\mathcal{RF}}_h\left(\tilde{G}_h, \Omega_h\right)$, $\tilde{\mathcal{RF}}_k\left(\tilde{G}_k, \Omega_k\right)$ are model modifications (12), (13); $g_h = \max_i g_{h,i}$, $i = \overline{1, \#\Omega_h}$, $g_{h,i}$ is element G_h; $g_k = \max_i g_{k,i}$, $i = \overline{1, \#\Omega_k}$. Designate g_p ($p = k, h$) the generalized input amplitude.

The condition (14) transformed into the form

$$g_h \tilde{\mathcal{RF}}_h\left(\tilde{G}_h, \Omega_h\right) \neq g_k \tilde{\mathcal{RF}}_k\left(\tilde{G}_k, \Omega_k\right).$$

As the input $u_h \in \mathrm{S}$, then $g_h \geq g_k$. This follows from

$$\begin{gathered} D_h\left(\mathcal{S}\left(u_h\right)\right) \geq D_k\left(\mathcal{S}\left(u_k\right)\right) \Rightarrow \\ \left|\tilde{\mathcal{RF}}_h\left(\tilde{G}_h, \Omega_h\right)\right| \geq \left|\tilde{\mathcal{RF}}_k\left(\tilde{G}_k, \Omega_k\right)\right|, \end{gathered}$$

and the model $\tilde{\mathcal{RF}}_h\left(\tilde{G}_h, \Omega_h\right)$ approximates the input ensured S-synchronization of the system (1).

Obtain the $d_{h,\Sigma}$-optimal the diameter $D_h\left(S_{ey}^h\right)$ from $\mathcal{S}\left(u_h\right) \Rightarrow S_{ey}^h$. As follows from (13), the framework S_{ey}^k does not have this property. Hence, $D_k\left(S_{ey}^k\right)$ gives the input $u_k \notin \mathrm{S}$, which has a smaller generalized amplitude.

Theorem 6. Let 1) the input u_k satisfies the condition (12a); 2) the framework S_{ey}^k corresponds to the input u_k; 3) such input $u_h \in \mathrm{S}$ exists that the condition (13a) held; Conditions (14) and (15) held; 4) $g_h \geq g_k$. Then (i) the S_φ-system is structurally non-identifiable by input u_k; (ii) structural parameters of the S_φ-system do not correspond to the system (1) with the identified framework S_{ey}^h.

Example. Consider a system with Bouc-Wen hysteresis (system S_{BW}) [12]

$$m\ddot{x} + c\dot{x} + F(x, z, t) = f(t), \quad (17)$$

$$F(x, z, t) = \alpha k x(t) + (1 - \alpha) k d z(t), \quad (18)$$

$$\dot{z} = d^{-1}\left(a\dot{x} - \beta|\dot{x}||z|^n \, sign(z) - \gamma \dot{x}|z|^n\right), \quad (19)$$

where $m > 0$ is weight, $c > 0$ is damping, $F(x, z, t)$ is the restoring force, $d > 0$, $n > 0$, $k > 0$, $\alpha \in (0,1)$, $f(t)$ is exciting force, a, β, γ are some numbers, $y = x$ is the system output. Denote by the system (17) - (19) as S_{BW}.

Let $n = 1.5$, $c = 2$, $m = 1$, $\beta = 0.5$, $\alpha = 0.7$, $k = 0.6$, $d = a = 1$. Set of data have the form $\mathrm{I}_o = \{u(t), y(t), t \in [0; t_e]\}$, $t_e > 0$. Consider inputs

$$\begin{gathered} u_0(t) = 2 - 2\sin(0.15\pi t), u_1(t) = 2 - 2\sin(0.35\pi t), \\ u_2(t) = 2 - 2\sin(0.5\pi t), \\ u_3(t) = 2 - 2\sin(0.15\pi t) + 0.2\sin(0.35\pi t). \end{gathered} \quad (20)$$

Design phase portraits S_i for the S_{BW}-system for each u_i and determine the diameters

$$D_{y,0}(S_0)=3.75, D_{y,1}(S_1)=1.728, \\ D_{y,2}(S_2)=1.08, D_{y,3}(S_3)=3.967. \quad (21)$$

The analysis gives $u_0(t)\in \mathrm{S}$. Therefore, S_0 is the reference and $d_{h,y}=D_{y,0}(S_0)$. Calculate the non-identifiability degree of the S_{BW}-system for various u_i

$$SI_1=0.549,\ SI_2=0.718,\ SI_3=-0.035. \quad (22)$$

Inputs u_1,u_2 give the structurally unidentifiable S_{BW}-system. S_{BW}-system with u_3 is structurally indistinguishable (Fig. 1).

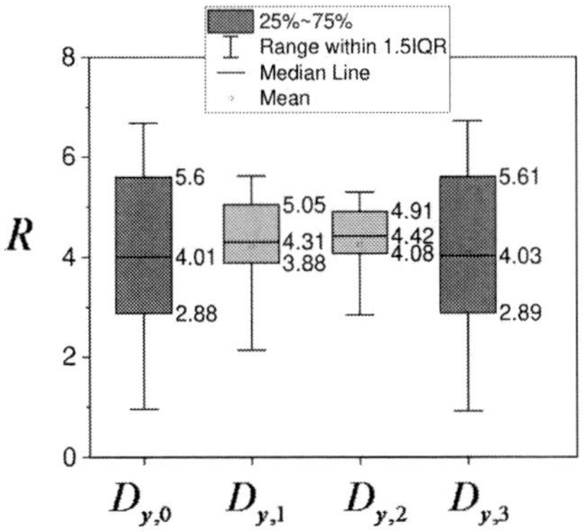

Fig. 1. Diameter S_i estimates in the ranked space

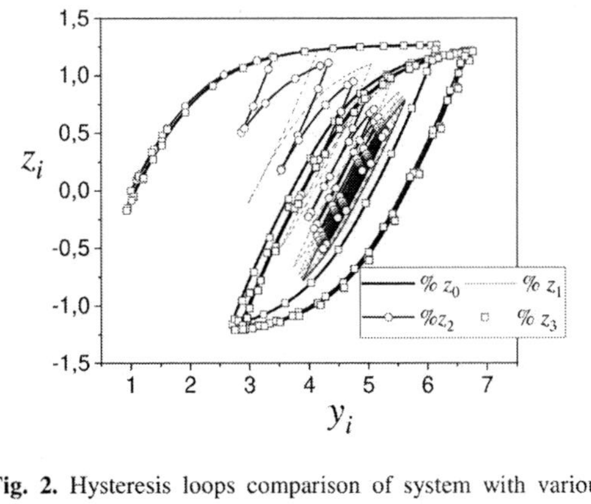

Fig. 2. Hysteresis loops comparison of system with various inputs

Analysis of (22) and Fig. 1 shows that structures $S_{ey,1}(u_1)$, $S_{ey,2}(u_2)$ are structures $\mathcal{NS}_{ey}$, and $S_{ey,3}(u_3)$ structure belongs to class S_{ey}^{LSI}. R is the range of the diameter distribution in Fig. 1

We see that the frequency properties of the input significantly affect the identifiability of the system. It is for the nonlinear part of the system where the change in input properties affects the estimate to structural parameters of the hysteresis. This effect presented in Fig. 2, where the output of the Bouc-Wen model (19) shown for various u_i.

The identifiability area Q_D showed in Fig. 3 for the S_{BW}-system. It confirms the made conclusions.

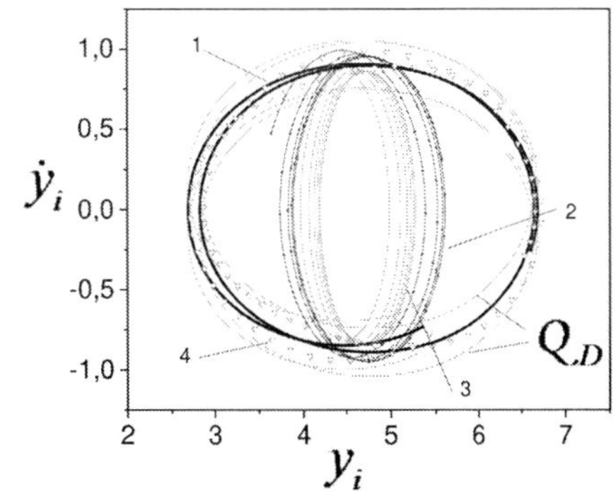

Fig. 3. Area Q_D of structural identifiability

So, we propose the method for quantifying estimate the degree identifiability of the system (1). The area of system identifiability obtained.

6 Conclusion

The concept of S-synchronizability of the input was introduced. Meeting the input S-synchronizability condition is the basis for structural identifiability and structural identification of the nonlinear system. The input influence on the estimating possibility of the structural parameters nonlinear system has been studied and conditions for the local identifiability and non-identifiability of the system have been obtained. The identifiability degree the system under uncertainty has been introduced. The influence of the constant excitation of input on the possibility of the system structural identifiability has been studied. The method for calculating the identifiability domain under uncertainty has been proposed.

References

1. N.A. Balonin *Theorems of identifiability* (St. Petersburg: Politekhnika publishing house, 2010)
2. J.D. Stigter, R.L. M.Peeters, Proceedings of the ECC 2007 Kos, **WeC09.4**, 3450-3456 (2007) ISBN: 978-960-89028-5-5
3. O.-T. Chis, J.e R Banga, E. Balsa-Canto, PLOS ONE, **6**(4) (2011)
4. M.P. Saccomani, K. Thomaseth, *Dynamics of mathematical models in biology. Ed. A. Rogato, V. Zazzu, M. Guarracino* (Springer, 2010) ISBN 978-3-319-45722-2

5. A. Villaverde, A. Barreiro, A. Papachristodoulou, PLOS Computational Biology, **12**(10) (2016).
6. N. Karabutov, Global Journal of Science Frontier Research: (A) Physics and Space Science, **18**(11) (2018).
7. N. Karabutov, International journal of intelligent systems and applications, **8**(7) (2016).
8. N. Karabutov, *In Nonlinearity problems, solutions and applications. Volume 1. Ed. L.A. Uvarova, A. B. Nadykto, A.V. Latyshev* (New York. Nova Science Publishers, Inc., 2017)
9. N. Karabutov, International journal of intelligent systems and applications, **7**(9) (2015)
10. G. Choquet, *L'enseignement de la geometrie* (Paris: Hermann, 1964)
11. N.N. Karabutov, Mekhatronika, Avtomatizatsiya, Upravlenie, **21**(6) (2020)
12. M. Ismail, F. Ikhouane, J. Rodellar, Arch Comput Methods Eng, **16** (2009)

Analysis of Tear Fluid for Evaluation on the Concentration of the Generic Anti-Glaucoma Drug Betoptik using Multisensory Stripping Voltammetry and Multivariate Statistics

I. I. Kolesnichenko[1], *L. M. Balashova*[2,3]

[1]Frumkin Institute of Physical Chemistry and Electrochemistry, Russian Academy of Sciences, RU-119071, Moscow, Russia
[2]Pirogov Russian National Research Medical University, RU-117997, Moscow, Russia
[3]International Research and Practice Center for Tissue Proliferation, Non-commercial Partnership, RU-119034, Moscow, Russia

Abstract. It is important to monitor the concentration of drugs after their instillation: this helps to find the right drug dosages. The proposed method is based on recognizing multidimensional images through the use of multidimensional statistics; the images are produced by multisensory electrochemical systems that provide reliable information about the tested objects. The measurements were carried out using a planar solid-state electrode. The test-system solution was a 0.05 M KCl solution, which contained the Zn2+, Cd2+, Pb2+, Cu2+ Co2+, and Hg2+ metal cations at the concentration of 5 10–5 M. A new electrochemical method for multisensory stripping voltammetry has been used to determine whether it is appropriate for determination of generic forms of the drug Betoptic, that is, Xonef, Betoftan, and Betalink EU in tear fluid. Measurements were carried out on a planar three-electrode interdigitated electrode. It has been shown that this method is effective for determination of the generics. The dynamics of changes in their concentrations over time were examined. After instillation of the generics in glaucoma patients, their concentrations in the Tear fluid remained unchanged within 12 h. The list of organic substances that can be determined by the method of multisensory stripping voltammetry has been extended.

1 Introduction

Composition of blood is always critically important. Its resilience has been studied in some detail including the active reaction (Ph), the ratio of electrolytes, glucose, and number of the formal elements and others. Due to the intensive development of high-tech methods for research in the modern world, there already exist large diagnostic capabilities. However, a large amount of information we always need to get is often unavailable. This raises the question about meaningful etiopathogenic using these techniques, especially in ophthalmology, because as of today there is no simple, informative and pathognomonic method for early diagnosis of ophthalmological diseases. Data needed for the studying of the effectiveness of these techniques are quite limited, and their efficiency in diagnosing different ophthalmic pathologies practically is not well studied that makes it difficult to determine their role. The medical and social importance of such pathologies as glaucoma, myopia, retinopathy and AMD (agerelated maculardegeneration) are indisputable so as and the early diagnostic of these diseases. Despite the numerous methods of diagnosis and treatment of these deceases, blindness and a high level of disability resulting from these ophthalmic pathologies are still a global problem. In this work, we propose an innovative method which is neither expensive and nor time consuming and easy to use. As a primary method, we use multisensory stripping (thick-film electrode) voltammetry (IVA) to determine the status of the homeostasis of the organism as a whole, assessing whole blood or serum and local homeostasis , appreciating the tear fluid of the anterior chamber of the eye and the vitreous. This allows tracng the evolution of biological fluids depending on the evolution of the pathological process. Built on the express-diagnostics of biological fluids, the method will be used for the preliminary assessment of various conditions. A method for recognition of multidimensional images is derived from analysis of multisensory stripping voltammetry electrochemical test system in the format of a "language".

Effective diagnostics through early disease detection implies simple procedures for initial patient examination, thereby reducing time and labor costs. In the line of the concept of metabolic profile, clinical chemistry is mostly concerned with analysis of body fluids. When conducting a comprehensive analysis of clinical chemistry is mainly limited to studies of the composition of the fluids of the human body [1].

A new approach to the analysis of volatile compounds is based on the use of "electronic nose" devices [2-4]. When using "electronic nose" and "electronic tongue" [5] each analyzed substance corresponds to some pattern of feedback system that enables multidimensional discriminate against various substances (or mixtures of substances) by comparing them to images formed in advance N-dimensional image database [6]. A fairly complete list of models "electronic nose and electronic tongue" is presented in [7], [8]. A number of foreign and domestic firms made bids for Annex developed analytical tools for evaluating the quality of foodstuffs [9] Wednesday, environmental monitoring [10], as well as for use in medical diagnosing [11].

The effective dosage of a new ophthalmic medicine is determined from the concentration-time profile of the drug in lacrimal fluid. Many analytical problems are solved by electrochemical methods which are quite simple, quick, and effective techniques for reliable quantitative analysis. The method of multisensory stripping voltammetry (MSV) for analysis of organic substances was developed at the Frumkin Institute of Physical Chemistry and Electrochemistry, Russian Academy of Sciences. This method is used to detect changes in the electrochemical activity of metal cations in solutions; the changes are due to their interaction with organic substances [12-19]. Metal cations (components of the test system) that are able to form complex compounds with organic substances are introduced into a 0.05 solution of KCl. The introduction of these metal cations is accompanied by peak dissolution currents on voltammograms: metal cations interact with organic compounds and resulting changes in metal dissolution currents are reflected on stripping voltammograms. These changes are evaluated either individually for each metal of the test system or integrally. Measurements are made on one indicator electrode, which has a definite advantage over those multisensor methods that use a set of electrodes. The method involves using a pre-created database to identify organic substances. It proved to be effective when determining the concentration of ophthalmic betaxolol [20-22]. In tear fluid. It was determined using a calibration function. We used the MSV method to build the concentration-time profile of betaxolol dissolved in tear fluid. It was shown that the concentration of betaxolol in tear fluid remains constant for 12 hours after instillation.

Glaucoma is one of the most severe forms of ocular pathology. After cataracts, glaucoma is the worldwide leading cause of blindness and low vision. In the Russian Federation, the share of glaucoma is 29% in the nosological structure of blindness and poor vision. The prevalence of glaucoma among the Russian population has increased by 2.8 times in recent years [23]. Glaucoma ranks first in the structure of primary visual disabilities in most regions of the Russian Federation. The main effect of most anti-glaucoma drugs is associated with reduction of intraocular pressure (IOP). One of the common IOP-lowering medications used as monotherapy or in combination with other drugs is Betoptic (betaxolol). Its chemical name is (2RS)-3-(2-propylamino)-1-[4-[2-cyclopropylmethoxy) lopropylmethoxy)ethyl]phenoxy]-propane-2-ol hydrochloride. The pharmacotherapeutic group is antiglaucoma medication; it is a selective beta-1-adrenoblocker. The pharmacological action of the betaxolol drug is selective beta-1-adrenoblockade without internal sympathomimetic activity; the drug does not have a membrane-stabilizing (local anesthetic) effect. With topical application, betaxolol reduces both elevated and normal intraocular pressure by reducing the production rate of ocular fluid. The onset of the hypotensive effect is observed in 30 min after instillation, and the maximum ophthalmotonus decrease occurs in approximately 2 h, according to the instructions for use of the drug. After a single instillation, the effect on ophthalmotonus persists for 12 h. Betaxolol does not decrease blood flow in the optic nerve compared toother beta-blockers. In addition, it does not cause miosis, spasm of accommodation, hemeralopia, or the effect of a "veil" over the eyes (in contrast to miotics). Regarding its pharmacokinetics betaxolol is highly lipophilic, resulting in good penetration through the cornea into the anterior chamber. The maximum concentration in the anterior chamber is observed 20 min after instillation. With topical use systemic absorption is low, the plasma concentrations is below the detection threshold (2 ng/mL). The drug is excreted mainly through the kidneys.

The popularity of betoptic has led to a set of its generics. The advantages of generics are low cost (due to the lack of costs on the elaboration, clinical trials and licensing fees), as well as the available results of clinical trials and accumulated experience of practical application of the drug (active ingredient) in the form of the original drug, which is copied by the generic versions. Despite the fact that generics are not original drugs, they must comply with requirements, i.e., they contain a ratio of active pharmaceutical ingredients, that of the qualitative and quantitative composition of their brand-name counterparts. It is necessary that the generic meets the requirements of the established standards. The production must comply with proper manufacturing practices. The generic must be equivalent to the original product in its degree and rate of absorption. A drug frequently has several analogues, but only those produced by different pharmaceutical companies. Therefore, the composition, and hence, their efficacy may differ. In addition, not all manufacturers of generic drugs comply to the requirements for generic production [24]. The Frumkin Institute of Physical Chemistry and Electrochemistry has developed a version of an electrochemical method for the analysis of organic substances: a multisensory stripping voltammetry method. This method measures the change in the electrochemical activity of metal cations in solution upon their interaction with organic substances [25, 26]. Metal cations, which have the ability to form complex compounds with organic substances, are introduced in solution of 0.05 M KCl (test system). These metals reveal the peaks of the dissolution currents on voltammograms. When a sample with the analyzed substance is introduced into the solution of a test system, metal cations interact with organic compounds, and their influence on the metal dissolution currents is recorded on a stripping voltammogram. These changes are determined for each of the metals of a test system or an integral evaluation of a change of the entire voltammogram is performed. Measurements are performed on a single indicator electrode, which has a significant advantage compared to multi-sensor methods using a range of electrodes. This method allows the identification of organic substances using a preliminary database.The effectiveness of this method has been shown for several ophthalmic medications: Visomitin, Betoptic, and Lanomax [27–30]. The concentration of Visomitin in the lacrimal fluid was measured based on a calibration relationship. The dynamics of changes in the Visomitin concentration in tear fluid over time were studied by multisensory stripping voltammetry; it was shown that its concentration significantly decreased in 2 h. Similarly, a possible change in the concentration of the Betoptic in the lacrimal fluid was investigated; it was shown that its concentration did not change 12 hours after instillation. The purpose of this study was to investigate the pharmacokinetics of generic forms of Betoptic, that is, Xonef, Betoftan, and Betalink EU using a multisensory stripping voltammetry method.

2 Materials and methods

The analyzer embodies a new approach using a solution containing various metal ions as an electrochemical test-system, the ions being able interact with organic substances to form complex compounds [21-31]. All known E-tongue devices use a set of electrodes as sensors, therefore the reproducibility of characteristics of the multi-sensor systems is problematic, which affects the reliability and pattern recognition in a negative way. We propose enhancing the reliability and pattern recognition of our multi-sensor stripping voltammetry method through the use of metal cations, dissolved in the background electrolyte, as sensors. One electrode is used to record voltamperograms by the stripping voltammetry technique to obtain multidimensional patterns of analytes.

The test-system (background electrolyte) is a 0.05 M KCl solution containing the metal cations such as Zn2+, Cd2+, Pb2+, Cu2+ Co2+, Hg2+ as sensors. Thus an array of sensors is formed on one electrode allowing the deviceto be used as both E-tongue. The measurements have been done using planar electrodes (Autocom, Moscow), which represent polypropylene plates with working and auxiliary electrodes (carbon material) and silver-chloride reference electrode (Brown electrode) The voltamperograms (sensor outputs) recorded by the method of stripping voltamperometry are characterized by specific features and good reproducibility. The appearance of the voltamperograms change when different organic substances are added to the test-system. Each of the organic substances (or mixture of substances) produces specific changes, which allows their discrimination.

Measurements were made on a planar electrode. The test system solution consisting of 0.05 M KCl contained cations of metals Zn2+, Cd2+, Pb2+, Cu2+ Co2+, and Hg2+ in a concentration of 5·10−5 M. First, the test system solution was applied to the indicator electrode and a background stripping voltammogram was obtained. The potential of cathodic metal deposition was set to −1.55 V (with respect to the silver chloride electrode), which was followed by a potential sweep of up to 0.3 V (Fig. 1, curve 1). In this work, the change in concentration has been studied in the tear fluid after instillation of the following Betoptic generic drugs: Xonef (Sentiss Pharma Pvt. Ltd., India), Betoftan (ROMPHARM Company, Romania), and Betalink EU (Unimed Pharma, the Slovak Republic).

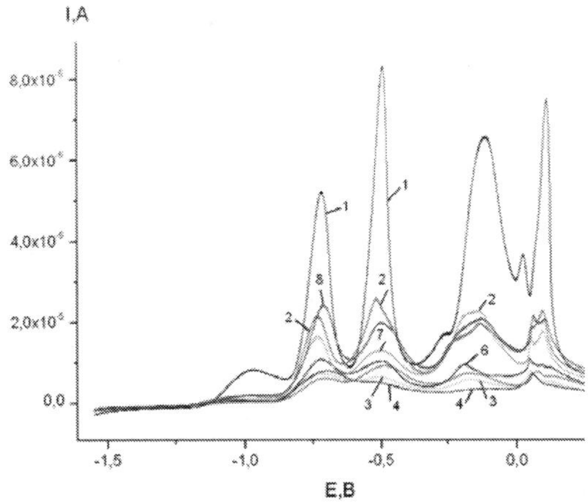

Fig.1. Stripping voltammograms of the test system with Betoftan in different concentrations: (*1*) test system; (*2*) lacrimal fluid of healthy people; (*3*) 4 × 10–3% of Betoftan; (*4*) 3 × 10–3% of Betoftan; (*5*) 2 × 10–3% of Betoftan; (*6*) 1 × 10–3% of Betoftan; (*7*) 0.5 × 10–3% of Betoftan; (*8*) 0.1 × 10–3% of Betoftan

Betoptic has a much more complex range of excipients, although the concentration of the active ingredient in all drugs is the same (0.5 mg of betaxolol hydrochloride). To conduct tests, 50 μL of a test system consisting of 0.05 M KCl and a range of metal cations (Zn^{2+},Cd^{2+}, Pb^{2+},Cu^{2+} ,Co^{2+}, Hg^{2+}) at a concentration of 5 × 10–5 M for each cation were loaded at planar electrodes (OOONPP Avtokom, Moscow); stripping voltammetry at a athodic potential of –1.55 V to deposit the metal ion on the electrode was performed with respect to a silver chloride reference electrode potential, in the positivegoing sweep of 0.3 V; the resulting voltammogram was the background (Fig. 1, curve *1*). The test system was removed from the electrode surface using filter paper. To study the effect of generics on the test system, 50 μL of fresh solution of the test system and a disk of porous material (filter paper) with the generic drug in the required concentration were placed on the electrode and a stripping voltammogram was again recorded. To study the effect of tear fluid on the test system, an 8-mm diameter disc of pre-treated porous material (filter paper, which had been thoroughly rinsed in alcohol and distilled water and dried at a temperature of 80°C for 3 h) was placed into the conjunctival sac of the eye. Next, 50 μL of the test system was applied again onto the electrode and a paper disk with tear fluid extracted from the conjunctival sac was placed on the electrode. A stripping voltammogram of the lacrimal fluid was recorded under the same conditions as the background curves. While investigating the effect of generics on the lacrimal fluid after instillation, the generic solution was administered into the conjunctival sac and a disk of filter paper was then placed there. After a certain time, the paper disc was removed from the conjunctival sac and placed on an electrode with a test system applied to it. Measurements were carried out after 30 min, 1 h and then every hour for 12 h after instillation of the drug.

3 Results and discussion

The effect of betoftan in different concentrations on the test-system. Figure 1 presents stripping voltammograms recorded when subjecting the test system to different concentrations of Betoftan. It is seen that with increasing drug concentration the lead currents decreased: from 19 to 12 μA (φ = –0.53 V) and cadmium from 20 to 11 μA (φ = -0.75 V). The currents of zinc, cobalt, and mercury almost did not change. Other generics with the same content of the active ingredient (betaxolol hydrochloride, 5 mg): Xonef and Betalink EU, demonstrated a similar effect on the test system. Thus, it has been shown that the method of multisensory stripping voltammetry can be used to measure the concentration of the generics in a test system.

The effect of the lacrimal fluid from healthy people on the test system. After recording the background curves using the above method (Fig. 2, curve *1*) a voltammogram of tear fluid was recorded under the same conditions. Figure 2 (curve *2*) shows the corresponding averaged voltammogram of tear fluid obtained from ten healthy people. The individual spread of voltammograms was 12%.

The time-dependent content of betoftan in the lacrimal fluid after instillation. One drop of Betoftan was in stilled into the conjunctival sac of the eye and an 8 mm-disk of filter paper was placed there. After certain periods of time, the paper disk was removed from the conjunctival sac and placed on the electrode with the test system applied to it. Measurements were carried out after 30 min, 1 h, and then every hour for 12 h after instillation of the drug. It has been found that the dissolution currents of metals did not change at the time of exposure to Betoftan from 30 min to 12 h. The corresponding voltammogram is shown in Fig. 2 (curve*3*). The results show that the concentration of Betoftan in the tear fluid did not change over 12 h.

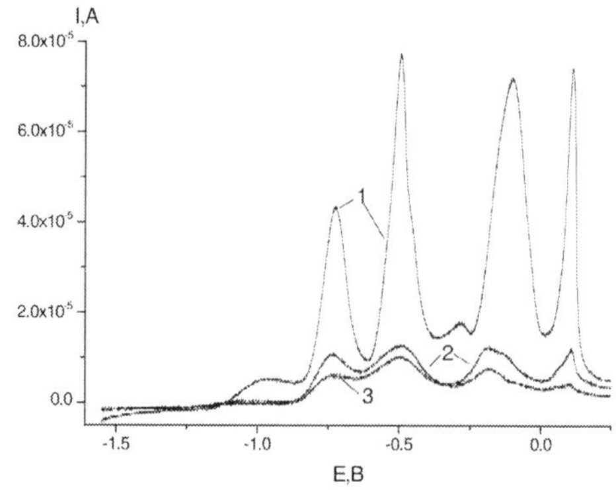

Fig. 2. Stripping voltammograms showing the effect of lacrimal fluid and Betoftan on the test system on the test system: (1) test system, (2) lacrimal fluid, (3) tear fluid after instillation of Betoftan.

Therefore, the drug is quite stable over time and continues to have an effect on the tear fluid 12 h after instillation, as well as on the cornea of the eye. The time-dependent concentrations Xonef, and Betalink EU were investigated in a similar way in the lacrimal fluid after instillation.

Table 1. The effects of the Betoptic and its generics with the same concentration of the active ingredient on the test system after instillation

Lacrimal fluid	Peaks of metal dissolution currents, μA					
	Zn	Cd	Pb	Cu	Co	Hg
Healthy people	1,6	14,6	14,0	13,2	13,0	12,8
Patients with glaucoma after instillation of Betoptic	1,6	11,2	12,4	10,0	5,6	4,6
Patients with glaucoma after instillation of Betoftan	2	10	13	7,3	6,1	5,5
Patients with glaucoma after instillation of Xonef	2	9,5	11,5	10,4	5,0	4,2
Patients with glaucoma after instillation of Betalink	2	11	12	9,5	6.0	5

Table 1 shows the effect of the Betoptic and its three generics with the same concentration of the active ingredient on the test system (betaxolol hydrochloride, 5 mg) 12 h after instillation. The same effect of the generics with the same concentration of the active ingredient on the test system indicates that betaxolol, rather than auxiliary components included in the formulations, has the major effect on the test system. It is noted that use of the generics caused burning and tingling sensations in the conjunctiva in the studied subjects upon instillation.

The dissolution currents of metals decrease 4–7 times. This is due to the fact that organic substances of tear fluid are complexing agents. .

4 Statistical Data Processing

A Voltamperogram is Multidimensional and divided into N-dimensional image, the amount of which depends on the complexity of the investigated system. Thus formed N-dimensional image object.

Data Dimensionality Reduction. The next step is to group processing correlated sensors into clusters by reducing the dimensionality of data with the subsequent allocation of M bulk of characterizing the cluster information. M varied to achieve maximum discriminatory ability.

There are many algorithms to reduce the dimensionality of the data, some of which were tested in the study: **Image of Primary Data** The result of the operation is voltamperogram (Fig 2)—current dependence of electrode potential, changing on linear time (sweep rate capacity). The peaks correspond to oxidation of metals test systems. Digital values (453 points) coming into the program for further processing in the form of a file from which to retrieve the value of the current. A voltamperogram factor analysis, principal component analysis, cluster analysis, Fourier transform [32].

5 Discrimination Algorithm

The purpose of discriminatory analysis is attributing the object under study to a certain group/class by its characteristics in an optimal way, minimizing expected losses and the probability of misclassification. Discriminatory analysis, as part of statistical multi-parameter analysis, comprises methods for classification of multidimensional objects through the use of available training samples.

Training samples were obtained by measuring parameters of known analytes to be subsequently stored in the database. A similarity measure is the square of the Mahalanobis distance, which shows how close the object is to the centroids of the groups of known samples in the space of canonical roots. The results of the discrimination are visually presented as a graph in the space of the first and second canonical roots, the known analytes being depicted as clouds of points surrounded by ellipses and the sample under study as a point.

Thus, the discriminatory analysis allows the objects under study to be classified by a set of group/class attributes The values of current were recorded in the database, assigned to one of three groups: "TEAR", "BET", or "TEST", indicating presumably healthy patients, or glaucoma patients, or the test-system (reference sample), respectively.

The statistics software package Statistica 7.0 was used for data processing.

6 Conclusions

Here we have shown that

(1) It has been shown that compared to a voltammogram of Tear fluid from a healthy person, the effects of Betoptic, Xonef, Betoftan, and Betalink were mainly observed for mercury and cobalt: the dissolution currents of these metals decreased from 12 to 6 μA for cobalt and from 12 to 5 μA for mercury.

(2) The studied drugs are quite stable over time and continue to exert an effect on the Tearl fluid 12 h after instillation.

(3) The use of the generic forms of Betoptic (Xonef, Betoftan, and Betalink EU) caused some minor discomfort upon instillation in the form of the burning or tingling sensation in the area of the conjunctiva.

(4) The method of multisensory stripping voltammetry allows one to analyze the time-dependent effect of various ophthalmic drugs on the Tear fluid in the Electronic language format.

References

1. A. D'Amico, C. Di Natale, R. Paolesse, et al., Sensors and Actuators B: Chemical, **130**, 458-465 (2008) http://dx.doi.org/10.1016/j.snb.2007.09.044
2. D. Pickel, G. Manucy, D. Walker, Applied Animal. Behaviour Science, **89**, 107-116 (2004) http://dx.doi.org/10.1016/j.applanim.2004.04.008
3. C.M. Willis, S.M. Church, C.M. Guest, et al., British Medical Journal, **329**, 712 (2004) http://dx.doi.org/10.1136/bmj.329.7468.712
4. M. McCulloch, T. Jezierski, M. Broffman, et al., Integrative Cancer Therapies, **5**, 30-39 (2006) http://dx.doi.org/10.1177/1534735405285096
5. P. Pelosi, K.C. Persaud, In Sensors and Sensory Systems for Advanced Robotics. In: Dario, P., Ed., Springer-Verlag, Berlin, **361** (1988)
6. S. Ampuero, J.O. Bosset, Sensors and Actuators B: Chemical, **94**, 1-12 (2003) http://dx.doi.org/10.1016/S0925-4005(03)00321-6
7. D. James, S.M. Scott, Z. Ali, W.T. O'Hare, Microchimica Acta, **149**, 1-17 (2005) http://dx.doi.org/10.1007/s00604-004-0291-6
8. N.V. Dolgopolov, M. Yablokov, Peace and Security, **3**, 54 (2007) [in Russian]
9. M. Aleixandre, J. Lozano, J. Gutirerrez, et al., Sensorsand Actuators B: Chemical, **131**, 71-76 (2008) http://dx.doi.org/10.1016/j.snb.2007.12.027

10. O. Helli, M. Sia, Sensors and Actuators B, **103**, 403-408 (2004) http://dx.doi.org/10.1016/j.snb.2004.04.069
11. E.R. Thaler, Laryngoscope, **112**, 1533-1542 (2002) http://dx.doi.org/10.1097/00005537-200209000-00002
12. D. James, S.M. Scott, Z. Ali, W.T. O'Hare, Microchimica Acta, **149**, 1-17 (2005) http://dx.doi.org/10.1007/s00604-004-0291-6
13. N.V. Dolgopolov, M. Yablokov, Peace and Security. **3**, 54 (2007) [in Russian]
14. M. Aleixandre, J. Lozano, J. Gutirerrez, et al., Sensors and Actuators B: Chemical, **131**, 71-76 (2008) http://dx.doi.org/10.1016/j.snb.2007.12.027
15. O. Helli, M. Sia, Sensors and Actuators B, **103**, 403-408 (2004) http://dx.doi.org/10.1016/j.snb.2004.04.069
16. E.R. Thaler, Laryngoscope, **112**, 1533-1542 (2002) http://dx.doi.org/10.1097/00005537-200209000-00002
17. W.A. Groves, E.T. Zellers, G.C. Frye, Analytica Chimica Acta, **371**, 131-143 (1998) http://dx.doi.org/10.1016/S0003-2670(98)00294-3
18. C. Di Natale, A. Macagnano, E. Martinelli, et al., Biosensors and Bioelectronics, **18**, 1209-1218 (2003) http://dx.doi.org/10.1016/S0956-5663(03)00086-1
19. W. Cao, Y. Duan, ClinicalChemistry, **52**, 800-811 (2006) http://dx.doi.org/10.1373/clinchem.2005.063545
20. V.P. Lukovtsev, A.N. Doronin, N.V. Lukovtseva, V.A. Semenov, V.M. Ganshin, Electrochemistry, **45**, 869-872 (2009) [in Russian]
21. V.N. Andreev, V.M. Ganshin, A.N. Doronin, et al., Patent of the Russian Federation No. 2375705 (2008) [in Russian]
22. I.I. Kolesnichenko, V.M. Ganshin, A.L. Kluev, et al., Proceeding of 4th International Scientific and Practical Conference. Publishing House of the Polytechnic University, Saint-Petersburg, **189** (2013) [in Russian]
23. .Nesterov, Glaucoma (Mediform, Moscow, 2014)
24. I. B. Bondareva, Klin. farmokinet., **1**(2), 7 (2005)
25. I.I. Kolesnichenko, A.L. Klyuev, V.M. Ganshin,et al., Prot. Met. Phys. Chem. Surf., **50**(4), 543 (2014)
26. L.M. Balashova, I.I. Kolecnichenko, E.P. Kantarzhi, Klin. Gepontol., **23**(9–10), 6 (2017)
27. I.I. Kolesnichenko, A.N. Doronin, E.P. Kantarzhi, L.M. Balashova, Vestn. MGTU im. N.E. Baumana, Ser. Estestv. Nauki, **2**, 107 (2018)
28. L.M. Balashova, V.A. Namiot, I.I. Kolecnichenko,et al., Biophysics, **63**(4), 655 (2018)
29. I.I. Kolesnichenko, V.M. Ganshin, et al. Proceeding of 3th International Scientific and Practical Conference. Publishing House of the Polytechnic University, Saint-Petersburg, 164-173 (2011) [in Russian]
30. I.I. Kolesnichenko, L.M. Balachova, E.P. Kantarzhi, American Journal of Analitical Chemistry, **7**, 588-596 (2016)
31. L.M. Balashova, V.A. Namiot, I.I. Kolesnichenko etc. Biophysics, **65**(6), 1211-1218 (2020)
32. .A.A. Khalafyan, Statistica 6. Statistical Data Analysis. 3rd Edition, Moscow. (2007) [in Russian]

Simulation of the Processing Process in a Glow Discharge Plasma Generator

Vladimir A. Logvin[1,*], *Ludmila A.* Yvarova[2]

[1]Interstate educational institution of the higher education "Belarusian-Russian University", BL-212000, Mogilev, Republic of Belarus doctoral candidate Institute of design-technology informatics Russian Academy of Sciences Moscow, Russia
[2]Moscow State Technological University "STANKIN", RU-127055, Moscow, Russia

Abstract. A theory has been developed that explains the structure formation in a thin surface layer of a material after low-energy exposure to a glow discharge plasma. This makes it possible to design new methods, technologies and automated devices for the development of an automated technological environment for strengthening various types of tools from various materials. Based on the developed theory, a self-learning system with a developing database was created for monitoring all stages of processing in a glow discharge plasma generator. Any crystal structure can be classified as a complex nonlinear system. When studying the dynamic response of such systems to an external low-energy effect, it was shown that by the time when the nonlinear vibrations stop, the atoms of the crystal lattice are stabilized in new positions. Based on the purpose of the products to be processed in the glow discharge plasma generator, with taking into account their operating conditions, it is possible to increase their production life up to 2 times or more. These technologies should be used at the final stage of the production process because they do not lead to distortion of the shape and size of the working surfaces of products being manufactured. In this case, the residual stresses are redistributed in the surface layer of the material with the formation of an equal structure that improves the running-in conditions of mating parts of devices and mechanisms. For cutting and forming tools, the run-in stage is reduced to the stage of uniform wear.
Keywords: plasma, glow discharge, control system, automated technological environment, modeling.

1 Introduction

An urgent problem of mechanical engineering is to ensure stable growth in the quality of manufactured products and, thus, to increase their competitiveness. The development of new technologies that ensure the formation of the necessary physical and mechanical properties of the surface layer of tool materials, adapted for use in an automated technological environment, has become an increasingly urgent problem of modern tool production.

Energy-efficient methods of ion-plasma processing that improve the functional properties of tools and other crystal structures include the action of glow discharge plasma formed in controlled technological environments [1, 2]. Structural changes appear in the surface layer of materials under the influence of a wide range of energy flows. The energy flux of glow discharge plasma radiation particles that initiate structural transformations causes changes in the physical and mechanical properties of the surface material of the products being strengthened. Structural transformations occur in the surface layer of materials under the influence of a wide range of glow discharge plasma radiation energies, which cause changes in their properties.

The formation of the desirable functional properties is ensured by using self-learning systems with a developing database for monitoring all stages of product processing in the glow discharge plasma generator. Based on the analysis and evaluation of the quality of functioning of the real object to be processed, it is possible to ensure the stability of the quality of operational management and, thus, the formation of the specified properties of products being processed. The energy of the particles formed in the discharge must correspond to the energy that ensures the active formation of point defects in the material surface layers. Such technologies are based on nonlinear effects inherent in the potential energy of interaction between structural elements of crystal lattices. Based on this, their structuring is provided under the low-energy effect of glow discharge plasma radiation. It is important to establish the nonlinear characteristics of the processed materials. This makes it possible to provoke internal rearrangements of systems with the formation of new metastable stable structures with minimal energy impact. These structures can be spatial, temporal, or spatio-temporal. The tendency of open systems to form autonomous time and space-time structures as a result of external energy influence is identified as self-organization of systems [3-8].

The treatment of ready-to-use products with glow discharge plasma radiation makes it possible to optimize the advantages of the processes of interaction of incoming charged particles with the surface of various materials. This allows achieving the required properties

* e-mail: logvinvladim@yandex.ru

by varying particle energy, radiation dose, temperature of irradiated objects, materials of electrodes and medium formed in the interelectrode gap [1, 2]. The radiation of the plasma glow discharge has a wide range of ion flux in ions and their energy. This does not require special formation and concentration of ion fluxes. This process can be classified as irradiation by a chaotic flow of incoming ions of the crystal lattice of materials. Since the development of technologies based on active, long-range nanostructuring by glow discharge plasma irradiation depends on the processes occurring in the interelectrode gap, the physical nature of these processes should be taken into account. Thus, it is important to take into account the physics behind these processes, especially for those occurring in the cathode dark space.

The main objective of this work is to develop a theory that can simulate the functioning of a control system for fast-flowing processes in a glow discharge plasma generator. This will allows designing new methods, technologies and automated devices that can be used in automated technological environments.

2 Method of research

Controlled formation and burning of glow discharge in the plasma generator in controlled technological environments, used for processing various materials for the purpose of changing in the surface layer physical and mechanical properties is an important condition for efficient reproducibility of processing results. To simulate these processes let's use the position of quantum mechanics which consists in the, that any system can be described by setting in the general case a complex wave function of the form. Ability to detect a charged particle at the moment time t at some point near-cathode space closed volume the plasma generator with the radius vector is determined by the probability density, which is represented by the square of the modulus of the wave function. Probability of finding a particle in an elementary volume working near-cathode space the plasma generator at a point with a radius vector at time t can be represented as

$$P(\vec{r},t)d^2\vec{r} = C^2|\psi(\vec{r},t)|^2 d^3\vec{r}, \qquad (1)$$

where C – is the constant accepted from the condition

$$\int_{-\infty}^{+\infty} \psi^*(\vec{r},t)\psi(\vec{r},t)\ d^3\vec{r} = 1, \qquad (2)$$

where $\psi^*(\vec{r},t)$ – function complex-conjugate $\psi(\vec{r},t)$.

Changes in function time $\psi(\vec{r},t)$ on condition, that the particle is moving in potential $U(\vec{r},t)$ they are reflected by the non-stationary Schrodinger equation [9] of the following form

$$i\hbar\frac{\partial\psi(\vec{r},t)}{\partial t} = -\frac{\hbar^2}{2m}\Delta\psi(\vec{r},t) + U(\vec{r},t)\psi(\vec{r},t), \qquad (3)$$

where m – is the mass of the particle, $\hbar$ – Planck constant divided by 2π

$$\Delta = \frac{\partial^2}{\partial x^2} + \frac{\partial^2}{\partial y^2} + \frac{\partial^2}{\partial z^2} \text{ – operator Laplace.}$$

In the one-dimensional case, the Schrödinger equation is represented in the following form

$$i\hbar\frac{\partial\psi(x,t)}{\partial t} = -\frac{\hbar^2}{2m}\frac{\partial^2}{\partial x^2}\psi(x,t) + U(x,t)\psi(x,t). \qquad (4)$$

Provided that the potential does not depend on time, the solution of equation (4) can be performed in the following form

$$\psi(x,t) = \varphi(x,t)e^{-\frac{iEt}{\hbar}}. \qquad (5)$$

If we assume that the particle arrives in the state described by the wave function (5), it must have a certain amount of energy E. When substituting (5) in (4) and performing transformations, we have a stationary Schrodinger equation of the form

$$\left[-\frac{\hbar^2}{2m}\frac{\partial^2}{\partial x^2} + U(x)\right]\varphi(x) = E\varphi(x), \qquad (6)$$

where $-\frac{\hbar^2}{2m}\frac{\partial^2}{\partial x^2} + U(x)$ energy operator $\hat{H}$ or the Hamilton operator [7-10],

$\varphi(x)$ – eigenfunction of the Hamiltonian,

E_n – operator's own value $\hat{H}$.

As a result, equation (6) can be written as

$$\hat{H}\varphi(x) = E\varphi(x). \qquad (7)$$

Due to the fact that the operator $\hat{H}$ it can have n proper functions φ_n and n corresponding energy values E_n, if they coincide, we get degenerate States, and if there are many continuous values, a continuous spectrum is formed. The stable position of any particle is considered to be the state of equilibrium or when the particle has the lowest energy reserve E_1 [9-14].

When the potential is stable *U(x) =const*

$$\left[-\frac{\hbar^2}{2m}\frac{\partial^2}{\partial x^2} + U(x)\right]\phi(x) - E\phi(x) = 0$$

$$\frac{\partial^2\phi(x)}{\partial x^2} - \frac{2m}{\hbar^2}(U(x) - E)\phi(x) = 0 \qquad (8)$$

After performing the appropriate transformations, we get:

$$\lambda^2 - \frac{2m}{\hbar^2}(U(x) - E) = 0.$$

When solving the presented equation, two directions are possible:

When

$U(x)<E$, $U(x)-E < 0$

and

$$\lambda_{1,2} = \frac{\sqrt{2m}}{\hbar}i\sqrt{E - U(x)}, \qquad (9)$$

and when $U(x)>E$, $U(x)-E > 0$ in this case, the necessary effect may not be because the impact of a charged particle on the material of the working surface of the tool being strengthened will be minimal or close to zero. As a result, we will analyze the first direction in solving the problem and:

$$\phi(x) = C_1 \cos\frac{\sqrt{2m}}{\hbar}\sqrt{E - U(x)}x + +C_2 \sin\frac{\sqrt{2m}}{\hbar}\sqrt{E - U(x)}x,$$

at $x_{min} = 0$ $\phi(x_{min}) = (C_1 \cos 0 + C_2 \sin 0\ ) = 0, C_1 = 0,$

at $x_{max} = 0$

$$\phi(x_{max}) = \left(C_2 \sin\frac{\sqrt{2m}}{\hbar}\sqrt{E - U(x)}x_{max} = 0\,, \frac{\sqrt{2m}}{\hbar}\sqrt{E - U(x)}x_{max} = \pi n, E_n = U(x) + \frac{\pi^2 n^2 \hbar^2}{2m x_{max}^2}\right)$$

$$\phi_n(x) = C_2 \sin\frac{\pi n x}{x_{max}}$$

$$\Psi = \sum_{n=1}^{\infty} B_n e^{\frac{-i\left(U(x)+\frac{\pi^2 n^2 \hbar^2}{2m x_{max}^2}\right)t}{\hbar}} \sin\frac{\pi n x}{x_{max}},$$

where $B_n = \frac{2}{x_{max}}\int_0^{x_{max}} \Psi(0)\sin\frac{\pi n x}{x_{max}}dx$ – the Fourier coefficients.

$$\Psi(0) = \Psi(t = 0).$$

Taking into account the fact that the mass of incident ions involved in interaction with the processed tools in the glow discharge plasma generator is different and the mass distribution is also different, we will use the exponential distribution. Then the total mass of the particles involved in the interaction with the treated surface will be

$$\Psi_\Sigma \int_{m_{min}}^{m_{max}} \psi(m) f(m) dm.$$

After performing substitutions and transformations, we get

$$\Phi(x) = \frac{\sqrt{2m}}{\hbar}\sqrt{E - b - ax}(C_2 J_{\frac{1}{3}}(-\frac{\sqrt{2m}}{\hbar a}(E - b - -ax)^{\frac{3}{2}}) + C_1 J_{\frac{-1}{3}}(-\frac{\sqrt{2m}}{\hbar a}(E - b - ax)^{\frac{3}{2}})$$

where $J_{\frac{1}{3}}$ even and $J_{\frac{-1}{3}}$ odd - Bessel functions.

Thus

$$\Phi(x) = \frac{\sqrt{2m}}{\hbar}\sqrt{E - b - ax}(C_2 J_{\frac{1}{3}}(\frac{\sqrt{2m}}{\hbar a}\left(E - b - ax)^{\frac{3}{2}}\right) + C_1 J_{\frac{-1}{3}}(\frac{\sqrt{2m}}{\hbar a}(E - b - ax)^{\frac{3}{2}})$$

Taking into account the boundary conditions, we perform the substitution and get

$$\begin{cases} C_1 J_{\frac{-1}{3}}(\frac{\sqrt{2m}}{\hbar a}(E - b)^{\frac{3}{2}}) + C_2 J_{\frac{1}{3}}(\frac{\sqrt{2m}}{\hbar a}(E - b)^{\frac{3}{2}}) = 0 \\ C_1 J_{\frac{-1}{3}}(\frac{\sqrt{2m}}{\hbar a}(E - b - a x_{max})^{\frac{3}{2}} + C_2 J_{\frac{1}{3}}\left(\frac{\sqrt{2m}}{\hbar a}(E - b - a x_{max})^{\frac{3}{2}}\right) = 0 \end{cases} \quad (10)$$

If the solution is trivial, the main determinant of the system must be zero

$$\Delta = J_{\frac{-1}{3}}(\frac{\sqrt{2m}}{\hbar a}(E - b)^{\frac{3}{2}}) J_{\frac{1}{3}}(\frac{\sqrt{2m}}{\hbar a}(E - b - a x_{max})^{\frac{3}{2}}) - -J_{\frac{-1}{3}}\left(\frac{\sqrt{2m}}{\hbar a}(E - b - a x_{max})^{\frac{3}{2}}\right) J_{\frac{1}{3}}\left(\frac{\sqrt{2m}}{\hbar a}(E - b)^{\frac{3}{2}}\right) = 0. \quad (11)$$

Equation (11) is a condition for determining the E_n and its solution can be represented as

$$\Psi \sum_{n=1}^{\infty} e^{\frac{-iE_n t}{\hbar}} (B_{1n} J_{\frac{-1}{3}}(\frac{\sqrt{2m}}{\hbar a}(E_n - b - ax)^{\frac{3}{2}}) + +B_{2n} J_{\frac{1}{3}}(\frac{\sqrt{2m}}{\hbar a}(E_n - b - ax)^{\frac{3}{2}})$$

Then

$$E_n = (\frac{\mu_n \hbar a}{\sqrt{2m}})^{\frac{2}{3}} + b$$

For each E_n set the possible sets x_{max}

$$x_{max\,j} = E_n - b - (\mu_j(E_n)\frac{\hbar a}{\sqrt{2m}})^{\frac{2}{3}}.$$

This allows you to match xmaxj and their corresponding possible values En. Due to the fact that the arguments of the Bessel function are much larger than 1, it is more convenient to operate with dimensionless quantities, and for this we will use asymptotic formulas, which for our case will take the form

$$J_{\frac{1}{3}}(x) \approx \sqrt{\frac{2}{\pi x}} cos(x - \frac{5\pi}{12})$$

$$J_{\frac{-1}{3}}(x) \approx \sqrt{\frac{2}{\pi x}} cos(x - \frac{\pi}{12})$$

From equation (11), after substitutions, we obtain

$$\frac{cos(\frac{\sqrt{2m}}{\hbar a}(E-b)^{\frac{3}{2}} - \frac{\pi}{12})}{cos(\frac{\sqrt{2m}}{\hbar a}(E-b)^{\frac{3}{2}} - \frac{5\pi}{12})} = \frac{cos(\frac{\sqrt{2m}}{\hbar a}(E-b-a x_{max})^{\frac{3}{2}} - \frac{\pi}{12})}{cos(\frac{\sqrt{2m}}{\hbar a}(E-b-a x_{max})^{\frac{3}{2}} - \frac{5\pi}{12\hbar a})} \quad (12)$$

As a result of solving equation (12), we obtain the valuesE_n.

In the general case, we denote

$$Z = \frac{\sqrt{2m}}{\hbar a}(E_n - b)^{\frac{3}{2}}$$

$$\alpha = \frac{(E - b - a x_{max})^{\frac{3}{2}}}{(E - b)^{\frac{3}{2}}}$$

Making substitutions from (10), we obtain

$$\frac{J_{-\frac{1}{3}}(Z)}{J_{\frac{1}{3}}(Z)} = \frac{J_{-\frac{1}{3}}(\alpha Z)}{J_{\frac{1}{3}}(\alpha Z)} \qquad (13)$$

Solving equation (13) we find the first few roots for different α and Z. Then we can determine different values E_n for n=0, 1, 2....

From equation (12), performing transformations, we obtain

$$\frac{\sqrt{2m}}{\hbar a}(E-b)^{\frac{3}{2}} = \frac{\pi n (E-b)^{\frac{3}{2}}}{(E-b)^{\frac{3}{2}} - (E-b-ax_{max})^{\frac{3}{2}}}$$

As a result of the transformations carried out, a dependence was obtained for determining the energy of ions formed in the plasma generator of a glow discharge in accordance with their mass andE_n

$$(E-b)^{\frac{3}{2}} - (E-b-ax_{max})^{\frac{3}{2}} = \frac{\hbar a \pi n}{\sqrt{2m}}$$

3 Results and discussion

Controlled formation of the ion flux in the glow discharge plasma generator was made possible due to the obtained analytical dependences for determining the energy of charged particles in accordance with their mass. The formation of charged particles in the interelectrode gap with a certain mass and energy is random. Using the Schrodinger equation to solve this problem is most appropriate. As a result of modeling, it is established that depending on the technological medium used and the rate of its pumping, the necessary ion flow can be formed. This flow has a wide range of masses and energies, which is able to perform appropriate transformations in the surface layer of the processed products. The use of self-learning systems with a developing database for monitoring all stages of product processing in a glow discharge plasma generator ensures its controlled formation and Gorenje. In controlled technological environments in a plasma generator, glow discharge combustion is characterized by the highest stability of reproducible results.

As a result of electron microscopic study of the fine structure of materials exposed to glow discharge plasma radiation, it was found that low-energy exposure to metal samples leads to an increase in the dislocation density at a depth exceeding the projective range of ions. This effect is called long-range [15, 16].

A significant increase in the dislocation density is observed in well-annealed samples with a low initial dislocation density (α-Fe). In fig. 1, a shows the dislocation structure of α-Fe before irradiation, and Fig. 1b after low-energy exposure to glow discharge plasma. It was found that the effect of a glow discharge plasma causes an increase in the dislocation density to a depth of 8.19 mm from the irradiation surface, and the structure corresponds to a deeply deformed state of the material [17].

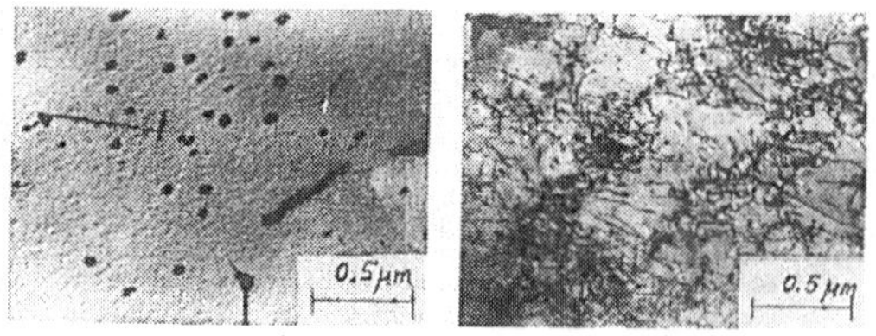

a) b)

Figure: 1. Dislocation structure of α-Fe: a - unirradiated state; b - after irradiation in a glow discharge at a voltage of 1 kV at a distance of 8.19 mm from the irradiation surface

Crystalline atomic structures can be classified as complex nonlinear systems. In the study of the dynamic response of such systems to an external low-energy impact, it was shown that by the time the nonlinear oscillations stop, the atoms of the crystal lattice are stabilized in new positions. The time period required for this stabilization is several orders of magnitude higher than the time of ordinary atomic relaxation. In this case, long-lived metastable structural states of the crystal lattice are formed. The type of the formed heterogeneous structures depends on the potential describing interatomic bonds and the magnitude of the external influence. The nonlinearity coefficients and additional states in multiwell potentials characterizing interatomic bonds depend on the type of potential or the properties of the irradiated material. Thus, for the Morse potential with the nonlinearity parameters established for α-Fe, it was shown that an external low-energy action causes the formation of solitons in nonlinear lattices moving along certain atomic chains [16]. Undamped pulsations during the motion of a soliton are transmitted to a depth exceeding the value of the projective path from the irradiated surface. In real crystal structures, dislocations are analogous to solitons. An increase in the dislocation density at sufficiently large distances from the irradiation surface is experimentally confirmed within the framework of the nonlinear theory. The displacement of atoms and their stabilization in the final state is the result of nonlinear collective vibrations of all atoms of the lattice after a low-energy impact. Therefore, by means of a weak external influence, it is possible to initiate internal processes in systems with the formation of new long-lived structural states. Thus, as a result, the crystal lattice represents an active zone for a long time.

4 Conclusion

The quality of the performed simulations was assessed when manufacturing products in a glow discharge plasma generator. It is shown that depending on the technological environment used two effects, short-range and long-range, were observed after the exposure to plasma. The short-range effect is caused by ion current transfer and leads to changes in the chemical and phase composition of the surface layer, modification and refinement of the crystal structure, and amorphization of the thin surface layer.

The long-range effect is manifested in the fact that the process of generating defects during processing in a glow discharge shows the features of a dissipative process with elements of self-organization and has a quasi-oscillating nature. Obviously, there is a variable amplifying effect of either the surface layers of the material or its volume. Defect density waves propagate through the volume of the material, which characterizes the process as a synergistic one. In this case, there is an oscillatory change in the dislocation density over time. Early fragmentation confirms the presence of oscillations in the dislocation subsystem and a multiparametric oscillatory process occurs.

The treatment of metals and alloys with low-energy glow discharge plasma leads to an increase in the dislocation density to a depth significantly exceeding the projective range of ions from the target surface. This phenomenon can be classified as a long-range effect.

Acknowledgments

This work was partially supported by the RSF project No. 18-11-00247. This work was carried out using equipment provided by the Center of Collective Use of MSUT "STANKIN".

References

1. I.V. Tereshko, V.A. Logvin, V.M. Tereshko, S.A. SHeptunov, Bulletin of the Bryansk State Technical University, **3**(51) (2016) DOI: 10.12737/22066
2. I.V. Tereshko, V.A. Logvin, V.M. Tereshko, S.A. SHeptunov, V.P. Red'ko, Proc. 6th Int. Konf. «Design and technological informatics», 21–29 (2017)
3. G. Haken, *Secrets of nature. Synergetics: the study of interaction* (Moscow-Izhevsk. Institute for Computer Research, 2003)
4. V.K. Vanag, *Dissipative structures in reaction-diffusion systems* (Experiment and theory, 2008)
5. Yu.P. Haraev, Polzunovsky almanac, **4**, 47–51 (2004)
6. W. Möller, *Fundamentals of Ion-Surface Interaction* (Technical University of Dresden, 2004)
7. I.A. Hamad, D. Robb, P.A. Rikvold, Computer Simulation Studies in Condensed-Matter Physics XIX. Proceedings of the Nineteenth Workshop, 89–93 (2006) https://doi.org/10.1007/978-3-540-85625-2_14
8. W. Eckstein, *Computer Simulation of Ion-Solids Interaction* (Springer-Verlag, Berlin, Heidelberg, New York, 1991)
9. L.D. Landau, E.M. Lifshic, *Quantum mechanics* (Non-relativistic theory. Moscow. 2004)
10. V.F. Zajcev, A.D. Polyanin, *Directory. Ordinary differential equations* (Moscow., 2001).
11. G. Korn, *The Mathematics Handbook for Scientists and Engineers: Definitions, Theorems, Formulas* (2014)
12. N.M. Chernavskaya, D.S. Chernavskii, L.A. Uvarova, Mathematical Models of Non-Linear Excitations, Transfer, dynamics, and Control in Condensed Systems and Other Media, 181-194, (1999) https://doi.org/10.1007/978-1-4615-4799-0_16
13. B.N. Aleksić, L.A. Uvarova, Opt Quant Electron, **52**, 175 (2020). https://doi.org/10.1007/s11082-020-02271-2
14. L.A. Uvarova, Y. Burenok, International Journal of Pure and Applied Mathematics, **109**(3), 691-707 (2016) doi:10.12732/ijpam.v109i3.17
15. I.V. Tereshko, Proc. 4 th all Russian Conf. with international participation of the scientific and technical confirence 2005, 210–215 (2005)
16. I.V. Tereshko, V.A. Logvin, V. Red'ko, Bulletin of the Belarusian-Russian University, **1**, 59-70 (2013) [in Russian]
17. I.V. Tereshko,. In: Application of Particle and Laser Beams in Materials Technology, 595–602 (1995) https://doi.org/10.1007/978-94-015-8459-3_41

Rhythm Drivers, Physical Properties, Mechanisms of Formation

Mikhail Mazurov[1,*]

[1]Russian University of Economics. G.V. Plekhanov, RU-117997, Moscow, Russia

Abstract. A mathematical model of the pacemaker is presented in the form of a nonlinear system of ordinary differential equations and in the form of a system of partial differential equations for distributed pacemakers. For the numerical study of the properties of the pacemaker, a modified axiomatic Wiener-Rosenbluth method was used using the properties of uniform almost periodic functions. Physical foundations, mechanisms of formation, properties of point and distributed pacemakers are described in detail.

1 Introduction

Rhythm generators are widely represented in physics, biology, physiology, nature, ecology and economics. The entire human body is permeated with rhythm generators. Some of the rhythm generators in the human body are shown in fig. 1.

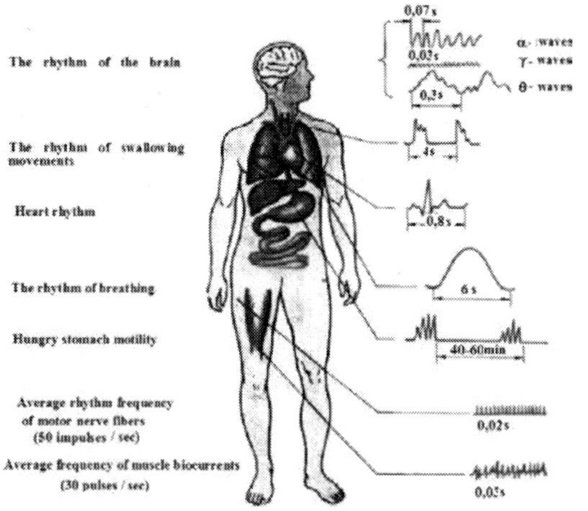

Fig. 1. Rhythm generators in the human body.

The entire human body is permeated with pacemakers. There are two types of neural networks responsible for rhythm generation. The first, using interconnected neurons, pacemakers or pacemakers, called a pacemaker (DR). The second type, a neural network based on the interaction of neuronal cells, devoid of their own pacemaker activity, but connected in a network. Such oscillators are called network or central rhythm generators (CGR). The emergence of oscillatory, rhythmic activity is due to the emergence of positive feedback in the network system. CGRs have a variety of structures and complexity.

In this work, we will consider DR based on relaxation oscillators that have their own pacemaker activity, called pacemakers; CGRs are not considered.

Let's define the physics of processes during the functioning of DR and the problems that arise during their study. These are the following problems:

1) Physical processes during the formation of a single rhythm or synchronization of partial rhythms of separate relaxation oscillators that make up DR. This is a complex problem in the physics of mutual synchronization of relaxation self-oscillations; 2) If the VR consists of several independent parts, then there is a need for mutual synchronization of individual parts. For example, the BP of respiration, located in the human medulla oblongata, consists of several parts that control various objects - the intercostal muscles, the abdominal diaphragm, which must work synchronously and must be rigidly synchronized; 3) The problem of synchronization of macroscopic VR with other subsystems of the body and the external environment. For example, the rhythm of the pacemaker of the heart or respiration as a whole should be determined by the internal state of the body and external load; 4) The problem of synchronization of macroscopic DR with the external environment of the organism. For example, a single DR rhythm located in the suprachiasmatic nucleus of the human medulla oblongata, which has a period of 23-25 hours, should be synchronized with a 24-hour circadian rhythm; 5) Cognitive intelligent regulation of the pacemaker depending on information from the external environment. For example, the formation of the □ rhythm in the hippocampus to realize the state of attention in the brain. Another striking example is the suppression of the alpha-rhythm generator in certain parts of the brain neocortex that are not involved in the processing of current essential information; 6) Real biological oscillators are a self-oscillating relaxation system (RSOS). This is due to the fact that effective mutual synchronization of the partial oscillators and components of the DR and synchronization of the DRs

* e-mail: mazurov37@mail.ru

themselves with each other and with the environment is possible if these are relaxation oscillators. This is due to the fact that the synchronization band of harmonic oscillators is much smaller than the synchronization band of the RSOS. A DR composed of mutually synchronized harmonic oscillators will not be a reliable, stable macroscopic system; 7) Influence of the spatial distribution of partial oscillators. The fact is that the nearest oscillators have the greatest influence on a particular oscillator, while those located at a great distance have less influence. Thus, real DR should be viewed as a distributed active environment; 8) The formation of a single rhythm can be significantly influenced by autowave interactions in a distributed active environment; 9) DR must transmit a stable signal of sufficient electrical power to the environment; Currently, the mechanisms of the formation of a single rhythm in the heart have been studied to a greater extent, this problem has been more than a century. Approaches to understanding the formation of a single rhythm in the distributed active environment of the heart have been found. DR and CGR have been studied to a lesser extent in other parts of the human body. In the future, the study of the physical properties of DR, the formation of their rhythm will be investigated on the basis of the DR of the heart.

2 Mathematical models for studying rhythm drivers

To study the nonlinear dynamics of the formation of a unified rhythm in the general case, when the VR is a distributed active medium, a system of nonlinear partial differential equations of parabolic type is used.

$$\frac{\partial E_i}{\partial t} = F_i(\vec{E}) + D_i \Delta E_i + f_i(x, y, z, t); \quad (i = 1, ..., n) \quad (1)$$

For some cases, it is sufficient to use a system of two equations.

$$\frac{\partial E_1}{\partial t} = f_1(E_1, E_2) + D\Delta E_1; \quad \frac{\partial E_2}{\partial t} = f_2(E_1, E_2) + D\Delta E_2 \quad (2)$$

The point system for partial differential equations (1) is the system nonlinear differential equations with a small parameter at the highest derivatives [1]

$$\varepsilon \frac{d\vec{x}}{dt} = \vec{f}(\vec{x}, \vec{y}, \vec{\beta}_1(t)); \quad \frac{d\vec{y}}{dt} = \vec{g}(\vec{x}, \vec{y}, \vec{\beta}_2(t)) \quad (3)$$

where,

$$\vec{x} = (x_1, ..., x_k), \quad \vec{y} = (y_1, ..., y_l),$$

$$\vec{f} = (f_1, ..., f_k), \quad \vec{g} = (g_1, ..., g_l), \quad \vec{\beta}_1(t), \ \vec{\beta}_2(t)$$

is the synchronizing effect. The phase space of system (3) R^n $(n = k + l)$ naturally splits into the direct sum of k – measured a dimensional subspace X^k and l – measured a dimensional subspace Y^l. The phase picture of Eq. (3) is characterized by the presence of "fast" and "slow" motions. If, as a result of the alternation of slow and fast motions, a closed trajectory arises, then the corresponding periodic solution (3) represents a relaxation oscillation. It was found that the control of relaxation oscillations can be achieved by controlling the individual phases of the oscillations by slow or fast ones. The most effective is the control of the slow phase.

To study the sinoatrial node from a set of relaxation oscillators - pacemakers, one can use the operator equation of electrical processes in the form

$$\vec{u} = \vec{L}\vec{u} + \vec{R}(\vec{u}).$$

The connection of the i-th pacemaker with the rest in the first approximation can be considered countably additive. If we consider the connection with the nearest neighbors to be the most important, then the equations can be used to describe the sinoatrial node

$$u_i(t) = \vec{L}u_i(t) + \sum_{j \in \{N\}} \alpha_{ij}(u_j - u_i), \quad (i = 1, ..., n),$$

where $\{N\}$ is the set of all pacemakers of the nearest environment. To build a model of competitive interaction of pacemakers in the future, we use the Fitzhue mathematical model [2]

$$\frac{du_1}{dt} = u_1 - \frac{u_1^3}{3} - u_2 + I, \quad \frac{du_2}{dt} = \varepsilon(a - u_1 + bu_2) \quad (4)$$

where, I, a, b, ε - parameters: $a = 0.7$, $b = 0.8$. At I = 0.14, the system is potentially self-oscillating; at I = 0.4, the equations describe relaxation oscillations. To study competition in a distributed system, we use the Fitzhu - Nagumo mathematical model of the type of equation (2) [3]

$$\frac{dx_1}{dt} = u_1 - \frac{u_1^3}{3} - u_2 + I + D\Delta u_1; \quad \frac{du_2}{dt} = \varepsilon(a - u_1 + bu_2) \quad (5)$$

where: u_1, u_2- variables, D- diffusion coefficient characterizing the propagation of autowave processes in a two-dimensional distributed active medium of pacemakers.

3 Formation of a single rhythm of the synoatrial node of the heart. Analytical results

There are no exact solutions to equations (3), therefore, the study of self-oscillating relaxation systems caused significant difficulties. A breakthrough in this area was

 https://doi.org/10.1051/epjconf/202124801007

the work of N. Wiener and A. Rosenbluth, who formulated the axiomatic theory of active excitable media [4]. On the basis of this theory, I.M. Gel'fand proposed a hypothesis about the formation of a single rhythm in a system of relaxation oscillators with different periods of oscillations. According to this theory, in an axiomatic excitable environment, a single rhythm of self-oscillations is equal to the highest rhythm for all excitable elements [5], that is

$$f_0 = \max_k f_k \qquad (k=1, \ldots, n)$$

However, the experimental results of physiologists did not confirm the Wiener-Gelfand theory [6-10]. This circumstance caused a heated debate among experimental physiologists, which is reflected in many publications. It was experimentally established that a single rhythm of the sinoatrial node is formed mainly at an intermediate level closer to the frequency of the high-frequency pacemaker.

4 Analytical results on the formation of a single heart rhythm

For the value of the total period T_0 of excitation in a theoretical study, the ratio was obtained [11-15]

$$T_0 = T_1 + \alpha_{12} K_2 (T_2 - T_1) / (K_1 + \alpha_{12}(K_2 - K_1))$$

where T_0 is the period of the formed rhythm, T_1, T_2 are the periods of interacting pacemakers, α_{12} is the coupling coefficient, K_1, K_2 is the slope of the slow phase of the pacemakers. If $K_1 \square K_2$, $\alpha_{12} << 1$, then, as follows from the last formula

$$T_0 = T_1 + T_{12}(T_2 - T_1),$$

$$f_0 = f_{\max} - \alpha_{12}(f_{\max} - f_{min}).$$

Thus, the results of calculations, confirmed by experimental studies, showed that the single rhythm of the pacemaker is not equal to the maximum possible in the pacemaker system, but it is less than the maximum by an amount $\alpha_{12}(f_{\max} - f_{min})$. N. Wiener's ingenious statement was generally confirmed with a small correction for the frequency.

For n pacemaker cells, an approximate expression is obtained for T_0 in the form

$$T_0 \square (1/k_1)(K_1 T_1 + \sum_{j=2}^{n} \alpha_{ij}(K_j T_j - K_1 T_1))$$

A qualitatively approximate location of a single rhythm between the rhythms of the original pacemakers is illustrated in Fig. 2.

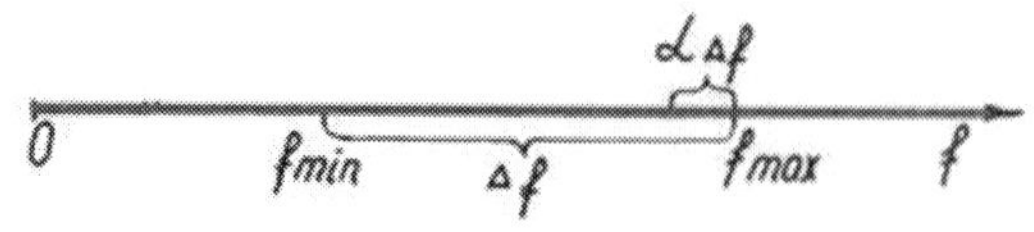

Fig. 2. The approximate location of a single rhythm between the rhythms of the original pacemakers.

5 Basic experimental data on the formation of a single heart rhythm

Consider the main types of experiments to study the formation of a single rhythm. Convincing results were obtained in experimental studies of the interaction in cell cultures of cardiomyocytes. The sinoatrial node was divided into separate cardiomyocyte cells using trypsin. It turned out that each isolated myocyte beats at its own rhythm. When contact is established between two myocardial cells, they synchronize after a while, that is, they begin to beat in a common rhythm. The authors believed that when synchronized, the faster cell should impose its rhythm on the slow one. However, synchronous beats usually occurred at an intermediate level between the frequencies of both pacemakers.

The study of the formation of a single rhythm of pacemaker cells was studied using the microelectrode technique during the interaction of fragments of the pacemaker tissue, and later in the interaction of single pacemaker cells. The works [7, 8] describe the interaction of individual fragments of the pacemaker tissue of the frog heart. The typical most often achieved result of interaction: the establishment of a single rhythm occurs at an intermediate level. The typical most often achieved result of interaction: the establishment of a single rhythm occurs at an intermediate level. In fig. 3, left, shows the results of one of the experiments in this series.

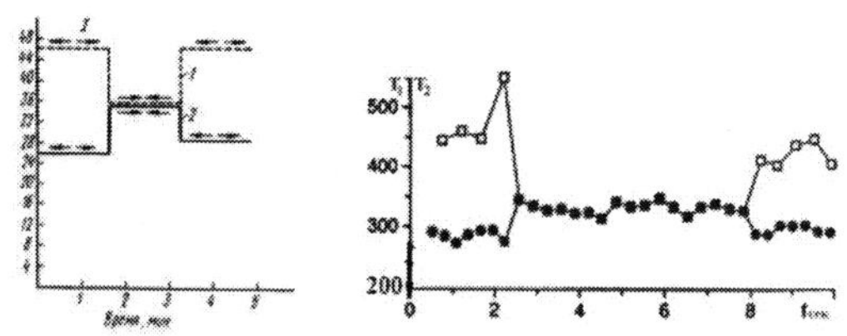

Fig. 3. Interaction of fragments of the pacemaker tissue of the frog's heart on the left; the rhythms of the preparations are 46 and 26 beats / min, the general rhythm is 35 beats / min; the right shows the interaction of two single pacemaker cells. A single rhythm is formed at an intermediate level closer to the rhythm of a cell with a higher rhythm.

The results of the initial work were brilliantly confirmed in experiments to study the interaction of two isolated pacemaker cells obtained from a rabbit tissue culture. and with computer fixation of the constancy of intercellular communication. With an increase in communication, rhythmic activity occurs in a single rhythm. These results are illustrated in fig. 3 on the right. Thus, at medium and large values of the connection, complete synchronization takes place, a single rhythm is formed at an intermediate level close to the level of a higher-frequency pacemaker, as shown in fig. 3 on the right. These experiments confirm the results of theoretical studies obtained in works [11-14].

6 Rhythm drivers hierarchy

An essential property of pacemakers developed by evolution is the formation of hierarchies. In the heart, the pacemaker is the sinoatrial node. In pathological conditions, the role of the pacemaker can be played by other underlying parts of the heart. The sinus-atrial node suppresses with more frequent impulses all the lower parts of the conduction system, but if it is damaged, the atrioventricular node, which generates impulses with a frequency of 40-50 per minute, can become a pacemaker. In the event of damage to this node, the fibers of the atrioventricular bundle (His bundle) can become a pacemaker, taking on its function. The frequency of the generated impulses and heart rate will be about 30-40 per minute. If these pacemakers do not work, then Purkinje Fibers can become them, setting the heart rate at about 20 per minute. Accordingly, the sinus node is the first order pacemaker, the atrio-gastric node is the second order pacemaker, and the Giss bundles are the third order pacemaker. Fibers of the ventricle are a fourth-order pacemaker. An illustration of the location of pacemakers of various orders in the heart is shown in fig. 4.

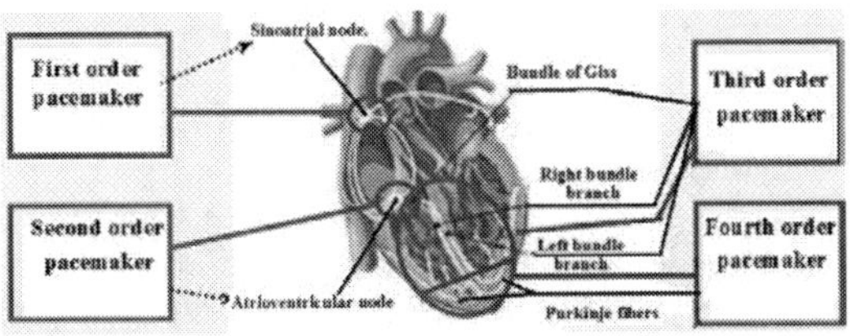

Fig. 4. Illustration of the location of pacemakers of various orders in the heart

7 Alignment of the synoatrial node and atrials

The DR must transmit a stable signal of sufficient electrical power to the environment. A two-dimensional model of the sinoatrial node was investigated, including pacemaker and atrial cells. The possibility of effective synchronization of atrial cells has been established. Transitional forms of cell potentials from pacemaker to atrial were found. In [15], the electrophysiological properties of the mosaic model of the sinoatrial node were investigated using a mathematical model. H. Zhang et al. a mosaic model of 20x20 dimension was investigated, including 400 cells of the sinoatrial node and atria [16]. A schematic representation of the model is shown in fig. 5.

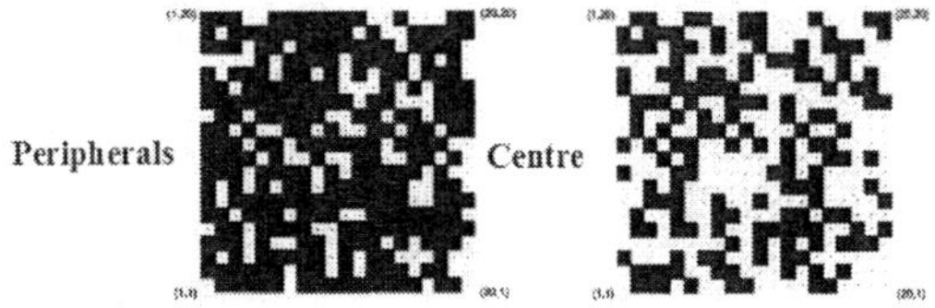

Fig. 5. Mosaic model of the center and periphery of the sinoatrial node. The lattice dimension is 20x20. The total number of nodal and atrial cells is 400. Nodal cells are shown in white squares, atrial cells in black. a) mosaic model of the periphery, b) the central part of the node.

8 Dynamics of competition and formation of a single rhythm of a two-dimensional synoatrial node

A numerical experiment was performed to test the existing hypotheses. Fitzhu – Nagumo equations (5) [3] were used as a mathematical model of DR. For the numerical solution of nonlinear partial differential equations of parabolic type, which can be reduced to a system of point nonlinear differential equations, the author's "scanning method" was used, which allows finding solutions of partial differential equations in inhomogeneous domains with complex geometry of boundaries [17]. The idea of a computational experiment is schematically illustrated in Fig. 5. The interaction of 10 thousand pacemakers is considered; this number of pacemakers is characteristic of the rabbit sinoatrial node. The system is described by 20 thousand nonlinear differential equations of Van der Pol - Fitzhju. The pacemakers of the node have a rhythm of about 60 beats / min, that is, a two-dimensional self-oscillating active medium is considered. There are also 5 groups of higher frequency pacemakers: 1) a pool of 9 pacemakers with a frequency of 70 beats / min; 2) a pool of 25 pacemakers with a frequency of 65 bpm; 3) a pool of 4 pacemakers with a frequency of 70 beats / min; 4) two pacemakers with a frequency of 75 beats / min; 4) one pacemaker with a frequency of 75 beats / min; pacemakers with a frequency of 75 beats / min have the highest rhythm. The scheme of the computational experiment for studying the interaction of the pacemakers of the sinoatrial node is shown in fig. 6 on the left, on the right shows the results of a computational experiment.

In fig. 6 clearly shows that the rate of capture of the excitation region by a group of pacemakers substantially depends on the frequency of oscillations of pacemakers in the group and on the number of pacemakers in the group, that is, on the power of the group. The higher the frequency and the number of pacemakers in the group, the faster the capture of the surrounding territory. At the same time, the frequency of pacemakers has a greater impact on the rate of capture of the surrounding territory than the number of pacemakers in the group. The dynamics of the wave movement at certain moments is interrupted by the excitation of the surrounding pacemakers. This destroys the wave periphery. It can be seen that a single rhythm is formed by a group of nine pacemakers. A more numerous group of twenty-five pacemakers, having a lower rhythm, fails to form a common rhythm; the rhythm of a group of nine pacemakers is imposed on it.

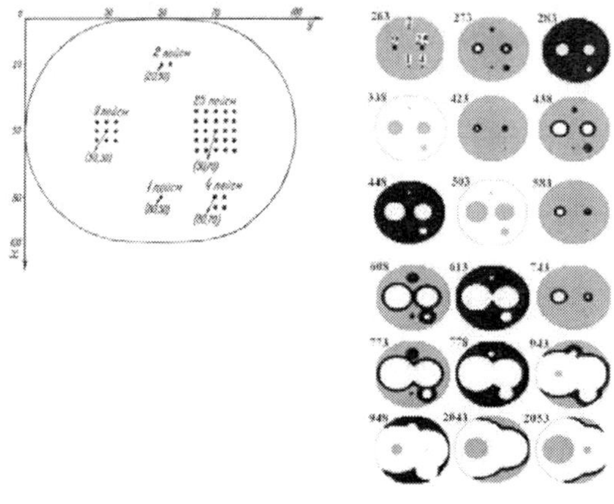

Fig. 6. The scheme of the computational experiment for studying the interaction of the pacemakers of the sinoatrial node is shown on the left, the results of the computational experiment are shown on the right.

Fig. 6 it can be seen that one pacemaker, even the highest-frequency one, cannot capture the rhythm of the entire node, nor can groups of two and four pacemakers do this, although as the number of pacemakers in a group increases, the size of their zone of influence increases. The results of a numerical experiment made it possible to visually confirm the main provisions on rhythmogenesis in the sinoatrial node of the heart: 1) rhythmogenesis is carried out by the pool with the highest rhythm; 2) the fastest cell alone cannot excite the entire node; 3) a single rhythm is established at an intermediate level close to the level of the highest-frequency cells; 4) a small number of high-frequency cells can create only a small focus of excitation (active vesicle), the size of which increases as the number of cells in the pool grows; 5) shows the efficiency of excitation of the sinoatrial node from its periphery. A further study of synchronization, competitive dynamics in neural ensembles was carried out in [18-20].

9 Driver rate control

Controlling the formed rhythm of DR and CGR is fundamentally different in terms of the physical mechanisms used for this. DR control is achieved due to the frequency of self-oscillations of each partial pacemaker, in connection with which the frequency of a single formed rhythm changes. In the CGR, the control of the formed rhythm is achieved by changing the parameter in the positive feedback loop. If the network is formed by oscillators, then the frequencies of these oscillators do not change.

There are constant tonic influences on the heart from the bulbar nuclei of the vagus nerve: with its bilateral transection, the frequency of palpitations increases 1.5-2.5 times. Stimulation of the vagus nerve causes a decrease or cessation of cardiac activity due to inhibition of the automatic function of the pacemakers of the sinoatrial node. The severity of this effect depends on the strength and frequency of stimulation of the vagus nerve. The negative chronotropic effect of irritation of the vagus nerve is associated with a slowdown in the generation of impulses in the pacemaker of the sinus node. Since, when the vagus nerve is irritated, a mediator, acetylcholine, is released at its endings, its interaction with the muscarinic receptors of the heart increases the permeability of the surface membrane of pacemaker cells for potassium ions. As a result, membrane hyperpolarization occurs, which slows down the development of slow spontaneous diastolic depolarization, and therefore the membrane potential later reaches a critical level. This leads to a decrease in the DR rhythm of the sinoatrial node. An illustration of the control of the rhythm of the sinoatrial node from the central nervous system - the medulla oblongata is shown in fig. 7.

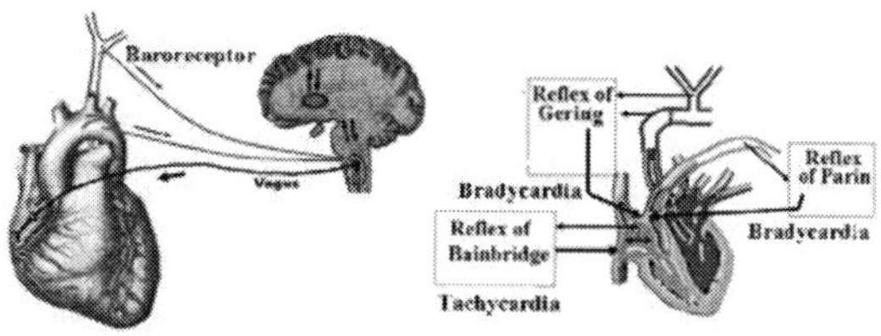

Fig. 7. Control of the rhythm of the sinoatrial node from the medulla oblongata of the central nervous system on the left, on the right, the control structure is shown using one's own cardiac reflexes. With an increase in blood pressure, the rhythm decreases. With a decrease in blood pressure, the rhythm increases.

Own cardiac reflexes form the basis of neurogenic regulation of the heart. Own reflexes of the heart are also manifested in response to mechanical stimulation of the heart chambers, in the walls of which there are a large number of baroreceptors. These include the Bainbridge reflex, which manifests itself as tachycardia in response to the rapid intravenous administration of a certain volume of blood. Among their own reflexes include the Hering reflex, which manifests itself in the form of bradycardia. The unloading reflex of V.V. Parin deserves a description: when the receptors of the pulmonary artery and pulmonary veins are irritated, hypotension is caused in the systemic circulation, and bradycardia develops. This reflex is used to unload the pulmonary circulation. Control of the rhythm of the sinoatrial node of the heart is possible by changing the duration of the phase of slow diastolic depolarization. The mechanism of this influence is described in more detail in [11-14].

10 Conclusion

It is shown that the pacemaker is an ensemble of many relaxation auto-generators with mutual synchronization. A mathematical model of a pacemaker in the form of a non-autonomous nonlinear system of ordinary differential equations with small parameters before derivatives is presented. For the numerical study of the properties of the pacemaker, a modified axiomatic Wiener-Rosenbluth method was used with the properties of uniform, almost periodic functions. It has been shown that the biological pacemaker is usually a hierarchical system. It has been shown that the human body and brain are permeated with many rhythm generators.

References

1. U.F. Mishchenko, N.Kh. Rozov, *Differential equations with a small parameter and relaxation oscillations* (M.: Science. 1975)
2. R.A. FitzHugh, Biophys. J., **1**, 445 (1961)
3. J. Nagumo, S. Arimoto, S. Yoshizawa, Proc. IRE, **50**, 2061 (1962)
4. N. Wiener, A. Rosenbluth, *Conducting impulses in the heart muscle. Mathematical formulation of the problem of impulse conduction in a network of connected excitable elements, in particular in the heart muscle* (M.: IL. Cybernetic. Sat., 3, 1961)
5. I.M. Gelfand, M.L. Tsetlin, DAN SSSR., **131**(6), 1242-1245 (1960)
6. I.M. Gelfand, S.A. Kovalev, L.M. Chailakhyan, DAN SSSR., **148**(4), 973-976 (1963)
7. Said Abu El Ela Kawser, M.G., Udelnov, G.S. Sukhova, Part II. Biological Sciences, **6**, 46-49 (1979)
8. M.E. Mazurov, G.S. Sukhova, M .: NDVSH. Biological Sciences, **1**(325), 140-149 (1991)
9. P.P. Aliev, L.V. Rozenshtraukh, Russian physiology. Magazine them. Sechenov, **92**(9), 1069-1077 (2006)
10. E.E. Verheijck., R. Wilders, R. Joiner et all., J. Gen. Physiol., **111**, 95-112 (1998)
11. M.E. Mazurov, Biophysics, **35**(6), 494-498 (1990)
12. M.E. Mazurov, Biophysics, **35**(6), 1001-1006 (1990)
13. M.E. Mazurov, Biophysics, **51**(6), 1092 (2006)
14. M.E. Mazurov, ZhVM and MF AS USSR, **31**(11), 1619-1636 (1991)
15. M.R. Boyett, H. Dobrzynski, M.K. Lancaster, S.A. Jones, H. Honjo, I. Kodama, J Cardiovasc Electrophysiol, **14**, 104-106 (2003)
16. H. Zhang, A.V. Holden, M.R. Boyett. Circulation, **103**, 584-588 (2001)
17. M.E. Mazurov, I.M. Kalyuzhny, Third International Conference "System Analysis and Information Technologies" (SAIT), 419-424 (2009)
18. M.E, Mazurov, Izvestiya RAN Phys. Series, **82**(1), 83-87 (2018)
19. M.E. Mazurov, Izvestia RAN. Series nat., **82**(11), 1564–1570 (2018)
20. M.E. Mazurov, Izvestia RAN. Series nat., **84**(3), 451-456 (2020)

The Nature of CVC Nonlinearity in Low-Voltage Scanning Tunneling Spectroscopy of Semiconductors

A.I. Loskutov*, *A.M.* Mandel, *E.E.* Karpova, *V.B.* Oshurko, *S.G.* Veselko, *A.A.* Sharts and *S.A.* Loskutov

*Moscow State Technological University “STANKIN”, RU-127055, Moscow, Russian Federation;

Abstract. A new model of field emission in a scanning tunnelling microscope was developed. The model describes the tunnelling current from a surface of semiconductor (semimetal) and allows estimating the pre-exponential factor in the expression for the tunneling probability. It is shown that this factor is directly related to the degree of localization of the electron density and determines the shape of the local tunnel current-voltage characteristics (LTCVCs) at low voltages. The model allows separating the contributions of surface electronic states of different symmetry (dimension) of the tunnelling current. The practical application of the model is demonstrated by the example of mathematical processing of the LTCVCs of HOPG surface containing different structural defects.
Keywords: charge tunneling, STM, STS, CVC, surface defects, monographite.

1 Introduction

The processes of assembling two-dimensional monolayers of organic molecules constantly attract the attention of researchers. They are observed in both liquid and solid surfaces. However, for their practical applications, such as functional elements of organic nano and microelectronics one must use substrate surface acts as a template for self-assemble two-dimensional monolayers and determines their final structure and electrical properties.

It is well known that the defects on the surface of solids are active sites in catalysis, adsorption and nucleating centers of a new phase [1-3], because it is on them free electrons are localized [4]. For example, randomly distributed charges, which are presented in the silicon oxide layer causes large potential fluctuations and substantially affect the structure of graphene deposited on the silicon surface [5].

There are different ways of visualizing structural defects on the surface of solids: by destructive mode (chemical, electrochemical, ion-plasma etching) and nondestructive mode (selective adsorption, electron and probe microscopies). However, local analysis of their electronic properties and their distribution over the surface are a more complicated problem.

The main objective of this work is experimentally test a new methodology for analyzing the LTCVCs, which allows one to separate the contributions of different electronic states to the tunneling current. This will enable separating "normal" surface electrons delocalized in the plane of the sample surface, from electrons, localized at surface defects, and thereby estimate the local density and type of the latter. This study is not only of theoretical interest, but also has practical value. Especially it is necessary for monitoring semiconductors surface properties in micro and nanoelectronic technologies, because defects disturb lateral mobility of the electrons.

At low electric fields electron emission from the semiconductor surface, when is measured by probe microscopy techniques, is a particular case of the tunnelling effect [6]. Such processes are usually studied with the semi-classical approximation that determines the amplitude of the transition up to an exponential factor.

In this voltage range electric field only slightly reduces the average height of the potential barrier at solid surface and its strength disappears from the exponent of the estimated CVC. In the same time, the pre-exponential factor in the transition amplitude is not taken into account at all. Since the density of electronic states, participating in the transition and "liberated" by the electric field on the surface of the probe, is proportional to the voltage U, the theoretical CVC of the tunnel current usually is purely ohmic at low voltages: I~U.

At the same time, numerous experiments show that the real CVC deviates from the linear law even in the vicinity of U=0. In this article we discuss the possible physical nature of this deviation and demonstrate that taking into account the above-mentioned pre-exponential factor it is possible to explain the tunneling probability and physically interpret such deviations. Unfortunately, the exact solution of the three-dimensional Schrodinger equation, describing electron tunneling from the real surface (i.e. containing various types of defects) of the semiconductor, is impossible for various reasons. Practically at low fields, it has no influence on the exponential factor of the tunnelling probability, but it is critically important for the pre-

* Corresponding author: ailoskutov@yandex.ru, tel.+79164919418

exponential factor, which determines the transition probability and, therefore, forms the tunneling current. To solve the problem, we have to use approximate methods. For example, in [7-9] there was used the approximation of electron tunneling from a potential well with an attractive δ-potential. In this case such approximation is adequate, because the surface electronic state released after tunneling was almost instantly filled by another electron. As a result, the calculated CVCs of the tunneling current in low electric fields were compiled to form: $I \sim U^{(4-n)/2}$ (the index $n = 0, 1, 2$ is determined by the dimension of the initial electronic state). In the particular case of free lateral surface electronic states (such as Tamm levels) $n = 2$ the result being to the standard [6, 10].

The traditional theory of tunneling current in STM [6] did not take into account the influence of the structure of the initial surface electronic states on the CVC. The reason is that the tunneling probability in an electric field is estimated rather crudely, in a purely exponential approximation, in fact repeating for a long time Fowler and Nordheim approximation. However, exactly the wave functions of the surface electrons included in the matrix elements of the tunnel junctions determine the structure of the pre-exponential factors in the tunneling current formula [11]. These factors can be identified by making the appropriate tunnel CVC normalization. It turns out that such a "normalized" CVC has the simple form of a power function, and the exponent is the number of degrees of freedom of the surface electron. Since surface electrons localize on structural defects, the degree of local CVC deviation from linearity (deviation of $n = 2$) can serve to a certain degree as an indicator of the quality of the local surface structure. In fact, a set of surface electronic states determines the fractal structure of the surface. This has, in one way or another, impact on all surface and emission properties of the body.

2 Theory

As mentioned in the introduction, the exact solution of the electron tunneling from the real surface of the semiconductor is almost impossible. However, this process relates to a broad class of phenomena induced by an external electric field. Consistent calculation of the tunneling current in STM with the pre-exponential factor in the tunneling probability is given in [11]. Here we briefly describe only the main points of this calculation, based on analogy with other areas of physics. Similar phenomena have been studied in almost all sections of quantum physics - nuclear physics, solid state physics, quantum field theory, etc. [8, 12-13]. Despite many differences, all such processes have a number of common features. The most important feature, for example, is a possibility to express the cross section and the probability of the relevant processes in terms of bilinear combinations of Airy functions. The argument of these functions is the ratio of two-characteristic length parameter: $u = \pm\lambda/l$. The first parameter is similar to the length of de Broglie wavelength (in relativistic - Compton wavelength) and characterizes the electron mobility. For this situation, the order of $\lambda = \hbar\,(2mW)^{-1/2}$, where $\hbar$ - Planck constant, m - mass of the electron (in a semiconductor - effective mass), W- height of the potential barrier between the tip and the sample (Fig. 1). The second parameter - the distance at which the electric field does work specific for this process $l = W/eE$, where e - electron charge, E - electric field. Positive values of the parameter u correspond to the energy range in which such process may occur without an external field. External electric field causes oscillations of the cross sections and probabilities due to the interference of de Broglie waves incident and reflected from the barrier created by the electric field [9].

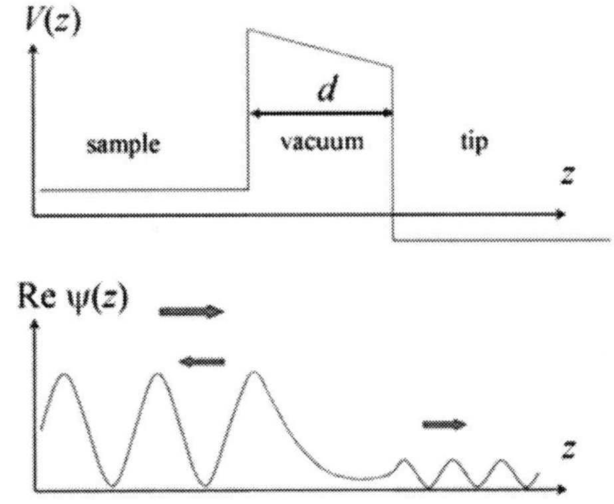

Fig. 1. The trapezoidal potential barrier energy structure and the real part of the electron wave function near the boundary of the sample and the probe.

If u<0 in the absence of the field the energy of the particle for transmission process is not enough, so it is only the electric field that increases the energy of electrons (no matter real or virtual). In this case, either cross sections or probabilities decrease monotonically with increasing energy deficit. If also $|u|$<<1 the probability is suppressed exponentially and the process has the tunneling nature. We are considering the field emission at low voltages that is precisely this type. Only in electric field (this corresponds to the length l of the triangular potential barrier, i.e., the potential barrier on the right side in Figure 1 is the probe potential) the tunneling probability p can be written in a universal form, where D – is a constant that does not depend on the external field [7, 11]

$$p = Du^{(2-n)/2}\exp(-4/3u) \qquad (1)$$

Above-described scheme is also consistent with quantum field processes: photo-generation of electron-positron pairs, nuclear processes, induced by beta decay and the splitting of the deuteron, internal photoelectric effect in semiconductors (Franz-Keldysh effect), auto-ionizing atoms and a number of similar threshold processes, induced by an electric field. The main issue for us here is the factors that determine the n in eq. (1). We can simplistically assume that n is determined mainly by the symmetry of the initial state. In this respect, the process of field emission from a semiconductor to a metal probe is rather typical. If electron tunneling came from (zero-dimensional) s-state to a free state with a continuous spectrum, n=0 (the same as potential of s-electron ionization from delta-quantum well).

Due to the fact, that the transition goes to the final state on the metal probe the exponent increases to two. Indeed, the electron transfer is possible only into the free state with zero Fermi function, and the number of such states, as noted above, is proportional to the external field. Since $u \sim E \sim U$ then the contribution of zero-dimensional surface states in the CVC of the tunnelling current depends on the external field as $I \sim U^2$. This is fully consistent with the well-known classical result of Fowler and Nordheim. In one-dimensional states occurring in linear structural defects (grain boundaries, edges of the steps, etc.); there is an additional degree of freedom. This gives an additional integral over the momentum variable in the calculation of the contribution to the current. So it enables calculating the contribution to a current over the momentum variable. This integral reduces the degree of u in the tunnelling probability and U respectively by ½, when determining the contribution to the tunnelling current. Respectively, in the lateral delocalized states with two degrees of freedom there would be two integrals and the degree of U is reduced to one: $I \sim U$ [6, 10, 11].

In analytical form, the density of the tunneling current from the surface of the semiconductor can be written as [10]

$$j = e\int_0^{W_m} p(W_z)\xi(W_z, eU)dW_z \quad (2)$$

where $p(W_z)$ - the probability of electron tunneling, and the integration should be made over the electron energy W_z along the axis normal to the surface; W_m - its maximum value. Value $\xi(W_z, eU)$ being usually referred to as the density of surface electronic states. However, the dimension of this magnitude (SI $m^{-2} * c^{-1} * J^{-1}$) shows that it is the flux density of particles per unit energy interval. It is proportional to the density of occupied states in the semiconductor and the density of free states in the metal (the probe). After applying the voltage to the probe, the electron states close to the Fermi level in the metal are "liberated". The result is the current from the surface of the semiconductor to the metal. Probability $p(W_z)$ has the form eq.(1) only for the triangular barrier in the absence of the other fields except the electric field. In this case, the barrier for electron tunnelling extends the entire length l. For a thin trapezoidal barrier (length $d \ll l$) sub-barrier electron trajectory is much shorter (Fig. 1). Calculating the tunneling exponent in the usual manner [11], we obtain:

$$-2\int_0^d dz\,(W_0 - eEz)^{\frac{1}{2}} = = -\frac{4}{3eE}\left[W_0^{3/2} - (W_0 - eEd)^{3/2}\right] \approx \approx -2W_0^{1/2} - eEd^2/2W_0^{1/2} \quad (3)$$

where W_0 – the height of the barrier on the left side. In low fields, the second term is usually neglected, so that the field strength in the tunnelling exponent reduces. As a result, if we use the approximation of n-dimensional binding δ-potential for low external fields $eU \ll W_0$, the tunneling probability of an electron with binding energy of $W = W_0 + W_m - W_z$ (the distance of electron - vacuum levels on the energy scale) from the semiconductor surface can be represented as

$$p \propto \left[\frac{eU}{dW_0^{3/2}}\right]^{(2-n)/2} \exp\left(-HdW_0^{\frac{1}{2}}\right) \quad (4)$$

where H - a dimensional constant is not of interest to us. Function $\xi(W_z, eU)$, appearing in eq.(2), can be estimated as

$$\xi(W_z, eU) \propto eU(W_m - W_z)^{\beta-1}. \quad (5)$$

In fact, the number of states in the probe "liberated" by applying voltage is proportional to eU. The constant β characterizes uniform filling of allowed states in areas such as the probe material and the studied surface. For example, for quasi-continuously filled conduction band β=1, for solitary or discharged surface electronic states $\beta \gg 1$, etc. Substituting eq. (3) - (5) in eq. (2), we obtain for the current density

$$j \propto \frac{eU^{(4-n)/2}}{d^{(2-n)/2}} \exp\left(-HdW_0^{\frac{1}{2}}\right) \cdot \int_0^{W_m} dW_z$$

$$(W_m - W_z)^{\beta-1}(W_0 + W_m - W_z)^{-3(2-n)/4} \cdot$$

$$exp\left[-Hd(W_m - W_z)/2W_0^{\frac{1}{2}}\right] \quad (6)$$

This is the Laplace integral, which can easily be estimated [11]. The main contribution to the integral comes from the region near the upper limit. As a result, one can get a final expression for the tunneling current to probe at low voltages $eU \ll W_0$

$$I(U) \approx KU^{(4-n)/2} \cdot W_0^{\beta/2-3(2-n)/4} \cdot d^{(n-2)/2-\beta} \exp(-HdW_0^{1/2}) \quad (7)$$

In special case n = 2, β = 1 this expression reduces to the standard formula [10]. It is not casually, since it is well known "ohmic" Simmons result obtained for the "metal-to-metal" tunneling by integrating over all possible values of the lateral moment. To get rid of extra constants normalize eq. (7) by an arbitrary data point (I_0, U_0)

$$I/I_0 = (U/U_0)^{(4-n)/2} \quad (8)$$

Sum up briefly the theoretical part of the work. The contribution of surface electronic states of various dimensions (symmetry) gives as a result the following asymptotic for the local CVC for small *U*, where the constants *A*, *B* and *C* give an idea of the contribution to the tunneling current: two-dimensional, one-dimensional and zero-dimensional states, respectively.

$$I = AU + BU^{3/2} + CU^2 \qquad (9)$$

Clearly, when $U \to 0$, the contribution of the second and the third term tends to zero, what explains the "vitality" of the traditional approach.

3 Experiment

It is well known that the electronic structure properties of bulk materials are influenced by crystallographic defects of different origin. Understanding of the degree of their influence on the electronic structure properties of poly- or mono-crystalline materials, such as carbon-based materials, is a necessary requirement if these materials are to be used for possible future electronic devices.

In a number of recent studies it was established the formation of three main types of structural defects on the surface of semiconductor materials based on carbon - point, flat and volumetric [14-15]. The other structural defects such as linear quantum wires or extended nanometer height steps were less common.

All of them formed quasi-localized electronic states with the energy near the Fermi surface. Since the density of electronic states on the surface of intrinsic semiconductor near the Fermi level is usually quite low, the "added" electronic states of defects significantly affected the local electronic structure of the semiconductor surface and therefore the tunneling current to and from the surface. For example, usually on the graphite surface there are many extended steps of different heights similar one-dimensional quantum wires. In this case, it was recommended to measure the electronic structure of graphite no closer than 5-20 nm (depending on temperature) from the edges of the steps [16-17]. These distances depend on temperature: the higher the temperature, the shorter the distance. Mono-crystalline graphite due to the peculiarities of its crystal structure and the possibility to obtain sufficiently simply an atomically clean surface is a model material for the study of the influence of structural defects on the electronic properties of the surface. Therefore, it makes sense to explore the more CVCs of the tunneling current from the graphite surface at different distances from the edge of the extended steps.

The highly oriented pyrolytic graphite (HOPG) ZYH type with misorientation angles (mosaic range) 3,5° - 5° (NT-MDT, Russia) was used as a test sample. Clean surface of HOPG was obtained by cleaving with the adhesive tape. Then the sample was placed quickly in a high vacuum scanning tunneling microscope (STM Solver HV-MFM, NT-MDT, Russia), evacuated to a pressure of not more than 10^{-6} Pa. Before STM measuring the sample was annealed in a high vacuum at 523 K to remove any surface contaminants, primarily water. As received the HOPG sample, like any other material, is not produced defect-free. Although the density of structural defects on flat atomic terraces is rather low, the existence of steps of different heights on the HOPG surface substantially increases their number. It is not always possible to observe such structural defects in the STM images even with atomic resolution. So getting independent information on the nature of structural defects and their location by using other methods is of practical importance.

When scanning and imaging the surface, the tunnelling parameters *I* and *U* determine the tip-to-substrate distance (*d*), the exact value of which is not well known. For the scanning tunneling spectroscopy (STS) the STM tip is moved to the position of interest with the feedback still activated. Therefore, *d* is adjusted according to the selected bias voltage and the set point for the tunnel current, before the feedback loop was interrupted. The vertical and lateral position of the tip is kept fixed, whereas a bias ramp changes from —1.5 to + 1.5. The measurement of LTCVCs conducted in the voltage range of ± 0,5 V, which was about an order less than the work function of graphite (4.3 eV). At these voltages, the electric field can still be considered rather low, compared to the internal field. During the CVCs data collecting the tip-to-substrate distance and the selected tunneling current set point of 0.5 nA were fixed according to the sample bias of —1.5 V and. Since all CVCs were measured for a single scan, we could assume that in the process of measuring *d* was not changed. At each selected point of the surface 10-15 CVCs were measured, the results were averaged and analyzed.

Unfortunately, the absolute value of *d* is not accessible experimentally and remains unknown. The tunneling current is contributed due to emission from the surface electronic states of the region, the exact size of which is difficult to assess. These states include also the states, localized on different kinds of structural surface defects. Since the graphite samples are misoriented with respect to basal plane on HOPG, the formation of steps and flat terraces on the clean surface is inevitable and well seen in Fig. 2.

Two series of measuring were conducted. The first consisted of four CVCs, taken on the different points along the edge of the step (Fig.2a), and the second - of five CVC measured at flat terrace at different distances from the step edges (Fig. 2b).

 https://doi.org/10.1051/epjconf/202124801008

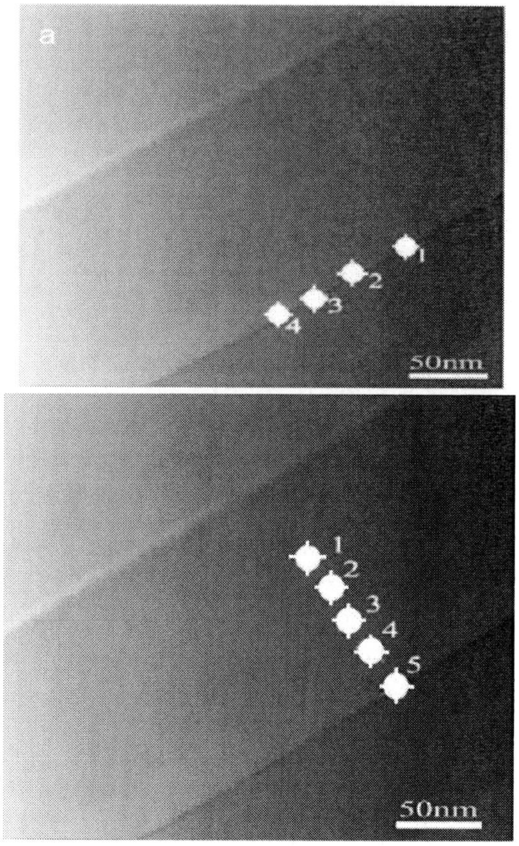

Figure 2. Points on the surface of graphite, in which the LTCVCs were measured: along the edge of the step (a) and perpendicular to the edge of the step (b) (point 5 is common).

These distances are given in Table 1.

Table 1. Position of the points of CVC measured on a flat terrace

The point at which the CVC is measured on the terrace	The distances along the step from the point 5, nm	The distance from the edge of the upper step, nm
1	94	52
2	72	74
3	48	97
4	26	117
5	0	144

Corresponding local tunnel CVCs and their corresponding dependencies of differential conductance versus applied voltage are presented in Fig. 3a and 3b.

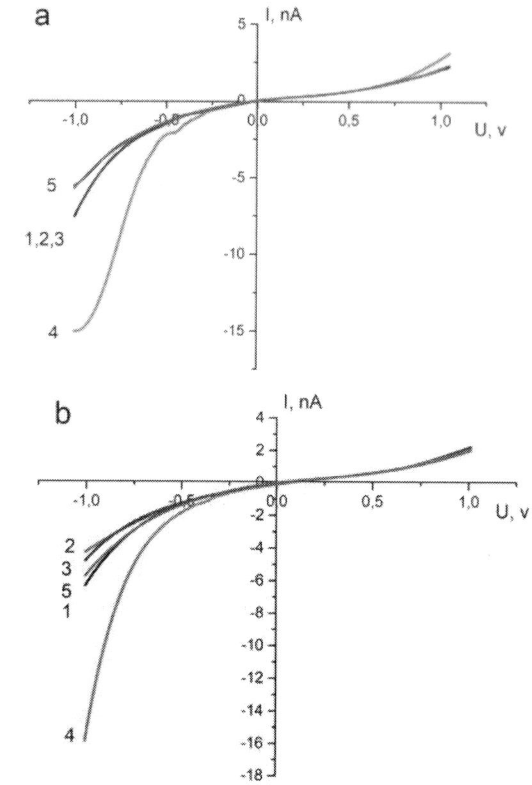

Figure 3. LTCVCs, measured at points along the edge of the step (a) and along the normal to the edge of the step (b).

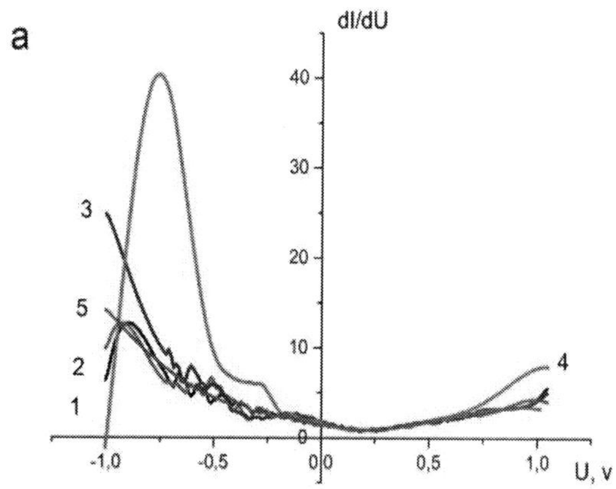

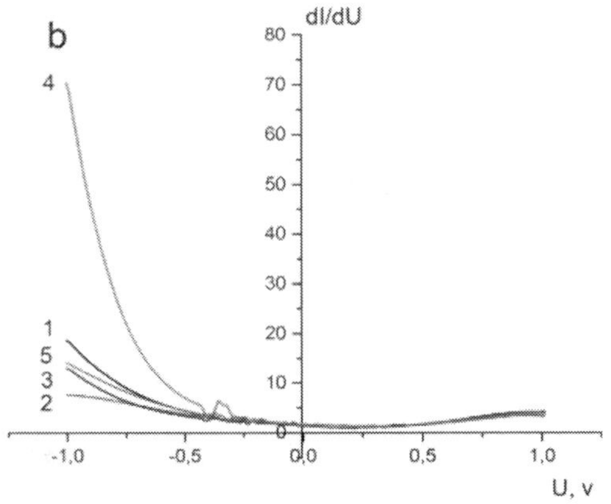

Figure 4. Differential conductance versus applied voltage, measured at points along the edge of the step (a) and along the normal to the edge of the step (b).

The results of mathematical processing of CVCs are shown in Tables 2 and 3.

Table 2. The coefficients of the approximation curve for the points along the edge of the step. Sign of the branches corresponds to the sign of the probe potential

Point	Branch	A	B	C
1	+	0,2895	0,3158	0,3947
1	-	0,9821	0,0060	0,0120
2	+	0,3241	0,2990	0,3769
2	-	0,9743	0,0069	0,0294
3	+	0,4984	0,3271	0,1745
3	-	0,9791	0,0004	0,0246
4	+	0,1216	0,0946	0,7838
4	-	0,9923	0,0010	0,0069

Table 3. The coefficients of the approximation curve for points across the step surface. Sign of the branches corresponds to the sign of the probe potential

Point	Branch	A	B	C
1	+	0,5650	0,0003	0,4346
1	-	0,8730	0,0397	0,0873
2	+	0,5224	0,0003	0,4773
2	-	0,8196	0,0074	0,1730
3	+	0,7127	0,0034	0,2840
3	-	0,8561	0,0227	0,1212
4	+	0,2937	0,0093	0,6970
4	-	0,8769	0,0077	0,1154
5	+	0,3069	0,2354	0,4577
5	-	0,9789	0,0017	0,0231

Each CVC was approximated by eq. (9) for the forward and backward branches, and optimal values of the coefficients were selected by the standard method of least squares.

4 Discussion and conclusion

Let us first consider the data in Fig. 4, which shows the distribution of the local density of electronic states (LDOS) as a function of energy. It is seen that, depending on the point chosen for measurements on the graphite surface, LDOS can vary significantly. In this case, the edge of the valence band (a sharp change of LDOS in the $E < E_F$ region) manifests itself in most of the spectra obtained. At $E > E_F$ in the region of free (vacant) states, the dependence of LDOS on energy in measurements along the edge of the step (Fig. 4a) has a different form compared to practically indistinguishable curves in measurements along the normal to the edge of the step (Fig. 4b). In the region of negative and positive voltages, oscillations of the tunneling current are observed, which have a regular character, in no way associated with noise during measurements and are not eliminated by smoothing. The strongest oscillations are observed in the range of negative voltages. According to [18-20], such oscillations can phenomenologicaly be attributed to localized π-states.

Turn first to the LTCVCs measurements at the points along the edge of the step. According to [16], in this case, one should expect a significant contribution to the tunneling current the one-dimensional states. The results of such treatment are presented in Table 2. As can be seen from the data of Table 2, there is a striking difference of forward (the flow of electrons from the sample to the probe) and the reverse CVC branches. Reverse branch - almost straight ohmic curve (factor A is very close to unity, and B and C are negligible). At the same time the forward branch has a significant curvature (all coefficients are comparable). This is consistent with the above model, as the shape of the CVC curve is determined basically by the initial symmetry of the electronic states (at the sample in the case of forward branch and the probe in the case of the reverse branch). On a graphite surface electrons are transferred from the states of various dimensions, which results in noticeable values of the coefficients B and C, while the surface state of the metal probe, as shown in Table 2, are not localized inside the metal.

This is consistently well with the conclusions of [13], where the theoretical model was proposed for the metal-metal tunneling (about the nature of differences in the forward and the reverse branches of CVC see also [7]). At points 1-3, the contribution of all three types of states is comparable. The point 4 on the forward branch markedly knocks out of this series, which may indicate the local concentration of defects in its vicinity. For example, this may be defects on the step edge such as "zigzag" or "chair" type, described in [17]. The differences of the CVC forward branch in point 4 of the other points can be observed visually from Fig. 4. To confirm our model it is important that on the reverse branch of CVC there is practically no effect at point 4.

Consider the next series of measurements on the flat surface of the terrace. The summation of approximation coefficients in Table 3 shows that the coefficients for point 5 along the line perpendicular to edge of the step on the forward branch as well as on the reverse branch of the CVC are quite consistent with similar coefficients for points 1-3 of the previous series. In other words, point 5 is quite a typical point on the edge of the step.

Secondly, in the CVCs, measured in 1-4 points, the coefficient B sharply decreases (almost to zero), i.e. the contribution of one-dimensional initial states decreases.

It is also quite understandable, because the above-mentioned states are localized on the linear defects. As mentioned before, the reverse branch of the CVC is much closer to a straight ohmic line than the forward branch. In this series of measurements deviations from Ohm's law (or more precisely, the values of the C coefficients) are more significant than in the previous ones. Concerning the forward CVC branches of 1-4 points on a flat terrace surface, its substantial nonlinearity may be associated not only with it, but with the existence of point surface defects, which are difficult to visualize. However, in general, the experimental data are consistent with the proposed model.

Of course, it would be possible to draw a conclusion about one correspondence of B and C coefficients to the local density of surface defects corresponding symmetry only after comparing them with the results of defect measurements by independent methods.

Nevertheless, these results give a basis for possible practical applications of the obtained parameters. First, if the measured degree of localisation of the surface electron is significantly greater than 1, then the presence of defects on the surface structure should be supposed. These hidden defects cannot always be visualized even with probe microscopy.

Secondly, in the STM constant tunneling current mode the topography of the studied surface depends not only on d, but also on the electron density distribution on the surface of the sample. Therefore, the situations may exist, when some corrections of the measured topography should be made. Of course, the value of n is not the only factor that determines a deviation of the observed CVC from a linear law or other simple dependences. However, if other factors are excluded, the value of n should be taken into account as a parameter treating a topology of surface defects.

Acknowledgements

This study was supported by the Ministry of Higher Education and Science of the Russian Federation grants NFSFS-2020-0025 and carried out using equipment provided by the Center of Collective Use of MSUT "STANKIN".

List of symbols

I - tunneling current
U – voltage at the probe
n - dimension of the initial electronic state on the surface
u - universal parameter characterizing the influence of an external electric field on a process
λ - de Broglie wave (or Compton) of the tunneling electron
l - electrical length or distance at which the electric field does the work required for process execution (in pure electric field coincides with the length of the triangular potential barrier).
$\hbar$ - Planck constant
m - mass of the electron (in a semiconductor - effective mass)
W - electron binding energy (for tunneling from the surface is equal to the local work function).
j - tunneling current density of (from) the semiconductor surface.
e - elementary charge.
E –electric field
p - quantum mechanical probability of the process (eg, tunneling from a potential well at the semiconductor surface).
D, H – constants, containing the studied variables
W_z- kinetic energy of the electron motion along the axis z, normal to the surface.
W_m- maximum value of the kinetic energy of the electron along the axis z.
$\xi(W_z, eU)$- so-called energy density of surface electronic states, to be exact - the electron flux density of the energy surface
d - distance between the probe and the surface, or the width of the trapezoidal potential barrier
β- parameter of degree of uniformity of filling surface electronic states.

References

1. M. Celestin, S. Krishnan, Sh. Bhansali, E. Stefanakos, Nano Research, **7**(5), 589 (2014)
2. A.I. Loskutov, O.Ya. Uryupina, S.N. Grigoriev, V.B. Oshurko, V.I. Roldughin, Colloid Journal, **75**(3), 301 (2013)
3. H. Luth, *Solid Surfaces, Interfaces and Thin Films*.(Springer, Berlin, 2010)
4. M. Grundmann, *The Physics of Semiconductors* (Springer, Berlin, 2006)
5. J. Martin, N. Acerman, G.Ulbricht, T.Lohmann, J.H. Smet, K. von Klitzing, A. Yacoby, Nat. Phys., **4**(144),. 46-89 (2008).
6. J. Tersoff, D. Hamann, Phys. Rev. B, **31**(2), 805 (1985).
7. V.N. Rodionov, G.A. Kravtsova, A.M. Mandel, JETP Letters, **75**(4), 363 (2002)
8. V.G. Kadyshevskii, G.A Kravtsova, A.M. Mandel,V.N. Rodionov, Theoretical and Mathematical Physics, **134**(2), 198 (2003)
9. V.N. Rodionov, A.M. Mandel, E.V. Arbuzova, arXiv: hep-ph/0510357v1, 11 (2005)
10. J. G.Simmons, J. Appl. Phys., **34**, 1793 (1963)
11. M. Mandel, S.N. Grigoryev, A.I. Loskutov, V.B. Ohurko, G.I. Solomakho, Journal of Computational and Theoretical Nanoscience, **12**(11), 1 (2015)
12. N.L. Manakov, M.V. Frolov, A.F. Starace, I.I. Fabrikant, J. Phys. B: At. Mol. Opt. Phys., **33**, R141 (2000)
13. V.N. Rodionov, G.A. Kravtsova, A.M. Mandel, JETP Letters, **78**(4), 218 (2003)
14. A.A. Kelly, K.M. Knowls, *Crystallography and Crystal Defects* (Willey&sons, London, 2012)
15. D.D.L. Chung, J. of Mater. Sci., **37**, 1475 (2002)

16. M. Fujita, K. Wakabayashi, K. Nakada, K. Kusakabe, J. Phys. Soc. Jpn, **65**, 1920 *(1996)*
17. Y. Niimi, T. Matsui, H. Kambara, K. Tagami, M. Tsukada, H. Fukuyama, Phys. Rev. B, **73**, 085421 *(2006)*
18. R.M. Feenstra, J.A. Stroscio, A.P. Fein. Phys. Rev. Lett., **58**, 1192 (1987)
19. R.M. Feenstra, J.A. Strościo. J. Vac. Sci. Technol.B, **5**(4), 923 (1987)
20. R.M. Feenstra, Phys. Rev. B, **50**(7), 4561 (1994)

Identification of the Oculo-Motor System in the Form Volterra Model Based on Eye-Tracking Data

Vitaliy D. Pavlenko[1,*], *Tetiana V.* Shamanina and *Vladislav V.* Chori

[1] Computerized Control Systems Department, Odessa National Polytechnic University, UK-65044, Odessa, Ukraine

Abstract. Instrumental computing and software tools have been developed for constructing a nonlinear dynamic model of the human oculo-motor system (OMS) based on the data of input-output experiments using test visual stimulus and innovative technology. Volterra model in the form of multidimensional transition functions of the 1st, 2nd and 3rd orders, taking into account the inertial and nonlinear properties of the OMS was used as the identification tool. Eye-tracking data developed in the Matlab environment are tested on real datasets from an experimental study of OMS.

1 Introduction

The study of human eye movements and the trajectory of their movement allows us to reveal the structure of the relationship of an individual with the environment, a person with the world. Knowledge about eye movement is of great theoretical and applied importance, expanding the possibilities of studying the specifics of many professions in order to improve the efficiency of the subject of labor activity [1-4].

The process of mastering knowledge is a central part of the learning process. Managing this process implies the existence of effective objective indicators for assessing an individual's intellectual abilities. The methods of psychological identification of an individual proposed in the project based on obtaining experimental data using the innovative Eye-tracking technology and computing means of processing them allow monitoring and diagnostics of the state of cognitive processes during the educational activities of students [5-9].

The aim of this work is to develop instrumental software tools for constructing a nonparametric dynamical model of the OMS human, taking into account its inertial and nonlinear properties, based on data from experimental input-output studies using test visual stimulus and innovative eye-tracking technology; implementation of the received information models in practice diagnostics of states cognitive processes.

1.1 The scope of application

The developed software enables support of the following tasks:

– The relationship study of mental states and cognitive processes in educational activities, a post-traumatic stress disorder, the diagnosis of the Parkinson's and Alzheimer's disease stage, checking the psychophysiological state of pilots and drivers, the professional suitability, the fatigue syndrome [10-16].

– The interaction of mental states and cognitive processes during the educational activities of students, an objective assessment of their cognitive development level, assessment of the effectiveness of training to improve mental processes and for psychological correction of personality [16].

– Extension of the individual's creative life due to the early diagnosis of degenerative processes of cognitive functions of the brain. Identification of a gifted personality (building a psychological model of the personality) and evaluation of its abilities. Professional selection (the identification and education of leaders) [9].

– The assimilation of scientific knowledge and their respective skills serves as the main goal and the main result of educational activities. The process of mastering knowledge is the central part of the learning process. Managing this process implies the existence of effective objective indicators for assessing an individual's intellectual abilities [7].

The methods of psychological identification of an individual proposed in the project, based on obtaining experimental data using eye tracking technology and computing means of processing them, allow monitoring and diagnostics of the state of cognitive processes during the educational activities of students.

1.2 Diagnostics neuronal processes

An intelligent information technology for diagnosing the states of neural processes based on nonparametric identification of OMS in the form of nonlinear dynamic Volterra models is proposed [17, 18]. The technology involves a consistent solution of the following tasks:

– *Identification of the OMS.* The *goal* is to obtain an OMS information model in the form of MTF. *Stages of implementation*: the supply of test signals to the inputs of the OMS (horizontally, vertically, diagonally); measurement of OMS responses to test signals with the

* e-mail: pavlenko_vitalij@mail.ru

help of an eye-tracker; MTF calculation based on input-output experiment data.

– *Building a diagnostic model of OMS.* The *goal* – the formation of the feature space. *Stages of implementation*: MTF compression; determination of the diagnostic value of symptoms; selection of the optimal system of signs – reduction of the diagnostic model.

– *Building a classifier of the psycho-physiological state of the individual based on the OMS model.* The *goal* is to build a family of decision rules for optimal classification. *Stages of implementation:* building decision rules based on OMS identification results – training; assessment of the accuracy of the classification – the exam; optimization of the diagnostic model.

– *Diagnostics psycho-physiological state of the personality.* The *goal* is to assess the condition of the individual. *Stages of implementation:* OMS identification; evaluation of diagnostic signs; classification – the assignment of the investigated individual to a particular class.

2 Approximation Volterra model of the nonlinear dynamical system

The input-output ratio for a nonlinear dynamical system (NDS) with an unknown structure (such as a "black box") with a single input and a single output can be represented by a discrete cubic Volterra polynomial in the form [17]:

$$y[m]=\sum_{n=1}^{3}y_n[m]=\sum_{k_1=0}^{m}w_1[k_1]x[m-k_1]+$$
$$+\sum_{k_1=0}^{m}\sum_{k_2=0}^{m}w_2[k_1,k_2]x[m-k_1]x[m-k_2]+ \qquad (1)$$
$$+\sum_{k_1=0}^{m}\sum_{k_2=0}^{m}\sum_{k_3=0}^{m}w_3[k_1,k_2,k_3]x[m-k_1]x[m-k_2]x[m-k_3],$$

where $w_1[k_1]$, $w_2[k_1,k_2]$, $w_3[k_1,k_2,k_3]$ – discrete weight functions (Volterra kernels) of the 1st, 2nd and 3rd orders; $x[m]$, $y[m]$ – input (stimulus) and output (response) function (signals) of the system, respectively; $y_n[m]$ – partial components of the response (convolution of n-th order sequences); m is a discrete time variable.

The problem of identification is to choice test signals $x[m]$ and develop an algorithm that allows based on the responses received $y[m]$ to identify partial components $y_n[m]$, (n = 1, 2, 3) and determine on their basis multidimensional Volterra kernels: $w_1[k_1]$, $w_2[k_1,k_2]$, $w_3[k_1,k_2,k_3]$ [18].

Taking into account the specifics of the studied OMS, test step signals are used for identification. If the test signal $x[m]=\theta[m]$, where $\theta[m]$ is a unit function (Heaviside function), then the partial components of the response $y_1[m]$, $y_2[m]$, $y_3[m]$ will be equal to the transient function of the first order $h_1[m]$ and diagonal sections of the transient functions of the second and third orders $h_2[m,m]$, $h_3[m,m,m]$, respectively [19-22]:

$$y_1[m]=h_1[m]=\sum_{k_1=0}^{m}w_1[m-k_1],$$
$$y_2[m]=h_2[m,m]=\sum_{k_1=0}^{m}\sum_{k_2=0}^{m}w_2[m-k_1,m-k_2], \qquad (2)$$
$$y_3[m]=h_3[m,m,m]=\sum_{k_1=0}^{m}\sum_{k_2=0}^{m}\sum_{k_3=0}^{m}w_3[m-k_1,m-k_2,m-k_3].$$

Determination of subdiagonal intersections of transient functions is based on the NDS test using L test step signals with given amplitudes a_i, $i=1,2,\ldots,L$ ($L>=N$, N is the degree of the Volterra polynomial). In this case the responses of the NDS are denoted by $y_1[m]$, $y_2[m]$, …, $y_L[m]$. Reviews of the Volterra model will be view

$$\tilde{y}_i[m]=a_i\hat{y}_1[m]+a_i^2\hat{y}_2[m]+a_i^3\hat{y}_3[m],\ i=\overline{1,L}, \qquad (3)$$

where $\hat{y}_1[m]=\hat{h}_1[m]$, $\hat{y}_2[m]=\hat{h}_2[m,m]$, $\hat{y}_3[m]=\hat{h}_3[m,m,m]$ – obtained estimates of the partial components of the model – multidimensional transition functions.

To determine the transition functions $h_1[m]$, $h_2[m,m]$, $h_3[m,m,m]$ is used the method of least squares (LSM), which provides the minimum standard error of the deviation of the model responses from the responses of the OMS to the same stimulus:

$$J_N=\sum_{j=1}^{L}\left(y_j[m]-\sum_{n=1}^{N}a_j^n\hat{y}_n[m]\right)^2\to\min. \qquad (4)$$

The minimization of criterion (4) is reduced to solving a system of normal Gaussian equations, which in vector-matrix form can be written as

$$\mathrm{A}'\mathrm{A}\hat{\mathrm{y}}=\mathrm{A}'\,\mathrm{y}, \qquad (5)$$

where $\mathrm{A}=\left\|\alpha_{ij}\right\|$, $\alpha_{ij}=a_i^j,\ i,j=\overline{1,N}$.

3 Computing of transient functions OMS

Information technology of the constructing a nonparametric dynamic model of the human OMS taking into account its inertial and nonlinear properties based on the data of experimental studies input-output was developed. As a basic OMS model – the Volterra model is used in the form of multidimensional transition functions.

Methods and tools for the identification of OMS have been developed using the help of eye tracking technology, and building a features space and optimal classification human states using machine learning. In the Laboratory of Motion Analysis and Interface

Ergonomics at the Lublin University of Technology (Lublin, Poland), joint studies of the human OMS were performed to obtain diagnostic information for solving urgent problems in the neuro informatics and the computational neuroscience. Experimental research was carried out using eye tracking technology with the use of the video based Tobii TX300 (300 Hz sampling rate) eye tracker and appropriate software [16].

The following instrumental algorithmic and software tools are developed to achieve the goal of the research:

– Formation of test signals in the form of bright dots on the computer monitor screen at different distances from the initial position horizontally, vertically and diagonally.

– Preprocessing (bringing the OMS responses to a common start and rationing to one) and analyzing the data obtained from the eye tracker.

– Constructing an identification model of OMS in the form of multidimensional transitional functions (integral transformations of Volterra kernels).

– Visualization of data and processing results of experimental research.

3.1 Experimental research of the OMS

When conducting experimental studies, such actions are carried out:

– The test subject is placed in front of the computer so that his eyes are at the center of the monitor at a distance of 40-50 cm from him.

– The subject's head is fixed in order to prevent its movements during the study and to ensure the same experimental conditions.

– On the subject's readiness, the Signal Manager of the test visual stimulus program is launched.

– A red circle appears in the center (or from its edge) – of the screen in the starting position.

– After a short pause (2-3 sec.), the circle in the starting position disappears and a circle of a different color appears at the point with the specified coordinates – a visual stimulus (test signal), which is displayed in this position for a specified duration (1-2 sec.) – the action makes the eye move in the direction of the visual stimulus.

– Then this stimulus circle disappears and a red circle appears in the starting position – this makes the eye move in the opposite direction to the starting position, after these actions the experiment ends.

– Using the eye tracker, the coordinates of the pupil of the eye are determined during its movement (reaction to the visual stimulus) in the period between the starting positions and the coordinate values are stored in the xls-file.

In the studies of each respondent, three experiments were successively implemented for three amplitudes of test signals in the horizontal direction. The distance between the starting position and the test incentives is equal to: 0.33 l_x, 0.66 l_x, 1.0 l_x, where l_x is the length of the monitor screen. Coordinates of the starting position (x = 0, y = 0.5 l_y), l_y – mean the width of the monitor screen.

The obtained results of measurements of the OMS responses at L=3 obtained with using the Tobii TX300 eye tracker in one study cycle ("Horizontally") are shown in Figure 1. Transient process in the OMS response to the test signal a_1 = 0.33 are illustrated on Figure 2.

4 Results

The experiments were organized in order to classify subjects by the state of fatigue. The data for constructing the model – the OMS responses to the same test signals, were obtained using the Tobii Pro TX300 eye tracker at different times of the day: "In the Morning" (before work) and "In the Evening" (after work). The average values of the OMS responses obtained from the eye tracker at various amplitudes of the test signals "In the Morning" and "In the Evening" are shown in Figure 3.

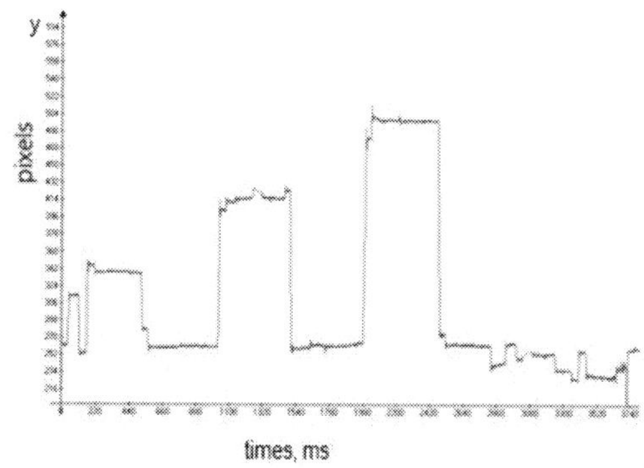

Fig. 1. OMS responses at L=3 obtained using the Tobii TX300 eye tracker

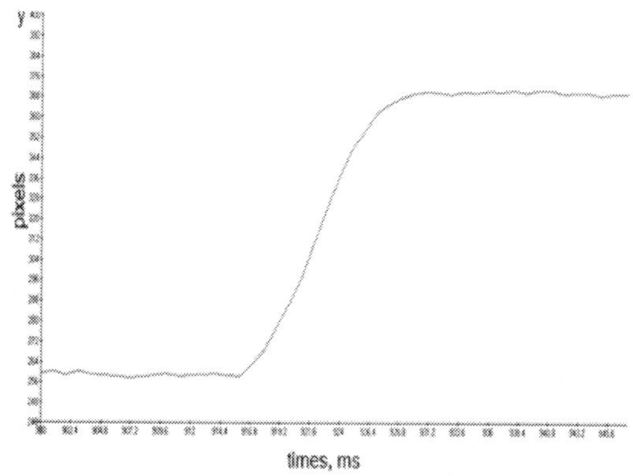

Fig. 2. Transition process in the OMS response to the test signal: a_1 = 0.33

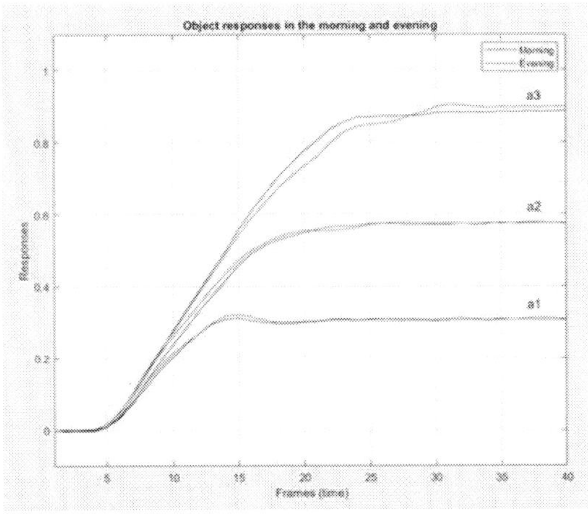

Fig. 3. Averaged OMS responses at various amplitudes of test signals "In the Morning" and "In the Evening"

According to averaged data of OMS responses on visual stimuli with a different distance from the start position on the basis of formula (5) the functions of the OMS were defined when approximation models of degrees $N = 3$ were used. Graphs of the transition functions estimates for the "In the Morning" and "In the Evening" states of the subject based on model (1) are shown in Figure 4.

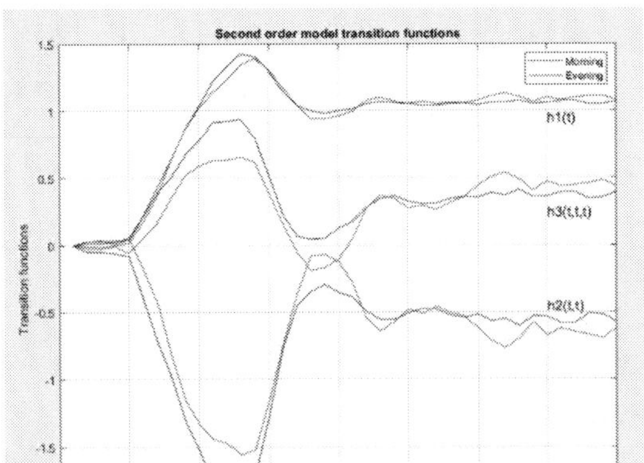

Fig. 4. Transition function estimates of the 1st, 2nd, and 3rd orders ($N = 3$) for states of the test subject "In the Morning" and "In the Evening"

Received responses with the help of calculations on models at $N = 3$ from various amplitudes of test signals "In the Morning". Graphs these are presented in comparison with similar responses OMS in Figure 5. Graphs of responses of the model OMS at $N = 3$ at various amplitudes of the test signals "In the Morning" and “In the Evening” are illustrated in Figure 6.

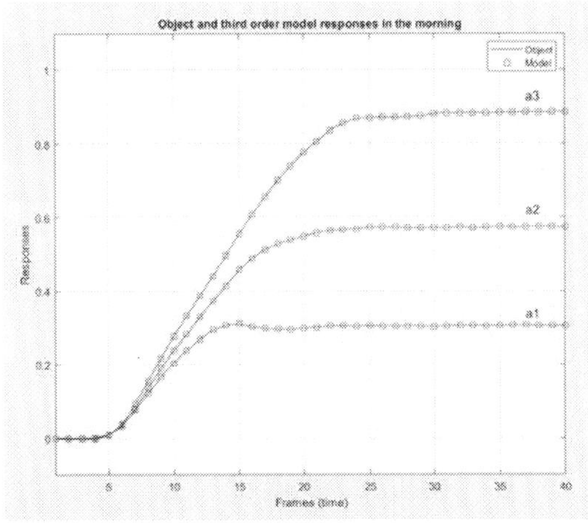

Fig, 5. Responses of the OMS and the model at $N = 3$ at various amplitudes of the test signals "In the Morning"

4.1 Deviation of the transient functions

The variability (deviation) of the MTF of different orders n of the approximation model of OMS for the states of the respondent "In the Morning" and "In the Evening" is quantified using the indicator of ε_{nN} – normalized standard deviation (6). The indicators deviation of the MTF of different orders n of the OMS approximation model for respondent states "In the Morning" and "In the Evening" are given in Table 1 and are represented by diagram in Figure 7.

$$\varepsilon_{nN} = \left(\frac{\sum_{m=0}^{M} (\hat{y}_{ne}[m] - \hat{y}_{nm}[m])^2}{\sum_{m=0}^{M} (\hat{y}_{nm}[m])^2} \right)^{1/2}, \quad n = \overline{1, N}. \quad (6)$$

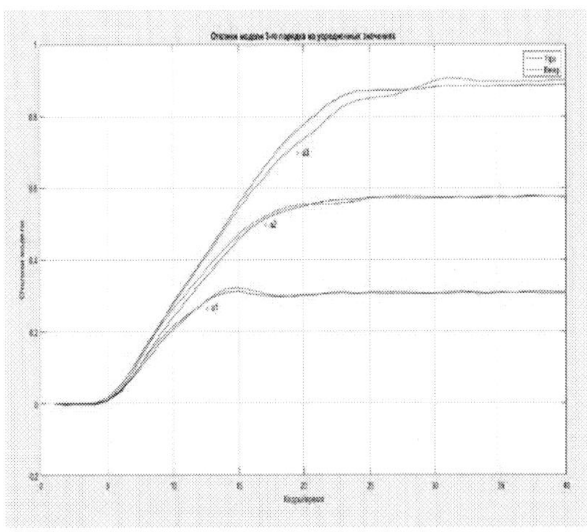

Fig. 6. The responses of the model at $N = 3$ at various amplitudes of the test signals "In the Morning" and "In the Evening"

Table 1. The deviation indicators of multidimensional transient functions

N	ε1N	ε2N	ε3N
1	0.019	–	–
2	0.051	0.232	–
3	0.04	0.199	0.322

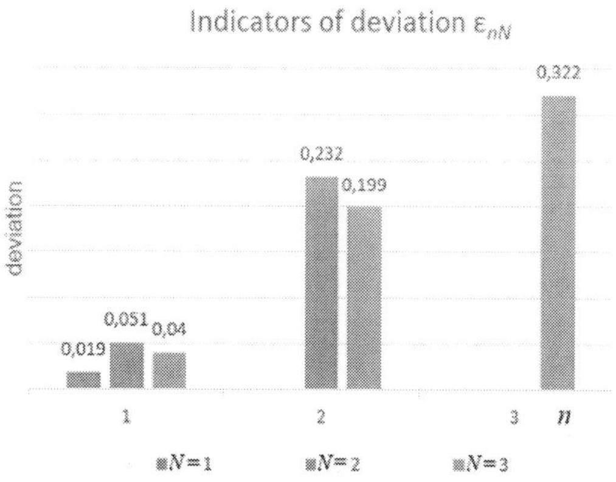

Fig. 7. Diagram of deviations indicators ε_{nN}

As can be seen from figure 5, the obtained transition function of the 1st order for the "In the Morning" and "In the Evening" are virtually independent of the status of the subject. However, the diagonal cross section of the transition functions of the second and third order change significantly in magnitude and, therefore, can be effectively used as the primary data source when building models of classifiers of psychophysiological conditions of the person using machine learning

4.2 Building a classifier of the fatigue

For estimate psycho-physiological state of the individual based on the OMS model conducted researchers:

– Building a feature space for designing the status classifier of a human using machine learning.

– Classifiers construction using deterministic and statistical methods of learning the pattern recognition based on the data obtained using eye tracking technology.

On base of training sets of data for an object's classes **A** ("In the Morning"), **B** ("In the Evening") there successively calculate discriminant function $d(\mathbf{x})$. To separate the two classes (dichotomy case) it uses discriminant function [23-25] of the form:

$$d(\mathbf{x}) = \frac{1}{2}\mathbf{x}'(\mathbf{S}_2^{-1} - \mathbf{S}_1^{-1})\mathbf{x} + (\mathbf{S}_1^{-1}\mathbf{m}_1 - \mathbf{S}_2^{-1}\mathbf{m}_2)'\mathbf{x} + \\ + \frac{1}{2}(\mathbf{m}_1'\mathbf{S}_1^{-1}\mathbf{m}_1 - \mathbf{m}_2'\mathbf{S}_2^{-1}\mathbf{m}_2 + \ln\frac{|\mathbf{S}_2|}{|\mathbf{S}_1|}) + \lambda_{\max} \quad (7)$$

where $\mathbf{x}=(x_1,x_2,\ldots,x_n)'$ – features combination, n – features space dimensionality, $\mathbf{m}_i$ – mathematical expectation vector for a features of class i, i=1, 2; $\mathbf{S}_i$=M[$(\mathbf{x}-\mathbf{m}_i)(\mathbf{x}-\mathbf{m}_i)'$] – covariance matrix for class i (M[] – mathematical expectation operation), $\mathbf{S}_i^{-1}$ – matrix inverse to $\mathbf{S}_i$, $|\mathbf{S}_i|$ – matrix determinant $\mathbf{S}_i$, $\lambda_{\max}$ – classification threshold that provides the highest criterion probability of the correct recognition of objects in the training sample.

The analysis of the quality of the combination of various features on base of the criterion probability of the correct recognition (P) is made. Quality of selected features combination from considered features set is evaluated by the result of classification on examination sample of data. A Bayesian classifier of a person's fatigue state in the space of featurs x_1 and x_2:

$$x_1 = \arg\max_t h_1'(t), \quad x_2 = \arg\min_t h_1'(t). \quad (8)$$

The estimation of the indicator of recognition reliability is $P = 0{,}9375$.

The classification of this data by the support vector method (SVM) [26] with a quadratic separating function in the space of features x_1 and x_2 gives the same value of the probability of correct recognition (P = 0,9375). The location of the objects of the training set in the space of features x_1 and x_2 is shown in Figure 8.

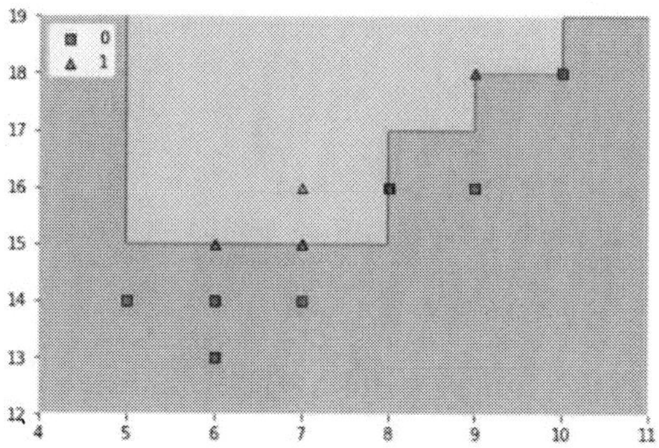

Fig. 8. The location of the objects of the training set in the space of features x_1 and x_2

5 Conclusions

Method and software have been developed for constructing a nonparametric dynamical model of the OMS human, taking into account its inertial and nonlinear properties, based on experimental input-output data using test visual stimuli and innovative eye-tracking technology. A nonlinear dynamic model in the form of a Volterra polynomial was used. The developed model was tested on real datasets from an experimental study of OMS.

Test visual stimulus in the form of bright points that are consistently displayed at different distances from the starting position are used. This formally corresponds to the different amplitudes of the step test signals. The transition functions of the 1st, 2nd, and 3rd orders are determined using LSM.

The developed software tools for data processing of the eye-tracking are tested on real data from an experimental study of the OMS. Verification of the constructed model confirms the adequacy model of the investigated OMS – a practical coincidence (within an acceptable error) of the responses of the OMS and its model at the same test signal.

The revealed variability of the transition functions of the 2nd and 3rd orders for different psychophysiological states of the respondent (level of fatigue) has observed. Thus, they can be used in diagnostic studies in the field of the neuroscience and psychology.

References

1. Y. Wamain, X. Corveleyn, L. Ott, Y. Coello, Neuropsychologia, **132**, 107121 (2019)
2. M. Daoudi, M. Coello, P. Desrosiers, L. Ott, Proc. FG 2018 (Xi'an, China, 2018)
3. F. Quesque, A.Mignon, Y. Coello, Consciousness and Cognition (2017)
4. L.S. Lanata, F.Di Gruttola, S.Di Modica, E.P. Scilingoб A. Greco1, Frontiers in Neuroscience, **13**, 1431 (2020)
5. N. Gueugneau, L. Crognier, C. Papaxanthis, Brain Research, **1187** (2008)
6. A. Guillot, C. Collet, Journal of Motor Behaviour, **37** (2005)
7. A. Guillot, C. Collet, Int. Review of Sport and Exercise Psychology, **1** (2008)
8. A. Guillot, C. Collet, V.A. Nguyen, F. Malouin, C. Richards, J. Doyon, Neuroimage, **41** (2008)
9. A. Guillot, M. Louis, C. Collet, In A. Guillot, C. Collet (Eds), *The neurophysiological foundations of mental and motor imagery* (Oxford University Press, 2010)
10. M. El Haj, Y. Coello, D. Kapogiannis, K. Gallouj, P. Antoine, J. of Alzheimer's Disease, **61**(2) (2018)
11. X. Corveleyn, J. Blampain, L. Ott, I. Lavenu, C. Delayen, A.Di Pastena, Y. Coello, Current Alzheimer Research, **15**(3) (2018)
12. D. Jansson, A. Medvedev, H Axelson, D. Nyholm, Advances in Experimental Medicine and Biology, **823** (2015)
13. D. Jansson, O. Rosén, A. Medvedev, IEEE Trans. control system technology, **23** (2015)
14. V. Bro, A. Medvedev, IEEE 56th Annual Conference on Decision and Control (Melbourne) (2017)
15. I. Rigas, O. Komogortsev, R. Shadmehr, ACM Trans. on Applied Perception, **13**(2) (2016)
16. V.D. Pavlenko, M. Milosz, M. Dzienkowski, J. of Physics: Conference Series, **1603** (IOP Publishing, 2020)
17. F.J. Doyle, R.K. Pearson, B.A. Ogunnaike, *Identification and control using Volterra models* (Springer Publ, 2002)
18. V. Pavlenko, S. Pavlenko, V. Speranskyy, In.V. Haasz, K.. Madani (Eds.), *Advanced Data Acquisition and Intelligent Data Processing* (River Publishers, 2014)
19. V. Pavlenko, D. Salata, M. Dombrovskyi, Y. Maksymenko, AIP Conf. Proc., **1872** (2017)
20. V. Pavlenko, I. Ivanov, E. Kravchenko, Proc. IDAACS'2017, **2** (Bucharest, Romania, 2017)
21. V. Pavlenko, D. Salata, H. Chaikovskyi, Int. J. of Biology and Biomedical Engineering, **11** (2017)
22. A. Fomin, M. Masri, V. Pavlenko, A. Fedorova, Eastern Eur. J. of Enterprise Technologies, **2/9**(74) (2015)
23. O. Fomin, V. Pavlenko, MMAR-2015 (Międzyzdroje, Poland) (Publisher IEEE, 2015)
24. A. Medvedev, O. Fomin, V. Pavlenko, V. Speranskyy, Proc. IDAACS'2017, **2** (Bucharest, Romania, 2017)
25. O. Fomin, A. Medvedev, V. Pavlenko, Proc. IDAACS'2015 (Warsaw, Poland, 2015)
26. V. Vapnik, *The Nature of Statistical Learning Theory* (Springer-Verlag New York Inc., 2010)

Non-Euclidean Biosymmetries and Algebraic Harmony in Genomes of Higher and Lower Organisms

Sergey Petoukhov[1, 2*], *Elena* Petukhova[1], and *Vitaly* Svirin[1, 2]

[1]Mechanical Engineering Research Institute of RAS, RU-101990, Moscow, Russia
[2]Moscow State Conservatory, Russia, RU-125009, Moscow, Russia

Abstract. The article is devoted to the study of the relationship of non-Euclidean symmetries in inherited biostructures with algebraic features of information nucleotide sequences in DNA molecules in the genomes of eukaryotes and prokaryotes. These genomic sequences obey the universal hyperbolic rules of the oligomer cooperative organization, which are associated with the harmonic progression 1/1, 1/2, 1/3,.., 1/*n*. The progression has long been known and studied in various branches of mathematics and physics. Now it has manifested itself in genetic informatics. The performed analysis of the harmonic progression revealed its connection with the cross-ratio, which is the main invariant of projective geometry. This connection consists in the fact that the magnitude of the cross-ratio is the same and is equal to 4/3 for any four adjacent members of this progression. The long DNA nucleotide sequences have fractal-like structure with so called epi-chains, whose structures are also related to the harmonic progression and the projective-geometrical symmetries. The received results are related additionally to a consideration of DNA double helix as helical antenna. This fact of the connection of genetic informatics with the main invariant of projective geometry can be used to explain the implementation of some non-Euclidean symmetries in genetically inherited structures of living bodies.

1 Introduction

The phenomena of the symmetric inherited structure of biological bodies have been the subject of research since ancient times. The importance of studying these phenomena is described, for example, in classic books [1, 2]. At the same time, until recently, all attention was paid to the biosymmetries of similarity, which are described using a group of geometric similarity transformations. According to F. Klein's Erlangen program, this group of transformations underlies Euclidean geometry and includes transformations of translations, rotations, specular reflections and scaling. At the same time, the hypothesis by V.I. Vernadsky on the possible important morphological significance of non-Euclidean symmetries in living bodies is widely known.

Clarification of the geometric foundations of morphology entails shifts in various areas of biomechanics and physiology: biomechanics of postures, biomechanics of growth and motor movements, psychophysics, etc. In the variety of organic forms, multiblock structures are of particular interest, the components of which are mutually conjugated on the basis of certain algorithms, repeated on the most diverse lines and branches of biological evolution. The general rule of such varieties, consisting of blocks or motivic units, is that the previous motivic unit is translated into the subsequent one by some fixed Euclidean or non-Euclidean transformation. In other words, so organized segmented varieties have a cyclic group of automorphisms, and their motivic units are arranged along the orbit of the corresponding cyclic group. Earlier, in the works of the Institute of Mechanical Engineering Research Institute of the Russian Academy of Sciences, a wide representation in the inherited configurations of the set of living bodies of non-Euclidean symmetries was revealed, which are based on the groups of projective and Möbius transformations, which generalize the group of similarity transformations and contain it as their subgroup [3].

Genetically inherited morphological structures must be structurally linked to the molecular genetic system for their genetically coding and transmission along the chain of generations. Therefore, it is important to look for traces of symmetries and invariants of non-Euclidean transformations in the genetic informatics of living bodies, including in the information texts of DNA heredity molecules carrying genetic information in the form of long nucleotide sequences. As a result of the work of the laboratory for research of biomechanical systems of the Mechanical Engineering Research Institute of the Russian Academy of Sciences in 2020, universal hyperbolic rules for the cooperative organization of genomes of higher and lower organisms were discovered [4, 5]. These rules are associated with harmonic progression (1), which has long been known in various fields of science and has now received the status

* Corresponding author: spetoukhov@gmail.com

of an important participant in genetic informatics, capable of influencing the characteristics of inherited biosymmetries:

$$1/1, 1/2, 1/3, 1/4, \ldots, 1/n \quad (1)$$

In this article, we consider the mathematical connections of this progression with the groups of projective and Möbius transformations, on which non-Euclidean symmetries are based. Besides we consider fractal-like features of long nucleotide sequences in helical DNA, which are also related to the harmonic progression (1) and give new materials to the known hypothesis about DNA double helix as a helical antenna. The article contains 5 sections and conclusions.

2 Computational approaches, models

Our main computational approaches and models of biological structures are based on the mathematical properties of harmonic progression (1), which were studied by Pythagoras, Orem, Leibniz, Newton, Euler, Fourier, Dirichlet, Riemann and many other researchers in connection with various problems of physics and mathematics. Nowadays this list is supplemented with the tasks of studying its role in the genetic information fundamentals of living matter and the structuring of inherited biological systems. Its historically known name - "harmonic progression" - sequence (1) was received due to its connection with a number of harmonics in music (or a number of lengths of standing waves when vibrating a stretched string). This article presents the results of an analysis of the conjugation of this progression with the main invariant of projective geometry, called the double ratio (as well as the wurf and anharmonic ratio). In the one-dimensional case, the group of projective transformations and the group of Möbius transformations coincide, and the double relation is simultaneously an invariant of both projective and Möbius transformations. In our research on non-Euclidean biosymmetries, we actively use the provisions of the famous Erlangen program of F. Klein, which defined geometry as the science of invariants of the transformation group. In accordance with this program, each group of transformations has its own geometry, which has its own individual set of invariants. In particular, the groups of projective and Möbius transformations define projective and Möbius geometries (the latter is also called the geometry of reciprocal radii or conformal geometry).

The cross-ratio (or the double ratio or the wurf) of the four numbers a, b, c, d is determined by the expression (2):

$$(ab, cd) = \{(c-a)(d-b)\} / \{(c-b)(d-a)\} \quad (2)$$

If the four numbers a, b, c, d correspond to the coordinates of four points on the number line, then expression (2) consists of the lengths of the corresponding segments between the indicated pairs of points. Under projective transformations of such a straight line, the lengths of the segments and the simple relations between them change, but the value of the cross-ratio remains unchanged.

3 Some results: projective symmetries in the harmonic progression

One of the properties of the harmonic progression (1) is that the value of the cross-ratio of any of its four neighboring numbers is equal to 4/3. Any four adjacent numbers in a series of natural numbers 1, 2, 3,..., n also have the same value of the cross-ratio. This means that the harmonic progression (1) and the natural series of numbers are projectively equivalent to each other (Fig. 1).

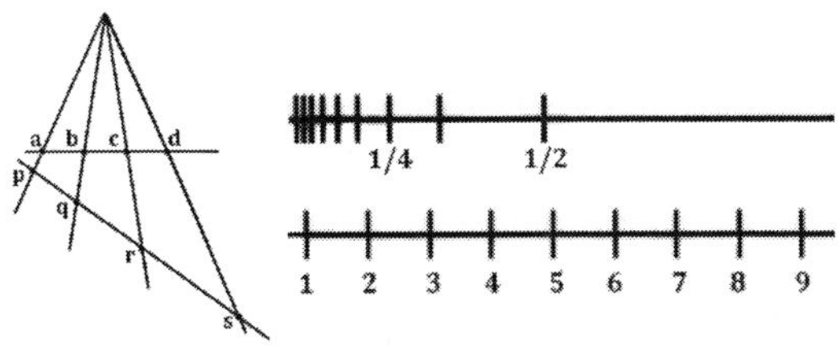

Fig. 1. Examples of projective transformation. Left: a central projective transformation changes the lengths of the line segments and their simple ratios, but retains the magnitude of the cross-ratio. Right: the location of the numbers of the harmonic progression (1) and a series of natural numbers on the number line.

The connection of the harmonic progression (1) with the cross-ratio leads to the following conclusions:

1. Various fragments of a harmonic progression can be considered as projectively symmetric to each other;
2. In the genomes of eukaryotes and prokaryotes, various hyperbolic sequences of oligomeric sums corresponding to the classes of A-, T-, C- and G-oligomers [5] are projectively similar to each other with the high accuracy with which hyperbolic rules are fulfilled in them;
3. All investigated genomes of eukaryotes and prokaryotes with respect to their sequences of oligomeric sums are projectively similar to each other (with the high accuracy, with which hyperbolic rules are fulfilled in them).

Similar hyperbolic rules associated with harmonic progression (1) are fulfilled not only in genomic DNA sequences of nucleotides, but also in the studied long proteins (such as the Titin protein containing 33423 amino acids and presented at https://www.ncbi.nlm.nih.gov/protein/ ACN81321.1) in relation to oligomer sums in their 20 classes of amino acids. The point is that in the case of protein sequences of 20 types of amino acids, 20 equivalence classes can be analyzed, each of which is determined by the corresponding type of amino acid and combines all n-oligomers (fragments of length n) of a given protein that begin with a given amino acid. For example, the amino acid Ala defines a class of Ala-oligomers, which includes all n-oligomers starting with this amino acid: the subset of Ala-doublets contains all 20 doublets that start with Ala (AlaAla, AlaArg, AlaAsn, ..., AlaCys); the subset of Ala-triplets contains all 400 triplets that start with Ala (AlaAlaAla, AlaAlaArg,...., AlaCysCys), and so on. The quantitative compositions of these subsets of n-oligomers of each of these amino acid classes for different n = 1, 2, 3, ... (here n is much less than the protein length) are mutually related with high accuracy and obey hyperbolic rules associated with harmonic progression (1). Accordingly, taking into account the mentioned connection of this progression with the cross-ratio, we can talk about the projective-geometric (as well as the Möbius-geometric) similarity of the named quantitative characteristics of the structure of long proteins such as the Titin protein [5].

Taking into account that the numerical characteristics of genetic sequences are associated with the main invariant of projective geometry, then it can be expected that projective-geometric and Möbius-geometric symmetries will appear in genetically inherited physiological structures. This allows us to consider in a new light the existence of the non-Euclidean symmetries previously established in the works of our Mechanical Engineering Research Institute of RAS in a wide variety of morphogenetic structures and growth processes in living bodies [3].

For example, the cross-ratio was applied to the analysis of growth changes in the longitudinal proportions of three-membered kinematic blocks in the human body: three-phalangeal fingers, three-membered limbs (shoulder-forearm-hand and thigh-calf-foot) and the entire three-membered body (in anthropology, the body is subdivided into the upper, trunk and lower segments). The value of the cross-ratio of the lengths of the three members in all these trebled blocks, at least during the entire postnatal period, is grouped around the reference value of 1.3. This is all the more interesting because the growth of the human body is essentially non-linear (Fig. 2): for example, during a person's life from birth, the upper part of the body grows 2.4 times, the trunk - 2.8 times, and the lower part - 3.8 times. This invariance of the cross-ratio value turns out to be largely true not only for normally developed people, but also for dwarfs and giants. This is also true for a wide range of highly organized animals.

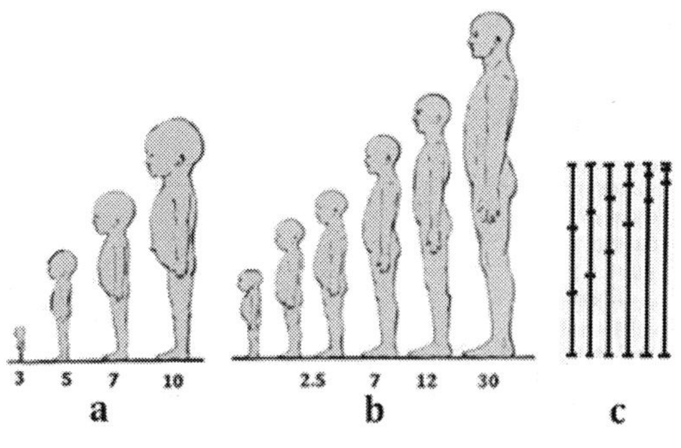

Fig. 2. Age-related changes in the proportions of the human body (from [3]). a, b - antenatal and postnatal stages: a - in the lunar months; b - in years (first newborn on the left); c - examples of projectively equivalent three-term segments, the cross-ratio values of which are equal to 1.3.

We note additionally that projective-geometric transformations and the cross-ratio as their main invariant are mathematical tools, on which the study of visual perception and the creation of methods of computer vision, pattern recognition, and artificial intelligence are largely based. A huge number of publications are devoted to this topic (see, for example, [6, 7]. The study of the relationship between the genetic system and genetically inherited biological structures, which are connected with non-Euclidean symmetries and harmonic progression, continues, including for the development of genetic biomechanics and algebraic biology. These issues, in particular, are considered at the International Interdisciplinary Seminar "Algebraic Biology and Systems Theory" (https://www.youtube.com/channel/UC8JLsuRzzPsRiHwrwEjMCtw).

The creators of quantum mechanics P. Jordan and (later) E. Schrödinger pointed out the key difference between living bodies and inanimate ones: inanimate objects are controlled by the average random motion of their millions of particles and the motion of an individual particle is not essential for the whole; on the contrary, in a living organism, the chosen - genetic - molecules have a dictatorial (!) influence on the entire organism (see the history of quantum biology [8]). In accordance with this statement, in our research we focus on the properties of DNA information sequences and study their influence on inherited biological structures.

Jordan in his work on quantum biology claimed that life's missing laws were the rules of chance and probability of the quantum world [8]. From the standpoint of Jordan's statement, the study of probabilities or frequencies of n-plets in long DNA sequences is important for discovering hidden biological laws and for developing quantum biology. Our study is closely connected with this direction of thoughts and, in particular, gave the following model.

The alphabet of 4 nucleotides of DNA is endowed with binary opposition traits "purine-or-pyrimidine" and "strong-or- weak hydrogen bonds". Taking this binary-oppositional indicators into account, one of the coauthors of this article created a quantum-information model of long DNA-texts, where notions of quantum qubits and their tensor products are used [4, 5].

In this model, 2n-qubit representations of long DNA texts were obtained as quantum chains from the corresponding n-plets. This quantum information model makes it possible to predict many non-trivial numerical characteristics of long DNA texts, again associated with the harmonic progression 1, 1/2, 1/3, 1/4, 1/5,..., 1/n. For example, the model predicts that in long DNA, the sums of n-ples starting with any of the 16 doublets are mutually related: the ratios of the sums of triplets, tetraplets, etc. to the sum of the selected doublet will be equal to the ratio of the second term 1/2 of the harmonic progression (1) to its subsequent members 1/3, 1/4, 1/5, ..., i.e. are equal to members of the following sequence:

$$3/2, 4/2, 5/2, 6/2, 7/2, ... \quad (3)$$

Similarly, in long DNA, the sums of n-ples beginning with any of 64 triplets are mutually related: the ratios of the sums of tetraplets, pentaplets and so on to the sum of the selected triplet are equal to the ratio of the third term of 1/3 of the harmonic progression to its subsequent members 1/4, 1/5, 1/6, ... that is, equal to members of the the following sequence:

$$4/3, 5/3, 6/3, 7/3, 8/3, ... \quad (4)$$

These predictions of the quantum information model were confirmed with high accuracy under analyzing long DNA texts in genomes [4, 5].

One can note that the cross-ratio of any of four neighboring numbers in numeric sequences (3, 4) and similar them is again equal to 4/3 as in the harmonic progression (1).

The effectiveness of the quantum-information model is evidenced by numerical data in favor of the fact that living organisms are quantum-information entities. This echoes the assumptions of many authors about the quantum-information basis of living things.

4 Other results: DNA epi-chains and helical antennas

Long DNA-texts have another previously unknown property associated with the harmonic progression and the so-called epi-chains of DNA. This property, which is described below, speaks of a fractal-like structure of these texts.

If in the DNA-texts of genomes we take only each kth letter (k = 2, 3, ..., 10, ..., 50, ..., 100, ...), then new shortened DNA-texts are formed with new sequences of letters A, T, C, G (Fig. 3). These shortened DNA-texts are called DNA epi-chains of corresponding order k [9]. It turns out that - for each such DNA-text - the same hyperbolic rules, connected with the harmonic progression, are fulfilled with high precision (it is supposed that k is much less than the length of the considered long DNA-text).

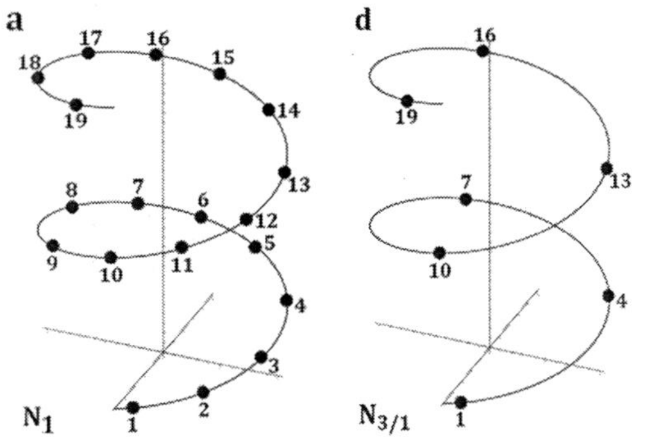

Fig. 3. Schematic representations of a single-stranded DNA and one od the examples of epi-chains of nucleotides; **a**, the sequence N_1 of numerated nucleotides of the DNA strand; **d**, the epi-chain of the third order $N_{3/1}$, which contains nucleotides with numbers 1-4-7-10-... .

The harmonic progression since the time of Pythagoras is associated with harmonics of music and standing waves in resonators. This allows us to assume that the connection between the harmonic progression and the numerical fractal structure of helical DNA-texts is of a resonant nature and reflects the existence of a certain system of standing waves. This is consistent with the known hypotheses that helical DNA is the helical antenna of electromagnetic waves, and therefore, like other helical chiral biomolecules, emits and absorbs electromagnetic waves of a certain circular polarization, which provides biomolecules with the possibility of exchanging radio waves of selective polarization. Many types of organisms are endowed with polarizing vision.

Helical antennas, including fractal ones, are important in space communications, radars, and cellular telephony. They - unlike other antennas - generate circularly polarized radio waves. They are insensitive to manufacturing defects and disturbances in the relative orientation of the transmitting and receiving antennas. Chirality of bio-structures and circular polarization of their electromagnetic waves are important for pharmacology, physiotherapy, etc.

An analog of the helical antenna was recently discovered by Swedish scientists in the tail of the spermatozoid. It has been suggested that these spirals act as a CCTV camera for the spermatozoids to bypass obstacles and search for the egg. [10].

Antenna-like helical structures are involved in the biomechanics of coordinated movements. For example, the unicellular organism Mixotricha paradoxa moves due to the 250 thousand spiral bacteria Treponema spirochetes on its surface, whose helical flagella are been twisting cooperatively, providing purposeful movement.

Electrical and mechanical vibrations in living bodies are closely related, since many of their structures are piezoelectric: nucleic acids, actin, dentin, tendons, bones, etc. Thus, electromagnetic phenomena are accompanied in biological bodies by the phenomena of vibration mechanics.

5 Some concluding remarks

Gr. Mendel discovered in his experiments on the crossing of organisms that the inheritance of their traits occurs according to algebraic rules, despite the colossal complexity and heterogeneity of the molecular structure of living bodies. In this regard, he can be considered the founder of algebraic biology.

According to Mendel's law of independent inheritance of traits, information from the level of DNA molecules dictates the macrostructure of living bodies through many independent channels, despite strong noise and interference. For example, hair, eye, and skin colors are inherited independently of each other. This determinism is provided by unknown algorithms of multichannel noise-immune coding. Accordingly, each organism is a multichannel noise-immune coding machine (in which DNA plays a dictatorial role).

Our studies of nucleotide DNA sequences in the genomes of higher and lower organisms reveal the existence in them of a system of quantitative hyperbolic rules of a universal nature, which are related to the harmonic progression (1).

The mathematical properties of this historically known progression, used in various fields of science and now found in the quantitative characteristics of molecular genetic informatics, indicate that the algebraic structures of genetic informatics can be a basis of non-Euclidean symmetries in the inherited morphology and biomechanics of movements [11]. On this path, new classes of models in biomechanics, psychophysics of sensory perception and algebraic biology, as well as new approaches in biotechnology and biotechnical systems, including artificial intelligence systems, are being developed.

The discovery of the relationship between genomes and harmonic progression also raises important questions about the relationship between the genetic harmony of organisms and musical harmony. These issues are now being dealt with by the special "Center for Interdisciplinary Research of Musical Creativity" of the Moscow State Tchaikovsky Conservatory.

Additional materials were obtained to the well-known statement based on the achievements of molecular biology: life is a partnership between genes and mathematics [12].

6 Conclusions

The present work leads us to the following conclusions :

1. The harmonic progression (1) is connected with the cross-ratio, which is the main invariant of projective geometry. This fact gives pieces of evidence that long DNA nucleotides sequences, whose structures are related to the harmonic progression, consist of fragments which are projective symmetric to each other. It leads to a new - genetic - consideration of the known phenomena of non-Euclidean symmetries in genetically inherited biological structures ;
2. The fractal-like structure of long DNA nucleotide sequences and their epi-chains of different orders is connected with the harmonic progression and the cross-ratio and projective symmetries;
3. The received results give new materials to a consideration of DNA double helix as a helical antenna, which emits and absorbs electromagnetic waves of circular polarization.

The authors are grateful to G. Darvas, E. Fimmel, I. Stepanyan, and G. Tolokonnikov for assistance in the work and discussion of the results.

References

1. H. Weyl, *Symmetry* (Princeton University Press., 2016) ISBN-13:978-0691173252, ISBN-10:0691173257
2. A.V. Shubnikov, V.A. Koptsik, *Symmetry in Science and Art, 3rd ed.* (Moscow: RChD, 2004).
3. S.V. Petoukhov, Computers & Mathematics with Applications, **17**(4-6), 505-534 (1989)
4. S.V. Petoukhov, Biosystems, **198**, 104273 (2020)
5. S.V. Petoukhov, Preprints 2020, 2020050471 (2020) DOI:10.20944/preprints202005.0471.v2.
6. R. Gonzalez, R. Woods, *Digital Image Processing* (Prentice Hall, 3rd ed, 2007)
7. D.A. Forsyth, J. Jean Ponce, *Computer Vision: A modern Approach* (Pearson, 2nd ed, 2011)
8. J. McFadden, J. Al-Khalili, The origins of quantum biology. Proceedings of the Royal Society A, **474**(2220), 1-13 (2018)
9. S.V. Petoukhov, Preprints 2019, 2019040011 (2019), DOI:10.20944/preprints201904.0011.v2.
10. D. Zabeo, J.M. Heumann, C.L. Schwartz, A. Suzuki-Shinjo, G. Morgan, P.O. Widlund, J.L. Höög, Sci.Reports, **8**(1), 2727 (2018)
11. N.A. Bernstein, *Essays on the physiology of movements and the physiology of activity* (Moscow, Medicine Publ., 1966).
12. I. Stewart, *Life's Other Secret: The New Mathematics of the Living World* (New York, Wiley, 1999).

Harmonic Progression in Bioinformatics and Recurrent Series in Inherited Biostructures

Vladimir Verevkin, *Sergey* Petoukhov*

Mechanical Engineering Research Institute of RAS, RU-101990, Moscow, Russia

Abstract. The article is devoted to the study of new approaches to the development of mathematical models in genetic biomechanics, which studies the structural relationships of the genetic coding system with genetically inherited biological forms. More specifically, we are talking about models based on the recurrent harmonic progression whose connection with the information sequences of DNA molecules in the genomes of higher and lower organisms was recently revealed. In particular, the article describes previously unknown connections of the function of natural logarithms with the structures of the molecular genetic system, which allow modelling the main psychophysical logarithmic law by Weber-Fechner and also many other logarithmic structures in genetically inherited biological systems. In physics, the harmonic progression is traditionally considered, first of all, as related to standing waves in resonators. Our results are correlated with Frohlich's vibration-resonant theory about collective quantum effects and long-range communication in biological systems.

1. Introduction

Investigations of symmetries in inherited biological structures are an important component of biomechanics, biophysics, biochemistry, and other sciences of living matter. Many monographs, international forums, including the Nobel Symposium on Symmetries, etc. are devoted to them. The continuation of these studies is an urgent task, which is associated with the disclosure of many secrets of living nature and the creation of new biotechnologies.

In particular, these studies have long drawn attention to the recurrent structure of many inherited biological configurations, which consist of recurrently repeating blocks related, for example, by similarity symmetries; these configurations are abundant in flowers, shells of mollusks, metameric bodies of animal organisms, etc. [1]. The inheritance of configurations relies on the work of the genetic coding system, which allows the reproduction of such forms in a series of generations. Recently, in the genomes of higher and lower organisms, an important rule for the recurrent organization of DNA nucleotide sequences was discovered; these sequences serve for the transfering genetic information [2, 3]. This hyperbolic rule for the relationship of oligomer sums is associated with a long-known harmonic progression (1):

$$1/1,\ 1/2,\ 1/3,\ 1/4,\ \ldots,\ 1/n \qquad (1)$$

The use of properties of this progression allows building new approches and models in genetic biomechanics, algebraic biology and biotechnologies.

2. Computanional approaches, models

Our computational approaches and models of biological structures are based on the mathematical properties of the harmonic progression (1). Its historically renowned name - "harmonic progression" - derives from its association with a series of harmonics in music (or with a series of standing wavelengths in a vibrating string). It is also indirectly associated with a set of trigonometric functions in the Fourier series. Pythagoras, Orem, Leibniz, Newton, Euler, Fourier, Dirichlet, Riemann and many others were engaged in research of mathematical properties, generalizations and applications of the harmonic progression in various problems. The unexpected discovery of this recurrent harmonic progression in genomes, which are the genetic information basis of living organisms, entails many new scientific issues. The aim and objectives of this study are to study how the mathematical properties of this recurrent genetic progression (1) are able to influence the inherited biomechanical structures bearing the stamp of the structural features of genetic coding. This study leads to the emergence of new model approaches in genetic biomechanics, developing the theme of structural relationships between genetic coding and inherited biomechanical phenomena.

Organisms are cooperative aggregates of a huge number of various molecules that have the ability to reproduce themself in their descendants. Gr. Mendel in experiments on crossing organisms showed that the inheritance of traits to descendants obeys certain algebraic rules. Thus, he actually became the founder of

* Corresponding author: spetoukhov@gmail.com

algebraic biology, which uses algebraic methods to model inherited biological phenomena. At present, in this regard, an international interdisciplinary seminar "Algebraic biology and systems theory" (https://www.youtube.com/channel/UC8JLsuRzzPsRiHwrwEjMCtw) is functioning in Moscow. According to Mendel's law of independent inheritance of traits, information from the level of genetic DNA molecules dictates the macrostructure of living bodies through many independent channels, despite strong noises and interference. For example, hair, eye, and skin colors are inherited independently of each other. This determinism is provided by unknown algorithms of multichannel noise-immune coding. Accordingly, each organism is a multichannel noise-immune coding machine. Our studies of the properties of genetic inheritance of biological structures and the possibilities of their algebraic modeling are based on the well-known mathematical formalisms of matrix and tensor analysis, recurrent sequences, the theory of information noise-immune coding, as well as on known academic information about genetic and other biological structures.

3. Results and discussion

Let us present some of the possibilities we have studied of using the mathematical properties of harmonic progression (1), which is associated with the organization of genomes of higher and lower organisms, for modeling inherited biological structures and processes.

Harmonic progression is related to the well-known concept of harmonic mean and is a recurrent sequence based on it: knowing two adjacent members of this progression, you can build the entire given progression. Indeed, the harmonic mean x_{n+1} for two numbers x_n and x_{n+2} is expressed by the relation (2):

$$x_{n+1} = ((x_n^{-1} + x_{n+2}^{-1})/2)^{-1} = 2\,x_n x_{n+2} /(x_n + x_{n+2}) \quad (2)$$

Knowing two neighboring numbers x_n and x_{n+1} in this triplet of numbers, the third number x_{n+2} is determined, respectively, by the relation (3):

$$x_{n+2} = x_n x_{n+1} /(2x_n - x_{n+1}) \quad (3)$$

Successively applying this recurrence relation (3) to the first terms $x_1 = 1$ and $x_2 = 1/2$ of the harmonic progression (1), we obtain the entire series of its terms. It seems interesting and useful to investigate in the future the harmonic mean ratios in inherited biological structures and processes.

Pairs of numbers n and $1/n$ define points on the plane with coordinates [n, $1/n$] in the Cartesian coordinate system belonging to the hyperbola y = $1/n$, which is closely related to the natural logarithm function ln (x). As you know, the natural logarithm can be defined for any positive real number a >1 as the area under the hyperbola y = $1/x$ from 1 to a (Fig. 1) [4].

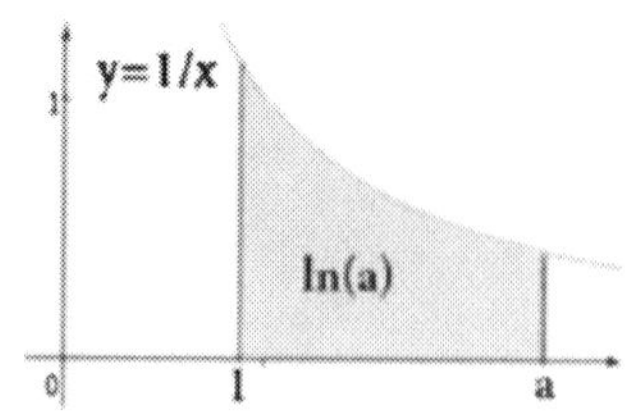

Fig. 1. Illustration of the natural logarithm ln (x) as the area under the hyperbola y = 1/x.

The natural logarithm function is widely used in mathematical biology to model inherited biological phenomena. For example, various types of inherited sensory perceptions - sight, hearing, smell, tactile sense, taste - are subject to the basic psychophysical Weber-Fechner law, which is logarithmic. This law states that the intensity of perception is proportional to the logarithm of the intensity of the stimulus. By virtue of this law, for example, the power of sound in engineering is expressed in a logarithmic decibel scale. The logarithmic nature of perception provides the body with a huge expansion of the range of perceived stimuli, which can differ in intensity by many millions or even billions of times: for example, hammer strikes on a steel plate generate noise that is a hundred billion times louder than the quiet rustle of leaves, and the brightness of a volt arc trillions of times the brightness of a faint star, barely visible in the night sky [5]. It can be assumed that organisms use natural logarithms as the basis for their internal counting.

Taking into account the named connection of the natural logarithm function with the hyperbola y = $1/x$, we note that the recurrent harmonic progression (1) is conjugate, respectively, with a recurrent sequence of natural logarithms (4):

$$\ln(1), \ln(2), \ln(3), \ldots, \ln(n) \quad (4)$$

Fig. 2 shows this recurrent sequence (4) in graphical form in a polar coordinate system, where a n integer positive variable n plays the role of the angular coordinate θ, and ln (n) plays the role of the radius vector r.

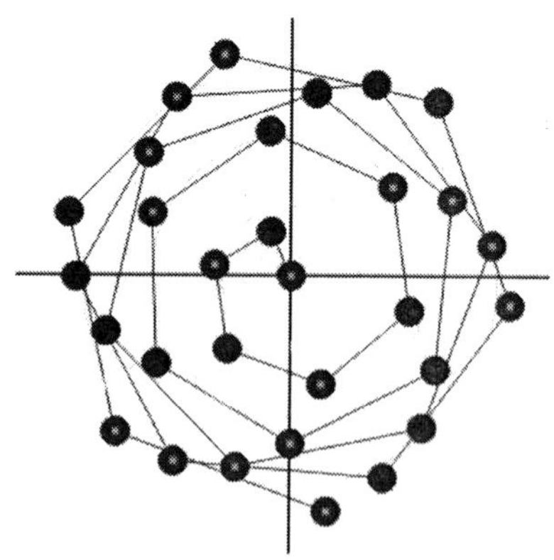

Fig. 2. The graph of a sequence of discrete values of the natural logarithm ln (n) in a polar coordinat system for n = 1, 2,..., 30.

Turning to discrete values of functions in modeling biological phenomena seems to be fundamentally important due to the discrete nature of the organization of both the genetic coding system (discrete DNA alphabets, etc.) and a variety of biomechanical processes (for example, the seemingly continuous movement of a human hand is provided due to the total effect of discrete contractions of a huge variety of muscle units, each of which has its own threshold of excitability and is contracted out of step with the rest). As is known, for all excitable cells of the body, the genetically inherited universal law of "all or nothing" of a discrete nature is valid: the cell does not respond at all to stimuli of a subthreshold value, but reacts to above-threshold stimuli with full amplitude. These facts indicate that biomechanics, if it takes into account these fundamental features of living tissue, should be developed as the mechanics of discrete sequences and processes, taking into account their relationship with discrete genetic coding.

The use of a polar coordinate system for the graphic representation of biological structures as configurations of a spiral type is justified by the fact that spirals have long been considered as "*the curves of life*" [6, 7]: spiral and helical configurations are characteristic of DNA molecules, proteins, bones, tendons, muscles, etc. etc. For example, the heart is a holistic spiral muscle that pushes blood by its action of twisting and untwisting, similar to how a washerwoman squeezes water out of laundry by twisting it. Spiral motions (nutation) a re observed during the growth of roots and shoots, tendrils of plants are spirally wrapped, a tissue in the trunks of trees g rows spirally, e tc. B ecause o f spiral configurations, all the fluids in the body (blood, lymph, and urine) are spiral.

Spirals and helices in living matter are quite different from spirals and helices in inanimate nature:

1. They ar e g enetically i nherited i n t he c hain o f generations and are associated with structures of a molecular-genetic system that is full of spirals and super spirals at different levels of its organization, from DNA to chromosomes;
2. Characteristics o f b iological spirals ar e as sociated with Fibonacci numbers at different levels and branches o f t he b iological e volution, from a lpha-peptides t o t he s tructure o f t he i ntegral b odies o f plants and animals, which is reflected in the famous bio-mathematical la ws o f p hyllotaxis, to which thousands of pu blications have be en de voted worldwide [8].

The amounts of the initial members of the harmonic progression (1) are termed harmonic numbers H_n. These numbers a ppear s o often i n the a nalysis of a lgorithms that computer scientists need a special notation for them. The series (5) presents first harmonic numbers H_n:

$$1,\ 3/2,\ 11/6,\ 25/12,\ 137/60, \ldots \qquad (5)$$

Harmonic numbers are related to the harmonic mean in th at th e n-th harmonic n umber i s al so n times th e reciprocal of the h armonic m ean of t he f irst n positive integers 1, 2, 3, …, n.

Harmonic n umbers ar e co nnected with t he f unction of n atural lo garithm: a s it is k nown, th e harmonic number H_n is t he d iscrete analog o f t he co ntinuous function l n(x). F ig. 3 shows a n a pproximation of t he function of the natural logarithm in Fig. 3 by the sum of areas 1, 1/2, 1/3, 1/4, … , 1/n of the shown rectangles; sums of these areas form the series of harmonic numbers H_n (4).

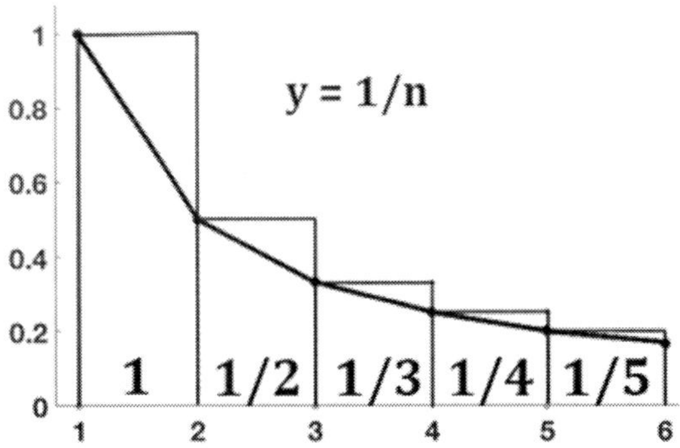

Fig. 3. An approximation of the area under the hyperbola y = 1/x by the sum of areas 1, 1/2, 1/3, 1/4, … , 1/n of the shown rectangles, whose areas form the series of harmonic numbers Hn (4).

In 1740, Euler obtained an asymptotic expression for the harmonic numbers H_n:

$$H_n = \ln(n) + \gamma + \varepsilon \qquad (6)$$

where γ = 0.5772… is the Euler-Mascheroni co nstant, and the value ε tends to zer o w ith i ncreasing n. Therefore, with increasing n, the harmonic number H_n is more and more accurately expressed by the sum (6):

$$H_n \approx \ln(n) + 0.5772 = \ln(n) + \ln(1.7810) \qquad (7)$$

To improve an approximation of the logarithmic function ln (x) by the total area of the rectangles (Fig. 4), you can c onsider - instead of the classic harmonic progression 1, 2, … , 1/ k - more fractional s equences, called t he " compressed h armonic p rogression" (or briefly, the C-harmonic progression):

$$1,\ 1+1/m,\ 1+2/m, \ldots,\ 1+k/m, \qquad (8)$$

where the additional factor m can take integer values 1, 2, 3,.... The a mounts o f th e initial members o f th e C-harmonic p rogression (8) a re t ermed b y au thors as "compressed h armonic numbers" (or C-harmonic numbers). F ig. 4 s hows a n i mproved a pproximation of the c ontinuous f unction l n (x) b y C -harmonic nu mbers $H_n^{[m]}$ at m = 3: the area under the hyperbola, representing the natural logarithm, is better approximated by the total area o f t he r ectangles equal t o t he C -harmonic nu mber $H_n^{[3]}$. T he l arger t he va lue m, t he b etter t he approximation o f t he continuous function l n (x) b y discrete C-harmonic numbers $H_n^{[m]}$ is provided.

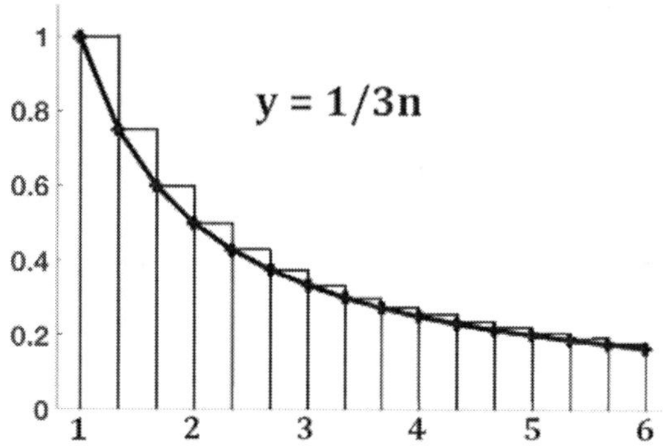

Fig. 4. The approximations of continuous function ln (x), by compressed harmonic numbers Hn[m] under m = 3.

The natural logarithm function $y = \ln(x)$, which is important for modeling inherited biological phenomena, is mutually inverse to the exponent function $x = \exp(y)$, which are also used for a long time to model morphogenetic and other bio-phenomena. Accordingly, the recurrent sequence (4) is mutually inverse to the recurrent sequence (9) of exponentials exp (n):

$$\exp(1), \exp(2), \exp(3), \ldots, \exp(n) \quad (9)$$

Fig. 5 shows a continuous exponential function exp(x) and a discrete recurrent exponential sequence in a polar coordinate system. These graphs correspond to continuous and discrete logarithmic spirals.

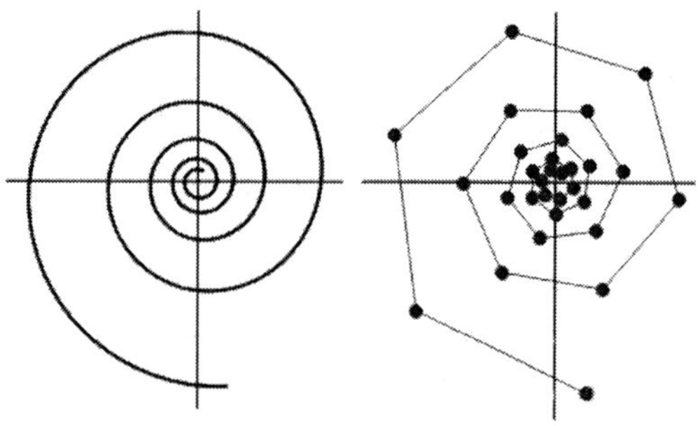

Fig. 5. Graphs of a continuous exponential function exp(0.1x) and discrete sequence exp (0.1n) for n = 1, 2,..., 30.

Logarithmic functions and logarithmic spirals model many genetically inherited biological phenomena, including morphogenetic configurations of phyllotaxis laws [8], which are associated with another recurrent sequence - a series of Fibonacci numbers related to the golden ratio.

Let us add that many cases of biomechanical movements - growth and motor - of living bodies can be modeled as a transition from one discrete recurrent logarithmic spiral to another. This corresponds to the statement of the classic of biomechanics N.A. Bernstein: "The idea that movement is in many respects like an organ seems extremely fruitful" [9].

Mendel discovered in his experiments on the crossing of organisms that the inheritance of their traits occurs according to algebraic rules, despite the colossal complexity and heterogeneity of the molecular structure of living bodies. In this regard, he can be considered the founder of algebraic biology. We believe that algebraic biology is connected with quantum biology, whose history is presented in [10]. The human adult body contains about 100 trillion cells, the ensemble of which behaves as a whole. It seems natural to make a consideration of such complex ensembles from the standpoint of quantum mechanics, to which the carriers of genetic information - DNA molecules - belong.

An important step in the development of quantum biology, to which our study of genetically inherited biological structures belong, was the creation of the theory by H. Frohlich for collective quantum effects in biological systems [11, 12]. This theory testifies in favor that living bodies are a kind of analog of quantum Bose-Einstein condensate. Nobel laureate R. Penrose uses Frolich's theory to substantiate his ideas about the brain as a quantum computer. He writes about it in his book [13, pp. 352 and 367]: «The distinguished physicist Herbert Frohlich suggested a possible role for collective quantum effects in biological systems. ... Frohlich was led to propose, in 1968, that there should be vibrational effects within active cells, which would resonate with microwave electromagnetic radiation, at 1011 Hz, as a result of a biological quantum coherence phenomenon. Instead of needing a low temperature, the effects arise from the existence of a large energy of metabolic drive. There is now some respectable observational evidence, in many biological systems, for precisely the kind of effect that Frohlich had predicted in 1968 ... He argued that so long as the energy of metabolic drive is large enough, and the dielectric properties of the materials concerned are sufficiently extreme, then there is the possibility of large-scale quantum coherence similar to that which occurs in the phenomena of superconductivity and superfluidity - sometimes referred to as Bose-Einstein condensation – even at the relatively high temperatures that are present in biological systems... In a Bose-Einstein condensate, large numbers of particles participate collectively in a single quantum state... There is a coherence on a large scale, where many of the strange features of quantum wavefunctions hold at a macroscopic level".

According to Fröhlich, under appropriate conditions a phenomenon quite similar to a Bose condensation may occur in substances, which possess longitudinal electric modes. If energy is fed into these modes and thence transferred to other degrees of freedom of the substance, then a stationary state will be reached in which the energy content of the electric modes is larger than in the thermal equilibrium. This excess energy is found to be channelled into a single mode - exactly as in Bose condensation - provided the energy supply exceeds a critical value. Under these circumstances a random supply of energy is thus not completely thermalized but partly used in maintaining a coherent electric wave in the substance. This theory is supported by many authors, for example [14-17].

Our results about hyperbolic rules of cooperative genomic organization connected with the harmonic progression, which is known in physics of standing waves in resonators, are correlated with ideas of the article "Genetic code as Language" (see in [11]). It was written by F.Frohlich, the son of H. Fröhlich. This article was devoted to quantum coherence states and long-range communication in genomes. Its author noted that the genome as a whole must contain some sort of long-distance communication. Using H. Frohlich's hypothesis, he wrote that long-range coherent vibrations will lead to resonance between a differentiated cell with its own characteristic vibrations and the chromosome such that the chromosome-particular region responding to this characteristic frequency will be activated or opened up so it can produce the appropriate proteins. Such a resonance could transport the embryological, already partially induced cells to their target and there they would be further fixed into producing the correct proteins for this organ through superimposed resonance and why if it were to o mature, it would not so adapt; there might be degrees of resonance. The conception of multi-resonance genetics was represented also in [18].

4. Some concluding remarks

The work of the laboratory for the research of biomechanical systems of the IMASH RAS (http://eng.imash.ru/) revealed fundamental hyperbolic rules for the organization of DNA information texts in the genomes of higher and lower organisms (eukaryotes and prokaryotes), associated with harmonic progression (1), which is well-known in science in connection with various mathematical and physical problems. The discovery of the important role of the harmonic progression in genetic informatics leads to the development of new mathematical models in genetic biomechanics, which studies the relationship of inherited biomechanical structures with genetic DNA texts, since DNA molecules play a dictatorial role in living organisms, determining their genetically inherited physiological structures [10]. Additionally, we note that harmonic progression plays a prominent role in aesthetics. This progression (1) is important not only for music with its systems of musical harmony and musical overtones. At least since the time of the Pythagorean doctrine of the aesthetics of proportions, there has been an idea that the composer R. Schumann expressed as follows: "*The aesthetics of one art is at the same time the aesthetics of another art; only the material is different*" [19]. In light of this, architecture has long been interpreted as frozen music and music as dynamic architecture. In connection with our research, assumptions arise about the genetic basis of a number of aesthetic parallels in various arts.

The above features of harmonic progression (1) do not exhaust its rich mathematical features, which can serve to create new approaches to the construction and substantiation of models in genetic biomechanics, as well as contribute to the development of algebraic and quantum biology [20].

5. Conclusions

The present study leads us to conclude that

(1) The discovery of the connection of cooperative genomic organization with the recurrent harmonic progression (1) gives new approaches for developing mathematical models of genetically inherited biological structures and also for searching hidden genetic rules;

(2) The connection of the genomic harmonic progression with functions of natural logarithms gives us a new understanding the main psychophysical law by Weber-Fechner and other genetically inherited logarithmic structures in living bodies;

(3) The received results are correlated with Frohlich's vibration-resonant theory about collective quantum effects and long-range communication in biological systems.

The authors are grateful to G. Darvas, E. Fimmel, V. Svirin, I. Stepanyan, and G. Tolokonnikov for assistance in the work and discussion of the results.

References

1. A.V. Shubnikov, V.A. Koptsik, *Symmetry in Science and Art, 3rd ed.* (RChD, Moscow, 2004)
2. S.V. Petoukhov, Preprints 2020, 2020050471 (2020)
3. S.V. Petoukhov, Biosystems, **198**, 104273 (2020)
4. J.H. Conway, R.K. Guy, *The Book of Number* (Copernicus, N.-Y., 1995)
5. N. Ya. Vilenkin, *Functions in nature and technology* (Prosveschenie, Moscow, 1985) [in Russian]
6. T.A. Cook, *The Curves of Life* (Constable and Co., London, 1914)
7. S.V. Petoukhov, V.I. Svirin, L.V. Khazina, Journal of Machinery Manufacture and Reliability, **44**(3), 249–253 (2015)
8. R.V. Jean, *Phyllotaxis: A Systematic Study in Plant Morphogenesis* (Cambridge University Press, 1994)
9. N.A. Bernstein, *Essays on the physiology of movements and the physiology of activity* (Moscow, Medicine Publ., 1966).
10. J. McFadden, J. Al-Khalili. Proceedings of the Royal Society A, **474**(2220), 1-13 (2018)
11. H. Fröhlich editor, *Biological Coherence and Response to External Stimuli* (Springer, 1988)
12. H. Fröhlich, F. Kremer, *Coherent Excitations in Biological Systems* (Springer-Verlag, 1983)
13. R. Penrose, *Shadows of the mind* (Oxford University Press, 1994)
14. G.J. Hyland, Electromagn. Med. Biol., **28**, 316-329 (2009)
15. V.N. Kadantsev, A.N. Goltsov, Russian technolog. journal, **6**(2) (2018)
16. I.V. Lundholm, H. Rodilla, W. Wahlgren, A. Duelli, G. Bourenkov, J. Vukusic, R. Friedman, J. Stake, T. Schneider, Structural Dynamics, **2**(5) (2015)
17. A.R. Vasconcellos, F.S. Vannucchi, S. Mascarenhas, R. Luzzi, Information, **3**(4) (2012)

18. S.V. Petoukhov, Biosystems, **139** (January 2016)
19. R. Schumann, *On Music and Musicians, ed. Konrad Wolff* (New York, Pantheon, 1969)
20. S.V. P etoukhov, M . H e, *Symmetrical Analysis Techniques for Genetic Systems and Bioinformatics: Advanced Patterns and Applications* (IGI G lobal, USA, 2010)

Gradient-Based Algorithm for Tracking the Activity of Neural Network Weights Changing

Anton Starodub[1,*], *Natalia* Eliseeva[1], *Milen* Georgiev[1]

[1]Moscow State University of Technology "STANKIN", RU-127055, Moscow, Russia

Abstract. The research conducted in this paper is in the field of machine learning. The main object of the research is the learning process of an artificial neural network in order to increase its efficiency. The algorithm based on the analysis of retrospective learning data. The dynamics of changes in the values of the weights of an artificial neural network during training is an important indicator of training efficiency. The algorithm proposed in this work is based on changing the weight gradients values. Changing of the gradients weights makes it possible to understand how actively the network weights change during training. This knowledge helps to diagnose the training process and makes an adjusting the training parameters. The results of the algorithm can be used to train an artificial neural network. The network will help to determine the set of measures (actions) needed to optimize the learning process by the algorithm results.

1. Introduction

Artificial neural networks are the most promising method of machine learning and are widely used for solving various applied problems such as image [1], text [2] and speech [3] recognition, semantic analysis [4], forecasting [5] and others.

One of the long-standing priority problems for the authors is monitoring and forecasting the implementation of priority areas of scientific and technological development in Russia. To solve this problem, experts need to analyze large amounts of information including collections of scientific publications and patent documents.

The use of artificial neural networks allows building in an automated mode semantic models of subject areas based on the extraction of actual objects of scientific research from textual information and the detection of their relationship. However, the result of such an application depends on the quality of training an artificial neural network, which requires consideration of the optimization of the training process to increase its efficiency and, thus, the overall accuracy of the network. That is why the work is largely focused on the learning process. The algorithm proposed in this work makes it possible to expand the information used in the analysis of learning an artificial neural network, which greatly simplifies the introduction of adjustments and increases both learning efficiency and accuracy of the artificial neural.

2. Training an artificial neural network

In the course of training an artificial neural network, it is necessary to change the weights of the network to minimize the error in its operation. Incorrectly chosen parameters of training an artificial neural network can lead to situations in which the global minimum of the error function will not be reached. It will negatively affect the accuracy of the network. The gradient property is usually used for training networks. It allows determining the direction of maximum decrease in the value of the error function. The proposed algorithm is also based on the values of the gradients of the weights of an artificial neural network and allows you to track its change during the training process.

2.1 Artificial neural network learning process

Neural network training is the search for the best set of weights to obtain the minimum network error and, as a consequence, the maximum solution accuracy. During training, it is required to change the weighting coefficients of the network so as to minimize the error of its operation on the sequence of training data.

Idealized training of an artificial neural network boils down to choosing the optimal direction of movement and step at each iteration. After a some number of iterations, the global minimum of the loss function (error) must be reached. The loss function characterizes the loss due to poor decision making on input data. The direction of movement during ideal learning should not change significantly. However, this cannot be guaranteed because of the complex relief of the loss function and the fact that adjustments to the weights during training are often made based on the operation of the network not on the full set of data for training, but only on a part of it (batch). The data batch has own global minimum, which in most cases differs from the global minimum of all training data.

* Corresponding author: starodub_a_v@mail.ru

Along with determining the direction of movement, there is the question of choosing a step size. If the step is too large, then there is a high probability of missing the global minimum of the loss function. As a result the direction of movement will need to be reversed. If the value is too small, then the chance of stopping at one of the local minima increases, instead of the global minimum. This will have a negative impact on the result of training, and subsequently the results of solving problems by the network.

2.2 Training with gradient

To train artificial neural networks, the concept of a gradient function is used. The gradient of a function is a vector indicating the direction of greatest growth of the scalar function. According to the definition, if we move in the direction opposite to the gradient, the loss function will decrease as much as possible.

One of the most common learning algorithms for neural networks is the so-called Gradient Descent. Using the previously described gradient properties, Gradient Descent allows you to optimize the network parameters by moving in the direction opposite to the gradient of the loss function.

To improve the work of learning using Gradient Descent, the following methods have been developed: Momentum gradient descent [6], Nesterov accelerated gradient [7], Adadelta [8], Adam [9] and others. Most of them are actively used and allow avoiding the main disadvantage of Gradient Descent – slow convergence rate.

3. Algorithm

The proposed algorithm makes it possible to track the dynamics of changes in the weights of an artificial neural network. This approach is based on the fact that the weights of an artificial neural network do not change uniformly. This suggests that part of the network learns worse than the other, perhaps not learning at all. Based on the calculation of the relative values of the gradients of the network weights, this can be controlled and the learning process adjusted accordingly. This is facilitated by the fact that the values of all gradients are calculated relative to their changes in past iterations, which allows them to be compared with each other on an equal basis. It is possible to reduce the consumption of fewer resources, in exchange for the detail of the algorithm readings. It should be noted that the algorithm has a drawback associated with a rare change in some weights of the neural network.

3.1 Algorithm Description

The purpose of the algorithm is to track the dynamics of changes in the parameters of the neural network. This will allow understanding and determining the uniformity of network training. Thus, if one part of the network learns significantly worse than the other, then this can be identified and the learning plan can be adjusted accordingly.

An increase in the dynamics of changes in weights in the process of training the network can occur in the area of local minima. This is usually associated with an increase in the curvature of the loss function when approaching the local minimum. So, for example, the Stochastic Gradient Descent method has problems with navigating through "ravines", that is, along areas where the surface bends much more strongly in one dimension than in another [10], which are usually found at local optima. If we determine the period of passage of the local minimum, then it is possible to apply an appropriate series of measures for the most effective network training in this area. In the future, this will increase the accuracy of the network, since with the passage of each area ofthe local minimum, the chance of getting into the area ofthe global minimum increases.

A significant decrease in the dynamics of changes in weights in the process of training a neural network can indicate a decrease in the efficiency of the process of its training. If the dynamics decreases to a minimum, this means that during the current course of training, the network will no longer change significantly. Thus, a decrease in dynamics can be one of the reasons for changing the training parameters.

A chaotic change in the weights of a neural network and not a decrease in the error, or a significant degradation of its decrease in an interval that is long for a specific task at the beginning of training, can mean:

- an error in the developed architecture of the neural network (the number of layers, their types, etc.),
- incorrectly selected training parameters (speed, impulse, etc.),
- problems with training data (quantity, quality, incorrect processing or transformation).

If the network undergoes chaotic changes during a significant number of iterations for a specific task in the conditional middle or end of the training process, then this is also a reason to change the training settings.

Using the results of the algorithm, it is also possible to train an artificial neural network that will help in classifying the readings of the algorithm. This will automate the process of using the algorithm by excluding the analysis of the results by a person.

3.2 The role of the gradient in the algorithm

As mentioned above, the proposed algorithm operates on the basis of calculating the values of gradients. The gradient change characterizes how the parameters of the neural network change. If the gradients of the weights change significantly more or less than their previous values, then the changes in the weights of the neural network will also be much more or less.

The algorithm captures the change in the values of the gradients of the network relative to their changes in the previous iterations. This allows considering equally the changes in the gradients of all network weights and accurately determining the training intervals where the gradient changes significantly more or less.

3.3 Algorithm stages

The algorithm can be divided into several stages. At the first stage, the values of the gradients of the current iteration are obtained and compared with those of the previous iteration. The second stage includes the formation of indicators by which the dynamics of changes in the weights of the neural network will be evaluated, and their recording into arrays. The third stage consists in adjusting the average values of changes in the gradients of the neural network weights, taking into account the changes in the gradients of the network weights of the given training iteration, and saving the current values of the network gradients. Let's consider these stages in more detail.

At the first stage, the values of the gradients of the weights of the artificial neural network for the current training step are obtained ($\nabla_\theta J(\theta)$). Then they are compared with the gradients obtained in the previous training step ($\nabla_Z J(Z)$), by finding their difference and calculating the modulus (1).

$$I = |\nabla_\theta J(\theta) - \nabla_Z J(Z)| \quad (1)$$

The result of comparison (I) is an array of values characterizing the change in the gradients of the weights at the current training iteration.

The second stage includes calculating the difference between the change in the values of the gradients obtained in the previous step (I) and the arithmetic mean of the changes in the gradients of the weights (M). Calculation of the modulus of the obtained values and, then, calculation of its ratio to the average values of changes in the gradients of the neural network weights (2).

$$dI = (|I - M|) / M \quad (2)$$

The result (dI) is an array of relative values characterizing the change in the weight gradients at the current training iteration. This, as mentioned earlier, allows you to operate with all the values of the gradient changes equally. If you do not perform these steps, then changes in small gradients can be overlapped by changes in the values of higher gradients in further calculations. Next, the arithmetic mean of the gradient changes for each layer is calculated. This allows you to have an idea of changing the gradient values of each layer separately. A general indicator is also formed, which is represented by the arithmetic mean of changes in the gradients of the entire network. After finding the average values of changes in gradients across layers and the network as a whole, they are recorded to save the history of changes.

The third stage is final. It includes adjusting the mean values of changes in the neural network weights. This is an array of values that includes the mean of the changes in the gradients of the network weights. They are corrected by summing the fraction (k_1) of the current average values of the gradient changes and the fraction (k_2) of the changes in the gradients of the neural network weights of the current iteration (3).

$$\nabla_M J(M) = k_1 * \nabla_M J(M) + k_2 * \nabla_I J(I) \quad (3)$$

Fractions are defined by values from zero to one, and should be summed to one. Adjusting the mean values of gradient changes allows you to react to changes in weight gradients by adapting to new values. This allows you to more accurately determine the moments of increase and decrease in activity. Also at this stage, the current values of the network gradients are saved for the first stage of the next iteration of the algorithm.

3.4 The possible algorithm corrections

3.4.1 Selecting gradients for processing

The algorithm provides the ability to process all the gradients of the network, but if there is a need to optimize the process, then you can resort to combining the gradient values. It is possible to summarize some gradient groups, gradient values of a layer, or the entire network. This, of course, gives an increase in the speed of the algorithm, due to the fact that the algorithm operates with fewer elements. The downside of this approach is that changes in large gradients overlap changes in smaller gradients, which negatively affects statistics. At the moment, the application is recommended for convolution layers that are used to work with images. This is due to the clear separation of filter weights, represented by matrices, which allows gradients to be grouped based on the similarity of changes within the filters.

3.4.2 Selection of correction coefficients for mean values of changes in gradients of network weights

Correction of the average values of changes in the gradients of the neural network weights occurs by summing the proportions of the corresponding elements of the arrays of the current average values of changes in the weight gradients and changes in the gradients of the weights of the network of the current iteration. The greater the proportion of the average values of changes in the gradients of the weights of the network, the less will be influenced by the values of changes in the gradients of the weights of the current iteration and, conversely, the smaller the proportion of the average values of changes in the gradients of the weights of the neural network, the more the values of changes in the gradients of the weights of the current iteration will affect them. It is recommended to choose a small fraction for the values of changes in the gradients of the network weights of the current iteration, otherwise the algorithm will be too sensitive to changes.

3.4.3 Algorithm Behavior During at the beginning

In order for the algorithm to start working, it is necessary to set the values for the gradients of the weights of the previous training iteration and the average change in the gradients of the network weights. To obtain these values, you should either obtain the necessary results in advance, or collect data at the start of training and then run the algorithm.

Initially, it is necessary to determine the number of iterations of the neural network operation, the averaging of changes in the weight gradients of which will serve as the first values for the average changes in the gradients of the network weights. This choice mainly affects the initial results of the algorithm, so it is allowed not to use large values for it.

3.4.4 Saving the dynamics changes in the gradients of the neural network weights

It is recommended to keep the dynamics of changes in the weights of layers and the network as a whole. This allows understanding why the overall network indicator has increased or decreased and whether this is due to a strong increase or decrease in the activity of a particular layer of the neural network or all layers. Nevertheless, keeping only the dynamics of changes in the weights of the network is admissible, but, accordingly, shows only changes in the network as a whole. Also, if computing resources allow saving the dynamics in more detail than changes in the layer weights, then this will give additional information about the network training process. This information can be useful, for example, in case of uneven changes in layer gradients.

3.5 Identified weaknesses of the algorithm

The algorithm displays the dynamics of changes in the gradients of the weights of the neural network. But if some weights are updated rarely, then the result of their changes can greatly affect the dynamics of changes in the weights of the current iteration. This problem can be solved by scaling values. It can be done by extracting the root of the results of the algorithm, the value of the degree of which makes it possible to adjust the level of scalability. Also, the values of such weight gradients can be reduced or excluded from the calculations altogether. Weights that are rarely updated have a mean value of gradient changes close to zero. Thus, by setting a low threshold for the average values of gradient changes that does not affect regularly changing weights, we can find weights that are rarely updated and perform the necessary operations with them.

4. Algorithm application

This section presents the graphs obtained using the proposed algorithm. The description of the graphs reveals the logic of applying the algorithm in the process of training a neural network.

Graph 1 (Fig. 1) shows a uniform change in the gradients of the network. If in this case the network error gradually decreases, then this serves as a reason for confidence in the absence of errors in the architecture of the neural network, the correctness of the training process and data for training.

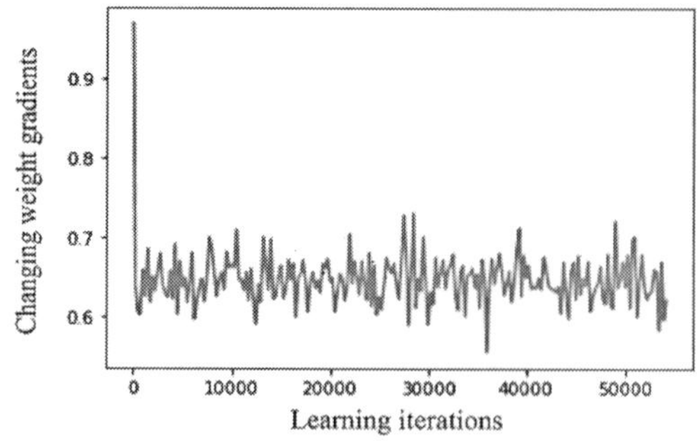

Fig. 1. Graph of uniform changes in the gradients of weights of an artificial neural network.

Graph 2 (Fig. 2) shows a chaotic change in the gradients of the artificial neural network. As it mentioned, this speaks of errors in the neural network architecture, problems in the training process and training data.

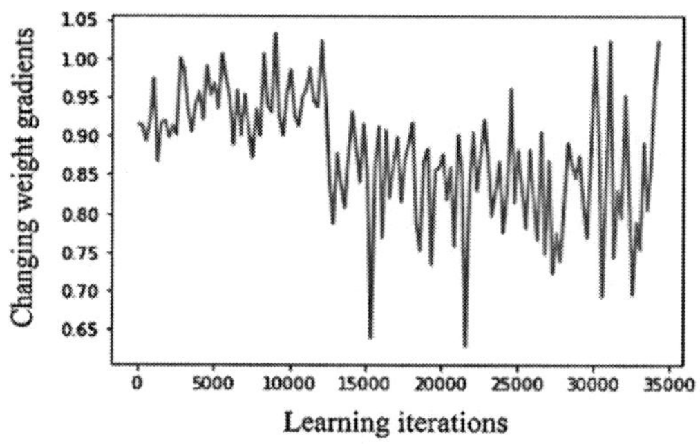

Fig. 2. Graph of uneven changes in the gradients of weights of an artificial neural network.

Graph 3 (Fig. 3) shows an increase in the activity of changes in neural network gradients. This may mean hitting a local minimum. In this case, the assumption was confirmed. After overcoming the area of the local minimum due to an increase in the learning rate, the error of the neural network on the test sample was reduced by 52 percent. Results are not guaranteed to be unambiguous - they can be either better or worse. It also cannot be ruled out that the assumption about the region of the local minimum may be false.

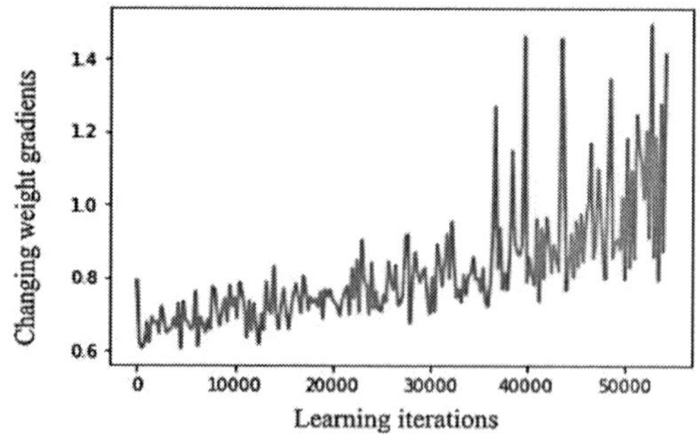

Fig. 3. Graph of increasing changes in the gradients of weights of an artificial neural network.

Graph 4 (Fig. 4) shows a decrease in the activity of changes in neural network gradients, which indicates a decrease in the efficiency of the learning process. In such cases, it is recommended to adjust the learning process, for example, by changing its speed.

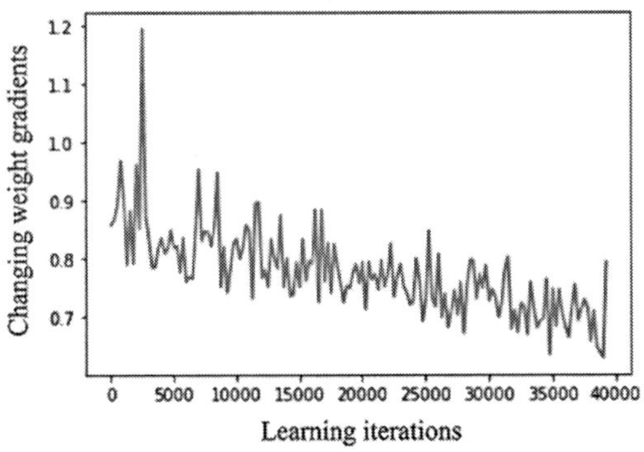

Fig. 4. Graph of decreasing changes in the gradients of weights of an artificial neural network.

Diagrams 5-7 (Fig. 5-7) show an uneven change in the parameters of the neural network. Its first layer is noticeably less subject to changes than the second and third.

This is because the gradient does not sufficiently extend to the last layer. This behavior may indicate problems in the architecture of the neural network.

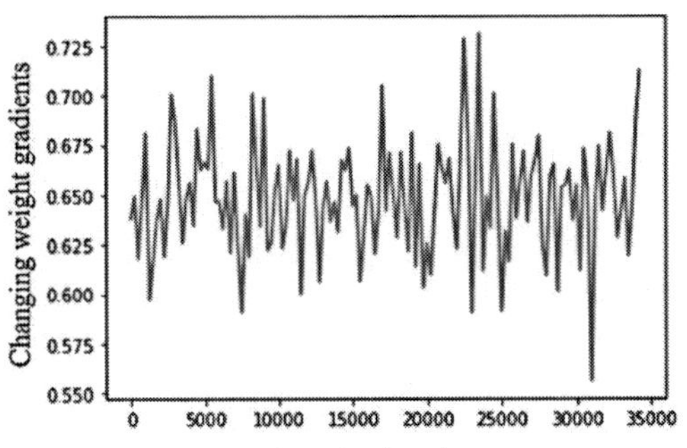

Fig. 5. Graph of changes in the gradients of weights of the first layer of an artificial neural network.

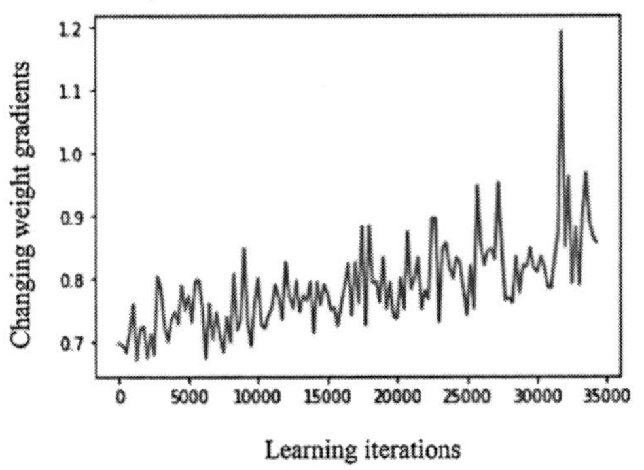

Fig. 6. Graph of changes in the gradients of weights of the second layer of an artificial neural network.

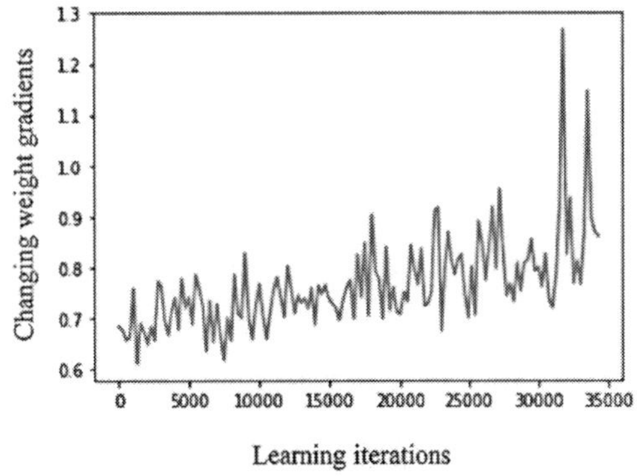

Fig. 7. Graph of changes in the gradients of weights of the third layer of an artificial neural network.

If we use only the general graph of changes in the gradients of the network weights (Fig. 8), then such behavior cannot be detected.

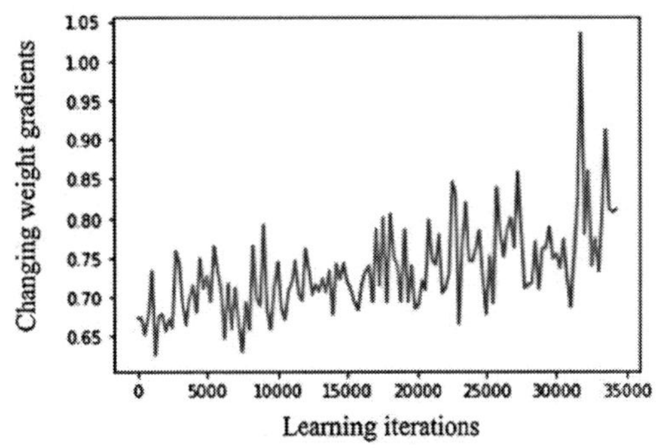

Fig. 8. Graph of the average changes in weights gradients in the first, second, and third layers of an artificial neural network.

5. Conclusion

Training an artificial neural network is critically important for obtaining high-quality solutions during the neural network operation. The developed algorithm allows obtaining retrospective data on its implementation at each iteration of the learning process. In case of unsatisfactory results, it is possible to analyze the stored dynamics of changes in the neural network weights, which serves as the basis for analyzing and adjusting training, which ultimately increases the probability of obtaining planned results. In the future, it is possible to implement an optimization scheme based on the results obtained by the algorithm, which will make it possible to quickly respond to changes in the dynamics of the weights of an artificial neural network.

Acknowledgments

We would like to thank the Ministry of Science and Higher Education of the Russian Federation for supporting this work via grant NO. 05.601.21.0019 (identification number RFMEFI60119X0019).

References

1. W. Rawat, Z. Wang, Neural Comp., **29**(9), 2352-2449 (2017) https://doi.org/10.1162/neco_a_00990
2. M. Yousef, K.F. Hussain, U.S. Mohammed, Pattern Recognit., **108**, 107482 (2020).
3. A.B. Nassif, et al., IEEE ACCESS, **7**, 19143-19165 (2019)
4. Y. Shen, et al., WWW '14, 373-374 (2014)
5. P. Yu, X. Yan, NEURAL COMPUT APPL, **32**(6), 1609-1628 (2020)
6. N. Qian, Neural Netw., **12**, 145–151 (1999)
7. Y. Nesterov, *A method for unconstrained convex minimization problem with the rate of convergence o(1/k2)* (Doklady an USSR, **269**, 543-547, 1983)
8. D. Zeiler, *ADADELTA: An Adaptive Learning Rate Method* (2012) https://arxiv.org/abs/1212.5701v1
9. P. Kingma, J. Lei, *Adam: a Method for Stochastic Optimization,* (ICLR, 1–13, 2015) https://arxiv.org/abs/1412.6980v9
10. S. Sutton, *Two problems with backpropagation and other steepest-descent learning procedures for networks* (Proceedings of the Eighth Annual Conference of the Cognitive Science Society, 1986)

Computer Visualization of Julia Sets for Maps beyond Complex Analyticity

Alexey Toporensky[1,*] *and Ivan Stepanyan* [2]

[1]Sternberg Astronomical Institute, RU-119234, Moscow, Russian Federation
[2]Mechanical Engineering Research Institute of the Russian Academy of Sciences, RU-101990, Moscow, Russian Federation

Abstract. Using the computer program creating Julia sets for two-dimensional maps we have made computer animation showing how Julia sets for the Peckham map alters when the parameter of the map is changing. The Peckham map is a one-parameter map which includes the complex map z=z^2+c, and is non-analytical for other values of the parameter. Computer animation of Julia fractal sets allows seeing how patterns typical for complex maps gradually destroy.

1 Introduction

Iterations of complex analytic maps appears to be a wonderful source of fractal structures, interesting from both mathematical and aesthetic viewpoints. Already the simplest map

$$z=z^2+c, \quad (1)$$

where c is a constant complex number contains very complicated dynamics, and Julia sets for certain values of the parameter c represent a true computer art (see examples in [1]).

There are several methods to generalize the map (1). One way is to retain the complex analyticity and consider polynomial or transcendent maps. There is a well developed mathematical theory for such kind of maps (see, for example, Milnor's lectures [2] and references therein). Julia sets for such maps has roughly the same structure, though can get additional symmetries in the polynomial case. Particular case of Zhukovski map have been considered in [3], for more general forms of maps within complex analyticity see, for example, [4]. Among recent results we can indicate the paper [5] where it was shown that the Julia set of a transcendental meromorphic map with at most finitely many poles cannot be contained in any finite set of straight lines and the paper [6] which describes the conditions of existence of Sierpiński carpet Julia sets and estimation of it's Hausdorff dimension.

We should also note that in some cases Julia set can bee smooth, see [7] which gives the examples of smooth Julia sets, among them: a circle, a segment, an infinite interval, a straight line, and the complex plane. It is shown that the functions studied in [7] are chaotic on their Julia sets. An interesting generalization of Mandelbrot and Julia sets to hyperbolic dynamics can be found in the paper [8].

Abandoning complex analyticity gives rise to new events, and the corresponding mathematical theory is developing intensively at the present time [9]. Visual patterns of analogs of the Mandelbrot set usually differ qualitatively from the fractal structure of the "true" Mandelbrot set, which can be seen clearly for the "burning ship" map [10]. This map differs from (1) only by the presence of moduli, and it is enough to convert the Mandelbrot set into amorphous structure with thin linear bands. In a broader contexts, both complex analytic maps and the "burning ship" map represent particular cases of two-dimensional real maps, which can be derived from the complex form by extracting real and imaginary parts. In principle, one-dimensional real maps also can have a complicated chaotic nature (see, for example, [11]), though they are less interesting from a viewpoint of a computer art.

In this form the complex analytic map (1) has the form

$$x=x^2 - y^2 + R(c), \quad (2)$$

$$y=2xy+I(c) \quad (3)$$

and in the "burning ship" map the equation (2) remains the same, while (3) becomes

$$y=2(x)(y) + I(c). \quad (4)$$

2 Results

For Mandelbrot and Julia sets for two-dimensional real maps these long quasi-parallel bands and other linear structures appear to be rather typical (see examples in [1]). This means that complex analytic maps have some special properties. Formally, they demand Cauchy-Riemann condition to be satisfied, which leads to absence of saddle points and absence of appearance or disappearance of fixed points with change of the

* e-mail: atopor@rambler.ru

https://doi.org/10.1051/epjconf/202124801013

parameter of the map (1) [9]. The letter follows directly from the fundamental theorem of algebra, the former follows from the fact that a complex multiplication is a combination of stretching and rotation, while a map in a saddle point stretches in one dimension and contracts in another one.

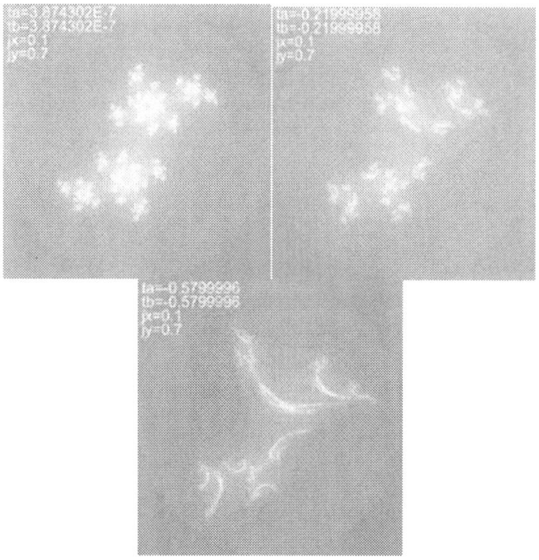

Fig.1. Three frame from the animation showing gradual change of Julia set with the deviation of complex analyticity. Parameters of the map is shown in the pictures. Here jx is real, and jy is the imagery parts of c, ta and tb represents a and b (they are equal for the Packham map). Left panel is the analytic map

In the paper [9] B. Peckham described how this properties have been lost when a map becomes non-analytical. For this purpose the map

$$z=z^2+c+a \quad (5)$$

have been used. It coincides with (1) for a=0, and is not analytic for any nonzero a. The goal of our paper is to describe typical visual forms of the Peckham map (gradual developing of linear structures through the classical Mandelbrot set for the Peckham map have been considered in [12].

Strictly speaking, the results presented below are related to filled Julia sets, i.e. sets of initial points such that iterations of the map (5) remake bounded from above. They are shown in white. Initial points from other regions leave the vicinity of the coordinate origins (we set it to a radius 10 disc in our numerical studies), the color encodes the escape time needed.

Keeping in mind possible generalizations of the Packham map, we realize more general map

$$x=x^2-y^2+R(c)+ax, \quad (6)$$

$$y=2xy+I(c)-by. \quad (7)$$

However, for the animations we have chosen the original Packham map, with a=b. Animations show gradual change of Julia sets with changing parameter a. Visual patterns of Julia sets rapidly loose features known from the complex analytic dynamics, becoming rather prolonged (and, usually, less aesthetically attractive). Figure 1 represent three frames from our animation, showing the transition from complex analytic pattern (left figure) to quasi-linear structures (right figure).

Some examples of Julia sets for the map with a not equal to be are presented in Fig.2. To our mind, current results indicate that "aesthetic maximum" is realized usually for the complex analytic map with a=b =0 in comparison with non-analytic maps.

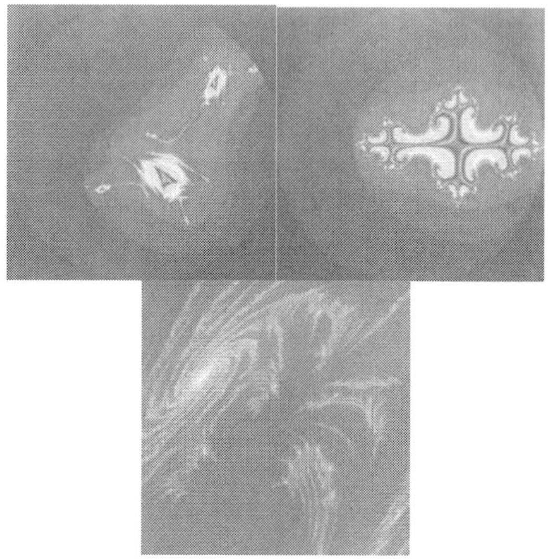

Fig.2. Examples of diagrams for general case of different parameters a and b. From left to right: a=- 0,75, b=1,6, x=0,74, y=0,97; a=1,14, b=1,39, x=0,28, y=0,01; a=- 0,6, b=0,3, x=-0,7, y=0,3

3 Discussion and Conclusions

We should however note that complex analytic dynamics is not unique in producing visually attractive Julia sets among two-dimensional real maps. Another example is given by the "burning ship" map. It's Mandelbrot set is rather typical for real maps, however, Julia sets represent unexpectedly interesting constructions, though probably less attractive than for analytic maps.

Fig.3 represents three frame from our visualization of Julia sets of the "burning ship" map. Adding the term proportional to $\bar{z}$ into the right hand side of the equations for "burning ship" map should destroy such structures. This is illustrated in Fig.4 created for the map

$$x=x^2-y^2+R(c)+ax, \quad (8)$$

$$y=2(x)(y)+I(c)-by. \quad (9)$$

The Fig.4 shows four frames from the corresponding animation.

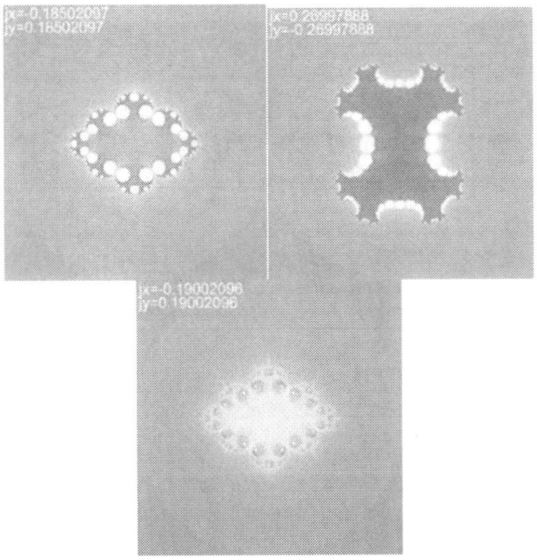

Fig.3. Examples of Julia sets for the "burning ship" map

We h ave co nsidered t he i nfluence o f co mplex non-analyticity up on visual forms o f J ulia s ets o f corresponding maps. For the Packham map the Julia sets rapidly evolve from known "classical" forms of the map (1) to forms typical for two-dimensional real maps with increasing of modulus of the parameter a. Already for |a| about 0 .2 t races o f co mplex an alyticity have b ecome hardly d etectable. W e can e ven c laim t hat a g eneral aesthetic attraction falls rapidly with deviating from complex analyticity (recognizing t hat such statement i s inevitably subjective a nd nonformalizable) c onfirming a special r ole o f co mplex a nalytical maps i n co mputer graphics.

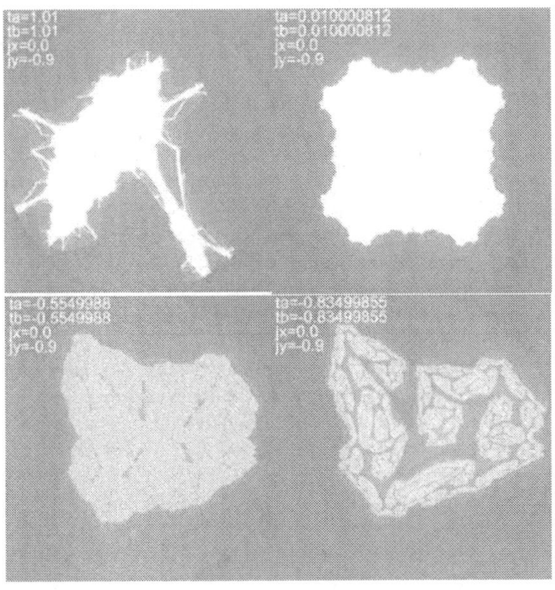

Fig.4. Diagrams for "deformed burning ship" map (8)-(9)

Nevertheless, examples of Julia sets for non-analytic map (2,4) show that the Cauchy - Riemann conditions are not strictly necessary for Julia sets to be aesthetically attractive (which is also interesting since the Mandelbrot set for the map (2,4) has totally different visual forms). This means t hat e ven for r ather s imple cas e o f t wo-dimensional maps the connection between mathematical struture a nd aesthetic a ttraction is n ot so ti ght, a nd the search for aesthetically attractive Julia sets should not be bound only to complex analytic maps, despite their special role in computer art.

References

1. H.-O. Peitgen, P.H. Richter, *The Beauty of Fractals* (Springer-Verlag, 1986)
2. J. M ilnor, *Dynamics i n One C omplex V ariable* (Vieweg, 1999)
3. J.B. M ugiraneza, I .J. I mage, G raphics an d S ignal Processing, **5**, 61-70 (2012)
4. Sh. Agarwal, I .J. C omputer N etwork a nd Information Security, **4**, 1-9 (2017)
5. C. C ao, Y . W ang, Acta M ath S ci, **40**, 903 –909 (2020) https://doi.org/10.1007/s10473-020-0401-5
6. Y. F u, F . Y ang, Math. Z ., **294**, 144 1–1456 (2020) https://doi.org/10.1007/s00209-019-02319-4
7. V.S. Sekovanov, J Math Sci, **245**, 202–216 (2020)
8. V. Blankers, et al., Fractal and Fractional, **3**(1) (2019) https://doi.org/10.3390/fractalfract3010006
9. B. P eckham, Int. J. Bifurcation a nd C haos, **8**(73) (1997)
10. M. M ichelitsch, O . R ossler, C omputers an d Graphics, **16**(435) (1992)
11. Sh. X u, Y . W ang, Y . G uo, C. W ang, I .J. I mage, Graphics and Signal Processing, **1**, 61-68 (2010)
12. A. T oporensky, *Quasi-Mandelbrot s ets f or perturbed c omplex analytic maps:visual patterns* (2008) arXiv:0807.1667

"Future Earth": Nigmatulin Hypothesis and Dynamic Model of Radiation Field of Ocean-Atmosphere System

Tamara Sushkevich[1,*], *Sergey* Strelkov[1], *and Svetlana* Maksakova[1]

[1]Keldysh Institute of Applied Mathematics of RAS, Department Dynamic systems, RU-125047, Moscow, Russia

Abstract. The United Nations has proclaimed a Decade of Ocean Science for Sustainable Development (2021-2030) to support efforts to reverse the cycle of decline in ocean health and gather ocean stakeholders worldwide behind a common framework that will ensure ocean science can fully support countries in creating improved conditions for sustainable development of the Ocean. The marine realm is the largest component of the Earth's system that stabilizes climate and support life on Earth and human well-being. Scientific understanding of the ocean's responses to pressures and management action is fundamental for sustainable development. Planet Earth is a natural example of a dynamic system with nonlinear processes that is in continuous change. The Earth's radiation field is a single physical field (electromagnetic radiation) and the unifying factor of the Earth dynamical system. The Earth's climate system is a natural environment that includes the atmosphere, the hydrosphere (oceans, seas, lakes, rivers), the cryosphere (land surface, snow, sea and mountain ice, etc.), and the biosphere that unites all living things. According to the hypothesis of R.I. Nigmatulin "Ocean is a dictator of climate". H2O and CO2 are competing climate influences. In this article, we propose original author's mathematical models for radiation blocks with hyperspectral data on absorption by atmospheric components. The new models are based on the development of the theory of the optical transfer operator and the method of influence functions in the theory of radiation transfer and Boltzmann equations, as well as the iterative method of characteristics with iteration convergence accelerations.

1 Introduction

The United Nations has proclaimed a **Decade of Ocean Science for Sustainable Development (2021-2030)** to support efforts to reverse the cycle of decline in ocean health and gather ocean stakeholders worldwide behind a common framework that will ensure ocean science can fully support countries in creating improved conditions for sustainable development of the Ocean [1]. The marine realm is the largest component of the Earth's system that stabilizes climate and support life on Earth and human well-being. Scientific understanding of the ocean's responses to pressures and management action is fundamental for sustainable development. Ocean observations and research are also essential to predict the consequences of change, design mitigation and guide adaptation. As mandated by the UN General Assembly the Intergovernmental Oceanographic Commission (IOC) of UNESCO will coordinate the Decade's preparatory process, inviting the global ocean community to plan for the next ten years in ocean science and technology.

From November 30 to December 12, 2015, The 21st Conference of the Parties Climate Summit (COP21) was held in Paris, where the heads of more than 150 States and about 40 thousand researchers. On December 12, 2015, the "Paris Agreement under the United Nations Framework Convention on Climate Change" ("**Paris Agreement**") was adopted, which, after signing from April 22, 2016 to April 21, 2017, entered into force instead of "Kyoto Protocol".

The global challenge to the Convention "**Agenda 21**", adopted at the United Nations Conference on Environment and Development, Rio de Janeiro, June 3-14, 1992, the Resolution "Transforming our world: the **2030 Agenda** for Sustainable Development", adopted by the General Assembly on 25 September 2015, and the super-large project, which has no analogues in scale and significance for all of humanity in world history is the World Global Scientific Program "Future Earth" (Program).

The fundamental basis of the Program to ensure the sustainable development and existence of human society and States on the planet were laid in the 20th century: this is the unprecedented flowering of mathematics, the invention of the computer, the Internet and the extraordinary earlier information, communication and digital technologies, as well as the exit of man into space, exploration of space, the development of the global observation systems and remote sensing of the Earth, objects and processes on the basis of the "light technologies" when creating the "Nuclear missile shield" and implementation of the "Atomic" and "Space" projects.

1951 was a key year in the life of M.V. Keldysh [2]. From the Resolution of the USSR Council of Ministers No. 1552-774ss / op of May 9, 1951 "On works on RDS-

* Corresponding author: tamaras@keldysh.ru

6T" (Atomic project: documents and materials. Vol. 3. Book 1. 2008. pp. 396-403.):

"4. Oblige the First Main Directorate under the USSR Ministers Council to organize a **mathematical section** (section 7) as part of the Scientific and technical council for scientific guidance on the development of designs, high-speed computers, as well as methods of their operation in the composition of: academician M.V. Keldysh – section president; section members: academician I.G. Petrovsky, academician S.L Sobolev, corresponding member N.N. Bogolyubov, corresponding member A.N. Tikhonov; section members (for computers): academician M.A. Lavrent'ev, corresponding member S.A. Lebedev, engineer Yu. Bazilevsky, engineer M.A. Lesechko.

Assign to Section 7 the consideration of plans for research, experimental and design work, as well as projects of mathematical machines and work plans of organizations performing calculation work on the subject of the First Main Directorate under the USSR Ministers Council. "

In the year of the 75th anniversary of the Soviet "Atomic Project" (1945-2020), it is important to note the creation of the world's strongest Russian mathematical school, which has survived and still occupies a leading position in world science. The foundations of applied mathematics were laid in the middle of the 20th century under the leadership of M.V. Keldysh – the Chief Mathematician responsible for computers and calculations in the "Atomic Project", "Space Project" and the "Nuclear Missile Shield" project ("Three Projects"). In 1953, the Chief Cosmonautics Theorist M.V. Keldysh [3,4] created the world's first Institute of applied mathematics of the USSR Academy of Sciences (Keldysh Institute) for the implementation of "Three Projects" using a computer.

At the stage of space exploration and the formation of space research and Earth remote sensing from space, the United States invited mathematicians and specialists in the theory of radiation transfer from different countries to NASA. However, the mathematical scientific school in the United States was not formed. In recent decades, in the centers of NASA and the United States, the EU, China, France, Japan and other countries, with the presence of supercomputers, researchers are passionate about information technology and data processing ("computer sciences").

During the "nuclear" threats in the 20th century, the scientists saved the planet for humanity. And this experience is now in great demand, since the threat of the global changes on the planet is already obvious. The Program "Future Earth" is a natural, comprehensive stage in the development of science, technology and society, aimed at ensuring life on planet Earth. The priceless lessons must be drawn from these successful projects. They have contributed to the development of mathematics, physics, chemistry, theory and applications of Earth sciences, and "computer sciences". The USSR pursued a real peace-loving policy and did not rush to develop means for delivering "nuclear strikes".

The recent UNESCO General Conference has adopted a resolution to proclaim 2022 as the International Year of Basic Sciences for Development (IYBSD). This is great news, as the Year would help highlight the crucial role of basic sciences for sustainable development, and emphasize their contributions to the implementation of the 2030 Agenda and achievement of the Sustainable Development Goals (SDGs).

This article focuses on the problems of the planet protection associated with the Earth's radiation field, the radiation transfer in natural environments, the role of solar and intrinsic radiation in the global Earth Climatic System (ECS) and Earth remote sensing (ERS). The radiation forcing, the role of the humidity in different phase states, the contribution of the ocean, and the radiation exchange between the atmosphere and the ocean are highlighted. These problems require deep study.

2 Nigmatulin hypothesis and the radiation factor

In the works of the great domestic scientists V.I. Vernadsky and N.N. Moiseev actually formulated the main provisions of the idea of sustainable development. N.N. Moiseev, perhaps, was the first to move from the geographical concepts of "climate" and "ecology" to the planetary problems of climate and sustainable development of the planet and created the scientific basis for the study of ECS and the co-evolution of nature and society.

"The ocean and short-period climate variability" – on such problems since the end of 1970 and in the 1980s the large-scale studies were conducted in the framework of the Russian program "Cuts" under the general scientific leadership of academicians G.I. Marchuk and A.S. Sargsyan. Initially, this program was only of Russian affiliation, but then part of it became part of the International Program World Ocean Circulation Experiment (WOCE).

The Earth Climate System is a natural environment, including the atmosphere, the hydrosphere (oceans, seas, lakes, rivers), the cryosphere (land surface, snow, sea and mountain ice, etc.), the biosphere that unites all living things. For quantitative assessments of the significance of different climate-forming factors, depending on solar and intrinsic radiation, we introduced a special characteristic of ECS – radiation exposure (forcing). According to expert estimates, from 40% to 60% of the impact of various factors on the evolution of climate falls on radiation forcing.

On the one hand, the solar radiation – a source of energy on Earth – is one of the integral factors of life support for the human, animal and plant world on Earth, as well as one of the determining components of the Earths ecosystem and biosphere, the behavior of which is characterized by the interaction of the individual components with the manifestation of synergism, i. e. feedbacks, which sometimes lead to mutual reinforcement of various processes.

On the other hand, the solar radiation field affects the mechanisms of variability (dynamic processes: circulation, convection, turbulent transport; radiation and photochemical processes) of the geophysical, meteorological, climatic conditions of the Earth, which have complex nonlinear relationships that make it difficult to predict possible effects, estimate their magnitude and significance.

On the third hand, the electromagnetic radiation detected by various means is the main source of information about the structure and physical properties of planetary atmospheres, surfaces, the ocean, the biosphere during conservative remote sensing. In active systems, a laser or spotlight can be used as an insolation source.

The physical picture of the climate of the ocean-land-atmosphere system was presented by A.S. Monin in the review [5]. R.I. Nigmatulin called the expert article "Ocean – the dictator of climate" [6]. More study of the H2O and CO2 cycle is needed. To test his realistic hypotheses for explaining the mechanisms of the most complex interaction of the atmosphere with clouds and the ocean and their influence on climate, it is important to be able to model the formation of clouds and to separate the contributions of the atmosphere and the ocean to the total radiation field of the Earth, recorded from space.

Nigmatulin hypothesis:

"The world ocean occupies more than 70% of the planet's surface. Solar radiation is the source of energy on Earth. It is absorbed and accumulated by the ocean, then re-emitted into the long-wave spectrum. This radiation is already absorbed by greenhouse gases - carbon dioxide and water vapor – and warms the atmosphere. And water vapor in the atmosphere is two orders of magnitude more than carbon dioxide. However, nature provides a compensatory mechanism... So far, this is a hypothesis: as the vapor content in the air increases, cloud cover should increase, which means that the reflection of solar radiation by clouds increases. There is a negative feedback..."

This is the most complex natural radiation process on a global scale, on which the future of the planet depends. To research radiation influence of the oceans on the Earth's evolution requires international cooperation with the involvement of integrated observing systems of radiation processes under water, at sea and from space with the support of supercomputer simulation [7].

The study of the hypothesis is carried out on the basis of a "scenario" approach. The complex model contains four main blocks:

1) solar radiation transfer in the ocean-atmosphere system,

2) evaporation from the ocean surface and formation of a humidity field in the atmosphere,

3) the formation of clouds,

4) transfer of solar radiation in the ocean-atmosphere system with clouds.

In this article, we propose original author's mathematical models for radiation blocks [8-10] with hyperspectral data on absorption by atmospheric components [11, 12]. The new models are based on the development of the theory of the optical transfer operator and the method of influence functions in the theory of radiation transfer and Boltzmann equations [8, 9], as well as the iterative method of characteristics with iteration convergence accelerations [10].

At present, numerical models of radiation transfer in the ocean-atmosphere system (OAS) and mechanisms of forming radiation fields in OAS with allowance for short-wavelength radiation transfer between the media continue to attract interest. Seven foreign models were compared to one another with the support of international foundations [13]. Five of the models were implemented by the Monte Carlo method; another one, by the method of invariant imbedding; and the remaining one, by the method of discrete coordinates, or more exactly, by the method of spherical harmonics (DISORT program) [14]. The latter two models used a preliminary Fourier expansion in terms of azimuth. All the models are one-dimensional in space and include an air–water Fresnel interface. The numerical models that most researchers use to calculate radiation characteristics are presented in report of the "Radiation commission" [15].

3 Model of radiation transfer in the ocean–atmosphere system and the influence functions method

Solar-radiation propagation in the ocean-atmosphere system are modeled as radiation transfer in a two-medium plane-parallel layer non-uniform in height and with the air–water interface reflecting and transmitting radiation by the Fresnel law.

3.1 Formulation of the mathematical problem

A one-dimensional planar model is considered of the atmosphere-ocean system illuminated by a mono-directional parallel flux of solar radiation. A new approach is proposed to radiation transfer modeling and daylight background formation for the atmosphere that is represented as a heterogeneous two-layer system each layer of which is described by different optical characteristics. The influence functions of each layer are determined by solutions of the radiation transfer boundary problem with an external mono-directional wide flux while the contribution of multiple scattering and absorption in the layer is taking into account.

The direction of radiation transfer in the atmosphere is described by the vector $s=(\mu,\varphi)$, $\mu=\cos\vartheta\in[-1,1]$, on the unit sphere $\Omega=[-1,1]\times[0,2\pi]$, where $\vartheta\in[0,\pi]$ is the zenith angle measured from the positive direction of the z-axis and $\varphi\in[0,2\pi]$ is the azimuth. The value $\varphi=0$ is attributed to the plane of the sun's vertical passing through the z-axis and the direction $s_0=(\mu_0,\varphi_0)$ of solar flux incidence upon the upper boundary of the layer $z=0$; $\mu_0=\cos\vartheta_0\in(0,1]$; $\vartheta_0\in[0,\pi/2)$; $\varphi_0=0$; and $s_0^-=(-\mu_0,\varphi_0)$.

If $s=(\mu,\varphi)$, then $-s=(-\mu,\varphi+\pi)$.

For the downward radiation

$$s = s^+ = (\mu^+, \varphi^+) \in \Omega^+ = \{(\mu, \varphi) : \mu \geq 0\};$$

for the upward radiation

$$s = s^- = (\mu^-, \varphi^-) \in \Omega^- = \{(\mu, \varphi) : \mu \leq 0\};$$

and $\Omega = \Omega^+ \cup \Omega^-$.

The direction of transfer in the ocean is defined by the vector $\tilde{s} = (\tilde{\mu}, \tilde{\varphi})$, $\tilde{\mu} = \cos\tilde{\vartheta} \in [-1,1]$, $\tilde{\varphi} \in [0, 2\pi]$, $\tilde{\varphi} = \varphi_0 = 0$, $\tilde{s} \in \Omega$; $\tilde{s}^+ = (\tilde{\mu}^+, \tilde{\varphi}^+) \in \Omega^+$; and $\tilde{s}^- = (\tilde{\mu}^-, \tilde{\varphi}^-) \in \Omega^-$.

The direction of the ray s^+ that is incident upon the interface from the side of the atmosphere and travels into the ocean after refraction is described by the vector

$$\overline{s} = \overline{s}^+ = (\overline{\mu}^+, \overline{\varphi}^+) \in \Omega^+_{crit} = [\mu_{crit}, 1] \times [0, 2\pi].$$

The direction

$$\overline{s} = \overline{s}^- = (\overline{\mu}^-, \overline{\varphi}^-) \in \Omega^-_{crit} = [-1, -\mu_{crit}] \times [0, 2\pi]$$

corresponds to the ray that is incident upon the interface from the side of the ocean and assumes the direction s^- in the atmosphere after refraction.

The interaction of radiation with the Fresnel boundary is governed by the geometrical-optics laws. The reflection law determines the relationship between the directions of incident $s = (\vartheta, \varphi)$ and reflected $s' = (\vartheta', \varphi')$ rays: $\mu' = -\mu$, $\varphi' = \varphi$. The refraction law controls the relationship between the directions of incident s and refracted $\overline{s} = (\overline{\vartheta}, \overline{\varphi})$ rays:

$$n_1 \sin\vartheta = n_2 \sin\overline{\vartheta}, \ \sin\overline{\vartheta} = \sin\vartheta / n, \ n = n_2/n_1, \ \varphi = \overline{\varphi},$$

where n_1 and n_2 are the refractive coefficients of the media. In accordance with the refraction law, we have

$$\overline{\mu} = k\mu / n, \ k = \sqrt{1 + (n^2 - 1)/\mu^2}, \ \mu_{crit} = \sqrt{1 - 1/n^2}.$$

If $n = 1.33$, then $\vartheta_{crit} \approx 48.75^\circ$.

The Fresnel coefficient is

$$r(\vartheta) = r(\overline{\vartheta}) = \frac{1}{2}\left[\frac{\sin^2(\overline{\vartheta} - \vartheta)}{\sin^2(\overline{\vartheta} + \vartheta)} + \frac{\mathrm{tg}^2(\overline{\vartheta} - \vartheta)}{\mathrm{tg}^2(\overline{\vartheta} + \vartheta)}\right].$$

For the normal incidence of radiation, $\overline{\vartheta} = \vartheta = 0$ and

$$r(\vartheta) = r(\overline{\vartheta}) = (n-1)^2 / (n+1)^2.$$

The directions $\tilde{s}^- \in \Omega^- \setminus \Omega^-_{crit}$ with $\tilde{\mu}^- \in (-\mu_{crit}, 0)$ in the ocean form the region of total internal reflection when $r(\tilde{\vartheta}) = 1$. The directions $\tilde{s}^+ \in \Omega^+ \setminus \Omega^+_{crit}$ with $\tilde{\mu}^+ \in (0, \mu_{crit})$ in the ocean lie in the shadow region, which is not illuminated by the light arriving from the atmosphere.

The spatial and angular distribution of solar-radiation intensity $\Phi(z, \mu, \varphi)$ in the two-medium OAS is described by the boundary-value problem

$$K\Phi = 0, \ \Phi|_t = \pi S_\lambda \delta(s - s_0), \ \Phi|_b = R_H \Phi,$$

$$\Phi|_{d1} = R_1\Phi + T_{21}\Phi, \ \Phi|_{d2} = R_2\Phi + T_{12}\Phi \qquad (1)$$

with the linear operators of transfer and collisions:

$$D \equiv \mu\frac{\partial}{\partial z} + \sigma_{tot}(z),$$

$$S\Phi \equiv \sigma_{sc}(z)\int_\Omega \gamma(z, s, s')\Phi(z, s')ds',$$

$$K \equiv D - S,$$

where $ds' = d\mu' d\varphi'$; $\sigma_{tot}(z)$ and $\sigma_{sc}(z)$ are the vertical profiles of the extinction and scattering coefficients, respectively; and $\gamma(z, s, s')$ is the scattering phase function. The source of radiation is an external parallel solar flux to the upper bound $z = 0$. At the level $z = h$ within the layer $z \in [0, H]$, the interface between two media is located.

For the Fresnel air–water interface, the reflection operators R_1, R_2 and the transmittance operators T_{12}, T_{21} act according to the following rules:

$$[R_1\Phi](s^-) = \Phi(h, s^+)\ r(\mu^+),$$
$$[R_2\Phi](\tilde{s}^+) = \Phi(h, \tilde{s}^-)\ r(|\tilde{\mu}^-|), \qquad (2)$$

$$[T_{12}\Phi](\overline{s}^+) = \Phi(h, s^+)[1 - r(\mu^+)]\ \mathrm{n}^2,$$
$$[T_{21}\Phi](s^-) = \Phi(h, \overline{s}^-)[1 - r(|\overline{\mu}^-|)]\ /\mathrm{n}^2. \qquad (3)$$

For the reflecting bottom,

$$[R_H\Phi](\tilde{s}^-) = \int_{\Omega^+} \Phi(H, \tilde{s}^+)\ q_H(\tilde{s}^-, \tilde{s}^+)\ d\tilde{s}^+.$$

The boundary conditions are written using the sets:

$$t=\{(z,s):\ z=0,\ s=s^{+}\in\Omega^{+}\},$$
$$b=\{(z,\bar{s}):\ z=H,\ \bar{s}=\bar{s}^{-}\in\Omega^{-}\},$$

$$d1=\{(z,s):\ z=h,\ s=s^{-}\in\Omega^{-}\},$$
$$d2=\{(z,\bar{s}):\ z=h,\ \bar{s}=\bar{s}^{+}\in\Omega^{+}\},$$

whose designations correspond to the first letters of the words top, bottom, and dividing.

For brevity and clarity, we will use the indices "*s*" (singular), "*c*" (continued), "*d*" (diffuse), the upper index "0" for the direct attenuated radiation, "*a*" (atmosphere), and "*ok*" (ocean).

We will solve the problem (1) by the method of influence functions on the basis of an OTO model. Consider several ways of structuring the solution of the initial problem (1), which correspond to different OTO approximations and make it possible to thoroughly study the mechanisms of forming the radiation field in the OAS.

The problem that describes the transfer of the direct solar flux in the OAS and includes only attenuation in the atmosphere and ocean and only one act of interaction with the interface:

$$D\Phi^{0}=0\ ,\ \Phi^{0}\big|_{t}=\pi\ S_{\lambda}\delta\ (s-s_{0})\ ,\ \Phi^{0}\big|_{b}=0\ ,$$
$$\Phi^{0}\big|_{d1}=R_{1}\Phi^{0},\ \Phi^{0}\big|_{d2}=T_{12}\Phi^{0}$$

can be solved in an explicit form if the Fresnel law is assumed to be valid at the interface. In the atmosphere,

$$\Phi^{0}(s_{0};z,s^{+})=\Phi^{0}(s_{0};z,s_{0})\delta\ (s^{+}-s_{0})\ ,$$
$$\Phi^{0}(s_{0};z,s_{0})=\pi\ S_{\lambda}\exp[-\tau(z)/\mu_{0}]. \tag{4}$$

The direct radiation reflected from the interface into the atmosphere in the direction s_{0}^{-} can be written as

$$\Phi^{0}(s_{0};z,s^{-})=$$

$$=\Phi^{0}(s_{0};h,s^{-})\ \exp\{-[\tau(h)-\tau(z)]/\mu_{0}\}, \tag{5}$$

$$\Phi^{0}(s_{0};h,s^{-})=E_{a}^{s}(s_{0};s^{-})\equiv R_{1}\Phi^{0}=$$
$$=\Phi^{0}(s_{0};h,s_{0})\ r(\mu_{0})\ \delta(s^{-}-s_{0}^{-}).$$

The direct radiation that penetrates into the ocean after refraction is written as

$$\Phi^{0}(\bar{s}_{0};z,\bar{s}^{+})=$$

$$=\Phi^{0}(\bar{s}_{0};h,\bar{s}^{+})\ \exp\{-[\tau(z)-\tau(h)]/\bar{\mu}_{0}\}, \tag{6}$$

$$\Phi^{0}(\bar{s}_{0};h,\bar{s}^{+})=E_{ok}^{s}(\bar{s}_{0};\bar{s}^{+})\equiv T_{12}\Phi^{0}=$$
$$=\Phi^{0}(s_{0};h,s_{0})\ [1\text{-}r(\mu_{0})]\ n^{2}\ \delta(\bar{s}_{0}-\bar{s}^{+}).$$

The optical thickness of the layer $[0,z]$ is described by the formula

$$\tau(z)=\int_{0}^{z}\sigma_{tot}(z')\ dz'.$$

3.2 The optical transfer operators and method of influence functions

For estimation of the influence of ocean radiation in the layer $z\in[0,H]$, we will consider the general boundary-value problem for the OAS (GBPOAS) when at least one of the conditions $T_{12}\neq 0$, $T_{21}\neq 0$ is satisfied:

$$K\Phi_{ok}=0,\ \Phi_{ok}\big|_{t}=0,\ \Phi_{ok}\big|_{b}=0,$$

$$\Phi_{ok}\big|_{d1}=\eta(R_{1}\Phi_{ok}+T_{21}\Phi_{ok}+E_{a}),$$
$$\Phi_{ok}\big|_{d2}=\eta(R_{2}\Phi_{ok}+T_{12}\Phi_{ok}+E_{ok}). \tag{7}$$

Here, $E_{a}=E_{a}^{s}+E_{a}^{d}$ and $E_{ok}=E_{ok}^{s}+E_{ok}^{d}$ are the source functions of radiation, which generally involve the singular E_{a}^{s}, E_{ok}^{s} and diffuse E_{a}^{d}, E_{ok}^{d} components, and the parameter η takes into account one act of radiation transmission through the interface.

Let us introduce the two-component algebraic vectors

$$\mathbf{\Theta}=\{\Theta_{a},\Theta_{ok}\},\quad \mathbf{\Phi}_{ok}=\{\Phi^{a},\Phi^{ok}\},$$

$$\mathbf{\Phi}_{n}=\{\Phi_{a,n},\Phi_{ok,n}\},\quad \mathbf{E}=\{E_{a},E_{ok}\}.$$

The solution of the GBPOAS (7) is sought in the form of a series of regular perturbations

$$\mathbf{\Phi}_{ok}=\sum_{n=1}^{\infty}\eta^{n}\mathbf{\Phi}_{n}, \tag{8}$$

whose components satisfy the recurrence system of the first boundary-value problems for the atmospheric layer $z\in[0,h]$ (FBPA) and the oceanic layer $z\in[h,H]$ (FBPO):

$$K\Phi_{a,n}=0,\ \Phi_{a,n}\big|_{t}=0,\ \Phi_{a,n}\big|_{d1}=F_{a,n-1};$$

$$K\Phi_{ok,n}=0,\ \Phi_{ok,n}\big|_{b}=0,\ \Phi_{ok,n}\big|_{d2}=F_{ok,n-1}$$

with the initial source approximations $F_{a,0} \equiv E_a$, $F_{ok,0} \equiv E_{ok}$.

The solutions in all n-approximations are found as linear functionals:

$$\Phi_{a,n} = \left(\Theta_a, F_{a,n-1}\right),\ \Phi_{ok,n} = \left(\Theta_{ok}, F_{ok,n-1}\right);$$

the sources are found:

$$F_{a,n} = R_1\Phi_{a,n} + T_{21}\Phi_{ok,n}\ ,\ F_{ok,n} =_2 \Phi_{ok,n} + T_{12}\Phi_{a,n}.$$

If $E_a\left(s^-\right)$ is the irradiance of the interface from the side of the atmosphere, then for $z \in [0,h]$, $s \in \Omega$ we have the functional

$$\left(\Theta_a, E_a\right)(z,s) = \frac{1}{2\pi}\int_{\Omega^-} \Theta_a\left(s_h^-; z, s\right) E_a\left(s_h^-\right) ds_h^-.$$

If $E_{ok}\left(\tilde{s}^+\right)$ is the irradiance of the interface from the side of the ocean, then for $z \in [h,H]$, or $\tilde{z} = z - h \in [0, H-h]$, $\tilde{s} \in \Omega$ we have the functional

$$\left(\Theta_{ok}, E_{ok}\right)(\tilde{z},\tilde{s}) = \frac{1}{2\pi}\int_{\Omega^+} \Theta_{ok}\left(\tilde{s}_h^+; \tilde{z}, \tilde{s}\right) E_{ok}\left(\tilde{s}_h^+\right) d\tilde{s}_h^+.$$

The kernels of the functionals are the atmosphere IF $\Theta_a\left(s_h^-; z, s\right)$, which represents the solution of the FBPA with the parameter $s_h^- \in \Omega^-$:

$$K\Theta_a = 0,\ \Theta_a\big|_t = 0,\ \Theta_a\big|_{d1} = \delta\left(s - s_h^-\right) \quad (9)$$

and the ocean IF $\Theta_{ok}\left(\tilde{s}_h^+; \tilde{z}, \tilde{s}\right)$, which represents the solution of the FBPO with the parameter $\tilde{s}_h^+ \in \Omega^+$:

$$K\Theta_{ok} = 0,\ \Theta_{ok}\big|_b = 0,\ \Theta_{ok}\big|_{d2} = \delta\left(\tilde{s} - \tilde{s}_h^+\right). \quad (10)$$

These problems (9) and (10) are similar to problems where the radiation source is a wide unidirectional (for example, solar) flux.

For the atmosphere, a set of atmosphere IFs with the parameter

$$s_h^- = \left\{\left(\mu_h^-, \varphi_h^-\right):\ \varphi_h^- = 0,\ \mu_h^- \in [-1,0)\right\}$$

are calculated as a superposition: $\Theta_a = \Theta_a^s + \Theta_a^d$. The singular component describes the transfer of direct radiation from the interface:

$$\Theta_a^s\left(s_h^-; z, s^-\right) =$$

$$= \exp\left\{-\left[\tau(h) - \tau(z)\right] / \left|\mu_h^-\right|\right\} \delta\left(s^- - s_h^-\right).$$

The continued component represents the solution of the FBPA with allowance for multiple scattering

$$K\Theta_a^d = \hat{S}\Theta_a^s,\ \Theta_a^d\big|_t = 0,\ \Theta_a^d\big|_{d1} = 0. \quad (11)$$

For the ocean, a set of ocean IFs with the parameter

$$\tilde{s}_h^+ = \left\{\left(\tilde{\mu}_h^+, \tilde{\varphi}_h^+\right):\ \tilde{\varphi}_h^+ = 0,\ \tilde{\mu}_h^+ \in (0,1]\right\}$$

as a superposition: $\Theta_{ok} = \Theta_{ok}^s + \Theta_{ok}^d$.

The singular component corresponds to the transmission of direct radiation from the interface:

$$\Theta_{ok}^s\left(\tilde{s}_h^+; \tilde{z}, \tilde{s}^+\right) = \exp\left[-\tau(\tilde{z}) / \tilde{\mu}_h^+\right] \delta\left(\tilde{s}^+ - \tilde{s}_h^+\right).$$

The continued component represents the solution of the FBPO with allowance for multiple scattering

$$K\Theta_{ok}^d = \hat{S}\Theta_{ok}^s,\ \Theta_{ok}^d\big|_b = 0,\ \Theta_{ok}^d\big|_{d2} = 0. \quad (12)$$

The diffuse components of the IFs of the atmosphere and ocean are calculated as the solutions of problems (11) and (12) obtained by the known methods in the approximations corresponding to the optical parameters of the two media.

Using the algebraic vector of the IFs of the atmosphere and ocean, we introduce a vector operation describing one act of radiation–interface interaction and taking into account multiple scattering in both media:

$$[\mathbf{GE}](s) \equiv P(\mathbf{\Theta}, \mathbf{E}) =$$

$$= \begin{bmatrix} R_1\left(\Theta_a, E_a\right) + T_{21}\left(\Theta_{ok}, E_{ok}\right) \\ T_{12}\left(\Theta_a, E_a\right) + R_2\left(\Theta_{ok}, E_{ok}\right) \end{bmatrix},$$

$$P \equiv \begin{bmatrix} R_1 & T_{21} \\ T_{12} & R_2 \end{bmatrix}, \quad (13)$$

where $(\mathbf{\Theta}, \mathbf{E}) = \left\{\left(\Theta_a, E_a\right), \left(\Theta_{ok}, E_{ok}\right)\right\}$ is the vector functional. Two successive n-approximations of series (8) are related by the recurrence formula

$$\mathbf{\Phi}_n = \left(\mathbf{\Theta}, P\mathbf{\Phi}_{n-1}\right).$$

An asymptotically exact solution of the GBPOAS (7) is obtained in the form of a linear functional, optical transfer operator of the OAS,

$$\mathbf{\Phi}_{ok} = (\mathbf{\Theta}, \mathbf{Z}),$$

where the "scenario" $\mathbf{Z} = \left\{Z_a, Z_{ok}\right\}$ of the irradiance pattern at the interface

$$\mathbf{Z} \equiv \sum_{n=0}^{\infty} G^n \mathbf{E} = \sum_{n=0}^{\infty} P\Phi_n$$

is the sum of the Neumann series in multiplicity of radiation transmission through the interface with the contribution of multiple scattering in both media taken into account.

The following identical representations are valid:

$$R_1(\Theta_a, E_a) = (R_1\Theta_a, E_a),$$

$$R_2(\Theta_{ok}, E_{ok}) = (R_2\Theta_{ok}, E_{ok}),$$

$$T_{12}(\Theta_a, E_a) = (T_{12}\Theta_a, E_a),$$

$$T_{21}(\Theta_{ok}, E_{ok}) = (T_{21}\Theta_{ok}, E_{ok}).$$

According to these identities, a more rapid calculation of the elements of the operator (13) can be performed by preliminarily determining $R_1\Theta_a$, $R_2\Theta_{ok}$, $T_{12}\Theta_a$, $T_{21}\Theta_{ok}$ from formulas (2) and (3).

If the ocean is regarded as the reflecting underlying surface, the contribution of radiation multiply reflected from the interface and multiply scattered in the atmosphere is described by the general boundary-value problem (GBPA) with $R_1 \not\equiv 0$ (the parameter ε takes into account an event of reflection from the interface):

$$K\Phi_{ar} = 0, \quad \Phi_{ar}|_t = 0, \quad \Phi_{ar}|_{d1} = \varepsilon(R_1\Phi_{ar} + E_{ar}). \tag{14}$$

The solution of the GBPA (14) is sought as the series

$$\Phi_{ar}(z,s) = \sum_{n=1}^{\infty} \varepsilon^n \Phi_n(z,s), \tag{15}$$

whose terms satisfy the recurrence system of the FBPA

$$K\Phi_n = 0, \quad \Phi_n|_t = 0, \quad \Phi_n|_{d1} = F_{n-1},$$
$$F_{n-1} = R_1\Phi_{n-1}, \quad F_0 = E_{ar}$$

and are found as linear functionals with kernels representing the atmosphere IFs:

$$\Phi_n = (\Theta_a, F_{n-1}).$$

If $E_{ar}(s^-)$ is a fixed irradiance of the interface, then using the operation

$$QE_{ar} \equiv R_1(\Theta_a, E_{ar}) = (R_1\Theta_a, E_{ar})$$

one can describe both the corresponding contribution of multiple scattering in the atmosphere, which is included through the atmosphere IFs, and one act of radiation reflection from the underlying surface. An asymptotically exact solution of the GBPA (14) is determined as a linear functional — the optical transfer operator of the underlying surface

$$\Phi_{ar} = (\Theta_a, Z_{ar}). \tag{16}$$

The full "scenario" of the irradiance pattern at the water surface is

$$Z_{ar}(s^-) \equiv \sum_{n=0}^{\infty} Q^n E_{ar} = \sum_{n=0}^{\infty} R_1\Phi_n = E_{ar} + R_1 Y_{ar}. \tag{17}$$

This Neumann series in multiplicity of radiation reflection from the water with allowance for multiple scattering in the air is expressed through the diffuse "irradiance" of the interface, which is caused by the backscattering of reflected radiation in the atmosphere:

$$Y_{ar}(s^+) = \sum_{n=1}^{\infty} \Phi_n(h, s^+). \tag{18}$$

3.3 Functional calculation

In the general case, it is necessary to separate singular and continued diffuse components, because each functional can involve four components with different properties:

$$(\Theta_a, E_a) = (\Theta_a^s, E_a^s) + (\Theta_a^d, E_a^s) + (\Theta_a^s, E_a^d) + (\Theta_a^d, E_a^d), \tag{19}$$

$$(\Theta_{ok}, E_{ok}) = (\Theta_{ok}^s, E_{ok}^s) + (\Theta_{ok}^d, E_{ok}^s) + (\Theta_{ok}^s, E_{ok}^d) + (\Theta_{ok}^d, E_{ok}^d). \tag{20}$$

The first three terms are expressed in an explicit form, and the fourth terms represent the continued components, which are calculated by the quadrature method. Let us present expressions for the components of the functionals (19) and (20).

If $z \in [0,h]$ and $s \in \Omega$ component of the functional (19)

$$(\Theta_a^s, E_a^s)\,(z, s^-) = E_a^s(s_0; s^-)\,\exp\{-[\tau(h) - \tau(z)]/\mu_0\}$$

is identical to the solution of the problem (5) for the direct solar flux reflected in the atmosphere from the interface:

$$\Phi_0(s_0; z, s^-) = (\Theta_a^s, E_a^s).$$

The second component of the functional (19) is a continued diffuse component:

$$(\Theta_a^d, E_a^s)(z, s) = E_a^s(s_0; s_0^-)\Theta_a^d(s_0^-; z, s).$$

The third component of the functional (19)

$$(\Theta_a^s, E_a^d)\ (z, s^-) =$$

$$= E_a^d(s^-)\ \exp\{-[\tau(h) - \tau(z)]/|\mu^-|\}$$

represents the direct radiation traveling from the boundary $z = h$ (a continued function).

The fourth component of the functional (19) is calculated taking into account the azimuth symmetry of the integrands, which are continued functions of angular arguments

$$E_a^d(\varphi) = E_a^d(2\pi - \varphi) = E_a^d(-\varphi),$$

$$\Theta_a^d(\varphi) = \Theta_a^d(2\pi - \varphi) = \Theta_a^d(-\varphi)$$

and using the quadrature-method formula

$$(\Theta_a^d, E_a^d)\ (z, s) =$$

$$= \frac{1}{2\pi}\int_0^\pi d\varphi_h^- \int_{-1}^0 [\Theta_a^d(\mu_h^-, 0; z, \mu, \varphi - \varphi_h^-) +$$

$$+ \Theta_a^d(\mu_h^-, 0; z, \mu, \varphi + \varphi_h^-)]\ E_a^d(\mu_h^-, \varphi_h^-)\, d\mu_h^-.$$

If $z \in [h, H]$ and $\tilde{s} \in \Omega$ the singular component of the functional (20)

$$(\Theta_{ok}^s, E_{ok}^s)\ (\tilde{z}, \tilde{s}^+) =$$

$$= E_{ok}^s(\tilde{s}_0; \tilde{s}^+)\ \exp[-\tau(\tilde{z})/\tilde{\mu}_0]$$

represents the direct solar flux in the ocean (6):

$$\Phi^0(\tilde{s}_0; z, \tilde{s}^+) = (\Theta_{ok}^s, E_{ok}^s).$$

The second component of the functional (20) is a continued component describing the contribution of the direct solar flux multiply scattered in the ocean:

$$(\Theta_{ok}^d, E_{ok}^s)(\tilde{z}, \tilde{s}) = E_{ok}^s(\tilde{s}_0; \tilde{s}_0^-)\Theta_{ok}^d(\tilde{\mu}_0, 0; \tilde{z}, \tilde{s}).$$

The third component of the functional (20)

$$(\Theta_{ok}^s, E_{ok}^d)(\tilde{z}, \tilde{s}^+) = E_{ok}^d(\tilde{s}^+)\exp[-\tau(\tilde{z})/\tilde{\mu}_h^+]$$

represents the direct radiation that travels from the boundary and is caused by the diffuse irradiance component. The fourth component of the functional (20) is calculated by the quadrature method with allowance for the azimuth properties:

$$(\Theta_{ok}^d, E_{ok}^d)\ (\tilde{z}, \tilde{s}) =$$

$$= \frac{1}{2\pi}\int_0^\pi d\tilde{\varphi}_h^+ \int_0^1 [\Theta_{ok}^d(\tilde{\mu}_h^+, 0; \tilde{z}, \tilde{\mu}, \tilde{\varphi} - \tilde{\varphi}_h^+) +$$

$$+ \Theta_{ok}^d(\tilde{\mu}_h^+, 0; \tilde{z}, \tilde{\mu}, \tilde{\varphi} + \tilde{\varphi}_h^+)]\ E_{ok}^d(\tilde{\mu}_h^+, \tilde{\varphi}_h^+)\, d\tilde{\mu}_h^+.$$

3.4 Structure of the OAS radiation field

The solution of the initial problem (1) is usually sought as a superposition

$$\Phi = \Phi^0 + \Phi_a + \Phi_d + \Phi_q.$$

The direct solar flux Φ^0 in the atmosphere is determined by the solution to the Cauchy problem

$$D\Phi^0 = 0\ , \quad \Phi^0\big|_t = \pi\, S_\lambda \delta\ (s - s_0)\ , \quad \Phi^0\big|_{d1} = 0$$

in the explicit form (4).

The background (diffuse) solar radiation multiply scattered in the atmosphere Φ_a is found as the solution of the FBPA with nonreflecting boundaries

$$K\Phi_a = S\Phi^0, \quad \Phi_a\big|_t = 0, \quad \Phi_a\big|_{d1} = 0.$$

The effect of the bottom in the layer $z \in [0, H]$ is described by the GBPOAS with the source $E_H = R_H \Phi_d$:

$$K\Phi_q = 0, \quad \Phi_q\big|_t = 0, \quad \Phi_q\big|_b = R_H\Phi_q + E_H,$$

$$\Phi_q\big|_{d1} = R_1\Phi_q + T_{21}\Phi_q, \quad \Phi_q\big|_{d2} = R_2\Phi_q + T_{12}\Phi_q.$$

If the bottom is Lambertian and its albedo is q, then

$$\Phi_q = qE_H\Psi/(1 - qc_0)\ , \quad c_0 = R_H\Psi,$$

where the influence function of the bottom $\Psi(z, s)$ is the solution of the problem

$$K\Psi = 0, \quad \Psi\big|_t = 0, \quad \Psi\big|_b = 1,$$

$$\Psi\big|_{d1} = R_1\Psi + T_{21}\Psi, \quad \Psi\big|_{d2} = R_2\Psi + T_{12}\Psi,$$

which reduces to a problem similar to (1.7).

The component Φ_d, which takes into account the contribution of the ocean as both a reflecting underlying surface and a scattering medium, can be written in different ways depending on the model chosen for the structuring and estimation of mechanisms involved in the formation of the OAS radiation field.

4 Mathematical model of radiation transfer in the ocean-atmosphere system with clouds

Boundary-value problem for multi-layer heterogeneous system is considered

$$K\Phi = F^{in},\quad \Phi\big|_{t\downarrow} = F_t^{\downarrow},\quad \Phi\big|_{b\uparrow} = R_b^{\uparrow}\Phi + F_b^{\uparrow},$$

at internal boundaries h_m for $m = 2 \div M$

$$\Phi\big|_{d\uparrow,m} = \varepsilon\ (R_m^{\uparrow}\Phi + T_m^{\uparrow}\Phi) + F_{m-1}^{\uparrow},$$

$$\Phi\big|_{d\downarrow,m} = \varepsilon\ (R_m^{\downarrow}\Phi + T_m^{\downarrow}\Phi) + F_m^{\downarrow};$$

integro-differential operator $K \equiv D - S$ with the linear operators: transfer operator D and the collision integral S; $m = 1 \div M$ – number of layer.

Radiation propagation through the internal boundaries between the layers is described by equal-measure restricted operators of reflectivity $R_m^{\downarrow}$, $R_m^{\uparrow}$ and transmittance $T_m^{\downarrow}$, $T_m^{\uparrow}$.

We seek solution in the form of a regular perturbation series

$$\Phi = \sum_{n=0}^{\infty} \varepsilon^n \Phi^{(n)}.$$

Let's introduce the algebraic vectors with the dimension $2M$:
n - approximation of the solution

$$\Phi^{(n)} = \{\Phi_1^{\downarrow(n)}, \Phi_1^{\uparrow(n)}, \Phi_2^{\downarrow(n)}, \Phi_2^{\uparrow(n)}, \ldots, \Phi_m^{\downarrow(n)}, \Phi_m^{\uparrow(n)}, \ldots, \Phi_M^{\downarrow(n)}, \Phi_M^{\uparrow(n)}\};$$

the complete solution

$$\Phi = \{\Phi_1^{\downarrow}, \Phi_1^{\uparrow}, \Phi_2^{\downarrow}, \Phi_2^{\uparrow}, \ldots, \Phi_m^{\downarrow}, \Phi_m^{\uparrow}, \ldots, \Phi_M^{\downarrow}, \Phi_M^{\uparrow}\};$$

n - approximation of the sources

$$F^{(n)} = \{F_1^{\downarrow(n)}, F_1^{\uparrow(n)}, F_2^{\downarrow(n)}, F_2^{\uparrow(n)}, \ldots, F_m^{\downarrow(n)}, F_m^{\uparrow(n)}, \ldots, F_M^{\downarrow(n)}, F_M^{\uparrow(n)}\};$$

the influence functions of the layers

$$\Theta = \{\Theta_1^{\downarrow}, \Theta_1^{\uparrow}, \Theta_2^{\downarrow}, \Theta_2^{\uparrow}, \ldots, \Theta_m^{\downarrow}, \Theta_m^{\uparrow}, \ldots, \Theta_M^{\downarrow}, \Theta_M^{\uparrow}\};$$

the initial approximation of the sources

$$E = \{E_1^{\downarrow}, E_1^{\uparrow}, E_2^{\downarrow}, E_2^{\uparrow}, \ldots, E_m^{\downarrow}, E_m^{\uparrow}, \ldots, E_M^{\downarrow}, E_M^{\uparrow}\};$$

the "scenario" on boundaries

$$Z = \{Z_1^{\downarrow}, Z_1^{\uparrow}, Z_2^{\downarrow}, Z_2^{\uparrow}, \ldots, Z_m^{\downarrow}, Z_m^{\uparrow}, \ldots, Z_M^{\downarrow}, Z_M^{\uparrow}\}.$$

We produce decomposition of the original problem at $2M$ problems with their own boundary conditions. The initial approximation – radiation from sources without radiation interchange between layers for $m = 1 \div M$:

$$K\Phi_m^{\downarrow(0)} = F_m^{\downarrow in},\quad \Phi_m^{\downarrow(0)}\Big|_{d\downarrow,m} = F_m^{\downarrow},\quad \Phi_m^{\downarrow(0)}\Big|_{d\uparrow,m+1} = 0;$$

$$K\Phi_m^{\uparrow(0)} = F_m^{\uparrow in},\quad \Phi_m^{\uparrow(0)}\Big|_{d\downarrow,m} = 0,\quad \Phi_m^{\uparrow(0)}\Big|_{d\uparrow,m+1} = F_m^{\uparrow}.$$

The approximation $n \geq 1$ – system of $2M$ equations for layers $m = 1 \div M$:

$$K\Phi_m^{\downarrow(n)} = 0,\quad \Phi_m^{\downarrow(n)}\Big|_{d\downarrow,m} = F_m^{\downarrow(n-1)},\quad \Phi_m^{\downarrow(n)}\Big|_{d\uparrow,m+1} = 0;$$
$$K\Phi_m^{\uparrow(n)} = 0,\quad \Phi_m^{\uparrow(n)}\Big|_{d\downarrow,m} = 0,\quad \Phi_m^{\uparrow(n)}\Big|_{d\uparrow,m+1} = F_m^{\uparrow(n-1)}$$

with sources at the boundaries h_m for $m = 1 \div M + 1$:

$$F_m^{\downarrow(n)} = T_m^{\downarrow}\Phi_{m-1}^{\downarrow(n)} + T_m^{\downarrow}\Phi_{m-1}^{\uparrow(n)} + R_m^{\downarrow}\Phi_m^{\downarrow(n)} + R_m^{\downarrow}\Phi_m^{\uparrow(n)};$$
$$F_m^{\uparrow(n)} = R_{m+1}^{\uparrow}\Phi_m^{\downarrow(n)} + R_{m+1}^{\uparrow}\Phi_m^{\uparrow(n)} + T_{m+1}^{\uparrow}\Phi_{m+1}^{\downarrow(n)} + T_{m+1}^{\uparrow}\Phi_{m+1}^{\uparrow(n)}.$$

Solution is assumed in the form of linear functionals for each layer with $m = 1 \div M$:

$$\Phi_m^{\downarrow(n)} = \left(\Theta_m^{\downarrow}, F_m^{\downarrow(n-1)}\right);\quad \Phi_m^{\uparrow(n)} = \left(\Theta_m^{\uparrow}, F_m^{\uparrow(n-1)}\right).$$

Functionals kernels – influence functions of layers with $m = 1 \div M$ are determined from boundary-value problems

$$K\Theta_m^{\downarrow} = 0,\quad \Theta_m^{\downarrow}\Big|_{d\downarrow,m} = f_{\delta,m}^{\downarrow},$$

$$\Theta_m^{\downarrow}\Big|_{d\uparrow,m+1} = 0;\quad f_{\delta,m}^{\downarrow} = \delta\left(s - s_m^{\downarrow}\right);$$

$$K\Theta_m^{\uparrow} = 0,\quad \Theta_m^{\uparrow}\Big|_{d\downarrow,m} = 0,$$

$$\Theta_m^{\uparrow}\Big|_{d\uparrow,m+1} = f_{\delta,m}^{\uparrow};\quad f_{\delta,m}^{\uparrow} = \delta\left(s - s_m^{\uparrow}\right).$$

The components of vectorial linear functional are computed for $z \in \left[h_m, h_{m+1}\right]$ by formulas:

$$\left(\Theta_m^{\downarrow}, f_m^{\downarrow}\right)(z,s) = \int_{\Omega^+} \Theta_m^{\downarrow}\left(s_m^{\downarrow}; z, s\right) f_m^{\downarrow}\left(h_m, s_m^{\downarrow}\right) ds_m^{\downarrow};$$

$$\left(\Theta_m^{\uparrow}, f_m^{\uparrow}\right)(z,s) = \int_{\Omega^-} \Theta_m^{\uparrow}\left(s_m^{\uparrow}; z, s\right) f_m^{\uparrow}\left(h_{m+1}, s_m^{\uparrow}\right) ds_m^{\uparrow},$$

where $f_m^{\downarrow}\left(h_m, s_m^{\downarrow}\right)$ is the brightness source distribution at boundary $z = h_m$ when $s_m^{\downarrow} \in \Omega^{\downarrow}$; $f_m^{\uparrow}\left(h_{m+1}, s_m^{\uparrow}\right)$ is the brightness source distribution at boundary $z = h_{m+1}$ when $s_m^{\uparrow} \in \Omega^{\uparrow}$.

5 Results

To calculate the total radiation inside or on the boundaries of the system with radiation exchange between the layers the vectorial optical transfer operator (VOTO) is constructed by the influence functions method. The kernels of operators are the influence functions of the layers. Operators of radiation transmittance and reflectance on the boundaries between the layers are formulated based on the collision integrals, and the separate layers are united in a system by these operators.

The n-approximation of solution is stated in form of vectorial linear functional

$$\Phi^{(n)} = \left(\Theta, F^{(n-1)}\right),$$

where sources in $(n-1)$-approximation $F = P\Phi^{(n-1)}$. Two sequential n-approximations are connected by recursion relation

$$\Phi^{(n)} = \left(\Theta, P\Phi^{(n-1)}\right),$$

where P – matrix of band type with layers reflection and trasmission characteristics. The notion can be found

$$\Phi^{(n)} = \left(\Theta, G^{n-1}E\right),$$

where E – initial approximation. Matrix-vector operation, which describe one act of radiation interaction with internal and external boundaries and take into account multiply scattering in the layers through influence functions, is determined:

$$GF = P(\Theta, F) =$$

$$\begin{bmatrix} 0 \\ R_2^{\uparrow}(\Theta_1^{\downarrow}, F_1^{\downarrow}) + R_2^{\uparrow}(\Theta_1^{\uparrow}, F_1^{\uparrow}) + \\ +T_2^{\uparrow}(\Theta_2^{\downarrow}, F_2^{\downarrow}) + T_2^{\uparrow}(\Theta_2^{\uparrow}, F_2^{\uparrow}) \\ \mathrm{M} \\ T_m^{\downarrow}(\Theta_{m-1}^{\downarrow}, F_{m-1}^{\downarrow}) + T_m^{\downarrow}(\Theta_{m-1}^{\uparrow}, F_{m-1}^{\uparrow}) + \\ +R_m^{\downarrow}(\Theta_m^{\downarrow}, F_m^{\downarrow}) + R_m^{\downarrow}(\Theta_m^{\uparrow}, F_m^{\uparrow}) \\ R_{m+1}^{\uparrow}(\Theta_m^{\downarrow}, F_m^{\downarrow}) + R_{m+1}^{\uparrow}(\Theta_m^{\uparrow}, F_m^{\uparrow}) + \\ +T_{m+1}^{\uparrow}(\Theta_{m+1}^{\downarrow}, F_{m+1}^{\downarrow}) + T_{m+1}^{\uparrow}(\Theta_{m+1}^{\uparrow}, F_{m+1}^{\uparrow}) \\ \mathrm{M} \\ T_M^{\downarrow}(\Theta_{M-1}^{\downarrow}, F_{M-1}^{\downarrow}) + T_M^{\downarrow}(\Theta_{M-1}^{\uparrow}, F_{M-1}^{\uparrow}) + \\ +R_M^{\downarrow}(\Theta_M^{\downarrow}, F_M^{\downarrow}) + R_M^{\downarrow}(\Theta_M^{\uparrow}, F_M^{\uparrow}) \\ R_b^{\uparrow}(\Theta_M^{\downarrow}, F_M^{\downarrow}) + R_b^{\uparrow}(\Theta_M^{\uparrow}, F_M^{\uparrow}) \end{bmatrix}$$

Asymptotical exact solution is obtained in the form of vectorial linear functional – optical transfer operator:

$$\Phi = \left(\Theta, Z\right).$$

"Scenario" – vector Z distribution of brightness on internal and external boundaries $m = 1 \div M + 1$:

$$Z \equiv YE \equiv \sum_{n=0}^{\infty} G^n E = \sum_{n=0}^{\infty} P\Phi^{(n)}$$

is sum of Neyman series by multiplicity passing of radiation through boundaries with consideration of multiply scattering contribution by the help of influence functions of each layer

Original mathematical tools are proposed for the first time to model radiation transfer in multi-layer non-homogenous heterogenic natural media with different radiation regimes within separate regions of the system. "Free atmosphere – multi-layer cloudiness – near surface layer of the atmosphere – ocean" system is heterogenic, in particular. The proposed approach is based on construction of generalized solutions in the form of matrix functionals; vectors of influence functions for each layer of the system serve as kernels of the functionals. The influence functions of the layers with different aerosol and molecular characteristics of scattering and absorption and radiation regimes can thus be calculated by various methods in different approaches of radiation transfer theory.

This approach enables to simulate radiation fields in a wide range of variations of the optical characteristics of these layers and to analyze mechanisms of radiation characteristics formation inside and outside the layers as well as to estimate any contribution of each region. Reflected radiation and distributions of radiation characteristics inside the layer near to the border illuminated by an external flux are thus calculated with higher accuracy than in common-used techniques.

The boundary problem for each layer can be solved depending on optical thickness, scattering and absorption characteristics by one of the following techniques:

i) as a solution of transfer equation with an azimuth dependence;

ii) as a solution of the problem with azimuth symmetry;

iii) as a solution in two-flux approach;

iv) as an approximate solution in asymptotic approach.

Operators of radiation transmittance and reflectance on the boundaries between the layers are formulated based on the collision integrals and the separate layers are united in a system by these operators. To calculate the total radiation inside or on the boundaries of the system with radiation exchange between the layers a matrix-vector operation is constructed for the operators the kernels of which are given by the influence functions of the layers.

The representation of the solution to boundary-value problem as a functional is the optical transfer operator of the radiation transfer system which establishes the explicit relationship between the recording radiation and the "scenarios" (the optical image) at the dividing boundaries of media. In turn, by the use of the influence functions, the "scenarios" is described clearly through the characteristics of the reflection and transmission of the dividing boundaries at the given its illumination. The influence functions are invariant about the conditions of the illumination and the properties of the dividing boundaries.

6 Conclusion

On a planetary scale, there is an urgent problem of creating international global monitoring of the Earth with the aim of studying its evolution and predicting natural disasters and anthropogenic-technogenic catastrophic processes, as well as environmental and climate changes on a local and global scale.

The theoretical basis for estimating radiation forcing is the solution of direct and inverse problems of the radiation transfer theory [8, 9]. Hyperspectral remote sensing of the Earth is the most important technology for monitoring and evaluating the radiation impact on the planet Earth. In recent years, close attention of researchers have focused on improving and developing methods and tools for more thorough and reliable accounting of the absorption of major and minor gases in the atmosphere and assessing the influence of the absorbing factor on radiation forcing.

The Table shows data on the number of spectral lines of absorbing gases in the most complete and reliable version of the HITRAN-2016 database (USA).

Table of spectral lines

Molecule	Sum of lines for all isotopes
H_2O	313787
CO_2	559874
O_3	449570
N_2O	160287
CO	5381
CH_4	450332
O_2	14085
Total	1953316

Scientists from different countries are involved in the formation of these databases [11]. Our team members are also engaged in analysis, "reference calculations" and applications of these databases to calculations [12]. According to 2016 data, about 2 million absorption lines are already known, and this does not take into account the continuous absorption of water in different phases (water vapor, droplets, ice crystals) and aerosols of different origin. To calculate the absorption coefficients, profiles of meteorological data (temperature, pressure) and component concentrations are required. And, of course, important times of day, seasons, regions, etc. And, of course, you need to take into account the diversity of underlying surfaces and the influence of oceans and clouds.

Acknowledgements

The article is dedicated to the memory of our teachers. Three Russian Geniuses – "Three K" – Mstislav Vsevolodovich Keldysh (10.02.1911-24.06.1978) – Chief Cosmonautics Theorist, Three times Hero of Socialist Labor (1956, 1961, 1971); Sergey Pavlovich Korolev (12.01.1907-14.01.1966) – Chief Designer of Cosmonautics, Twice Hero of Socialist lLbor (1956, 1961); Igor Vasilyevich Kurchatov (12.01.1903-07.02.1960) – "Father of the Russian atomic bomb", Three Times Hero of Socialist Labor (1949, 1951, 1954) – they headed and made the scientific and technological revolution in the Union of Soviet Socialist Republics (USSR) in the middle of the XX-th century, and their names forever entered the history of the world mankind civilization , the USSR and Russia. It is thanks to their scientific exploits and merits that the strategic parity of the two political systems and military blocs led by USSR)and USA was ensured, and for almost 75 years there are no global world wars on the planet.

This research was supported by the State project 0017-2019-0002.

References

1. V.E. Ryabinin, V.I. Sychev, Uchenye zapiski RSHU, **54**, 73-79 (2019)
2. M.V. Keldysh, *Creative portrait on the memoirs of contemporaries* (Nauka, Moscow, 2001)
3. T.A. Sushkevich, Jr. Rse. Cosmos, **8**(1), 9-25 (2011)
4. T.A. Sushkevich, Jr. Rse. Cosmos, **8**(4), 9-22 (2011)
5. A.S. Monin, Yu.A. Shishkov, UFN, **170**(4), 419-445 (2000)
6. R.I. Nigmatulin, Expert, **34**, 46-51 (2018)

7. T.A. Sushkevich, Fundam. Prikl. Gidrofiz., **13**, 103-109 (2020)
8. T.A. Sushkevich, S.A. Strelkov, *Characteristics Method in Atmospheric Optics Problems* (Nauka, Moscow, 1990)
9. T.A. Sushkevich, *Mathematical Models of Radiation Transfer* (BINOM, Laboratory of Knowledge, Moscow, 2005)
10. T.A. Sushkevich, A.A. Ioltukhovsky, Preprint KIAM RAS, **9** (1986)
11. I.E. Gordon, L.S. Rothman, C. Hill *et al.*, JQSRT **203**, 3-69 (2017)
12. B.A. Fomin, G.E. Kolokutin, Jr. Rse. Cosmos **16**(3), 17-24 (2019)
13. C.D. Mobley, B. Gentili, H.R. Gordon, Z. Jin, G.W. Kattawar, A. Morel, Appl. Opt., **32**, 7484-7504 (1993)
14. Z. Jin, K. Stanmes, Appl. Opt., **33**, 431-443 (1994)
15. J. Lenoble, Radiative transfer in scattering and absorbing atmospheres: standard computational procedures (A. DEEPAK Publishing, Hampton, Virginia, USA, 1985).

Modelling Biomolecular Structures in Categorical Systems Theory

Georgy Tolokonnikov[1,*]

[1]Federal Scientific Agroengineering Center VIM, RU-109428, Moscow, Russia

Abstract. In the systemic movement there exist numerous approaches to systems, the most profound of which is the theory of functional systems by Anokhin, which remained largely intuitive science until his pioneering works. The basic principles of functional systems are formalized with the help of the convolutional polycategories in the form of categorical systems theory, which embraced the main systemic approaches, including the traditional mathematical theory of systems. Convolutional polycategories can be built using categorical splices that directly model the external and internal parts of systems. For an algebraic biology using the categorical theory of systems in relation to systemic constructions, the main task of which is to predict the properties of organisms from the genome using strict algebraic methods, new categorical methods are proposed that are widely used in categorical systems theory. These methods are based on the theory of categorical splices, with the help of which the behaviour of quantum-mechanical particles is modelled, in particular, within the framework of the proposed representation of molecules, including RNA and DNA, as categorical systems. Thus, new algebraic and categorical methods (associative algebras with identities, PROP, categorical splices) are involved in the analysis of the genome. The paper presents new results on these matters.

1 Introduction

Systems theory was born in biology, there are numerous systems approaches, which include the direction of mathematical systems theory, with the concept of a system given by M. Mesarovich, V.M. Matrosov and other scientists (see [1], review in [2]), ergatic systems "man-machine" of engineering psychology Lomov [3] and others. There are approaches initiated in biology and medicine by L. Bertalanffy, P.K. Anokhin, G.N. Kryzhanovsky [4] and other researchers. The most profound of the approaches is the theory of functional systems by P.K. Anokhin [5], which remained an essentially intuitive science until the author's works (see [6-9] and references therein). The basic principles of functional systems are formalized with the help of (generalizing polycategories of Szabo (1975)) convolutional polycategories in the form of categorical systems theory, which embraced the main system approaches, including the traditional mathematical theory of systems. Convolutional polycategories can be built using categorical splices that directly model the external and internal parts of systems. For the algebraic biology using the categorical theory of systems [10,11], the main task of which is to predict the properties of organisms from the genome using strict algebraic methods, new categorical methods used in categorical systems theory are proposed. These methods are based on the theory of categorical splices, with the help of which the behaviour of biomolecules, including RNA and DNA, as categorical systems is modelled. Thus, new algebraic and categorical methods are involved in genome analysis. From the point of view of the systemic paradigm, all concepts must be reduced to systems and their aggregates, just as it is implemented in the set-theoretic paradigm according to N. Burbaki, when all concepts are reduced to sets and subsets. The functional systems of P.K. Anokhin do not describe biomolecules as systems. This paper proposes a strictly mathematical description of biomolecules as categorical systems. Unlike the genetic code, the properties of DNA and RNA molecules in algebraic biology have not yet been rigorously studied mathematically. Thus, from the point of view of applications in algebraic biology, the work is a definite step in the study of the properties of DNA and RNA molecules by algebraic methods. The work is also devoted to the study of the properties of categorical splices, in particular, duality. Duality in category theory is known to be a fundamental property. In the theory of categorical splices, there is not one, but several types of duality, including the traditional duality (generated by a change in the direction of the arrows). In particular, the stated duality theorem turns out to be valid, generalizing to gluings and new types of duality in them the fundamental theorem of Hatcher [12] substantiating the application of the duality principle in category theory [13]. The listed results are new and original.

2 Categorical splices

Let's introduce a four-sorted first-order logic language

* Corresponding author: admcit@mail.oru

with variables $p, p', \bar{p}, \bar{p}'$ responsible for splices, convolutions, inner splices and inner convolutions, respectively. Convolutions are represented using: specified letters; functional symbols of co-regions $C_i, C'_i, \bar{C}_i, \overline{C'}_i$; names $\pi_i, \pi'_i, \bar{\pi}_i, \overline{\pi'}_i$; values $ev_\alpha, ev'_\alpha, \overline{ev}_\alpha, \overline{ev'}_\alpha$; predicates M_b, M_r, M_t, M_l,; using a graphic representation of formulas similar to that introduced by Hetcher [12] when constructing the first-order language of category theory, for example

$$x_i \stackrel{\text{def}}{=} \pi_i(p), b_i \stackrel{\text{def}}{=} C_i(p), x_i = \pi_i(p) \wedge b_i = C_i(p) \wedge$$

$$\wedge\, x_j = \pi_j(p) \wedge b_j = C_j(p) \wedge x_i = x_j$$

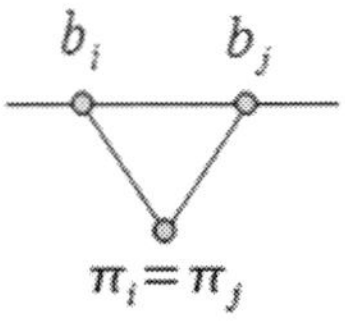

The properties of predicates and other letters of the signature are determined by axioms. Formulas $x_i = \pi_i(p), b_i = C_i(p)$ will be denoted by π_i, C_i. The convolution S is defined by the ratio, the symbol ev has the opposite and the opposite convolution is determined by the $S = M_b ev M_l$. The complete splices can be represented as a conjunction formula $a \wedge b' \wedge \bar{c} \wedge \bar{d}'$, divided into four subformulas in accordance with the sort of variables, located in the 2x2 quadrants of the table

a	*b*
c	*d*

Equivalences between co-regions, between names that correspond to the equal sign, are depicted by connecting the right and left letters inequality with dashed lines. The predicates M_t, M_r, M_l, M_b are determined by the following formula transformations. M_b moves to the lower quadrants from the upper quadrants only those pairs $C_k\pi_k$ that have co-area connections, M_t moves to the upper quadrants from the lower quadrants only those pairs $C_k\pi_k$ that have co-area connections, M_l moves vertically and inside a pair from right quadrants to left ones, M_r moves connected pairs from left quadrants to rights.

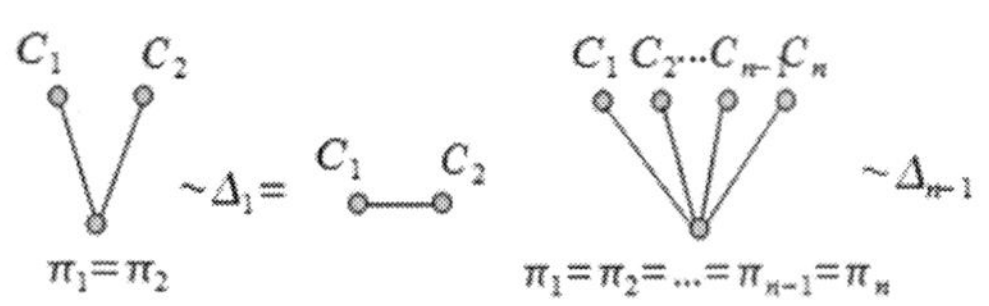

Fig. 1. Modeling simplices by splices

The function symbol *ev* translates π_i, C_i to $ev(\pi_i), ev(C_j)$ and places the = sign in the appropriate places. If you do not impose restrictions on the choice of splices and convolutions, then we speak of free general splices. Special categorical splices are constructed from a fixed set of convolutions and a fixed set of splices, while the original splices are called generators, and all the rest are obtained by all possible applications of these convolutions to splices. The splices model the set of affine simplexes (Fig. 1). For clarity, consider the following formula as an example.

$$C_k \wedge \pi_k \wedge C_i \wedge \pi_i \wedge C_m \wedge \pi_m \wedge C_n \wedge \pi_n \wedge C'_j \wedge \pi'_j$$
$$\wedge\, C'_w \wedge \pi'_w \wedge C'_v \wedge \pi'_v \wedge \pi_k$$
$$= \pi_i \wedge \pi_m = \pi_n \wedge \pi'_j = \pi'_v \wedge C_i$$
$$= C'_j \wedge C_m = C_v$$

We will omit the conjunction sign, replace equalities with a line connecting the right and left sides of the equalities. The table for the formula is

$C_k\pi_kC_i\pi_iC_m\pi_mC_n\pi_n$	$C'_j\pi'_jC'_w\pi'_wC'_v\pi'_v$
$\emptyset$	$\emptyset$

For convolution, we have a table

$C_k\pi_kC_n\pi_n$	$C'_w\pi'_w$
$C_i\pi_iC_m\pi_m$	$C'_j\pi'_jC'_v\pi'_v$

Applying, M_l we get

$C_k\pi_kC_n\pi_n$ $C_w\pi_w$	$\emptyset$
$C_i\pi_iC_m\pi_m$ $C_j\pi_jC_v\pi_v$	$\emptyset$

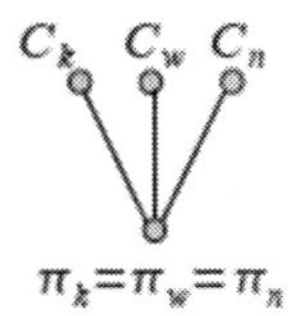

A complete restoration of the original splice is possible as follows. We apply a transfer M_r to the final table in relation to $C_w\pi_w\bar{C}_j\bar{\pi}_j\bar{C}_v\bar{\pi}_v$, we will get a table that coincides with the previous final table. Then we

apply M_t to it, which gives the original table and, thereby, the original splice.

For a given splice, it is possible to construct a new Δ'-dual splice by transferring the contents in the right parts of the splice table to the left and at the same time vice versa (splice-convolution duality). Similarly, you can construct a new $\bar{\Delta}$-dual splice by transferring the content in the upper parts of the splice table to the lower parts and at the same time vice versa (duality "outside-inside"). If the names in the splice formula are replaced by the corresponding co-domains and simultaneously vice versa, then again the splice is obtained, Δ_π^C-dual of the original splice, in which the equalities of the co-domains turn into the equal names and vice versa (the duality "name-co-domain").

3 Higher categorical k-splices and general systems

Let there be k-splices with sets of variables of four sorts, functional symbols, and predicates. It is not difficult to construct a unified formal theory of higher categorical k-splices. Let's go to a table with 2k columns, in the first k columns, in order, insert $a_i, c_i, i = 1, \dots, k$, in the second k columns, insert $b_i, d_i, i = 1, \dots, k$, as indicated below

a_1	...	a_k	b_1	...	b_k
c_1	...	c_k	d_1	...	d_k

The above types of duality (Δ',$\bar{\Delta}$,Δ_π^C) are naturally carried over to the theory of k-splices. Here a form of duality $\hat{\Delta}$, already known from category theory, arises, associated with the replacement of the direction of the arrows. It takes the form of interchange by places in the formula of a general k-splices or any other formula of the first-order language of the formal theory of k-splices of one type of splices to another. There can be several such replacements in k-splices.

Let formulas be given in the language of categorical splices. The transformations of formulas, respectively, Δ',$\bar{\Delta}$,Δ_π^C,$\hat{\Delta}$-types defined above are called their transformation Δ',$\bar{\Delta}$,Δ_π^C,$\hat{\Delta}$-duality. The fundamental role played by duality in the categories is well known. It is logically substantiated in [12]. For the listed dualities in the theory of categorical k-splices, an analogue of the theorem proved in [12] holds.

Theorem. Let in the formal theory of categorical k-splices be given a provable well-constructed formula A. Then dual - formulas are also provable in this theory.

We considered systems in convolutional polycategories and in 1-splices [6-9], having higher k-splices, we can naturally define k-systems or higher systems on them. At the same time, k-systems are used for modeling neurons with connections other than spike connections.

In the last decade, research has been expanding on the non-spike connections of neurons (see, for example, [14] and references therein). Traditional neural networks in physiological terminology realize only the level of reflexes, which, as taught by modern physiology and the theory of functional systems, is not enough for the implementation of cognitive functions. A possible way out is to complicate the models of neurons [15] and to model new types of connections between neurons [14]. More formally, we can say that the main idea we propose with regard to neural networks is the transition from associative compositional convolutional polycategories with corona-type convolutions, which simulate traditional artificial neural networks of arbitrary topology [7,9], to networks based on higher categorical splices, modeling also other types of neuronal connections besides the usual communication carried out by spikes.

4 Higher Convolutional polycategories and their gluing representation

In convolutional polycategories, poly-arrows have inputs and outputs. Their modeling is possible using 2-splices. Let's look at an example of traditional categories. Two arrows, when the beginning of one arrow coincides with the end of the other arrow, and the convolution (corresponding to the composition of the arrows), corresponds to the formula (between the letters the conjunction sign is omitted)

$$C_1^{(1)}\pi_1^{(1)}C_2^{(1)}\pi_2^{(1)}C_3^{(1)'}\pi_3^{(1)'}C_1^{(2)}\pi_1^{(2)}C_2^{(2)}\pi_2^{(2)}C_3^{(2)'}\pi_3^{(2)'}$$

$$\left(\pi_1^{(1)} = \pi_1^{(2)}\right)\left(\pi_2^{(1)} = \pi_2^{(2)}\right)\left(\pi_3^{(1)'} = \pi_3^{(2)'}\right)$$

$$\left(C_1^{(2)} = C_2^{(1)}\right)\left(C_2^{(1)} = C_3^{(1)'}\right)$$

$$\left(C_1^{(2)} = C_3^{(2)'}\right)\left(C_3^{(1)'} = C_3^{(2)'}\right).$$

The second pairs $C^{(2)}, \pi^{(2)}$ will be denoted by D, ω. We have a table

$C_1\pi_1C_2\pi_2$	$D_1\omega_1D_2\omega_2$	$C_3'\pi_3'$	$D_3'\omega_3'$
$\emptyset$	$\emptyset$	$\emptyset$	$\emptyset$

Apply M_b from convolution $S = M_l \mathrm{ev}\, M_b$, we get

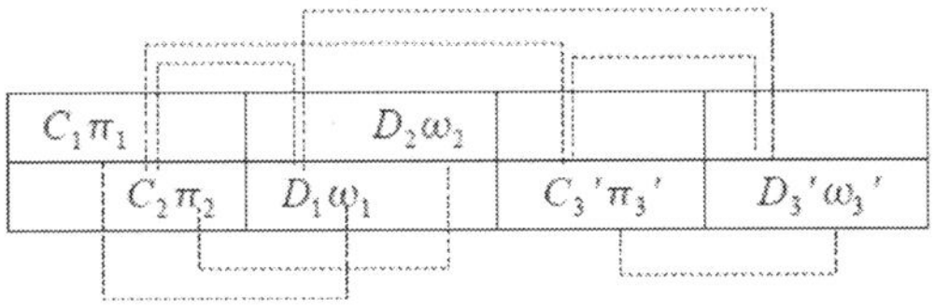

Now we apply ev, $ev\ C_i = \check{C}_i, ev\ \pi_i = \check{\pi}_i, ev\ D_i = \check{D}_i$ which changes the parameters available in the models, with equal parameters equates (we connect with an equal sign, in the diagram it is highlighted in bold dotted line) the names. For each model (category) ev must be explicitly set. We get

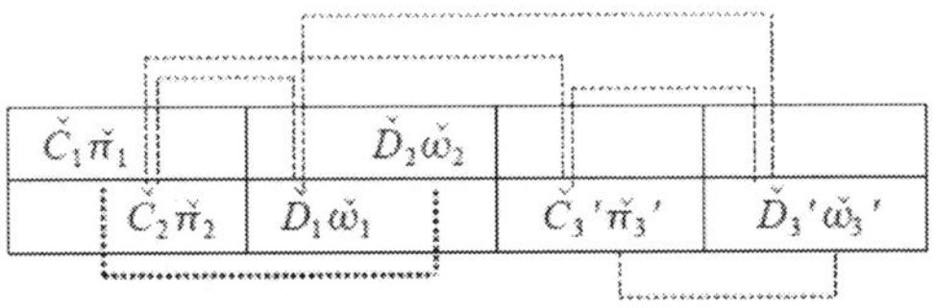

Convolutions of several arrows for a category in a Set form a nerve of this category, the nerve construction is transferred to convolutional polycategories and categorical splices.

We will explicitly take into account the parameters of the general case using an easily generalized example of a simplicial category, which is one of the main examples of categories used, for example, in algebraic topology. Its objects are finite ordinals $r_j \in \bar{m} = \{\bar{0}, \bar{1}, \dots, \bar{m}-1\}$, arrows are of non-decreasing functions between ordinals. We construct arrows from elementary splices

$$C_i\pi_i = r_i\pi_i(r_i, r_i, id(i,i)), D_j\omega_j = r_j\pi_j(r_j, r_j, id(j,j))$$

as follows

$$r_i\pi_i(r_i, r_j, \alpha_{ij})r_j\omega_j(r_i, r_j, \alpha_{ij})\left(\pi_i(r_i, r_j, \alpha_{ij}) = \omega_j(r_i, r_j, \alpha_{ij})\right), \alpha_{ij} \in Hom(r_i, r_j).$$

For clarity, the lines between the names and objects will be dashed rω, the arrow will take a graphical form

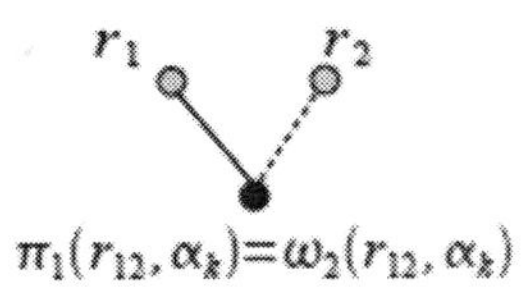

The indices i, j are omitted from the mapping $\alpha_k \in Hom(r_i, r_j)$ (they are restored, if necessary, from the indices of the objects, k numbers the arrows from $om(r_i, r_j), (r_i, r_j) = r_{ij}$. We conduct the convolution $\gamma = \alpha^\circ\beta$

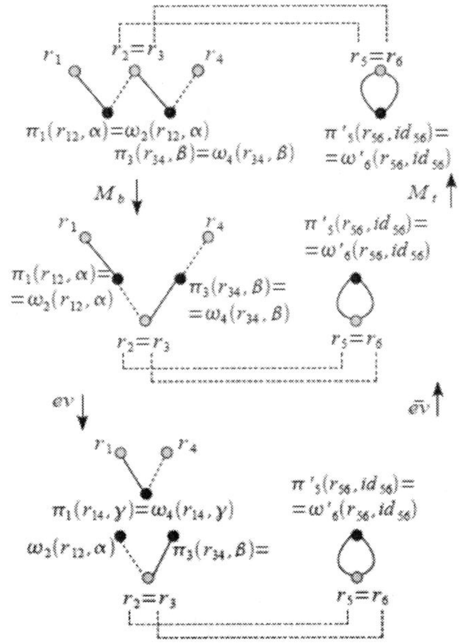

We got the result of the convolution along with the "history" from which the original splice is restored. To restore, we apply the convolution with $\overline{ev}$ and M_t to the last scheme (as indicated in the diagram on the right).

5 Categorical models of chemical bond

The adiabatic approximation in quantum chemistry consists in a separate consideration of the motions of nuclei and electrons. It can be assumed that chemical bonds are determined by the density of the distribution of the electron charge in the electron cloud, specified by the wave function of the molecule, which ensures the equilibrium position of the nuclei (the Gelman-Feynman theorem). An electron cloud in different molecules can have the same substructures, each of these substructures can correspond to one or another type of chemical bond. After some refinements, the structural chemical fomrules become equivalent to the corresponding splices and their convolutions (covalent strokes are associated with splices co-regions). For example, methane Fig. 2. is obtained by simple convolutions.

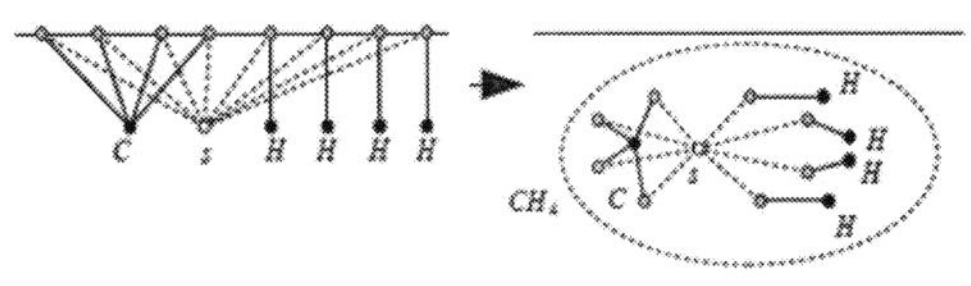

Fig. 2. Convolutions for methane

Alkane models are represented by associative algebras with identities, the multiplication operation is ternary. Hydrocarbons with double and triple bonds are modeled taking into account different types of convolutions for bonds (Fig.3)

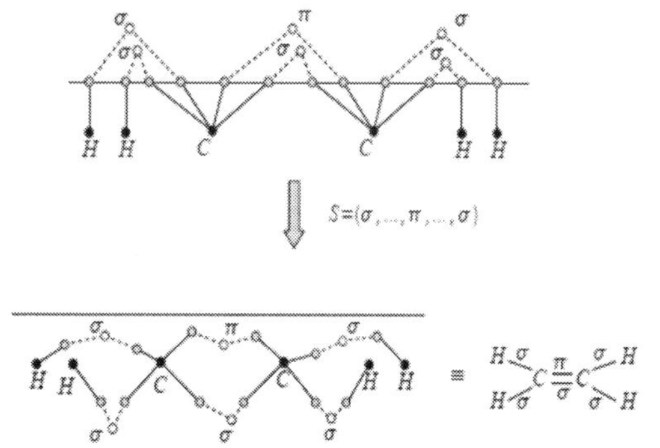

Fig. 3. Convolutions for ethylene

For the aromatic bond, a convolution with six (for benzene) and a large number of H atoms (for other aromatic hydrocarbons) is selected.

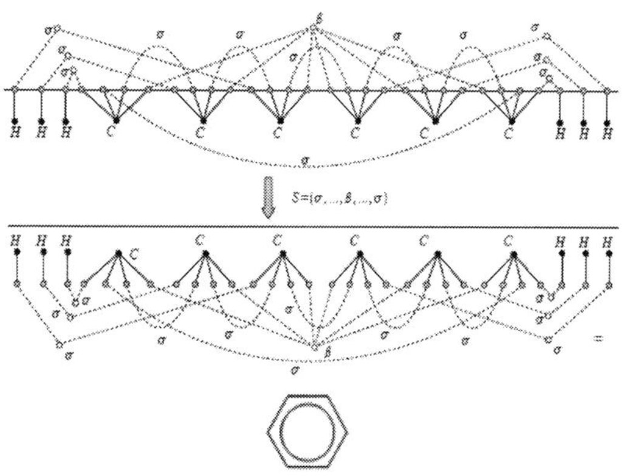

Fig. 4. Modeling a benzene molecule by splices

Cyclic hydrocarbons are modeled naturally. The properties of the algebras, PROPs and splices that appeared in this way, studied by strictly mathematical methods, give the properties of biomolecules. The most important biomolecules for algebraic biology, RNA and DNA, are also modeled by splices. It is not difficult to write out splices corresponding to nucleotides and nucleosides. Further construction of RNA and DNA molecules from them is reduced to algebraic convolutions. We emphasize the strictly mathematical nature of these models and the operations in them, which equips algebraic biology with a number of categorical algebraic research methods. The DNA molecule uses a phosphodiester bond. The polynucleotide chain is built through the formation of phosphodiester bonds between the third and fifth carbons of sugars (3C′ and 5C′). Phosphodiester bond is modeled by the following splices

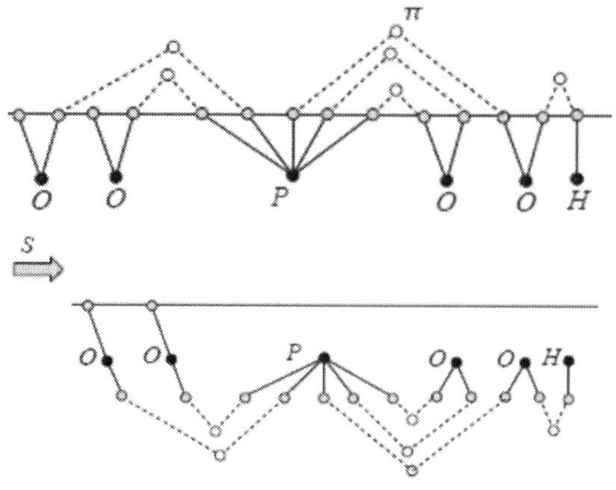

Fig. 5. Modeling a phosphodiester bond by splices

It is possible to simulate hydrogen bonds in DNA and RNA that are responsible for the secondary structure by adding new types of elements (marked with yellow circles) into splices co-regions and convolutions between them. We also note another option for taking into account the differences in bonds using higher splices, when certain types of bonds are modeled by the corresponding splices included in the higher splices, which completely model the molecule. Let us illustrate the formation of water dimers and their models

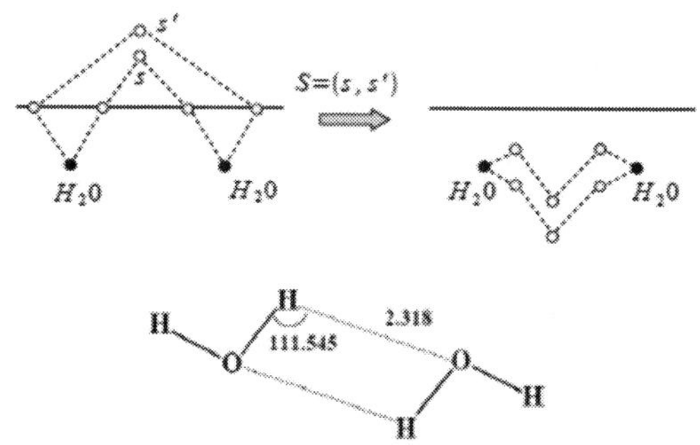

Fig. 6. Modeling of water dimer isomers by splices

The geometry of molecules, in particular the angles and positions of atomic nuclei for dimers, can also be modeled by adding parameters to the corresponding convolutions. The disassembled types of communication and their models are sufficient to represent the entire DNA (or RNA) molecule, including various levels of its structure. The examples given make it possible to construct formulas and their graphic designations for DNA and RNA, due to their cumbersomeness, they are not presented here.

6 Conclusion

One of the main tasks of algebraic biology is to predict the properties of organisms from the genome using rigorous algebraic methods. The paper proposes new categorical methods used in categorical systems theory. These methods are based on the theory of categorical splices, the elements of which are presented in the work. With their help, the behavior of quantum-mechanical objects, in particular, RNA and DNA biomolecules, as categorical systems, is modeled. New algebraic and categorical methods (associative algebras with identities, PROP, categorical splices) are involved in the analysis of the genome. The presented results are new and original.

References

1. Yi Lin, *General systems theory: A Mathematical Approach* (Kluwer Academic Publishers New York, 2002)
2. S.N. Vasiliev, A.K. Zherlov, E.A. Fedosov, B.E. Fedunov, *Intelligent control of dynamic systems* (Moscow, Phys.-math. literature, 2000)
3. B.F. Lomov, *Man and technology* (M., Soviet radio, 1966)
4. G.N. Kryzhanovsky, *Fundamentals of General Pathophysiology* (Moscow, Medical Information Agency, 2011)
5. P.K. Anokhin, *Fundamental questions of the general theory of functional systems, Principles of systemic organization of functions* (M., Science, 1973)
6. G.K. Tolokonnikov, Biomachsystems. Theory and applications, **1**, 31-213 (2016)
7. G.K. Tolokonnikov, *Biomachsystems*, **2**(1), 127-174 (2018)
8. G.K. Tolokonnikov, *Biomachsystems*, **2**(4), 41-144 (2018)
9. G.K. Tolokonnikov, Advances in Intelligent Systems and Computing, **938**, 259-267 (2020)
10. S. Petoukhov, *Matrix genetics, algebra of genetic code, noise immunity* (M., RHD, 2008)
11. G.K. Tolokonnikov, S.V. Petoukhov. Advances in Intelligent Systems and Computing, **1126**, 55-64 (2020)
12. W. Hatcher, *The logical foundations of mathematics* (Perg.Pr., 1982)
13. S. Mac Lane, *Categories for a working mathematician* (M., 2004)
14. O. Kuznetsov, AI and decision making, **2**, 3-20 (2018)
15. A.A. Zhdanov, *Autonomous artificial intelligence* (M., BINOM, 2012)

Different-Scale Simulation of Flows in Porous Media

Marina Trapeznikova*, *Natalia* Churbanova, and *Antonina* Chechina

*Keldysh Institute of Applied Mathematics RAS, RU-125047 Moscow, Russia

Abstract. The paper considers the development of algorithms for an adequate description of processes of different scales in porous media. The choice of a computational technique is determined by the reference size of the problem being solved. Models of porous medium flow under Darcy's law, neglecting the medium microstructure, are used for the simulation at macro-scale. While at micro-scale, a direct description of fluid flow in porous channels with complex geometry by means of gas dynamic equations is used. In the first case the proposed model of non-isothermal multiphase multicomponent flow in a porous medium includes the mass balance and total energy conservation equations modified by analogy to the known quasi-gas dynamic equations. The model features are the introduction of minimal reference scales in space and in time and the change of the system type from parabolic to hyperbolic to increase the stability of explicit difference schemes applied for approximation. In the second case the dimensionless form of the quasi-gas dynamic system with pressure decomposition, developed by the authors earlier, is adapted to the simulation of flows in the pore space. The fictitious domain method is proposed to reproduce the core microstructure. The developed approaches have been verified by test predictions.

1 Introduction

The numerical simulation of fluid flows in porous media is one of urgent research topics all over the world. These studies are of great practical importance for the petroleum and metallurgical industries, since modeling the hydrocarbon recovery and minerals mining, calculations of groundwater dynamics, modeling operations in various technological units, for example, in catalytic reactors, as well as the digital analysis of small rock samples can reduce the number of expensive natural experiments. The flows under consideration are essentially multi-scale due to a large scatter of physical quantities and a variety of physical processes, such as convection, diffusion, phase transitions and reactions [1]. Therefore, models of porous media flows need further improvement for an adequate description of processes of different scales.

This paper does not propose a single universal approach to modeling multiscale flows. The choice of a model and computational technique should be determined by the reference size of the applied problem being solved.

To simulate flows at macro-scale (e.g. modeling flows in the subsurface or in technological installations with reference sizes of the order of meters and kilometers) models of porous medium flow under Darcy's law are used [2-4].

To simulate flows at micro-scale (e.g. investigation of core samples with reference sizes of the order of micrometers and less) a direct description of fluid flow in the pore space of rocks by means of gas dynamic equations can be used [5, 6].

In the both cases the present research is based on further development of the known quasi-gas dynamic (QGD) system of equations [7]. Governing models are transformed in accordance with physical peculiarities of flows under consideration. The developed approaches have been verified by a number of test predictions.

2 Macro-scale flow modeling

2.1 QGD-based hyperbolic model of porous medium flow aimed at explicit algorithms

We propose two approaches for modeling flows on a macro-scale within the framework of Darcy's law when the medium microstructure can be negligible. First, the original mathematical model of multiphase fluid flow has been constructed by analogy to the QGD system of equations [8, 9]. The phase continuity equations were modified and got additional dissipative terms for the solution smoothing. The equations' type was transformed from parabolic to hyperbolic to increase the stability of explicit difference schemes used for the approximation.

The second approach is based on the classical model of compressible fluid flow in a porous medium [2-4] and the introduction of mass flux relaxation [10] into the continuity equations. Thus a hyperbolized system of equations has been also obtained [11]. The model allows for separation of the pressure and saturation variables and for the use of explicit schemes to determine the average pressure and the phase saturation. The both approaches apply logically simple computational

* Corresponding author: mtrapez@yandex.ru

algorithms aimed at efficient adaptation to massively parallel processing architectures.

The current version of the QGD-based hyperbolic model considers non-isothermal multiphase multicomponent flow through a non-deformable isotropic porous medium [12]. By a component a single chemical compound (a pure component) or a mixture of compounds (a pseudo-component) is assumed. Under phases we assume mobile phases, their number in the system is no more than three: these are water, Non-Aqueous Phase Liquid (NAPL) and gas, further subscripts w, n and g denote them respectively. The number of components n_κ can be arbitrary, the same component may be present in different phases. Liquids are considered as slightly compressible, gas is ideal, the rock skeleton is incompressible. The temperature of all phases and the rock is considered identical. The mass and energy transfer between the phases occurs.

The mass concentration of component κ in phase α is presented as follows:

$$C_\alpha^\kappa = \frac{m_\alpha^\kappa}{m_\alpha}, \quad \kappa = 1, \ldots, n_\kappa, \quad \alpha = w, n, g \qquad (1)$$

where m_α^κ is the mass of component κ in phase α, m_α is the mass of phase α.

The model consists of:

- the mass balance equation for each component;
- the extended Darcy's law for each phase;
- the total energy conservation law;
- state equations accounting for the multicomponent composition of phases;
- capillary pressures as functions of saturations;
- constitutive relations on the sum of saturations and the sum of concentrations;
- constansts of phase equilibrium (so-called "*K*-values") specified for all components to be condensed.

As the model naturally follows from the hyperbolized QGD model the second time derivatives and regularizers with small parameters τ and l are present in equations. In porous media the minimal reference length l is the distance at which the rock microstructure is negligible, $l : 10^2$ rock grain sizes. The minimal reference time τ is the time interval for inner equilibrium establishing in the volume of linear size l. The estimation $\tau : h/c$ is valid at numerical implementation [8] (h is the computational grid step in space, c is the sound speed in fluid).

A computational algorithm of the explicit type has been developed for numerical implementation of the model. The number of primary variables depends on the problem statement. For example, for the simulation of two-phase two-component flow the set of primary variables consists of the temperature, one phase pressure, the other phase saturation and the concentration of one component in a phase.

The algorithm specific feature is that equations for calculating the primary variables are not derived directly. The equations of conservation of mass and total energy are approximated by explicit difference schemes, and some auxiliary quantities are found from them on each time level. And then, solving the system of algebraic equations with these auxiliary quantities in the right-hand sides, one can obtain the primary variables locally in each computational point.

Derivation and substantiation of the model as well as detailed description of the computational algorithm and its verification via test predictions can be found in previous works by the authors [8, 9, 11, 12].

2.2 Test predictions

One of macro-scale test problems concerning phase transitions is the heat pipe effect simulation [13]. A horizontal column filled by a porous medium saturated with water and air is considered. This is a two-phase two component system: the water phase consists of the water component only while the gas phase consists of two components – water (as steam) and air. We choose the gas pressure P_g, the water saturation S_w, the temperature T and the concentration of water in the gas phase C_g^w (1) as primary variables. Figure 1 illustrates the problem statement.

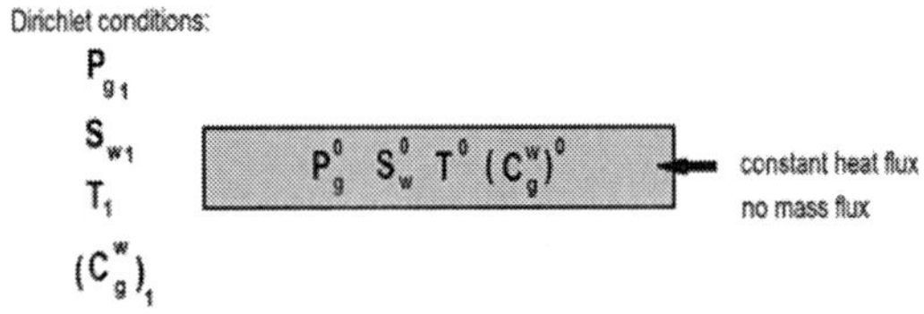

Fig. 1. The heat pipe problem statement.

The initial and boundary conditions are as follows:

$$P_g^0 = 1\,\text{atm}, \quad S_w^0 = 0.5, \quad T^0 = 343.15\,\text{K}, \quad \left(C_g^w\right)^0 = 0.05,$$

$$P_{g\,1} = 1\,\text{atm}, \quad S_{w\,1} = 0.98, \quad T_1 = 343.15\,\text{K}, \quad \left(C_g^w\right)_1 = 0.29.$$

The constant heat flux of 1000 W into the column and zero fluxes for all mass components are given at the right-hand boundary. In comparison with this test calculation from [12] the conditions are changed, in particular, the heat flux is reduced.

As the water component may evaporate at heating and condense at cooling, the *K*-value is specified for it as follows [14]:

$$K_{gw}^w = \frac{C_g^w}{C_w^w} = \exp\left[5.373\left(1+\omega_w\right)\left(1-\frac{T_c}{T}\right)\right]\frac{P_c}{P_g} \qquad (2)$$

where ω_w is the water molecule acentric factor, T_c and P_c are critical values of the temperature and the pressure.

Van Genuchten relationships [4] are used to describe relative phase permeabilities and capillary pressures.

The following process is reproduced as a result of computations: due to the heat flux the system is heated and some amount of the water component passes from the water phase to the gas phase. Figures 2-5 show the obtained numerical results at different time moments. One can observe the expected water-gas phase transition: the water saturation is decreased at the right boundary (see Fig. 4) while the concentration of water steam in the gas phase is increased respectively (see Fig. 5). As $C_w^w = 1$ during the process, C_g^w is completely determined by the exponent from formula (2).

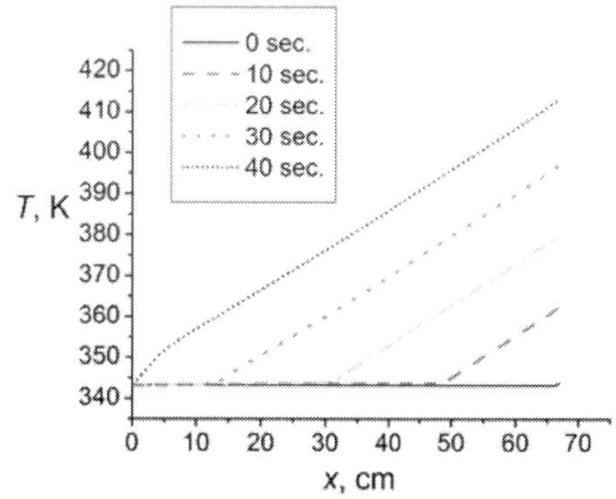

Fig. 2. Temperature in the heat pipe at different time moments.

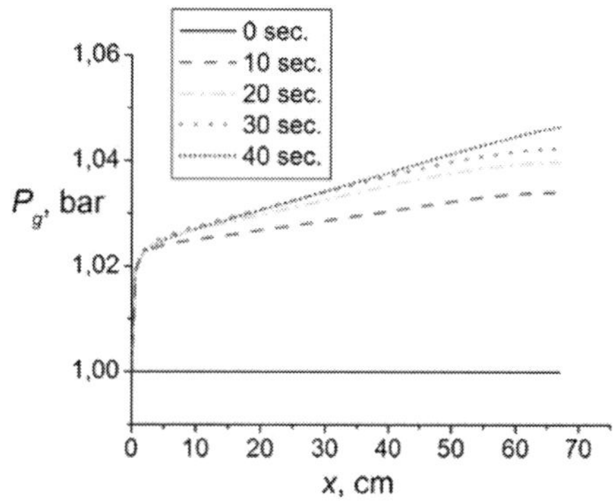

Fig. 3. Gas phase pressure in the heat pipe at different time moments.

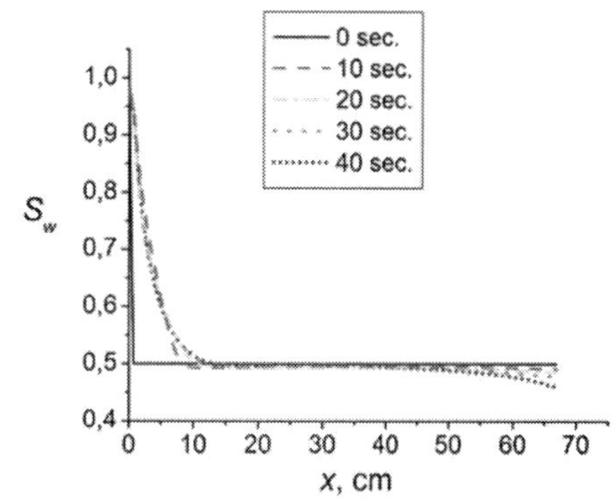

Fig. 4. Water saturation in the heat pipe at different time moments.

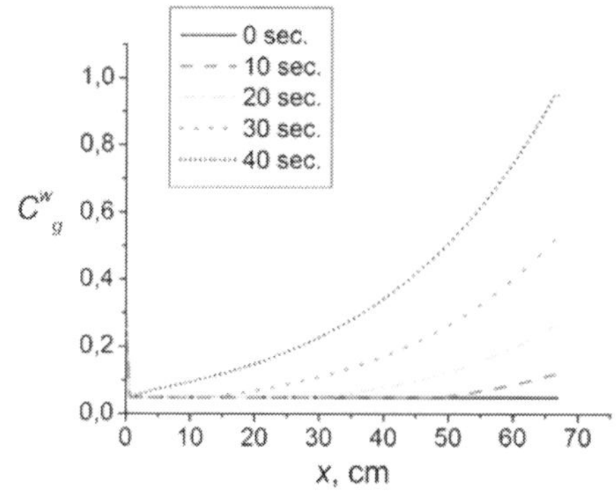

Fig. 5. Concentration of water steam in the gas phase in the heat pipe at different time moments.

The steam flows away from the high temperature region (see Fig. 2) under the gas pressure gradient (see Fig. 3). It condenses after reaching a cooler region near the left boundary. One can conclude that the proposed computational approach ensures correct predictions of multiphase multicomponent porous media flows.

3 Micro-scale flow modeling

3.1 Adaptation of QGD equations with decomposed pressure for modeling flow in pores

When modeling processes at micro-scale, in particular, in problems of core research, one of the main modern trends is direct numerical simulation of fluid flow in porous channels with complex geometry using gas dynamic equations (see, for example, [15]). The paper proposes to adapt for these purposes the system of equations with pressure decomposition developed earlier by the authors [16].

A well-known problem arises when simulating flows of weakly compressible fluids, which often include flows in porous media. Pressure changes are proportional to density changes times the Mach number squared. At low Mach numbers this leads to the pressure-density uncoupling. To overcome this problem, the authors developed an algorithm that treated the pressure instead of the density as the primary variable: the pressure was obtained via a system of equations, and the density was found from the equation of state [16].

In view of specific pressure behavior the pressure is decomposed into a sum of two components – the volume-averaged part $\overline{p}(t)$ and the dynamic part $\Pi(t,x)$ with reference values P_0 and $\rho_0 U_0^2$ correspondingly:

$$P(t,x) = P_0\,\overline{p}(t) + \rho_0 U_0^2\,\Pi(t,x). \tag{3}$$

Then we get the dimensionless pressure:

$$p(t,x) = \overline{p}(t) + \gamma M^2\,\Pi(t,x). \tag{4}$$

As a result, the system of QGD equations with pressure decomposition in the dimensionless form looks as follows (for brevity, the spatially one-dimensional isothermal case is given):

$$\frac{\partial \rho}{\partial t}+\frac{\partial \rho u}{\partial x}=\frac{\gamma M^2}{\mathrm{Re}}\frac{\partial}{\partial x}\frac{\tau}{2}\frac{\partial}{\partial x}(\rho u^2+\Pi) \quad (5)$$

$$\frac{\partial \rho u}{\partial t}+\frac{\partial}{\partial x}(\rho u^2+\Pi)= =\frac{\gamma M^2}{\mathrm{Re}}\frac{\partial}{\partial x}\frac{\tau}{2}\frac{\partial}{\partial x}(\rho u^3+3\Pi u)+ +\frac{1}{\mathrm{Re}}\frac{\partial}{\partial x}\frac{\tau}{2}\frac{\partial}{\partial x}(3\bar{p}u) \quad (6)$$

The system does not contain singularities at $M \to 0$ and therefore it can be used even for $M=0$.

One can introduce the pressure increments (n is the time level number)

$$\delta\bar{p}=\bar{p}^{n+1}-\bar{p}^n, \quad \delta\Pi=\Pi^{n+1}-\Pi^n \quad (7)$$

and derive finite difference equations for obtaining the intermediate values of these increments [16]. The equation of elliptic type for the intermediate value of the dynamic pressure increment has the next form:

$$\mathrm{div}_h\,\mathrm{grad}_h(\delta\Pi)^*-\frac{\gamma M^2}{(\Delta t)^2 T^{n+1}}(\delta\Pi)^*= \frac{1}{\Delta t}\mathrm{div}_h(\rho\mathbf{u})^{n+1/2}+ +\frac{1}{(\Delta t)^2}\frac{(\delta\bar{p})^*}{T^{n+1}}+\frac{p^n}{(\Delta t)^2}\left(\frac{1}{T^{n+1}}-\frac{1}{T^n}\right)- -\frac{\gamma M^2}{\Delta t\,\mathrm{Re}}\left[\left(\frac{\tau}{2}(\rho u^2+\Pi)_{\bar{x}}\right)_x+\left(\frac{\tau}{2}(\rho v^2+\Pi)_{\bar{y}}\right)_y\right]^n \quad (8)$$

and the equation for the intermediate value of the averaged pressure increment is as follows:

$$(\delta\bar{p})^*= =-\left(\sum_{\omega}\frac{1}{T^{n+1}}h_1h_2\right)^{-1}\left\{\Delta t\sum_{\partial\omega}((\rho\mathbf{u})^{n+1},\mathbf{n})h_{\partial\omega}+\right. +\sum_{\omega}p^n\left(\frac{1}{T^{n+1}}-\frac{1}{T^n}\right)h_1h_2- -\Delta t\frac{\gamma M^2}{\mathrm{Re}}\sum_{\omega}\left[\left(\frac{\tau}{2}(\rho u^2+\Pi)_{\bar{x}}\right)_x+\right. \left.\left.+\left(\frac{\tau}{2}(\rho v^2+\Pi)_{\bar{y}}\right)_y\right]^n h_1h_2\right\}. \quad (9)$$

The final values of the average, dynamic and total pressure are calculated using the following formulas:

$$\delta\bar{p}=(\delta\bar{p})^*+\frac{\gamma M^2}{V_\Omega}\int_\Omega(\delta\Pi)^*dv, \quad (10)$$

$$\delta\Pi=(\delta\Pi)^*-\frac{1}{V_\Omega}\int_\Omega(\delta\Pi)^*dv, \quad (11)$$

$$p^{n+1}=p^n+\delta\bar{p}+\gamma M^2\delta\Pi. \quad (12)$$

Since it is supposed to investigate flow in a porous medium, the algorithm should be adapted to computations in regions of complex shape. For this, it is proposed to use the method of fictitious domains. The entire computational domain is enclosed in a rectangle, in which a rectangular computational grid is set. On this grid, a map of the area is compiled: the indicator map_i in the grid cell equals 1 if the cell belongs to the computational domain, that is, to the flow area, and 0 otherwise. Then the difference derivatives of the convective terms can be written in the form:

$$\left.\frac{\partial(\rho u^2)}{\partial x}\right|_i : 0.5\left(\frac{map_{i+1}\,map_i\left((\rho u^2)_{i+1}+(\rho u^2)_i\right)}{h}-\right. \left.-\frac{map_i\,map_{i-1}\left((\rho u^2)_i+(\rho u^2)_{i-1}\right)}{h}\right) \quad (13)$$

and the pressure gradient in the form:

$$\left.\frac{\partial\Pi}{\partial x}\right|_i : 0.5\left(\frac{map_{i+1}\,map_i\,(\Pi_{i+1}-\Pi_i)}{h}+\right. \left.+\frac{map_i\,map_{i-1}\,(\Pi_i-\Pi_{i-1})}{h}\right). \quad (14)$$

Thus, difference equations with indicators are written uniformly throughout the domain so that the no-flow condition is automatically satisfied on a wall.

3.2 Test predictions

The developed algorithm was approved by the problem of flow around an unconnected complex-shaped body (see Fig. 6).

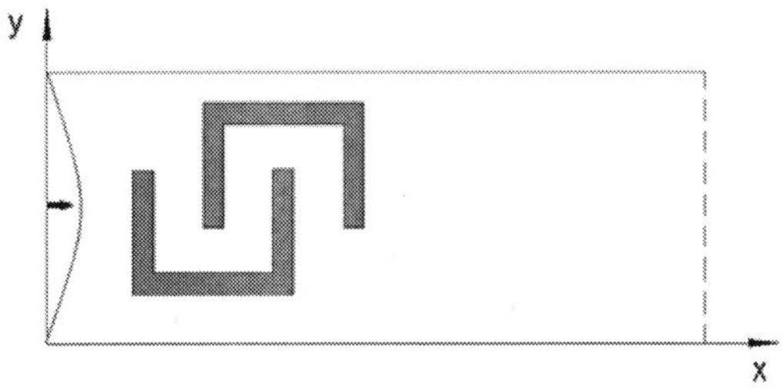

Fig. 6. Statement of the problem of flow around the body of complex shape.

The computational domain is a rectangular channel with rigid impermeable walls. On the left, a stream flows into the channel, the horizontal velocity has a parabolic profile with the maximum equal to 1.5. Inside the channel there is an impermeable area, consisting of two unconnected segments. The no-flow conditions are realized on all walls using the fictitious domain method described above, the grid size is 300×100. For the outgoing stream, the condition $\partial u / \partial x = 0$ is set.

Figure 7 shows the results of calculations – the velocity fields at different Reynolds numbers. At small Re flow is more laminar, at larger Re the signs of turbulence are clearly manifested, the vorticity corresponds to the physics of the process.

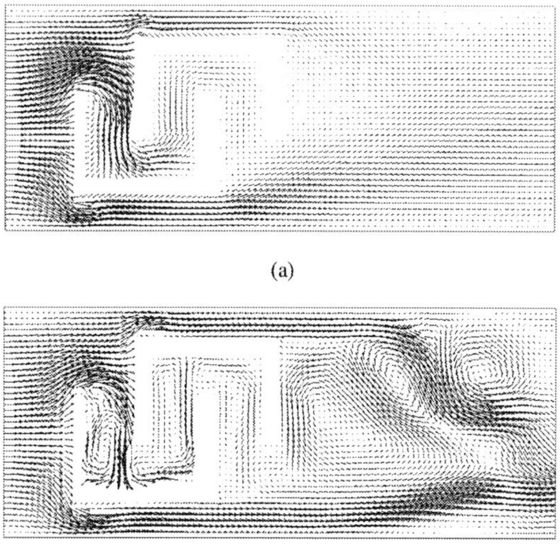

(a)

(b)

Fig. 7. Flow around the body of complex shape, Re = 30 (a) and Re = 400 (b).

Currently, calculations are carried out in domains with a complex internal structure imitating rock samples.

4 Conclusions

The developed computational techniques have a wide range of applications. At macro-scale they include, for example:

- the simulation of thermal methods of oil recovery to increase the production rate of high-viscosity oil;
- the solution of ecology problems on the thermally enhanced remediation of contaminated soils;
- the simulation of processes in technological units at the purification of crude oil or natural gas.

At micro-scale , they include the digital core analysis becoming an alternative to field and laboratory experiments in the oil and gas industry to obtain reliable data on reservoir properties and, as a result, to improve the macro-level models.

References

1. *Simulation of Flow in Porous Media. Applications in Energy and Environment.* Radon Series on Computational and Applied Mathematics, **12**, ed. by P. Bastian et al. (De Gruyter, Berlin, 2013)
2. K. Aziz, A. Settari, *Petroleum Reservoir Simulation* (Applied Science Publ., London, 1979)
3. Z. Chen, *Reservoir Simulation: Mathematical Techniques in Oil Recovery* (SIAM, Philadelphia, 2007)
4. G.F. Pinder, W.G. Gray, *Essentials of Multiphase Flow and Transport in Porous Media* (John Wiley & Sons, Hoboken, NJ, 2008)
5. P. Meakin, A.M. Tartakovsky, Rev. Geophys., **47**(3), RG3002 (2009)
6. M. Rolle, D. Hochstetler, G. Chiogna, P.K. Kitanidis, P. Grathwohl, Transp. Porous Med., **93**(3), 347–362 (2012)
7. B.N. Chetverushkin, *Kinetic Schemes and Quasi-Gas Dynamic System of Equations* (CIMNE, Barcelona, 2008)
8. B. Chetverushkin, N. Churbanova, A. Kuleshov, A. Lyupa, M. Trapeznikova, Rus. J. Numer. Anal. Math. Modelling, **31**(4), 187–196 (2016)
9. M. Trapeznikova, N. Churbanova, A. Lyupa, J. Math. Chem., **58**(3), 595–611 (2020)
10. A.A. Davydov, B.N. Chetverushkin, E.V.Shil'nikov, Comp. Math. Math. Phys., **50**(12), 2157–2165 (2010)
11. M. Trapeznikova, N. Churbanova, A. Lyupa, In: *Proc. of ECCM6&ECFD7*, ed. by R. Owen et al., pp. 637–647 (CIMNE, Barcelona, 2018)
12. M. Trapeznikova, N. Churbanova, A. Lyupa, In: *IV International Conference "Modeling of Nonlinear Processes and Systems" (MNPS-2019).* EPJ Web of Conferences, **224**, ed. by A. Nadykto et al., 02010 (2019)
13. H. Class, R. Helmig, P. Bastian, Advances in Water Resources, **25**, 533–550 (2002)
14. A.I. Brusilovsky, *Phase Transitions in the Development of Oil and Gas Fields* (Graal Publ., Moscow, 2002) [In Russian]
15. V. Balashov, E.B. Savenkov, In: *Physical and Mathematical Modeling of Earth and Environment Processes (2018).* Springer Proceedings in Earth and Environmental Sciences, ed. by V. Karev et al., 374–379 (Springer, Cham, 2019)
16. B.N. Chetverushkin, N.G. Churbanova, M.A. Trapeznikova, In: *Parallel Computational Fluid Dynamics 2004 - Multidisciplinary Applications*, ed. by G. Winter et al., 297–304 (Elsevier B.V., Amsterdam, 2005)

Improving the Learning P ower of Artificial I ntelligence Usi ng Multimodal Deep Learning

Eugene Yu. Shchetinin[1], *Leonid* Sevastianov[2]

[1]Financial University under the Government of Russian Federation, Department of Mathematics, RU-125993, Moscow, Russia
[2]Russian University of Peoples Friendship, Department of Informatics, RU-117198, Moscow, Russia

Abstract. Computer paralinguistic analysis is widely used in security systems, biometric research, call centers and banks. Paralinguistic models estimate different physical properties of voice, such as pitch, intensity, formants and harmonics to classify emotions. The main goal is to find such features that would be robust to outliers and will retain variety of human voice properties at the same time. Moreover, the model used must be able to estimate features on a time scale for an effective analysis of voice variability. In this paper a paralinguistic model based on Bidirectional Long Short-Term Memory (BLSTM) neural network is described, which was trained for vocal-based emotion recognition. The main advantage of this network architecture is that each module of the network consists of several interconnected layers, providing the ability to recognize flexible long-term dependencies in data, which is important in context of vocal analysis. We explain the architecture of a bidirectional neural network model, its main advantages over regular neural networks and compare experimental results of BLSTM network with other models.

1 Introduction

Paralinguistics is a branch of linguistics, main scope of which is analyzing of non-verbal aspects of language – such as tempo, pitch, tone and intonation. The main goal is to analyze not what was said, but how it was said. The field of computational paralinguistics deals with analysis and synthesis of paralinguistic phenomena and research in this area has become very active in recent years [1,2].

Computer emotion classification aims to extract and classify human emotions from audio source. Different models, which estimate different physical properties of voice, such as pitch, volume and harmonics, are used. Such classifiers usually serve as parts of security systems, mobile assistants and biometric research algorithms [3,4].

The main difficulty is finding such features that would be robust to outliers and will retain variety of human voice properties at the same time [5]. Moreover, the model used must be able to estimate features on a time scale for an effective analysis of voice variability. Usually this is done by extracting features based on a sliding-window scheme, which solves the problem of data normalization and helps to reduce overfitting [6-9].

Another challenge is absence of uniform standard of human emotion types. One of possible solutions is MPEG4 standard, which divides human emotions in 6 groups: aggression, disgust, happiness, sadness and surprise. The last 6th group is called "neutral" and is used in case of absence of any emotions.

With development of computational power, neural networks have become widely used in field of speech recognition and emotional classification. In this paper a paralinguistic model based on Bidirectional Long Short-Term Memory (BLSTM) neural network is described, which was trained for vocal-based emotion recognition. The main advantage of this network architecture is that each module of the network consists of several interconnected layers: output layer, input layer and forget gate. Modules of BLSTM network are interconnected, so model can use context both from the past and from the future, thus providing the ability to recognize flexible long-term dependencies in data, which is important in context of vocal analysis. In the next paragraph we describe the architecture more thoroughly.

2 Bidirectional ne ural ne twork architecture

Given an input sequence $X = (x_1, \dots, x_T)$, a standard RNN computes a sequence of hidden vectors $h = (h_1, \dots, h_T)$ and output vectors $y = (y_1, \dots, y_T)$, by recursively evaluating the following equations from time steps $t = 1$ to $t = T$:

$$h_t = f_{act}(W_{xh}x_t + W_{hh}h_{t-1} + b_h)$$

$$y_t = W_{hy}h_t + b_y$$

Where W denote weight matrices, b – bias vectors and fact is the activation function of the hidden layer.

A typical LSTM layer consists of a number of recurrently connected memory blocks [10]. Memory block is a structure, consisting of a several memory cells, sharing the same input and output gate. The main goal of memory block is to store information within the network. Since each memory block has as many gate units, as a single memory cell, it's more efficient to use block structure (fig.1).

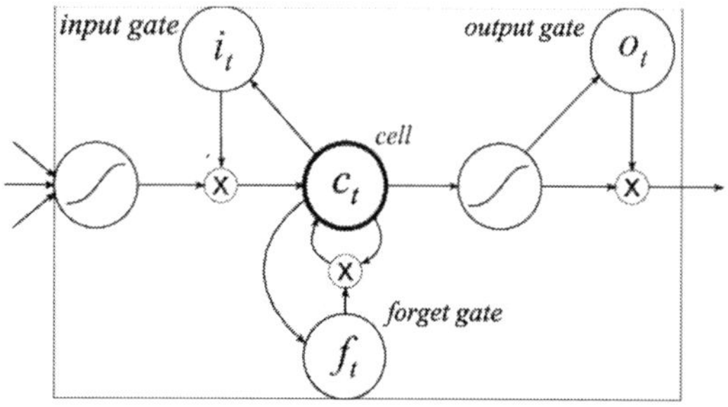

Fig. 1. Long Short-Term Memory Block

2.1 Memory Cells

Memory cell is a singular multiplicative unit of a LSTM network. Each memory cell is built around a central linear unit with a fixed self-connection, called nets. In addition to this, memory cell gets input from a multiplicative unit outj, called the output gate, and multiplicative unit inpj, called the input gate. Shortly, we have:

$$y^{out_j}(t) = f_{out_j}\left(net_{out_j}(t)\right);\ y^{in_j}(t) = f_{in_j}\left(net_{in_j}(t)\right);$$

$$net_{out_j}(t) = \sum_u w_{out_j u} y^u(t-1)$$

$$net_{in_j}(t) = \sum_u w_{in_j u} y^u(t-1)$$

$$net_{c_j}(t) = \sum_u w_{c_j u} y^u(t-1)$$

All these inputs convey useful information about the current state of the net. For example, an input gate may use inputs from other memory cells, to decide which information to store in its memory cells.

2.2 Input and output gates

Error signals within a memory cell cannot change, but they may get superimposed. The output gate has to learn which errors to trap, by appropriately scaling them. The input gate has to learn when to release errors, again by appropriately scaling them.

Distributed output representations usually require output gates, but not always both gates are necessary. Even in this case, however, output gates prevent the net's attempts at storing long time lag memories from perturbing activations.

2.3 Forget gate

One of the main advantages of a LSTM model is ability to store data across the time lags, and carry error signals through time. However, in certain conditions this may contribute to a weakness. In a situation when an input is a continuous stream, the cell states may grow in unbounded fashion, causing saturation of output function and making the cell to degenerate into an ordinary BPTT unit. In order to overcome this problem LSTM network was modified with an addition of forget gate, which learns to reset memory blocks once their contents are useless.

A forget gate activation is calculated like any other gate activation function, and squashed with a sigmoid function:

$$y^{\varphi_j}(t) = f_{\varphi_j}\Big(\sum_u w_{\varphi_j u} y^u(t-1)\Big)$$

Where y^{φ_j} functions as a weight of self-recurrent connections of the internal state $s_{c_j^n}$in the next equation:

$$s_{c_j^n} = y^{\varphi_j}(t)s_{c_j^n}(t-1) + y^{in_j}(t)g(net_{c_j^n}(t))$$

Where $s_{c_j^n}(0) = 0$. Bias weights for LSTM gates are initialized with negative values for input and output gates, and positive for forget gates. This way the initial forget gate activation is almost 1.0, which implies the cell will behave almost as a standard LSMT cell until it has learned to forget.

2.4 Vanishing gradient problem

If memory cell c_j inputs are mostly positive or mostly negative, then its internal state s_j will tend to drift away over time. This is potentially dangerous, for the $h_0(s_j)$ will then adopt very small values, and the gradient will vanish. One way to circumvent this problem is to choose an appropriate function h. But h(x) = x, for instance, has the disadvantage of unrestricted memory cell output range. A simple method might have been using ReLu as activation functions for the gates, but this approach leads to model diverge, due to the nature of LSTM gates [11].

A simple but effective way of solving drift problems at the beginning of learning is to initially bias the input gate in_j towards zero. Although there is a tradeoff between the magnitudes of$h_0(s_j)$ on the one hand and of y_{inj} andon the other, the potential negative effect of input gate bias is negligible compared to the one of the drifting effect [12].

3 Feature extraction and dataset

In order to perform any kind of machine learning, we first need to collect data and extract features from it. In voice-based emotion classification data used is recordings, presented in audio format. Features that can be extracted from audio source, can be classified into two big groups: low-level descriptors (LLD) and high-level descriptors (HLD). In broader terms, low-level descriptorsare those related to the signal itself and have little or no meaning to the end-user. In other words, and thinking in terms of the audio domain, these descriptors cannot be heard. On the other hand, high-level descriptors are meaningful and might be related to semantic or syntactic features of the sound. They will be used to classify sound objects into the class they belong.

Here are some of the commonly used LLD's, which we have also used in our experiment:

- Pitch -the perceived frequency of sound. The autocorrelation method for pitch estimation was used to determine changes in speaking behavior in response to factors relating to stress, intonation and emotional changes.
- Formants and their bandwidths.A formant is the spectral shaping that results from an acoustic resonance of the human vocal tract. They play a vital role in identifying human emotions. A 13th order linear prediction (LP) filter was calculated on each voiced frame of speech. Next, the first three formant frequencies and bandwidths were calculated from the roots of the polynomial LP predictor.
- Energy - the area under the squared magnitude of the considered signal.
- Jitter & Shimmer - Frequency perturbation also called jitter refers to the short-term (cycle to cycle) fluctuations in pitch (F0). It is obtained by measuring the fundamental frequency (pitch) of each cycle of vibration, subtracting it from the previous F0 values, and dividing it by the average F0. Shimmer on the other hand is calculated in similar fashion; however, the period to period variability of the signal peak to peak amplitude is calculated instead.

All of these features were extracted by using an open-source audio analysis Python toolkit Librosa [13]. For each of the feature, we computed a delta coefficient and 10 statistic functionals: mean, standard deviation, kurtosis, skewness, minimum and maximum value, relative position, and range, all in all resulting in 160 total features for each recording.

For our experiment we used AIBO dataset, which contains 18216 recordings, classified by following types of emotions: neutral, calm, happy, sad, angry, fearful, disgust and suprised. It's important to outline that there is a big class imbalance: neutral class contributes to more than 60% of the dataset, while anger and positive classes contribute to 8.1% and 4.8% accordingly. This dataset was made of children interactions with a pet robot called "Aibo". The children were led to believe that the Aibo was responding to their commands, whereas the robot was actually controlled by a human operator. Operator caused the Aibo to perform a fixed, predetermined sequence of actions,sometimes disobediently, thereby provoking emotional reactions. The data was collected from 51 children, and all recordings were cut into 1 second long parts.

4 Experiment results

The model itself was implemented with Keras deep-learning library, using Python programming language. The first model we trained was a BLSTM neural network, with a 160-100-50-5 topology: an input layer with size 160, two hidden layers with 100 and 50 nodes accordingly, and an output layer of size 5. For comparison, we have also trained a one-directional LSTM model, and a regular RNN model with same topologies. In order to cope with the class imbalance, we have tried and compared two different methods: SMOTE – synthetic oversampling and random neutral-class undersampling [14].

The results of computer experiments can be seen on table 1: It shows the values of the emotion classification accuracy metrics (average accuracy, F1-metric, and average AUC) obtained after applying the trained models to the test sampleOn the Fig.2 the graphs of ROC analysis functions and values of AUC indicators for all classes of emotions are given. Fig.3 shows the processess of training and tesitng the BLSTM network.

Table 1. Comparison of different neural network architectures

Models	Model performances		
	Type of class balancing	**Accuracy Train**	**Accuracy Test**
LSTM	No class balancing	46.25	36.3
	SMOTE	44.9	40.5
	Undersampling	44.55	42.2
BLSTM	No class balancing	54.4	50.2
	SMOTE	56.1	48.6
	Undersampling	56.6	52.3
RNN	No class balancing	43.7	41.0
	SMOTE	40.9	36.2
	Undersampling	42.4	39.8

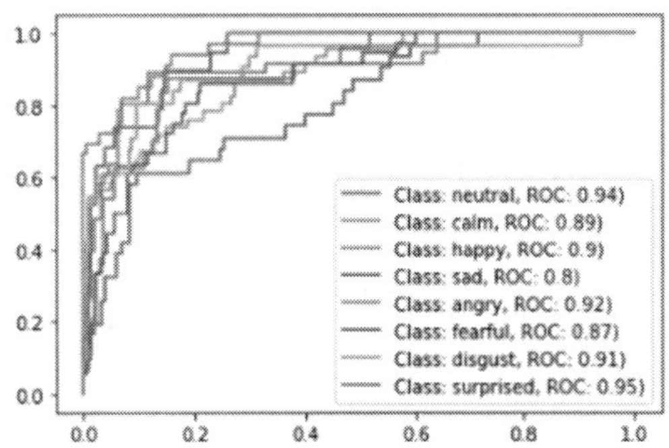

Fig. 2. ROC-AUC curves for emotion classes and their AUC.

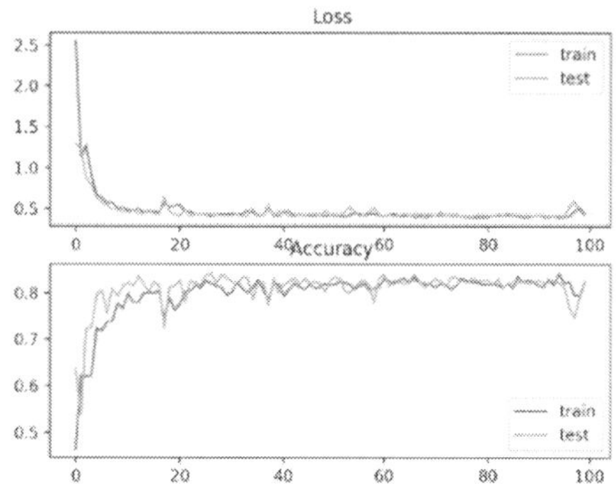

Fig. 3. Training graphs for the BLSTM neural network ensemble model. Upper graph: loss function-number of epochs; lower graph: accuracy-number of epochs

5 Conclusion

In this paper we presented a BLSTM neural network in application to voice-based emotion classification. The model shown much better results, compared to one-directional recurrent neural network, and slightly outperformed a one-directional LSTM network. This increase in accuracy was able by using memory blocks, which is a distinct feature of LSTM architecture. Further improvement of these results may lay in increasing the amount of data and features, used for training, and in stacking or blending different types of neural network architectures.

Based on the results of the research, the following conclusions are made. It is obvious that speech alone is not enough for high accuracy of emotion recognition, it is necessary to use additional data sources (videos, facial expressions, gestures, etc.). in many ways, the success of the algorithm depends on the quality of the training database. It should include all the main types of emotions displayed by experts, preferably in equal proportions. For this purpose, it is necessary to replenish and expand existing databases by creating new records, for example, using generative neural networks, as well as using transfer learning.

References

1. B. Schuller, IEEE Signal Processing Magazine, **29**(4), 97-101 (2012)
2. B. Schuller, S. Steidl, A. Batliner, A. Vinciarelli, F. Burkhardt, R. van Son, Computer Speech and Language, Special Issue on Next Generation Computational Paralinguistics, (2014)
3. B. Schuller, A. Batliner, *Computational Paralinguistics: Emotion, Affect and Personality in Speech and Language Processing*, (Wiley, 2013) DOI:10.1002/9781118706664
4. B. Schuller, F. Eyben, G. Rigoll, *Static and Dynamic Modelling for the Recognition of Non-Verbal Vocalisations in Conversational Speech*, Perception in Multimodal Dialogue Systems: 4th IEEE Tutorial and Research Workshop on Perception and Interactive Technologies for Speech-Based Systems, **5078** of Lecture Notes on Computer Science (LNCS), (Springer, Berlin/Heidelberg, 2008)
5. J. Wagner, F. Lingenfelser, E. Andre, Proc. of Interspeech, Lyon, France, **2013**, 168-172 (2013)
6. A. Janicki, Proc. of Interspeech, Lyon, France, **2013**, 153-157 (2013)
7. A. Stuhlsatz, C. Meyer, F. Eyben, T. Zielke, G. Meier, B. Schuller, Proc. of ICASSP, Prague, Czech Republic, (2011) DOI: 10.1109/ICASSP.2011.5947651
8. R. Brueckner,B. Schuller, Proc. of Interspeech, Portland, OR, USA, **2012**, 290-293 (2012)
9. R. Brueckner, B. Schuller, Proc. of ASRU, Olomouc, Czech Republic, (2013) DOI: 10.1109/ASRU.2013.6707757
10. S. Hochreiter, Y. Bengio, P. Frasconi, J. Schmidhuber, A Field Guide to Dynamical Recurrent Neural Networks, Kremer and Kolen, Eds. IEEE Press, **14**, 237-243 (2001)
11. F. Gers, N. Schraudolph, J. Schmidhuber, Journal of Machine Learning Research, **3**, 115-143 (2002)
12. M. Schuster, K. Paliwal, IEEE Transactions on Signal Processing, **45**(11), 2673 - 2681 (1997) DOI: 10.1109/78.650093
13. Librosa reference manual. https://www.Librosa.org,
14. L.A. Sevastianov, E.Yu. Shchetinin, Inform. Primen., **14**(1), 63–70 (2020) https://doi.org/10.14357/19922264200109

Plastic Flow in a Thin Layer: the Theory, Formulations of the Boundary- Value Problems and the Applications

Vagid Kadymov [1], *Elena* Yanovskaya[2]

[1]Technological University- MIREA, RU-119484, Moscow, Russia
[2]Moscow State Technological University "STANKIN", RU-127055, Moscow, Russia

Abstract. Two-dimensional, averaged over the thickness of the layer, mathematical theory of the spreading of a plastic layer on the plane has been studied. General and simplified mathematical formulations of boundary value problem were presented. The problem of plastic stretching of a strip by forces applied on its "clamped" ends was investigated. The analysis of various modes of the process was carried out, which are determined by both the total compression force of the ends and the total tensile force. Mathematical analogy between the process of the free spreading of a plastic layer on the plane and the process of heat transfer was studied. For known forms of a domain occupied by a thin plastic layer at the initial time and for a given law of convergence of the plates, the evolution of the boundary of a plastic layer spreading was described. The exact particular solutions of the aforementioned problem was obtained.

1 Introduction

Problems of plastic flow of the material in a relatively thin layer between the converging surfaces of the tool bodies form a fairly wide class. This class includes most of the technological problems of material treatment by pressure, such as stamping and pressing of thin-walled structural elements, thin-sheet rolling and many others. These are complex space problems of plastic shaping with a variety of parameters that determine the properties of the material. In these processes high pressures develop, which are an order of magnitude higher than the shear properties of the layer material, so that in the initial approximation, the properties of the material in the layer are close to those of a hydrodynamic liquid [1-3]. Almost along the entire contact surface, slippage is observed, in which the contact surfaces coincide with the sliding surfaces and the friction forces on them reach a maximum value equal to the yield strength of the layer material [2]. The accuracy requirements of the final part define another feature of these processes. As known [2], high contact pressures cause normal elastic movements of the working surfaces of the tool bodies, commensurate with the thickness of the plastic layer. Therefore, tin this case, ignoring elastic movements of the tool bodies becomes unjustified. In high-speed processes of material treatment by pressure the viscosity and inertia forces play an essential role, which greatly complicate modeling of the physical processes. In these processes, temperature and heat flows play an important role. Thus, heat generation due to the sliding friction forces on the contact surface and the dissipation of mechanical energy should also be considered.

Based on the analysis of the solution of L.Prandtl's problem, A.A. Ilyushin has formulated a number of hypotheses regarding the kinematics and the friction forces on the contact. He then utilized them to develop an effective two-dimensional theory of flow in a thin plastic layer that allowed to efficiently explore these processes. Within the framework of this model, a general boundary value problem for a "viscous" liquid has been formulated. This problem is described by nonlinear partial differential equations of the second order with respect to three unknown functions: contact pressure and two components of the flow velocity [1, 2, 5, 6]. It should be noted that until recently, all studies in this area were conducted on the basis of a partial simplified formulation of the boundary value problem in the model of an "ideal" fluid described by nonlinear partial differential equations of the first order, in which tangent stresses are neglected. In the framework of the simplified formulation, the kinematics of the spreading of the plastic layer was investigated and the evolutionary equation determining the free boundary of the spreading layer was obtained. In [3] other forms of representation of the evolutionary equation have been received. What's more, it has been established that this equation is reduced to a special case of the nonlinear equation of thermal conductivity.

This theory has been extended to the case of plastic flow on elastically deformable surfaces [2] and on surfaces with a pronounced texture (anisotropy of the properties of friction forces on contact). The theory has also been applied for the flows in high-speed processes and in the treatment processes of material volumetrically compressible and material with structural dependence (powder, sintered material).

2 Equations and mathematics

General statement of the boundary-value problem of the spreading of a plastic layer. The basic equations of the boundary-value problem of plastic layer flow on the plane have the form [1, 2, 5]:

$$\frac{\partial p}{\partial x}=\frac{\sigma_s}{3}\frac{dt}{d\lambda}\Delta u-\frac{\tau_s}{h}\left(\frac{u-u_1}{\sqrt{(u-u_1)^2+(v-v)^2}}+\frac{u-u_2}{\sqrt{(u-u_1)^2+(v-v)^2}}\right), \quad (2.1)$$

$$\frac{\partial p}{\partial y}=\frac{\sigma_s}{3}\frac{dt}{d\lambda}\Delta v-\frac{\tau_s}{h}\left(\frac{v-v_1}{\sqrt{(u-u_1)^2+(v-v)^2}}+\frac{v-v_2}{\sqrt{(u-u_1)^2+(v-v)^2}}\right), \quad (2.2)$$

$$\frac{\partial u}{\partial x}+\frac{\partial v}{\partial y}-\frac{d\lambda}{dt}=0, \quad (2.3)$$

where p,u,v are the contact pressure and components of flow velocity; $\vec{v}_1=(u_1,v_1)$ and $\vec{v}_2=(u_2,v_2)$- the known speeds of internal movements of surfaces of tool bodies relative to a fixed geometric surface; $\sigma_S=\sqrt{3}\tau_S$ is the yield strength of plastic material; $h=h(t)$ is the known law of change of the layer thickness; $\lambda(t)=\ln(h(t_0)/h(t))$ is the degree of deformation, where t_0 is some initial point of time. In particular, if there are no internal motions of physical surfaces ($\vec{v}_1=\vec{v}_2=0$), then the differential equations of quasi -static equilibrium are simplified:

$$\frac{\partial p}{\partial x}=\frac{\sigma_s}{3}\frac{dt}{d\lambda}\Delta u-\frac{2\tau_s}{h}\frac{u}{\sqrt{u^2+v^2}} \quad (2.1)'$$

$$\frac{\partial p}{\partial y}=\frac{\sigma_s}{3}\frac{dt}{d\lambda}\Delta v-\frac{2\tau_s}{h}\frac{v}{\sqrt{u^2+v^2}} \quad (2.2)'$$

On a free previously unknown boundary the two scalar dynamic conditions are specified:

$$\vec{\sigma}^{(n)}=\tilde{\sigma}\cdot\vec{n}=\left(\sigma_{xx}n_x+\sigma_{xy}n_y;\sigma_{xy}n_x+\sigma_{yy}n_y\right)=\vec{0}, \quad (2.4)$$

and the kinematic condition is given:

$$v_n\equiv\vec{v}\cdot\vec{n}=-\frac{\frac{\partial F}{\partial t}}{|gradF|} \quad (2.5)$$

In (2.4)" $\vec{n}=(n_x,n_y)=\nabla F/|\nabla F|$ is the normal vector to the boundary, and the components of the stress tensor $\tilde{\sigma}$ have the form:

$$\sigma_{xx}=-p+\frac{2}{3}\sigma_s+\frac{2}{3}\sigma_s\frac{dt}{d\lambda}\frac{\partial u}{\partial x},$$

$$\sigma_{xy}=\frac{1}{3}\sigma_s\frac{dt}{d\lambda}\left(\frac{\partial u}{\partial y}+\frac{\partial v}{\partial x}\right),$$

$$\sigma_{yy}=-p+\frac{2}{3}\sigma_s+\frac{2}{3}\sigma_s\frac{dt}{d\lambda}\frac{\partial v}{\partial y}.$$

At the tool contact boundary (usually known) $\Gamma_\sigma: F(x,y,t)=0$, the following two conditions are set: one dynamic condition with respect to the shear stress

$$|\vec{\sigma}^{(n)}\cdot\vec{\tau}|=|\sigma_{xx}n_xn_y+\sigma_{[y}(n_x^2+n_y^2)-\sigma_{yy}n_xn_y|=\mu\tau_s,\ 0\le\mu\le 1, \quad (2.6)$$

and one kinematic condition of the impermeability

$$v_n\equiv\vec{v}\cdot\vec{n}=-\frac{\partial G/\partial t}{|gradG|}, \quad (2.7)$$

where $\vec{\tau}$ and $\vec{n}$ are the unitary tangent and normal vectors with respect to Γ_c. In particular, at the fixed boundary, the normal velocity is zero.

Let's move on to dimensionless variables. Let $\sigma_s, L_0, h_0, (dh(t_0)/dt)$ be the characteristic values of the yield strength of the material, the linear size and thickness of the layer, and the rate of convergence of the plates. We introduce a small parameter $\varepsilon=h_0/L_0<<1$, characterizing the fineness of the plastic layer. The analysis of incompressibility equation leads to: a) $t_0=L_0/v_0$; b) $v_0=-(dh(t_0)/dt)/\varepsilon$. The latter means that the characteristic value of the flow velocity along the layer is much greater than the rate of convergence of the plates, which is commonly expected. Let's consider p_0 as the characteristic value of the contact pressure in the plastic layer. The analysis of the quasistatic equilibrium equations shows that the characteristic value of the contact pressure is much higher than the yield strength of the layer material, i.e. $p_0=\sigma_s/\varepsilon$. We now have the following system of equations in dimensionless form:

$$\frac{\partial\overline{p}}{\partial\overline{x}}=\varepsilon\frac{d\overline{t}}{d\lambda}\Delta\overline{u}-\frac{2}{\sqrt{3}}\frac{1}{\overline{h}}\frac{\overline{u}}{\sqrt{\overline{u}^2+\overline{v}^2}}, \quad (2.1)''$$

$$\frac{\partial\overline{p}}{\partial\overline{y}}=\varepsilon\frac{d\overline{t}}{d\lambda}\Delta\overline{v}-\frac{2}{\sqrt{3}}\frac{1}{\overline{h}}\frac{\overline{v}}{\sqrt{\overline{u}^2+\overline{v}^2}}, \quad (2.2)''$$

$$\frac{\partial \overline{u}}{\partial \overline{x}}+\frac{\partial \overline{v}}{\partial \overline{y}}-\frac{d\lambda}{d\overline{t}}=0\,, \qquad (2.3)\text{''}$$

where dimensionless quantities are marked by a dash at the top, $\overline{h}=h/h(t_0)$.

The conditions at the free boundary $\Gamma_\sigma : \overline{F}(\overline{x},\overline{y},\overline{t})=0$:

$$\left[-\bar{p}+\left(\frac{2}{3}+\frac{2}{3}\frac{d\bar{t}}{d\lambda}\frac{\partial \bar{u}}{\partial \bar{x}}\right)\varepsilon\right]n_x+\left[\frac{1}{3}\frac{d\bar{t}}{d\lambda}\left(\frac{\partial \bar{u}}{\partial y}+\frac{\partial \bar{v}}{\partial x}\right)\varepsilon\right]n_y=0, \qquad (2.4)\text{''}$$

$$\left[\frac{1}{3}\frac{d\bar{t}}{d\lambda}\left(\frac{\partial \bar{u}}{\partial y}+\frac{\partial \bar{v}}{\partial x}\right)\varepsilon\right]n_x+\left[-\bar{p}+\left(\frac{2}{3}+\frac{2}{3}\frac{d\bar{t}}{d\lambda}\frac{\partial \bar{v}}{\partial y}\right)\varepsilon\right]n_y=0,$$

$$\bar{v}_n=-\frac{\partial\bar{F}/\partial\bar{t}}{|grad\bar{F}|} \qquad (2.5)\text{''}$$

The conditions at the tool contact boundary Γ_C:

$$\left[-\bar{p}+\left(\frac{2}{3}+\frac{2}{3}\frac{d\bar{t}}{d\lambda}\frac{\partial \bar{u}}{\partial \bar{x}}\right)\varepsilon\right]n_x n_y+\left[\frac{1}{3}\frac{d\bar{t}}{d\lambda}\left(\frac{\partial \bar{u}}{\partial y}+\frac{\partial \bar{v}}{\partial x}\right)\varepsilon\right](n_y^2-n_x^2)-\left[-\bar{p}+\left(\frac{2}{3}+\frac{2}{3}\frac{d\bar{t}}{d\lambda}\frac{\partial \bar{v}}{\partial y}\right)\varepsilon\right]n_x n_y=\pm\frac{\mu}{\sqrt{3}}\varepsilon, \qquad (2.6)\text{''}$$

$$\bar{v}_n=-\frac{\partial\bar{G}/\partial\bar{t}}{|grad\bar{G}|} \qquad (2.7)\text{''}$$

The system of differential equations (2.1)"-(2.3)" together with the boundary conditions (2.4)" - (2.7)" constitutes the general boundary-value problem of the spreading of a "viscous fluid".

In the main approximation ($\varepsilon = 0$), the original equations are simplified by decreasing the order of the system:

$$\frac{\partial\bar{p}}{\partial\bar{x}}=-\frac{2}{\sqrt{3}}\frac{1}{h}\frac{\bar{u}}{\sqrt{\bar{u}^2+\bar{v}^2}},$$

$$\frac{\partial\bar{p}}{\partial\bar{y}}=-\frac{2}{\sqrt{3}}\frac{1}{h}\frac{\bar{v}}{\sqrt{\bar{u}^2+\bar{v}^2}}, \qquad (2.8)$$

$$\frac{\partial\bar{u}}{\partial\bar{x}}+\frac{\partial\bar{v}}{\partial\bar{y}}-\frac{\partial\lambda}{\partial\bar{t}}=0\,.$$

It is clear that the number of boundary conditions for the simplified problem of the "ideal fluid" also decreases.

Solution of the contact problem of plate tension (in a simplified formulation). Let's consider the problem of plastic stretching of a strip by forces applied on its "clamped" ends (Fig.1). In this case, different modes are possible, which are determined by the values of both the total compression force of the ends and the total tensile force.

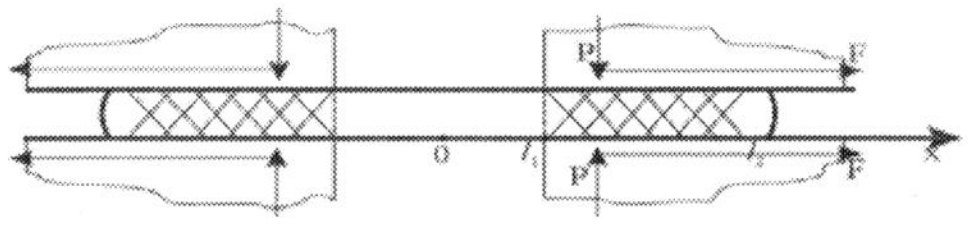

Fig.1. On the problem of complex plate loading.

3 Solution

Due to the symmetry of initial conditions let us formulate the boundary value problem for the area $0<x<l_2(t)$: in the first part $0<x<l_1(t)$ the strip material is in a state of tension, and in another part $l_1(t)<x<l_2(t)$ the material subjected to complex loading in the form of pressing with stretching (or compression, depending on the direction of the applied force $F(t)$. Let's write out the differential equations of the boundary value problem in the domain (l_1,l_2):

$$\frac{\partial u}{\partial x}-\frac{d\lambda}{dt}=0,\quad \frac{\partial p}{\partial x}=-\frac{2\tau_s}{h}\frac{u-u_0}{|u-u_0|}, \qquad (3.1)$$

where $u=u_0(t)$ - the unknown speed of movement of tools.

The boundary condition on the right free end:

$$x=l_2(t):\quad p(l_2,t)=\sigma_s\,, \qquad (3.2)$$

Integrating the equations (3.1), we get:

$$u(x,t)=-\frac{1}{h}\frac{dh}{dt}(x-x_0)+u_0(t), \qquad (3.3)$$

$$p(x,t)=\begin{cases}-\dfrac{2\tau_s x}{h}+c_2, & x\in(x_0,l_2(t))\\ \dfrac{2\tau_s x}{h}+c_1, & x\in(l_1(t),x_0)\end{cases}, \qquad (3.4)$$

where $x=x_0(t)$ - the unknown branch line of the plastic flow in the band relative to the tool:

$$x=x_0(t):\quad u(x_0,t)=u_0(t).$$

To determine the three unknown values, two other conditions must be specified, in addition to the condition (3.2):

- the condition of continuity of the contact pressure at the point of branching of the flow;
- the some unknown, generally speaking, condition at the left end at $x=l_1(t)$.

From (3.2) we get:

$$-\frac{2\tau_s}{h}l_2+c_2=\sigma_s, c_2=\sigma_s+\frac{2\tau_s}{h}l_2\,. \qquad (3.5)$$

From the condition of continuity of the contact pressure at the branch point of the flow $x = x_0(t)$, taking into account the condition (3.5), we find:

$$c_1 = c_2 - \frac{2\tau_s}{h}(2x_0) = \sigma_s + \frac{2\tau_s}{h}(l_2 - 2x_0) \qquad (3.6)$$

Thus, we found a solution with an accuracy of up to $x_0(t)$. Let's look at a few cases.

A) Let the following condition be met

$$u(l_1,t) = -\frac{1}{h}\frac{dh}{dt}(l_1 - x_0) + u_0(t) = 0, \qquad (3.7)$$

meaning that the flow rate of the layer material is zero on the left (stationary!) end of the contact area (i.e., the left end of the contact area moves at a speed $u_0(t)$, but there is no band flow rate at the point $x = l_1(t)$. In other words, there is a plastic pressing of the material in the capture area, and the Central part of the strip remains in its natural state (not deformed). It is clear that in this case we have a problem on the plastic sediment of the end part $(l_1(t); l_2(t))$, which corresponds to the boundary condition of the left free end:

$$x = l_1(t): \quad p(x,t) = \sigma_s \qquad (3.8)$$

from which we get that

$$c_1 = -\frac{2\tau_s}{h}l_1 + \sigma_s \qquad (3.9)$$

The condition (3.7) relates the rate of pressing (dh/dt) and the rate of translational movement of the tool ($u_0(t)$). In particular, it follows from (3.7) that if $u_0(t) = 0$ we obtain that $x_0 \equiv l_1$ (i.e., the left end of the part of the strip being deposited remains stationary). Comparing (3.6) and (3.9), we get:

$$x_0(t) = \frac{l_1(t) + l_2(t)}{2}$$

Now let's return to the condition (3.7), from which, taking into account the last ratio, we find such a value of the tool speed (in our case, the same as the speed of expansion of the middle part of the strip), at which the left end of the contact area is free of stresses:

$$u_0(t) = \left(-\frac{1}{h}\frac{dh}{dt}\right)\frac{l_2 - l_1}{2},$$

or taking into account the incompressibility condition written in integral form

$$(l_1 - l_2)h = (l_{10} - l_{20})h_0\,,$$

we get:

$$u_0(t) = -\frac{dh}{dt}\frac{l_{20} - l_{10}}{2}h_0 \qquad (3.10)$$

We need to find the laws of change $l_1(t)$ и $l_2(t)$:

$$\frac{dl_1}{dt} = u_0(t), \quad l_1(t_0) = l_{10} \Rightarrow l_1(t) = \int_0^t u_0(t)dt + l_{10};$$

$$\frac{dl_2}{dt} = u(l_2,t) = -\frac{1}{h}\frac{dh}{dt}\left(l_2 - \frac{l_2 + l_1}{2}\right) + u_0)t), \; l_2(t_0) = l_{20}.$$

From the last system we get:

$$\frac{d(l_2 - l_1)}{dt} = -\frac{1}{h}\frac{dh}{dt}\frac{1}{2}(l_2 - l_1) \Rightarrow l_2 - l_1\frac{\bar{c}_1}{\sqrt{h}}$$
$$\Rightarrow \; l_2 = l_1 + \frac{\bar{c}_1}{\sqrt{h}}. \qquad (3.11)$$

If we take into account the conditions (3.10), we get

$$\frac{dl_1}{dt} = u_0(t) = \left(-\frac{1}{h}\frac{dh}{dt}\right)\frac{1}{2}(l_2 - l_1) = \left(-\frac{1}{h}\frac{dh}{dt}\right)\frac{1}{2}\frac{\bar{c}_1}{\sqrt{h}},$$

$$l_1(t) = \frac{\bar{c}_1}{\sqrt{h}} + \bar{c}_2; \; l_2(t) = \frac{2\bar{c}_1}{\sqrt{h}} + \bar{c}_2. \qquad (3.12)$$

Now let's substitute the initial data:

$$t = t_0 \;:\; l_{10} = \frac{\bar{c}_1}{\sqrt{h_0}} + \bar{c}_2\,;\, l_{20} = \frac{2\bar{c}_1}{\sqrt{h_0}} + \bar{c}_2 \Rightarrow$$

$$\bar{c}_1 = (l_{20} - l_{10})\sqrt{h_0}; \; \bar{c}_2 = 2l_{10} - l_{20}.$$

The solution (3.12) takes the final form:

$$\begin{cases} l_1(t) = (l_{20} - l_{10})\sqrt{\dfrac{h_0}{h}} + (2l_{10} - l_{20}) \\ l_2(t) = 2(l_{20} - l_{10})\sqrt{\dfrac{h_0}{h}} + (2l_{10} - l_{20}) \end{cases}. \qquad (3.13)$$

B) Let the condition now be met

$$u(l_1,t) = -\frac{1}{h}\frac{dh}{dt}(l_1 - x_0) + u_0(t) > 0, \qquad (3.14)$$

that is, the left end of the (plastic) layer material in the contact area moves to the right, and $q(l_1) = q_0 = -\sigma_s$. This condition necessarily leads to plastic stretching of the middle part of the strip $(0 < x < l_1(t))$, and, according to the condition of full plasticity [1]

$$\sigma_{yy} = -p; \sigma_{xx} = -p + \sigma_s,$$

we have:

$$p(l_1,t) = 0, \qquad (3.15)$$

thus, we got the missing boundary condition of the problem.

Rewrite (3.15) with (3.4), (3.6):

$$\frac{2\tau_s}{h}l_1 + c_1 = 0 \Rightarrow \frac{2\tau_s}{h}l_1 + \left[\sigma_s + \frac{2\tau_s}{h}(l_2 - 2x_0)\right] = 0,$$

$$x_0(t) = \frac{\sigma_s}{(4\tau_s/h)} + \frac{l_1 + l_2}{2} > \frac{l_1 + l_2}{2}, \quad (3.16)$$

that is, the branching line moves slightly towards the free-flowing end $x = l_2(t)$, and we can accept it with an accuracy up to infinitesimal (the theory of flow in a thin plastic layer [1] is characterized by an accuracy of order $\delta = h/l << 1$) that:

$$x_0 = \frac{l_1 + l_2}{2} \quad (3.16)'$$

If we substitute (3.16)' in (3.14), we get a condition for $u_0(t)$, when performing which simultaneously with the plastic pressing of the "clamped" end parts, the plastic stretching of the central part of the strip occurs:

$$u_0(t) > -\frac{1}{h}\frac{dh}{dt}(x_0 - l_1) = \left(-\frac{1}{h}\frac{dh}{dt}\right)\left(\frac{l_2 - l_1}{2}\right)$$

4 Results

Differential equation of the plastic layer free spreading on a plane and its particular solutions. Let us consider the process of free spreading of a thin plastic layer enclosed between two approaching surfaces of tool bodies. Such flows can be identified in a wide range of technological processes of material treatment by pressure, such as stamping and pressing of thin-walled structural elements, sheet rolling and others. In [1] using a two-dimensional averaging over the layer thickness, an efficient mathematical model to describe the flow process of a thin plastic layer is proposed.

Below, we will focus on the study of the kinematics of plastic layer spreading on a plane (which as a rule constitutes the main difficulty in application of boundary value problem), present a mathematical model of its description, and write out some particular solutions to this problem in analytical form.

It is known [3,4] that the quasilinear parabolic equation

$$\frac{\partial w}{\partial \tau} = \frac{\partial}{\partial x}\left(K(w)\frac{\partial w}{\partial x}\right) \quad (4.1)$$

allows to describe fairly well various physical processes such as diffusion of heat in a continuous medium, heat and mass transfer, combustion and filtration. In particular, when in (4.1)

$$K(w) = cw^m, \quad (4.2)$$

the equation is often used to describe the non-stationary heat transfer in a stationary medium, provided that the coefficient of thermal diffusivity is a power function of temperature. There are many exact solutions of equation (4.1), provided (4.2). For the cases, when $m = 1, -1, -2, -4/3, -2/3$ such solutions can be found in [4]. Note, that the problem of the plastic layer free spreading on a plane is reduced to the study of the equation (4.1), in which $K(w) = \frac{1}{2}w^2$.

Let suppose that a thin layer of plastic material at the initial moment of time occupies a plane domain bounded by a piecewise smooth convex contour $y = \varphi(x, t_0) \equiv \varphi(x)$, symmetrical with respect to the horizontal axis x. Assume that the layer shrinks by the mean of two parallel rough planes that are converging along the common normal. In [2, 3] the following evolution equation for finding the contour $F(x, y, t) = y - \varphi(x, t) = 0$ of the domain occupied by spreading plastic layer is obtained:

$$\frac{\partial \varphi}{\partial t_1} - \varphi\left[1 + \left(\frac{\partial \varphi}{\partial x}\right)^2\right] - \frac{\varphi^2}{2}\frac{\partial^2 \varphi}{\partial x^2} = 0, \quad (4.3)$$

where $t_1(t) = \ln(h(t_0)/h(t))$ is a dimensionless modified time ($\frac{dt_1}{dt} = -\frac{1}{h}\frac{dh}{dt} > 0$), and $h = h(t)$ is a well-known variation law of the layer thickness. In [3] the equation (4.3) is reduced to some particular form of nonlinear equation of heat conduction (4.1), with respect to the function $w(x, \tau)$, $(K(w) \equiv \frac{1}{2}w^2)$:

$$\frac{\partial w}{\partial \tau} = \frac{1}{2}\frac{\partial}{\partial x}\left(w^2\frac{\partial w}{\partial x}\right), \quad \varphi(x, \tau_1) \equiv e^{t_1}w(x, \tau), \tau \equiv e^{2t_1}/2 \quad (4.4)$$

Let us consider another type of the evolution equation of spreading obtained from (4) using the substitution $q = w^2$:

$$4\frac{\partial q}{\partial \tau} = 2q\frac{\partial^2 q}{\partial x^2} + \left(\frac{\partial q}{\partial x}\right)^2. \quad (4.5)$$

In this paper all previously known exact solutions of the nonlinear equation (4.4) have been presented, namely:

1. $(x, \tau) = f(\tau)g(x)$, with:

1a) $f(\tau) = C_1, g(x) = C_2$- the solution of the problem on the spreading of a strip parallel to the axis x;

1b) $f(\tau) = c_1$, $g = \sqrt[3]{c_3x + c_4}$ - the solution of the problem on spreading of the domain bounded by cubic parabola;

1c) $2f^2(\alpha_1 + c_0\tau) = -1$, $\int \frac{g^2 dg}{\sqrt{c_0 g^4 + \alpha_2}} = \pm x + \alpha_3$ - the solution of the problem on spreading of the domain bounded by curve in the form of an elliptic integral;

2. $4(\tau - \gamma x) = C_2 + (\gamma w)^2 - 2C_1\gamma w + C_1^2 \ln(C_1 + \gamma w)^2$ - the solution of the problem on the spreading of the domain bounded by the parabola;

3. $q \equiv w^2 = -\frac{x^2}{2(\tau + \tau_0)} + \frac{A_3}{\sqrt{\tau + \tau_0}}$ - the solution of the problem on the spreading of a plastic layer in the domain bounded by the ellipse whose major semi-axis lies on the x -axis;

4. $q \equiv w^2 = -\frac{x^2}{2(\tau + \tau_0)} + \frac{A_3 x^{2/3}}{(\tau + \tau_0)^{7/9}}$ - the generalized solution of the problem on the spreading of a plastic layer in the domain bounded by the ellipse;

5. $w(x, \tau) = \pm\sqrt{q^2} = \pm\frac{x}{\sqrt{2(\tau_0 - \tau)}}$ - the solution of the problem on the spreading of a plastic layer in the domain in the form of a wedge;

6. $q \equiv w^2 = \frac{x^2}{2(\tau_0 - \tau)} + \frac{A_3}{\sqrt{\tau_0 - \tau}}$ - the solution of the problem on the spreading of the domain bounded by the hyperbola, whose real or imaginary axis, depending on the sign of A_3, is coincided with the axis x;

7. $q \equiv w^2 = \frac{x^2}{2(\tau_0 - \tau)} + \frac{A_3 x^{2/3}}{(\tau_0 - \tau)^{7/9}}$ - the generalized solution of the problem on the spreading of a plastic layer bounded by a hyperbola.

5 Conclusion

The paper focuses on the theory of flow in a thin plastic layer, which was developed by A. A. Ilyushin and served as a guide in the calculations of technological processes of material treatment by pressure, such as stamping and pressing of thin-walled structural elements, thin-sheet rolling, and others. This paper reveals the current state of development of the mathematical modeling of contact problems of plastic flow in a thin layer.

The paper presents a solution to the contact problem of plastic stretching of the strip by forces applied on the "clamped" ends of the strip. The possible modes of the process were studied, which are determined by both the total compression force of the ends and the total tensile force.

The mathematical analogy between the process of the free spreading of a plastic layer on the plane and the process of heat transfer has been investigated. The exact particular solutions of the considered problem were obtained. The obtained results may be of interest for researchers in the field of nonlinear problems of heat and mass transfer as well as for engineers in the field of designing and carrying out calculation of processes of metals treatment by pressure.

The obtained results have interesting practical applications. In particular, analytical solutions of various generalizations of the classical L. Prandtl problem given in section 3 are of interest in the technique of studying the problem of straightening plates using uniaxial stretching beyond the elastic limit. It should be noted that straightening by stretching allows to obtain a high degree of flatness of thin wide strips and sheets made of high-strength steels and special alloys [6], which straightening by other methods does not give satisfactory results. The results may be used to describe a wide class of processes such as heat and-mass exchange in the technologies, biology and medicine, which are modeled using nonlinear parabolic equations, and for which there exist a mathematical analogy with the problems of spreading plastic layers considered in section 4.

This work was carried out using equipment provided by the Center of Collective Use of MSUT "STANKIN".

References

1. A. Ilyushin, *Proceedings (1946-1966). V.2. Plasticity* (M.:Fizmatlit. 2004)
2. I.A. Kiyko, *Theory of plastic flow* (M.: MSU. 1978)
3. N. Belov, V. Kadymov, Mechanics of solids, **46**(1), 36-46 (2011)
4. A. Polyanin, V. Zaitsev, *Nonlinear equations of mathematical physics. Exact Solutions* (Moscow. Fizmatlit, 2002)
5. V. Kadymov, .Journ."Mathematical Methods in the Applied Sciences", **38**(16), 3421-3431 (2015) https://doi.org/10.1002/mma.3403
6. V. Kadymov, E. Yanovskaya, EPJ Web Conf., **224** (2019) https://doi.org/10.1051/epjconf/201922401002

Transmutation Operator Method for Solving Heat Conduction Problem

Oleg Yaremko[1,*], *Natalia* Yaremko[2]

[1]Moscow State Technological University "STANKIN ", RU-127055, Moscow, Russia
[2]Penza State University, RU-440026, Penza, Russia

Abstract. The transmutation operator method is extended to the case of functions of two variables. The transmutation operator flattens the function, i.e. the transmutation operator replaces a function with discontinuous partial derivatives on the coordinate axes by a continuously differentiable function. The work reveal the properties of the transmutation operator, and prove the commutativity of the transmutation operator and the Laplace operator. It was found that the Cauchy problem for the Laplace equation with internal conjugations in an unbounded domain can be replaced with the model Cauchy problem for the two-dimensional Laplace equation. As a result, a new analytical method for solving initial-boundary value problems for a two-dimensional heat equation has been developed. The factorization of the transmutation operator is established as a product of two one-dimensional transmutation operators. The form of the transmutation operator establishing the isomorphism of two mathematical models of heat conduction in unbounded media with different physical characteristics was found and descrfibed.

1 Introduction

The transmutation operator establishes a connection between the solutions of two problems of mathematical physics,one of the problems is considered as a model one.

Example 1. The Cauchy problem for the heat equation [6]

$$\begin{cases} \tilde{u}_t = \tilde{u}_{xx}, t>0, -\infty<x<\infty \\ \tilde{u}(0,x) = f(x), -\infty<x<\infty \end{cases} \quad (1)$$

Is model. Let's consider a new Cauchy problem

$$\begin{cases} u_t = a^2 u_{xx}, t>0, -\infty<x<\infty \\ u(0,x) = f(x), -\infty<x<\infty \end{cases} \quad (2)$$

An operator $J:\tilde{u} \to u$ that establishes an isomorphism of the solution space for problems (1) and (2) is called a transmutation operator. In this example, the transmutation operator is defined by the formula

$$u(t,x) \equiv J\left[\tilde{u}(t,x)\right] = \tilde{u}\left(a^2 t, x\right).$$

Example 2. Consider the initial-boundary value Cauchy problem for the heat equation

$$\begin{cases} u_{1t} = a_1^2 u_{1xx}, t>0, -\infty<x<0 \\ u_1(0,x) = f_1(x), -\infty<x<0 \end{cases}$$

$$\begin{cases} u_{2t} = a_2^2 u_{2xx}, t>0, 0<x<\infty \\ u_2(0,x) = f_2(x), 0<x<\infty \end{cases} \quad (3)$$

with internal conjugation conditions

$$\begin{cases} u_1(t,0) = u_2(t,0), t>0, \\ \lambda_1 u_{1x}(t,0) = \lambda_2 u_{2x}(t,0), 0<t. \end{cases} \quad (4)$$

Transmutation operator $J_1 : \tilde{f} \to f$,

$$f = f_1 H(-x) + f_2 H(x),$$

where $H(x)-$Heaviside step function, is defained by formula

$$f_2(x) = \tilde{f}(x/a_2), x>0, \quad (5)$$

$$f_1(x) = \left(\frac{1}{2}+\nu\right)\tilde{f}(x/a_1) + \left(\frac{1}{2}-\nu\right)\tilde{f}(-x/a_1), x<0,$$

here

$$\nu = \frac{1}{2}\frac{\lambda_2}{\lambda_1}\frac{a_1}{a_2}.$$

In the model problem, the initial condition is chosen in the form $\tilde{f} = J_1^{-1}[f]$. In more detail, the inverse operator J_1^{-1} with respect to the transmutation operator J_1 has the form

* Corresponding author: yaremki8@gmail.com

$$\tilde{f}(x)=f_2(a_2x), x>0,$$
$$\tilde{f}(x)=\frac{2}{1+2\nu}f_1(a_1x)-\frac{1-2\nu}{1+2\nu}f_2(-a_1x), x<0. \quad (6)$$

So, the solution to problem (3)-(4) has the form [6]

$$u(x,t)=\frac{1}{2\sqrt{\pi t}}\int_{R}J_1\left[\exp\left(-\frac{(x-y)^2}{4t}\right)\right]J_1^{-1}\left[f(y)\right]dy.$$

In the authors' work [1,2,21], the formula for solution of problem (3) - (4) is transformed to the form

$$u_1=\frac{1}{2a_1\sqrt{\pi t}}\int_{-\infty}^{0}\left(e^{-\frac{(x-\xi)^2}{4a_1^2t}}-e^{-\frac{(x+\xi)^2}{4a_1^2t}}\frac{1-2\nu}{1+2\nu}\right)f_1(\xi)d\xi+$$

$$+\frac{1}{a_2\sqrt{\pi t}}\int_{0}^{\infty}e^{-\frac{(a_1^{-1}x-a_2^{-1}\xi)^2}{4t}}(1+2\nu)^{-1}f_2(\xi)d\xi, t<0, x<0;$$

$$u_2=\frac{1}{2a_1\sqrt{\pi t}}\int_{-\infty}^{0}e^{-\frac{(a_2^{-1}x-a_1^{-1}\xi)^2}{4t}}2\nu(1+2\nu)^{-1}f_1(\xi)d\xi+ \quad (7)$$

$$+\frac{1}{2a_2\sqrt{\pi t}}\int_{0}^{\infty}\left(e^{-\frac{(x-\xi)^2}{4a_2^2t}}+e^{-\frac{(x+\xi)^2}{4a_2^2t}}\frac{1-2\nu}{1+2\nu}\right)f_2(\xi)d\xi, t, x>0.$$

Formula (7) is a generalization of the Poisson formula of heat conduction theory [6].

In works [3,4,5,7,14,16,18,20] the theoretical foundations of transmutation operator's method are created. Papers [8,10-13,15,17,19] are devoted to various applications of the method, including the solving of partial differential equations. The most important of the applications of the developed method is the solution of boundary value problems in areas with axial symmetry [3,11,12]. The transmutation operator method allows problems with axial symmetry to be reduced to boundary value problems in domains with plane symmetry.

Example 3. Let $\tilde{u}(t,x)$ be the solution of the initial-boundary value problem (3) - (4) with parameters $\tilde{a}_1,\tilde{a}_2,\tilde{\lambda}_1,\tilde{\lambda}_2$ and initial condition

$$\tilde{f}(x)=\tilde{f}_2(x)H(-x)+\tilde{f}_2(x)H(x).$$

and $u(t,x)$ is the solution to initial-boundary value problem (3)-(4) with parameters $a_1,a_2,\lambda_1,\lambda_2$ and initial condition

$$f(x)=f_2(x)H(-x)+f_2(x)H(x),$$

then the transmutation operator J is defined so that $J[\tilde{f}]=f; u=J[\tilde{u}]$. The transmutation operator J is a linear combination of stretched or compressed operators and reflections. Let's write out the expression of the transmutation operator J

$$f_2(x)=\tilde{f}_2\left(\frac{\tilde{a}_2}{a_2}x\right), x>0,$$

$$f_1(x)=\frac{1+k}{1+\tilde{k}}\tilde{f}_1\left(\frac{\tilde{a}_1}{a_2}x\right)-\frac{\tilde{k}-k}{1+\tilde{k}}\tilde{f}_2\left(-\frac{\tilde{a}_2}{a_1}x\right).$$

Remark 1. Inverse operator J^{-1} with respect to transmutation operator J can be calculated by formulas

$$\tilde{f}_2(x)=f_2\left(\frac{a_2}{\tilde{a}_2}x\right), x>0,$$

$$\tilde{f}_1(x)=\frac{1+\tilde{k}}{1+k}f_1\left(\frac{a_1}{\tilde{a}_2}x\right)-\frac{k-\tilde{k}}{1+k}f_2\left(-\frac{a_2}{\tilde{a}_1}x\right), x<0,$$

where

$$k=\frac{\lambda_2}{\lambda_1}\frac{a_1}{a_2}, \tilde{k}=\frac{\tilde{\lambda}_2}{\tilde{\lambda}_1}\frac{\tilde{a}_1}{\tilde{a}_2}.$$

The solution to initial-value problem (3)-(4) with parameters $a_1,a_2,\lambda_1,\lambda_2$ and initial condition

$$f(x)=f_1(x)H(-x)+f_2(x)H(x),$$

can be obtained by algorithm:

1. find the initial conditions for the model problem according to the rule $\tilde{f}(x)=J^{-1}[f(x)]$;
2. calculate the solution $\tilde{u}(t,x)$ to the model problem;
3. determine the solution of problem (3) - (4) by the formula $u(t,x)=J^{-1}[\tilde{u}(t,x)]$.

2 Transmutation operator's method for two-dimensional problems

2.1 Statement of the problem

Let us set the goal of obtaining the Poisson formula for solving the heat equation in an unbounded four-layer plate. Consider a separate system of equations describing thermal conductivity in a four-layer medium

$$u_{ijt}=a_i^2u_{ij,xx}+b_j^2u_{ij,yy}, t>0, x,y\in D_{ij}, \quad (8)$$

where

$$D_{11}=\{(x,y)/x<0,y<0\}, D_{12}=\{(x,y)/x<0,y>0\},$$
$$D_{21}=\{(x,y)/x>0,y<0\}, D_{22}=\{(x,y)/x>0,y>0\}$$

with initial conditions

$$u_{ij}(0,x,y)=f_{ij}(x,y), i,j=1,2 \quad (9)$$

with internal conjugation conditions on coordinate lines $x=0, y=0$

$$u_{1j}=u_{2j}, \quad \lambda_1\partial_x u_{1j}=\lambda_2\partial_x u_{2j}, x=0. \quad (10)$$

$$u_{i1}=u_{i2}, \quad \mu_1\partial_y u_{i1}=\mu_2\partial_y u_{i2}, y=0.$$

Consider the model equation of heat conduction in an unbounded medium

$$\tilde{u}_t=\tilde{u}_{xx}+\tilde{u}_{yy}, t>0, x,y\in R^2, \quad (11)$$

with initial condition

$$\tilde{u}(0,x,y)=\tilde{f}(x,y). \quad (12)$$

In order to solve the two-dimensional heat conduction problem (8), (9), (10), it is necessary to determine the transmutation operator. Transmutation operator J establishes isomorphism (8),(9),(10) and (11)-(12), so that $J:\tilde{f}\to f; J:\tilde{u}\to u$.

2.2 Construction of transmutation operator

It will be proved here that the transmutation operator J represents as linear combination of stretched or compressed transforms and reflections.

Theorem 1. Transmutation operator J has the form

$$u_{22}(t,x,y)=\tilde{u}(t,x/a_2,y/b_2),(x,y)\in D_{22},$$

$$u_{12}(t,x,y)=\left(\frac{1}{2}+\nu\right)\tilde{u}(t,x/a_1,y/b_2)+$$
$$+\left(\frac{1}{2}-\nu\right)\tilde{u}(t,-x/a_1,y/b_2),(x,y)\in D_{12}$$

$$u_{21}(t,x,y)=\left(\frac{1}{2}+\mu\right)\tilde{u}(t,x/a_2,y/b_1)+$$
$$+\left(\frac{1}{2}-\mu\right)\tilde{u}(t,x/a_2,-y/b_1),(x,y)\in D_{21} \quad (13)$$

$$u_{11}(t,x,y)=\left(\frac{1}{2}+\nu\right)\left(\frac{1}{2}+\mu\right)\tilde{u}(t,x/a_1,y/b_1)+$$
$$+\left(\frac{1}{2}+\nu\right)\left(\frac{1}{2}-\mu\right)\tilde{u}(t,x/a_1,-y/b_1)+$$
$$+\left(\frac{1}{2}-\nu\right)\left(\frac{1}{2}+\mu\right)\tilde{u}(t,-x/a_1,y/b_1)+$$
$$+\left(\frac{1}{2}-\nu\right)\left(\frac{1}{2}-\mu\right)\tilde{u}(t,-x/a_1,-y/b_1),(x,y)\in D_{11}$$

here

$$\nu=\frac{1}{2}\frac{\lambda_2}{\lambda_1}\frac{a_1}{a_2}, \mu=\frac{1}{2}\frac{\mu_2}{\mu_1}\frac{a_1}{a_2}.$$

Proof. Equations (1) are satisfied by construction. Check the conjugation conditions for the component u_{11}

$$u_{11}=u_{21}, \quad \lambda_1\partial_x u_{11}=\lambda_2\partial_x u_{21}, x=0,$$

$$u_{11}=u_{12}, \quad \mu_1\partial_y u_{11}=\mu_2\partial_y u_{12}, y=0.$$

Let's check the first of the four conditions. We have

$$u_{11}(0,y)=\left(\frac{1}{2}+\nu\right)\left(\frac{1}{2}+\mu\right)\tilde{u}(0,y/b_1)+$$
$$+\left(\frac{1}{2}+\nu\right)\left(\frac{1}{2}-\mu\right)\tilde{u}(0,-y/b_1)+$$
$$+\left(\frac{1}{2}-\nu\right)\left(\frac{1}{2}+\mu\right)\tilde{u}(0,y/b_1)+$$
$$+\left(\frac{1}{2}-\nu\right)\left(\frac{1}{2}-\mu\right)\tilde{u}(0,-y/b_1)=$$
$$=\left(\frac{1}{2}+\mu\right)\tilde{u}(0,y/b_1)+\left(\frac{1}{2}-\mu\right)\tilde{u}(0,-y/b_1).$$

$$u_{21}(0,y)=\left(\frac{1}{2}+\mu\right)\tilde{u}(0,y/b_1)+\left(\frac{1}{2}-\mu\right)\tilde{u}(0,-y/b_1).$$

The third condition is established similarly. To check the second condition, we calculate the derivatives in the direction of the normal

$$\lambda_1 u_{11x}(x,y)=\left(\frac{1}{2}+\nu\right)\left(\frac{1}{2}+\mu\right)\frac{\lambda_1}{a_1}\tilde{u}_x(0,y/b_1)+$$
$$+\left(\frac{1}{2}+\nu\right)\left(\frac{1}{2}-\mu\right)\frac{\lambda_1}{a_1}\tilde{u}_x(0,-y/b_1)-$$
$$-\left(\frac{1}{2}-\nu\right)\left(\frac{1}{2}+\mu\right)\frac{\lambda_1}{a_1}\tilde{u}_x(0,y/b_1)-$$
$$-\left(\frac{1}{2}-\nu\right)\left(\frac{1}{2}-\mu\right)\frac{\lambda_1}{a_1}\tilde{u}_x(0,-y/b_1)=$$
$$=2\nu\left(\frac{1}{2}+\mu\right)\frac{\lambda_1}{a_1}\tilde{u}_x(0,y/b_1)+2\nu\left(\frac{1}{2}-\mu\right)\frac{\lambda_1}{a_1}\tilde{u}_x(0,-y/b_1).$$

Then

$$\lambda_2 u_{21x}(0,y)=\left(\frac{1}{2}+\mu\right)\frac{\lambda_2}{a_2}\tilde{u}(0,y/b_1)+$$
$$+\left(\frac{1}{2}-\mu\right)\frac{\lambda_2}{a_2}\tilde{u}(x/a_2,-y/b_1).$$

Further, the fourth conjugation condition is proved similarly. The conjugation conditions for the functions $u_{12}, u_{21}u_{22}$.

Theorem 2. The inverse operator J^{-1} to the transfmutation operator J has the form

$$\tilde{u}(x,y)=u_{22}(a_2x,b_2y),(x,y)\in D_{22},$$

$$\tilde{u}(x,y)=\frac{2}{1+2\nu}u_{12}(a_1x,b_2y)-\frac{1-2\nu}{1+2\nu}u_{22}(-a_1x,b_2y),$$
$(x,y)\in D_{12}$,

$$\tilde{u}(x,y)=\frac{2}{1+2\mu}u_{21}(a_2x,b_1y)-\frac{1-2\mu}{1+2\mu}u_{22}(a_2x,-b_1y),$$

$$(x,y)\in D_{21}, \qquad (14)$$

$$\tilde{u}(x,y)=\frac{4}{(1+2\nu)(1+2\mu)}u_{11}(a_1x,b_1y)-$$
$$-\frac{1-2\mu}{1+2\mu}\left(\frac{2}{1+2\nu}u_{12}(a_1x,-b_2y)-\frac{1-2\nu}{1+2\nu}u_{22}(-a_1x,-b_2y)\right)$$
$$-\frac{1-2\nu}{1+2\nu}\left(\frac{2}{1+2\mu}u_{21}(-a_2x,b_1y)-\frac{1-2\mu}{1+2\mu}u_{22}(-a_2x,-b_1y)\right)-$$
$$-\frac{1-2\mu}{1+2\mu}\frac{1-2\nu}{1+2\nu}u_{22}(-a_2x,-b_2y)\tilde{u}(-x,-y),(x,y)\in D_{11}.$$

Proof. From the first equation of (13) we determine the function $\tilde{u}(x,y)$ in the domain D_{22} . Then we move on to the second equation. Considering the first step, we get the expression of function $\tilde{u}(x,y)$ in the area D_{12}. We finish the calculations by considering the fourth equation from (13), using the previously obtained values of the function $\tilde{u}(x,y)$ in the domains D_{22},D_{21},D_{12}.

In the next section, for a better understanding of the construction of the transfmutation operator, we will establish the possibility of its decomposition as a product of simple transfmutation operators J_1 and J_2.

2.3 Factorization of the transmutation operator

The transfmutation operator J_1 acts on a function of the variable x. Similarly, we define a transformation operator J_2 that acts on a function of the variable y. Let a function $\tilde{h}(y)$ be continuously differentiable on the real axis, and let the function

$$h(y)=h_1(y)H(-y)+h_2(y)H(y)$$

was defined as follow

$$h_2(y)=\tilde{h}(y/b_2),y>0,$$
$$h_1(y)=\left(\frac{1}{2}+\mu\right)\tilde{h}(y/b_1)+\left(\frac{1}{2}-\mu\right)\tilde{h}(-y/b_1),y<0$$

where

$$\mu=\frac{1}{2}\frac{\mu_2}{\mu_1}\frac{b_1}{b_2},$$

then the function $h(y)=h_1(y)H(-x)+h_2(y)H(x)$ is continuously differentiable in $R\setminus\{0\}$ and the conjugation condition

$$h_1(0)=h_2(0),$$
$$\mu_1h_{1y}(0)=\mu_2h_{2y}(0)$$

are hold. So, for transmutation oherator J_2 the following formula is true:

$$h(y)=J_2\left[\tilde{h}(y)\right].$$

Inverse operator J_2^{-1} takes form

$$\tilde{h}(y)=h_2(b_2y),y>0,$$
$$\tilde{h}(x)=\frac{2}{1+2\nu}h_1(b_1y)-\frac{1-2\nu}{1+2\nu}h_2(-b_1y),y<0$$

Theorem 3. The transmutation operator J is factorized as the product of an operator J_1 that acts on the variable x and an operator J_2 that acts on y, i.e. $J=J_1J_2$. The inverse operator J^{-1} to the transmutation operator also admits factorization in the form $J^{-1}=J_2^{-1}J_1^{-1}$.

The proof follows from the definition of the operators J_1 and J_2.

Theorem 4. The transmutation operator J and the Laplace operator Δ are commutative

$$\Delta=\left(a_1^2\partial_{xx}+b_1^2\partial_{yy}\right)H(-x)H(-y)+$$
$$\left(a_2^2\partial_{xx}+b_1^2\partial_{yy}\right)H(x)H(-y)+$$
$$+\left(a_1^2\partial_{xx}+b_2^2\partial_{yy}\right)H(-x)H(y)+$$
$$\left(a_2^2\partial_{xx}+b_2^2\partial_{yy}\right)H(-x)H(-y),$$

i.e. $J\Delta=\Delta J$.

The proof follows from Theorem 1.

2.4. Extending the Results to the Vector Case

Let a vector-function

$$f(x)=f_1(x)H(-x)+f_2(x)H(x),$$

where

$$f_i(x)=\left(f_{i1}(x)\ \ \dots\ \ f_{in}(x)\right)^*,i=1,2$$

be continuous in, continuously differentiable in R , and continuously differentiable in $R\setminus\{0\}$. The parameters

$a_1, a_2, \lambda_1, \lambda_2$ are matrices of $n\times n$ size for each of which all eigenvalues are different and positive.

The definition of a matrix argument function can be found in the book [9]. We present an unknown definition of the vector function of a matrix argument.

Definition 1. Suppose, that all eigenvalues of matrices A and B of dimensions $n\times n$ are different and positive, and the vector -function $f(x)$ is given in R and has $n\times 1$-size, then the vector -function $f(Ax)$ of the matrix argument is determined by the rule

$$f(Ax)=\sum_{k=1}^{n} f_k(Ax)e_k,\quad e_k=\begin{pmatrix}0 & \dots & \underset{k}{\underset{\smile}{1}} & \dots & 0\end{pmatrix}^{*}.$$

and the vector-function $Bf(Ax)$ is defined by the rule

$$Bf(Ax)=Bf(x)(x\to Ax).$$

Let us explain the last formula. First, you need to calculate the product $Bf(x)$, and then to apply the definition of a vector- function of a matrix argument. However, the order of operations cannot be changed.

Example 4. If we have matrices

$$A=\begin{pmatrix}-1 & 3\\ -2 & 4\end{pmatrix}, B=\begin{pmatrix}1 & 2\\ 0 & 3\end{pmatrix}$$

then

$$Bf(Ax)=\begin{pmatrix}-5x+29x^2\\ -6x+30x^2\end{pmatrix},$$

$$Bf(x)(x\to Ax)=\begin{pmatrix}-x+17x^2\\ -2x+18x^2\end{pmatrix},$$

and

$$Bf(Ax)\neq Bf(x)(x\to Ax).$$

Note, if matrices A and B are permute the equality is true.

The vector analogs of transformation operators have the following form:

Transmutation operator $J_1:\tilde{f}\to f$, where

$$f=f_1H(-x)+f_2H(x),$$

$H(x)-$ Heaviside step function, and f is defined by equality

$$f_2(x)=\tilde{f}(a_2^{-1}x), x>0,$$

$$f_1(x)=\left(\left(\frac{E}{2}+\nu\right)\tilde{f}(x)+\left(\frac{E}{2}-\nu\right)\tilde{f}(-x)\right)(x\to a_1^{-1}), x<0,$$

where

$$\nu=\frac{1}{2}a_1\lambda_1^{-1}\lambda_2 a_2^{-1}.$$

The inverse vector transmutation operator J_1^{-1} for the operator J_1 has the form

$$\tilde{f}(x)=f_2(a_2x), x>0,$$

$$\tilde{f}(x)=\left(2(E+2\nu)^{-1}f_1(x)-(E-2\nu)(E+2\nu)^{-1}f_2(-x)\right)$$
$$(x\to a_1x), x<0.$$

The vector analogue of the initial-boundary value Cauchy problem (3) - (4) for the heat equation is posed as follows: find a solution to the system of differential equations

$$\begin{cases}u_{1t}=a_1^2u_{1xx}, t>0, -\infty<x<0,\\ u_1(0,x)=f_1(x), -\infty<x<0,\end{cases}$$

$$\begin{cases}u_{2t}=a_2^2u_{2xx}, t>0, 0<x<\infty,\\ u_2(0,x)=f_2(x), 0<x<\infty,\end{cases}$$

satisfied the conditions of internal conjugation

$$\begin{cases}u_1(t,0)=u_2(t,0), t>0,\\ \lambda_1u_{1x}(t,0)=\lambda_2u_{2x}(t,0), 0<t,\end{cases}$$

where matrices λ_1,λ_2 $n\times n$ - size eigenvalues of wich are different and positive. The solution of this vector analogue of problem (3) - (4) can be determined by the formula with help of transmutation operator J_1 and its inverse operator J_1^{-1} :

$$u(x,t)=\frac{1}{2\sqrt{\pi t}}\int_{\mathbf{R}} J_1\left[\exp\left(-\frac{(x-y)^2}{4t}\right)\right]J_1^{-1}\left[f(y)\right]dy.$$

3 Solution of the initial-boundary value problem for the heat equation in an unbounded piecewise-homogeneous plate

We find the solution to the initial-boundary value problem (1) - (3) using Theorem 1. We apply the transmutation operator J from theorem 1 to problem (1) - (3). In the images, we obtain problem (4) - (5), in which the initial state of the temperature field is determined by the formula $\tilde{f}=J[f]$. We use the Poisson formula for the two-dimensional Poisson equation

$$\tilde{u}(t,x,y)=\frac{1}{2\pi t}\int_{R^2} e^{\frac{(x-\xi)^2+(y-\eta)2}{4t}}\tilde{f}(\xi,\eta)d\xi d\eta$$

On the base of theorem 1 we have

$$u(t,x,y)=\frac{1}{2\pi t}\int_{R^2} J\left[e^{-\frac{(x-\xi)^2+(y-\eta)2}{4t}}\right]\tilde{f}(\xi,\eta)d\xi d\eta.$$

In each of the four areas $D_{22}, D_{21}, D_{12}, D_{11}$, we obtain a solution to problem (1) - (3)

$$u_{22}(t,x,y)=\frac{1}{2\pi t}\int_{R^2} e^{-\frac{(x/a_2-\xi)^2+(y/b_2-\eta)2}{4t}}\tilde{f}(\xi,\eta)d\xi d\eta,$$

$(x,y)\in D_{22}$,

$$u_{12}(t,x,y)=\frac{1}{2\pi t}\int_{R^2}\left(\left(\frac{1}{2}+\nu\right)e^{-\frac{(x/a_1-\xi)^2+(y/b_2-\eta)2}{4t}}+\right.$$
$$\left.+\left(\frac{1}{2}-\nu\right)e^{-\frac{(x/a_1+\xi)^2+(y/b_2-\eta)2}{4t}}\right)\tilde{f}(\xi,\eta)d\xi d\eta,(x,y)\in D_{12},$$

$$u_{21}(t,x,y)=\frac{1}{2\pi t}\int_{R^2}\left(\left(\frac{1}{2}+\mu\right)e^{-\frac{(x/a_2-\xi)^2+(y/b_1-\eta)2}{4t}}+\right.$$
$$\left.+\left(\frac{1}{2}-\mu\right)e^{-\frac{(x/a_2-\xi)^2+(y/b_1+\eta)2}{4t}}\right)\tilde{f}(\xi,\eta)d\xi d\eta,(x,y)D_{21},$$

$$u_{11}(t,x,y)=\frac{1}{2\pi t}\int_{R^2}\left(\left(\frac{1}{2}+\nu\right)\left(\frac{1}{2}+\mu\right)e^{-\frac{(x/a_1-\xi)^2+(y/b_1-\eta)2}{4t}}+\right.$$
$$+\left(\frac{1}{2}+\nu\right)\left(\frac{1}{2}-\mu\right)e^{-\frac{(x/a_1-\xi)^2+(y/b_1+\eta)2}{4t}}+$$
$$+\left(\frac{1}{2}-\nu\right)\left(\frac{1}{2}+\mu\right)e^{-\frac{(x/a_1+\xi)^2+(y/b_1-\eta)2}{4t}}+$$
$$\left.+\left(\frac{1}{2}-\nu\right)\left(\frac{1}{2}-\mu\right)e^{-\frac{(x/a_1+\xi)^2+(y/b_1+\eta)^2}{4t}}\right)\tilde{f}(\xi,\eta)d\xi d\eta, in\ D_{11}.$$

Further, we can simplify the found formulas. To do this, we use the expression of the function $\tilde{f}(\xi,\eta)$ from theorem 2. As a result of simplifications for each of the four components, we obtain formulas in which only the initial data (9) participate. For example, we present the formula for the component u_{11}

$$u_{11}=\frac{1}{4\pi a_1 b_1 t}\int_{D_{11}}\left(e^{-\frac{(x-\xi)^2}{4a_1^2 t}}-e^{-\frac{(x+\xi)^2}{4a_1^2 t}}\frac{1-2\nu}{1+2\nu}\right)\cdot$$
$$\cdot\left(e^{-\frac{(y-\eta)^2}{4b_1^2 t}}-e^{-\frac{(y+\eta)^2}{4b_1^2 t}}\frac{1-2\mu}{1+2\mu}\right)f_{11}(\xi,\eta)d\xi d\eta+$$

$$+\frac{1}{2a_1 b_2\pi t}\int_{D_{12}}\left(e^{-\frac{(x/a_1-\xi/a_1)^2}{4t}}-e^{-\frac{(x/a_1+\xi/a_1)^2}{4t}}\frac{1-2\nu}{1+2\nu}\right)\cdot$$
$$\cdot(1+2\mu)^{-1}e^{-\frac{(y/b_1-\eta/b_2)^2}{4t}}d\eta f_{12}(\xi,\eta)d\xi d\eta+ \tag{15}$$
$$+\frac{1}{2a_2 b_1\pi t}\int_{D_{21}} e^{-\frac{(x/a_1-\xi/b_1)^2}{4t}}(1+2\nu)^{-1}\cdot$$
$$\cdot\left(e^{-\frac{(y/b_1-\eta/b_1)^2}{4t}}-e^{-\frac{(y/b_1+\eta/b_1)^2}{4t}}\frac{1-2\mu}{1+2\mu}\right)\cdot$$
$$\cdot f_{21}(\xi,\eta)d\xi d\eta+$$
$$+\frac{1}{\pi a_2 b_2 t}\int_{D_{22}} e^{-\frac{(x/a_1-\xi/a_2)^2}{4t}}e^{-\frac{(y/b_1-\eta/b_2)^2}{4t}}\cdot$$
$$(1+2\nu)^{-1}(1+2\mu)^{-1}f_{22}(\xi,\eta)d\xi d\eta, x,y<0.$$

Similarly, the components u_{12}, u_{21}, u_{22} can be found.

4 Discussion

The method of Fourier integral transforms, well-known in classical mathematics, is an alternative to the method of transmutation operators developed in this article. Let's compare both methods. Let us associate problem (9) - (11) with two one-dimensional Sturm-Liouville problems. The first problem is to determine a bounded on the real axis solution to the equation with piecewise constant coefficients

$$\begin{aligned}&a_1^2\varphi_{1xx}+\alpha^2\varphi_1, x<0,\\&a_2^2\varphi_{2xx}+\alpha^2\varphi_2, x>0\end{aligned} \tag{16}$$

with internal conjgation conditions on the stright line $x=0$

$$\varphi_1=\varphi_2,\quad \lambda_1\varphi_{1x}=\lambda_2\varphi_{2x}, x=0. \tag{17}$$

The solution to problem (16) - (17) defines the kernel of the inverse Fourier integral transform F_{1x}^{-1} with a point of division on the real axis [21]

$$f(x)\equiv F_{1x}^{-1}\left[\tilde{f}(\alpha)\right]=\int_{-\infty}^{\infty}\varphi(x,\alpha)\tilde{f}(\alpha)d\alpha.$$

Then consider the dual Sturm-Liouville problem: find a bounded on the real axis solution of the equation with piecewise constant coefficients

$$\begin{aligned}&a_1^2\varphi^*_{1xx}+\alpha^2\varphi^*_1=0, x<0,\\&a_2^2\varphi^*_{2xx}+\alpha^2\varphi^*_2=0, x>0\end{aligned} \tag{18}$$

with internal conjgation conditions on the stright $x=0$

$$\lambda_2\varphi^*_1=\lambda_1\varphi^*_2,\qquad \varphi^*_{1x}=\varphi^*_{2x}, x=0. \tag{19}$$

The solution to this dual problem (18) - (19) defines the kernel of the direct Fourier integral transform F_{1x} with a point of division on the real axis

$$\tilde{f}(\alpha)\equiv F_{1x}\left[f(x)\right]=\int_{-\infty}^{\infty}\varphi^*(x,\alpha)f(x)dx.$$

The expressions for the kernels of integral transforms are written out in [21].

The direct F_{1y} and inverse F_{1y}^{-1} Fourier transforms with the division point in the variable y are defined similarly. Moreover, in the Sturm-Liouville problem, it is necessary to replace $\alpha\to\beta, a_i\to b_i;\lambda_i\to\mu_i, i=1,2$.

To solve problem (9) - (11), we apply the Fourier integral transforms F_{1x} and F_{1y}. According to the scheme proposed in [21], we obtain a formula for solving problem (9) - (11):

$$u(t,x,y)=F_{1x}^{-1}F_{1y}^{-1}\left[e^{-(\alpha^2+\beta^2)t}F_{1x}F_{1y}\left(f(x,y)\right)\right]. \tag{20}$$

Comparing formulas (15) and (20), we conclude the advantage of transmutation operator's method. In formula (15), the solution is given by a double integral, and in formula (20) it is need to use four -times integral. Thus, the transmutation operator's method has all the advantages of the integral transforms method and is more efficient from a computational point of view.

5 Conclusion

The article has solves a number of problems. In particular, a solution is found for a two-dimensional problem of heat conduction in an unbounded thin plate with four sections with different thermophysical parameters. The solution is found in terms of the transmutation operator. The construction of the transmutation operator and its inverse is found, the factorization of the transmutation operator is found in the form of the product of transmutation operators with respect to the variables x and y. The connection between the transmutation operator method and the method of integral transforms is established. It is shown that from the computational point of view the method of transmutation operators is more efficient than the method of integral transforms.

The method of transmutation operators makes it possible to predict the structure of the temperature field of an unbounded piecewise homogeneous thin plate from the known temperature field of a model piecewise homogeneous (or homogeneous) thin plate.

Just like the method of integral transforms, the method of transmutation operators is universal. It can be used to solve differential equations of hyperbolic and elliptic types. The technique of applying the method is similar to the method of Fourier integral transforms:

1.From the problem in the originals, we pass to the problems in the images,

2. we solve the problem in images,
3. back to the originals space.

The difference lies in the fact that the image space coincides with the space of the originals of a certain model problem.

Example 4. Consider the initial-boundary value Cauchy problem for the equation of vibrations of a infinite string with two sections of different dencity

$$\begin{cases} u_{1tt}=a_1^2u_{1xx}, t>0, -\infty<x<0,\\ u_1(0,x)=f_1(x), -\infty<x<0,\\ \partial_t u_1(0,x)=0, -\infty<x<0, \end{cases}$$

$$\begin{cases} u_{2tt}=a_2^2u_{2xx}, t>0, 0<x<\infty\\ u_2(0,x)=f_2(x), 0<x<\infty\\ \partial_t u_2(0,x)=0, 0<x<\infty \end{cases} \tag{21}$$

with internal conjgation conditions

$$\begin{cases} u_1(t,0)=u_2(t,0), t>0,\\ \lambda_1u_{1x}(t,0)=\lambda_2u_{2x}(t,0), 0<t. \end{cases} \tag{22}$$

The Cauchy problem for the equation of string vibrations is a model problem

$$\begin{cases} \tilde{u}_{tt}=\tilde{u}_{xx}, t>0, -\infty<x<\infty,\\ \tilde{u}(0,x)=\tilde{f}(x), -\infty<x<0,\\ \partial_t\tilde{u}(0,x)=0, -\infty<x<\infty, \end{cases}$$

The solution to the model problem is determined by the d'Alembert formula

$$\tilde{u}(t,x)=\frac{\tilde{f}(x-t)+\tilde{f}(x+t)}{2}.$$

Thus, the solution to problem (21) - (22) has the form

$$u(t,x)=J_{1x}\left[\frac{\tilde{f}(x-t)+\tilde{f}(x+t)}{2}\right].$$

In the expanded form, the aforeentioned formula is given in the work [18].

The work is supported by the Scientific Research by Educational Organizations in 2020-2022 Project under Grant No FSFS- 2020-0031 (075-03-2020-079/2).

References

1. I.I. Bavrin, O.E. Yaremko, Differ. Equ., **40**(8), 1149–1160 (2004) https://doi.org/10.1023/B:DIEQ.0000049832.37277.95

2. I.I. Bavrin, O.E. Yaremko, Doklady Mathematics, **68**(3), 371-375 (2003)
3. R. Carroll, *Transmutation and Operator Differential Equations* (North Holland, 1979)
4. R. Carroll, *Transmutation, Scattering Theory and Special Functions* (North-Holland, 1982)
5. R. Carroll, *Transmutation theory and applications* (North-Holland, 1985)
6. R. Courant, D. Hilbert, *Methods of Mathematical Physics, vol* II (Wiley) New York, 1962)
7. M.M. Dzrbashian, *Harmonic Analysis and Boundary Value Problems in the Complex Domain* (Birkhauser, 1993)
8. A. Fitouhi, I. Jebabli, E.L. Shishkina, S.M. Sitnik, EJDE, **130**, 1-27 (2018)
9. F.R. Gantmacher. *The Theory of Matrices* (Chelsea, New York, 1959)
10. V.V. Katrakhov, S.M. Sitnik, Doklades of the Russian Academy of Sciences, **337**(3), 307-311, (1994)
11. V.V. Kravchenko, S.M. Torba, CAOT, **9**(2), 379-429 (2015)
12. V.V. Kravchenko, V.A. Vicente-Benítez, Math. Meth. Appl. Sci., **43**(16), 9455-9486 (2020)
13. A.A. Kilbas, H.M. Srivastava, J.J. Truhillo, *Theory and Applicationsof Fractional Differential Equations* (Elsevier, 2006)
14. A.A. Kilbas, M. Saigo, *H - transforms. Theory and applications* (Chapman and Hall, CRC.2004)
15. N.H. Mahmoud, Transactions of the AMS, **352**(8), 3687–3706 (2000)
16. S.G. Samko, A.A. Kilbas, O.I. Marichev, *Integrals and Derivatives of Fractional Order and Some Their Applications* (Nauka i Tekhnika, Minsk, 1987) [in Russian]
17. E. Shishkina, S. Sitnik, *Transmutations, Singular and Fractional Differential Equations with Applications to Mathematical Physics*, (Elsevier, 2020)
18. S.M Sitnik, O. Yaremko, N. Yaremko, *Transmutation Operators and Applications* (Springer 2020)
19. Ta Li, Proc. AMS, **11**(2), 290–298 (1960)
20. B.M. Levitan, *The theory of generalized shift operators* (Nauka, Moscow, 1973)
21. O.E. Yaremko, JMAG, **9**(4), 594-603 (2013)

Excitation of Some Nitrogen-Like Ions: Calculations and Comparison with Measurements

O.B. Nadykto[1], *B.A.* Nadykto[2], *A.B.* Nadykto[1]

[1]Department of Applied Mathematics, Moscow State Technological University "STANKIN", RU-127055, Moscow, Russia
[2]Russian Federal Nuclear Center – All-Russian Scientific Research Institute of Experimental Physics (RFNC-VNIIEF), Sarov, Nizhny Novgorod Region, RU-607188, Russia

Abstract. Electronic energies of one-electron excited states $1s^22s^22p^2ns\ ^4P$, $1s^22s^22p^2ns\ ^2P$, $1s^22s^22p^2np\ ^4S$, $1s^22s^22p^2np\ ^2S$, $1s^22s^22p^2np\ ^4P$, $1s^22s^22p^2np\ ^2P$, $1s^22s^22p^2np\ ^4D$, $1s^22s^22p^2np\ ^2D$, $1s^22s^22p^2nd\ ^4P$, $1s^22s^22p^2nd\ ^2P$, $1s^22s^22p^2nd\ ^4D$, $1s^22s^22p^2nd\ ^2D$, $1s^22s^22p^2nd\ ^4F$, $1s^22s^22p^2nd\ ^2F$ of nitrogen-like ions with n=3-12 and Z=7-50 have been calculated. For all the aforementioned electronic configurations, fine-structure levels have been determined for states with different total momentum J. The comparison of theoretical results with the available experimental data shows that theory and measurements agree well. In particular, relative accuracy of 10^{-4} has been achieved in the electronic energy of N-like at Z > 10, while the typical deviation in the calculated fine-structure levels from experimental data is in the order of 10^{-2}

1 Introduction

Knowledge of the energy levels of multiply charged ions with large nuclear charge Z is critically important for solving multiple emerging problems of plasma physics and astrophysics. Obtaining accurate energy levels is not an easy task because it requires accurate simultaneous accounting for relativistic and radiation effects. In the literature, there exist relevant quantum-mechanical studies concerning atomic energy levels and other related properties, for example [1-11]. The studies are dedicated to both the general theoretical approach [1-5] and specific methods for calculating energy levels of multielectron ions with varying Z [6-11]. A large fraction of complexity in getting accurate energy level is related to the fact that the Schrödinger equation is nonrelativistic, and, thus, the theoretical description of ions with $Z>>1$ requires obtaining relativistic corrections. The Dirac's approach [3] is considered by many to be more suitable for spectroscopic calculations. In both scenarios, taking into account the radiation effects is needed.

The commonly used handbooks [12]- [13] contain a compilation of experimental data on the energy levels of multiply charged ions with variable Z. Due to the complexity of the spectroscopic measurements of such ions, the amount of existing experimental data is quite small. On the other hand, direct quantum-mechanical simulations have very large, sometimes prohibitively high, computational costs and encounter significant difficulties related with the use of a number of approximations. The modified Bohr model [14] allows obtaining a large amount of data on the energies of various configurations of multiply charged ions with varying Z at nearly zeros computational costs. In this work, the modified Bohr model [14] is used to obtain energy levels of multiply charged N-like ions. In particular, energies of one-electron excited states $1s^22s^22p^2ns\ ^4P$, $1s^22s^22p^2ns\ ^2P$, $1s^22s^22p^2np\ ^4S$, $1s^22s^22p^2np\ ^2S$, $1s^22s^22p^2np\ ^4P$, $1s^22s^22p^2np\ ^2P$, $1s^22s^22p^2np\ ^4D$, $1s^22s^22p^2np\ ^2D$, $1s^22s^22p^2nd\ ^4P$, $1s^22s^22p^2nd\ ^2P$, $1s^22s^22p^2nd\ ^4D$, $1s^22s^22p^2nd\ ^2D$, $1s^22s^22p^2nd\ ^4F$, $1s^22s^22p^2nd\ ^2F$ of N-like ions with n=3-12 and Z=7-50 have been calculated and compared with available experimental data.

2 Model and Method

As for the ground state of boron-like and carbon-like ions [15], [16], [17], the energy of the ground state of nitrogen-like ions, 1s22s22p3, is described assuming the formation of a hybrid state of all (five) 2s22p3 outer electrons with the identical wave function parameter. The interaction potential of 2s electrons and 2p electrons with internal 1s electrons is described in the same way as in lithium-like ions [18-22]. The parameter of the hybridized wave function for ground state is determined by minimizing the sum of electronic energies of outer electrons. In this way, we get accurate energies of the ground state of ions with varying Z. An empirical parameter in this model is the screening coefficient of the nuclear charge due to the interaction of electrons in the 2s22p3 configuration. Excited states 1s22s2p4 are also considered as hybridized states with the identical wave function parameter of outer electrons.

* e-mail: anadykto@gmail.com

3 Results and discussion

After the accurate theoretical description of hybridized energy states of carbon-like ions is achieved, we begin modeling of multiple one-electron excitations of N-like ions, where the outer electron is in the state with $n > 2$. The ionic core $1s^22s^22p^2$ yields multiple series of one-electron excited states of N-like ions $1s^22s^22p^2nl$ with n>2 and varying l=0,1,2,...,n-1. Different sets of similar series $1s^22s2p^3nl$, $1s^22p^4nl$ are realized in the case, when the carbon-like core is in excited $1s^22s2p^3$ or $1s^22p^4$ states. The parameters for calculating the energy of one-electron excited states of nitrogen-like ions are given in Table 1.

Table 1. Parameters of one-electron excited states of N-like ions

State	SIGM3	A_{13}	B_{13}	A_{23}	B_{23}
Third shell					
$1s^22s^22p^2(^3P)ns^4P$	0.0	0.8045	0.15	0.359	0.135
$1s^22s^22p^2(^3P)ns^2P$	0.0	0.8045	0.15	0.333	0.118
$1s^22s^22p^2(^3P)np^4P$	0.0	0.2455	-0.084	0.288	0.07
$1s^22s^22p^2(^3P)np^2P$	0.0	0.2455	-0.084	0.248	0.05
$1s^22s^22p^2(^3P)np^4D$	0.0	0.2455	-0.084	0.300	0.075
$1s^22s^22p^2(^3P)np^2D$	0.0	0.2455	-0.084	0.27	0.055
$1s^22s^22p^2(^3P)np^4S$	0.0	0.2455	-0.084	0.252	0.075
$1s^22s^22p^2(^3P)np^2S$	0.0	0.2455	-0.084	0.322	0.081
$1s^22s^22p^2(^3P)nd^4P$	0.0	0.0105	-0.003	0.087	-0.055
$1s^22s^22p^2(^3P)nd^2P$	0.0	0.0105	-0.003	0.130	-0.105
$1s^22s^22p^2(^3P)nd^4F$	0.0	0.0105	-0.003	0.130	-0.075
$1s^22s^22p^2(^3P)nd^2F$	0.0	0.0105	-0.003	0.093	-0.065
$1s^22s^22p^2(^3P)nd^4D$	0.0	0.0105	-0.003	0.088	-0.045
$1s^22s^22p^2(^3P)nd^2D$	0.0	0.0105	-0.003	0.033	-0.045
$1s^22s2p^3(^5S)3s^4S$	0.0	0.8045	0.150	0.265	0.218
$1s^22s2p^3(^5S)3s^6S$	0.0	0.8045	0.150	0.35	0.130
$1s^22s2p^3(^5S)3p^4P$	0.0	0.2455	-0.084	0.22	0.103
$1s^22s2p^3(^5S)3p^6P$	0.0	0.2455	-0.084	0.272	0.103
$1s^22s2p^3(^5S)3d^4D$	0.0	0.0105	-0.003	0.004	-0.022
$1s^22s2p^3(^5S)3d^6D$	0.0	0.0105	-0.003	0.096	-0.035

4 Energy levels of $1s^22s^22p^2(^3P)ns$ ^4P states of N-like ions.

The constants of interaction of 2s and 2p electrons with 1s electrons are same as in the states of lithium-like ions $1s^22s$ ^2S and $1s^22p$ ^2P, respectively, or in $1s^22s^22p^2$ ^3P states of carbon-like ions. Electrons in ns states interacts with the $1s^2$ shell in the same way as in lithium-like ions. For ns states, the parameters of interaction of electrons of the second and third shells were chosen. The calculated values of the energy $1s^22s^22p^2(^3P)3s$ $^4P_{5/2}$ – $1s^22s^22p^3$ $^4S_{3/2}$ and fine splitting $1s^22s^22p^23s$ $^4P_{5/2}$ – $1s^22s^22p^23s$ $^4P_{1/2}$ for ions with Z = 7-20 are shown in Table. 2. The same table shows a comparison of the calculation with the experimental data the handbooks [12]- [13]. As seen from the comparison, the theoretical and measured values of electronic energies and fine splitting are in good agreement. The fine splitting 1s22s22p23s 4P5/2 –1s22s22p23s 4P1/2 is equal to the fine splitting $1s^22s^22p^2$ $^3P_2$ –$1s^22s^22p^2$ $^3P_0$ calculated taking into account the effect of the third shell electron on the energy levels of the second shell. As the principal quantum number n of the third shell electron increases, $1s^22s^22p^2ns$ $^4P_{5/2}$ –$1s^22s^22p^2ns$ $^4P_{1/2}$ increases, approaching the value $1s^22s^22p^2$ $^3P_2$ –$1s^22s^22p^2$ $^3P_0$ for carbon-like ions. The comparison of data shows that $1s^22s^22p^2ns$ $^4P_{5/2}$ –$1s^22s^22p^2ns$ $^4P_{1/2}$ at Z = 18 was probably measured with a large error.

Table 2. Comparison with experiment of the calculated energies (in eV) of $1s^22s^22p^2(^3P)3s$ $^4P_{5/2}$ states of N-like ions and fine splitting $1s^22s^22p^23s$ $^4P_{5/2}$ – $1s^22s^22p^23s$ $^4P_{1/2}$

Z	$1s^22s^22p^2(^3P)3s$ $^4P_{5/2}$– $1s^22s^22p^3$ $^4S_{3/2}$			$1s^22s^22p^23s$ $^4P_{5/2}$–$1s^22s^22p^23s$ $^4P_{1/2}$		
	E_{th}	E_{exp}	$E_{th} - E_{exp}$	E_{th}	E_{exp}	$E_{th} - E_{exp}$
7	10.2754	10.3362	-0.06073	0.01335	0.00998	0.00337
8	22.9071	22.9992	-0.09210	0.03171	0.03271	-0.00101
9	39.3130	39.3328	-0.01984	0.06472	0.06549	-0.00077
10	59.4422	59.4696	-0.02738	0.11870	0.11779	0.00091
11	83.4252	83.4119	0.01329	0.20102	0.20098	0.00004
12	111.176	111.146	0.03021	0.32021	0.30996	0.01025
13	142.709	142.697	0.01227	0.48589	0.47362	0.01226
14	178.048	178.058	-0.01004	0.70881	0.69556	0.01326
15	217.199	217.203	-0.00494	1.00091	0.97328	0.02762
16	260.143	260.178	-0.03502	1.37523	1.36384	0.01139
17	306.913	306.955	-0.04234	1.84603		
18	357.497	357.611	-0.11390	2.42873	2.14494	0.28379
19	411.889			3.13996		
20	470.116			3.99758		

Comparison with experiment of the calculated energies of $1s^22s^22p^2(^3P)3p$ $^4D_{7/2}$, $1s^22s^22p^2(^3P)3p$ $^4P_{5/2}$ states of N-like ions and fine splitting $1s^22s^22p^23p$ $^4D_{7/2}$ – $1s^22s^22p^23p$ $^4D_{1/2}$ and $1s^22s^22p^23p$ $^4P_{5/2}$ – $1s^22s^22p^23p$ $^4P_{1/2}$ values is given in Tables 3 and 4. The fine splitting $1s^22s^22p^23p$ $^4D_{7/2}$ – $1s^22s^22p^23p$ $^4D_{1/2}$ is equal to the $1s^22s^22p^23p$ $^4P_{5/2}$ – $1s^22s^22p^23p$ $^4P_{1/2}$ calculated taking into account the effect of the third shell electron on the state on the second electronic shell. Its slight difference from $1s^22s^22p^23s$ $^4P_{5/2}$ – $1s^22s^22p^23s$ $^4P_{1/2}$ is explained by the different influence of 3p and 3s electrons on the states of $22s^22p^2$ electrons. The fine splitting $1s^22s^22p^23p$ $^4P_{5/2}$ – $1s^22s^22p^23p$ $^4P_{1/2}$ value is equal to half of $1s^22s^22p^2$ $^3P_2$ – $1s^22s^22p^2$ $^3P_0$

calculated with accounting for the effect of the third shell electron on the second shell.

Table 3. Comparison with experiments of the calculated energies (in eV) of $1s^22s^22p^2(^3P)3p\ ^4D_{7/2}$ states of nitrogen-like ions and fine splitting $1s^22s^22p^23p\ ^4D_{7/2} - 1s^22s^22p^23p\ ^4D_{1/2}$

Z	$1s^22s^22p^2(^3P)3p\ ^4D_{7/2}- 1s^22s^22p^3\ ^4S_{3/2}$			$1s^22s^22p^23p\ ^4D_{7/2}-1s^22s^22p^23p\ ^4D_{1/2}$		
	E_{th}	E_{exp}	$E_{th} - E_{exp}$	E_{th}	E_{exp}	$E_{th} - E_{exp}$
7	11.7395	11.7639	-0.02447	0.01374	0.01375	-0.00002
8	25.6715	25.6652	0.00628	0.03270	0.03369	-0.00099
9	43.3078	43.3036	0.004197	0.06657	0.06987	-0.00329
10	64.6054			0.12165		
11	89.7067			0.20534		
12	118.533			0.32618		
13	151.100			0.49382		
14	187.432			0.71906		
15	227.531			1.01383		
16	271.376			1.39123		
17	318.993			1.86552		
18	370.363			2.45217		
19	425.474			3.16782		
20	484.344			4.03036		

Table 4. Comparison with experiments of the calculated energies (in eV) of $1s^22s^22p^2(^3P)3p\ ^4P_{5/2}$ states of nitrogen-like ions and fine splitting $1s^22s^22p^23p\ ^4P_{5/2} - 1s^22s^22p^23p\ ^4P_{1/2}$

Z	$1s^22s^22p^2(^3P)3p\ ^4P_{5/2}- 1s^22s^22p^3\ ^4S_{3/2}$			$1s^22s^22p^23p\ ^4P_{5/2}-1s^22s^22p^23p\ ^4P_{1/2}$		
	E_{th}	E_{exp}	$E_{th} - E_{exp}$	E_{th}	E_{exp}	$E_{th} - E_{exp}$
7	11.8377	11.8447	-0.00697	0.00689	0.00703	-0.00014
8	25.8834	25.8490	0.03446	0.01641	0.01712	-0.00071
9	43.6356	43.5829	0.05272	0.03340	0.03509	-0.00168
10	65.0483	65.0943	-0.04608	0.06102	0.07761	-0.01659
11	90.2631			0.10298		
12	119.201			0.16353		
13	151.880			0.24753		
14	188.321			0.36035		
15	228.531			0.50798		
16	272.485			0.69696		
17	320.210			0.93444		
18	371.688			1.22814		
19	426.908			1.58640		
20	485.886			2.01816		

Comparison of the calculated energies of $1s^22s^22p^2(^3P)3p\ ^2S_{1/2}$ and $1s^22s^22p^2(^3P)3p\ ^4S_{3/2}$ states of nitrogen-like ions with experiment data is shown in Table 5. Experimental data [12]-[13] are available for ions with Z = 7-9 only. As seen from the comparison, the relative deviation of the theory from measurements is in the order of 10^{-4} - 10^{-3}.

Table 5. Comparison with experiments of the calculated energies (in eV) of $1s^22s^22p^2(^3P)3p\ ^4P_{5/2}$ states of nitrogen-like ions and fine splitting $1s^22s^22p^23p\ ^4P_{5/2} - 1s^22s^22p^23p\ ^4P_{1/2}$

Z	$1s^22s^22p^23p\ ^2S_{1/2}- 1s^22s^22p^3\ ^4S_{3/2}$			$1s^22s^22p^23p\ ^4S_{3/2}- 1s^22s^22p^3\ ^4S_{3/2}$		
	E_{th}	E_{exp}	$E_{th} - E_{exp}$	E_{th}	E_{exp}	$E_{th} - E_{exp}$
7	11.5665	11.6027	-0.03619	11.9957	11.9958	-0.00011
8	25.2895	25.2858	0.00371	26.3050	26.3049	0.00001
9	42.7118	42.7053	0.00658	44.3600	44.3219	0.03817
10	63.7976			66.0871	66.0814	0.00572
11	88.6902			91.6207		
12	117.310			120.879		
13	149.674			153.879		
14	185.803			190.641		
15	225.701			231.171		
16	269.346			275.446		
17	316.763			323.491		
18	367.934			375.290		
19	422.847			430.830		
20	481.519			490.129		

Comparison of the calculated energies of $1s^22s^22p^2(^3P)3p\ ^2D_{5/2}$ states of nitrogen-like ions and fine splitting $1s^22s^22p^23p\ ^2D_{5/2}$ – $1s^22s^22p^23p\ ^2D_{3/2}$ with experimental data is given in Table 6. Fine splitting $1s^22s^22p^23p\ ^2D_{5/2}$ – $1s^22s^22p^23p\ ^2D_{3/2}$ is equal to 2/3 of $1s^22s^22p^2\ ^3P_2$ –$1s^22s^22p^2\ ^3P_0$ calculated taking into account the effect of the third shell electron on the electrons of the second shell.

Table 6. Comparison with experiments of the calculated energies (in eV) of $1s^22s^22p^2(^3P)3p\ ^2D_{5/2}$ states of nitrogen-like ions and fine splitting $1s^22s^22p^23p\ ^2D_{5/2}$ – $1s^22s^22p^23p\ ^2D_{3/2}$

Z	$1s^22s^22p^2(^3P)3p\ ^2D_{5/2}$– $1s^22s^22p^3\ ^4S_{3/2}$			$1s^22s^22p^23p\ ^2D_{5/2}$–$1s^22s^22p^23p\ ^2D_{3/2}$		
	E_{th}	E_{exp}	$E_{th} – E_{exp}$	E_{th}	E_{exp}	$E_{th} – E_{exp}$
7	12.0203	12.0097	0.01063	0.00920	0.00947	-0.00026
8	26.2501	26.2492	0.00088	0.02194	0.01124	0.01070
9	44.1840	44.1846	-0.00063	0.04467	0.04840	-0.00373
10	65.7740			0.08159		
11	91.1626			0.13767		
12	120.271			0.21858		
13	153.119			0.33078		
14	189.727			0.48147		
15	230.102			0.67862		
16	274.221			0.93096		
17	322.110			1.24802		
18	373.752			1.64011		
19	429.134			2.11834		
20	488.275			2.69464		

A comparison of the calculated energies of $1s^22s^22p^2(^3P)3p\ ^2P_{3/2}$ states of nitrogen-like ions and fine splitting $1s^22s^22p^23p\ ^2P_{3/2}$ – $1s^22s^22p^23p\ ^2P_{1/2}$ with measurements is given in Table 7. The fine splitting $1s^22s^22p^23p\ ^2P_{3/2}$ – $1s^22s^22p^23p\ ^2P_{1/2}$ is 1/3 of $1s^22s^22p^2\ ^3P_2$ –$1s^22s^22p^2\ ^3P_0$ calculated taking into account the effect of the third shell electron on the electrons of the second shell.

Table 7. Comparison of the calculated energies (in eV) of $1s^22s^22p^2(^3P)3p\ ^2P_{3/2}$ states of nitrogen-like ions and fine splitting $1s^22s^22p^23p\ ^2P_{3/2}$ – $1s^22s^22p^23p\ ^2P_{1/2}$ with experiments.

Z	$1s^22s^22p^2(^3P)3p\ ^2P_{3/2}$– $1s^22s^22p^3\ ^4S_{3/2}$			$1s^22s^22p^23p\ ^2P_{3/2}$–$1s^22s^22p^23p\ ^2P_{1/2}$		
	E_{th}	E_{exp}	$E_{th} – E_{exp}$	E_{th}	E_{exp}	$E_{th} – E_{exp}$
7	12.1506	12.1265	0.02418	0.00461	0.00443	0.00019
8	26.5607	26.5613	-0.00057	0.01101	0.00741	0.00360
9	44.6888	44.6884	0.00040	0.02241	0.01077	0.01164
10	66.4770			0.04094		
11	92.0647			0.06906		
12	121.372			0.10964		
13	154.419			0.16588		
14	191.225			0.24140		
15	231.798			0.34019		
16	276.114			0.46661		
17	324.200			0.62543		
18	376.038			0.82181		
19	431.618			1.06132		
20	490.955			1.34991		

The energies of $1s^22s^22p^2(^3P)3d\ ^{4,2}F$, $1s^22s^22p^2(^3P)3d\ ^{4,2}D$, $1s^22s^22p^2(^3P)3d\ ^4P$ states of nitrogen-like ions were calculated. The constants of interaction of 2s and 2p electrons with 1s electrons were the same as in the case of the states of lithium-like ions $1s^22s\ ^2S$ and $1s^22p\ ^2P$, respectively or $1s^22s^22p^2\ ^3P$ states of carbon-like ions. The electron in nd states interacts with the $1s^2$ shell in the same way as in lithium-like ions. For nd states, the parameters of interaction of electrons of the second and third shells were chosen. The calculated values of the energy 1s22s22p2 (3P) 3d 4F9 / 2– 11$s^22s^22p^2\ ^3P$ s22s22p3 4S3 / 2 and fine splitting 1s22s22p23d 4F9 / 2–1s22s22p23d 4F3 / 2 for ions with Z = 7-20 are given in Table. 8. The same table shows a comparison of the calculated values with experimental data taken from the handbooks [12], [13]. The calculated values are in good agreement with the experiment. The fine splitting $1s^22s^22p^23d\ ^4F_{9/2}$–$1s^22s^22p^23d\ ^4F_{3/2}$ is equal to the fine splitting $1s^22s^22p^2\ ^3P_2$ –$1s^22s^22p^2\ ^3P_0$ calculated with accounting for the impact of the third shell electron on the second shell electrons. Its slight difference from $1s^22s^22p^23s\ ^4P_{5/2}$ –$1s^22s^22p^23s\ ^4P_{1/2}$ is explained by the different influence of 3d and 3s electrons on the state of $2s^22p^2$ electrons.

Table 8. Comparison of the calculated energies (in eV) of $1s^22s^22p^2(^3P)3d\ ^4F_{9/2}$ states of nitrogen-like ions and fine splitting $1s^22s^22p^23d\ ^4F_{9/2}$–$1s^22s^22p^23d\ F_{3/2}$ with experiments.

Z	$1s^22s^22p^2(^3P)3d\ ^4F_{9/2}$– $1s^22s^22p^3\ ^4S_{3/2}$			$1s^22s^22p^23d\ ^4F_{9/2}$–$1s^22s^22p^23d\ ^4F_{3/2}$		
	E_{th}	E_{exp}	$E_{th}-E_{exp}$	E_{th}	E_{exp}	$E_{th}-E_{exp}$
7	13.1289	12.9896	0.13938	0.01401	0.01265	0.00136
8	28.7120	28.6939	0.01808	0.03357	0.01664	0.01693
9	48.0746	48.0722	0.00240	0.06844	0.05805	0.01039
10	71.1253			0.12498		
11	97.9919			0.21067		
12	128.591			0.33412		
13	162.941			0.50505		
14	201.064			0.73432		
15	242.968			1.03395		
16	288.635			1.41711		
17	338.093			1.89813		
18	391.329			2.49255		
19	448.335			3.21711		
20	509.133			4.08975		

Comparison with experiments of the calculated energies of $1s^22s^22p^2(^3P)3d\ ^4D_{7/2}$– $1s^22s^22p^3\ ^4S_{3/2}$, $1s^22s^22p^2(^3P)3d\ ^4P_{5/2}$– $1s^22s^22p^3\ ^4S_{3/2}$ states of nitrogen-like ions and fine splitting $1s^22s^22p^23d\ ^4D_{7/2}$–$1s^22s^22p^23d\ ^4D_{1/2}$ and $1s^22s^22p^23d\ ^4P_{5/2}$ – $1s^22s^22p^23d\ ^4P_{1/2}$ is given in Tables 9 and 10. The fine splitting interval $1s^22s^22p^23d\ ^4D_{7/2}$–$1s^22s^22p^23d\ ^4D_{1/2}$ is 0.33 of the fine splitting calculated with accounting for the impact of the third shell electron on the second shell electrons. The fine splitting $1s^22s^22p^23d\ ^4P_{5/2}$ – $1s^22s^22p^23d\ ^4P_{1/2}$ is equal to half of the fine splitting $1s^22s^22p^2\ ^3P_2$–$1s^22s^22p^2\ ^3P_0$.

Table 9. Comparison of the calculated energies (in eV) of $1s^22s^22p^2(^3P)3d\ ^4D_{7/2}$– $1s^22s^22p^3\ ^4S_{3/2}$ and $1s^22s^22p^3\ ^4S_{3/2}$ states of nitrogen-like ions and fine splitting $1s^22s^22p^23d\ ^4D_{7/2}$–$1s^22s^22p^23d\ ^4D_{1/2}$ with experiments.

Z	$1s^22s^22p^2(^3P)3d\ ^4D_{7/2}$– $1s^22s^22p^3\ ^4S_{3/2}$			$1s^22s^22p^23d\ ^4D_{7/2}$–$1s^22s^22p^23d\ ^4D_{1/2}$		
	E_{th}	E_{exp}	$E_{th}-E_{exp}$	E_{th}	E_{exp}	$E_{th}-E_{exp}$
7	13.1004	13.0209	0.07945	0.00467	0.00409	0.00058
8	28.8158	28.8580	-0.04222	0.01120	0.00523	0.00597
9	48.3931	48.3801	0.01301	0.02285	0.01116	0.01169
10	71.7037			0.04175		
11	98.8576			0.07040		
12	129.762			0.11167		
13	164.429			0.16880		
14	202.878			0.24544		
15	245.115			0.34558		
16	291.119			0.47362		
17	340.917			0.63436		
18	394.496			0.83296		
19	451.847			1.07502		
20	512.992			1.36653		

5 Conclusion

Electronic energies of one-electron excited states $1s^22s^22p^2ns\ ^4P$, $1s^22s^22p^2ns\ ^2P$, $1s^22s^22p^2np\ ^4S$, $1s^22s^22p^2np\ ^2S$, $1s^22s^22p^2np\ ^4P$, $1s^22s^22p^2np\ ^2P$, $1s^22s^22p^2np\ ^4D$, $1s^22s^22p^2np\ ^2D$, $1s^22s^22p^2nd\ ^4P$, $1s^22s^22p^2nd\ ^2P$, $1s^22s^22p^2nd\ ^4D$, $1s^22s^22p^2nd\ ^2D$, $1s^22s^22p^2nd\ ^4F$, $1s^22s^22p^2nd\ ^2F$ of nitrogen-like ions with n=3-12 and Z=7-50 have been calculated. For all the aforementioned electronic configurations, fine-structure levels have been determined for states with different total momentum J. The comparison of theoretical results with the available experimental data shows that theory and measurements agree well. In particular, relative accuracy of 10^{-4} has been achieved in the electronic energy of N-like at Z > 10, while the typical deviation in the calculated fine-structure levels from experimental data is in the order of 10^{-2}.

The dependences of the fine structure on the nuclear charge Z and the principal quantum number n for a number of configurations of nitrogen-like ions have been determined. The fine splitting $1s^22s^22p^23s\ ^4P_{5/2}$ – $1s^22s^22p^23s\ ^4P_{1/2}$ of the energy levels of N-like ions is equal to the fine splitting of $1s^22s^22p^2\ ^3P_2$–$1s^22s^22p^2\ ^3P_0$ of the energy levels of carbon- like ions calculated taking into account the influence of the third shell electron on the state of the second shell electrons. With the increase in the principal quantum number n of the third shell electron, the fine splitting $1s^22s^22p^2ns\ ^4P_{5/2}$ – $1s^22s^22p^2ns\ ^4P_{1/2}$ slightly increases, reaching the value of the fine splitting $1s^22s^22p^2\ ^3P_2$ –$1s^22s^22p^2\ ^3P_0$ for carbon-like ions. The fine splitting $1s^22s^22p^23p\ ^4P_{5/2}$ –

* e-mail: anadykto@gmail.com

$1s^22s^22p^23p\ ^4P_{1/2}$ is equal to 0.5 of the fine splitting $1s^22s^22p^2\ ^3P_2$ – $1s^22s^22p^2\ ^3P_0$. The fine splitting $1s^22s^22p^23p\ ^2P_{3/2}$ – $1s^22s^22p^23p\ ^2P_{1/2}$ is equal to the fine splitting $1s^22s^22p^2\ ^3P_2$ –$1s^22s^22p^2\ ^3P_0$. The fine splitting $1s^22s^22p^23p\ ^2D_{5/2}$ – $1s^22s^22p^23p\ ^2D_{3/2}$ is equal to 2/3 of $1s^22s^22p^2\ ^3P_2$ –$1s^22s^22p^2\ ^3P_0$. The fine splitting $1s^22s^22p^23p\ ^2P_{3/2}$ – $1s^22s^22p^23p\ ^2P_{1/2}$ is equal to 0.33 of the fine splitting $1s^22s^22p^2\ ^3P_2$ –$1s^22s^22p^2\ ^3P_0$. The fine splitting $1s^22s^22p^23d\ ^4F_{9/2}$–$1s^22s^22p^23d\ ^4F_{3/2}$ is equal to the fine splitting $1s^22s^22p^2\ ^3P_2$ –$1s^22s^22p^2\ ^3P_0$.Its slight difference from the $1s^22s^22p^23s\ ^4P_{5/2}$ —$1s^22s^22p^23s\ ^4P_{1/2}$ is explained by the different influence of 3d and 3s electrons on the state of $2s^22p^2$ electrons. The fine splitting $1s^22s^22p^23d\ ^4D_{7/2}$–$1s^22s^22p^23d\ ^4D_{1/2}$ is equal to 0.33 of the fine splitting $1s^22s^22p^2\ ^3P_2$ –$1s^22s^22p^2\ ^3P_0$. The fine splitting $1s^22s^22p^23d\ ^4P_{5/2}$ – $1s^22s^22p^23d\ ^4P_{1/2}$ is equal to 0.5 of the fine splitting $1s^22s^22p^2\ ^3P_2$ – $1s^22s^22p^2\ ^3P_0$. The fine splitting $1s^22s^22p^23d\ ^2D_{5/2}$–$1s^22s^22p^23d\ ^2D_{3/2}$ is equal to 0.2 of the fine splitting $1s^22s^22p^2\ ^3P_2$ –$1s^22s^22p^2\ ^3P_0$. In the case of the fine splitting of $1s^22s^22p^23d\ ^2P$, the energy $1s^22s^22p^23d\ ^2P_{3/2}$ is lower than that of $1s^22s^22p^23d\ ^2P_{1/2}$ and fine splitting $1s^22s^22p^23d\ ^2P_{1/2}$ –$1s^22s^22p^23d\ ^2P_{3/2}$ is equal to 0.33 of $1s^22s^22p\ ^2P_{3/2}$– $1s^22s^22p\ ^2P_{1/2}$.

Experimental data for the energy levels of outer electrons in N-like ions with the principal quantum number n> 3 are very limited. However, taken together, they confirm the correctness of the proposed theoretical approach, which was validated in more detail at n = 3. In many cases, the fine splitting of N-like ions is associated with the splitting of the energy of the $1s^22s^22p^2$ state of carbon-like ions and does not decrease with increasing n. Instead, it slightly increases, approaching the value of the fine splitting $1s^22s^22p^2\ ^3P_2$ –$1s^22s^22p^2\ ^3P_0$ for carbon-like ions. The fine splitting of the energy levels of nl electrons is small compared to the interval $1s^22s^22p^2\ ^3P_2$ –$1s^22s^22p^2\ ^3P_0$ and decreases sharply with increasing n.

Acknowledgments

This work was partially supported by the RSF project No. 18-11-00247. This work was carried out using equipment provided by the Center of Collective Use of MSUT "STANKIN".

References

1. L.D. Landau, E.M. Lifshitz, *Quantum mechanics: non-relativistic theory (Vol. 3)* (Pergamon Press Ltd. 1977)
ISBN:0-08-020940-8
2. H.A. Bethe, E.E. Salpeter, *Quantum mechanics of one-and two-electron atoms* (Springer Science & Business Media, 2012)
ISBN-13: 978-0-306-20022-9
3. P.A.M. Dirac, *The principles of quantum mechanics (No. 27)* (Oxford University Press, 1981)
ISBN: 0-19-852011-5
4. E.U. Condon, E.U. Condon, G.H. Shortley, *The theory of atomic spectra.* (Cambridge University Press., 1951)
ISBN: 0-521-09209-4
5. I.I. Sobelman. *Introduction to the theory of atomic spectra* (Moscow: Nauka, 1977) [in Russian]
6. C. Froese-Fischer, *The Hartree-Fock method for atoms: a numerical approach* (Wiley, 1977)
7. R.D. Cowan, *The theory of atomic structure and spectra (No. 3)* (Univ of California Press., 1981)
ISBN: 978-0-520-03821-9
8. A.A. Nikitin, Z.B. Rudzikas. *Fundamentals of the theory of spectra of atoms and ions* (Moscow: Nauka, 1983) [in Russian]
9. U.I. Safronova, V.S. Senashenko, *Theory of spectra of multiply charged ions* (Moscow: Energoatomizdat, 1984) [in Russian]
10. M.A. Brown, A.D. Gurchumelia, W.I. Safronov. *Relativistic theory of the atom* (Moscow: Nauka, 1984) [in Russian]
11. L.A. Weinstein, V.P. Shevelko, *Structure and characteristics of ions in hot plasma* (Moscow: Nauka, 1986) [in Russian]
12. G.I. Tachiev, C.F. Fischer, Astronomy & Astrophysics, **385**(2), 716-723 (2002)
DOI: 10.1051/0004-6361:20011816
13. L. Radžiūtė, J. Ekman, P. Jönsson, G. Gaigalas, Astronomy & Astrophysics, **582**(A61), 6 (2015)
http://dx.doi.org/10.1051/0004-6361/201526708
14. E. Träbert, Atoms, **2**(1), 15-85 (2014)
doi:10.3390/atoms2010015
15. K. Wang, P. Jönsson, J. Ekman, R. Si, Z.B. Chen, Y.G. Li, ... & J. Yan, Journal of Quantitative Spectroscopy and Radiative Transfer, **194**, 108-112 (2017)
https://doi.org/10.1016/j.jqsrt.2017.03.014
16. K. Wang, S. Li, P. Jönsson, N. Fu, W. Dang, X.L. Guo, ... & R. Si, Journal of Quantitative Spectroscopy and Radiative Transfer, **187**, 375-402 (2017)
https://doi.org/10.1016/j.jqsrt.2016.10.011
17. K. Wang, R. Si, W. Dang, P. Jönsson, X.L. Guo, S. Li, ... & D.F. Li, The Astrophysical Journal Supplement Series, **223**(1), 3 (2016)
http://dx.doi.org/10.3847/0067-0049/223/1/3
18. C.E. Moore, *Atomic Energy Levels, Circular of the National Bureau of Standards* (Superintendent of Documents, US Government Printing Office, Washington, 25, 1949)
19. S. Bashkin, J.O. Stoner, *Atomic energy levels and Grotrian diagrams: Hydrogen I-phosphorus XV* (North-Holland Publ. Comp., 1975)
ISBN: 0-7204-0322-7
20. B.A. Nadykto, Physics-Uspekhi, **36**(9), 794 (1993) [in Russian]
https://doi.org/10.1070/PU1993v036n09ABEH002307

21. B.A. Nadykto, O.B. Nadykto, Questions of atomic science and technology. Series Theoretical and Applied Physics, **1**, 3-12 (2007) [in Russian]
22. B.A. Nadykto, O.B. Nadykto, Questions of atomic science and technology. Series Theoretical and Applied Physics, **1**, 13-50 (2007) [in Russian]
23. B.A. Nadykto, O.B. Nadykto, Questions of atomic science and technology. Series Theoretical and Applied Physics, **1**, 51-99 (2007) [in Russian]

 https://doi.org/10.1051/epjconf/202124801021

Mathematical Modeling of Working Memory in the Presence of Random Disturbance using Neural Field Equations

Pedro M. Lima[1,*], *Wolfram* Erlhagen[2,**], *Gennady Yu.* Kulikov[1,***], and *Maria V.* Kulikova[1,****]

[1]Centro de Matemática Computacional e Estocástica, Instituto Superior Técnico, University of Lisbon, PR-1049 001 Lisboa, Portugal
[2]Centro de Matemática, Universidade do Minho, PR-4710-057, Guimarães, Portugal

Abstract. In this paper, we describe a neural field model which explains how a population of cortical neurons may encode in its firing pattern simultaneously the nature and time of sequential stimulus events. Moreover, we investigate how noise-induced perturbations may affect the coding process. This is obtained by means of a two-dimensional neural field equation, where one dimension repre- sents the nature of the event (for example, the color of a light signal) and the other represents the moment when the signal has occurred. The additive noise is represented by a Q-Wiener process. Some numerical experiments reported are carried out using a computational algorithm for two-dimensional stochastic neural field equations.

1 Introduction

In recent years, significant progress has been made in understanding the brain electrodynamics using mathematical techniques. Neural field models represent the large-scale dynamics of spatially structured networks of neurons in terms of nonlinear integro-differential equations. Such models are becoming increasingly important for the interpretation of experimental data, including those obtained from EEG, fMRI and optical imaging [2].

In Robotics, NFE are a powerful tool for modeling and analyzing the dynamic behavior of neural populations in cognitive tasks [4]. They explain the existence of stimulus-specific, persistent population activity which bridges gaps between sensation and action in order to mediate higher-order cognitive functions like working memory, planning and decision making. The neural activity is maintained in the absence of external stimuli due to strong excitatory and inhibitory feedback loops within the population. Since NFE are defined over continuous perceptual or motor variables such as orientation, color, or movement direction, the neural interactions can be characterized by their distance in feature space. Typically, neurons tuned to similar values of the continuous variable excite each other and those tuned to dissimilar values inhibit each other. The steady states that the field dynamics develops in response to a transiently presented cue are often called bump attractors since they represent bell-shaped activity patterns in metric space. Due to the assumed translation-invariant connection structure, the network can hold a continuous family of bumps, with each of the attractor states encoding a specific stimulus value. However, since bumps are neutrally stable, their positions in the continuous attractor manifold can be shifted by random noise, potentially leading to a deterioration of the encoded information over time (see, for example, [10]).

In numerical simulations, we test the robustness of a dynamic neural field model of serial order in the presence of additive noise in an experiment in which a sequence of color cues is presented. The two-dimensional field model implements the idea that the serial order of stimulus events can be represented in terms of their location along the feature dimension and the temporal dimension. The position of each individual bump of a multi-bump pattern encodes the memory of a specific color cue presented at a specific time after the sequence onset. Importantly, the joint cue-timing representation allows us to address the problem of cue repetitions which has been identified as a major challenge for models of serial order [7]. A deterministic model describing this process is presented in [16], where numerical simulations are also reported. Here, we investigate how noise-induced perturbations may affect the coding process.

Some numerical experiments reported are carried out using a computational algorithm for two-dimensional stochastic neural field equations. This numerical algorithm is MATLAB-based and presented in [14] in detail. The numerical results are discussed and their physical interpretation is explained.

1.1 Deterministic Neural Field Equation

The neural field equation was first introduced in the seminal work of Wilson and Cowan [23], as a model of spatiotemporal dependence of neural activity. Here, we apply the NFE proposed by Amari [1], as a biophysically mechanistic model of pattern formation in neural tissue:

* e-mail: plima@math.tenico.ulisboa.pt; **e-mail: wolfram.erlhagen@math.uminho.pt;
***e-mail: gennady.kulikov@tecnico.ulisboa.pt; ****e-mail: maria.kulikova@tecnico.ulisboa.pt

$$\frac{\partial}{\partial t}U(\mathbf{x},t) = I(\mathbf{x},t) - \alpha U(\mathbf{x},t) + + \int_{\Omega} F(\|\mathbf{x}-\mathbf{y}\|_2) S(U(\mathbf{y},t))d\mathbf{y}, \quad t \in [0,T], \quad \mathbf{x},\mathbf{y} \in \Omega \subset \mathbb{R}^2, \tag{1}$$

where

- U(x, t) (the unknown function) denotes the membrane potential in point X at time t;
- I(x, t) represents the external sources of excitation;
- α is the potential decay rate;
- S is the dependence between the firing rate of the neurons and their membrane potentials (sigmoidal or Heaviside function);
- F(x−y2) defines the coupling strength between any two neurons at the positions x and y which is assumed to depend on their distance, only. To support the existence of stable multi-bump solutions, the connectivity function is chosen to be of oscillatory type with several zero crossings. They define the distances where excitation or inhibition dominates [6, 15].

1.2 Working Memory Application

Working memory (WM) defined as the ability to actively retain stimulus information over short periods of time is crucial for cognitive control of behavior. Persistent, stimulus-selective activity observed in many brain areas is commonly believed to represent a neural substrate of WM. In neural field models, WM is implemented in a bi-stable regime of the field dynamics in which a brief localized input may switch from a homogeneous resting state below threshold to a self-stabilized bump. Fig.1 shows an example of a 1D field representing the continuous dimension color. When two or more transient inputs are presented simultaneously or sequentially, a multi-item bump pattern stabilizes. Fig.2 illustrates the formation of a twobump pattern in response to two simultaneously presented inputs. Notice that the population activity decays slightly after cessation of the inputs but the recurrent interactions are strong enough to maintain the pattern above the threshold.

The 2D-field model of serial order of sequential events adds a temporal dimension (X) to the color dimension (Y)[1]. This choice is motivated by the experimental observation of neurons that are tuned to the temporal intervals and/or the ordinal structure of sequential tasks [20]. We use a traveling wave (assumed to be triggered by a sequence onset signal) as a second input to the field to integrate the information about the event timing relative to the sequence onset. Traveling waves which are observed in many cortical areas provide a subthreshold depolarization to individual neurons and increase their firing probability to the presentation of external stimuli [21]. Consisting with this view, we assume that only at field positions receiving the traveling wave and the color input simultaneously, the combined input is strong enough to trigger the evolution of a bump. A projection of the bump on the X and Y axes reveals bell-shaped activity profiles which are consistent with the notion of broadly tuned neurons in the feature (color) dimension and the time dimension [3, 20]. Figs. 3 and 4 illustrate, respectively, the localized input and the wave input to the 2D-field. If a signal of color Y occurs at a certain time, we have a ridge-like input which is localized in the feature space Y but extends in the time dimension X. The wave has the form of a ridge which extends in the Y direction and propagates in the direction of X with elapsed time t since the sequence onset occurred at t = 0. One bump evolves in response to the combined input (see Fig.5), which persists after all inputs have been switched off (see Fig. 6).

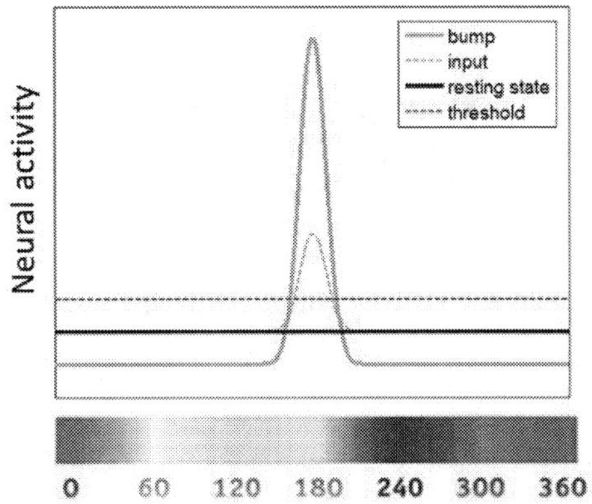

Figure 1. Stable bump triggered by a single color stimulus

1.3 Stochastic Models

There are several reasons why neural fields may be affected by noise:

- Irregularity of spikes;
- Non-homogeneous or irregular connectivity;

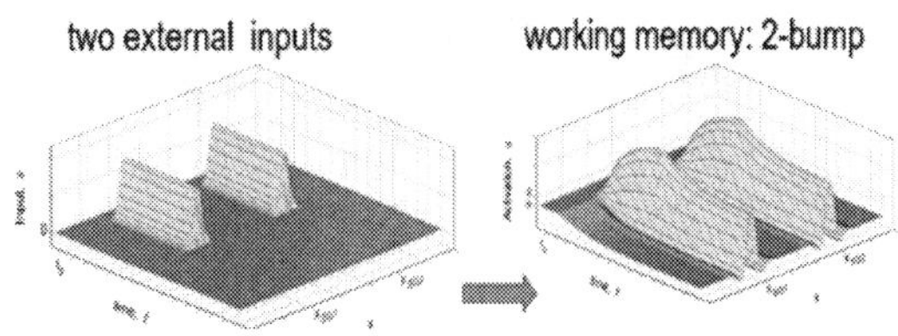

Figure 2. External stimuli and their representations in working memory

[1]When describing the working memory problems the coordinates are given by the vector x = (X, Y).

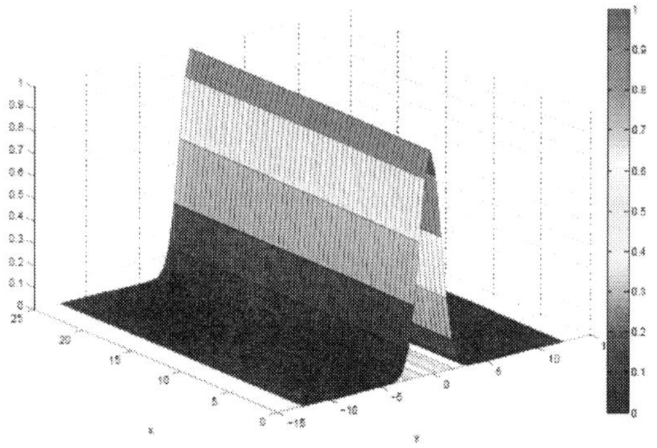

Figure 3. Surface graph of the input function corresponding to a light signal of the color $Y = 0$

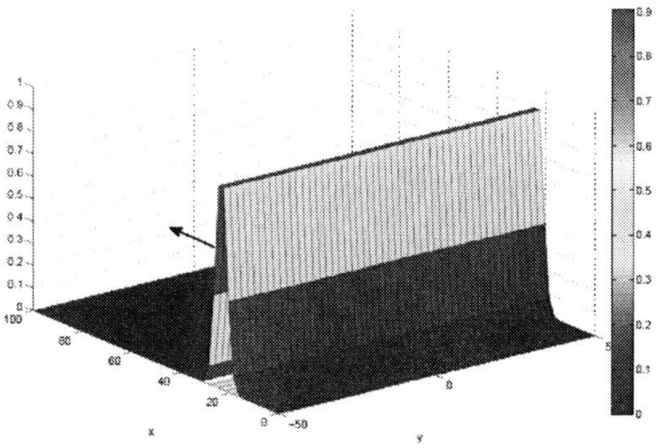

Figure 4. Input function corresponding to a traveling wave, moving in the direction of the X axis

- Perturbations of external stimulus;

By introducing stochastic models, we aim at answering questions as follows:

- Does noise interfere in the existence of stationary solutions?
- How far the stationary solutions can be modified as a result of noise?

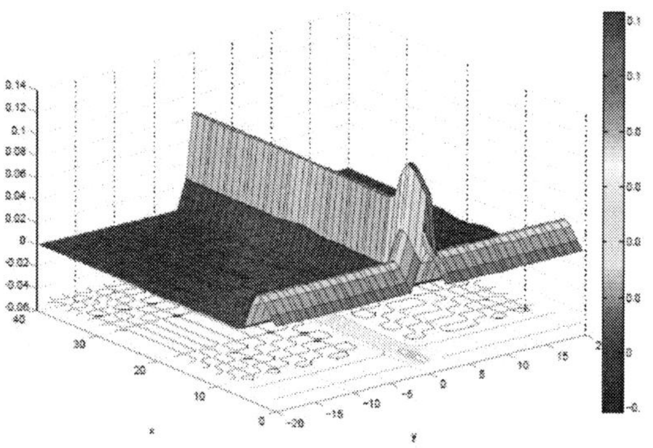

Figure 5. Combination of two inputs generating a bump state

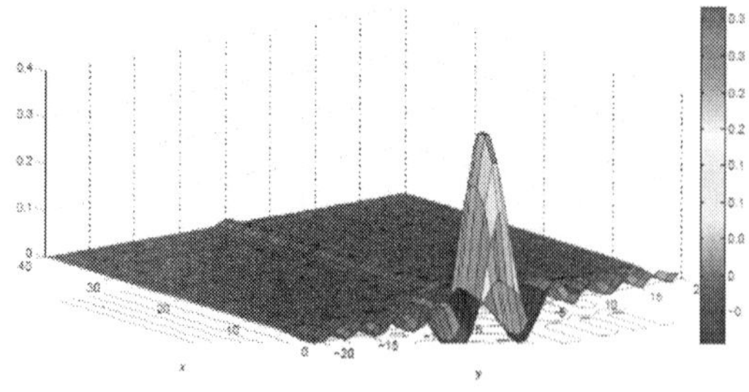

Figure 6. Example of a stable bump solution which remains after all the inputs have been switched off

- May a stationary solution be transformed into another one just as a result of noise? What is the probability of this event?

In [12], a stochastic version of the neural field equation with additive noise is introduced as follows:

$$dU(\mathbf{x}, t) = \left(I(\mathbf{x}, t) - \alpha U(\mathbf{x}, t) + \right.$$

$$\left. + \int_{\Omega} F(\|\mathbf{x} - \mathbf{y}\|_2) S(U(\mathbf{y}), t) d\mathbf{y} \right) dt + \epsilon dW(\mathbf{x}, t), \quad (2)$$

where $t \in [0, T]$, $x = (x_1, x_2) \in \Omega \subset R^2$, $W(x, t)$ is a Q-Wiener process. The initial condition

$U_0(x) = U(x, 0)$ is supposed to satisfy the periodic boundary conditions in space $x \in R^2$

2 Numerical Algorithm

Since NFE cannot in general be solved analytically, numerical methods for their approximation have been developed by many researchers. In [5], a computational method is introduced which applies quadrature rule in space to reduce the problem to a system of delay differential equations, which is then solved by a standard algorithm for this kind of equations. A more efficient approach is proposed in [8], [9], where the authors introduce a new algorithm to deal with the convolution kernel of the equation and use the Fast Fourier Transforms to reduce significantly the computational effort required by numerical integration. Recently, a new numerical method for the approximation of two-dimensional neural fields was introduced, with use of an implicit second order scheme for the integration in time and applying the Chebyshev interpolation to reduce the dimensions of the matrices [17]. Some applications of this algorithm to Neuroscience problems are discussed in [18].

In the stochastic case, the numerical approximation of NFE becomes a harder challenge.

Numerical algorithms for the stochastic NFE are proposed in [19] and [13].

The numerical integration scheme proposed in the present work is rooted in the Karhunen-Loève presentation in the following form:

$$U(\mathbf{x},t)=\sum_{k=0}^{\infty}\sum_{l=0}^{\infty}u_{kl}(t)v_{kl}(\mathbf{x}). \tag{3}$$

In series (3), the deterministic continuous-in-space factors $v_{kl}(x)$ defined in the square domain $[-L, L] \times [-L, L]$ of real numbers are the eigenfunctions of the covariance operator in the 2D-SNFE model (2) under examination. Here and below, these are chosen to satisfy the formula

$$v_{kl}(\mathbf{x}):=\begin{cases}\frac{1}{2L}, & \text{if } k=0,\ l=0,\\ \frac{1}{\sqrt{2}L}\cos(k\pi x_1/L), & \text{if } k\geq 1,\ l=0,\\ \frac{1}{\sqrt{2}L}\cos(l\pi x_2/L), & \text{if } k=0,\ l\geq 1,\\ \frac{1}{L}\cos(k\pi x_1/L)\cos(l\pi x_2/L), & \text{if } k\geq 1,\ l\geq 1.\end{cases} \tag{4}$$

We point out that functions (4) establish an orthonormal basis in the Hilbert space associated with the problem at hand.

Next, the uncorrelated random processes $u_{kl}(t)$ defined on the time interval [0, T] obey the following SDE:

$$du_{kl}(t) = \Big[\langle I(\mathbf{x},t),v_{kl}(\mathbf{x})\rangle - \alpha u_{kl}(t) + FS_{kl}(U(\mathbf{x},t))\Big]dt + \varepsilon\lambda_{kl}d\beta_{kl}(t) \tag{5}$$

whose nonlinear term refers to the double integral, that is

$$FS_{kl}(U(\mathbf{x},t)) := \int_{\Omega} v_{kl}(\mathbf{x})\left[\int_{\Omega} F(\|\mathbf{x}-\mathbf{y}\|_2) \times S\Big(\sum_{n=0}^{\infty}\sum_{m=0}^{\infty}u_{nm}(t)v_{nm}(\mathbf{y})\Big)d\mathbf{y}\right]d\mathbf{x}, \tag{6}$$

the inner product in SDE (5) enjoys the standard definition

$$\langle I(\mathbf{x},t),v_{kl}(\mathbf{x})\rangle := \int_{\Omega} I(\mathbf{x},t)v_{kl}(\mathbf{x})\,d\mathbf{x} \tag{7}$$

at any time t, and the scalars λ_{kl} and the stochastic process increments $d\beta_{kl}(t)$ come from the infinite series representation of the differential $dW(x, t)$ of the trace-class space-valued Q-Wiener process in 2D-SNFE (2). It is known that the mentioned representation has the following fashion:

$$dW(\mathbf{x},t)=\sum_{k=0}^{\infty}\sum_{l=0}^{\infty}v_{kl}(\mathbf{x})\lambda_{kl}d\beta_{kl}(t) \tag{8}$$

with the stochastic processes $\beta_{kl}(t)$ being the mutually independent Brownian motions with zero mean and unit variance and λ_{kl} standing for the eigenvalues of the covariance operator Q. The correlation in this noise obeys the formula

$$\mathrm{E}\{W(\mathbf{x},t)W(\mathbf{y},s)\} = \min\{t,s\}\frac{1}{2\xi}\exp\left(\frac{-\pi\|\mathbf{x}-\mathbf{y}\|_2^2}{4\xi^2}\right)$$

for any space points x, y $\in \Omega$ and time instants t, s > 0. Here and below, $\mathrm{E}\{\cdot\}$ refers to the expectation operator, and the fixed parameter ξ determines the spatial correlation length. Following [22], we impose the condition $\xi \ll 2L$ and, then, yield explicitly values of the eigenvalues λ_{kl} of the covariance operator Q in series (8). These are found to be:

$$\lambda_{kl} = \sqrt{\exp(-\xi^2(k^2+l^2)/(4\pi))}. \tag{9}$$

In what follows, the Galerkin-type approximation underlying our numerical scheme implies that the infinite series (3) is truncated to a finite summation formula. We choose a sufficiently large positive integer K and replace solution (3) to 2D-SNFE (2) with its approximation of the fashion

$$U_K(\mathbf{x},t) := \sum_{k=0}^{K}\sum_{l=0}^{K}u_{kl}(t)v_{kl}(\mathbf{x}) \tag{10}$$

where the deterministic functions $v_{kl}(x)$ are given by rule (4) and the stochastic processes $u_{kl}(t)$ satisfy SDE (5).

Taking into account the 2D-square-form of the domain $\mathbf{\Omega}$, we introduce then an equidistant 2D-space-mesh as follows:

$$\mathbf{x}_1 := \{x_1^0=-L, x_1^i = x_1^0 + ih_x, i=0,1,\dots,N, x_1^N = L\},$$
$$\mathbf{x}_2 := \{x_2^0=-L, x_2^j = x_2^0 + ih_x, i=0,1,\dots,N, x_2^N = L\}$$

with its step size $h_x = 2L/N$, where N is a user-supplied quantity of the subdivision steps fulfilled in each direction of the square $[-L, L] \times [-L, L]$ underlying 2D-SNFE (2). More formally, this mesh **X** consists of 2D-vectors in the domain Ω, that is X is the square matrix of size N+1 whose (i, j)th entry is the vector with the coordinates $x^i_1 = x^0_1 + ih_x$ $x^j_2 = x^0_2 + jh_x$. We recall that the number of summation terms in the approximate solution (10) and that of the discretization steps used in the domain Ω must obey the condition $N \gg K$, as explained in [14].

First, we substitute the Galerkin-type solution (10) into SDE (5). Second, we replace the continuous space-dependent functions in equations (5) and (6) with their values calculated at nodes of the mesh **X** and approximate the integrals arisen by means of the Trapezoidal Rule summations, as follows:

$$\langle I(\mathbf{x},t),v_{kl}(\mathbf{x})\rangle \approx h_x^2\sum_{i=0}^{N-1}\sum_{j=0}^{N-1}I(x_1^i,x_2^j,t)v_{kl}(x_1^i,x_2^j), \tag{11}$$

$$FS_{kl}(U(\mathbf{x},t)) \approx h_x^4 \sum_{i=0}^{N-1}\sum_{j=0}^{N-1} v_{kl}(x_1^i, x_2^j) \times \Big[\sum_{p=0}^{N-1}\sum_{q=0}^{N-1} F\Big(\Big\|\begin{bmatrix} x_1^i - x_1^p \\ x_2^j - x_2^q \end{bmatrix}\Big\|_2\Big) \times S\Big(\sum_{n=0}^{K}\sum_{m=0}^{K} u_{nm}(t) v_{nm}(x_1^p, x_2^q)\Big)\Big] \quad (12)$$

Further, we explain our effective MATLAB-based solution technique for attacking the space-discretized version of SDE (5)–(9) rooted in approximations (10)–(12). At first we need to vectorize all values and functions defined at entries of the space-discterized-meshmatrix **X**. Here, we employ the built-in MATLAB command reshape for converting any matrix into the requested vectorized fashion. For instance, the mesh **X** is translated into its column-wise vector representation by the command **X**(:). We note that formulas (11) and (12) imply that the last row and column of the mesh array **X** are not used and must be excluded from the calculation, as that explained in [14]. That is why we further remove them from the mesh. This entails that the (i, j)th entry of such a truncated matrix **X** and the $(j \times N + i + 1)$th entry in its vectorized representation **X**(:) coincide. Then, we vectorize the scalar stochastic processes $u_{kl}(t)$ in SDE (5) to the form

$$\mathbf{u}(t) := [u_{00}(t), u_{01}(t), \ldots, u_{0K}(t), u_{10}(t), u_{11}(t), \ldots, u_{1K}(t), \ldots, u_{K0}(t), u_{K1}(t), \ldots, u_{KK}(t)]^{\mathrm{T}}, \quad (13)$$

where each random variable $u_{kl}(t)$ is presented by the $(k\times K + l + 1)$th entry of vector (13) for any $k, l = 0, 1, \ldots, K$.

Similarly, we reshape and vectorize all the Fourier-basis-functions (4) evaluated at nodes of our truncated space mesh **X**. For that, we assemble the rectangular matrix **V** of size $(K + 1)^2 \times N^2$ whose rows are sorted at the same manner as the stochastic processes $u_{kl}(t)$ in vector (13). More formally, every eigenfunction $v_{kl}(\mathbf{x})$ is calculated first at space points (x^i, x^j), i, j=0, 1, . . . , N–1, in use. Second, the resulting matrix derived for the function $v_{kl}(\mathbf{X})$ evaluated at all nodes of our truncated mesh **X** is further reshaped to the requested vectorized fashion of size N^2 by the MATLAB command $v_{kl}(\mathbf{X}(:))$. Here and below, we follow the MATLAB notation in which $v_{kl}(\mathbf{X})$ refers to the $(N \times N)$-matrix whose entries are the values $v_{kl}(x^i, x^j)$, $i, j = 0, 1, \ldots, N - 1$. Having fulfilled the above procedure for each eigenfunction in the Fourier-basis-equations (4), we obtain a set of $(K + 1)^2$ reshaped vectors $v_{1:(K+1)^2}(\mathbf{X}(:))$ of size N^2. Eventually, the (k, l)th vector is saved in the form of the $(k \times K + l + 1)$th row in the Fourier-basis-matrix $\mathbf{V} \in R^{(K+1)^2 \times N^2}$. In other words, the rows of such a matrix **V** are sorted at the same manner as the unknown variables $u_{kl}(t)$ in vector (13), that is every row of this matrix **V** stores the values of a particular Fourier-basis-function (4) evaluated at the vectorized-truncated-space-mesh **X**(:). The above notation converts the inner product (11) to its MATLAB-oriented manner, that is

$$\langle I(\mathbf{x},t), v_{kl}(\mathbf{x})\rangle \approx h_x^2 \mathbf{V}(k \times K + l + 1, :) \times I(\mathbf{X}(:), t). \quad (14)$$

Eventually, with use of formulas (11)–(14), the stochastic processes $u_k(x)$ $k = 0, 1, \ldots, K$, employed in the numerical solution (10) are computed by solving the vectorized SDE

$$d\mathbf{u}(t) = [h_x^2 \mathbf{V} \times I(\mathbf{X}(:), t) - \alpha\mathbf{u}(t) + \mathbf{FS}(U_K(\mathbf{X}(:), t))]dt + \varepsilon d\mathbf{W} \quad (15)$$

where the term $\mathbf{FS}(U_K(\mathbf{X}(:), t))$ stands for the $(K + 1)^2$-dimensional column-vector with its $(k \times K + l + 1)$th entry determined by equation (12), and the random-disturbance-vector **dW** is the multivariate zero-mean white Gaussian process with the diagonal covariance of the fash-ion $\mathbf{\Lambda} := \mathrm{diag}\{\lambda^2_{00}, \lambda^2_{01}, \ldots, \lambda^2_{0K}, \lambda^2_{10}, \lambda^2_{11}, \ldots, \lambda^2_{0K}, \ldots, \lambda^2_{K0}, \ldots, \lambda^2_{KK}\}$ in which all nonzero entries come from formula (9) and sorted in the same way as the entries $u_{kl}(t)$ in vector (13).

Finally, we utilize the Euler-Maruyama scheme [11, section 10.2] for attacking SDE (15) on some equidistant mesh predefined by the user in the time interval [0, T] of 2D-SNFE (2). The time-integration step size ht is assumed to be sufficiently small. The effective MATLAB-based implementation of the above-presented-SNFE-solution-method is based on the computation of the term $\mathbf{FS}(U_K(\mathbf{X}(:), t))$ in SDE (15) by the formula

$$\mathbf{FS}(U_K(\mathbf{X}(:), t)) = h_x^4 \mathbf{V} \times \mathbf{F} \times \mathbf{s} \quad (16)$$

where the time-variant column-vector **s** is of size N^2 and, for any $i, j = 0, 1, \ldots, N - 1$, its $(j \times N + i + 1)$th entry is given by the value $S(\mathbf{u}^{\mathrm{T}}(t)\mathbf{V}(:, j \times N + i + 1))$ at each time instant t. We recall that the matrix **V** represents the values of Fourier-basis-functions (4) evaluated at the vectorized-truncated-space-mesh **X**(:), but the MATLAB command $\mathbf{V}(:, j \times N + i + 1)$ returns the $(j \times N + i + 1)$th column in the latter matrix. In other words, the mentioned column contains the values of all eigenfunctions $v_{kl}(\mathbf{x})$, $k, l = 0, 1, \ldots, K$, computed at the space position (x^i, x^j). Lastly, the notation **F** refers to the symmetric matrix of size N^2 whose $(j \times N + i + 1, q \times N + p + 1)$-entry means the following factor in calculation (12):

$$F\Big(\Big\|\begin{bmatrix} x_1^i - x_1^p \\ x_2^j - x_2^q \end{bmatrix}\Big\|_2\Big), \quad i, j, p, q = 0, 1, \ldots, N - 1.$$

3 Numerical Results

The numerical results presented in [16] support the conjecture that if the external input has appropriate intensity and duration, and if the connectivity kernel is of a certain type (see equation (17)) the neural activity can generate stable multi-bump solutions which contain the information carried by the external signals.

Here, we begin by validating those results with the new numerical algorithm described in Section 2 and, then, we introduce some noise and observe how this noise affects the behavior of the solutions, in particular, how the structure of the multi-bump solutions can be changed.

The connectivity kernel F is of the oscillating type

$$F(r) = A\exp(-lr)\,(l\sin(a_1 r) + \cos(a_1 r)), \quad (17)$$

where A, a_1, l are certain positive constants (see [16]), $r = \sqrt{X^2 + Y^2}$.

The activation function S is of the Heaviside type, that is,

$$S(U) = H(U - b),$$

where $b > 0$ is a certain threshold.

The function I represents the sum of the external inputs with several components of the form

$$I(X, Y, t) = \sum_{i=0}^{n} I_i(X, Y, t),$$

where I_0 is a traveling wave of the form

$$I_0(X, Y, t) = \alpha_0 \exp(-\gamma_0 (X - vt)^2), \quad (18)$$

$0 \le t \le T$, $(X, Y) \in [-L, L] \times [-L, L]$, $v > 0$ is a positive number (the propagation speed of the traveling wave). The remaining inputs are localized (have the centers at C_i) and can be written as:

$$I_i(X, Y, t) = \alpha_i \exp\left(-\gamma_i (Y - C_i)^2\right), \quad (19)$$

where $C_i \in [-L, L]$, α_i, γ_i, are given positive numbers, $i = 1, \ldots, n$.

As the initial condition, we take

$$U(\mathbf{x}, 0) \equiv 0.$$

We carry out a set of numerical experiments. In each of them, we have a certain sequence of external stimuli, and simulate three different cases, with Ε = 0 (deterministic case), Ε = 0.04 (weak noise) and Ε = 0.08 (strong noise). In all the cases where the stochastic equation is simulated, we perform $ns = 100$ Monte Carlo runs and display the average value (as an approximation of the mathematical expectation of solution).

In all the subsequent examples, the domain of discretization is [−20, 20] × [−20, 20], where the solution is plotted on the domain [0, 20]×[−20, 20]; the step size in time is $\tau = 0.1$, the number of basis functions (3) is set by $K = 50$, and the step size in space is $h = 0.04$.

Example 1. In this example, the external input consists of the traveling wave I_0, described by (18), and the localized signal I_1, given by (19), which are constant during the time interval $t \in [0, 1.5]$ and, then, vanish. Their parameters are taken to be $\alpha_0 = 0.12$, $\gamma_0 = 1$, $v = 1.5$; $\alpha_1 = 0.12$, $\gamma_1 = 1$, $C_1 = 0$. The threshold for the activation function is $b = 0.1$; the connectivity kernel is described by (17), with the constants $A = 0.06$, $l = 0.8$, $a_1 = 1$.

The snapshots of the solution at the time $t = 0.5$ are displayed in Fig. 7. In the deterministic case, we see that this solution reproduces the inputs I_0 (traveling wave) and I_1. Moreover, at the place where the graphics of the two inputs intersect, we observe suprathreshold activity ($U > b = 0.1$). In the case of the weak noise (Ε = 0.04), we observe some perturbation of the signals, but the main features of the solution are preserved. Finally, in the case of the strong noise (Ε = 0.08), the perturbation of the activity is even higher, but the bump is still well visible at the intersection of the stimuli.

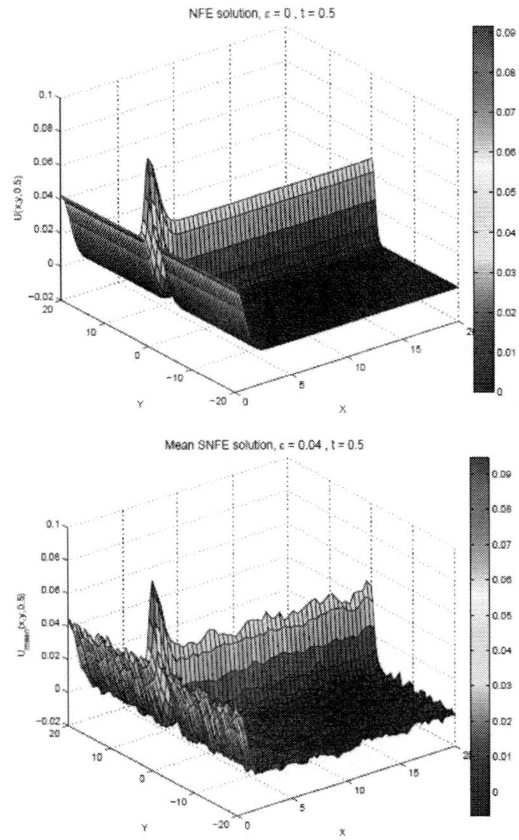

https://doi.org/10.1051/epjconf/202124801021

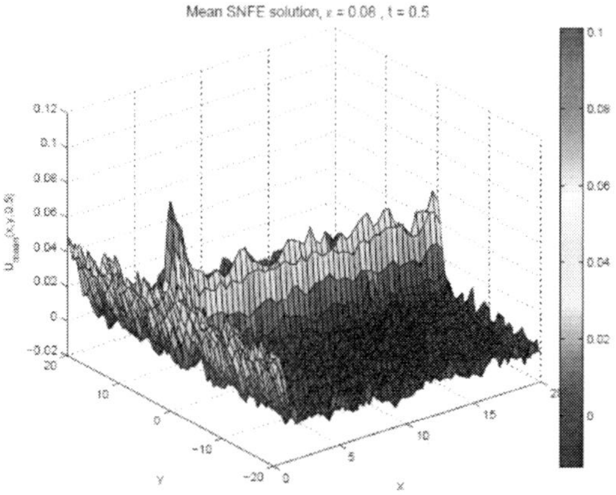

Figure 7. Snapshots of the solution of Example 1 at time $t = 0.5$: up- E = 0; down, left-hand side - E = 0.04; down, right-hand side - E = 0.08

Fig. 8 displays the snapshots of the solution at the time $t = 2.5$. Here, in the deterministic case, the effect of the input I_0 (traveling wave) is no more visible; the image of the input I_1 is considerably reduced (as expected, since the corresponding signals have been turned off at the time $t = 1.5$). However, in the place where the graphics of the two inputs intersect, the one-bump solution still exists and carries the information about the nature of the signal and the time at which it has occurred. In the case of the weak noise (E = 0.04), we observe some perturbation of this one-bump pattern, but the coordinates of the peak still reproduce the characteristics of the input. Finally, in the case of the strong noise (E = 0.08), we still observe suprathreshold activity in the place of the intersection, but the peak position deviates slightly from the location of the external input.

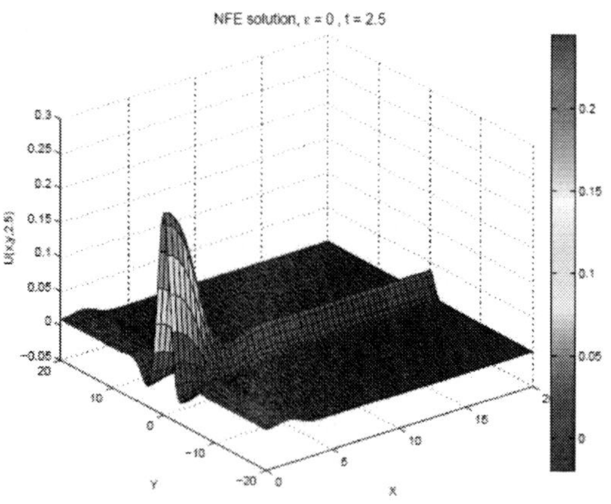

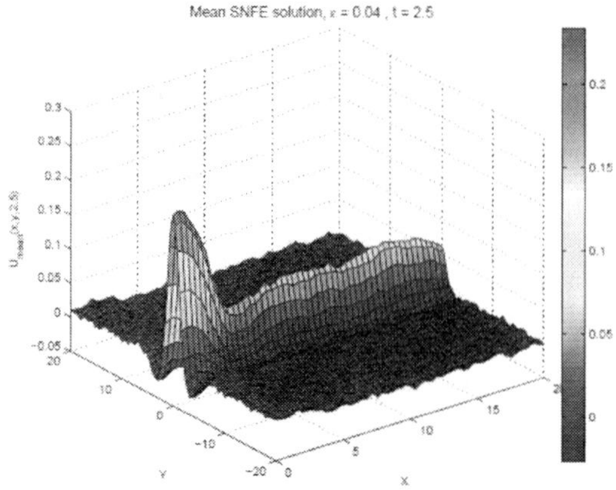

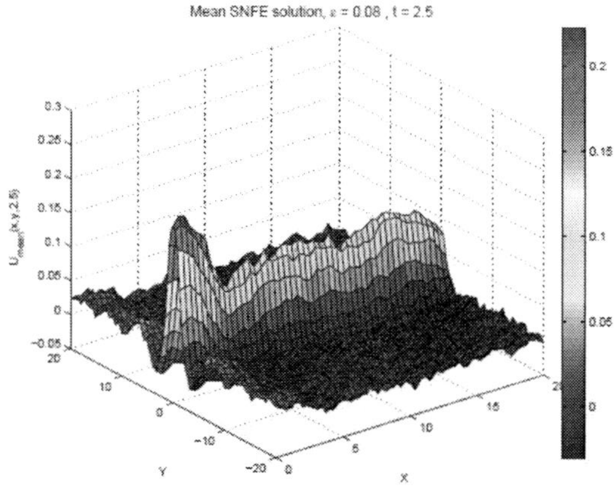

Figure 8. Snapshots of the solution of Example 1 at time $t = 2.5$: up- E = 0; down, left-hand side - E = 0.04; down, right-hand side - E = 0.08

Example 2. In this case, we have again the traveling wave I_0 and two localized inputs I_1 and I_2, with $\alpha_1 = 0.12$, $\gamma_1 = 1$, $C = -10$ and $\alpha_2 = 0.12$, $\gamma_2 = 1$, $C_2 = 10$, which correspond to the experiment with two different colors. As in the previous example, the inputs I_0, I_1 and I_2 are constant in the time interval $0 < t < 1$, and, then, are removed. The other parameters have the same values as in the previous example.

The snapshots of the solution at the time $t = 1$ are displayed in Fig. 9 In the deterministic case, we see that the solution reproduces the inputs I_0 (traveling wave), I_1 and I_2. Moreover, at the places where the graphic of I_0 intersects I_1 and I_2, we observe suprathreshold activity ($U > b = 0.1$). In the case of the weak noise (E = 0.04), we observe some perturbation of the patterns, but the main features of this solution are preserved. Finally, in the case of the strong noise (E = 0.08), the perturbation of the activity pattern is even higher, but the bump is still well visible at the intersections of the stimuli.

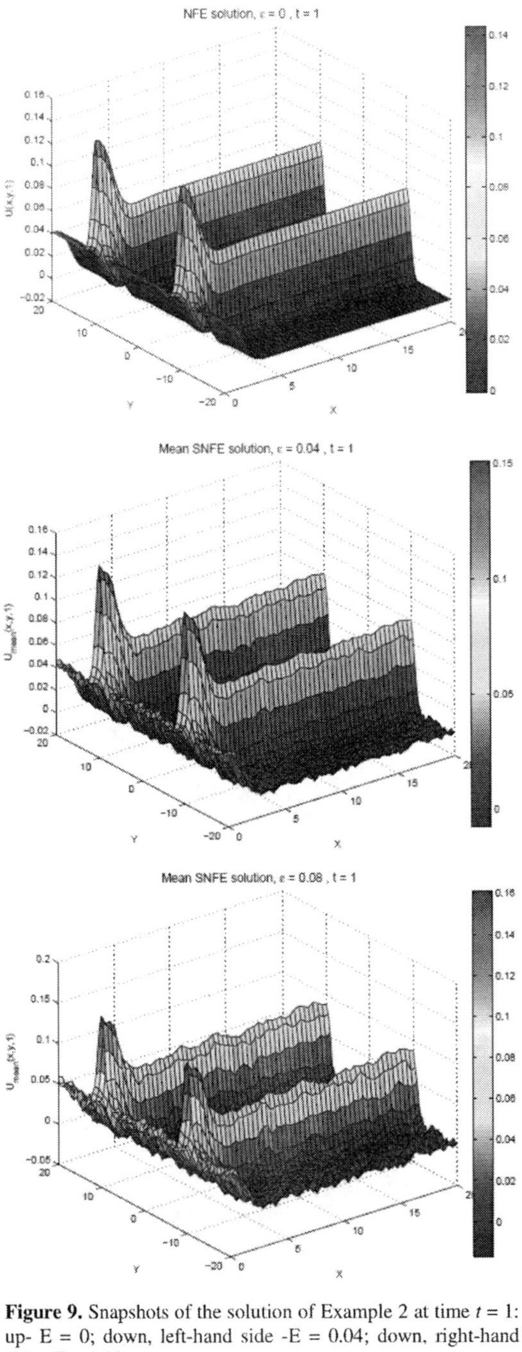

Figure 9. Snapshots of the solution of Example 2 at time $t = 1$: up- E = 0; down, left-hand side -E = 0.04; down, right-hand side - E = 0.08

Fig. 10 displays the snapshots of the solution at the time $t = 5$. Here, in the deterministic case, the effect of inputs I_0 (traveling wave), I_1 and I_2 are no more visible (since these inputs have been turned off at time $t = 1$.) However, similarly to what happens in the previous example, the suprathreshold activity remains at the places of the intersections. In this case, this suprathreshold activity takes the form of a two-bump solution, where the *Y*-coordinate of each peak corresponds to one color of the signal. Note that these two peaks have the same *X*-coordinate, which means that the two inputs occurred simultaneously. The effects of the weak noise (E = 0.04) and strong noise (E = 0.08) are less visible in this case in comparison with those observed in the previous example, because we consider the later time instant, that is, $t = 5$.

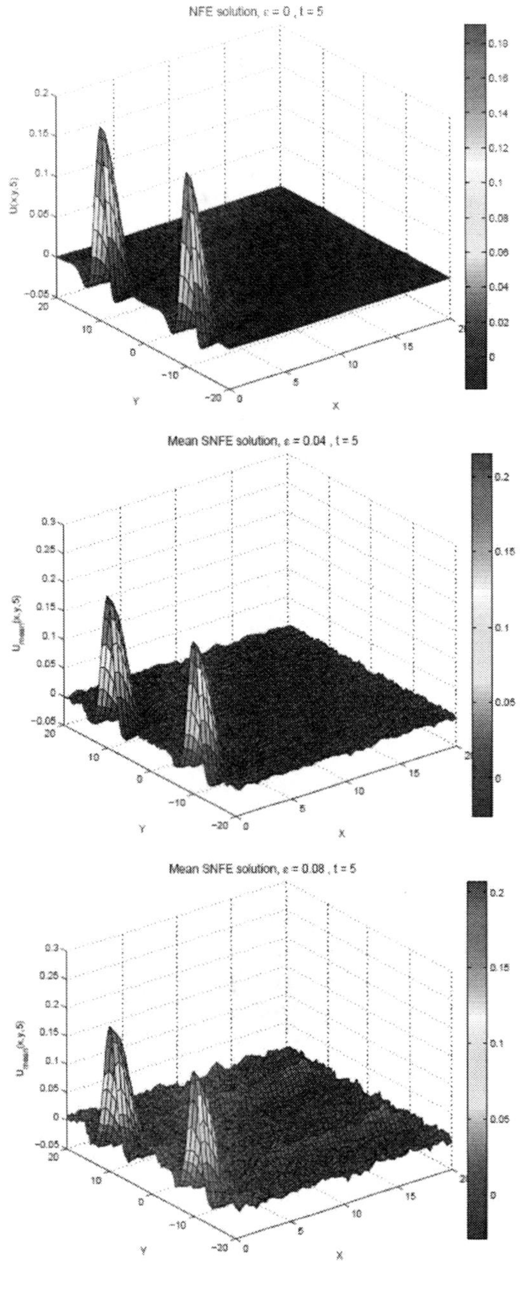

Figure 10. Snapshots of the solution of Example 2 at time $t = 5$: up- E = 0; down, left-hand side -E = 0.04; down, right-hand side - E = 0.08

Example 3. In this example, we have three series of inputs at different times. First, in the time interval [0, 1],

https://doi.org/10.1051/epjconf/202124801021

we employ I_0, I_1 and I_2 (with the same parameters as in Example 2); then, in [1, 5], we implement only the traveling wave I_0; finally, in [5, 6], we have I_0, I_3 and I_4, where $I_3 = 1.2I_1$ and $I_4 = 1.2I_2$. After $t = 6$, all the inputs are removed.

The output of this example is depicted in Fig.11 (t = 1), Fig.12 (t = 6), Fig. 13 (t = 8), Fig.14 (t = 12), and Fig. 15 (t = 15).

In Fig. 11, we observe the formation of two bumps, which correspond to the signals that occurred in the time interval [0, 1]. The graph in Fig. 12 shows the formation of a new set of two bumps. These relate to the signals in the time interval [5, 6]. In Fig. 13, in the deterministic case, we observe a solution with four bumps, as can be expected: two of them correspond to the first input signal, and two of them to the second one. This solution contains the whole information about the sequence of the stimuli in use.

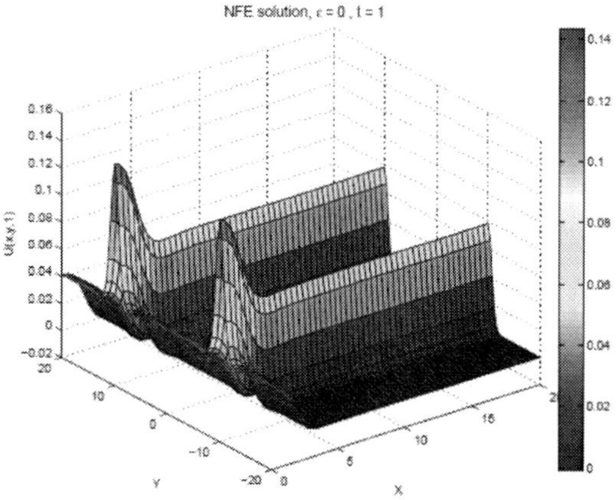

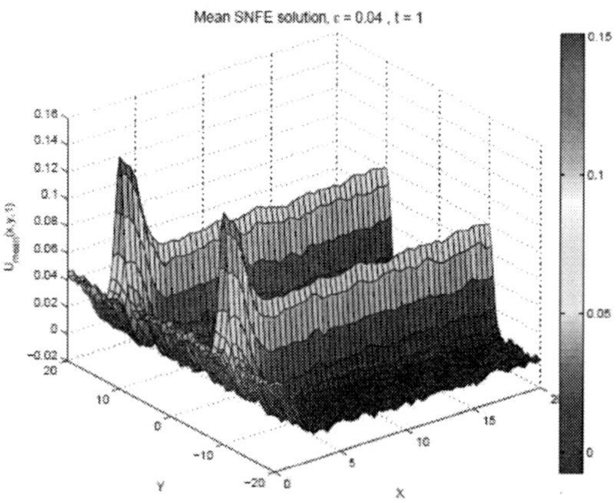

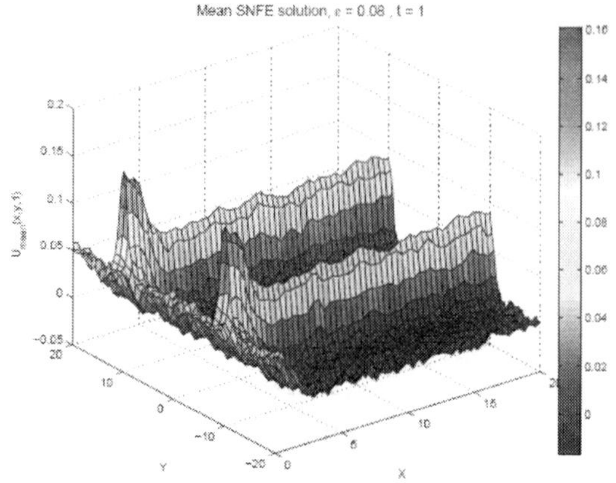

Figure 11. Snapshots of the solution of Example 3 at time t = 1: up- E = 0; down, left-hand side -E = 0.04; down, right-hand side- E = 0.08

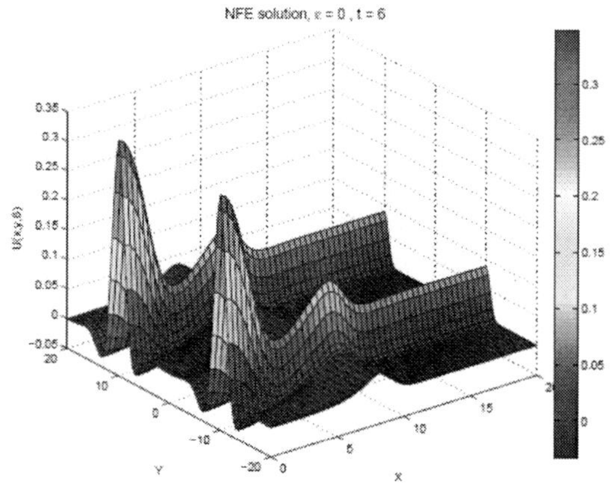

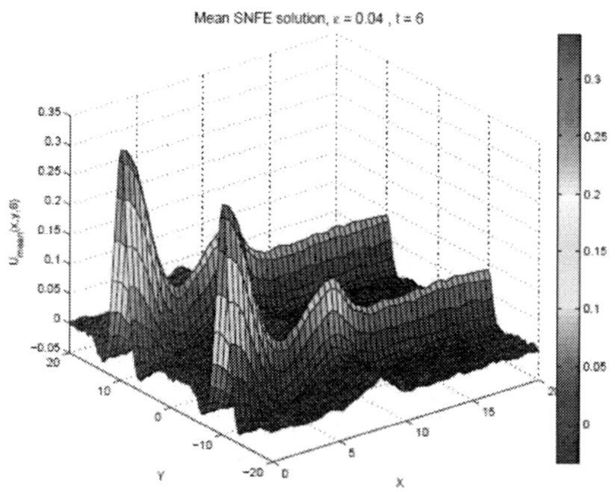

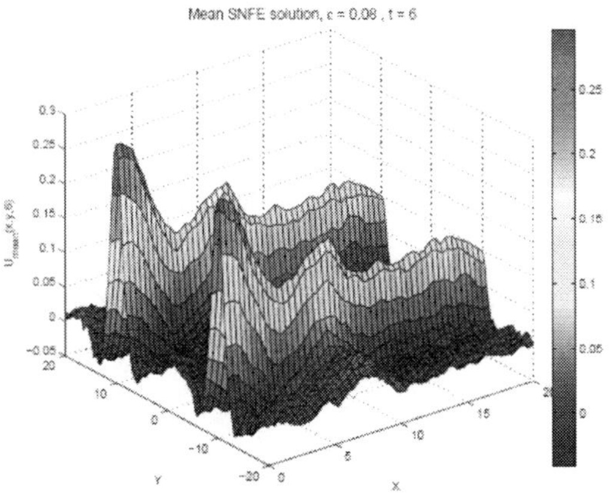

Figure 12. Snapshots of the solution of Example 3 at time t = 6: up- E = 0; down, left-hand side -E = 0.04; down, right-hand side - E = 0.08

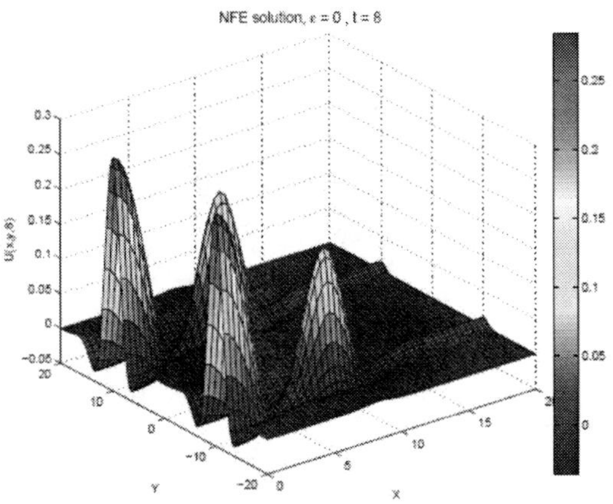

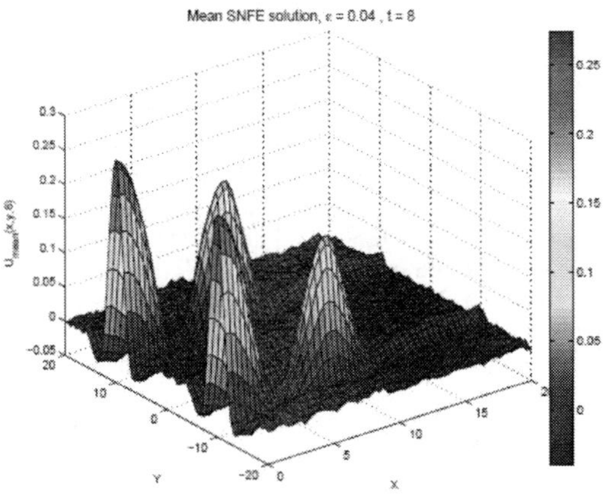

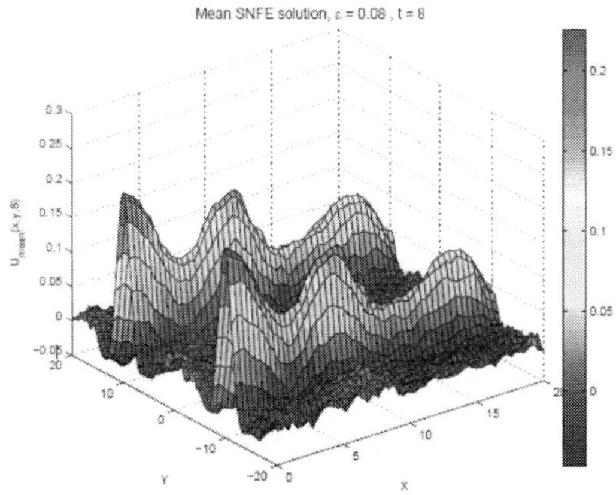

Figure 13. Snapshots of the solution of Example 3 at time t = 8: up- E = 0; down,left-hand side -E = 0.04; down, right-hand side - E = 0.08

In the case of the strong noise (E = 0.08), Fig.13 shows an interesting feature that is not observed in the previous examples. Instead of 4 bumps, the field activity displays 6 ones, that is, two additional bumps have been formed just as the effect of the implemented noise. Note that this does not happen in the deterministic case, nor in the case of the weak noise.

In order to observe how this picture changes in time, we have carried out computations over a longer time interval, that is, we show the results at the time t = 12 and t = 15. Such outputs are exhibited in Fig.14 and Fig.15. In these graphs, we observe that the two additional bumps, which appear as the result of the noise in use, do not disappear or even decrease in time.

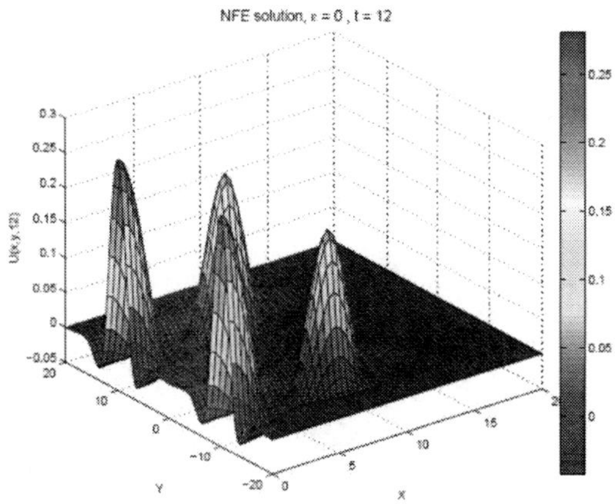

https://doi.org/10.1051/epjconf/202124801021

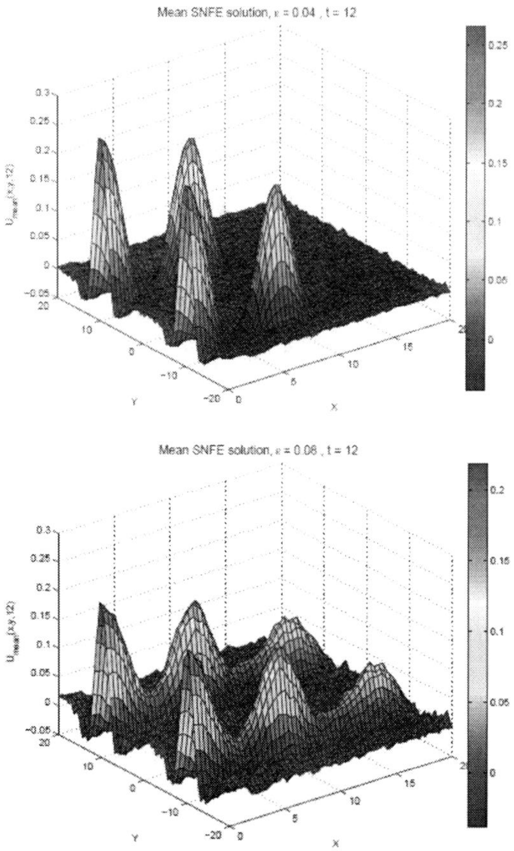

Figure 14. Snapshots of the solution of Example 3 at time t = 12: up- E = 0; down, left-hand side -E = 0.04; down, right-hand side - E = 0.08

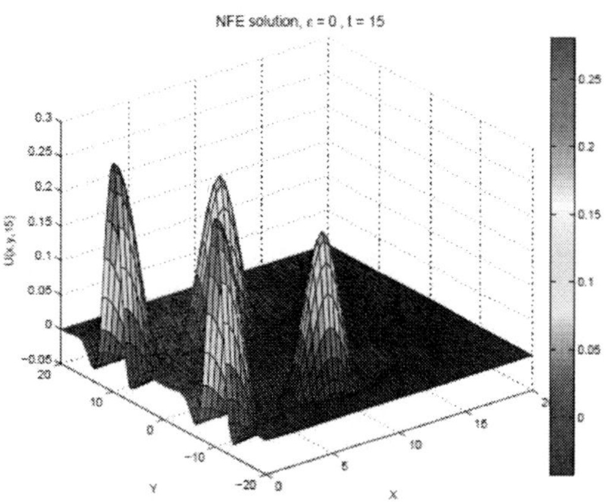

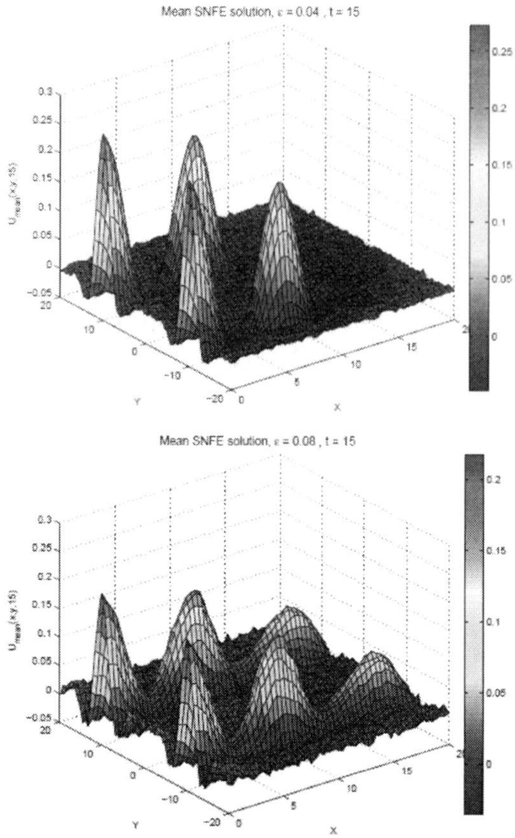

Figure 15. Snapshots of the solution of Example 3 at time t = 15: up- E = 0; down, left-hand side -E = 0.04; down, right-hand side - E = 0.08

This is an example of false memories that can be induced by random noise. Differently from the input-induced bumps, their activity level is subthreshold (less than b = 0.1). This can be explained as follows: only part of the paths of the stochastic solution contain these two additional bumps (whose height should be above the suprathreshold value), while the remaining ones are similar to the deterministic solution. Since what we show in the picture is the mean solution, the heights of these bumps are averaged on all the paths computed and, hence, these are subthreshold).

It is also worth to mention that the distance between a 'false bump' and the closest real memory representation is approximately the coordinate of one of the local maxima of the connectivity function, that is, the location of false memory receives excitation from existing bumps.

4 Conclusion

We have described a two-dimensional neural field model which explains how a population of cortical neurons may encode in its self-sustained firing pattern simultaneously the nature and time of sequential stimulus events.

The postulated wave mechanism explains how a nervous system lacking specific sensors for temporal perception may develop neurons that respond to specific interval durations.

The numerical results presented support the conjecture that if the external input has appropriate intensity and duration, and if the connection kernel is of the oscillatory type described here, the neural activity can generate stable multi-bump solutions which reliably contain the information carried by the external signal.

In the case of the weak noise, the structure of the stationary solutions is preserved in the sense that the mean solution of the stochastic equation has the same number of bumps as the solution of the deterministic one has.

In the case of the strong noise, the structure of the stationary solutions can be significantly changed, affecting the precision of the memorized information or even generating false memories.

Acknowledgements

The authors acknowledge the financial support of the portuguese FCT (Fundação para a Ciência e Tecnologia), through projects UIDB/04621/2020, UIDP/04621/2020 (IST), UIDB/00013/2020, UIDP/00013/2020 (UMinho) and PTDC/MAT-APL/31393/2017.

References

1. S.L. Amari, Biol. Cybernet., **27**(2), 77–87 (1977)
2. S. Coombes, NeuroImage, **52**(3), 731–739 (2010)
3. W. Erlhagen, A. Bastian, D. Jancke, A. Riehle, G. Schoner, Journal of Neuroscience Methods, **94**(1), 53–66 (1999)
4. W. Erlhagen, E. Bicho, Journal of Neural Engineering, **3**(3), R36 (2006)
5. G. Faye, O. Faugeras, Physica D, **239**, 561–578 (2010)
6. F. Ferreira, W. Erlhagen, E. Bicho, Physica D: Nonlinear Phenomena, **326**, 32–51 (2016)
7. F. Ferreira, W. Wojtak, E. Sousa, L. Louro, E. Bicho, W. Erlhagen, IEEE Transactions on Cognitive and Developmental Systems (2020) DOI:10.1109/TCDS.2020.2991789
8. A. Hutt, N. Rougier, Physical Review,E, **82**, 055701 (2010)
9. A. Hutt, N. Rougier, In: S. Coombes et al. (Eds.), Neural Fields The- ory and Applications, 175–183, Springer (2014) https://doi.org/10.1007/978-3-642-54593-1_6
10. Z. Kilpatrick, B. Ermentrout, SIAM J. Appl. Dyn. Syst., **12**, 61–94 (2013)
11. P. Kloeden, E. Platen, *Numerical Solution of Stochastic Differential Equations* (Berlin: Springer, 1999)
12. C. Kühn, M.G. Riedler, J. Math. Neurosci, **4**(1), 1–33 (2014)
13. G.Yu. Kulikov, P.M. Lima, M.V. Kulikova, Journal of Computational and Applied Mathematics, J. Comput. Appl. Math., **387**, 112563 (2021)
14. M.V. Kulikova, G.Yu. Kulikov, P.M. Lima, Proceedings of the 23rd International Conference on System Theory, Control and Computing (ICSTCC), 650-655 (2019)
15. C.R. Laing, W.C. Troy, B. Gutkin, G.B. Ermentrout, SIAM Journal on Applied Mathematics, **63**(1), 62–97 (2002)
16. P.M.Lima, W. Erlhagen, Proceedings of the 2018 European Control Con-ference (ECC), 2040-2045 (2018)
17. P.M. Lima, E. Buckwar, SIAM Journal of Scientific Computing, **37**, B962– B979 (2015)
18. P.M. Lima, E. Buckwar, Proceedings of the Second International Conference on Mathematics and Computers in Sciences and in Industry, 131-137, IEEE Conference Publications (2015)
19. P.M. Lima, Advances in Mathematical Methods and High Performance Computing: V.K. Sing , D. Gao, A. Fischer, Eds., 51-67 (2019)
20. H. Mechant et al., The Journal of Neuroscience, **33**(21), 9082–9096 (2013)
21. L. Muller, F. Chavane, J. Reynolds, T.J. Sejnowski, Nature Reviews Neuroscience, **19**(5), 255-268 (2018)
22. T. Shardlow, J. Comput. Appl. Math., **175**, 429–446 (2005)
23. H.R. Wilson, J.D. Cowan, Bipophys. J., **12**, 1–24 (1972)

Distributions of Two Atoms Collisions over the Surface of the Condensed Phase

Sergey Zheltov[1], *Leonid* Pletnev[2]

[1]Math Department, Tver State University, RU-170036, Tver, Russia
[2]Dep of General scientific disciplines, Tver State Technical University, RU-170036, Tver, Russia Russia

Abstract. The processes of heat and mass transfer are closely related to the evaporation of a substance from the surface of the condensed phase. The interaction of outgoing molecules from the surface of the condensed phase with condensed phase molecules plays a fundamental role. A simpler case of evaporation is the departure of atoms from the surface of the condensed phase, i.e. the atoms overcome the potential barrier on the surface of the condensed phase. Depending on the evaporation rate, a Knudsen layer appears above the surface of the condensed phase. In this paper, based on the model of rigid spheres, the density distributions of the collision distances and the average values of the collision distances of two atoms emitted simultaneously from the surface of the condensed phase above the surface are analyzed. Distributions of the collision distance depending on the surface temperature, the size of the potential barrier, and the size of the evaporation area are obtained. Computer experiments were performed using the Monte Carlo method. To obtain the results of numerical simulation, a parallel algorithm adapted to calculations on graphics processors with CUDA technology was developed.

1 Introduction

The processes of heat and mass transfer in systems are related both to the interaction of gas with the walls of the systems [1-5] and to the interaction of the evaporating substance with the surface of the condensed phase [6-11]. It is of considerable interest to consider the evaporation process from a microscopic point of view. In this case, it is necessary to know the interaction potentials of molecules or atoms of a substance with each other or the walls of systems. Analysis of the evaporation of a substance consisting of molecules is a more complex task compared to the evaporation of a substance consisting of atoms, since when interacting with molecules, the exchange between the internal degrees of freedom of the molecules must be taken into account. With low-intensity evaporation of a substance, atoms or molecules can travel a sufficiently long distance before colliding [12-15], so there is a Knudsen layer near the wall surface and the surface of the condensed phase. In this layer, there are significant changes in the parameters that characterize the gas - temperature, density, etc. Due to the lack of accurate data on the interaction of atoms with each other, jumps of these parameters on the surface are introduced.

To describe the gas in this layer, the Boltzmann equation or model equations derived from it is used. The problems of describing a gas in this layer arise because the interaction processes are non-equilibrium, i.e. the velocity distribution function of gas atoms differs from the equilibrium distribution function. The development of computer technology has made it possible to use Monte Carlo and molecular dynamics methods to analyze the evaporation of a substance. It should be noted that even the use of modern computers and software does not allow for a detailed analysis of the evaporation process, since in real conditions it is necessary to calculate a huge number of outgoing molecules and, especially, their interaction with each other in the gas phase.

2 Problem statement

This paper presents the results of calculations of model computer experiments to study the distributions of collisions of two atoms over the surface of a condensed phase. The atoms flew simultaneously into the vacuum from a limited area of a square surface with a side $a = 10^{-9}$ m. The model of atoms in the form of rigid spheres with a radius $R = 1.5\cdot10^{-10}$ m was used. The masses of the atoms were assumed to be equal to m = 40 AU. Fig. 1 shows a scheme of the departure of two atoms from the bounded surface of the condensed phase.

The coordinates of the atoms $(x_i;y_i)$ flying out from the surface of the condensed phase and colliding over its surface were played out using a sensor of uniformly distributed random variables. The distance a_{12} was defined as the distance between the centers of atoms and could be in the range $(3\cdot10^{-10}\text{m};38.18\cdot10^{-10}\text{m})$. The atomic velocity components $(v_{xi};v_{yi};v_{zi})$ were played out using a normal distribution sensor. If a collision between atoms occurred, the collision parameters were entered in the corresponding arrays; if not, a new pair of atoms was played.

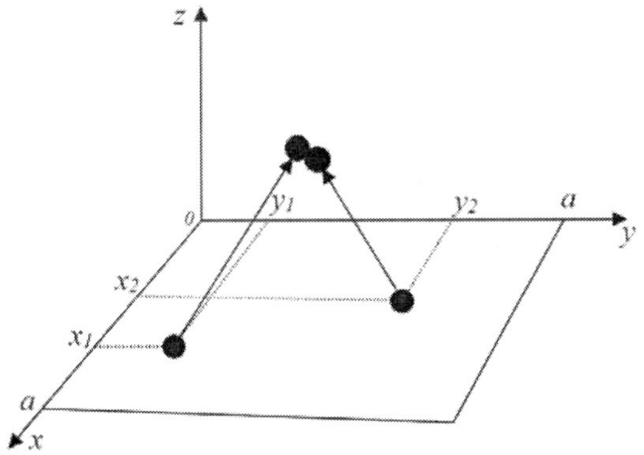

Fig. 1. Scheme of departure of two atoms from the surface of the condensed phase.

A special feature of the computer experiments is that the calculations are made taking into account the potential barrier on the surface of the condensed phase U formed by the interatomic forces of the condensed phase. From the z – component of each outgoing atom, a part of the energy equal to the value of the potential barrier U is subtracted. If the kinetic energy of the z component is greater than the value of the potential barrier U, then the atom was considered to have flown out. Otherwise, a new value of the z-component of the speed was played. It was found that the calculations depend on the dimensionless parameter $r = U / kT$, where T is the surface temperature of the condensed phase and k is the Boltzmann constant. As was found in [16,17], the average velocity component v_z of the ejected atoms perpendicular to the surface of the condensed phase increases with an increase in the parameter r. The other two components of the velocities of the ejected atoms v_x and v_y did not change.

Computer experiments were performed using the Monte Carlo method for several hundred million pairs of atoms. The original problem has a so-called data parallelism and belongs to the SIMD class. Thus, it is possible to simulate the collision of several million pairs of atoms simultaneously. To increase efficiency, an algorithm that was adapted to computing on graphics processors with CUDA technology was developed. The calculations were performed on the NVIDIA Tesla K80 GPU of the heterogeneous HybriLIT platform, which is part of the multifunctional information and computing complex of the JINR information technology Laboratory (Dubna).

3 Purpose of research

Because not all ejected atoms collided with each other above the surface, is of interest to determine the density distributions of the distance a_{12} between the colliding atoms at the time of departure on the surface of the condensed phase, i.e., surface distance for which collision over the surface is real. In addition to the densities of the a_{12} distance distributions, the average distances of a_{12} were obtained.

As a result of computer experiments, the density distributions of the distances of the collision of atoms on the surface of the condensed phase z and the average distance values of collisions of atoms z_{av} above the surface of the condensed phase were determined.

The densities of distributions of the total collision distances of l atoms and the average values of the total collision distances of l_{av} atoms from the surface of the condensed phase were determined. In fact, this meant that the lengths of free paths of atoms before collisions were determined.

All the above distribution densities and average values are obtained for various values of the surface temperature of the condensed phase T, the value of the dimensionless parameter r, and the size of the square evaporation area with side a.

Additional studies were performed to obtain the densities of velocity distributions v_z between colliding atoms.

4 Results of calculation

The densities of distance distributions a_{12} depending on the parameter r are shown in Fig. 2. Normalization of all distribution densities was performed by dividing each value of the number in the cell by the largest value for all distributions. The distributions have maxima that depend on the value of the parameter r, which shift towards smaller distances a_{12} as this parameter increases. This can be explained by the fact that as the parameter r increases, the average values of the v_z components increase, and the atoms may collide if they are closer to each other. Computer experiments were performed for the surface temperature $T = 50$ K. Calculations with other temperature values led to the same results. The temperature value changes the values of the average values of the components of the atomic velocities and, accordingly, the time before the collision, but not the point of collision of the atoms.

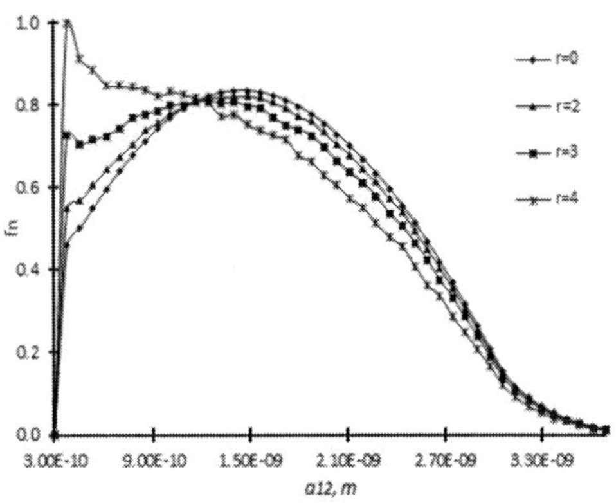

Fig. 2. Densities of distance distributions a_{12}.

It should be noted that the calculations were performed for the values of the parameter $r \in [0;12]$. As it was found, the results of calculations change slightly with an increase in the parameter r from 8 to 12. The results of calculations obtained for the value $r = 8$ can be considered as limiting results, i.e. atoms fly out from the surface according to the cosine law.

This is confirmed by the distribution of average distances a_{12} depending on the parameter r, shown in Fig. 3. The distribution of average distances a_{12} does not depend on temperature and decreases with increasing parameter r.

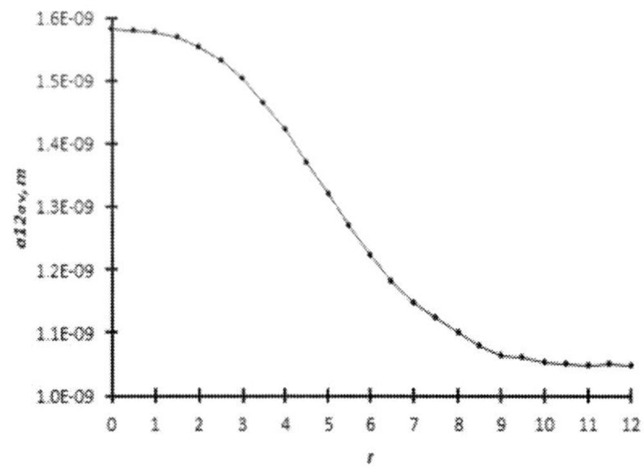

Fig. 3. Average distances a_{12} between colliding atoms on the surface of a condensed phase.

As a result of calculations, the distances z above the surface of the condensed phase at which the atoms collided above the surface of the condensed phase were determined. The value of z was defined as the product of the atom with the minimum velocity for the time of movement before the collision. Fig. 4 shows the results of calculations of the density distributions of collision distances z for several values of the parameter r. Calculations were performed for the surface of the condensed phase with a temperature $T = 50$ K. As the parameter r increases, the maxima of the distributions shift to the right, towards larger collision distances, and increase slightly. It is found that the distributions do not depend on the surface temperature, since the temperature affects the velocity components, i.e. the atoms collide faster, but at the same point.

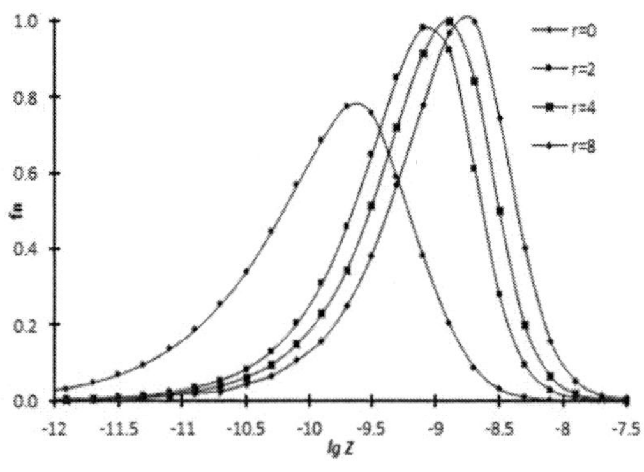

Fig. 4. Densities of distributions of collision distances z depending on the parameter r.

Colliding atoms, after leaving the surface of the condensed phase, could fly out of the evaporation area. The free path of the atoms from the point of departure from the surface of the condensed phase to the point of collision of the atoms l was determined. The value l was defined as the square root of the sum of the squares of the components of the atomic runs along the axes of the coordinate system. This value, on average, is almost twice as large as the corresponding z value. Fig. 5 shows the normalized densities of distributions of the distances of atomic runs to collisions l. It is found that for the value l, the regularities of the distribution densities are the same as for the distribution densities of the value z. As the parameter r increases, the maxima of distributions increase and shift to the area of large l. Just like the collision distribution densities z in Fig. 4, the distribution densities for l do not depend on the surface temperature of the condensed phase.

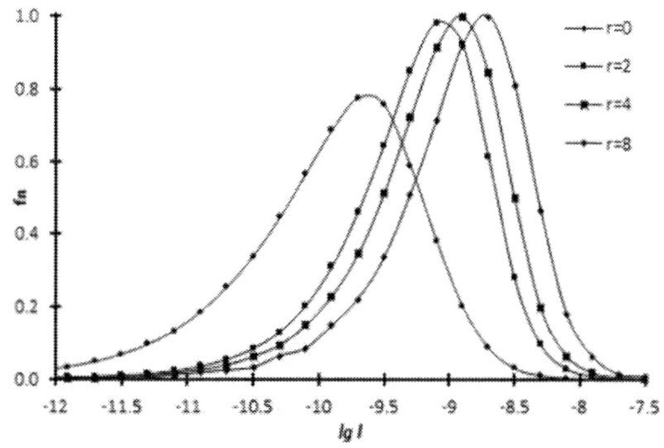

Fig. 5. Densities of distributions of collision distances l depending on the parameter r.

Figure 6 shows the distributions of the average collision distances z_{av} and l_{av} depending on the parameter r. It is found that the distributions of the average values of these values do not depend on the surface temperature for the same reason as the distribution densities. As the parameter r increases, the average distance z_{av} and l_{av} between colliding atoms increases.

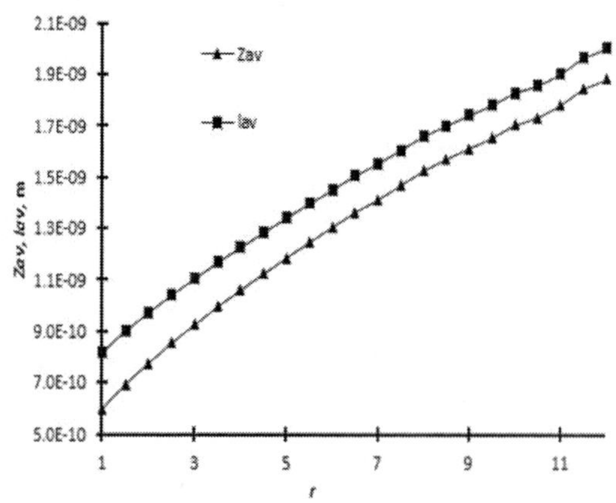

Fig. 6. Distributions of average z_{av} and l_{av} values depending on the parameter r.

A special series of computer experiments was performed to establish the dependences of the distribution densities and average collision values for the z_{av} and l_{av} values depending on the size of the evaporation area a. The results of calculations of distribution densities for the value z for the parameter $r = 2$ are shown in Fig. 7. With the increase in the size of the evaporation area, the maxima decrease in height and shift to the right, towards larger values of z. It was found that the densities of z collision distributions do not depend on the surface temperature.

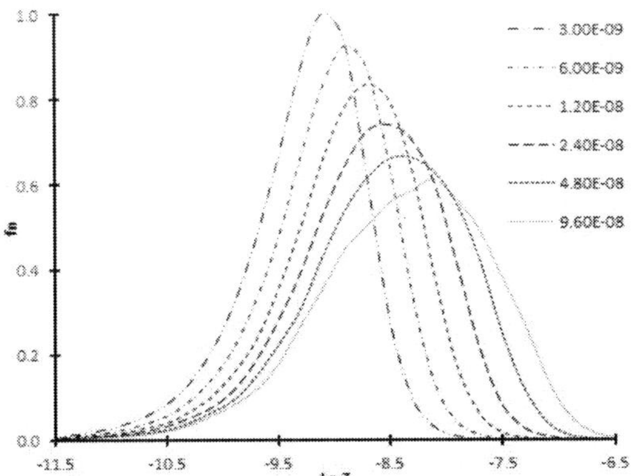

Fig. 7. Densities of distance distributions z depending on the evaporation region a. $r = 2$.

Fig. 8 shows the results of calculations of the distance distribution densities for the value z for the parameter $r = 8$. The obtained distributions do not depend on the surface temperature as well as for $r = 2$. The forms of the z distance distribution densities for $r = 2$ and $r = 8$ coincide, but are shifted to the left for $r = 8$.

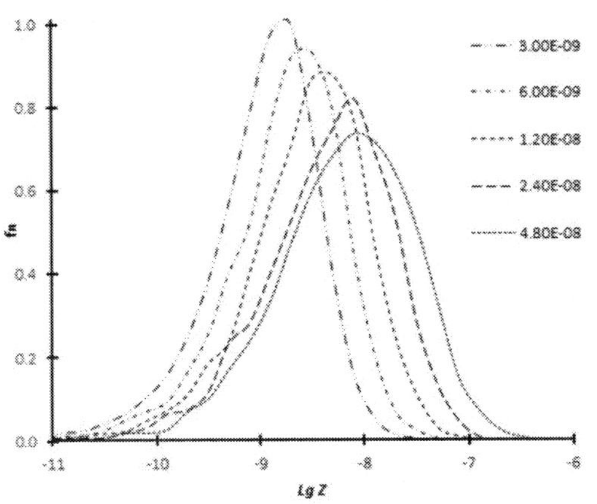

Fig. 8. Densities of distance distributions z depending on the evaporation region a. $r = 8$.

Densities of distributions of free path lengths l were determined depending on the size of the evaporation area a for temperatures $T = 50$ K and $T = 100$ K. It is established that there is no difference between the densities of distributions of free path lengths l. The calculation results are shown in Fig. 9.

Similar results are obtained for the value of the parameter $r = 8$. The densities of distance distributions l do not depend on the surface temperature T. The calculation results are shown in Fig. 10.

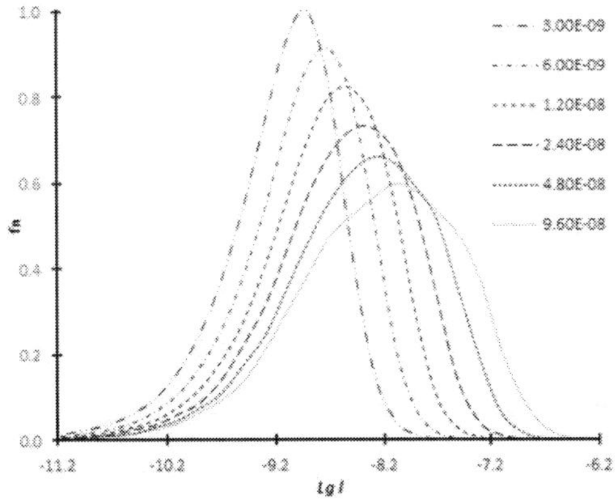

Fig. 9. Densities of distance distributions l depending on the evaporation region a. $r = 2$.

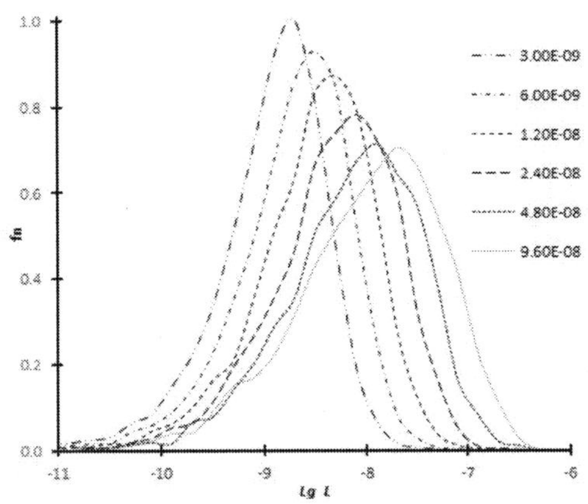

Fig. 10. Densities of distance distributions l depending on the evaporation region a. $r = 8$.

The distributions of the average distances z_{av} and l_{av} depending on the evaporation area a are shown in Fig. 11. As the size of the evaporation region a increases, the average values of these values change almost linearly.

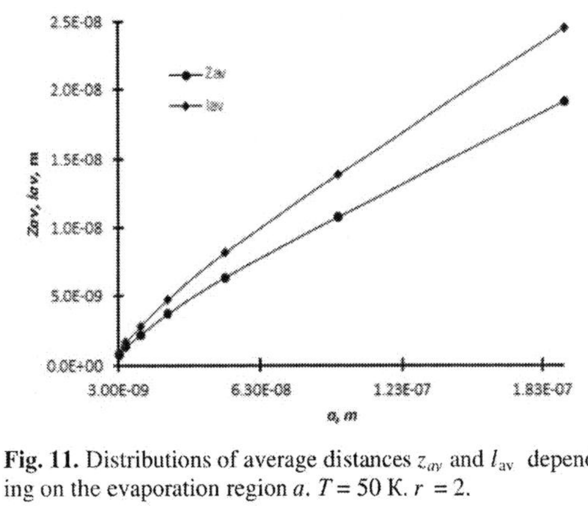

Fig. 11. Distributions of average distances z_{av} and l_{av} depending on the evaporation region a. $T = 50$ K. $r = 2$.

The ratio between the corresponding z_{av} and l_{av} values remains constant. Similar distributions are obtained for other values of the surface temperature T and the parameter r.

The distance between collisions of atoms depends on the velocity components of these atoms after overcoming the potential barrier on the surface of the condensed phase. The regularities of distributions of the average v_z components depending on the parameter r for two temperature values are shown in Fig. 12.

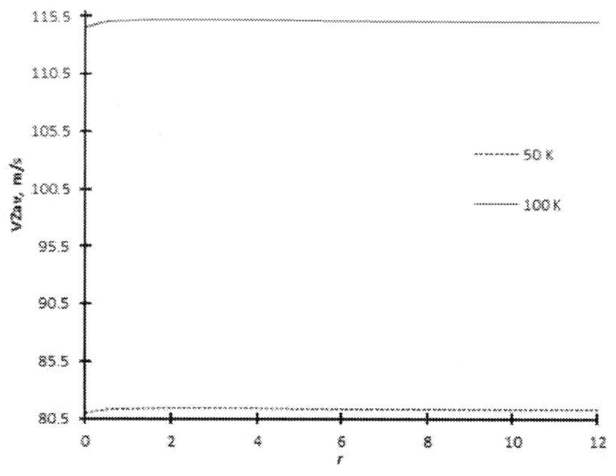

Fig. 12. Distributions of average values of atomic velocities v_z depending on the parameter r. $T = 50$ K.

Distributions are almost parallel lines that do not depend on the parameter r. The ratio of the average velocities of the components v_z is equal to the square root of the inverse ratio of temperatures.

5 Conclusions

Computer experiments made it possible to determine the structure of the Knudsen layer in the framework of the proposed model of the collision of two atoms flying from a limited area of the condensed phase surface. Taking into account the influence of the interaction potential on the conditions of atomic collisions allowed us to obtain a more accurate structure of the Knudsen layer. The obtained densities of distributions of atomic collision distances allowed us to establish the complex structure of this layer.

Analysis of these distributions has shown that the main number of collisions of atoms occurs in a wide range of distances. From a few nanometers to hundreds of nanometers. To determine the Knudsen layer, you can use the average values of z_{av} and l_{av}, which almost coincide with the maxima of the distributions.

The results of calculations for the collision of two atoms can be used as a basis for calculating the collisions of three atoms simultaneously flying from the surface.

References

1. J. Lee, T. Laoui, R. Karnik, Nat. Nanotech., **9** (2014)
2. Y. Li, M.A. Alibakhshi, Y. Zhao, C. Duan, Nano Lett., **17** (2017)
3. K.L. Wilke, B. Barabadi, Z. Lu, T. Zhang, E.N. Wang, Appl. Phys. Lett., **111** (2017)
4. R. Xiao, S.C. Maroo, E.N. Wang, Appl. Phys. Lett., **102** (2013)
5. A. Frezzotti, Phys. Fluids, **23** (2011)
6. A.H. Persad, C.A. Ward, Chem. Rev., **116** (2016)
7. M. Knudsen, Ann. Phys., **352** (1915)
8. F.G. Tcheremissine, Comput. Math. Phys., **46** (2006)
9. T. Ishiyama, S. Fujikawa, T. Kurz, W. Lauterborn, Phys Rev E, **88** (2013)
10. M. Kon, K. Kobayashi, M. Watanabe, Int. J. Heat Mass Transfer, **99** (2016)
11. K. Kobayashi, K. Sasaki, M. Kon, H. Fujii, M. Watanabe, Microfluid Nanofluid, **21** (2017)
12. S. Cheng, J.B. Lechman, S.J. Plimpton, G.S. Grest, J. Chem. Phys., **134** (2011)
13. M. Kon, K. Kobayashi, M. Watanabe, Phys. Fluids, **26** (2014)
14. M. Kon, K. Kobayashi, M. Watanabe, Eur. J. Mech. B Fluids, **64** (2017)
15. K. Kobayashi, K. Hori, M. Kon, K. Sasaki, M. Watanabe, Heat Mass Transfer, **52** (2016)
16. L.V. Pletnev, N.I. Gamayunov, V.M. Zamyatin, *Mathematical Models of Non-Linear Excitations, Transfer, Dynamics, and Control in Condensed Systems and Other Media. L.A. Uvarova (eds)* (Springer, Boston, MA. 1999) https://doi.org/10.1007/978-1-4615-4799- 0_13
17. L.V. Pletnev, J. Monte Carlo Methods and Applications, **6**, 3 (2000)

Probabilities of Atomic Departures from the Slit System

Leonid Pletnev

Dep of General scientific disciplines, Tver State Technical University, RU-170036, Tver, Russia Russia

Abstract. The theoretical description of experimental data on transport in open systems has been of interest for more than a hundred years. Boltzmann proposed a new view of the transfer of matter. Now it is possible to describe the transfer processes from a microscopic point of view. The solution of the Boltzmann equation and the equations derived from it is a complex problem related to the mathematical problems of solving such equations. On the other hand, the complexity of the solution is related to the geometry of the system in which the transfer process takes place. Fundamental physical calculations are made for systems: flat, slotted, cylindrical, rectangular, etc. In the free molecular mode of gas flow, collisions of molecules occur mainly with the walls of systems. In this connection, there was a direction related to the calculation of the probabilities of atomic transport in the system. In this paper, we propose an approach for determining the probabilities of atomic outcomes from slit systems depending on the relative height of the walls of H systems. Exact formulas are obtained for calculating the probabilities of atomic departures from systems without colliding with walls, the distribution of atomic collisions over the height of the system wall, the probabilities of atoms entering the condensed phase after a single collision with the system walls, and the probabilities of atomic departures from systems after a single collision with walls. The accuracy of the obtained formulas was compared with the data obtained from computer experiments using the Monte Carlo method.

1 Introduction

The problems of heat and mass transfer in systems have been of interest to researchers for many decades. The approach from the molecular point of view allows us to obtain different results depending on the parameters and conditions used. Clausing, using a probabilistic approach, determined the flow of matter in the cylindrical and slit systems [1]. On the basis of probable reasoning, Knudsen, Smolukhovsky, and Dushman obtained formulas for a constant gas flow in a free-molecular flow regime in long and short cylindrical pipes. In the mid-60s, Waldman did a lot of work on non-equilibrium thermodynamics in the boundary layer [2]. Using the first approximation of the iterative procedure for solving the Boltzmann equation, Fustoss obtained a formula for the molecular flow as a function of the observation angle [3]. Thus, regularities were determined for a larger number of system parameters. All the works used certain approximations, but there are no exact solutions. With the development of computer technology, it became possible to use the Monte Carlo method for analyzing the flows of matter in systems [4]. The results of calculations were analyzed depending on the number of hours played.

Modeling the process of heat and mass transfer in open systems is an urgent task, both from the theoretical and practical point of view. The peculiarity of the description of the free-molecular transport ($Kn > 1$) of the gas phase flow is that the gas dynamics equations and the Boltzmann equation cannot be used because of large density fluctuations in the gas and the insignificant number of collisions of molecules in the gas phase. The almost complete absence of collisions of atoms or molecules in the gas phase allows us to calculate the trajectories of a large number of particles. We can assume that all collisions of atoms occur with the walls of systems. The trajectories of molecules represent straight lines from one surface to another.

An important factor is the presence of a potential barrier on the surface of the condensed phase and the walls of systems. The potential barrier on the surface of the condensed phase and the walls of the system is a filter that cuts off slow particles, changes the velocity component perpendicular to the surface, and, accordingly, the trajectory of the molecules. This also applies to atoms. In the limiting case, when the potential barrier value $U = 0$, the atoms flew out from the surface of the condensed phase equally likely in all directions. This means that the integrand included in all integrals presented in the paper is $\rho = 1$.

In the present paper, the obtained probabilities are compared with the results obtained by the Monte Carlo method. It was assumed that the particles are point-like, and the results of calculations depend only on the relative height of the wall of the system H. The relative wall height is defined as the ratio of the wall height to its width. The influence of the number of N atoms played on the probability of atomic departures was established. The masses of the played atoms are equal to

40 a.u.m., but this did not affect the results of calculations. The calculation results depend on the parameter $r = U / kT$, where T is the system temperature and k is the Boltzmann constant [5,6]. Since in our case $U = 0$, the parameter $r = 0$. At $r \to \infty$, the atoms fly off the surface according to the cosine law, and the integrand will be different.

When solving such problems, three problems arise: the correctness of the program algorithm, the correctness of the random number sensors, and the number of particles played N to obtain results with a given accuracy.

2 Calculations

In this part of the work, on the basis of the proposed approach, an exact analytical formula was obtained for the probability of an atom leaving the surface of a condensed phase without colliding with the walls of the $W1(0;0)$ system. The first index determines the number of collisions of atoms with the walls of systems, and the second - the value of the parameter $r = 0$. The results of theoretical calculations were compared with the results of computer experiments.

Fig. 1 shows a cross-section diagram of a slit system with a relative wall height H. The atom is located on the surface of the condensed phase ($z = 0$) at the point x. An atom from the surface of the condensed phase can fly out of the slit system at an angle that falls within the interval $\theta \in (\theta_{cr1};\theta_{cr2})$. The value of the critical angle $\theta_{cr1} = arctg(x/H)$.

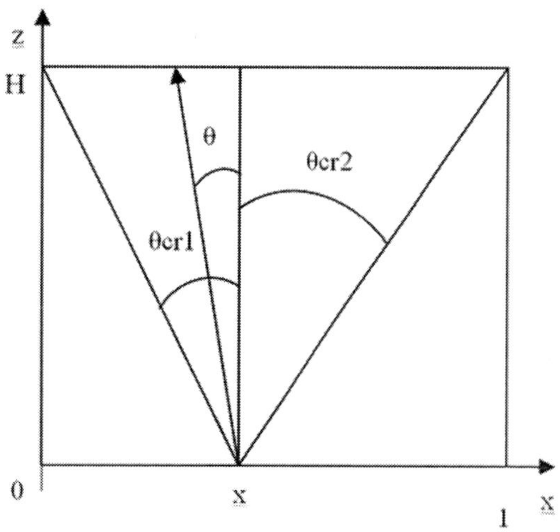

Fig. 1. The scheme of the slot system and the trajectory of departure from the system.

Find the probability of the atom's departure through the left part of the system. Due to the symmetry of the slit system, the probability of an atom flying through the right side of the system will be the same. Integrating the integral function with respect to the variable x from 0 to 1 and with respect to the variable θ from 0 to θ_{cr1}, taking into account the normalization of I, we obtain:

$$W1(0;0) = \frac{1}{I}\int_0^1 dx \int_0^{arctg\frac{x}{H}} d\theta, \qquad (1)$$

where

$$I = \int_0^1 dx \int_0^{\pi/2} d\theta. \qquad (2)$$

After calculating the double integrals, we obtain an exact formula for the probability of an atom leaving the slit system H without colliding with the walls of the system:

$$W1(0;0) = \frac{2}{\pi}(arctg\frac{1}{H} - \frac{H}{2}\ln(1+\frac{1}{H^2})) \quad (3)$$

When $H = 0$, we get that the probability of departure of atoms from such a system is 1. When $H \to \infty$ the probability of departure of atoms from a slit system tends to 0.

To compare the results of calculations obtained using this formula and the results obtained in computer experiments, computer experiments were performed with different numbers of N atoms. The calculation results are shown in table 1.

Table 1. The number of atoms ejected from the slot systems and the probability of atom departures.

N/H	0,1	0,7	2	10
The number of atoms that escaped from systems without colliding with walls				
10^5	79225	36540	15295	3216
10^6	790366	363838	152738	31390
10^7	7897575	3637825	1531312	317319
10^8	78971615	36349497	15317031	3177349
The probability of departure of atoms W1(0;0)				
10^5	0,792251	0,365401	0,152958	0,032161
10^6	0,790366	0,363838	0,152738	0,031392
10^7	0,789758	0,363783	0,153131	0,031732
10^8	0,789716	0,363495	0,153170	0,031773
Relative error				
10^5	0,328814	0,547619	0,104609	1,187504
10^6	0,091224	0,120658	0,243554	1,236381
10^7	0,014245	0,105420	0,013844	0,145595
10^8	0,009012	0,026402	0,039374	0,014509

The probability $W1(0;0)$ was defined as the ratio of the number of atoms that flew out of the system without colliding with the walls to the number of atoms played. In fact, this ratio determines the arithmetic mean, but for very large N it tends to the corresponding probability. For clarity, the first part of the table shows the results of calculations of the number of atoms that flew out of the system without colliding with the walls of the systems. It should be noted that the table shows the results of calculations of individual computer experiments. As the number of N atoms being played increases, the number of outgoing atoms increases, but the first digits practically do not change.

The results of flight probabilities obtained in computer experiments are shown in the middle part of the table.1. For comparison, the results of calculations obtained by the formula (3) are given. The last part of the table shows the corresponding relative errors of the results of computer experiments and theoretical calculations. We can note a tendency to decrease errors with an increase in the number of played N atoms. The dependence of the probability of atomic departures on the relative height of the walls of H systems is shown in Fig. 2.

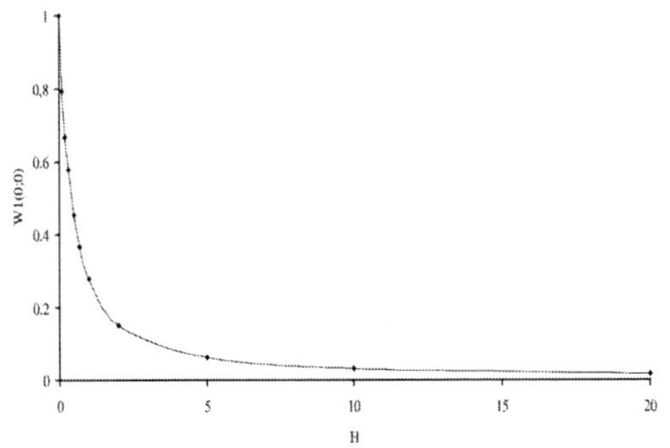

Fig. 2. Graph of the probability of departure from the systems without colliding with the walls.

Fig. 3 shows the dependences of relative errors ε as a percentage of the relative height of the walls of H systems. The relative errors obtained for large values of H have a large value, since the number of escaping atoms is less than for small slit systems. It can be noted that relative errors do not exceed 1.5%.

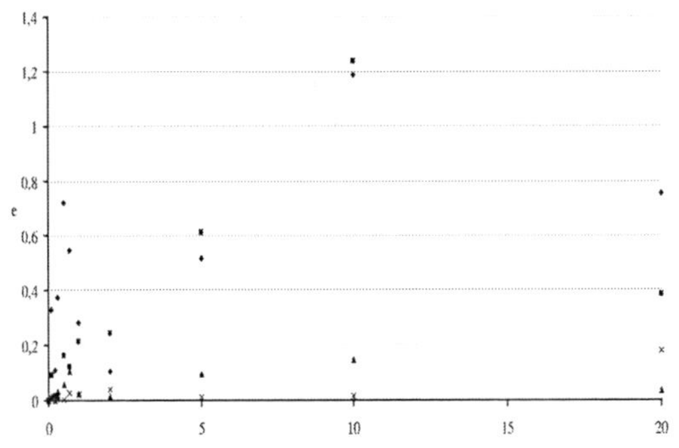

Fig. 3. Distributions of relative errors between theoretical values and computer experiments.

In another series of 12 computer experiments, data were obtained on the number of atoms that flew out of systems without colliding with the walls of systems to determine statistical deviations. The results of calculations are shown in table 2. The arithmetic mean of the corresponding data was calculated and divided by the number of played atoms $N = 10^7$. For comparison, the results of calculations using the formula (3) are presented. A comparison of the calculation results shows that the relative error of the average arithmetic probabilities of atomic departures and theoretical values does not exceed several hundredths of a percent.

Table 2. The number of atoms ejected from the slit systems, depending on the relative height of the walls of the systems.

N=10^7/H	0,1	2	10
1	7897575	1531312	317319
2	7896446	1530683	317484
3	7897084	1532382	316729
4	7896121	1533553	317127
5	7894780	1530533	317676
6	7895923	1531763	317572
7	7895626	1531524	318848
8	7897074	1532479	318318
9	7893911	1533045	318041
10	7894926	1529914	318632
11	7898377	1530485	316935
12	7896230	1530291	317641
Σ	94754073	18377964	3812322
$\overline{X}$	7896173	1531497	317693,5
$W1_{MC}$	0,789617	0,153149	0,031769
$W1_{theor}$	0,789645	0,15311	0,031778
ε%	0,003546	0,025472	0,028636

The analysis of the obtained results showed an excellent agreement between the data of computer experiments and the results obtained by formula (3).

We find formulas for determining the probability of an atom hitting the wall of the system and the density of the distribution of collisions of atoms with the wall after departure from the surface of the condensed phase. Fig. 4 shows a cross-section diagram of a slit system with a relative wall height H. The atom is located on the surface of the condensed phase ($z = 0$) at the point x.

An atom that has escaped from the surface of the condensed phase can hit the wall of the slit system at an angle that belongs to the interval $\theta \in (\theta_{cr1}; \pi / 2)$. The value of the critical angle $\theta_{cr1} = arctg(x / H)$.

Find the probability of an atom hitting the left wall of the system. Due to the symmetry of the slit system, the probability of an atom hitting the right wall of the system will be the same.

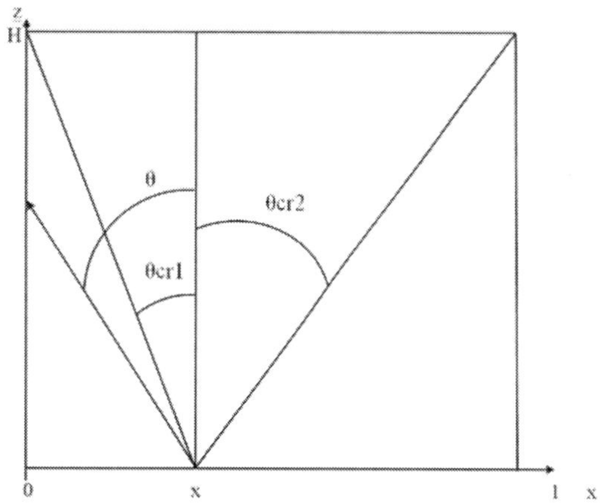

Fig. 4. Scheme of atom departure from the surface of the condensed phase and hit the wall of the system.

Integrating the integrand function with respect to the variable x from 0 to 1 and with respect to the variable θ from θ_{cr1} to $\pi/2$, taking into account the normalization of I, we obtain:

$$W2(0;0) = \frac{1}{I}\int_0^1 dx \int_{arctg\frac{x}{H}}^{\frac{\pi}{2}} d\theta, \qquad (4)$$

where

$$I = \int_0^1 dx \int_0^{\pi/2} d\theta. \qquad (5)$$

After calculating the double integrals, we obtain an exact formula for the probability of an atom hitting the walls of the slit system after leaving the surface of the condensed phase:

$$W2(0;0) = 1 - \frac{2}{\pi}(arctg\frac{1}{H} - \frac{H}{2}\ln(1+\frac{1}{H^2})). \qquad (6)$$

This formula could be obtained from the condition:

$$W1(0;0) + W2(0;0) = 1. \qquad (7)$$

The density of the probability distribution of collisions of atoms along the wall height is determined as the first derivative of the probability of atoms hitting the wall of the system after departure from the surface of the condensed phase. Replacing H with z in formula (6) and taking the derivative with respect to z, we obtain:

$$p2(0;0) = \frac{dW2(0;0)}{dz} = \frac{1}{2\pi}\ln(\frac{z^2+1}{z^2}). \qquad (8)$$

An exact formula is obtained for calculating the density of the probability distribution of collisions of atoms with one wall of a slit system after departure from the surface of the condensed phase. In the limiting case for $z \to 0$, we get that $p2(0;0) \to \infty$. In another limiting case, for $z \to \infty$ $p2(0;0) \to 0$.

The results of calculations obtained by formula (8) were compared with the results obtained in computer experiments. Computer experiments were carried out with different numbers of N atoms and systems with wall heights $H = 0.5$ and $H = 5$. The height of the wall of the system was divided into 20 intervals, and the number of atoms trapped in this cell was determined. The results of calculations for a slit system with $H = 0.5$ are shown in table 3. The density of atoms was determined as the ratio of the number of atoms trapped in a given cell to the number of atoms played and the length of the cell. The results of calculations of individual experiments are presented. The relative error ε was less than one percent. The largest deviation of the theoretical values from the results of computer experiments is observed in the first two cells near the surface of the condensed phase. This can be explained by the fact that the point $z = 0$ is the break point of the function (8).

Table 3. Densities of probability distributions of collisions of atoms over the height of the system wall.

z	0,0125	0,0625	0,1125	0,1625
M-K	1,49194	0,88363	0,69661	0,58422
теор	1,39487	0,88316	0,69744	0,58254
e,%	6,95907	0,05355	0,11986	0,28804
z	0,2625	0,3125	0,3625	0,4125
M-K	0,43518	0,38606	0,34160	0,30709
теор	0,43634	0,38507	0,34264	0,30687
e,%	0,26630	0,25709	0,30351	0,0701

Table 4. Densities of probability distributions of collisions of atoms over the height of the system wall.

z	0,125	0,625	1,125	1,625	2,125
M-K	0,7627	0,2046	0,0932	0,0513	0,0319
теор	0,6643	0,2020	0,0926	0,0511	0,0318
e,%	14,804	1,2687	0,5784	0,4863	0,3831
z	2,625	3,125	3,625	4,125	4,625
M-K	0,0216	0,0155	0,0116	0,0091	0,0072
теор	0,0215	0,0155	0,0116	0,0091	0,0072
e,%	0,1664	0,2642	0,2253	0,7798	0,3565

Table 4 shows the results of calculations for determining the densities of atomic probability distributions along the system wall for $H = 5$. The same regularities are established as for the system with $H = 0.5$. Table 3 shows the results of calculations for the same system with $H = 5$, but with an order of magnitude greater number of played atoms. The absolute error gets better for some z values, but the relative error, if any, improves only slightly.

Fig. 5 shows the density of the probability distribution of atoms over the height of the slit system wall with $H = 0.5$.

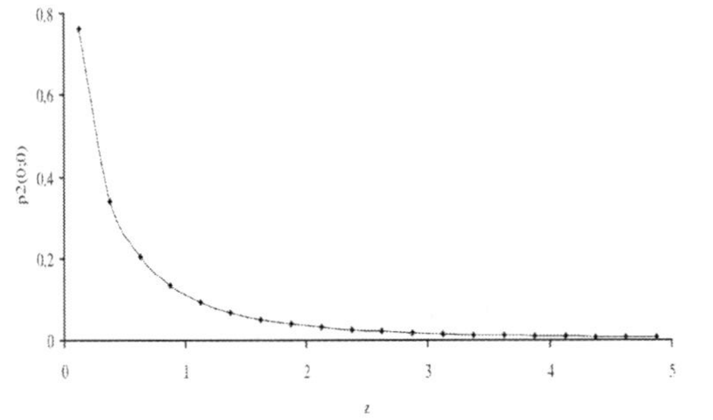

Fig. 5. Density of the probability distribution of collisions of atoms with the system wall. $H = 5$.

The distribution graphs obtained using the Monte Carlo method and the formula (8) practically do not differ from each other.

Analysis of the calculation results showed a good agreement between the data of computer experiments and the results obtained by formula (8). The obtained data can be used for comparison with the data obtained in real experiments on sputtering on glass plates.

We find the probability of atoms entering the condensed phase after leaving the system wall. In previous calculations, it was found that the density of the distribution of the collision of atoms over the height of the wall of the slit system is determined by the exact formula (8).

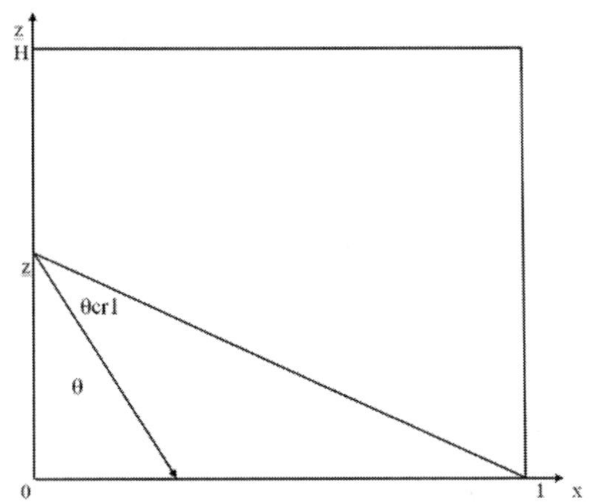

Fig. 6. Diagram of an atom entering the condensed phase after a collision with a wall.

An atom that has escaped from the wall surface can enter the condensed phase at an angle corresponding to the interval $\theta \in (0;\theta_{cr1})$. The value of the critical angle $\theta_{cr1} = arctg(1/z)$. Due to the symmetry of the slit system, the probability of an atom entering the condensed phase from the right wall of the system will be the same. Integrating the integrand (8) over the variable z from 0 to H and over the variable θ from 0 to θ_{cr1}, taking into account the normalization of I, we obtain the double integral:

$$W3(1;0) = \frac{1}{2\pi I}\int_0^H dz \ln(\frac{z^2+1}{z^2}) \int_0^{arctg\frac{1}{z}} d\theta = \\ = \frac{1}{2\pi I}\int_0^H dz \ln(\frac{z^2+1}{z^2}) arctg\frac{1}{z} \quad (9)$$

where

$$I = \frac{\pi}{2}. \quad (10)$$

The integral over the variable z can be expressed in terms of special functions, but it is advisable to solve it by numerical method.

The calculation results obtained by formula (9) were compared with the results obtained in computer experiments. Computer experiments were carried out with different numbers of N atoms and systems with several heights of H walls. The results of calculations for slit systems are shown in table 5. This table shows the results of calculations of individual computer experiments. The third part of the table shows the corresponding relative errors ε.

Table 5. The probability of atoms entering the condensed phase W3(1; 0).

H	0,1	2	5
Number of atoms trapped in the condensed phase			
$N=10^5$	10171	32235	32964
$N=10^6$	102054	321949	331802
$N=10^7$	1022759	3217659	3314116
$N=10^8$	10230987	32178188	33130734
Probabilities W3(1;0)			
$N=10^5$	0,10171	0,32235	0,32964
$N=10^6$	0,10205	0,32194	0,33180
$N=10^7$	0,10227	0,32176	0,33141
$N=10^8$	0,10231	0,32178	0,33130
Теор.	0,10233	0,32182	0,33134
Relative errors			
$N=10^5$	0,61462	0,16188	0,51306
$N=10^6$	0,27848	0,03728	0,13943
$N=10^7$	0,06165	0,01960	0,02160
$N=10^8$	0,02846	0,01464	0,00985

For the number of played atoms $N = 10^6$, the relative error does not exceed one percent. as the number of atoms played increases, the relative error decreases. for the case $N = 10^8$, the relative error does not exceed a few tenths of a percent.

Fig. 7 shows the probability distribution of atoms entering the condensed phase $W3$ (1;0) after departure from the surfaces of the walls of systems, depending on N.

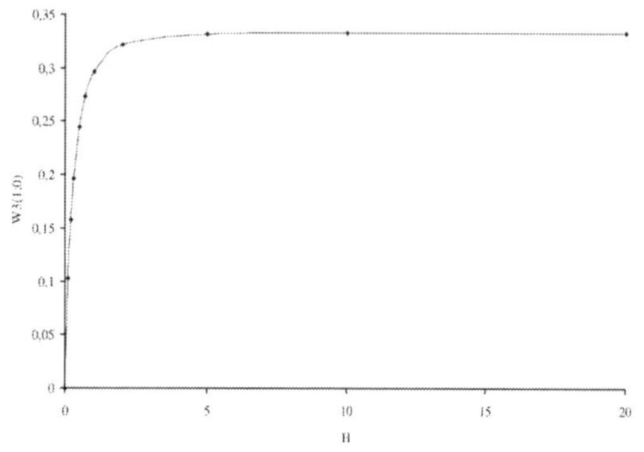

Fig. 7. The probability of atoms entering the condensed phase after a single collision with the walls.

The distribution graphs obtained using the Monte Carlo method and the formula (9) practically do not differ from each other, as can be seen from table 5. Analysis of the probability graph $W3$ (1; 0) allowed us to establish that up to the value $H = 2$ there is a rapid growth, which is replaced by a slight increase in this value. It can be assumed that the curve approaches the asymptotic value, which is equal to 1/3. Relative errors are related both to the number of atoms played N, and to the accuracy of calculating a certain integral.

Using the previously obtained formula for the density distribution of collisions of atoms with the system wall (8), we find the probability of the atoms leaving the slit system after one collision with the system wall. The scheme of the atom's departure is shown in Fig. 8.

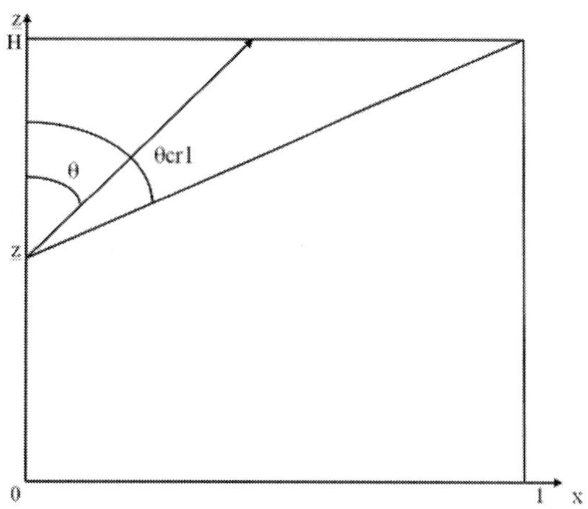

Fig. 8. Scheme of atom departure from the system after colliding with the wall.

An atom that has escaped from the wall surface can escape from the system at an angle that belongs to the interval $\theta \in (0;\theta_{cr1})$. The value of the critical angle $\theta_{cr1} = arctg(1/(H-z))$. Due to the symmetry of the slit system, the probability of an atom flying out of the system from the right wall of the system will be the same. By integrating the integrand (8) over the variable z from 0 to H and over the variable θ from 0 to θ_{cr1}, taking into account the normalization of I, we obtain the double integral:

$$W1(1;0) = \frac{1}{2\pi I}\int_0^H dz \ln(\frac{z^2+1}{z^2}) \int_0^{arctg\frac{1}{H-z}} d\theta =$$

$$= \frac{1}{2\pi I}\int_0^H dz \ln(\frac{z^2+1}{z^2}) arctg\frac{1}{H-z} \quad (11)$$

where

$$I = \frac{\pi}{2}. \quad (12)$$

The calculation of the integral with respect to the variable θ is not difficult, but the integration with respect to the variable z is quite complex and its solution for various values of H is obtained numerically. The results of the calculations are shown in the second part of table 6.

Table 6. Probabilities of atomic departures from slit systems W1(1; 0).

H	0,1	0,7	2	5
Number of atoms ejected from the system				
$N=10^5$	9955	23122	16544	7524
$N=10^6$	100874	231482	166827	75363
$N=10^7$	1012771	2322137	1670716	753142
$N=10^8$	10130587	23239929	16697251	7529611
The probability of departure of atoms $W1$(1; 0)				
$N=10^5$	0,09955	0,23122	0,16544	0,07524
$N=10^6$	0,10087	0,23148	0,16682	0,07536
$N=10^7$	0,10127	0,23221	0,16707	0,07531
$N=10^8$	0,10130	0,23239	0,16697	0,07529
Teop.	0,10133	0,23247	0,16697	0,07532
Relative errors				
$N=10^5$	1,75760	0,53770	0,91811	0,10621
$N=10^6$	0,45099	0,42500	0,08743	0,05709
$N=10^7$	0,05319	0,11025	0,05905	0,0077
$N=10^8$	0,0248	0,03041	0,00029	0,03171

For comparison with theoretical calculations, the results of calculations of computer experiments carried out using the Monte Carlo method are presented. Computer experiments were carried out with different numbers of N atoms and systems with several heights of H walls. In this table, the results of calculations of individual computer experiments are presented. The third part of the table shows the corresponding relative errors ε. For the number of atoms played $N = 10^6$, the relative error exceeded one percent in only two cases due to the small number of atoms that flew out. As the number of atoms played increases, the relative error decreases. For the case $N = 10^8$, the relative error does not exceed a few hundredths of a percent.

Fig. 9 shows the probability distributions of atomic departures from slit systems without colliding with the walls $W1(0;\ 0)$ and after one collision with the walls $W1(1;\ 0)$, depending on the relative height of the walls of systems H. The distribution graphs obtained using the Monte Carlo method and the formula (2) practically do not differ from each other, as can be seen from table 6. Analysis of the probability distribution graphs allowed us to establish qualitatively different types of dependencies. For atoms emerging from systems without collisions with the system walls, the distribution is monotonically decreasing. The graph of the probability distribution of atomic departures from systems after a single collision with walls has a maximum. The presence of the maximum can be explained by the fact that with an increase in the relative height of the walls of H systems, the probability of falling on them of atoms that have escaped from the surface of the concentrated phase increases, but at the same time the probability of departure after a collision with the walls of the system decreases. Relative errors are related both to the number of atoms played N, and to the accuracy of calculating a certain integral. Similar distributions of the probabilities of atomic departures from slit systems have been established for a larger number of collisions of atoms with the walls of systems. There was a decrease in the magnitude of the maxima and a slight shift of the maxima to the right.

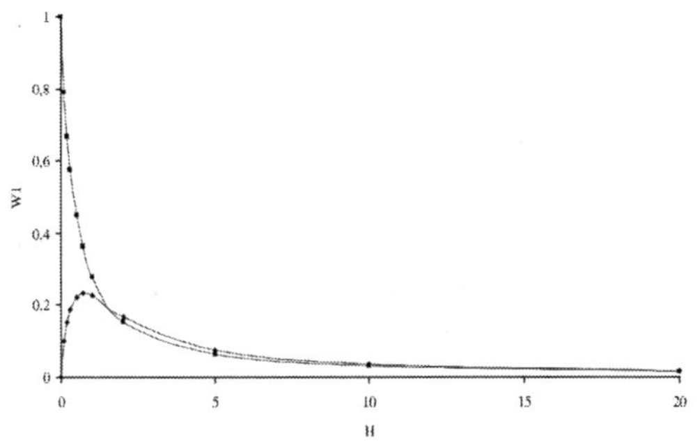

Fig. 9. The probability of atomic departures from the slit systems without collisions with the walls and after a single collision with the walls. ■ - $W1(0;0)$, ♦ - $W1(1;0)$.

Analysis of the calculation results showed an excellent match between the data of computer experiments and the results of calculations obtained by formula (11).

3 Conclusion

An approach is proposed for obtaining precise formulas for determining the probabilities of atomic outcomes from slit systems. The results of computer experiments confirmed the data obtained by the formulas with high accuracy. This indicates both the correctness of the approach for obtaining theoretical probabilities of outcomes, and the correctness of the developed computer program, the correct operation of random number sensors, and the optimal number of atoms played.

The obtained formulas can be used for testing newly developed random number sensors. This indicates the correct operation of the developed program and the correct operation of the random variable sensors.

The obtained theoretical probabilities of outcomes and the results of computer experiments can be used for comparison with the data obtained in real experiments.

References

1. P. Clansing, Annalen der Physic., **12**, 5 (1932)
2. L.Z. Waldmann, Naturforsch., A**22** (1967)
3. L. Fustoss, Rep. polyt. Mech. Eng., **21**, 3-4 (1977)
4. G.A. Bird, *Molecular Gas Dynamics and the Direct Simulation of Gas Flows* (Oxford Engineering Science Series, 1994)
5. L.V. Pletnev, N.I. Gamayunov, V.M. Zamyatin, *Mathematical Models of Non-Linear Excitations, Transfer, Dynamics, and Control in Condensed Systems and Other Media. L.A. Uvarova (eds)* (Springer, Boston, MA., 1999) https://doi.org/10.1007/978-1-4615-4799- 0_13
6. L.V. Pletnev, J. Monte Carlo Methods and Applications, **6**(3) (2000)

Modeling Cognitive Activity of the Human Brain by the Mathematical Apparatus of Quantum Mechanics

Alexandr Petukhov[1,*]

[1]Keldysh Institute of Applied Mathematics RAS, RU-125047, Moscow, Russia

Abstract. This paper discusses the possible approaches to modeling the cognitive activity of the human brain using the mathematical apparatus of quantum mechanics (primarily – potential wells and virtual particles) in terms of the theory of information representations. The article briefly describes the proposed theory of information representations, draws analogies, and identifies common features of information representations of the human mind and Feynman's virtual particles. The human mind is represented as a one-dimensional potential well with finite walls of different sizes and internal potential barrier simulating the boundary between consciousness and subconsciousness. This creates a foundation for a mathematical apparatus that can make it possible to forecast particular cognitive functions of the human brain. The results of these studies can be used to create predictive models of various cognitive disorders (diseases) and to be used in diagnostics.

1 Introduction

Natural science aspects of how the environmental information factors impact living systems are still relatively poorly studied [1]. There has been quite a number of new electrophysiological methods created, primarily, the registration of evoked and event-related potentials recorded in experimental behavioral situations, that made it possible to approach the study of physiological mechanisms of individual stages of information processing: sensory analysis, attention mobilization, image formation, extraction of memory standards, decision making, etc. The study of time parameters of electrophysiological reactions to stimuli of different types and under different conditions made it possible for the first time to register time parameters, i.e. to estimate the duration of individual stages of information processing directly at the level of brain substrate [1-2]. Many fundamental facts were also obtained using classical methods of psychology, which were used to study and analyze the behavior of subjects in various social situations.

The problem of describing the processes of information transfer and processing by an individual is fundamental for modern cognitive science. Relatively recently, unique natural science models of information transfer from individual to individual [2-5], cognitive activity modeling [6-13], and others have appeared. However, many of the presented models are either poorly scalable or insufficiently formalized and therefore do not allow to fundamentally explain the processes of information transfer and its distortion resulting from interaction with the external communicative environment.

Moreover, one aspect of human cognitive activity is the fact that a person's thinking is not code-based (as compared to a computer), but is a result of the interaction of multiple images. Although these images have quite specific material structure at the heart of their functioning (namely, electrical and chemical activity in the human brain), from the point of view of conventional mathematical models their description is difficult for several reasons [11].

Therefore, this paper proposes new methods for describing the activity of information representations (which, in turn, simulate the cognitive activity of an individual) based on the mathematical apparatus of quantum physics (potential wells and virtual particles used in physics to describe fundamental interactions [14]).

2 A brief overview of the theory of information representations

The proposed theory is based on the idea of a universal cognitive unit [11] of information in the human mind - an information representation (or image), the space in which it exists, its topology and properties. Information representations (hereinafter referred to as IRs) can be defined as displays of objects and events in any feature space.

Correspondingly, the theory of information representations (hereinafter referred to as TIR) is a way to describe information interactions of individuals, as well as several cognitive functions of a person.

* Corresponding author: spetoukhov@gmail.com

TIR views the human mind as a huge structured number of interacting images, which are constantly affected by external factors.

Images with higher energy (we introduce the notion of energy E to describe the communicative activity of images) are located "above" and closer to the edge of the individual's information image space, which is why they interact with external elements much more often; meanwhile, images with lower energy and longer response rate are located closer to the center of the space and relatively rarely interact actively with external stimuli.

TIR gives an alternative perspective on some characteristic regularities in the human mind, allowing for a correct interpretation and explanation of some of them.

The information image can not be communicated between individuals without changes. Each IR is unique because each individual has specific individual experiences.

Besides, an individual is unable to convey the image that exists in their mind, in their IR space to another individual directly. For this purpose, they use various communication apparatuses formed within the social superstructure of communication or the communication field (CF). A communication field (CF) is an information community of individual experiences and collective unconsciousness formed as a result of an individual's presence in society. The communicative apparatus includes speech, visual, tactile, symbolic, and other ways of information transmission.

In previous works, the mathematical model of interaction of information representations was based on diffusion equations (first of all, on the Langevin equation). This approach allowed to describe several special cases of cognitive processes but also had some limitations related to the specificity of IRs. More details about TIR and its application to real cognitive processes and experimental testing are available in [15].

3 A brief overview of the theory of information images

What is the peculiarity of the theory of virtual particles and why should it be used to describe cognitive processes in the human brain? Virtual particle (VP) is quanta of relativistic wave fields participating in vacuum fluctuations. From the general quantum-mechanical point of view, VP can be considered as particles arising in intermediate states of transition processes and interaction of particles. VPs have the same quantum numbers as ordinary real particles, the quantum numbers of which, in turn, correspond to one of the real elementary particles (with mass m), for which, however, the usual connection between energy and momentum is not satisfied.

Which means that for them:

$$E^2 \neq m^2c^4 + p^2c^2$$

Where c is the speed of light in vacuum.

Therefore, a measure of "virtuality" of the particle was introduced, which is expressed as:

$$Q^2 = E^2 - m^2c^4 - p^2c^2 \quad (1)$$

Virtual particles cannot "travel infinitely"; they are generated and must either be absorbed by some particle or disintegrate [14]. Basically, virtual particles are the way the interaction takes place.

This is very close to the concept of IRs for human interaction, because they are also present only in the virtual world, but also have material properties, can have a concrete impact on the matter (on the human body), and are designed to describe the processes of exchange, generation, and processing of information by an individual. IRs also cannot "exist infinitely"; their existence is conditioned exclusively by the presence of a perceptive individual (imagine a picture that no one is looking at). Therefore, each information image must also be absorbed or embodied in a material form [15].

Additionally, VPs are responsible for the quantum mechanism of interaction of particles - they are the carriers of interactions. For example, the scattering of charged particles due to electromagnetic interactions between them on quantum field representations is carried out through the exchange of virtual photons.

Virtual particles exist in intermediate short duration states, for which the usual ratio between energy, pulse and mass is not met. Other characteristics of virtual particles are electric charge, spin, baryon charge, etc. - are the same as for corresponding real particles [14].

The concept of "Virtual Particles" and virtual processes is central to modern quantum field theory. In this theory, the interaction of particles and their mutual transformations are considered as the generation or absorption by one free particle of other (virtual) particles. Any particle continuously emits and absorbs VPs of different types. For example, a proton emits and absorbs virtual pi-mesons (and other VPs) and thus appears surrounded by a cloud. In this case, the number of VPs is uncertain.

Here we can also draw an analogy with an individual who generates a "cloud" of their information images, which essentially represent their mind and accumulated individual and social experience [15].

In addition, from the point of view of classical physics, a free particle (a particle which is not affected by external forces, i.e. resting or moving evenly and linearly) can neither generate nor absorb another particle (for example, a free electron can neither emit nor absorb a photon), since such an occurrence would violate either the law of conservation of energy or the law of conservation of momentum. Indeed, the resting electron has the minimum possible energy. That is why such an electron can not emit a photon that always possesses energy, as this would violate the law of energy conservation. If the electron moves at a constant speed, it also cannot (due to its kinetic energy) generate a photon, because this would violate the law of conservation of momentum: the electron's momentum loss caused by the energy loss from photon generation would be larger than the impulse of the photon corresponding to its energy (due to the difference in masses of these particles). The same applies to the process of absorption of the photon by a free electron.

However, the situation in quantum mechanics is quite different. According to the fundamental principle of quantum mechanics - the uncertainty principle (see Indeterminacy Relation), the energy of any particle "existing" in a small interval of time Δt is not fixed. The energy and momentum are continuously fluctuating, and during small intervals of time, the law of energy conservation may be "temporarily violated" (in the classical sense), while processes occurring at small scales may be accompanied by "local violations" of the law of conservation of momentum [14].

Moreover, it would not be accurate to refer to conservation laws in the usual sense in the case of IR, which, of course, are limited to certain energy indicators based on their material carriers, i.e. – human beings. However, if we are not talking about the entire IR system, but about its individual information images, it is obvious that the classical connection (1) will also be broken (in previous models IRs were presented as a higher number of interacting particles [15]).

Sometimes, for the sake of clarity, the concept of "virtual particles" is explained somewhat differently. Namely, it is mentioned that in the process of interaction, the law of energy conservation is satisfied with some permissions. This does not contradict quantum mechanics: according to the indeterminacy principle, an event lasting for a finite period of time does not allow to determine the energy with an accuracy exceeding a certain limit. Roughly speaking, intermediate particles "borrow energy" for a short time. In this case, ordinary particles can be generated and disappear in the process of interaction, only with a small violation of the law of energy conservation.

The interaction of ordinary, real particles in the vast majority of cases occurs via emission and absorption (exchange) of VPs. The energy and momentum of real particles before and after the reaction remain unchanged, and during the reaction, the laws of preservation of these values are not satisfied. The whole theory is constructed in such a way that any reaction can be represented as a result of various virtual processes occurring during a short reaction time [16].

Additionally, the interaction of individuals is a process of emission and absorption (processing and perception) of IRs. Thus, we can describe any information interaction as a result of virtual processes related to IR.

Thus, there are quite a few analogies between the processes of interaction of particles in quantum field theory and individuals in the theory of information images. Of course, such analogies cannot be interpreted as a sign of a complete concordance of the two theories; however, it should be taken into account that an individual is also a part of nature, and, therefore, those patterns, laws, phenomena that surround us (not only phenomena related to VPs but also diffusion, laws of conservation, etc.) must somehow manifest in individuals and their cognitive processes.

However, this can only be reliably verified by using the appropriate mathematical apparatus to predict specific cognitive processes and verify results in an experimental study.

4 Mathematical apparatus

The classical method for describing a virtual process (a process involving virtual particles) is the Feynman diagram method [17]. However, the direct application of such a device for IRs would hardly be possible. First, it is necessary to perform parameterization of IRs and compare these values with VP parameters.

Interactions of virtual particles are described by the propagator (distribution function) - a function that determines the amplitude of probability of distribution of a relativistic field (particle) between two stages of interaction. The propagator is determined through the evolutionary operator:

$$D(x,t;x',t') = \langle x|\hat{U}(t,t')|x'\rangle$$

It allows considering the influence of virtual particles. In fact, the propagator is a function of the Green's wave equation. In general, in elementary particle physics and statistical physics, Green's functions are often used as propagators in Feynman diagrams (and the expression "Green's function" is often applied in general to the correlation function in quantum field theory).

Within the model, the interaction of two individuals is represented as the interaction of two systems with the help of communication fields (which are quantized through IR - as an analog of VP) in the information environment. Also, of interest is the impact on one individual of a given information environment (e.g., media, Internet resources, social environment, etc.).

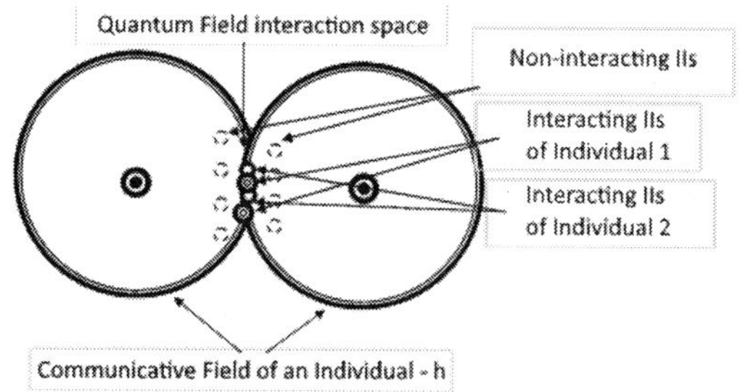

Fig. 1. Interaction of communicative fields of two individuals

The interaction function for these two individuals or an individual and the external action should be written accordingly.

Since we are talking about the IR space, it is obvious that this space will have certain coordinates; the particles simulating the IR movement in this space will also have a certain momentum, energy. At the same time, we should keep in mind the presence of local minimum potential particle energy in the space. Thus, an individual's mind can be represented as a potential well, inside of which information representations make vibrational movements. It is clear, that it is common for physics to solve the problem of quantization of the field inside the potential well to determine energy levels (by solving the stationary Schrödinger equation [18]). However, in this case, we know about the presence of particles in advance.

The potential well can be represented in the simplest form as - **Fig.2.**

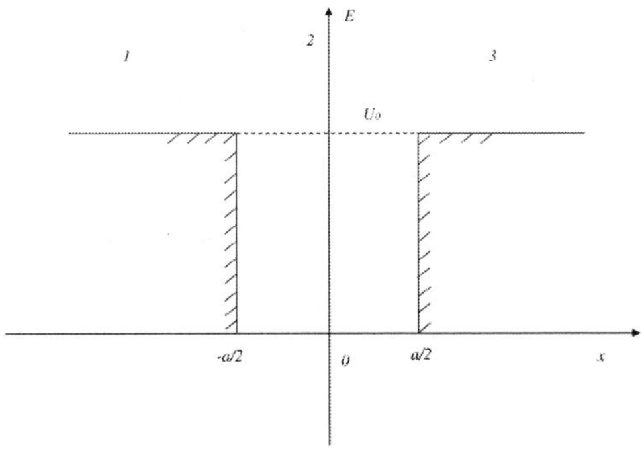

Fig. 2. The most basic one-dimensional potential well with finite walls.

$$U(x) = \begin{cases} 0, -\frac{a}{2} < x < \frac{a}{2}, \text{region 2} \\ U_0, \quad x \geq \frac{a}{2}, \text{region 3} \\ \quad\;\; x \leq -\frac{a}{2}, \text{region 1} \end{cases} \quad (2)$$

When a particle crosses the potential barrier in the form of a wall, it simulates the information interaction of an individual (i.e., particle disturbance, the transition from one energy level to a higher one).

A more interesting case of a potential well is the well uneven walls and an additional internal barrier. Such barrier and heterogeneity of walls simulate certain specific properties of the human mind, in particular, the conditional division into consciousness-subconsciousness, complexity of interaction with the external environment, etc. Theoretically, depending on the mental state of the modeled individual (or their individual cognitive function), such barriers may be more complex and repeated multiple times - Fig.3.

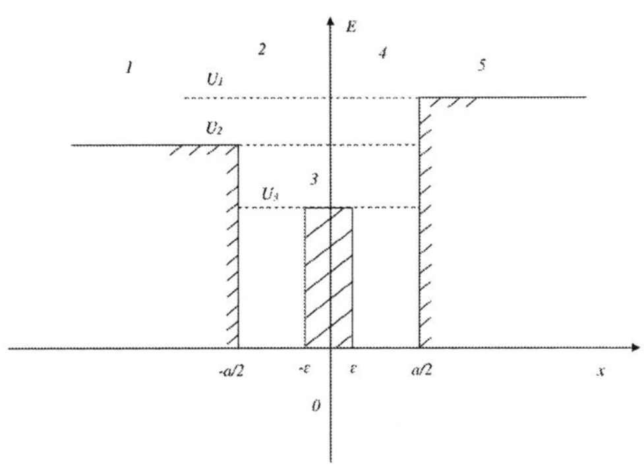

Fig .3. One-dimensional potential well with finite walls. A version with uneven walls and internal potential barrier.

$$U(x) = \begin{cases} U_1, x \geq \frac{a}{2} \\ 0, \varepsilon < x < \frac{a}{2} \\ U_2, -\varepsilon \leq x <\leq \varepsilon \\ 0, -\frac{a}{2} < x < -\varepsilon \\ U_0, x \leq -\frac{a}{2} \end{cases} \quad (3)$$

Thus, the impact of the external information environment on the mind of an individual can be described as the impact on particles (modeling information images) in a potential well. Let us record the equations for the case in Fig. 2.

The Schrödinger equation outside the potential well for the information image (IR) will look as:

$$\frac{d^2\varphi_{out}(x)}{dx^2} - \frac{2m}{\hbar^2}(U_0 - E)\varphi_{out} = 0 \quad (4)$$

Where,

φ_{out} is a wave function outside the potential well.
E is the total energy of an IR.
m is mass (complexity) of IR.
$\hbar$ is Planck constant.

By introducing the interaction coefficient k_1,

$$k_1 = \sqrt{\frac{2m}{\hbar^2}(U_0 - E)}$$

We obtain,

$$\frac{d^2\varphi_{out}(x)}{dx^2} - k_1{}^2\varphi_{out} = 0 \quad (5)$$

For region 1, the solutions of this equation will look as:

$$\varphi_1(x) = A_1 e^{k_1 x} + B_1 e^{-k_1 x}$$

$$\varphi_3(x) = A_3 e^{k_1 x} + B_3 e^{-k_1 x} \quad (6)$$

For the wave function to be limited (which is impossible otherwise for real information images), B_1 и $A_3 = 0$.

For region 2, located in the well (i.e. in the human mind), it will look as:

$$\frac{d^2\varphi_{out}(x)}{dx^2} + \frac{2m}{\hbar^2}(E)\varphi_{out} = 0 \quad (7)$$

Here, the interaction coefficient k_2 will be,

$$k_2 = \sqrt{\frac{2m}{\hbar^2}E} \quad (8)$$

And the solution will look as:

$$\varphi_2(x) = C\sin(k_2 x + \propto)$$

Thus, the wave functions for an information image in the mind of an individual will have a form:

$$\varphi_1(x) = A_1 e^{k_1 x}$$

$$\varphi_2(x) = C\sin(k_2 x + \alpha)$$

$$\varphi_3(x) = B_3 e^{-k_1 x} \quad (9)$$

The external impact will be recorded by disturbance V(x), which is added to the general form of the Schrödinger equation:

$$\frac{d^2\varphi(x)}{dx^2} - \frac{2m}{\hbar^2}(U(x) + V(x) - E)\varphi = 0 \quad (10)$$

Disturbance can be set in different ways, depending on its type, for example, by normal distribution or other methods.

This model should become the foundation for experimental research and practical testing of the modeling results in the future.

5 Conclusion

Thus, in this paper, we presented the fundamentals of the mathematical model of the human mind based on the mathematical apparatus of quantum mechanics (primarily, the potential wells and elements of virtual particles in the problem statement), defined the task of modeling the cognitive activity of the human brain based on classical mathematical models and physical analogies, gave a brief overview of the theory of information representations and its foundations, highlighted the general regularities and the phenomena connected with the behavior of information representations of the human mind and virtual particles and proposed options of potential wells for modeling cognitive brain functions.

This allowed us to propose a foundation for a mathematical model, which will be further tested experimentally (based on external information impact on the individual and his psychophysiological reactions).

The results of these studies can be used to create predictive models of various cognitive disorders (diseases) and to be used in diagnostics. First of all, this concerns the problems of perception of information, the processing of complex images (connecting different categories of information, for example, color and sound), etc.

References

1. Y.I. Aleksandrov, Proceedings of ISA RAS, **61**(3), 3-25 (2011)
2. D.S. Chernavsky, *Synergy and information. Dynamic information theory*. (URSS, 2009)
3. W. Gevers, R.C. Kadosh, W. Notebaert, Journal of Experimental Psychology: Learning, Memory, and Cognition, **37**(5), 1243–1249 (2011) DOI:10.1037/a0023550
4. T.M. Lee, H.L. Liu, C.C. Chan, (Eds.), Neuroimage. **28**(2), 305-313 (2005) https://doi.org/10.1016/j.neuroimage.2005.06.051
5. D. Griffith, F. Greitzer, Cognitive informatics and natural intelligence, **1**(1), 39-52 (2007) DOI:10.4018/jcini.2007010103
6. J. Vandekerckhove, Journal of Mathematical Psychology, **60**, 58–71 (2014)
7. O. Faugeras, J. Inglis, Journal of Mathematical Biology, **71**(2), 259-300 (2015)
8. B.W. Kooi, J. Math. Biol. **71**, 1575–1605 (2015) DOI: 10.1007/s00285-015-0869-0
9. P. Haazebroek, S. van Dantzig, B. Hommel, Cognitive Processing, **12**(4), 355-365 (2011) DOI:10.1007/s10339-011-0408-x
10. B.B. Velichkovsky, A.N. Gusev, V.F. Vinogradova, O.A. Arbekova, Experimental psychology, **9**(1), 5-20 (2016) [in Russian] DOI:10.17759/exppsy.2016090102
11. K.V. Anokhin. *Probes for Mapping Nervous Networks during Training. Principles and mechanisms of human brain activity* (L.: Nauka, 1989)
12. F. Pana, et al, Neuroscience Letters, **628**, 35–39 (2016) https://doi.org/10.1016/j.neulet.2016.05.062
13. A.R. Nikolaev, R.N. Meghanathan, C. van Leeuwen, Brain and Cognition, **107**, 55–83 (2016) https://doi.org/10.1016/j.bandc.2016.06.004
14. *Microcosm Physics* (1980)
15. A.Y. Petukhov, S.A. Polevaya, International Journal of Biomathematics, **10**(6), 1750092 (2017) DOI:10.1142/S1793524517500929
16. C. Wilson, G. Johansson, A. Pourkabirian, et al, Nature, **479**, 376–379 (2011) https://doi.org/10.1038/nature10561
17. R. Feynman, *Theory of Fundamental Processes* (M., 1978)
18. A. Bohm, *Quantum mechanics: foundations and applications* (M. Mir, 1990)

Application of Algorithms for Placement of Orthogonal Polyhedrons for Solving the Problems of Packing Objects of Complex Geometric Shape

Vladislav Chekanin[1,*] and *Alexander* Chekanin[1]

[1]Moscow State Technological University "STANKIN", RU-127055, Moscow, Russia

Abstract. The article is devoted to the development and research of algorithms for placing objects of complex geometric shapes. To solve the placement problem is proposed an approach which consists in transforming the shape of all objects and further application of the developed algorithm for placing orthogonal polyhedrons of arbitrary dimension to the resulting transformed objects. In the process of transforming the shape of the objects being placed, they are initially voxelized, after which the developed decomposition algorithm is applied to the resulting voxelized objects, which provides the formation of orthogonal polyhedrons consisting of the largest possible orthogonal objects. The proposed model of potential containers is used to describe the free space of containers as a set of orthogonal areas. The developed algorithm for the placement of orthogonal polyhedrons provides a fast solution to NP-hard problems of placing objects of complex geometric shapes without resorting to the use of time-consuming nonlinear programming methods. Examples of the practical application of the developed algorithms for modeling the dense layout of parts of complex geometric shapes on the platform of a 3D printer are given.

1 Introduction

The problem of packing objects of complex geometric shape has a large number of practical applications in various areas. Solving the problems of cutting industrial materials, rational usage of free spaces (for example, ships or aircraft), geometric surface coverage with specified shapes, modelling the microstructure of composite materials as well as a number of other optimization problems is reduced to packing problem [1-6].

All packing problems are NP-hard optimization problems for which there are no algorithms of polynomial complexity to solve them, which makes it urgent to develop algorithms that provide suboptimal solutions in an acceptable time. Modern methods for solving problems of packing objects of complex geometric shape are based on the use of the hodograph vector function of dense placement, which requires the subsequent application of nonlinear programming methods characterized by high computational complexity [7-12].

To reduce the complexity of the problem, we propose an approach that consists in packing of voxelized objects [13-15] by applying the developed algorithm of placement of orthogonal polyhedrons.

We will consider the problem of placement orthogonal polyhedrons in the general D-dimensional case. A container is specified in the form of D-dimensional parallelepiped with the dimensions $\{W^1;W^2;\ldots;W^D\}$ (the superscript in formulas means the number of the coordinate axis) as well as specified a set of n orthogonal polyhedrons $O_i, i \in \{1,\ldots,n\}$, each of which consists of m_i orthogonal objects in the form of D-dimensional parallelepipeds $o_{i,k}, k \in \{1,\ldots,m_i\}$ with the dimensions $\{w_{i,k}^1;w_{i,k}^2;\ldots;w_{i,k}^D\}$, the position of which relative to each other is specified using vectors $\{z_{i,k}^1;z_{i,k}^2;\ldots;z_{i,k}^D\}$ containing the coordinates of orthogonal objects in the local coordinate system associated with each orthogonal polyhedron O_i [16]. In the particular case, when all orthogonal polyhedrons consist of only one orthogonal object $m_i = 1\ \forall i \in \{1,\ldots,n\}$, the considered problem will be reduced to the classic D-dimensional orthogonal packing problem [1, 2].

2 Description of free container spaces

To describe a packing is used the developed model of potential containers [17-19]. Under a potential container (PC) placed in a container at some its point is understood an imaginary orthogonal object with the largest possible dimensions that can be placed at this point without overlapping with any packed into the container object and edges of the container. Each potential container h is described with a vector $\{p_h^1;p_h^2;\ldots;p_h^D\}$ containing its

* Corresponding author: vladchekanin@rambler.ru

dimensions and a vector $\{x_h^1; x_h^2; \ldots; x_h^D\}$ containing coordinates of its point which is nearest to the origin of the container containing it.

All existing free orthogonal spaces located in a container are described by a set of potential containers. When a new orthogonal object is put into a container it is necessary to verify the correctness of the placement. The model of potential containers guarantees the correct placement of an orthogonal object if it overlaps no borders of the potential container in which it is located. In this case when an object is put at some point of a container instead of checking on the intersection with all placed into the container objects is required to check only one condition of placement of this object entirely within the potential container, located at this point. This ensures a higher speed of formation the orthogonal packing.

When an orthogonal object i with the dimensions $\{w_i^1; w_i^2; \ldots; w_i^D\}$ is placed at a point $\{x_i^1; x_i^2; \ldots; x_i^D\}$ it divides the potential container h into a set of smaller potential containers from two sets:

1) set of D potential containers with the dimensions $\{p_h^1; p_h^2; \ldots; p_h^{d-1}; x_i^d - x_h^d; p_h^{d+1}; \ldots, p_h^D\}$ located in the origin $\{x_h^1; x_h^2; \ldots; x_h^d; \ldots; x_h^D\}$ of the potential container h produced under the following conditions: $x_i^d > x_h^d$ and $x_i^d < x_h^d + p_h^d \quad \forall d \in \{1, \ldots, D\}$;

2) set of D potential containers with the dimensions $\{p_h^1; p_h^2; \ldots; p_h^{d-1}; x_h^d + p_h^d - x_i^d - w_i^d; p_h^{d+1}; \ldots; p_h^D\}$ located at D points with coordinates $\{x_h^1; x_h^2; \ldots; x_h^{d-1}; x_i^d + w_i^d; x_h^{d+1}; \ldots; x_h^D\}$ produced under the following conditions: $x_i^d + w_i^d > x_h^d$ and $x_i^d + w_i^d < x_h^d + p_h^d \quad \forall d \in \{1, \ldots, D\}$.

Figure 1 presents all new potential containers which are formed in a three-dimensional potential container after placing an orthogonal object inside it.

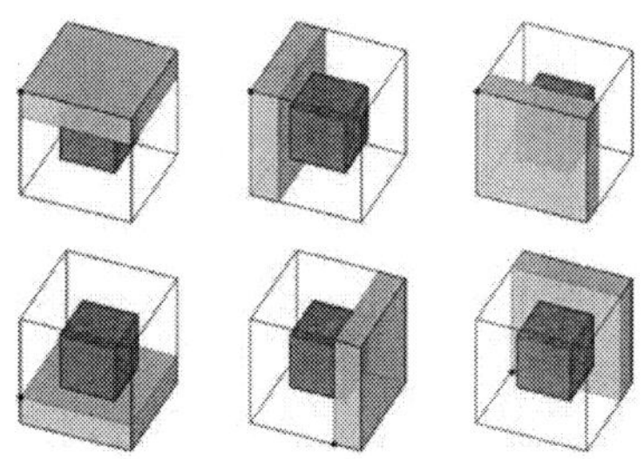

Fig. 1. New potential containers in a three-dimensional case.

To update a set of potential containers after placing an orthogonal polyhedron O_i at the point $\{X_i^1, X_i^2, \ldots, X_i^D\}$ of a D-dimensional container, the free spaces of which are described by a set of potential containers Ω_0, the following algorithm is performed.

Step 1. Create a set $\Omega_0' \subset \Omega_0$ of potential containers $h: \exists d \in \{1, \ldots, D\}: x_h^d \le X_i^d + S_i^d$, where $\{S_i^1, S_i^2, \ldots, S_i^D\}$ denotes the overall dimensions of a D-dimensional parallelepiped bounding the orthogonal polyhedron O_i: $S_i^d = \max\left(z_{i,k}^d + w_{i,k}^d\right)$, $\forall d \in \{1, \ldots, D\}$, $k \in \{1, \ldots, m_i\}$.

Step 2. Place an orthogonal polyhedron O_i in the specified position $\{X_i^1, X_i^2, \ldots, X_i^D\}$ of a new identical empty container, as a result of which a set of potential containers Ω will be formed in it. Placement of the orthogonal polyhedron is performed by sequentially placing of all its orthogonal objects $o_{i,k}$, $k \in \{1, \ldots, m_i\}$.

Step 3. Apply the intersection operation (Figure 2) to sets of potential containers Ω_0' and Ω to get a set of potential containers $\Omega_0'' = \Omega_0' \cap \Omega$ that describes all free spaces of the original container in the area of the placed orthogonal polyhedron O_i. During the intersection operation, each set of potential containers is considered as an orthogonal polyhedron, consisting of orthogonal objects whose parameters coincide with the parameters of the corresponding potential containers.

Step 4. Replace in the set Ω_0 all potential containers that are also in the set Ω_0' with potential containers from the set Ω_0''.

This algorithm is shown in Figure 3.

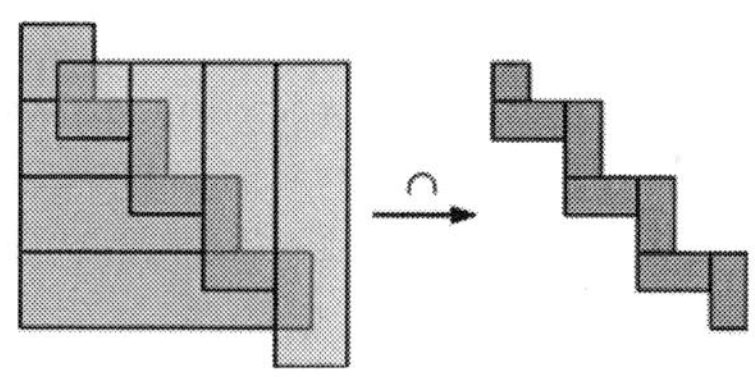

Fig. 2. Intersection operation.

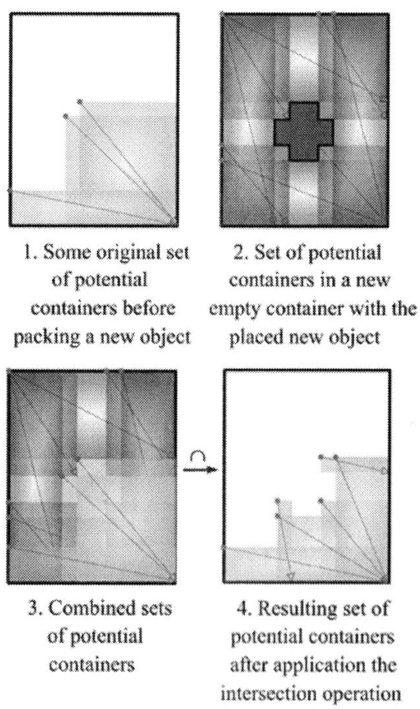

Fig. 3. Updating of potential containers.

3 Preparing objects for placement

As a result of converting objects to a voxel representation, orthogonal polyhedrons consisting of a large number of cubes are obtained. To reduce the number of orthogonal objects forming an orthogonal polyhedron, a decomposition algorithm has been developed. The algorithm provides the decomposition of a D-dimensional orthogonal polyhedron V into a set of large orthogonal objects includes steps 1–6.

Step 1. Create an empty D-dimensional orthogonal container 1 with the overall dimensions $\{W_1^1, W_1^2, \ldots, W_1^D\}$ matching the dimensions of a D-dimensional parallelepiped bounding the original orthogonal polyhedron V ($W_1^d = S^d, \forall d \in \{1,\ldots,D\}$, where S^d is the length of the packing measured along the axis d.

Step 2. Place the orthogonal polyhedron V into container 1.

As a result in the container 1, a set of potential containers Ω_1 with the overall dimensions $\{w_{k_1}^1, w_{k_1}^2, \ldots, w_{k_1}^D\}$, $k_1 \in \Omega_1$ located at points $\{p_{k_1}^1, p_{k_1}^2, \ldots, p_{k_1}^D\}$ will be formed. The set of potential containers Ω_1 describes the space of container 1, which does not belong to the placed orthogonal polyhedron V.

Step 3. Create an empty D-dimensional orthogonal container 2 with the overall dimensions that match the overall dimensions of the container 1 ($W_2^d = W_1^d$, $\forall d \in \{1,\ldots,D\}$).

Step 4. Place in the container 2 a set of D-dimensional orthogonal objects with parameters matching the parameters of potential containers from the container 1 (when placing objects, their mutual overlap is allowing): $x_i^d = p_{k_1}^d$ and $w_i^d = w_{k_1}^d$ $\forall d \in \{1,\ldots,D\}$, $i = 1\ldots|\Omega_1|$. As a result in the container 2, a set of potential containers Ω_2 with the overall dimensions $\{w_{k_2}^1, w_{k_2}^2, \ldots, w_{k_2}^D\}$, $k_2 \in \Omega_2$, located at points $\{p_{k_2}^1, p_{k_2}^2, \ldots, p_{k_2}^D\}$, will be formed. The set Ω_2 describes the space of container 2, which belongs to the orthogonal polyhedron V.

Step 5. Create a D-dimensional orthogonal polyhedron V', consisting of orthogonal objects with parameters matching the parameters of potential containers from the container 2: $z_i^d = p_{k_2}^d$ and $w_i^d = w_{k_2}^d$ $\forall d \in \{1,\ldots,D\}$, $i = 1\ldots|\Omega_2|$, so the orthogonal polyhedron V' will contain the set of all orthogonal objects into which it can be decomposed.

Step 6. Apply the addition operation to all orthogonal objects that are part of the orthogonal polyhedron V'. As a result, an orthogonal polyhedron O will be obtained, consisting of the largest orthogonal objects that do not overlap each other.

The operation of addition orthogonal objects (Figure 4) includes the following steps.

1. Place all objects from the set A into the set A^+. Create sets $B = \varnothing$, $C = \varnothing$.

2. Sort the set A^+ in descending order of volumes (areas) of objects.

3. Include the first object $o_1 \in A^+$ in the set C.

4. For each current object $o_k \in A^+$ $(k = 2\ldots|A^+|)$ check for overlapping with the object o_1. If the objects do not overlap each other, then place the current object o_k in the set B; otherwise, perform the cutting off procedure for all objects, and then place the obtained objects in the set B.

5. Clear the set A^+. If the set $B \neq \varnothing$, then move all the objects from the set B to the set A^+, and then go to step 2.

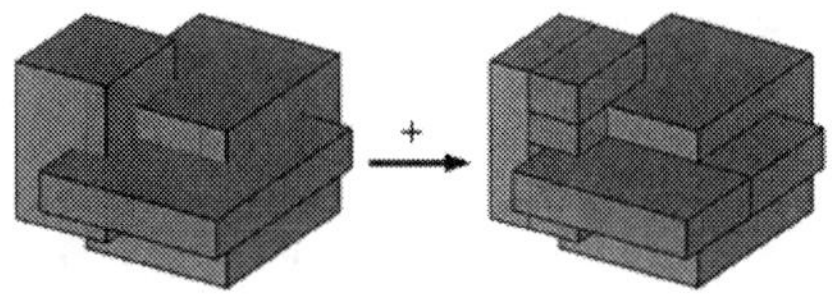

Fig. 4. Addition operation.

Unlike decomposition algorithms developed by other authors [20-22], the proposed algorithm is described and programmatically implemented invariantly with respect to the dimension of the problem to be solved.

The decomposition significantly increases the speed of placing objects. Figures 5 as an example shows original voxelized ball as well as the obtained decomposed orthogonal polyhedron. In this figure also shown time which was spent to place the orthogonal polyhedrons (here and later in this paper, the results of computational experiments performed on a personal computer with CPU Intel Core i5-8400, 2,8 GHz and 8 GB RAM are presented). An example of the decomposition of a complex geometric shape is shown in the Figure 6.

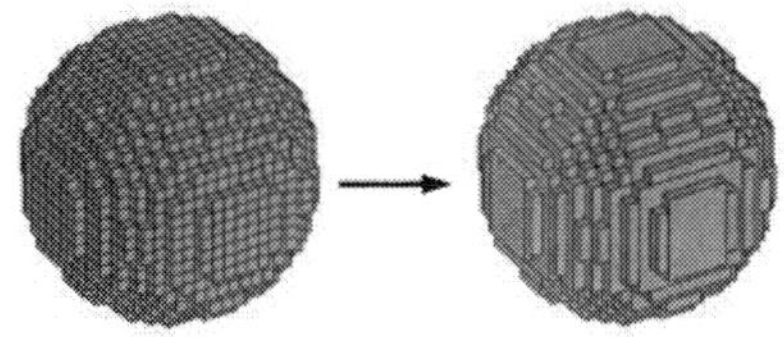

Fig. 5. Decomposition of a voxelized ball.

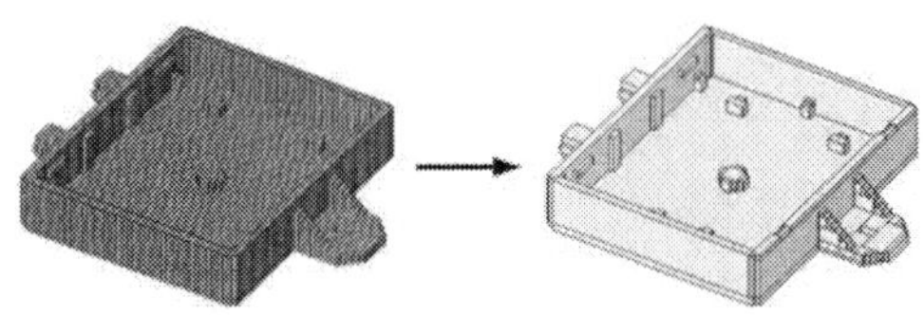

Fig. 6. Decomposition of a voxelized object of a complex geometric shape.

4 Placement of orthogonal polyhedrons

Figure 7 presents the developed algorithm for packing an orthogonal polyhedron (OP) O_i of arbitrary dimension D into a container.

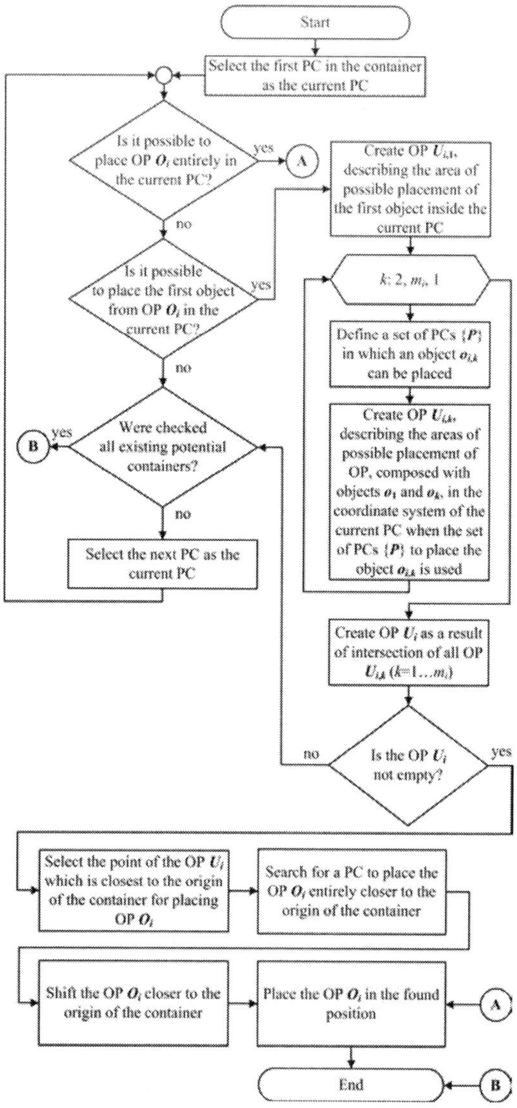

Fig. 7. Algorithm for placing an orthogonal polyhedron.

This algorithm is based on the creation of orthogonal polyhedrons that determine the possible placement areas for each object of complex shape. To reduce the complexity of the block diagram, in Figure 7 is presented a part of the algorithm which provides placing of an orthogonal polyhedron inside only one current container. Algorithm for packing orthogonal polyhedrons in a set of containers in presented in the paper [23].

The steps for sequentially determining the region of admissible placement of an orthogonal polyhedron in a given potential container are shown in Figure 7. The final orthogonal polyhedron is obtained after application the intersection operation to all orthogonal polyhedrons describing areas of admissible placement.

Figure 9 presents an example of placement orthogonal polyhedrons into a container.

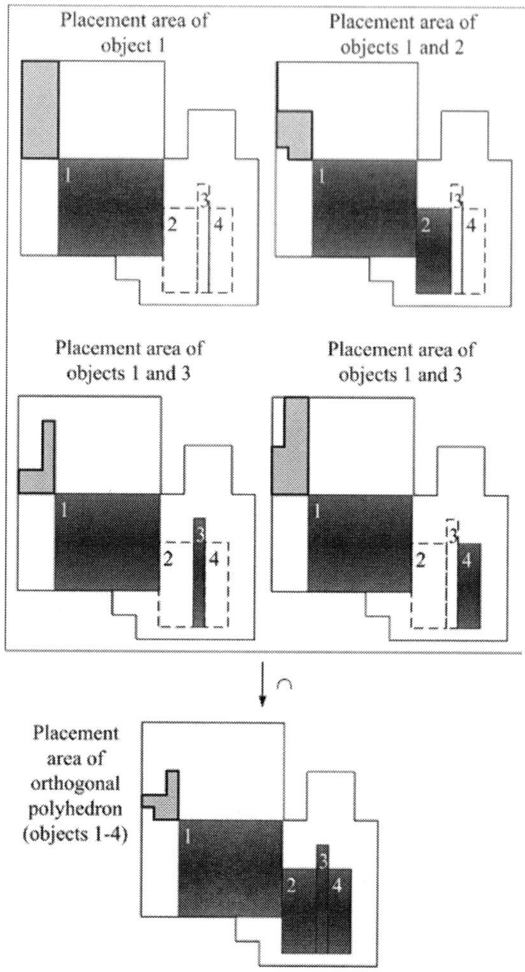

Fig. 8. Finding the admissible area of placement of an orthogonal polyhedron (filled with a gradient fill) in a potential container located in the upper left corner of the container.

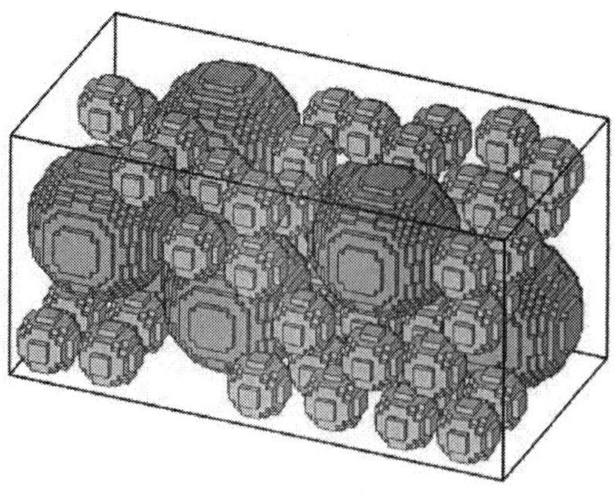

Fig. 9. Placement of voxelized balls into a parallelepiped.

5 Practical application

As an example of the practical application of the developed algorithms for the formation and placement of orthogonal polyhedrons, the problem of parts (details) layout on the platform of a 3D printer is considered.

Solving the problem of optimizing the layout of objects is of great importance in additive technologies, since it can significantly reduce the consumption of material, reduce the time spent on preparing the layout of objects inside the platform (container) of a 3D printer, as well as the time spent on their direct manufacture. The use of denser layouts additionally leads to a reduction in energy costs in the process of manufacturing objects, and also reduces the equipment depreciation used.

At present, the richest functionality for solving the problems of additive manufacturing is provided by the software Materialise Magics (Materialise NV, Lovaine, Belgium), which is used by the world's leading manufacturers of equipment for 3D printing [24]. The widespread recognition of this tool in the global market for additive technologies explains its choice for a comparative analysis of the obtained results.

We consider a three-dimensional container with the following parameters: length: 340 mm, width: 340 mm, height: 620 mm, the gap from the bottom of the container is 9 mm, the gap from the side faces of the container is 10 mm, minimum distance between placed objects is 6 mm. Parameters of details are given in Figure 10 (200 items of three types). All details can be rotated by multiples of 90° when placed.

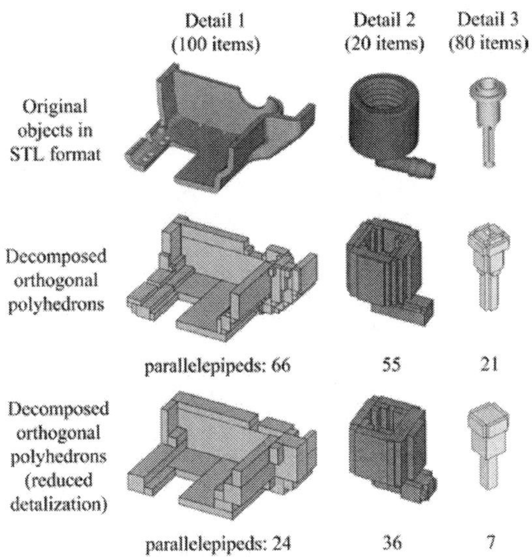

Fig. 10. Parameters of details.

Figure 11 shows the best solution obtained by the Sinter module of the software Materialise Magics, version 23.0.1.19 (layout height: 141.0 mm, solution time: 300 seconds). Figure 12 presents the denser placement of orthogonal polyhedrons found by the designed software Packer which implements all developed algorithms [25, 26] (height: 122.4 mm, solution time: 107 seconds).

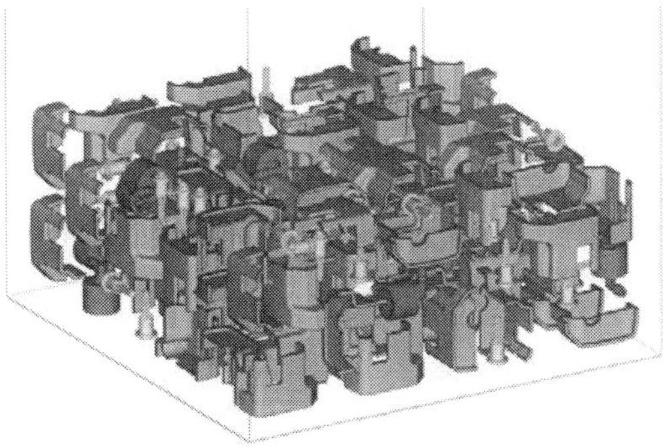

Fig. 11. The best placement obtained by Materialise Magics.

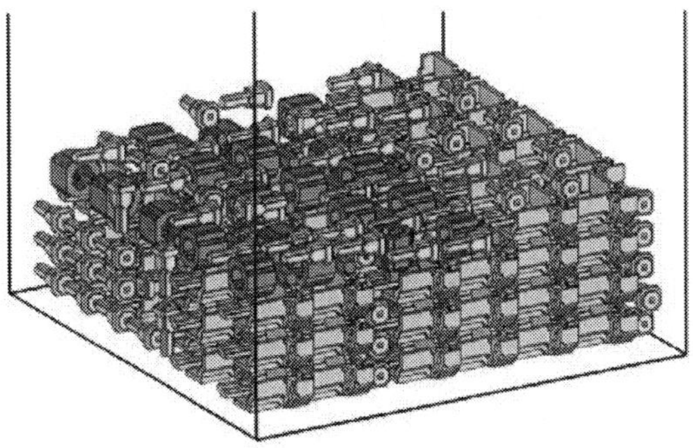

Fig. 12. Placement of orthogonal polyhedrons (decomposed voxelized objects) obtained by the developed algorithms.

Reducing the detalization on voxelization results in a significant increase in the placement speed with a slight loss of placement density. The resulting layout of details with reduced detalization is shown in Figure 13 (height: 124.4 mm, solution time: 18 seconds).

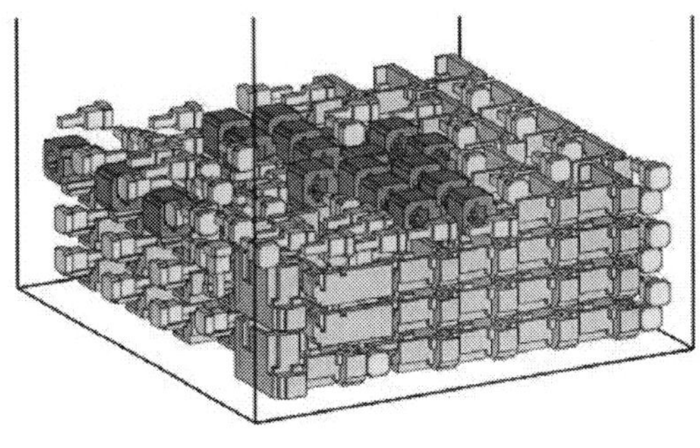

Fig. 13. Placement of orthogonal polyhedrons (decomposed voxelized objects with reduced object detalization) obtained by the developed algorithms.

6 Conclusion

The article proposes a method for solving the problem of placing objects of complex geometric shapes, which consists in transforming it to the problem of placing orthogonal polyhedrons.

All considered algorithms are developed and implemented independently from the dimension of the problem being solved, which makes them applicable for solving arbitrary types of cutting and packing problems.

The proposed algorithm for placing an orthogonal polyhedron of arbitrary dimension solves the problem of obtaining the area closest to the origin of the container which can be used for the placement of all orthogonal objects that are part of the considered orthogonal polyhedron. As a result of analyzing the set of all existing free orthogonal areas of the container, the most suitable area of the container is determined for each orthogonal polyhedron when forming a dense placement of objects.

The efficiency of application of the developed and software-implemented algorithms for the formation and placement of orthogonal polyhedrons is investigated by the example of solving the problem of modeling the layout of objects of complex geometric shape on the platform of a 3D printer. It is shown that the algorithms presented in the work provide an average of 10% denser layouts in a several times shorter than the layouts generated by Materialize Magics software which is the market leader in software solutions for additive technologies.

References

1. G. Wäscher, H. Haußner, H. Schumann, Eur. J. Oper. Res., **183**(3), 1109-1130 (2007)
2. A. Bortfeldt, G. Wäscher, Eur. J. Oper. Res., **229**(1), 1-20 (2013)
3. T.G. Crainic, G. Perboli, R. Tadei, INFORMS J. Comput., **20**(3), 368-384 (2008)
4. R.J. Mailloux, S.G. Santarelli, T.M. Roberts, D. Luu, Int. J. Antenn. Propag., **956524** (2009)
5. V.A. Chekanin, A.V. Chekanin, Comp. Math. Math. Phys+, **53**(10), 1457-1465 (2013)
6. V.A. Chekanin, A.V. Chekanin, Applied Mechanics and Materials, **598**, 377-381 (2014)
7. Y. Stoyan, T. Romanova, A. Pankratov, A. Chugay, Springer Optimization and Its Applications, **105**, 265-293 (2015)
8. N. Chernov, Y. Stoyan, T. Romanova, Computational Geometry, **43**(5), 535-553 (2010)
9. J. Bennell, G. Scheithauer, Y. Stoyan, Ann. Oper. Res., **179**(1), 343-368 (2010)
10. E.G. Birgin, L.H. Bustamante, H.F. Callisaya, J.M. Martínez, Int. T. Oper. Res., **20**(3), 365-389 (2013)
11. E.G. Birgin, R.D. Lobato, J.M. Martínez, J. Global. Optim., **65**(4), 709-743 (2016)
12. M. Verkhoturov, A. Petunin, G. Verkhoturova, K. Danilov, D. Kurennov, IFAC-PapersOnLine, **49**(12), 1-5 (2016)
13. A.V. Tolok, N.B. Tolok, Automat. Rem. Contr+, **79**(9), 1703-1712 (2018)
14. A.C.J. de Korte, H.J.H. Brouwers, Powder Technol., **233**, 319-324 (2013)
15. T. Byholm, M. Toivakka, J. Westerholm, Powder Technol., **196**(2), 139-146 (2009)
16. V.A. Chekanin, A.V. Chekanin, Lecture Notes in Electrical Engineering, **641**, 891-900 (2020).
17. V.A. Chekanin, A.V. Chekanin, *Advances in Mechanical Engineering. Lecture Notes in Mechanical Engineering, 33-38* (Springer, Cham, 2015)
18. V.A. Chekanin, A.V. Chekanin, *Advances in Mechanical Engineering. Lecture Notes in Mechanical Engineering, 27-35* (Springer, Cham, 2017)
19. V.A. Chekanin, A.V. Chekanin, *Advances in Mechanical Engineering. Lecture Notes in Mechanical Engineering, 87-92* (Springer, Cham, 2016)
20. S. Durocher, S. Mehrabi, Theor. Comput. Sci., **689**, 157-168 (2017)
21. T. Biedl, M. Derka, V. Irvine, A. Lubiw, D. Mondal, A. Turcotte, Lecture Notes in Computer Science, **10807**, 146-160 (2018)
22. P. Floderus, J. Jansson, C. Levcopoulos, A. Lingas, D. Sledneu, Algorithmica, **80**(1), 136-154 (2018)
23. V.A. Chekanin, A.V. Chekanin, *Advances in Mechanical Engineering. Lecture Notes in Mechanical Engineering, 41-48* (Springer, Cham, 2020)
24. S. Hällgren, L. Pejryd, J. Ekengren, Procedia CIRP, **50**, 518-523 (2016)
25. V.A. Chekanin, A.V. Chekanin, Journal of Theoretical and Applied Information Technology, **88**(3), 421-430 (2016)
26. V.A. Chekanin, A.V. Chekanin, Advances in Intelligent Systems Research, **123**, 379-382 (2015)

EPJ Web of Conferences **248**, 02002 (2021)
MNPS-2020
https://doi.org/10.1051/epjconf/202124802002

Driver Behaviour Algorithms for the Cellular Automata-Based Mathematical Model of Traffic Flows

Antonina Chechina*, *Natalia* Churbanova, and *Marina* Trapeznikova

Keldysh Institute of Applied Mathematics RAS, RU-125047, Moscow, Russia

Abstract. The work is devoted to the development of new algorithms describing the complex interaction of drivers for the parallel numerical implementation of the traffic flow model developed by the authors based on the cellular automata theory. As is known from observations, the behaviour of drivers largely determines how difficult it will be to pass a section with a bottleneck (a narrowing, a motionless obstacle, etc.) or even without it at the same flow values. In such conditions, cars can move in a synchronized flow at a more or less constant speed or go into "stop-and-go" mode. The authors have developed a set of algorithms for a large number of different road situations, which is featured in this article.

1 Introduction

Traffic flow modelling is an essential field of research nowadays, as traffic management problems require solving on everyday basis. Now, having high-performance supercomputers at hand, one can use developed traffic models and algorithms to create software for real time routing and infrastructural tasks that require immediate attention, as well as for long term road and city development tasks.

The model considered in this article was developed by the authors a few years ago and since then is being constantly modified and refined, mainly by building realistic algorithms of driver interactions in various situations and on different road elements (see, for example, [1]).

The main focus of this work are new algorithms for the model. The algorithm of driving around a wide obstacle (i.e., road maintenance works on a multilane highway) has been developed. The algorithm for boundary conditions exchange for the correct representation of a traffic jam spread across the network in case of parallel computations has been developed.

2 Cellular automata (CA) model for traffic flows

Traffic flow modelling allows for various approaches, though all models can be divided into two big groups: macroscopic models and microscopic models. As a rule, macro models are based on the similarity between a vehicle flow and a compressible fluid flow, so traffic is described by hydrodynamic equations and only average characteristics (flow density and velocity) are considered. Micro models on the other hand consider each vehicle separately. Usually the motion equations are written out, but it's not the only way of constructing a microscopic mathematical model of traffic flows. The approach based on cellular automata (CA) realizes this possibility in order to avoid dealing with differential equations and to ensure more flexibility in depicting drivers' interactions.

The cellular automata (CA) approach has a long and successful history of usage in various fields of research, including medical, biological and social sciences. It was introduced into the traffic field by K. Nagel and M. Schreckenberg in 1992 [2] and, since then, is widely used for modelling vehicular traffic all over the world [3-9].

The basics of the CA traffic flow model are as follows: the computational domain (the road, see Fig. 1) is divided into equal cells, each cell is 7.5 m long and one lane wide (the amount of space a single car takes up in the maximum density case – i.e., when being in a traffic jam). The distance in the model is measured in cells, a vehicle speed – in the number of cells per time step, time is measured in steps, time step is equal to 1 second.

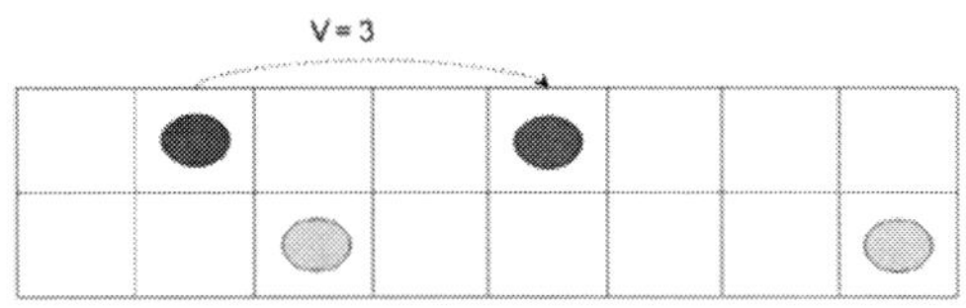

Fig. 1. The computational domain in the CA model.

Every time step the cell state update takes place. First, drivers make decision about changing lanes; second – vehicles move along the road by the rules of one lane traffic from Nagel-Schreckenberg model. The block-scheme for vehicular movement along the road is presented in Fig. 2.

* Corresponding author: chechina.antonina@yandex.ru

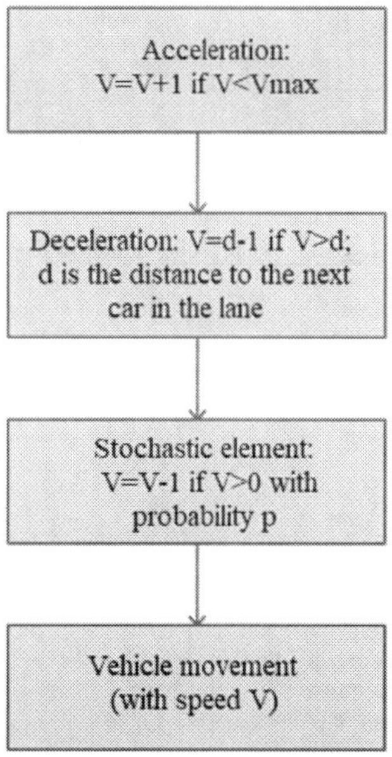

Fig. 2. Nagel-Schreckenberg model rules for movement along a road.

Here V is the speed of a vehicle, V_{max} is maximal speed of a vehicle, d is the distance to the next car in the lane, and p is the probability of additional speed decrease by a driver.

Basically, most complicated algorithms in the model are connected to driver behaviour. The decision making regarding a lane change depends on various factors. For each of the following situations the sequence of operations before reaching the decision will be different:

• closer/farther the car is from the intersection;
• narrowing/widening of a road;
• driver's goal (a turn, a certain exit, etc);
• presence of obstacles in a given and/or an adjacent lane;
• presence of a congestion in a given/adjacent lane;
• presence/absence of road markings prohibiting lane change;
• on-ramp or an entrance from a secondary road to a main road;
• crossing a non-signalized/signalized intersection;
• U-turn;
etc.

The authors have developed a set of algorithms for a large number of different road situations (see [10]).

3 Numerical realization

The model is realized as a program package written using C/C++, Glut library for visualizing the computation along with it and MPI library for boundary data exchange during parallel calculations on high-performance supercomputer systems. The program has a module structure, each module answers for modeling traffic on a certain road element (an intersection, a strait road fragment, a U-turn, etc.). From these elements a road network is constructed. Each element is computed on a separate CPU.

4 The algorithm for changing lanes for driving around a wide obstacle

One of the new algorithms developed for the CA model is the algorithm of drivers' decision making regarding changing lanes in presence of a wide obstacle that blocks several lanes on a multilane highway. This situation occurs fairly often when part of the road is closed for repairs. Lanes can also be blocked by slow-moving heavy duty vehicles that clean the road. The main difference between this case and the one where drivers need to move around small obstacles (car/cars that stopped due to an accident or breakage) is that the width and the position of the obstacle should be taken into consideration while choosing from which side to by-pass an obstacle to –the left one or the right one.

The algorithm of this decision making is shown in Fig. 3.

The Boolean type variable that determines if the lane change happens for the certain car at the current time step is at first set to *true* value if the step number is even (for moving to the right) or odd (for moving to the left). Then, it is determined whether the lane change would be advantageous to the driver (that is, if it results in getting into a faster and/or freer lane). For that, the speed and the distance to the next car in the current as well as in the neighbouring lane are checked and compared. If the lane change is indeed advantageous, the value of the variable stays *true*, if not – it is reset to *false*. Next, the presence of an obstacle is checked a few cells ahead, both in the current lane and in the target one, and the width of the obstacle is evaluated. To do so, the next free cell in the sequence of occupied ones is searched across the road in the direction of a supposed lane change and in the opposite one. These distances are compared to each other. If the obstacle is there, but not in the target lane, or it's there in the current lane and in the target one, but the distance to the first unoccupied cell is shorter in the desired lane change direction at the given time step, the lane change variable is set to *true*. After that, the safety is checked and lane change is cancelled if it can result in a road accident. Finally, if lane change variable is *true* after the safety check, the randomization rule is applied: the random value is received (*rand()* function in C/C++) and if it is greater than the given probability, lane change takes place, if not – it is cancelled.

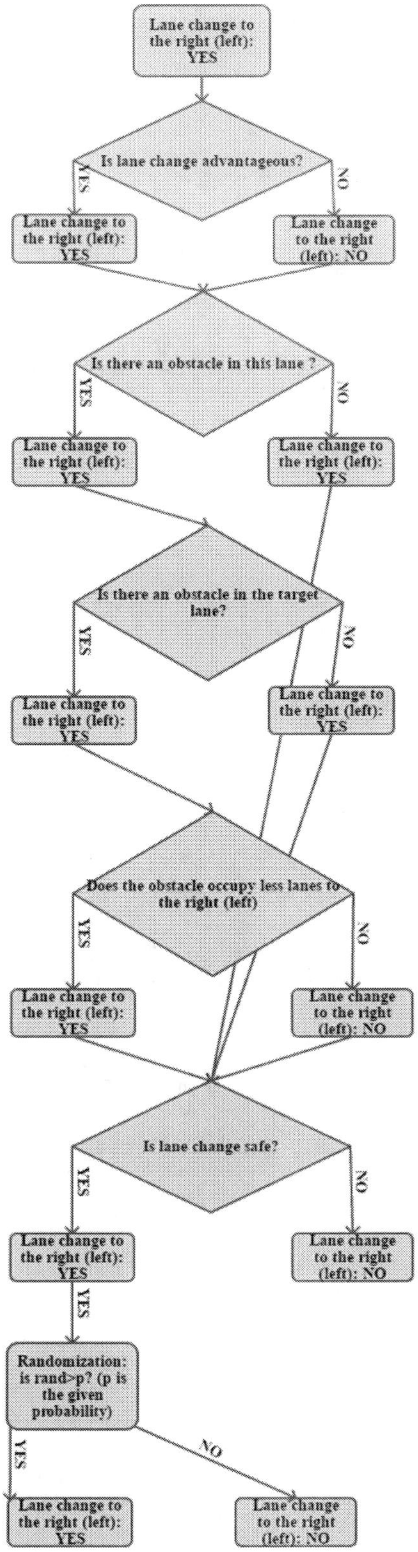

Fig. 3. Block-scheme of the algorithm of driving around a wide obstacle.

The algorithm is realized as a module for the program package developed by the authors. The output is shown in Fig. 4.

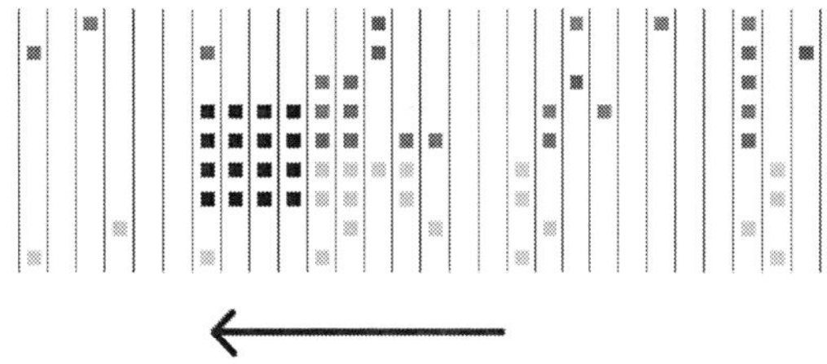

Fig. 4. Modelling driving around a wide obstacle.

Figure 4 represents the 9-lane highway road. The movement direction is shown with the arrow (right to left). The traffic here is right-hand side. A cluster of 4x4 black squares represents a wide obstacle. The cars are depicted by squares of different shades of gray. Darker cars enter this road fragment by first 5 lanes (slower ones), lighter cars use faster lanes - from 6 to 9.

Here we can see that darker cars choose to move to the right while lighter ones change lanes to the left, this way ensuring that obstacle by-passing takes less steps, which is the goal of the algorithm and reproduces real life driver behaviour.

5 The boundary conditions exchange algorithm for parallel computations

As computations for each road element are carried out on different processors and drivers don't "know" anything about traffic conditions on the next element, some probing needs to take place before a car might exit the current road element in order to see if there's the possibility to enter the next one. For that, flag variables are set to indicate the situation, and they are exchanged before actual boundary condition exchange takes place.

Block-scheme for the algorithm of car exiting a road element is shown in Fig. 5.

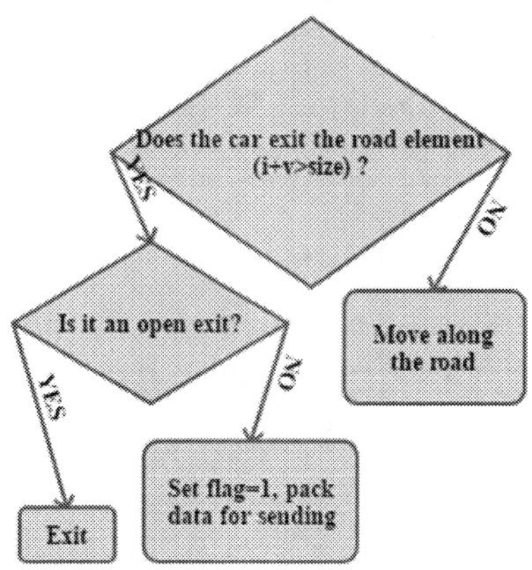

Fig. 5. Block-scheme of the algorithm of exiting a road element

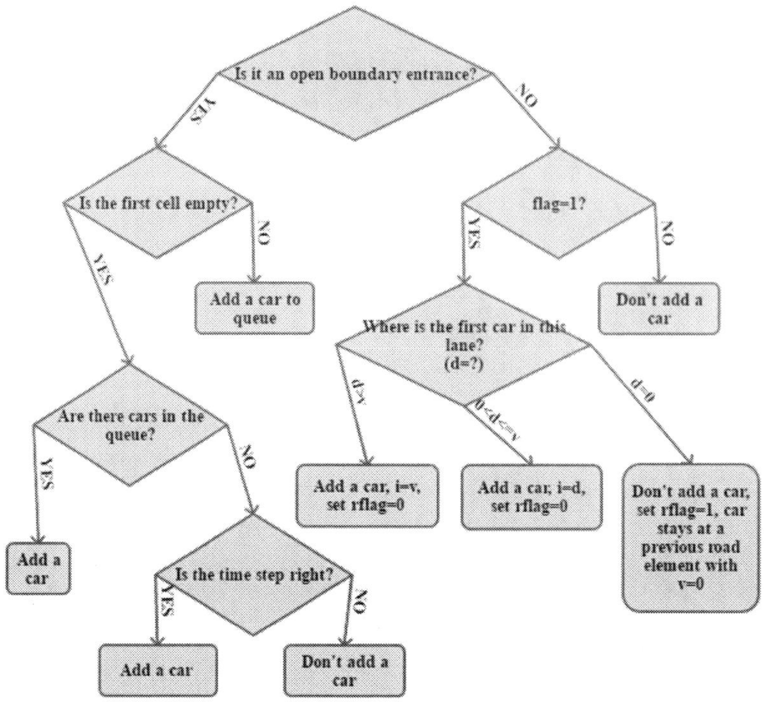

Fig. 6. Block-scheme of the algorithm of adding a car to a road element.

When a vehicle is reaching the end of the road on an element, we check if it's going to leave it at the current time step. If the number of cell i exceeds the size of the road after adding the current speed value, the car is about to exit. If it's a free exit with an open boundary, the vehicle just leaves the computational domain. If the road element has a neighbouring one, the data about the exiting car (its current speed, ID number, desired destination, maximal speed, the ability of the driver to become cooperative) is packed for sending and the flag variable that indicates that the there's data for receiving is set to 1 (*flag*=1).

When cars are added to the system, it happens differently whether it is a free boundary entrance or not (see Fig. 6). If it is free, we check if the first cell is empty. If it is, a car can be added, if not – it is added to the queue to be put on the road as soon as the first cell is empty. If a road element is connected to its neighbour, we check if there's data packed for transmission on the processor that is accountable for modelling a neighbouring road fragment. If said data exists, we check if there is a possibility of adding a car according to the position of its target cell. If adding is possible, we do so, setting the return flag variable to 0 (*rflag*=0) indicating that the car was added and thus can be removed from the previous element. If adding a car was not successful, the return flag is set to 1 and the car stays on the previous road element with its speed set to zero. In order to avoid high deceleration values, the situation is constantly checked for congestions at boundaries to slow down cars that approach the entrances to congested road elements.

To test the algorithm, the computations on different road networks were carried out. The results show it works correctly.

6 Conclusions

New algorithms for the CA model developed by authors earlier were presented in the article. The algorithms were realized as a part of the program package developed by authors and test problems were solved to confirm their adequate representation of real life traffic.

The developed program package for high performance computing systems can be used for traffic modelling on city road networks, taking into consideration different driving strategies and evaluating the efficiency of measures taken in order to decrease traffic jams. It can be used as a basis for Intelligent Transport System of a metropolis.

The calculations were performed with the use of supercomputers installed in the Collective Usage Centre of KIAM RAS [11].

References

1. A. Chechina, N. Churbanova, M. Trapeznikova, A. Ermakov, M. German, Int. J. of Eng. & Techn., **7**(2.28), 225-227 (2018)
2. K. Nagel, M. Schreckenberg, J. Phys. I France, **2**, 2221–2229 (1992)
3. S. Maerivoet, B. De Moor, Physics Reports, **419**(1), 1-64 (2005)

4. M. E. Lárraga, L. Alvarez-Icaza, Physica A: Statistical Mechanics and its Applications, **389**(23), 5425-5438 (2010)
5. B. Kerner, S. Klenov, G. Hermanns, M. Schreckenberg, Physica A: Statistical Mechanics and its Applications, **392**(18), 4083-4105 (2013)
6. H. Jiang, Zh. Zhang, Q. Huang, P. Xie, Safety Science, **82**, 182-189 (2016)
7. A.P. Buslaev, A.G. Tatashev, M.V. Yashina, Int. J. of Eng. & Techn., **7**(2.28), 351-356 (2018)
8. Kerner B., Klenov S., Schreckenberg M. Phys. Rev. E., **84** (2011)
9. J. Vasic, J. Heather, H.J. Ruskin, Physica A: Statistical Mechanics and its Applications, **391**(8), 2720-2729 (2012)
10. N.G. Churbanova, A.A. Chechina, M.A. Trapeznikova, P.A. Sokolov, Mathematica Montisnigri, **XLVI**, 72-90 (2019)
11. KIAM – The official site of Keldysh Institute of Applied Mathematics, https://www.kiam.ru/MVS/resourses/

Mathematical Modeling of Technological Modes of Operation of Liquid Stamping of Aluminum Alloy AK7

Vagid Kadymov[1], *Evgeny* Sosenushkin[2]

[1]Russian Technological University- MIREA, RU-119484, Moscow, Russia
[2]Moscow State Technological University "STANKIN", RU-127055, Moscow, Russia

Abstract. In this paper we examine a mathematical model of solidification of the melt in a two-phase zone during the formation of forgings through liquid stamping. From the simultaneous solution of the Fourier equations of thermal conductivity for the melt, the two-phase zone, and the solid crust, the solidification kinetics has been established and the effect of external pressure on the temperature fields and components of the solidification time of hollow thin-walled forgings has been thoroughly assessed.

1 Introduction

The technology of forming of crystallized metal (liquid stamping) successfully competes with the processes of casting [1] and metal processing by pressure [2] in the hot state in closed dies [3, 4]. It is necessary to expand the technological capabilities of metal processing by pressure in the manufacture of products made of hard-to-deform alloys and metals with low plastic properties. The metal utilization rate increases to 0.95 due to reduced allowances for further machining and the absence of Gating systems [5] and the deformation forces are reduced by 3-5 times when the technology is implemented. The metal of forgings acquires a fine-grained structure, and the physical and mechanical properties of products that are produced by stamping of crystallized metal increase by 15-20% [6-8]. All these trends are determined by the technological parameters of the liquid stamping process, which include, in particular, the deformation force, temperature intervals for pouring molten metal into the die cavity, the temperature of the die elements themselves, as well as the time of filling the die cavity and crystallization, the value of which is of particular importance when stamping thin-walled forgings. The term "thin-walled" is used in relation to hollow forgings, of which the wall thickness is less than one-tenth of the maximum forging size or less than one-tenth of the diameter in case of axisymmetric forgings. The formation of the forging structure depends on the external pressure, the intensity of heat exchange with the tool and the environment [9]. Modeling of heat exchange conditions allows us to estimate the thermal fields of contacting bodies, as well as the duration of the cooling stages of forgings [10-12]. However, it is necessary to draw attention to the problems, associated with this type of forming technology, among which the durability of the die tool [13, 14] remains one of the most important problems.

2 Problem formulation

To describe the modeling of heat exchange processes during solidification at the boundary of two contiguous bodies, we use the Fourier equation of thermal conductivity and the one-valuedness conditions that characterize the phenomenon of heat propagation under study. The geometric parameters and physical characteristics of contacting bodies are usually known from the problem statement. Time conditions are defined as temperature distribution in the bodies for a specific point of time. For example, at the initial point of time, the temperature is distributed evenly in the body, meaning if t=0, T=T0.

In the real technological process of crystallization under pressure, the liquid metal contacts the surface of the die, so as a boundary condition, thus we will use the condition of equality of heat fluxes on the surface of contact of the liquid phase of the crystallizing alloy with the die material:

$$-\lambda\left(\frac{\partial T}{\partial n}\right)_n = \alpha\left(T_n - T_c\right), \tag{1}$$

where λ is the thermal conductivity of the melt, W/(m×K); α is the coefficient of heat transfer by contact, W/(m2×K). $\left(\frac{\partial T}{\partial n}\right)_n$ is temperature gradient on the contact surface, K / m; Tn, Tc – temperatures of the contact surface and the medium, respectively, K. The diagram of process of hardening of metal of hollow forgings in the two-phase zone is shown in Fig. 1.

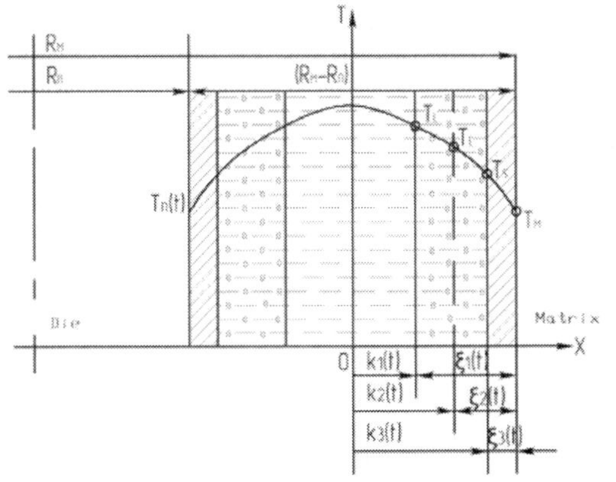

Fig. 1. The process of solidification of a hollow forging in the two-phase zone:
temperature fields: TL – liquidus temperature; TE - the temperature of the conventional solidification front; TS – solidus temperature; Tп(-x,t)=Tм(x,t)=Tш(t) is the temperature of the punch and die matrix; geometric parameters: Rм, Rп – the radii of the die and punch; s=Rм-Rп – the thickness of the forging wall; ξ(t) – the thickness of the solid crust; k1(t), k3(t) – the coordinates of the fronts of the beginning and the end of crystallization; k2(t) - the coordinate of the conditional crystallization front.

The technological process of forming the forgings with melt crystallization under pressure consists of the following stages:

- melting of the alloy in an electric furnace at the temperature of the furnace Tп, K;
- pouring into the die a portion of the melt with a volume of V (m3) at a temperature of Tзал (K) during t (s);
- removal of the heat of overheating T (K) for the time t1 (s);
- solidification of the forging in the range of liquidus and solidus temperatures [TL, TS] at the crystallization temperature Tкр, (K) under external pressure p, MPa during the removal of the latent heat of crystallization t2 (s);
- cooling of the forging in the die cavity and subsequent ejection.

Let's establish a system of differential equations that describe the heat exchange between the forging and the die during solidification.

For superheated melt in the non-solidified part of the forging

$$\frac{\partial T_1}{\partial t} = a_1 \frac{\partial^2 T_1}{\partial x^2}; \; 0 < x < k_1(t) \qquad (2)$$

3 Equations and mathematics

To mathematically describe the processes occurring in a two-phase zone, we can use the heat conduction equation in general form, taking into account the heat release from internal sources distributed over the volume [2]:

$$c_2\gamma_2 \frac{\partial T_2}{\partial t} = \lambda_2 \frac{\partial^2 T_2}{\partial x^2} + L\gamma_2 \frac{\partial \psi}{\partial t}; \; k_1(t) < x < k_2(t).$$

The relative amount of solid phase, crystallizing in the two-phase forging zone, is characterized by the function ψ, which is defined as a function of temperature, and the rate of formation of the solid phase is as follows:

$$\frac{\partial \psi}{\partial t} = \frac{\partial \psi}{\partial T} \cdot \frac{\partial T}{\partial t} = \mu \frac{\partial T}{\partial t},$$

where $\mu = \frac{\partial \psi}{\partial T}$ is the rate of crystallization of the alloy within the temperature range [TL, TS]. If taken as a function $\psi = \sqrt{xt}$, its partial time derivative is defined as $\frac{\partial \psi}{\partial t} = \frac{1}{2}\frac{\sqrt{x}}{\sqrt{t}}$. The simplest accounting for the fraction of the solid phase $\psi(T_2)$ during crystallization in the two-phase zone

$$\psi(T_2) = \frac{T_L - T_2}{T_L - T_S} \qquad (3)$$

For a hard crust

$$\frac{\partial T_3}{\partial t} = a_3 \frac{\partial^2 T_3}{\partial x^2}; \; k_3(t) < x < \frac{s}{2} \qquad (4)$$

The conditions of equality of heat fluxes at the front of the beginning of crystallization

$$-\lambda_1 \frac{\partial T_1(k_1,t)}{\partial x} = -\lambda_2 \frac{\partial T_2(k_1,t)}{\partial x} \qquad (5)$$

at the front of the end of crystallization in the two-phase zone

$$-\lambda_2 \frac{\partial T_2(k_2,t)}{\partial x} = -\lambda_3 \frac{\partial T_3(k_1,t)}{\partial x} \qquad (6)$$

Boundary conditions:

$$T_1(k_1,t) = T_2(k_1,t) = T_L$$
$$T_2(k_2,t) = T_3(k_2,t) = T_S \qquad (7)$$

To account for physical processes in the two-phase zone within the alloy crystallization interval let us choose the temperature $T_E = \frac{1}{2}(T_L - T_S)$ and assume that the crystallization process is almost complete in the two-phase zone by the time this temperature is reached. In this case, the solidification kinetics is characterized by the growth rate of the solid crust $U(t) = \frac{\partial \xi}{\partial t}$. Therefore, it is possible to replace the condition of equality of heat fluxes at the front of the end of crystallization in the two-phase zone with the differential equation of Stefan solidification, and replace the coordinate k3(t) with k2(t). Then, for a two-phase zone, the differential equation of solidification [2] is defined as follows:

$$-\lambda_2 \frac{\partial T_2(k_2,t)}{\partial x} + [1-\psi(k_2,t)]L\gamma_3 U = -\lambda_3 \frac{\partial T_3(k_2,t)}{\partial x};\quad k_1(t) < x < k_2(t). \quad (8)$$

The conditions for equality of heat fluxes:

$$-\lambda_i \frac{\partial T_i\left(\frac{s}{2},t\right)}{\partial x} = q_0(t);\ \frac{\partial T_i(0,t)}{\partial x} = 0;\ i=1 \text{ and } 3; \quad (9)$$

or

$$T_i\left(\frac{s}{2},t\right) = T_0(t);\ \frac{\partial T_i(0,t)}{\partial x} = 0;\ i=1 \text{ and } 3; \quad (10)$$

$$k_2(t_{зал}) = \frac{s}{2};\quad T_1(x,t_{зал}) = T_E = const;\quad T_1(k_2,t) = T_3(k_2,t) = T_E. \quad (11)$$

To reduce the intensity of heat exchange between the melt and the die, a layer of technological lubricant is applied to the surface of the die, from which (lubricant) the liquid phase evaporates upon contact with the melt, and its components partially burn out, forming an air-gaseous layer. Solid components, such as graphite, settle on the surface of the tool, on which an oxide film is formed at elevated temperatures. As a result, the gap between the forging and the die is filled with layers of these components. The gap in the thermal ratio creates a thermal resistance that is quantified:

$$R_{заз} = \frac{x_{заз}}{\lambda_{заз}}, \quad (12)$$

where Xзаз is the gap thickness, m; λзаз is the effective coefficient of thermal conductivity of the substance filling the gap, W/(m×K). The inverse value of the thermal resistance is called the coefficient of heat transfer through the gap. In a multi-layer gap, this coefficient is equal to:

$$\beta = \frac{1}{R_{заз}} = \frac{1}{R_1} + \frac{1}{R_2} = \ldots = \frac{1}{R_n}. \quad (13)$$

The relatively low heat exchange rate of the melt with the die and the die with the environment makes it possible to ignore the unevenness of the thermal fields in the cross sections of the forging and the die. Under these conditions, it is appropriate to independently consider the characteristic stages of the technological process. The constancy of the thermal resistance of the gap substance and the temperature head for the metal melt in the mobile and stationary state means that the value of the heat flow from the melt into the stamp through the gap remains unchanged in both cases. The flow of metal in the die at the filling stage is expressed in the fact that the stamp is additionally warmed up, and the melt loses some of the heat. The amount of heat that is transferred from the melt to the die through the punch and die matrix during the time dt is defined as the ratio:

$$d^2Q_1 = \beta(T_1 - T_n)dS_n dt + \beta(T_1 - T_м)dS_м dt. \quad (14)$$

Change in the heat content of the melt during the same time:

$$d^2Q_2 = -c_1\gamma_1 dVdT, \quad (15)$$

where $V = V_м - V_n$ -is the forging volume, m3. The heat balance equation is defined as:

$$\beta(T_1 - T_n)dS_n dt + \beta(T_1 - T_м)dS_м dt = -c_1\gamma_1(dV_м - dV_n)dT \quad (16)$$

Let us solve it with respect to time dt:

$$dt = -\frac{1}{\beta}\frac{c_1\gamma_1(dV_м - dV_n)dT}{T_1(dS_n + dS_м) - (T_n dS_n + T_м dS_м)}.$$

After integration we obtain:

$$t = -\frac{1}{\beta}\frac{c_1\gamma_1(dV_м - dV_n)}{(dS_n + dS_м)}\cdot \ln\left|(T_1 - T_n)dS_n + (T_1 - T_м)dS_м\right| + C,$$

where the value C is the integration constant, which is excluded under the condition t=0

$\ln\left|(T_1 - T_n)dS_n + (T_1 - T_м)dS_м\right|$=0, thus C=0,

$$t = -\frac{1}{\beta}\frac{c_1\gamma_1(dV_м - dV_n)}{(dS_n + dS_м)}\cdot \ln\left|(T_1 - T_n)dS_n + (T_1 - T_м)dS_м\right| \quad (17)$$

One of the main parameters when forming a crystallizing metal are the pressure and the duration of its impact on the solidifying metal. First of all, the pressure, applied to the liquid metal during its crystallization, eliminates gas porosity [15]. The release of gases from the metal during crystallization occurs due to the lower solubility of gases in the solid metal compared to its liquid state. Practice shows that the decrease in porosity occurs to a much greater extent than it would occur according to the Boyle-Mariotte law. The fact is that in addition to the compressing effect of pressure on the gas porosity, the pressure also increases the solubility of gases in the liquid metal, which leads to a decrease in the amount of gas released from the metal when it solidifies. Liquid metal under the influence of pressure during deformation solidifies much faster than when casting in a metal mold without applying pressure. This phenomenon is explained in terms of thermodynamics of the process. For a two-phase thermodynamically equilibrium system, temperature is a function of pressure, and this dependence is expressed by the Clausius-Clapeyron equation [5]:

$$\frac{dT_{кр}}{dp} = \frac{\Delta V T_{кр}}{L},\ \Delta T = \int_0^p \frac{\Delta V T_{кр}}{L} dp = \frac{\Delta V T_{кр}}{L} p$$

Next, the melt crystallization temperature is determined depending on the external pressure p=100-250 MPa:

$$T^p_{кр} = T_{кр} + \frac{(V_1 - V_3)T_{кр}}{L} p \quad (18)$$

where V3 is the volume of forging in the solid state, m3. The temperature of pouring the melt into the die depends on the temperature range of metal solidification. The wider the range, the lower can be the temperature of pouring the melt, as a part of the solution to the equation of solidification, and that is usually the minimum for the particular conditions of cooling during filling of the die, but greater than the metal crystallization temperature. Therefore, the calculation is carried out in accordance with the formula proposed by A.I. Veynik [1]:

$$T_{зал} = T^p_{кр} + \frac{(T^p_{кр} - T_4)b_4}{b_1} \quad (19)$$

where $b_i = \sqrt{\lambda_i c_i \gamma_i}$ - coefficient of heat accumulation (i=1,4): for die, index i=4.

Relative pour temperature is defined as follows:

$$\theta_{зал} = \frac{T_{зал} - T_4}{T_L - T_4} \quad (20)$$

Now the time when a portion of the melt is poured into the die can be defined [2]:

$$t_{зал} = \left[\frac{c_1\gamma_1(\theta_{зал} - 1)}{1{,}13 b_4 \theta_{зал}}\right]^2 R^2_{пр} \quad (21)$$

Melt superheat temperature is:

$$T_{пер} = T_{зал} - T^p_{кр} \quad (22)$$

Temperature head is:

$$\Delta T = T_{зал} - T_4 \quad (23)$$

The time of superheat removal is found from equation (17), taking into account the known volumes and contact areas, which, when decomposing the function y=ln(x) into a Maclaurin series, takes the form:

$$t_1 = \frac{R_{пр}\xi(T)c_1\gamma_1}{\lambda_1}\frac{T_{пер}}{\Delta T} \quad (24)$$

where $R_{пр} = \frac{V}{S_п + S_м}$ is the reduced forging radius, m. The temperature on the surface of the die in contact with the melt can be found [16]:

$$T_к = \frac{T_{зал}b_1 + T_4 b_4}{b_1 + b_4} \quad (25)$$

and when a solid crust is formed with a thickness of ξ (t), the temperature of its contact with the die can be found by analogy:

$$T_к = \frac{T^p_{кр}b_3 + T_4 b_4}{b_3 + b_4} \quad (26)$$

The height of the melt rise in the die channel depends on the fill temperature, die temperature, slider movement speed, and channel geometry [17]. For axisymmetric forgings in the coordinate system associated with the punch, under the condition of continuity of the jet, the ratio of speeds is obtained:

$$v_K = \frac{R^2_М}{R^2_М - R^2_П} v_П \quad (27)$$

where vK, vП are the speed in the die channel and the speed of the punch respectively; RM, RП – radii of the die matrix and punch.

4 Results

The results of modeling the process of solidification of an aluminum alloy in a two-phase zone during liquid stamping of a hollow forging are presented in the form of graphs. Fig. 2 shows the effect of external pressure applied to the meltduring crystallization on the temperature factors.

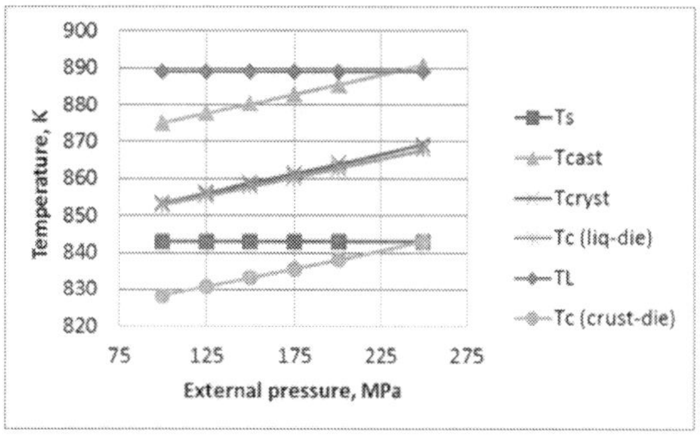

Fig. 2. Dependence of temperature factors on external pressure

Thus, the crystallization temperature is directly dependent on the pressure. The graph shows that as the external pressure developed by the press increases, the crystallization temperature increases. The law of temperature change is linear. Even at low pressures, the crystallization temperature does not cross the solidus line of this alloy, and the superheat temperature varies within small limits-3.01 K. This suggests that under the influence of external pressure, the alloy crystallizes earlier at temperatures above the solidus temperature. This feature is confirmed in the practice of foundry production, which aims at reducing the temperature of overheating in order to avoid hot cracks [1]. So, the temperature of crystallization under pressure becomes higher than the temperature of the melt subjected to deformation. In this case, the entire volume of the melt, which did not have time to solidify before applying pressure, is immediately supercooled relative to the new temperature of the beginning of crystallization, which naturally causes the appearance of crystallization centers in it. The greater the pressure applied to the melt or the lower the temperature of the melt to which the pressure is applied, the greater supercooling will be achieved relative to the new crystallization temperature and the more fine-grained structure will be formed during solidification. The higher the contact surface temperature, the longer it takes to remove the superheat temperature (Fig. 3, a). The superheat temperature also affects the proportion of the solid phase in the two-phase solidification zone. In accordance with the linear law of the change in the fraction of the solid phase depending on the temperature, the nature of this change is shown in Fig. 3, b.

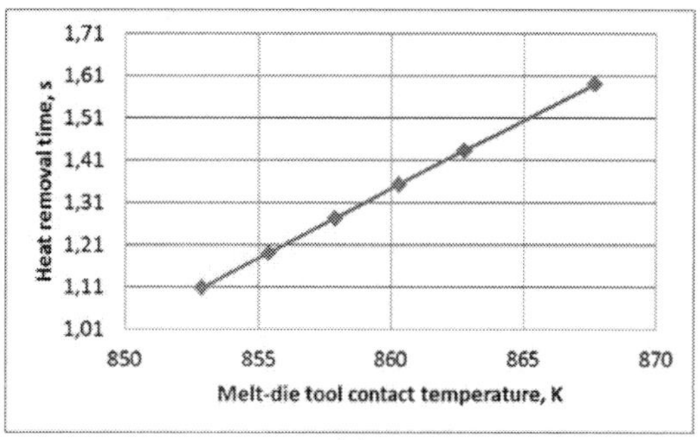

(a)

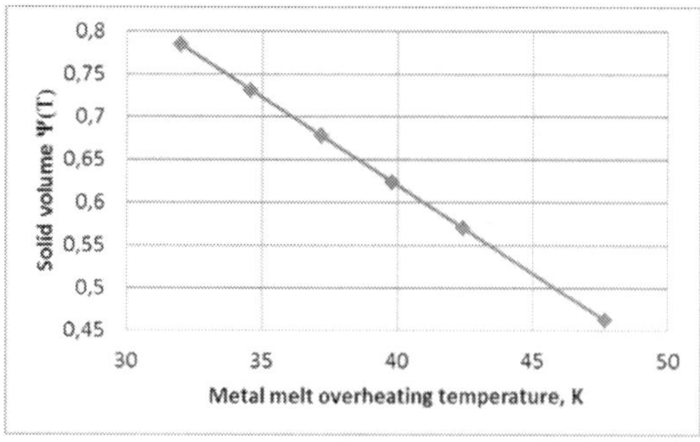

(b)

Fig. 3. Dependence of the superheat temperature removal time (a) on the contact surface temperature and the effect of the melt superheat temperature on the solid phase fraction during solidification in the temperature range (b).

With a decrease in the superheating temperature, the proportion of the solid phase in the two-phase solidification zone increases, which does not contradict the physics of the process. The graphs in Fig. 4 revealed the influence of the preheating temperature of the die on the temperature parameters of the melt. After pouring the melt and starting the deformation process, part of the heat passes into the die, the melt is supercooled and its temperature drops to the value of the crystallization temperature. Therefore, as the graph shows, the temperature on the contact surface is not much different from the crystallization temperature.

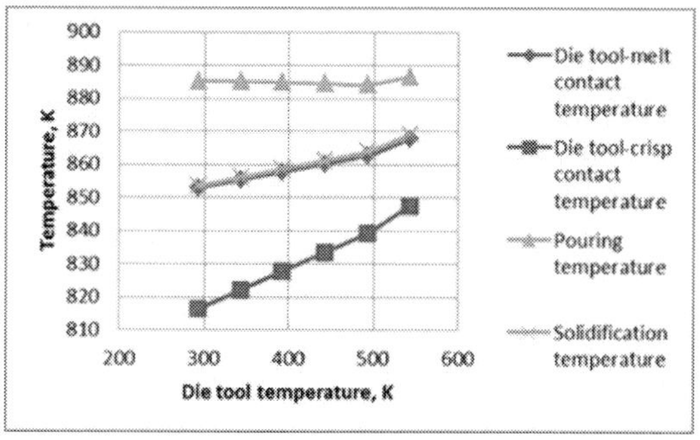

Fig. 4. Dependence of temperature fields on the die temperature

Since the forging material is constantly in contact with the die surface, part of the metal crystallizes to form a solid crust with a thickness of ξ (t), the heat exchange conditions change and the temperature on the contact surface decreases (see Fig. 4).

An important point that affects the quality of forging is filling of the die cavity with melt until it is completely solidified. The higher the pouring (filling) temperature, the more time it takes to remove the superheat temperature. You can control the process of filling of the die by adjusting the speed of the press slider. Low-speed hydraulic presses are usually used for the implementation of the liquid stamping process, which ensure that the slider stands without pressure relief. According to the graphs in Fig. 5 you can select the temperature of the melt pouring into the die and the speed of movement of the press slider to ensure the flow of the melt in the gap between the punch and the matrix, with a guaranteed rise to the desired height equal to the height of the wall of the thin-walled hollow forging until it is completely solidified.

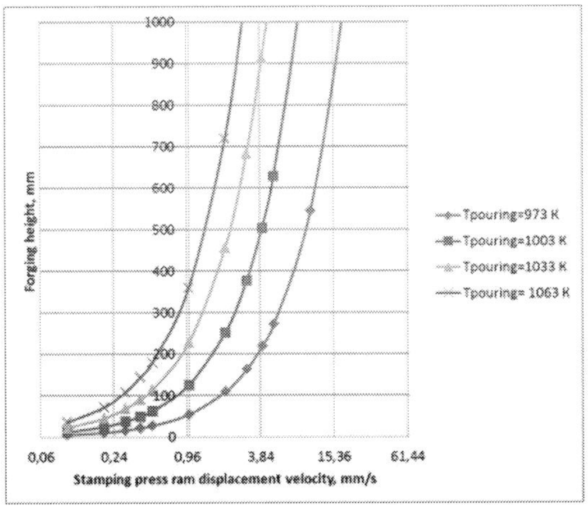

Fig. 5. Influence of the slider movement speed on the forging wall height at different filling temperatures

In the example, a forging with a wall height of 50 mm is used, the melt will rise to a height of this value before solidifying at a fill temperature of 973 K, at a tool movement speed of 1 mm/s. According to Fig. 5, the speed of movement of the slider can be reduced while increasing the pouring temperature, and vice versa to increase the working speed of the press with a longer standing of the slider at the end of the stroke to ensure complete solidification of the metal before ejection of the forging.

Experimental studies were carried out to verify the adequacy of theoretical calculations and the choice of preferred technological parameters for manufacturing thin-walled axisymmetric forgings by the crystallizing metal stamping method. Stamping was performed on a hydraulic press model DB2432A with a nominal force of 1600 kN. The aluminum alloy was smelted in an electric crucible furnace PP50, with a maximum heating temperature of 1,373 K. The melting time of 10 kg of charge was 1.5 hours.

5 Conclusion

The present study leads us to the following key conclusions:

1. Specific conditions of solidification of alloys in the temperature range with the formation of a two-phase zone are described by a mathematical model of solidification in relation to the technology of stamping metal crystallizing under pressure. The developed model was used to analyze the technological parameters of the process of liquid stamping of hollow thin-walled forgings made of AK7 casting alloy. The temperature and time parameters of the technological process of an aluminum alloy crystallizing under pressure were calculated, the control of which makes it possible to influence the process of solidification and formation of the structure of hollow thin-walled forgings.

2. The conducted experimental study confirmed the possibility of manufacturing hollow axisymmetric forgings with thin walls of the required height by stamping metal that crystallizes under pressure.

3. Control of theoretically justified technological parameters of the liquid stamping process allows for improved quality and uniformity of the structure of forgings made of AK7 alloy. When stamping other forgings that are similar in shape and size to the one under study, the process parameters can be recommended as a starting point, which, if necessary, are refined depending on the specific size, configuration and brand of the material.

This work was carried out using equipment provided by the Center of Collective Use of MSUT "STANKIN".

References

1. G.F. Balandin, *Theory of casting forming* (MSTU Publishers, Moscow, 1998)
2. H. Bayani, S.M.H. Mirbagheri, M. Barzegari, S. Firoozi, Journal of Materials Research and Technology, **3**(1), 55-70 (2014)
3. I.Ya. Belousov, A.A. Saprykin, B.A. Sivak Proceedings of the conference "Week of metals in Moscow", 423-428 (2008) [in Russian]
4. G.S. Dmitrienko, L.A. Uvarova, Vestnik MSTU "Stankin", **1**(19), 50-54 (2012) [in Russian]
5. L.G. Grishin, O.D. Solodukho, G.A. Ioffe, A.A. Saprykin, Forging and stamping production, **2**, 4-6 (1996)
6. Yu.P. Kirdeev, I.Ya. Belousov, A.I. Rakogon Forging and stamping production, **3**, 9-11 (2002)
7. S. Murali, M.S. Yong, Journal of Materials Processing Technology, **210**(10), 1276-1281 (2010)
8. E.N. Sosenushkin, V.A. Kadymov, E.A. Yanovskaya, A.A. Tatarencev, A.E. Sosenushkin,, Key Engineering Materials, **684**, 57–66 (2016) https://doi.org/10.4028/www.scientific.net/kem.684.57
9. E.N. Sosenushkin, *Progressive Processes of Bulk Stamping* (Moscow: Mashinostroenie. 2011)
10. E.N. Sosenushkin, L.S. Frantsuzova, V.V. Grishin, Zagot. Proizv. Mashinostr., **5**, 5–12 (2005) [in Russian]
11. E.N. Sosenushkin, Vestnik MSTU "Stankin", **2**, 12–20 (2010) [in Russian]
12. E.N. Sosenushkin, A.V. Khromenkov, Yu.A. Melnik, Friction and wear, **35**(6), 752-758 (2014)
13. E.N. Sosenushkin, E.A. Yanovskaya, K.N. Ivanov, T.A. Kinzhaev, non-Ferrous metals, **11**(911), 83-88 (2018) [in Russian]

14. E.N. Sosenushkin, L.S. Frantsuzova, E.A. Yanovskaya, T.A. Kinzhaev, Metallurg, **3**, 25-29 (2018) [in Russian]
15. V.N. Popov, A.A. Yushkanov, D.A. Ore, Vestnik MSTU "Stankin", **1**(24), 98-102 (2012) [in Russian]
16. M.R. Portnoy, V.V. Kuzin, Vestnik MSTU "Stankin", **2**(37), 21-26 (2016) [in Russian]
17. A.I. Veinik, *Thermal Principles of the Theory of Casting* (Mashgiz, Moscow, 1953) [in Russian]

The Integrated Statistical Control Method of Technical Condition of Radio-Electronic Equipment for Prototyping of Robotic Systems

Pavel Budko[1], *Vladimir* Goydenko[1], *Aleksei* Rozhnov[2,*] and *Aleksei* Vinogradenko[1]

[1]PJC "Inteltech", RU-197342, St. Petersburg, Russia
[2]V.A. Trapeznikov Institute of Control Sciences of Russian Academy of Sciences, RU-117997, Moscow, Russia

Abstract. In the article, the approach of integrated statistical control and diagnostics of the radio-electronic equipment, which is based on the analysis, integration of the measuring information arriving from the thermal imager, magnetic field sensors, tension and humidity of air and also statistical control of equipment parameters, is presented. The solution of a control and diagnostics problem is based on monitoring of changes in the average level of a multi-parameter process. Here, Hotelling control charts, integration and processing of the measured information are used to provide the analysis and decision-making on the influence of the operator. The algorithm implements a technique of integrated statistical control of the technical condition of the radio-electronic equipment and is capable to determine the level of its precritical condition. The proposed technique increases the reliability of the identification of technical conditions of radio-electronic equipment and expands the scope of technical means of control and diagnostics.

1 Introduction

Various Radio-Electronic Equipment (REE) is an important part of the structure of modern technological systems in various branches of industrial electronics and electrical equipment. The value and a relative share of volume of REE in technological systems over the past years continuously increase as well as demands in the creation of effecient systems of control and diagnostics at stages of tests and operation. Nondestructive control is especially important for the uninhabited (unattended) units which are located in remote areas where carrying out control and the REE diagnostic functions by servicemen is impossible. Collecting diagnostic information in similar units has to be carried out in the real-time, without allowing a transition of the precritical Technical Condition (TC) of REE to an accident. The purpose of this work is the development of a new approach of integrated statistical control and diagnostics of REE based on the statistical analysis and an integration of the *Measuring Information* (MI) received from the thermal imager, sensors of the magnetic field, tension and relative humidity.

Here, we consider a problem of assessment of the TC of objects under control. The developed approach of integrated statistical control of the TC of REE has to meet the following requirements:

1. To carry out an integration of sensor readings so that to define the TC of a unit under control in one parameter "invisible" to one type of sensors, but revealed by sensors of other type;
2. To reveal emergency signals at multidimensional statistical control;
3. To increase the application field of control and diagnostics.

The possibility of realization of the proposed approach relies on use of nondestructive methods of control and diagnostics applied in various branches of industrial electronics and electrical equipment [1-6].

Taking into account the conducted studies of modern approaches and controls for non-residential (sparsely populated) units for control, the task was to develop an approach to the integrated statistical control of the TC of REE, which allows diagnosing the pre-critical and critical state of controlled objects in the real time online regime [7-9].

The stage-by-stage description of process of integrated control of REE is given and the approach to reduction of redundancy of MI in the control system is shown in section two.

The third section describes the reason for low reliability in the existing control systems. The approach of statistical control of the TC of REE based on multidimensional methods of statistical control, which is carrying out correcting of values of the REE parameters during her operation by results of selective control, is established. Separate stages of the approach of integrated statistical control of the TC of REE are presented in the form of the algorithm implementing this approach. In the next sections, the level of the precritical state of the controlled unit is determined, which is identified by the rate at which its controlled parameter exceeds the

* Corresponding author: rozhnov@ipu.ru

permissible values by obtaining all the relevant measurable data (replenished data set) [10-12].

2 Integrated control of technical condition of REE

In the proposed approach the combined essence of control consists in obtaining measuring information on REE based on j hybrid sources (a thermogram, an electromagnetic response, humidity of the air, tension) received, respectively, from the thermal imager, sensors of the tension of magnetic field, the humidity of the air, the voltmeter. Considering that transfer of measuring information on the controlled equipment from sensors on blocks of processing of MI is carried out constantly, during the normal work to the equipment, its volume will be superfluous. For the elimination of redundancy, it is necessary to involve one (two) type of sensors removing information on the most critical for a certain type of equipment to parameters. In those timepoints when controlled parameters go beyond admissible limits, information arriving from sensors will confirm a critical (preemergence) condition of the equipment. In this case, it is important that the data arriving from each sensor supplemented each other gives a fuller picture.

On the basis of the statistical analysis of the measured parameters several samples of REE establish reliability range D, i.e. the area of operating states representing an interval of dispersion of values of parameters (proceeding from installation of admissible limits), corresponding to operating state of REE in general. On obviously REE efficient elements on versatile signs create them "a working profile" which is remembered in the form of reference matrixes of the TC. During the current operation (tests) of REE measure instant values of signals from exits of j of sensors, and results of the measured instant values of signals digitize and remember in the form of vectors of the digital sequence N length. From the received vectors of the digital sequence N length, form a matrix of the TC dimension of $N \times N$ which is compared to reference matrixes of the TC of "a working profile". Depending on rate of approach to reliability range limits technical condition of the REE is determined.

By results of comparison estimate the TC of the REE element, on the greatest number of coincidence of sectors of the compared reliability ranges to the level of precritical condition and also identify the place of refusal (malfunction) of an element of the equipment. Estimate REE condition at several stages, and at the first stage, use information on places of refusals and conditions of each REE element according to which find a violation of the set functioning mode (normal or critical condition) by comparison with reliability range. At the next stage, the level of the precritical state of the controlled unit is determined, which is identified by the rate at which its controlled parameter exceeds the permissible limits by measuring all available estimation information (data set).

The stages of implementation of assessment of REE, presented in the figure number one in a probabilistic state graph of system display: $P_1 = 1 - P_2$ is a prior probability of lack of an emergency; P_2 is a prior probability of emergence of an emergency; α is a false refusal (error of the first sort); β is an undetected refusal (error of the second sort); N is a normal work; A is an accident (refusal); k = 1, 2, ..., K are the REE levels. At the same time classes of the TC of REE are designated 1, 2, ..., 5_1, 6_1, where $1 = P_2\bar{\beta}_0\bar{\beta}_P...\bar{\beta}_k$; $2 = P_1\alpha_0\alpha_P...\alpha_k$; $3 = P_2\bar{\beta}_0\bar{\beta}_P...\beta_k$; $4 = P_1\alpha_0\alpha_P...\bar{\alpha}_k$; $5 = P_2\beta_0$; $6 = P_1\bar{\alpha}_0$; $5_1 = P_2\bar{\beta}_0\beta_P$; $6_1 = P_1\alpha_0\bar{\alpha}_P$. Next, we identify the place of refusal of the REE element.

Classes of the TC of REE, mean the following:

- "1" – the system is blocked, the refusal is found and distinguished;
- "2" – the system is efficient, false detection and recognition;
- "3" – the system is blocked, the refusal is found, but not distinguished;
- "4" – the system is efficient, false detection isn't distinguished;
- "5" – the system is blocked, the refusal isn't found;
- "6" – the system is efficient, recognized as efficient.

And, in the figure number one classes of the TC "1", "2", "3", "4" multilevel equipment's of REE belong to the final stage of control, and the classes "51", "61", etc. are presented at each of K levels.

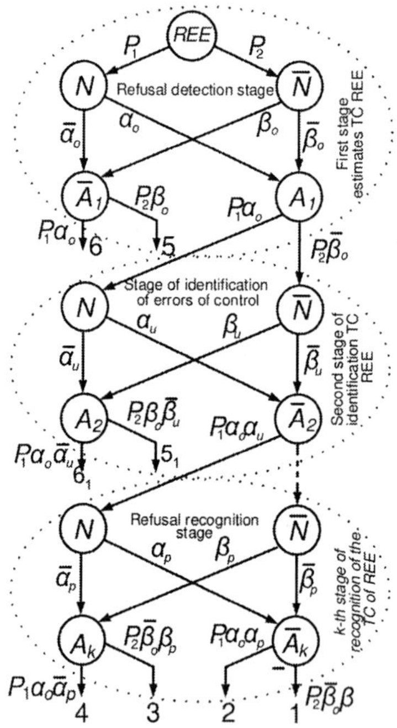

Fig. 1. The probabilistic approach for recognition of the technical condition of REE implementing the stage-by-stage principle of decision-making.

The most preferable is the class "6" when the system is efficient and recognized as efficient. The class of technical condition "1" goes the following on preference

– the system is blocked, the refusal is found and distinguished. A variation number of preferences of classes of technical condition of the REE can be written down in a look: "6" > "1" > "4" > "2" > "3" > "5".

3 Multidimensional statistical control of technical condition of REE

Control of the REE separate parameters without their interrelations doesn't provide the required size of reliability of control or excessively overestimates operational indicators; at the same time numerous false signals of accident are possible.

Multidimensional Statistical Analysis Methods can be used to control and manage REE for several correlated indicators in checking with *Data Envelopment Analysis* [7]. Their application means correcting of values of the REE parameters during her operation by results of selective control for maintenance of statistically operated and stable process of work of REE. The developed approach, is based on the analysis and processing of the statistical data arriving in real time from various sensors with REE during the current operation, taking into account tracking of the average level of the correlated indicators for their assessment and decision-making on the operating influence from the operator. So, researches in the interests of substantiating the components are performed in accordance with the *Hardwarein-the-Loop* (HIL) paradigm. In our case, a HIL simulation includes emulation of sensors and actuators infrastructure of highly automated (intelligent) transport systems [10].

Monitoring of change of the average level of multiple parameter process works of REE are carried out by *Hotelling's chart*, that is check of a zero hypothesis of equality of a vector of averages to in advance set vector [8]. Its application assumes selection of selections of constant volume at regular intervals and comparison of value of selective statistics with the critical value corresponding to position of control border on the card. Excess of critical value serves as a signal of change of level of control and demands intervention in process of work of REE for his stabilization.

Taking into account controlled j of indicators of multiple parameter process $X_1, X_2, \ldots, X_p$ results of observations of a controlled indicator are analyzed [5-6]: for each instant selection of $t(t=1,\ldots,m)$ the statistics are described

$$T_t^2 = u(\overline{X}_t - \mu_0)^{\mathrm{T}} S^{-1}(\overline{X}_t - \mu_0), \quad (1)$$

where u is a sample size; $\overline{X}_t$ is a vector of averages in selection of t (vector assessment μ), $\overline{X}_t = (\overline{x}_{t1} \ldots \overline{x}_{tp})^T$, $\overline{x}_{tp}$ is an average value in t instant selection on j to parameter ($j=1,\ldots,p$); μ_0 is a vector of averages, $\mu_0 = (\mu_1,\ldots,\mu_p)^T$; S is an assessment of a dispersion matrix Σ, defining dispersion of indicators of quality and degree of narrowness of their communication,

$$s_{jk} = \frac{1}{m(n-1)} \sum_{t=1}^{m} \sum_{i=1}^{u} (x_{ijt} - \mu_j)(x_{ikt} - \mu_k). \quad (2)$$

At stable process of removal of measuring information the condition $T_t^2 < T_{\kappa p}^2$ has to be satisfied, where $T_{\kappa p}^2$ is a border of critical area of range of reliability. At assessment a component of a dispersion matrix on a formula (2) position of control limit of the *Hotelling's* charts decides on use of the current instant selections by a formula

$$T_{\kappa p}^2 = \frac{j(m-1)(u-1)}{mu - m - j + 1} F_{1-\alpha}(j, mu - m - j + 1), \quad (3)$$

where $F_{1-\alpha}(k_1, k_2)$ is a quantile of Fischer distribution with numbers of degrees of freedom k_1 and k_2, where k_1 is in numerator, k_2 is in a denominator; α is a value of significance value (probability of false alarm).

The main criterion of violation of stability of process is an excess of *Hotelling's* statistics (1) for control border (3) [4]. However, the use of this criterion does not always provide a sufficiently rapid identification of significant changes in the process under study.

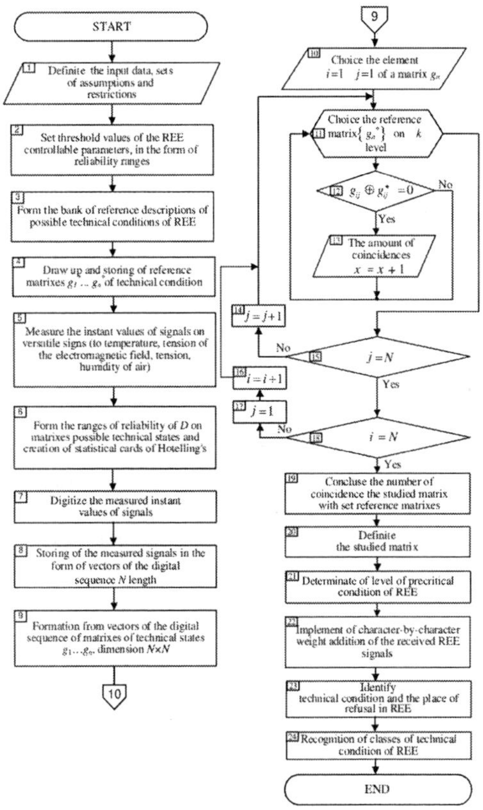

Fig. 2. The algorithm of integrated statistical control of technical condition in overall approach of REE prototyping.

If the deviation of controlled parameters is recorded only in one parameter (for example, humidity of air), and others process is in reliability range limits, then the general values of statistics T_i^2 cannot excess over. In this case, for obtaining information on an aberration of the most "critical" parameter in the course of control it is necessary to create not one, but four *Hotelling's* charts (depending on the number of sensors, controlled parameters), for pairs of parameters.

This gives the opportunity to determinate a deviation of "critical" parameter more clearly relative to others.

The work includes recommendations for system integration of newest vehicles and user services, info- and communication technologies and automation tools with smart infrastructure. The attendant system changes, aimed at providing a set of innovative DEA-services for various modes of vehicle, are presumably designed at achieving sustainable mobility through increasing the efficiency, safety and environmental friendliness of the transport process and improving the infrastructure of highly automated and/or self-driving transport systems.

4 Conclusion

The results of the present study show that various control instruments providing an increase in reliability of identification and sensitivity to detection of emergency (preemergency) situations can be used for control of the REE parameters of, especially, autonomous vehicles.

The idea of the complex statistical control of technical conditions of REE presented in the article is based on the integration of several types of sensors and can be used for development of the versatile automated integrated control systems for autonomous vehicles. The corroboration of the developed approach will include prototyping techniques, design approaches, practical applications and implications of robotic systems based on the developed methodologies. In the near future, technologies to support the implementation of such systems and units, as well as related economic, ethical and social considerations, will become a very important component of the intensive growth of multi-layered projects [13-14].

Acknowledgments

This work was partially supported by the RFBR according to the completed project No. 16-29-04326 (hybrid intelligence). The report was partially financially supported by RFBR project No. 19-29-09030 (hybrid intelligence components) and the RSF project No. 17-11-01353 (DEA and FDH application). This research was partially based by the Presidium of the Russian Academy of Sciences, Program No. 7 (30) "*Theory and Technologies of Multi-level Decentralized Group Control under Confrontation and Cooperation*" (investigation of smart infrastructure).

References

1. D.C. Montgomery, *Introduction to statistical quality control* (New York: John Wiley and Sons, 2009)
2. S. Ryvkin, A. Rozhnov, I. Lobanov, Proceedings IEEE 20th International Symposium on Electrical Apparatus and Technologies, (SIELA), 8447067 (2018) DOI:10.1109/SIELA.2018.8447067
3. F. Aparisi, J. Jabaloyes, A. Carrion, Comm. in Statistics – Theory and Methods, **28**(11), 2671–2686 (1999)
4. Y. K. Chen, K. L. Hsieh, European Journal of Operational Research, **182**(3), 1251–1262 (2007)
5. P. A. Budko, G. A. Zhukov, A. M. Vinogradenko, V.K. Goydenko, Sea radio electronics, **58**(4), 20–23 (2016)
6. P. A. Budko, A. M. Vinogradenko, A. I. Litvinov, Mechatronics, automation, control, **9**(162), 53–58 (2017)
7. A. Lychev, A. Rozhnov, I. Lobanov, Computer Vision in Control Systems-6, Springer, **182**, 127-139 (2020) doi:10.1007/978-3-030-39177-5_10
8. V. N. Klyachkin, I. N. Karpunina, M. K. Fyodorova, Automation of control processes, **45**(3), 58–64 (2016)
9. J. Katzel, Control Engineering, **54**(2), 50–54 (2007)
10. A.V. Rozhnov, I.A. Lobanov, Proceedings of 2017 10th International Conference Management of Large-Scale System Development (MLSD), 8109678 (2017) DOI:10.1109/MLSD.2017.8109678
11. S. Ryvkin, A. Rozhnov, I. Lobanov, L. Chernyshov, Proceedings 20th International Symposium on Electrical Apparatus and Technologies (SIELA), 8447069 (2018) DOI:10.1109/SIELA.2018.8447069
12. V. Abrosimov, V. Goncharenko, S. Ryvkin, A. Rozhnov, I. Lobanov, ACEMP 2017, IEEE, 905–910 (2017) doi: 10.1109/OPTIM.2017.7975085
13. A. Lychev, A. Rozhnov, Reshetnev-readings, **2**, 447-448 (2017) [in Russian]
14. L. C. Jain, M. N. Favorskaya, Computer Vision in Control Systems-6, **182**, Springer, 1-6 (2020) doi:10.1007/978-3-030-39177-5 1

EPJ Web of Conferences **248**, 02005 (2021) https://doi.org/10.1051/epjconf/202124802005
MNPS-2020

Information-Active Systems Design Tool: "IS-2" Integrated Environment

Andrey Gavrilov, *Galina* Volkova, and *Olga* Novoselova*

Department of information technologies and computing systems, Institute of information technology, Moscow State University of Technology "STANKIN", RU-127055, Moscow, Russia

Abstract. The article describes the tool for designing information-active systems - the integrated environment "IS-2": the concept of the methodology automation of intellectual labor (MAIL), the method of modeling the visual integrated environment, its structure, and features of the software solution. The Methodology Automation of Intellectual Labor (MAIL) offers an industrial way to create information-active, intelligent, automated and other systems. Their design involves the formation of models at three stages - initial modeling, conceptual modeling, and infological modeling. To automate the design process according to MAIL, a visual integrated environment (VIS) that supports this process is being developed at MSUT "STANKIN". A formal description of the structure of the models, the structure of creating models process, the method of modeling VIS has been completed. The features of the integrated environment "IS-2", obtained in the process of its development such as the formation of a universal (library of universal components used in the development of modules) component and specialized, are described. The composition of the versions is described including graphical editors for forming constituent models and their linking, means of intermediate documentation of models in the form of diagrams and specifications, and prototypes of modules for supporting analytical processing and synthesis of models.

1 Introduction

The methodology of the automation of intellectual labor (MAIL) defines an industrial way of creating various systems. This methodology involves dividing the creation process into stages such as initial modeling, conceptual modeling, infological modeling, datalogical modeling (logical and physical) and implementation. The implementation of any stage of the MAIL consists in the application of an invariant set of procedures for the formation of models at two levels of complexity: generalized, in the case, when the generated model describes a set of design tasks, and local, when the model describes just one design task. Since local and generalized models have a similar structure, their execution procedures are algorithmically identical [1].

Thus, in order to automate the design process of information-active, automated, informational and other systems in accordance with MAIL, a set of software tools is needed to support all the stages (initial, conceptual and infological modeling) of their implementation for any structure of each model. The approach to the automation of procedures by methodology used earlier made it possible to quickly obtain ready-to-use tools for supporting individual procedures at various stages of the design of information-active and automated systems in a short time. However, with the increase in their number, the development of complete projects containing models from different stages has become more difficult. This occurred due to the lack of interaction and information transfer between these software tools. In addition, for the developed tools, different implementation platforms were often used, which made it difficult to further debug and develop them.

To solve the problem of constructing a project of system (automated, informational, information-active, etc.), a visual integrated system/environent (VIS) is being developed, which will include the interface of support tools for various procedures of the methodology and an automated method for their formation such as visual modeling. To reduce the time of creation of both various VIS modules and the entire system, a method for modeling the environment has been developed, taking into account the structural features of models, their formation and execution [2].

2 The structure of information-active systems models

MAIL offers modeling methods for various stages of the life cycle of information-active or automated systems (AS). Each method contains a formal description of two model representations - universal and for subject problems. The model representation for subject tasks (ST) consists of models formed at two levels of abstraction - an object level model and a specific level model. Since the VIS is supposed to be used to form models of object-level subject tasks, its structure should be based on the structure of these models. [1-3]

* Corresponding author: olga.n10@yandex.ru

Since the structures of the models are identical, on their basis a unified model of the object level of the ***n***-th object problem of the ***i***-th stage of creation was compiled:

$$M2_i(n) =< EL2_i(n), ST2_i(n), DN2_i(n), FN2_i(n), R2_i^{SDF}(n) > \quad (1)$$

where $EL2_i(n)$ is the set of elements of the ST model, $ST2_i(n)$ is the set of static relations on the elements of the object-level model (static component), $DN2_i(n)$ is the set of dynamic constraints on the elements of the model (dynamic component), $FN2_i(n)$ is the set of functional relations on the elements of the model (functional component), $R2_i^{SDF}(n)$ is the linkage of the components. Each component includes two structures: the main and the derivative for the static component, a system of constraints on the elements of the main and derivative of static structures, and a system of functional relations on the elements of the main and derivative of static structures.

From the point of view of a graphical description of structures for object-level models of the ***n***-th subject task, all components can be formally represented by graphs: static, dynamic, functional and meaningful (component linking graph):

$$M2_i(n) \rightarrow \{G_i^{tr}(n)\}, \quad G_i^{tr}(n) = (V_i^{tr}(n), E_i^{tr}(n)), \quad (2)$$

where the indices ***t, r*** determine the type of the graph and its components:

$$t \in ('ST','DN','FN','R') \quad (3)$$

denote the type of the component represented by the graph, $r \in [1,2]$ determine the structure index.
From the point of view of the tabular description of structures, each model of object-level for the n-th subject task is fixed in the form of specifications, which can be formally represented by relational relations. Each structure of the model can contain specifications for describing structure elements, for describing binary and / or ternary relationships on structure elements.

$$M2_i(n) \rightarrow \{R_i^{tre}(n)\}, \quad (4)$$

$$R_i^{tre}(n) \subset D_i^{tre1} \times D_i^{tre2} \times \ldots \times D_i^{trem}, \quad (5)$$

$$R_i^{tre}(n) = \{r_i^{tren}: r_i^{tren} = (d_i^{tre1-x}, d_i^{tre2-y}, \ldots, d_i^{trem-z})\}, \quad (6)$$

where the indices ***t, r*** determine the type of specification describing the ***tr***-th structure, the index ***e*** defines the index of the specification: ***e*** = 1 for a table with elements, ***e*** = 2 for a table with binary links, ***e*** = 3 for a table with ternary links, ***e***> 3 for additional specifications.

3 Process structure

The structure of creating models process can be formally presented as follows:

$$Y = (P, TP), P = \cup_u P_u, \quad P_u \cap P_k = \emptyset, \; u \neq k, \quad (7)$$

$$P_u = \{p_{uv}\}, \quad (8)$$

$$TP \subset P \times P, \quad (9)$$

where *P* is a set of actions, *TP* is a set of links between actions that describe the structure of actions. P_u - a set of actions of the ***u***-th level of complexity, p_{uv} - the ***v***-th action at the ***u***-th level of complexity in general. In this case, actions at each difficulty level can be defined as follows:

$$\overline{p_{1i(s)}} = p_{1i}\&(p_{0s}, p_{1i}) \quad (10)$$

- ***i***-th stage of ***s***-th process,

$$\overline{p_{2j(si)}} = p_{2j}\&(p_{1i}, p_{2j})\&(p_{0s}, p_{1i}) \quad (11)$$

- ***j***-th procedure of the ***i***-th stage of the ***s***-th process,

$$\overline{p_{3k(sij)}} = p_{3k}\&(p_{2j}, p_{3k})\&(p_{1i}, p_{2j})\&(p_{0s}, p_{1i}) \quad (12)$$

- the ***k***-th operation of the ***j***-th procedure of the ***i***-th stage of the ***s***-th process,

$$\overline{p_{4l(sijk)}} = p_{4l}\&(p_{3k}, p_{4l})\&(p_{2j}, p_{3k})\&(p_{1i}, p_{2j})\&(p_{0s}, p_{1i}) \quad (13)$$

- ***l***-th action of the ***k***-th operation of the ***j***-th procedure of the ***i***-th stage of the ***s***-th process.

4 Visual Integrated Environment (VIS) modeling method

The method of modeling a visual integrated environment consists in its sequential presentation at various levels of detail as a set of fundamentally different structural elements. [2]

Taking into account the structure of models of information-active systems and procedures for their processing, the visual integrated environment can be represented as a set of models of various levels of abstraction:

$$VIS =< VIS1, VIS2, VIS3, R^{123} >, \quad (14)$$

Where $R^{123} = \{R^{12}, R^{23}\}$ - is a set of interrelationships between elements of VIS models of different levels of abstraction. (see fig. 1)

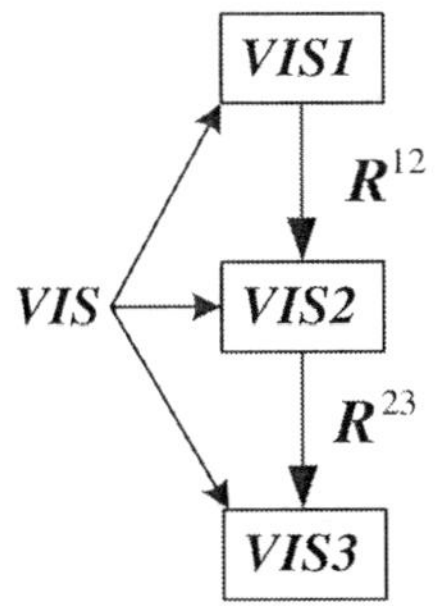

Fig. 1. Representation of the VIS implementation model taking into account the levels of abstraction

For each action $\overline{p_{4l(sijk)}}$, there can be many versions of the support tools implemented as functional modules

$$M^{sijkl} = \{m_f^{sijkl}\} \tag{15}$$

Based on this, the visual integrated environment can be represented as a system of functional modules:

$$\mathrm{VIS1} =< \mathrm{M}, \mathrm{R}^1 >, \tag{16}$$

$$\mathrm{M} = \bigcup_{s,i,j,k,l,f} \mathrm{M}^{\mathrm{sijkl}} \tag{17}$$

where M is a set of functional modules, $R1 \subset M \times M$ is a set of relationships between functional modules.

For a set of modules M that implement all processes, information, functional and interface components can be distinguished that implement the corresponding parts of the modules:

$$M^{sijkl} \rightarrow < \mathrm{INF}^{\mathrm{sijkl}}, \mathrm{FUN}^{\mathrm{sijkl}}, \mathrm{INT}^{\mathrm{sijkl}}, \mathrm{R}^{\mathrm{sijkl}} >, \tag{18}$$

$$\mathrm{R2}^{\mathrm{sijkl}} \subset \mathrm{INF}^{\mathrm{sijkl}} \times \mathrm{FUN}^{\mathrm{sijkl}} \times \mathrm{INT}^{\mathrm{sijkl}}. \tag{19}$$

Thus, at the second level of abstraction - the level of logical components, the visual integrated environment can be represented as a set of components and their relationships:

$$VIS2 =< INF, FUN, INT, R2 >. \tag{20}$$

Each component at the file system level is represented by its elements of the corresponding types.

$$INF^{sijkl} =< TI^{sijkl}, DI^{sijkl}, FI^{sijkl}, R2^{sijkl} > \tag{21}$$

for an information component, which is implemented by information volumes, directories, files. The set of files of the functional and interface components are formed in the same way.

Based on the foregoing, the VIS model at the third level of abstraction - the file system (physical), is formally defined as a set of volumes, directories and files:

$$VIS3 =< T, D, F, R3 >, \tag{22}$$

$$\mathrm{R3} \subset \mathrm{T} \times \mathrm{D} \times \mathrm{F} \times \ldots \times \mathrm{T} \times \mathrm{D} \times \mathrm{F}, \tag{23}$$

$$\mathrm{R3} = \mathrm{R3}_1 \bigcup \mathrm{R3}_2 \tag{24}$$

At the same time, $R3_1$ determines the storage structure of the VIS, i.e. hierarchical connections between the elements of the file system, and $R3_2$ - the structure of functioning, i.e. the relationship between files within a specific version of the VIS.

The identical structure of the models at each stage, an invariant set of modeling procedures made it possible to identify invariant parts in the implementation of functional modules - universal files that implement identical information, functional or interface parts of the modules. For example, for information component files of various modules:

$$FI^{sijkl} \bigcap FI^{smjkl} = UFI^{jkl}, \tag{25}$$

where UFI^{jkl} is the set of universal information files that implement the ***l***-th action of the ***j***-th procedure for the ***k***-th component of the model at various stages ***i*** and ***m***.

The set of such files that implement the universal part of the logical components of the VIS must be assembled into a separate library - a separate constructive element of the environment, which makes it possible to reduce the implementation time of its modules. The considered method for modeling VIS allows it to be represented in the matrix form shown in Fig. 2. Each area defines the class of the VIS module - its place within the entire system. The class of the module determines its functional purpose, input and output data. The colored sections show an example of module classes that can be built on generic components: red - modules that support a certain type of model, blue - support processing of a certain component, green - support a certain procedure. Data structures, processing algorithms and presentation forms can be invariant.

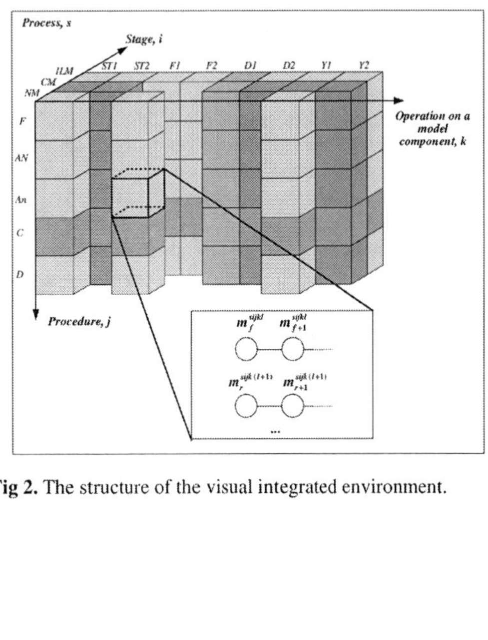

Fig 2. The structure of the visual integrated environment.

Symbols. Modeling procedures: F - formation, AN - analysis, An -analytical processing, C - synthesis, D - documentation. Stages: NM - initial modeling, CM - conceptual modeling, ILM - infological modeling; components of the models: ST1 - basic static structure, ST2 - derived static structure, F1 - functional structure of the first kind, F2 - functional structure of the second kind, D1 - dynamic structure of the first kind, D2 - dynamic structure of the second kind, Y1 - linking ST1, D1 and F1, Y2 - linking CT2, D2 and F2.

5 VIS implementation

When developing a software environment that ensures the integration of the means to support the procedures of the MAIL, the main task is to ensure information flows between the means that can operate simultaneously on different computers. It should also be borne in mind that a single modeling procedure can have multiple support tools, depending on the configuration the integrated environment itself. A schematic diagram of information interaction between classes of such tools within one stage of creating information-active and automated systems is shown in Fig. 3.

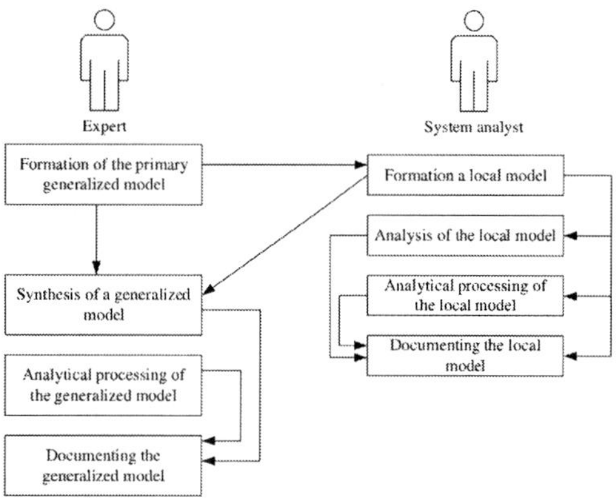

Fig 3. Information flows between classes of integrated environment modules

Based on this, a visual integrated environment (VIS) was developed in the form of the "IS-2" software package (see Fig. 4-5), which contains functional modules supporting the procedures of the MAIL stages. This software package "IS-2" provides a graphical method for developing models and automatic generation of specifications based on the constructed diagrams. The complex manages the operation of modules, as well as organizes their information interface. Since the models have a similar composition and are processed by an invariant set of procedures, during the development of the VIS, the unification of software solutions was made - the general information, functional and interface aspects of various modules were highlighted. Later they were moved to a single repository of universal components. [4, 5]

Created at the Department of Information Technologies and Computing Systems of the MSUT "STANKIN" implementation - the software complex" IS-2 "allows you to form models at various stages of system design (initial, conceptual, infological), to perform analytical processing of models, including restructuring the algorithm for solving the task. Also "IS-2" can form a generalized model for a complex of subject tasks, integrating algorithms for solving subject tasks into a single representation of the solution for the complex. [4, 6-9]

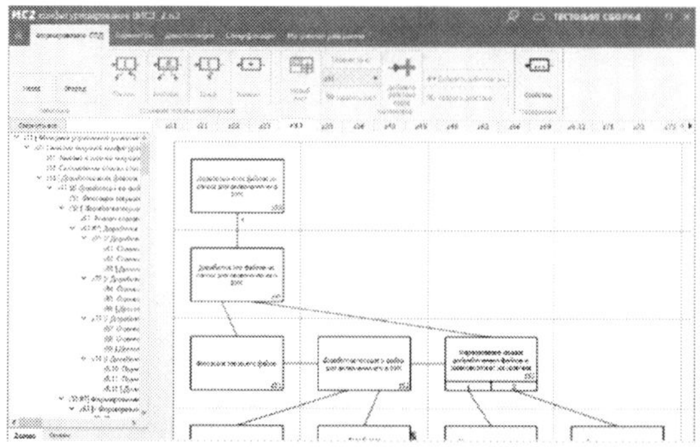

Fig. 4. Graphical interface of "IS-2": module for forming a system of subject actions

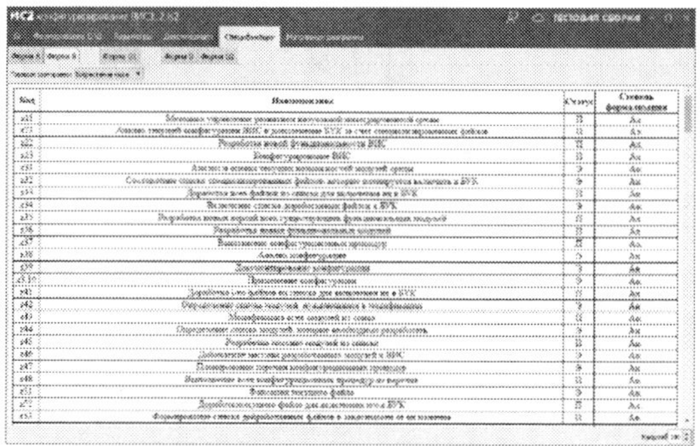

Fig. 5. Graphical interface of "IS-2": module for outputting specifications of the initial model

The integrated environment "IS-2" consists of two subsystems - a specialized one, which includes many modules of the environment, and a universal one, presented in the form of a library of universal components (LUC) used in the development of modules. LUC represents software tools for developers of "IS-2" modules, namely: data structures describing models as graph structures, software libraries containing the main processing algorithms for these structures, which are associated with the implementation of functional modeling procedures, as well as with various forms of information presentation reflected in the models. The library also includes tools for converting invariant structures to models of a certain type for further work with them in the environment modules. Thus, the design and development of modules is greatly simplified due to the repeated use of already implemented software structures.

A feature of the IS-2 is the presence of several modes of its operation. The main operating mode of the complex is the development of information-active and automated systems, their stage-by-stage modeling and subsequent implementation. But, since the environment itself is a complex information-active system, it can be used to develop new versions of itself.

6 Conclusion

At the present time, the "IS-2" versions are developed and used in the educational process and scientific activities of the Department of Information Technologies and Computing Systems of the MSUT "STANKIN" to support the initial and conceptual modeling. Each version contains graphic editors for forming components models and their linkage, means of intermediate documentation of models in the form of diagrams and specifications. Also, prototypes of modules supporting analytical processing and model synthesis were developed and included in the environment. The development of modules for conjugation of various adjacent stages of modeling and modules for supporting infological modeling is in the final stage [6, 10-12].

It should be noted that the "IS-2" includes a SDK (a Software Development Kit), a set of software tools used by third-party developers to create their own modules for the environment.

This article was prepared in the framework of the project No. 17-29-07057 "Development of methods of restructuring and integration for semantic and syntactic representations when creating automation systems for design and management processes" under the RFBR grant. This work was carried out using equipment provided by the Center of Collective Use of MSUT "STANKIN".

References

1. G.D.Volkova, *Methodology of intellectual labor automation. Monograph* (M.: Janus-K, 2013)
2. G.D. Volkova, A.G. Gavrilov, Vestnik MSUT "STANKIN", **4**(39), 85-89 (2016) [in Russian]
3. O.V. Novoselova, G.D. Volkova, A.G. Gavrilov, Proceedings of higher educational institutions. Volga region. Technical science, **1**(29), 81-91 (2014) [in Russian]
4. A.G. Gavrilov, G.D. Volkova, O.V. Novoselova, Scientific and technical journal "Information technologies in design and production", **2**, 30-36 (2018)
5. A.G. Gavrilov, G.D. Volkova, O.V. Novoselova, Digital economy: technology, management, human capital: materials of the III All-Russian scientific and practical conference, 10-13 (Vologda: LLC "Marker", 2020) [in Russian]
6. A.G. Gavrilov, O.V. Novoselova, G.D. Volkova, Vestnik MSUT "STANKIN". Scientific peer-reviewed journal, **2**(41), 83-88 (2017) [in Russian]
7. O.V. Novoselova, A.A. Ruzheynikov, Science today: challenges and prospects: materials of the international scientific and practical conference, 72-74 (Vologda: LLC "Marker", 2020) [in Russian]
8. O. Novoselova, A. Ruzheynikov, A. Gavrilov, EPJ. Web Conf., **224**, 06008 (2019) https://doi.org/10.1051/epjconf/201922406008
9. O. Novoselova, A. Sidorov, IOP Conf. Ser.: Mater. Sci. Eng., **709**, 044014 (2020) https://doi.org/10.1088/1757-899X/709/4/044014
10. A.A. Ruzheynikov, Materials of the student scientific and practical conference "Automation and Information Technologies (AIT-2019)", 105 (Moscow: MSUT "STANKIN", 2019) [in Russian]
11. A.A. Ruzheynikov, Information technologies in modeling and management: approaches, methods, solutions: Collection of scientific articles of the II All-Russian scientific conference with international participation, **1**, 546-552 (Togliatti: Publisher Kachalin Alexander Vasilievich, 2019) [in Russian]
12. I.Yu. Pavlov, Materials of the student scientific and practical conference "Automation and Information Technologies (AIT-2020)", **3**, 148-150 (Moscow: FGBOU VO "MSUT" STANKIN ", 2020) [in Russian]

Design of Water Fire Extinguishing System for Protection of Metro Rolling Stock Stagnation Zones

Ilya Filippov[1], *Vladimir* Yusupdzhanov[1], *Olga* Zybina[1], *Ilya* Klochihin[1], and *Ekaterina* Chalovskaya[1]

[1]Peter the Great St.Petersburg Polytechnic University, Higher school of technosphere safety, RU-195251, Saint-Petersburg, Russian Federation

Abstract. Metro rolling stock stagnation zones are quite long. This causes a number of difficulties in choosing and designing a water fire extinguishing system, as it directly affects the pressure loss in pipelines when water is supplied over such a long distance. That is why the task of this work is to design two different types of water fire extinguishing for metro dead ends and to perform hydraulic calculations, based on the results of which to choose the most suitable system for the object under study. The choice of a fire extinguishing system to ensure fire safety of dead ends of rolling stock sludge is made in the favor of the fire extinguishing system with high-pressure water spray.

1 Introduction

Fire extinguishing is designed for localization of ignition of the railway car within the zone of dead ends, which provides night rolling stock to prevent the spread of fire, protect structures, cable networks and equipment and to ensure safe conditions for the firefighting forces of EMERCOM of Russia and firefighters [1-7].

The purpose of the work is to develop a methodology for evaluating fire extinguishing systems in relation to various types of water fire extinguishing systems, further selecting the most suitable one for working in areas of sedimentation of rolling stock and ensuring fire protection of the object.

To achieve this goal, the following problems have to be solved:

1. Analysis of fire safety of the object;
2. Design of two different types of water fire extinguishing for the object/system being studied;
3. The hydraulic calculations of the designed fire suppression systems;
4. Selection of the most suitable fire extinguishing system for the object/system being studied based on the calculation results.

The object of research in this work is the fire protection system in the dead ends of the metro with areas of sedimentation of rolling stock. The subject of research pursued in this paper is the development of the project of the most suitable fire extinguishing system for operation in the conditions of the metro object being studied.

Metro dead ends are a complex object/system, both in terms of construction and technical issues [8-11]. In the field of fire fighting, there are a sufficient number of legal acts regulating the requirements for design, installation and operation [12]. However, objects that need to be protected from a possible fire have a striking difference from each other and have their own distinct features. That is why there exist additional conditions and features in the design, installation, commissioning and operation of such objects and it is also necessary to develop special recommendations for the installation and operation of fire extinguishing on them.

This work is important due to the need to prevent a fire in the dead ends of the underground areas of sediment rolling stock, to choose the right sprinkler system and to ensure all the necessary technical means for the exploitation of the fire-extinguishing system and effective extinguishing of a possible fire [12-15]. The results of the present study amy be interesting and useful for developers/designers of fire protection systems, specialists in the field of transport, fire and technosphere safety.

The source of fire load in dead ends is the eight-car rolling stock. According to the degree of fire hazard, the extinguishing zone belongs to the 2nd group of premises [16, 17]. Ignition of rolling stock by burning substances and class A materials, solid substances [18] is possible. According to the [19], it is a zone of class P-IIA in which solid combustible substances are handled [19, 20]. Between the distillation tunnels and dead ends of the night sediment, it is necessary to build two-thread drencher water curtains based on drencher sprayers [17, 20, 21]. General-purpose automatic fire systems are designed in accordance with [17], and automatic fire extinguishing systems with high-pressure thin-spray water are designed in accordance with internal fire safety standards and [17].

In accordance with the group of premises for fire hazard (group 2), fire classes (A, E), clause 5.1.4 and section 5.4 of SP 5.13130.2009, with table 5.1, the estimated fire extinguishing area is 120 sq. m. and the duration of the automatic fire extinguishing system opertaion is at least 60 minutes [17]. The irrigation intensity when using a general-purpose fire extinguishing system for the protected area is assumed to be 0.180 l/(s m2). The intensity of irrigation when using automatic fire extinguishing system with thinly sprayed high-pressure water is not strictly limited [17]

2 Methods

2.1 Methods of hydraulic calculation of fire extinguishing systems with high-pressure thin-spray water

Water fire extinguishing systems use various communications. An integral part of the aggregate water fire extinguishing system are pipelines that supply the extinguishing agent (water) to the sprayers. It is in connection with this function that hydraulic calculations are necessary.

In order to confirm (clarify) the selected diameters of the pipeline and pipeline fittings, a hydraulic calculation is performed by calculating the pressure drop during the flow of water from the pumping unit to the most remote sprayers of the calculated irrigation area using the Darcy-Weisbach method. [21]

To simplify the calculation, the pipeline is divided into sections so that within each section there is no change in the parameters of the liquid flow (flow rate). The velocity of the liquid flow is determined by the formula (1).

$$V=(4\cdot Q)/(\pi\cdot D_i^2) \quad (1)$$

where Q is the liquid flow rate, m^3/s

D_i – inner diameter of the pipe, m.

Pressure loss at each site consist of the pressure losses along the length of the pipe (due to wall friction), pressure losses in smooth pipe bends (with bending pipe is meant for a smooth change of flow direction along the radius not less than 2,5·D, where D is the diameter of the pipe) and the pressure drop with a sharp bend (a sharp turn of flow at 90°, such as a tee or elbow). Thus, the pressure drop in the section is determined by the formula (2):

$$\Delta\rho=\Delta\rho_\lambda+\Delta\rho_b+\Delta\rho_e, \quad (2)$$

where $\Delta\rho_\lambda$ is the pressure loss along the length of the pipe;

$\Delta\rho_b$ – pressure loss during smooth bending of the pipe;

$\Delta\rho_e$ – pressure loss when the pipe bends sharply.

To determine the pressure loss along the length of the pipe, the Reynolds number must be calculated using the formula (3):

$$Re= (\rho\cdot V\cdot Di)/\mu \quad (3)$$

where ρ is the density of the liquid;

μ is the dynamic viscosity of the liquid.

Depending on the Reynolds number, the fluid flow mode for this section is determined.

For laminar flow (Re < 2300), the pressure loss along the length of the pipe is determined by the Darcy-Weisbach formula (4):

$$\Delta\rho_\lambda = \lambda_L\cdot(L\cdot V^2)/(2\cdot D)\cdot\rho \quad (4)$$

where L is the length of the pipeline section;

D – diameter of pipeline;

λ_L – coefficient of pressure losses along the length of the;

ρ is the density of the liquid.

For laminar flow, the coefficient of pressure loss for friction along the length is determined by the Poiseuille formula (5):

$$\lambda_L = 64/Re \quad (5)$$

For turbulent flow (Re > 2300), the pressure loss along the length of the pipe is determined by the Darcy-Weisbach formula (6):

$$\Delta\rho_\lambda=\lambda T\cdot(L\cdot V^2)/(2\cdot D)\cdot\rho \quad (6)$$

where L is the length of the pipeline section;

λ_L – coefficient of pressure loss along the length.

In a turbulent flow, the coefficient of pressure loss for friction along the length is determined from the equation (7):

$$1/\sqrt{(\lambda_T)}=-2\cdot Lg\cdot[2{,}51/(Re\cdot\sqrt{(\lambda_T)})+e/(3{,}72\cdot D_i)] \quad (7)$$

where e is the relative roughness (which in laminar flow does not affect the flow structure, but only increases the streamlined surface, and also does not affect the friction resistance);

D_i – inner diameter of the pipe, m.

From this equation, the coefficient λ_L is determined using the iterative method.

Also, the Blasius formula (8) is often used to determine the coefficient of pressure loss for friction along the length during turbulent flow:

$$\lambda_L=0{,}316/\sqrt[4]{Re} \quad (8)$$

However, this paper uses the derivation of the coefficient λ_L from equation 7.

The pressure loss during a smooth bend of the pipe is determined by the formula (9):

$$\Delta\rho_b=(B\cdot\varepsilon_b\cdot\rho\cdot V^2)/2 \quad (9)$$

where B is the number of smooth bends in the selected section;

ε_b – coefficient of pressure loss during bending.

The total pressure loss is defined as the algebraic sum of the losses in each section by the formula (10):

$$\Delta\rho_\Sigma= \sum\Delta\rho_i \quad (10)$$

The initial data for the hydraulic calculation are situational plans and a block diagram of an automatic fire extinguishing system with fine water.

The calculation section is determined from the diagram. For the first approximation, the pipe diameters are selected depending on the characteristics of the designed fire extinguishing system. In this paper, the following pipe diameters are selected:
- from pump unit to sectional valves – D38 pipe;
- looped feed pipe – D38 pipe;
- taps to sprayers – D12 pipe.

Constant coefficients required for calculation: ρ=1000kg/m^3; μ=0,001Pa·s; e=3·10^{-5}.

The maximum pressure drop is determined from the calculation results. Then the result obtained after calculations is added to the minimum pressure on the sprayer equal to 6 MPa. If it is necessary to reduce the output pressure, it is necessary to increase the dimensions of the valve and pipeline diameters, and then perform the hydraulic calculation again.

2.2 Methods of hydraulic calculation of standard water fire extinguishing systems

Further, the paper presents the main mathematical dependencies of the calculation method [17].

The following formula (11) is used to determine the estimated water flow on the dictating sprinkler located within the dictating irrigation zone):

$$q=10\cdot K\cdot\sqrt{P} \quad (11)$$

where q is the flow rate of the extinguishing agent (in this case, water) on the dictating sprinkler, *l/s*;
K is the productivity coefficient of the sprinkler, which is set depending on the technical documentation for the product, l/(s·MPa$^{0.5}$);
P - pressure on the dictating sprinkler, *MPa*.

This formula (12) is used to determine pipe diameters in a section of the pipeline network

$$d=1000\cdot\sqrt{((4\cdot Q)/\pi\mu v)} \quad (12)$$

where d is the diameter of the pipeline sections in the intervals between sprinklers, mm;
Q – direct flow rate of sprinklers, l / s;
μ – coefficient of discharge;
v – the speed of water movement within the calculated sections of the pipeline network, m/s (the speed must be below 10 m/s).

The following formula (13) is used to determine the pressure loss in a section of the pipeline network:

$$P=(Q\cdot L)/(100\cdot K_T) \quad (13)$$

where Q is the total consumption of fire extinguishing agent (water) on the first and second sprinklers, l/s;
K_T – specific characteristic of the pipeline, l/s;
L is the length of the pipeline.

The pressure of the second sprinkler after the dictating one is determined by the formula (14):

$$P_2=P_1+P_{(1\text{-}2)} \quad (14)$$

The flow rate of the second sprinkler is obtained from the formula (15):

$$q_2=10\cdot K\cdot\sqrt{(P_2)} \quad (15)$$

The generalized characteristic of a series (branch) is calculated by the formula (16):

$$B=Q_2/P \quad (16)$$

For the calculation of automatic sprinkler fire extinguishing systems, the condition presented in the formula (17):

$$Q_n\leq Q_c \quad (17)$$

where Q_n is the standard flow rate of an automatic sprinkler fire extinguishing system;
Q_c - actual consumption of the automatic sprinkler fire extinguishing system.

To determine the number of sprinklers that can provide the actual flow rate of an automatic sprinkler fire extinguishing system Q_c with an irrigation intensity exceeding the standard (when taking into account the configuration of the selected protected area), the condition presented in the formula (18) is used:

$$n\geq S/\Omega \quad (18)$$

where n is the minimum number of sprinkler sprinklers that can provide the actual flow rate of an automatic sprinkler fire extinguishing system Q_c with an irrigation intensity exceeding the standard;
S is the minimum protected irrigated area, taking into account the data presented in table 5.1 of [17];
Ω is the conditional calculated area, the protection of which is provided by one sprinkler and is determined by the following formula (2.19):

$$\Omega=L^2 \quad (19)$$

where L is the set distance between sprinklers.

Due to the fact that the pressure on each Sprinter is different, and the lowest pressure is on the dictating Sprinter, it is necessary to take into account the flow rate of each Sprinter from the total amount.

To determine the total water consumption of a drencher automatic fire extinguishing system, a sequential summation of the costs of all sprinklers located in the protected area is carried out using the formula (20):

$$Q_d=\sum\nolimits^{n}_{n\text{-}1} q_n \quad (20)$$

where Q_d is the calculated flow rate of the drencher automatic fire extinguishing system, l/s;
q_n - direct flow rate of the nth sprinkler, l / s;
n is the total number of sprinklers located in the protected area.

The flow rate Q_s of an automatic sprinkler fire extinguishing system with a water curtain is determined by the formula (21):

$$Q_s=Q_c+Q_z \quad (21)$$

where Q_c is the flow rate of the automatic sprinkler fire extinguishing system;
Q_z - water curtain flow rate.

If a combined fire water supply system is designed on the site (i.e. combined lines from automatic fire extinguishing systems and internal fire water), then allowed to set one group of pumps, while ensuring that group flow Q, is equal to the summed costs for each vocopro-water, that represented in the formula (22):

$$Q=Qs+Qint \quad (22)$$

where Q_s, Q_{int} – respectively, the flow rate of the internal fire-fighting water supply and the flow rate in the pipeline network of an automatic fire extinguishing system.

To calculate the required pressure of a fire pump, this formula is generally used (23):

$$P_n=P_h+P_v+\sum P_L+P_{CU}+P_D+Z-P_{pf}=P_{req}-P_{fp} \quad (23)$$

where P_n is the required pressure on the fire pump, MPa;
P_h – pressure loss on the horizontal section of the AB pipeline, MPa;
P_v – pressure loss on the vertical section of the DB pipeline, MPa;
P_L – pressure loss in local resistances, MPa;
P_{CU} – the local resistance in the control units (at the gates, valves and alarm valve), MPa;
P_D – pressure on the dictating sprinkler, MPa;
Z – piezometric pressure (geometric height of the dictating sprinkler above the fire pump axis), MPa; Z=H/100;
P_{fp} – pressure at the inlet of the fire pump, MPa;
P_{req} – required pressure of the fire pump, MPa.

The following formulas (24), (25) are used to determine the pressure loss in the control units:
- in the sprinkler:

$$P_{CUS}=\xi_{CUS}\cdot\gamma\cdot Q^2=(\xi_{VS}+\xi_{sod})\cdot\gamma\cdot Q^2 \quad (24)$$

- in drencher:

$$P_{CUD}=\xi_{CUD}\cdot\gamma\cdot Q^2=(\xi_{VD}+2\cdot\xi_{sod})\cdot\gamma\cdot Q^2 \quad (25)$$

where ξ_{CUS}, ξ_{CUD}, ξ_{VS}, ξ_{VD}, ξ_{sod} - pressure loss coefficients, respectively, in the sprinkler and drencher control unit, in the sprinkler and drencher signal valve and in the shut-off device (accepted according to the technical documentation for the control unit as a whole or for each signal valve, gate or gate valve individually);
γ is the density of water, kg/m^3;
Q-estimated flow rate of water passing through the control unit, m^3/h.

In approximate calculations, local resistances are assumed to be equal to 20% of the resistance of the entire pipeline network, taking into account losses in control unit.

Unless otherwise specified in the technical specifications, the calculations are carried out so that the pressure at the control units does not exceed 1 MPa.

The duration of the fire extinguishing agent supply (in this case, water) is accepted, taking into account the selected group of the protection object according to table 5.1, Appendix B of [17].

After a proper study of the calculation methods used in the work, it is necessary to give the methodology itself, according to which the choice of a particular fire extinguishing system was made based on the results of calculations.

3 Results

3.1. Hydraulic calculation of the fire extinguishing system with high-pressure water spray

The maximum pressure drop is determined from the results obtained during the calculation. Then the result obtained after calculations is added to the minimum pressure on the sprayer equal to 6 MPa. If it is necessary to reduce the output pressure, increase the size of the valve and pipeline diameters, and then perform the hydraulic calculation again. The calculation results are shown in table 1.

Table 1. Results of calculation of the fire extinguishing system with high-pressure water spray

№	Q, l/min	L, m	E	B	V, m/s	Re	dPi, kg/sm2	dP, kg/sm2
1	34,2	0,5	2	2	8,96	80679	1,56	1,56
2	34,2	9	2	2	0,71	22691	0,03	1,59
3	68,4	9	2	2	1,42	45382	0,09	1,68
4	85,50	380	15	15	1,77	56728	4,73	6,41
5	171	5	8	8	3,55	113455	0,79	7,20
6	342	5	8	8	7,09	226911	3,12	10,33
7	171	3	8	8	3,55	113455	0,70	11,03
8	128,25	425	45	45	2,66	85092	12,50	23,54
9	256,5	3	6	6	5,32	170183	1,25	24,79

The following designations are used in the table:
- № – number of the pipeline section;
- L – length of the pipe section;
- Q-consumption;
- E – the number of sharp bends of the pipe in the section where the flow turns 90O
- B – number of smooth bends on the section;
- Δp – pressure loss on the pipe section;
- ∑Δp - total head loss.

The pipe diameter of 1 section of the pipeline is 12 mm, the pipe diameter of 2-9 sections of the pipeline is 38 mm. The wall thickness of the pipe of 1 section of the pipeline is 1.5 mm, the wall Thickness of the pipes of 2-9 sections of the pipeline is 3 mm.

Based on the calculation results, it can be concluded that at the maximum flow rate of 513.0 l/min (simultaneous operation of 10 sprinklers on an estimated area of 120 m2 – 342.0 l/min and a drainpipe curtain – 171.0 l/min), the system performance is not impaired. The minimum pressure on the sprayers does not fall below 60 Bar (6.0 MPa).

To ensure normal operation of the system, it is necessary to ensure a maximum pressure drop of no more than 25 Bar (2.5 MPa), since the guaranteed pressure at the outlet of the pump unit is 85 Bar (8.5 MPa), and the minimum operating pressure at the sprayers is 60 Bar (6.0 MPa).

As follows from the calculations, the total pressure drop and pipe velocity in all cases do not exceed the permissible values. Therefore, the selected pipe diameters meet the system requirement.

The pump unit is selected for a maximum design flow rate of 513.0 l/min and a 10% capacity margin according to the manufacturer's requirements. Based on the fact that the capacity of one pump unit is 100 l/min, the project adopted a pumping unit consisting of 7 pump units (6 working and 1 backup).

3.2 Hydraulic calculation of standard water fire extinguishing systems

Operation of sprinklers in dead ends of night sedimentation of rolling stock on 3 track (this section is selected as potentially dangerous due to the large number of sprayers and significant distance from the pumping station). Extinguishing is carried out at the expense of the pumping unit in the room of the pumping automatic firefighting system at the platform level of the station complex.

The initial data is given below:
- Height of placement of the dictating sprinkler – 5 m;
- Irrigation intensity of the protected area – 0.180 $l/(s*m^2)$;
- Standard consumption (not less than) – 45 l/s;
- The minimum area of the sprinkler automatic firefighting system is 120 m^2;
- The performance factor of the sprinkler – 0,71;
- Area protected by a single sprinkler with the required intensity – 12 m^2;
- Pressure of the dictating sprinkler – 0.5 MPa;
- The height of the section control unit is 1.5 meters;
- The estimated standard consumption of the section is 45 l/s.

The pipeline in the protected area was divided into 9 sections. Each section between two sprinklers is 4.5 meters long. The length of the supply pipeline is defined as 390 meters. The type of pipeline is circular.

To provide fire extinguishing of areas of sedimentation of rolling stock with a fire extinguishing agent, a pumping unit was selected with two pumps connected in series and one backup.

The maximum water consumption per section, according to the calculation results, is 46.158 l/s.

The maximum pressure on the section is 0.804 MPa.

The supply pipeline from the pump unit to the control unit is divided into two sections. The first section is 10 meters long-from the pump unit in the pump room of the automatic firefighting system to the exit outside the room. The second section is 425 meters long-from the premises of the pumping STATION located on the platform of the station complex to the protection zones of dead ends of night sediment of rolling stock. The pipeline type is a dead end.

Based on the calculation results, the following parameters of fire pumps that should be used in the pump unit are determined: water

Consumption – 46.158 l/s (166.168 m^3/h);

Pressure – 1,399 MPa (139,904 meters of water column).

After completing the hydraulic calculation of the designed general-purpose fire extinguishing system, you should proceed directly to the selection of the most suitable system for operation in the conditions of metro dead ends based on the results of the calculations.

3.3 Selecting the type of water fire extinguishing based on the results of calculations

The system of water fire extinguishing with thinly sprayed high-pressure water requires a water flow rate of 34.2 l/min to ensure the protection of the object. The maximum calculated flow rate with simultaneous operation of 10 sprinklers on an estimated area of 120 m2 is equal to 342.0 l/min. Taking into account the flow rate that is required for the operation of the drencher curtain (171.0 l/min), the total maximum flow rate is 513.0 l/min.

To ensure extinguishing during operation of a general-purpose water fire extinguishing system,a water flow rate of 5-5.5 l/s (300-330 l/min) is required on sprinklers. The total water consumption of the General-purpose system will be 46.158 l/s (2769.48 l/min).

Therefore, it is concluded that the fire extinguishing system is more economical in terms of the consumption of fire extinguishing agent (water), as opposed to a low-pressure system. For the required water reserves, a comparison is made in the work below.

Considering the designed systems by pipe diameters, it is logical to conclude that in a low-pressure system, pipes of a larger diameter are used than in a high-pressure thin-spray water system.

Namely, for a low-pressure system-pipes with a diameter of 100 mm in the section from the pump unit to the control units and pipes with a diameter of 78 mm in the sections of the supply pipeline.

System the water mist high pressure is applied, the pipe diameter of 38 mm on the plots from the pumping unit to control nodes and on sections of the supply line, pipe diameter 12 mm on carts to dispensers.

The linear meter of the DN78 pipe used in the low-pressure system has a mass of 7.8 kg. The running meter of the DN100 pipe is 9.7 kg.

Consequently, on the total mass mounted for the needs of fire communications, with the exception of the weight of the pumping unit and the sprinklers, we can conclude that the mass of a low pressure system will be (given the first two sections 10 and 425 m, respectively, 9 sections of the supply line with a length of 4.5 meters each and a section from the management node to the design area 390 meters) – 32160,5 kg.

The linear meter of the DN12 pipe made of corrosion-resistant steel has a mass of 0.27 kg. Linear meter of DN38 pipe made of corrosion-resistant steel – 2.62 kg.

Consider the total mass of the designed fire extinguishing system with high-pressure water spray. Taking into account sections equal to 0.5 meters with a pipe diameter of DN12 and other sections with a pipe diameter of DN38 with a total length of 839 meters (also excluding the mass of the pump unit and sprayers), the total mass of installed communications will be 19354.8 kg.

Having studied the cost of steel electric-welded pipes for general-purpose fire extinguishing systems, it should be concluded that the price of one meter of this pipe with diameters DN78 is approximately 350 rubles. One meter of electric-welded pipe with a diameter of DN100 will cost about 390 rubles.

The cost of one meter of stainless steel pipe made of steel grade 12X18N10T and a diameter of DN38 is approximately 1400 rubles. The cost of one meter of such a pipe with a diameter of DN12 is 260 rubles.

This leads to the conclusion that the total cost of pipeline communications for the installation of the designed low-pressure fire extinguishing system will be approximately 1421150 rubles.

The total cost of purchasing a pipeline for installing a fire extinguishing system with high-pressure water spray will be about 10377440 rubles.

Let's move on to comparing the designed systems with respect to the required supply of fire extinguishing agent to ensure extinguishing.

For a general-purpose system, based on the results of hydraulic calculations, a water flow rate of 46.158 l/s or 2769.48 l/min is required. Taking into account the fact that according to the standards given in this work, it is necessary to ensure extinguishing within 60 minutes, it should be concluded that the total amount of water for the general purpose fire extinguishing system will be 166168.8 liters.

However, it is necessary to take into account the mandatory 10% supply of fire extinguishing agent. Therefore, the total amount of water required for this fire extinguishing system is 182785.68 liters. For example, the volume of a standard pool for the Olympic games and world Championships is 2500,000 liters of water, the configuration of which is 50 meters long, 25 meters wide and 2 meters deep. To store the volume of water required for the fire extinguishing system in question, a tank 10 meters long, 10 meters wide and 2 meters high would be required.

This is quite a large volume, especially for metro facilities, the majority of which are located underground. It is simply not possible to place such a large reservoir within the space-planning solutions of the metro object. The only solution that can be taken to provide a fire extinguishing system with a fire extinguishing agent is to interact with the city water utility and request the necessary amount of water from them. However, when designing it is necessary to take into account the fact that the water utility does not always enter into a contract for the supply of water for the needs of fire extinguishing various objects.

Let's move on to the assessment of the necessary supply of fire extinguishing agent to ensure extinguishing by a high-pressure fire extinguishing system with thinly sprayed water.

Based on the calculation results, it is necessary to provide a flow rate of 513 l/min. This number already includes a 10% supply of fire extinguishing agent. Also, taking into account the condition of fire extinguishing within 60 minutes, 30780 liters of water will be required. This volume of water required a much smaller reservoir. Approximately 5 meters long, 4 meters wide and only 1.5 meters high. In principle, such a water tank, if divided into several separate vessels, can fit even in the pump room of the automatic firefighting system, given that according to the standards this room should be at least 60 m2. In the case of a city water utility, it is much easier to allocate and provide such a volume of water than the required volume for a general – purpose system.

4 Discussion

The results of comparing the hydraulic calculations of two different fire extinguishing systems are presented below. According to the consumption of fire extinguishing agent and total water supply, the fire extinguishing system with thinly sprayed high-pressure water has an advantage over the other one.

Based on the diameter and weight of the linear meter of pipes, it is concluded that the total mass of communications and the mass of individual elements is significantly lower for a fire extinguishing system with high-pressure water in contrast to a general-purpose system.

However, despite the advantage in weight, the high-pressure water spray system is much inferior in terms of cost. However, when designing a fire extinguishing system for a metro facility, the emphasis should be on operational efficiency rather than costs. That is why the choice is made in the favor of the high-pressure water spray system and , thus, a fire extinguishing system to ensure fire safety of dead ends of rolling stock sludge has to include high-pressure water spraying.

References

1. P. Zhu, W. Wei, J. Mei, J. Tu, R. Ding, Procedia Engineering, **205**, 87-94 (2017) https://doi.org/10.1016/j.proeng.2017.09.938
2. T. Chen, S.Y. Zhang, L.Z. Zhao, et al, IOP Conference Series: Earth and Environmental Science, **69,** 012179 (2017)
3. J. Yang, J. Liang, Q.-Z. He, Procedia Engineering, **211**, 897-900 (2018) https://doi.org/10.1016/j.proeng.2017.12.089

4. Y. Zhu, Y. Wang, X. Zhou, IOP Conference Series: Materials Science and Engineering, **423**, 012019 (2018)
5. E.Y.Y. Chan, et al, International Journal of Environmental Research and Public Health, **16**(2), 228 (2019) https://doi.org/10.3390/ijerph16020228
6. S. Moriyama, et al, Fire Safety Science, **8**, 1461-1472 (2005) doi:10.3801/IAFSS.FSS.8-1461
7. Z. Wang, et al, Procedia Engineering, **84**, 595-602 (2014)
8. V. Maslak, et al, Procedia Engineering, **165**, 214-223 (2016)
9. D. Gabay, Tun. And Und. Sp. Tech., **17**(2), 139-143 (2002) https://doi.org/10.1016/S0886-7798(02)00016-0
10. R. Gao, A. Li, Y. Zhang, N. Luo, Saf. Sc., **80**, 94-104 (2015) https://doi.org/10.1016/j.ssci.2015.07.015
11. L. J. Hong, C. J. Liu, Chin., Jour. of Und. Sp. and Eng., **1**(1), 149-154 (2005)
12. M. Gravit, A. Vaititckii, A. Shpakova, Procedia Engineering, **165**, 1667-1672 (2016).
13. W.K. Chow, International Journal on Architectural Science, **5**(2), 35-42 (2004)
14. J. Mao, et al, Journal of the China railway Society, **6** (2010)
15. R.E. Shaltout, M.A. Ismail, Sustainable Rail Transport. Springer, Cham., 209-228 (2020) https://doi.org/10.1007/978-3-030-19519-9_7
16. *Subways. The updated edition of SNiP 32-02-2003*, (SP 120.13330.2012, with Amendments No. 1-4)
17. *Systems of fire protection. Automatic fire-extinguishing and alarm systems. Designing and regulations rules*, (SP 5.13130.2009, with Change N 1)
18. *Fire engineering. Classification of fires* (GOST 27331-87)
19. *Rules for arrangement of electrical installations* (PUE 7th edition)
20. *Means for guarding. Fire and guarding fire signalization. Type, basic parameters and dimensions* (GOST 26342-84, with Changes N 1,2)
21. *Automatic fire extinguishing system with high-pressure spray water "ALFA-FOG". Design recommendations* (STO-20161283-001-2017)

Application of Digital Twin Technology for Modelling of Information Security Level of Industrial Plant

Natalia Mikhailovna. Kuznetsova.[1,*], *Tatyana Vladimirovna* Karlova[2,*]

[1]Candidate of technical science, docent, Automated Control Systems and Information Processing Department, Moscow State University of Technology "STANKIN", RU-127055, Moscow, Russia
[2]Doctor of sociological Science, candidate of technical science, Professor, Automated Control Systems and Information Processing Department, Moscow State University of Technology "STANKIN", RU-127055, Moscow, Russia

Abstract. Maintaining the high level of information security at all stages of production is one of the most important tasks of modern industrial plans. In this case, the complex (integrated) approach plays a special role in which information security is realized on maximum number of automated systems and communication channels. The article is devoted to the mechanism of modelling the realization of external and internal information security threats by means of digital twin application. The presented model is a generalized digital copy of all industrial automated systems.

1 Introduction

Digital twin technology is becoming more popular among industrial enterprises. The main purpose of the technology is creating the model (copy) that is closest to reality. These models should contain the complex of detailed digital copies of physical objects in all automated systems (further AS).

Digital twin technology is at the intersection of technologies "Industry 4.0" and "Industrial Internet of things" [1].

Detailed digital twins could be used for maintaining (and increasing) the level industrial information security. This requires:

- creating the detailed digital model of all AS and communication channels of enterprise:
  a) creating the digital models of computing complexes (further CC);
  b) creating the digital models of data stores (further DS);
  c) creating the digital models of local area networks (further LAN);
  d) creating the digital models of automated workstations (further AW) for employees;
- modelling the information security incidents:
  a) modelling (simulation) employee behavior;
  b) modelling (simulation) "behavior" of active equipment (technics);
  c) modelling (simulation) "behavior" of passive equipment (communication channels);
  d) modelling (simulation) of changes in environment parameters;
- integrated modelling of all systems and incidents:
  a) building incident trees (in particular for the study of avalanche effects);
  b) using Swiss Cheese Model (from risk theory) [2].

2 Detailed digital model of all enterprise automated systems

Architecture scheme of interaction between main enterprises AS elements is shown on Figure 1.

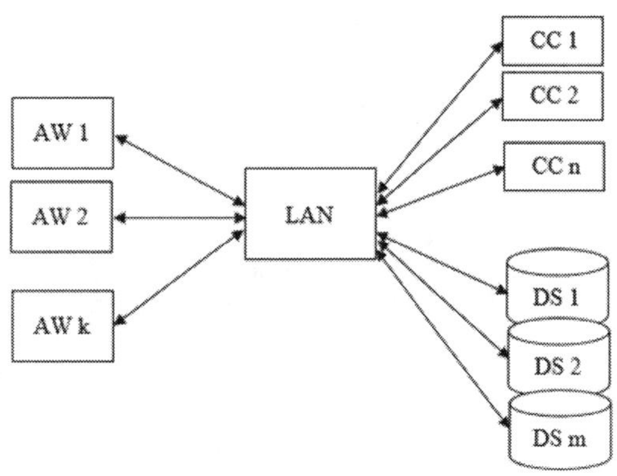

Fig. 1. Architecture scheme of interaction between main enterprise AS elements.

2.1 Digital Model of CC

Digital model of CC should contain:
- the logic of all information processing algorithms;
- the logic of interaction between clusters;
- technical characteristics of the equipment;

* Corresponding authors: knm87@mail.ru, karlova-t@yandex.ru

- technical characteristics of data transmission channels within CCi and between CC;
- data exchange interfaces (e.g. MPI – Message Passing Interface);
- features of the operation of RAM (Random Access Memory) and permanent memory;
- environmental conditions:
  a) wet;
  b) temperature;
  c) electromagnetic compatibility of equipment (including TEMPEST or "Compromising emanation") etc.;
- territorial location, etc.

2.2 Digital model of DS

Digital model of DS should contain:
- feature of applied storage technologies (access mode, backup mode, protocols used, etc.);
- technical characteristics of the equipment;
- technical characteristics of data transmission channels within DSj and between DS;
- territorial location;
- environmental conditions, etc.

2.3 Digital model of LAN

Digital model of LAN should contain:
- network topology;
- technical characteristics of data transmission channels;
- territorial location of all network parts;
- heterogeneity of data transmission medium (on physical level);
- environmental conditions, etc.

2.4 Digital model of AW

Digital model of AW should contain:
- first, all data "entry points":
  a) USB-ports;
  b) floppy drives;
  c) printers;
  d) faxes etc. [3-5]
- technical characteristics of the equipment.

It is important to note, that creating all kind of enterprise AS is needed for modelling (including all stages of life cycle of the enterprise target product) [6]:
- integrated monitoring and control design system:
  a) CAD (Computer-Aided Design) / CAM (Computer-Aided Manufacturing) / CAE (Computer-Aided Engineering);
  b) SCADA (Supervisory Control And Data Acquisition);
- integrated resource management system:
  a) ERP (Enterprise Resource Planning);
  b) MRP (Material Requirements Planning);
- document management system;
- tasks trackers;
- common programming environments.

3 Modelling the information security incidents

When the complete (detailed) digital copy of AS is created, development of event scenarios is needed. Event scenarios should depend on employees' behavior, equipment condition and the external environment.

If the behavior of equipment and environment could be predicted by changing a relatively small number of parameters, then employees' behavior modelling requires research on psychological, emotional, moral and motivation factors.

Modelling of equipment reliability can be carried out by mean of Monte-Carlo method [6]. For this, the technical characteristics AS equipment should be used as parameters. Also, this method could be used for modelling of external environment parameters.

It is important to note, that Monte-Carlo method assumes the presence of a large amount of computing resources.

When employee's behavior modelling, it is necessary to take into account:
- human factor:
  a) tiredness;
  b) tensity;
  c) neuropsychic stability;
  d) stress level etc.;
- psycho-emotional state:
  a) suffered shocks [7];
  b) love etc.;
- morals and principles:
  a) upbringing;
  b) education;
  c) origin;
  d) religiosity etc.

Based on the listed characteristics possible "personal portraits" of employees are formed. For correct modelling, "personal portraits" (which saved in database) are discussed and coordinated with experts-psychoanalysts.

4 Integrated modelling of all systems and incidents

4.1 Incident tree

While modelling of employee's, equipment and environment behavior the possibility of a series of interrelated events should be taken into account – avalanche effect, in which each of the events does not pose a serious threat, but their combination can lead to a decrease in the enterprise information security level.

4.2 Model "Cheese holes" (Swiss Cheese Model)

Model "Cheese holes" (Swiss Cheese Model) assumes that in any system there are vulnerability at different levels. However, the levels of protection ("cheese layers") overlaps, and therefore the general state of system is stable (Figure 2a). However, there is the possibility that "cheese holes" would be overlapped not by cheese layers, but by the "holes" (lined up), which would lead to the global (system) information security threat implementation (Figure 2b).

In order to minimize the likelihood of lowering the information security level, it is necessary to:

- reduce the number of cheese "holes";
- reduce in the size of "cheese holes".

"Cheese holes" could be any kind of vulnerabilities in any part of enterprise AS in any time. Thereby while creating the model of "Cheese holes" it is necessary to consider both special and temporal aspects of information security [8].

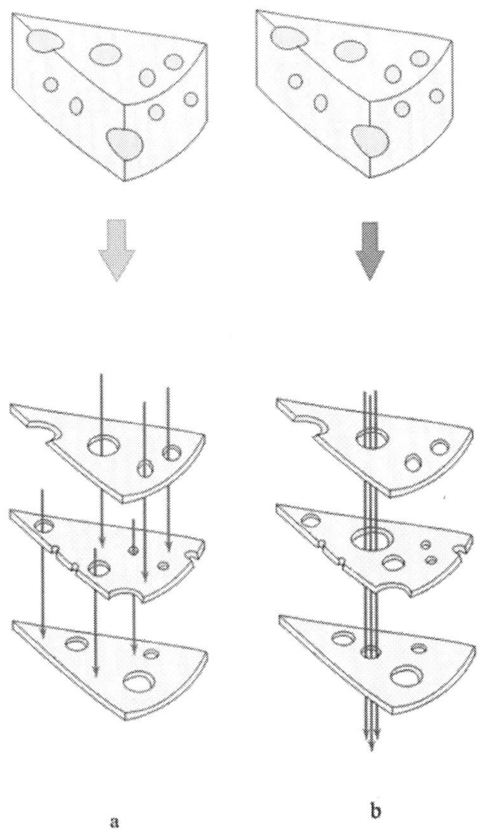

Fig. 2. Model «Cheese holes»:
a – «overlapping holes with layers of cheese»;
b – «overlapping holes».

5 Realization of modelling the digital twin

When using the concept of digital twins to simulate information security incidents and assess the protection level of all enterprise AS it is necessary to describe as accurate as possible all business processes and resources [9].

The cost of realization directly depends on degree of approximation of the digital model to reality. Also, the choice of modelling methods defines the realization cost.

6 Conclusion

Modelling physical systems using the conception of digital twins is a convenient tool for predicting the behavior of complex technical objects. The article is devoted to the methodology for using digital twins to estimate the information security level of the enterprise's automated systems. The methodology takes into account modelling of complex automated systems, communications, employees' actions, possible security incidents. In addition, methods of creating incident trees and "cheese holes" are presented as tools for integrate modelling. The methodology presented in the article allows to estimate the information security level as accurate as possible, to identify the "bottlenecks" and to take actions to eliminate vulnerabilities on time.

References

1. Digital twin. Available at: https://en.wikipedia.org/wiki/Digital_twin (accessed 15 October 2020)
2. Swiss Cheese Model. Available at: https://en.wikipedia.org/wiki/Swiss_cheese_model+ https://en.wikipedia.org/wiki/Chain_of_events_(accident_analysis) (accessed 15 October 2020)
3. N.M. Kuznetsova, T.V. Karlova, S.A. Sheptunov, Proceedings of the 2017 International Conference "Quality Management, Transport and Information Security, Information Technologies" (IT&QM&IS) – Proceedings Edited by S.Shaposhnikov 2017 St.Peterburg, Russia: Saint Petersburg Electrotechnical University "LETI", 199-202 (2017) ISBN 978-1-5386-0703-9 DOI:10.1109/ITMQIS.2017.8085797
4. Karlova, T.V., Bekmeshov, A.Y., Kuznetsova, N.M.. Proceedings of the 2019 IEEE International Conference "Quality Management, Transport and Information Security, Information Technologies" (IT&QM&IS) – Proceedings Edited by S. Shaposhnikov, St. Petersburg, Russia: Saint Petersburg Electrotechnical University "LETI", (2019) ISBN 978-1-7281-2594-7 DOI:10.1109/ITQMIS.2019.8928412
5. N.M. Kuznetsova, T.V. Karlova, S.A. Sheptunov, A.Y. Bekmeshov, IEEE Conference on Quality Management, Transport and Information Security, Information Technologies (IT&MQ&IS). Proceedings. – M.: "Quality", 23-27 (2016) ISBN 978-5-94768-071-3
6. A.M. Kolchuzhkin, N.M. Kuznetsova, Materials of XII scientific conference MSUT «STANKIN» and

«Education and scientific center for mathematical modelling MSUT «STANKIN» Collection of reports / Edited by O.A. Kazakov – M.: «STANKIN», 434 (2009)

7. Y.S. Shoigu, *Psychology of extreme situations* (S-Pb: Piter – 2019)
8. ICS cybersecurity. Available at: InfoWatch.ru (accessed 15 October 2020)
9. N.M. Kuznetsova, T.V. Karlova *Total Quality Management. Solving the problem of increasing the level of information security at an industrial enterprise as part of quality management system (course of lectures and laboratory workshops) Tutorial.* (M.: Janus-K, 2019) ISBN 978-5-8037-0779-0

Development of Experimental Prototypes for the Teaching of Science

Guillermo Terán[1], *Fausto* Ávalos[2], *Washington* Lomas[3,*] and *Iliana* Quintana[3]

[1]Universidad Central del Ecuador, Faculty of Filosofía, Letras y Ciencias de la Educación, EC-170521, Quito, Ecuador
[2]Universidad Central del Ecuador, Faculty of Ingeniería y Ciencias Aplicadas, EC-170521, Quito, Ecuador
[3]Universidad Central del Ecuador, Centro de Física, EC-170521, Quito, Ecuador

Abstract. The following document presents the development of educational experimental prototypes. The themes of the prototypes cover principles or concepts of physics. Like kinematics, oscillatory movements and shocks. Each prototype has sensors that allow measurements of the variables involved in the experimentation. At the same time, mobile applications are developed for devices with Android operating system, which allow the user to control the prototype from their cell phone and also view the variables measured by the experimentation prototypes. The applications also have a data export system that allows to obtain a file in csv format. In this way, the measurements are obtained in the form of a table to later be processed through office automation tools or other numerical computing software such as Matlab.

1 Introduction

To improve the learning experience of the students of the Universidad Central del Ecuador and in order to reinforce their knowledge through experimentation, experimental prototypes are developed which, through the use of technology, allow students to carry out practices that demonstrate physical principles or concepts.

The interaction with the laboratory prototypes will have a positive impact on the creativity and development of cognitive thinking of the students since the transfer of knowledge with acquisition and interpretation of data will be efficient, guaranteeing the consolidation of concepts of the different themes of Physics such as a branch of the Sciences.

2 Methodology

Starting from physical principles, variouscriteria are analyzed for the development ofexperimental prototypes. In this first stage, topics such as parabolic movement, shocks, capacitors, light composition have been chosen.

Each prototype consists of:

- Experimentation equipment: includes all the necessary materials to carry out the experimentation, within each of them there is an electronic system which contains one or more sensors necessary to carry out the measurements of variables that allow the completion of the experiment, a communication system that allows the information obtained to be sent to the mobile device and allows the user to interact with the equipment.
- Mobile application for devices with Android operating system: each experimentation equipment has a mobile application, which allows the user to manipulate the equipment and observe the measurements made. Depending on each experiment there are different options within the application, within which graphs of each of the variables measured with respect to time can be observed.

Below, the components of each prototype mentioned above are detailed in a better way.

2.1 Parabolic Movement.

Inside the prototype of the parabolic movement there is a system actuated by a spring, which allows a particle to be thrown, in this case a small sphere to carry out the experimentation. The launch angle is configured through the mobile application, by means of a servomotor. In the cannon we have a pair of infrared sensors that allow us to measure the initial velocity of the particle. Additionally, there is a small platform that has an impact sensor which allows obtaining the time of flight of the particle.

For experimentation purposes, the friction between the sphere and the barrel is considered to be zero.

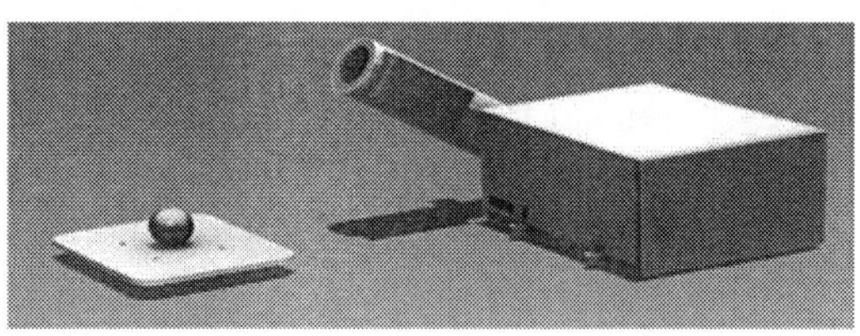

Fig. 1. Parabolic movement experimentation prototype.

* e-mail: wplomas@uce.edu.ec

The mobile application that allows interaction with the prototype is shown in Fig.

There are the following options within the application, among which we have:

- Connect button: allows you to select the bluetooth device with which to connect, in this case with the parabolic movement experimentation prototype.
- Disconnect button: allows you to disconnect the experimentation prototype.
- The status label shows if the prototype is connected or disconnected.
- Through the slider the launch angle of the particle is selected and once selected, the angle is configured through the OK button.The initial speed measured by the prototype is shown on the "Initial Speed" label.
- - The "Flight Time" label shows the flight time measured by the prototype.

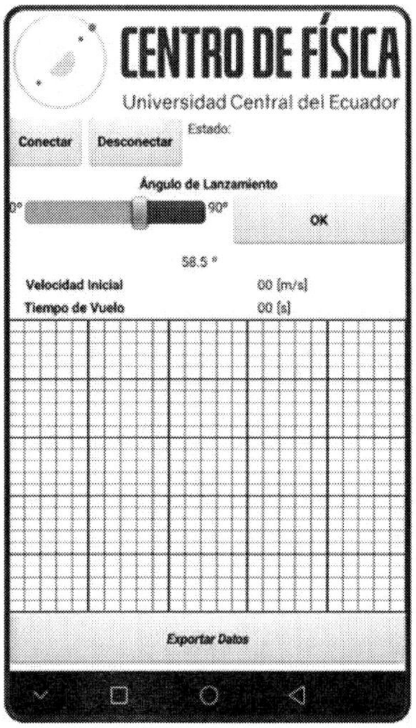

Fig. 2. Mobile application of the parabolic movement prototype.

2.2 One-dimensional shocks.

In the prototype of the one-dimensional collision, there is a system activated by a spring, which allows a particle to be thrown, which impacts with another sphere that is in the line of action of the first sphere. In this way both spheres impact, and through sensors placed at the initial end and at the end end, the initial conditions before the crash and the final conditions after the crash can be obtained. The sensors placed are a pair of infrared sensors at the initial end and at the end end to thereby measure the velocities of the particles before and after the collision. For the purposes of experimentation, it is considered that the friction between the test spheres and the track on which the experiment is carried out is zero.

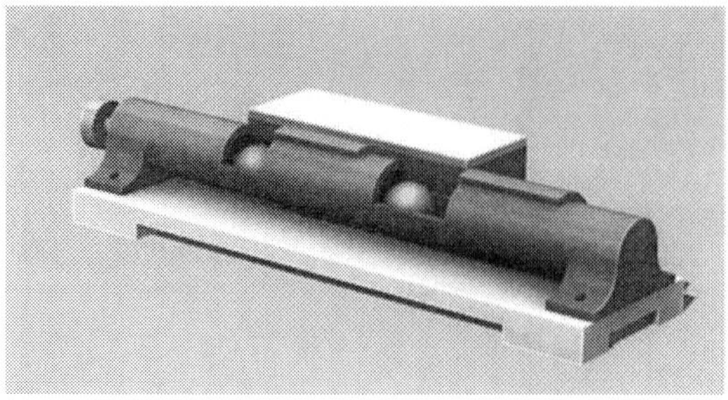

Fig. 3. One-dimensional shocks experimentation prototype

In Fig. 4 the application is shown to be able to interact with the prototype. Similar to the application described above, there are buttons, for example the prototype connection and disconnection buttons. In addition to the label that indicates if the prototype is connected or disconnected. This application additionally has:

- Two labels in which the initial conditions before the collision are observed, that is, the initial speeds. For the particle that is impacted in this case, a null initial velocity will be considered.
- Two labels showing the conditions after the crash, speeds after the crash.
- In case of placing spheres of different mass, there are two text boxes in which the masses of each of the spheres can be entered.
- In addition, it has three additional labels in which the kinetic energy is shown before and after the crash. It iscalculated from the velocity data and mass data of the test spheres.

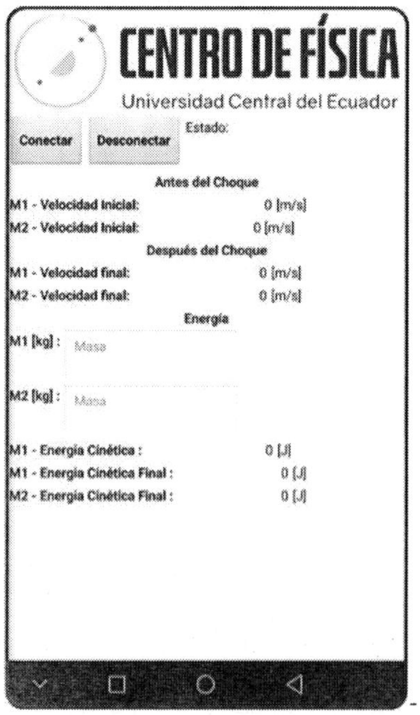

Fig. 4. One-dimensional shocks prototype mobile application

2.3 Charging and Discharging Capacitors.

The prototype developed for charging and discharging capacitors is shown in Fig. 5. It allows observing the charge and discharge graphs of an R-C set in series. Resistance and capacitance values can be changed as user substitutes. In the upper part it has a switch which allows you to choose if you want tocharge the capacitor or discharge it. In the application, the voltage-time graph is observed to see how the voltage changes according to the R-C set placed.

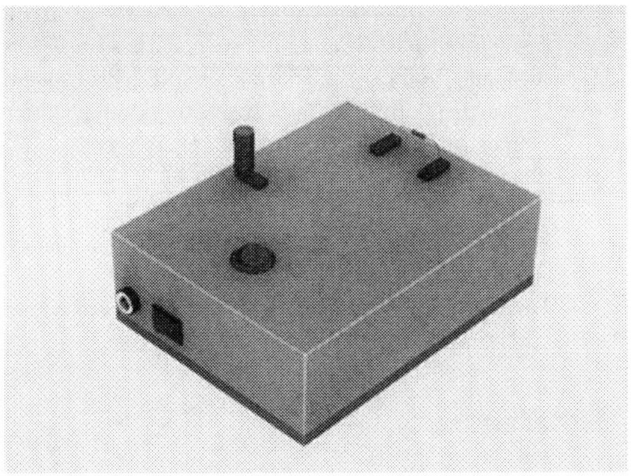

Fig. 5. Capacitor charging and discharging experimentation prototype

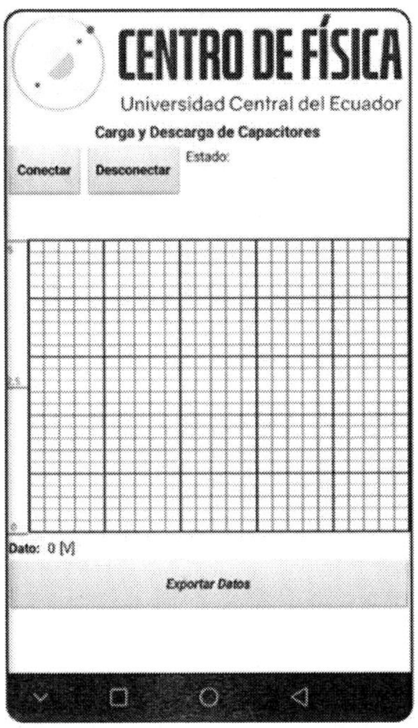

Fig. 6. Capacitor charging and discharging prototype mobile application

In Fig. 6 you can see the application developed for the prototype, the connection and disconnection buttons as well as the connection status label is like the previous ones described. If you have a squared area which allows you to see the graph of voltage - time of the charge or discharge process. In this case, the prototype uses a charging voltage of 5V. In addition, there is a label that allows you to see what the instantaneous voltage is.

The voltage-time data is recorded and can be exported to obtain a file in csv format. To obtain the file with the registered data, use the Export Data button.

2.4 Composition of light.

The following prototype allows us to observe the additive synthesis of colors. Different colors are obtained from a set of RGB LEDs. The prototype is shown in Fig. 7.

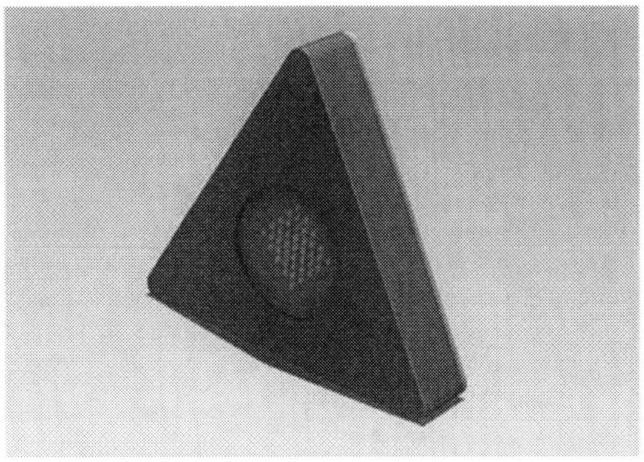

Fig. 7. Light composition experimentation prototype

The application that allows selecting the prototype light is seen in Fig. 8. The connection and disconnection buttons as well as the connection status label were previously detailed.

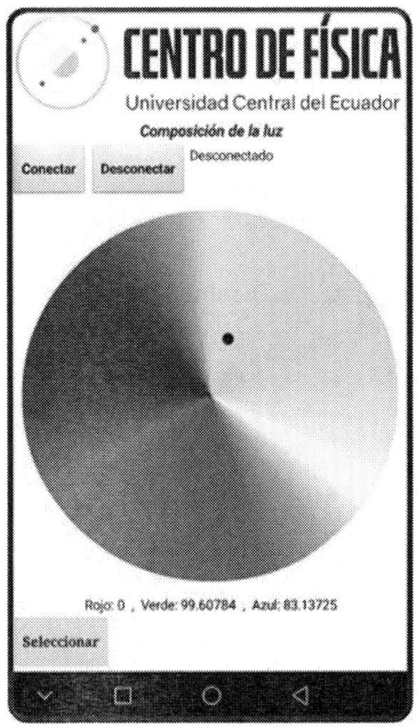

Fig. 8. Light Composition Prototype Mobile App

There is a wheel-shaped color palette from which you can select a color from it. Below the color wheel, you can see the composition percentage of each selected color: red, green and blue. Once the color has been selected tobe reflected in the prototype, the "Select" button must be selected.

3 Results

Each of the developed prototypes allows obtaining data which can be compared with data obtained from the respective theory.

Inside each of the prototypes there is a bluetooth module which was renamed so that it can be quickly identified within the pairing with the Android device.

3.1 Parabolic Movement.

The parabolic movement prototype is in operation, allowing the initial speed and time of flight of the particle to be obtained.

The application correctly shows the data obtained, for the prototype test an Arduino Nano development board is being used. At the end of the first tests, a printed circuit board of the prototype will be used, after this process the prototype will be next to go to tests and validation by the work team of the Centro de Física de la Universidad Central del Ecuador.

3.2 One-dimensional shocks.

When carrying out the experimentation, it is observed that sometimes the spheres bounce off the walls of the cylindrical track within which the experimental process is carried out. Because of this, experimentation is done a couple of times to get the data without these disturbances.

At the moment an Arduino development board is being used for prototype testing, the next step is to only use the microcontroller inside a printed circuit board to decrease the size of the components used. Once this process has been carried out, the prototype will go through tests and validation by the team of the Centro de Física de la Universidad Central del Ecuador.

3.3 Charging and Discharging Capacitors.

The prototype is in operation, thanks to the small graphic area of the application, it is possible to observe the voltage - time graphs, allowing to observe the charge and discharge graphs of the capacitor in the R-C assembly, similar to a small oscilloscope.

Knowing the time constant of an R-C series circuit (resistance - Capacitor), the exported data representing the voltage - time values can be used to calculate the capacitance values with which it was used for experimentation. Obtaining errors of 3%.

3.4 Composition of light.

This prototype is demonstrative, the selected color can be obtained through the mobile application. In the back it has holes which allow air circulation to reduce the temperature of the transistors used as drivers.

4 Conclusion

- Each prototype allows the measurement of the minimum variables to obtain a comparison with the theoretical data obtained, in this way the experimentation with the theory is associated.
- The interface of each application is friendly to the user, showing the data in a clear way. Each application has common elements which will allow students to quickly adapt to each of the prototypes.
- The use of applications makes the student show a greater interest in the experimentation and study of Physics.
- The percentage of error of the data obtained compared with the data of the respective theory is less than 5%.
- The prototypes are low cost, using materials easily accessible in the Ecuadorian market.

References

1. R.A. Serway, J.W. Jewitt, Física para ciencias e ingeniería, **10**, (2019)
2. F.P. Beer, E.R. Johnston, P.J. Cornwell, Mecánica Vectorial para ingenieros Dinámica, **9**, (2010)
3. H. Young, R. Freedman, University Physics (with Modern Physics), **12**, (2008)
4. R.F. Coughlin, F.F. Driscoll, Amplificadores Operacionales y Circuitos Integrado, **5**, (2000)
5. R.L. Boylestad, L. Nashelsky, Electrónica: Teoría de Circuitos y Dispositivos Electrónicos, **10**, (2009)

Mathematical Modeling of Vibration Dampers of Vibration-Insulated Structures

Evgeny Sosenushkin[1], *Oksana* Ivanova[1], *Elena* Yanovskaya[1], and *Yuliya* Vinogradova[1]

[1]Moscow State Technological University "STANKIN", RU-127055, Moscow, Russia

Abstract. Vibration dampers are installed on the machine foundations in order to reduce the vibration level. Such technological solutions are most expedient in the case of a harmonic load with a low instability of the vibration frequency. Unfortunately, dampers do not provide such a large reduction in the dynamic effect on the base, as vibration isolation, but in some cases their efficiency turns out to be quite sufficient with a relatively simple implementation and low manufacturing cost. The use of dynamic vibration dampers gives a great effect when an increased vibration of foundations occurs during the operation of equipment in metallurgical production, for example, when processing materials by pressure, reconstructing enterprises and replacing heavy equipment. During the operation of heavy forging equipment and manipulators for various purposes, the foundations of these devices can be considered as a rigid body. The model soil on which this foundation is installed can be considered a homogeneous elastic isotropic half-space. When calculating with such mathematical models, one can use solutions of the corresponding dynamic contact problems. A comparative analysis of the effectiveness of damping foundation vibrations using different foundation models, including the model of an elastic, homogeneous half-space and a system of semi-infinite rods, the modulus of elasticity of which increases with depth according to the quadratic law, shows a fairly close agreement.

1 Introduction

The classical equation of bending vibrations of a plate is applicable with sufficient accuracy as long as the bending wavelength is not less than five times the plate thickness [1]. For a rectangular plate loaded with lumped masses located periodically, the bending vibration equation has the form:

$$D\nabla^4 W + \rho h \frac{\partial^2 W}{\partial t^2} = h \sum_{i=1}^{m_1} \sum_{j=1}^{m_2} \rho_{ij} \delta(x - -x_i)\delta(y - y_j) \frac{\partial^2 W}{\partial t^2}(x_i, y_j, t) \quad (1)$$

where $0 \le x \le a, 0 \le y \le b,\ x_i = \frac{ia}{m_1+1}, y_j == \frac{jb}{m_2+1}$, wherein $i = 1,2, \dots m_1, j = 1,2, \dots m_2$.

$W(x, y, t)$ - deflection or deviation of the point (x, y) from the equilibrium position.

$D = \frac{Eh}{12(1-\nu^2)}$ – bending stiffness of the plate;

$\rho h = q$ – plate mass per unit surface.

This equation can be used if the distance from the edge of the plate to the point at which the vibrations are considered is greater than the thickness of the plate h, which in many cases can be considered as a constant value. [2- 5]

The boundary conditions for this problem can be formulated as follows:

1. For a supported edge, deflection and bending moment must be zero.

$$W = 0,\ \frac{\partial^2 W}{\partial x^2} + \nu \frac{\partial^2 W}{\partial y^2} = 0 \text{ for } x = 0 \text{ or } x == a,$$

or

$$W = 0,\ \frac{\partial^2 W}{\partial y^2} + \nu \frac{\partial^2 W}{\partial x^2} = 0 \text{ for } y = 0 \text{ or } y == b. \quad (2.1)$$

2. For the sealed edge

$$W = 0,\ \frac{\partial W}{\partial x} = 0, \text{for } x = 0 \text{ or } x = a,$$

$$W = 0,\ \frac{\partial W}{\partial y} = 0, \text{for } y = 0 \text{ or } y = b. \quad (2.2)$$

2 Equations and mathematics

Initial conditions:

$$W(x, y, 0) = \varphi(x, y), \qquad \frac{\partial W}{\partial t} == \psi(x, y). \quad (3)$$

An analytical solution to problem (1, 2, 3) is sought in the form

$$W(x,y,t) = \sum_{n=1}^{\infty}\sum_{k=1}^{\infty}(a_{kn}cos\frac{k\pi x}{a}cos\frac{n\pi y}{b} + +b_{kn}cos\frac{k\pi x}{a}sin\frac{n\pi y}{b} + c_{kn}sin\frac{k\pi x}{a}cos\frac{n\pi y}{b} + +d_{kn}sin\frac{k\pi x}{a}sin\frac{n\pi y}{b})e^{i\omega_{kn}t}, \quad (4)$$

$$\nabla^4 W = \frac{\partial^4 W}{\partial x^4} + 2\frac{\partial^4 W}{\partial x^2 \partial y^2} + \frac{\partial^4 W}{\partial y^4} = \sum_{n=1}^{\infty}\sum_{k=1}^{\infty}((\frac{k\pi}{a})^4 + 2\left(\frac{k\pi}{a}\right)^2\left(\frac{n\pi}{b}\right)^2 + +\left(\frac{n\pi}{b}\right)^4)(a_{kn}cos\frac{k\pi x}{a}cos\frac{n\pi y}{b} + +b_{kn}cos\frac{k\pi x}{a}sin\frac{n\pi y}{b} + c_{kn}sin\frac{k\pi x}{a}cos\frac{n\pi y}{b} + +d_{kn}sin\frac{k\pi x}{a}sin\frac{n\pi y}{b})e^{i\omega_{kn}t} = \sum_{n=1}^{\infty}\sum_{k=1}^{\infty}\left(\left(\frac{k\pi}{a}\right)^2 + \left(\frac{n\pi}{b}\right)^2\right)^2(a_{kn}cos\frac{k\pi x}{a}cos\frac{n\pi y}{b} + +b_{kn}cos\frac{k\pi x}{a}sin\frac{n\pi y}{b} + c_{kn}sin\frac{k\pi x}{a}cos\frac{n\pi y}{b} + +d_{kn}sin\frac{k\pi x}{a}sin\frac{n\pi y}{b})e^{i\omega_{kn}t}, \quad (5)$$

$$\frac{\partial^2 W}{\partial t^2} = \sum_{n=1}^{\infty}\sum_{k=1}^{\infty}(\omega_{kn})^2(a_{kn}cos\frac{k\pi x}{a}cos\frac{n\pi y}{b} + +b_{kn}cos\frac{k\pi x}{a}sin\frac{n\pi y}{b} + c_{kn}sin\frac{k\pi x}{a}cos\frac{n\pi y}{b} + +d_{kn}sin\frac{k\pi x}{a}sin\frac{n\pi y}{b})e^{i\omega_{kn}t}. \quad (6)$$

$$\sum_{i=1}^{m_1}\delta(x-x_i) = 2\sum_{m=1}^{\infty}\sum_{i=1}^{m}(cos\frac{m\pi x}{a}cos\frac{m\pi x_i}{a} + +sin\frac{m\pi x}{a}sin\frac{m\pi x_i}{a}) \quad (7)$$

$$\sum_{j=1}^{m_2}\delta(y-y_j) = 2\sum_{p=1}^{\infty}\sum_{i=1}^{p}(cos\frac{p\pi y}{b}cos\frac{p\pi y_j}{b} + sin\frac{p\pi y}{b}sin\frac{p\pi y_j}{b}) \quad (8)$$

Substitute (4,5,6,7,8) into equation (1):

$$D\sum_{n=1}^{\infty}\sum_{k=1}^{\infty}\left(\left(\frac{k\pi}{a}\right)^2 + \left(\frac{n\pi}{b}\right)^2\right)^2(a_{kn}cos\frac{k\pi x}{a}cos\frac{n\pi y}{b} + b_{kn}cos\frac{k\pi x}{a}sin\frac{n\pi y}{b} + c_{kn}sin\frac{k\pi x}{a}cos\frac{n\pi y}{b} + d_{kn}sin\frac{k\pi x}{a}sin\frac{n\pi y}{b})e^{i\omega_{kn}t} - \rho h\sum_{n=1}^{\infty}\sum_{k=1}^{\infty}(\omega_{kn})^2(a_{kn}cos\frac{k\pi x}{a}cos\frac{n\pi y}{b}$$

$$+b_{kn}cos\frac{k\pi x}{a}sin\frac{n\pi y}{b} + c_{kn}sin\frac{k\pi x}{a}cos\frac{n\pi y}{b} + d_{kn}sin\frac{k\pi x}{a}sin\frac{n\pi y}{b})e^{i\omega_{kn}t} = 4h\sum_{i=1}^{m_1}\sum_{j=1}^{m_2}\rho_{ij}\sum_{m=1}^{\infty}\sum_{i=1}^{m}(cos\frac{m\pi x}{a}cos\frac{m\pi x_i}{a} + sin\frac{m\pi x}{a}sin\frac{m\pi x_i}{a})\sum_{p=1}^{\infty}\sum_{i=1}^{p}(cos\frac{p\pi y}{b}cos\frac{p\pi y_j}{b} + sin\frac{p\pi y}{b}sin\frac{p\pi y_j}{b})$$

$$\sum_{n=1}^{\infty}\sum_{k=1}^{\infty}(\omega_{kn})^2(a_{kn}cos\frac{k\pi x_i}{a}cos\frac{n\pi y_j}{b} + +b_{kn}cos\frac{k\pi x_i}{a}sin\frac{n\pi y_j}{b} + +c_{kn}sin\frac{k\pi x_i}{a}cos\frac{n\pi y_j}{b} + +d_{kn}sin\frac{k\pi x_i}{a}sin\frac{n\pi y_j}{b})e^{i\omega_{kn}t}. \quad (9)$$

We denote

$$D_{kn}(\omega_{kn}) = D\cdot\sum_{n=1}^{\infty}\sum_{k=1}^{\infty}(\left(\left(\frac{k\pi}{a}\right)^2 + \left(\frac{n\pi}{b}\right)^2\right)^2 - \rho h\omega_{kn}^2)\cdot e^{i\omega_{kn}t},$$

$$H_{kn}(\omega_{kn}) = 4\sum_{i=1}^{m_1}\sum_{j=1}^{m_2}\rho_{ij}h\sum_{n=1}^{\infty}\sum_{k=1}^{\infty}\omega_{kn}^2\cdot e^{i\omega_{kn}t}\cdot t. \quad (*)$$

Using these notation, equation (9) takes the following form

$$\frac{D_{kn}(\omega_{kn})}{H_{kn}(\omega_{kn})}\left(a_{kn}cos\frac{k\pi x}{a}cos\frac{n\pi y}{b} + b_{kn}cos\frac{k\pi x}{a}sin\frac{n\pi y}{b} + c_{kn}sin\frac{k\pi x}{a}cos\frac{n\pi y}{b} + +d_{kn}sin\frac{k\pi x}{a}sin\frac{n\pi y}{b}\right) = \sum_{i=1}^{m_1}\sum_{j=1}^{m_2}\sum_{m=1}^{\infty}(cos\frac{k\pi x}{a}cos\frac{m\pi x_i}{a} + +sin\frac{m\pi x}{a}sin\frac{m\pi x_i}{a})\sum_{p=1}^{\infty}(cos\frac{p\pi y}{b}cos\frac{p\pi y_j}{b} + +sin\frac{p\pi y}{b}sin\frac{p\pi y_j}{b})(a_{kn}cos\frac{k\pi x_i}{a}cos\frac{n\pi y_j}{b} + +b_{kn}cos\frac{k\pi x_i}{a}sin\frac{n\pi y_j}{b} + +c_{kn}sin\frac{k\pi x_i}{a}cos\frac{n\pi y_j}{b} + d_{kn}sin\frac{k\pi x_i}{a}sin\frac{n\pi y_j}{b}) \quad (9.1)$$

After applying the summation operators and changing the places of the summation indices $k \to m$, $m \to k, n \to p, p \to n$, equation (9) takes the form:

$$D_{kn}(a_{kn}cos\frac{k\pi x}{a}cos\frac{n\pi y}{b} + b_{kn}cos\frac{k\pi x}{a}sin\frac{n\pi y}{b} + c_{kn}sin\frac{k\pi x}{a}cos\frac{n\pi y}{b} + d_{kn}sin\frac{k\pi x}{a}sin\frac{n\pi y}{b}) =$$

$$= H_{kn}\sum_{m=1}^{\infty}\sum_{p=1}^{\infty}\sum_{k=1}^{\infty}\sum_{n=1}^{\infty}(a_{mp}\cos\frac{k\pi x}{a}cos\frac{n\pi y}{b}\cdot \cdot\sum_{i=1}^{\infty}cos\frac{k\pi x_i}{a}cos\frac{m\pi x_i}{a}\cdot\sum_{j=1}^{\infty}cos\frac{n\pi y_j}{b}cos\frac{p\pi y_j}{b} + b_{mp}cos\frac{k\pi x}{a}sin\frac{n\pi y}{b}\sum_{i=1}^{\infty}cos\frac{k\pi x_i}{a}cos\frac{m\pi x_i}{a} \sum_{j=1}^{\infty}sin\frac{n\pi y_j}{b}sin\frac{p\pi y_j}{b} + c_{kn}sin\frac{k\pi x}{a}cos\frac{n\pi y}{b}\sum_{i=1}^{\infty}sin\frac{k\pi x_i}{a}sin\frac{m\pi x_i}{a}$$

$\sum_{j=1}^{\infty} cos\frac{n\pi y_j}{b} cos\frac{p\pi y_j}{b} +$
$d_{kn} sin\frac{k\pi x}{a} sin\frac{n\pi y}{b} \sum_{i=1}^{\infty} sin\frac{k\pi x_i}{a} sin\frac{m\pi x_i}{a}$ (9.2)

Equating the terms of the equation from its left and right sides with the same combinations at cos and sin, we obtain the following system of equations

$$D_{kn}(a_{kn} cos\frac{k\pi x}{a} cos\frac{n\pi y}{b}) - H_{kn}\sum_{m=1}^{\infty}\sum_{p=1}^{\infty}\sum_{k=1}^{\infty}\sum_{n=1}^{\infty} a_{mp} cos\frac{k\pi x}{a} \cdot$$

$$\cdot cos\frac{n\pi y}{b}\sum_{i=1}^{\infty} cos\frac{k\pi x_i}{a} cos\frac{m\pi x_i}{a} \cdot\sum_{j=1}^{\infty} cos\frac{n\pi y_j}{b} cos\frac{p\pi y_j}{b} = 0.$$

$$D_{kn}(b_{kn} cos\frac{k\pi x}{a} sin\frac{n\pi y}{b}) - H_{kn}\sum_{m=1}^{\infty}\sum_{p=1}^{\infty}\sum_{k=1}^{\infty}\sum_{n=1}^{\infty} b_{mp} cos\frac{k\pi x}{a}$$

$$sin\frac{n\pi y}{b}\sum_{i=1}^{\infty} cos\frac{k\pi x_i}{a} cos\frac{m\pi x_i}{a} \cdot \sum_{j=1}^{\infty} sin\frac{n\pi y_j}{b} sin\frac{p\pi y_j}{b} = 0.$$

$$D_{kn}(c_{kn} sin\frac{k\pi x}{a} cos\frac{n\pi y}{b}) - H_{kn}\sum_{m=1}^{\infty}\sum_{p=1}^{\infty}\sum_{k=1}^{\infty}\sum_{n=1}^{\infty} c_{mp} sin\frac{k\pi x}{a} cos\frac{n\pi y}{b}$$

$$\sum_{i=1}^{\infty} sin\frac{k\pi x_i}{a} sin\frac{m\pi x_i}{a} \cdot \sum_{j=1}^{\infty} cos\frac{n\pi y_j}{b} cos\frac{p\pi y_j}{b} = 0.$$

$$D_{kn}(d_{kn} sin\frac{k\pi x}{a} sin\frac{n\pi y}{b}) - H_{kn}\sum_{m=1}^{\infty}\sum_{p=1}^{\infty}\sum_{k=1}^{\infty}\sum_{n=1}^{\infty} d_{mp} sin\frac{k\pi x}{a}$$

$$sin\frac{n\pi y}{b}\sum_{i=1}^{\infty} sin\frac{k\pi x_i}{a} sin\frac{m\pi x_i}{a} \cdot \sum_{j=1}^{\infty} sin\frac{n\pi y_j}{b} sin\frac{p\pi y_j}{b} = 0.$$

After a slight simplification, the system of equations will be reduced to the following form

$$D_{kn}(a_{kn}) - H_{kn}\sum_{m=1}^{\infty}\sum_{p=1}^{\infty}\sum_{k=1}^{\infty}\sum_{n=1}^{\infty} a_{mp}\sum_{i=1}^{\infty} cos\frac{k\pi x_i}{a}$$

$$cos\frac{m\pi x_i}{a}\sum_{j=1}^{\infty} cos\frac{n\pi y_j}{b} cos\frac{p\pi y_j}{b} = 0.$$

$$D_{kn}(b_{kn}) - H_{kn}\sum_{m=1}^{\infty}\sum_{p=1}^{\infty}\sum_{k=1}^{\infty}\sum_{n=1}^{\infty} b_{mp}\sum_{i=1}^{\infty} cos\frac{k\pi x_i}{a}$$

$$cos\frac{m\pi x_i}{a}\sum_{j=1}^{\infty} sin\frac{n\pi y_j}{b} sin\frac{p\pi y_j}{b} = 0.$$

$$D_{kn}(c_{kn}) - H_{kn}\sum_{m=1}^{\infty}\sum_{p=1}^{\infty}\sum_{k=1}^{\infty}\sum_{n=1}^{\infty} c_{mp}\sum_{i=1}^{\infty} sin\frac{k\pi x_i}{a}$$

$$sin\frac{m\pi x_i}{a}\sum_{j=1}^{\infty} cos\frac{n\pi y_j}{b} cos\frac{p\pi y_j}{b} = 0.$$

$$D_{kn}(d_{kn}) - H_{kn}\sum_{m=1}^{\infty}\sum_{p=1}^{\infty}\sum_{k=1}^{\infty}\sum_{n=1}^{\infty} d_{mp}\sum_{i=1}^{\infty} sin\frac{k\pi x_i}{a}$$

$$sin\frac{m\pi x_i}{a}\sum_{j=1}^{\infty} sin\frac{n\pi y_j}{b} sin\frac{p\pi y_j}{b} = 0.$$

Since changes in the left and right sides of the equation should occur with the same frequencies, the modes of natural oscillations ξ_k are excited at the frequency of the acting force, i.e. the eigenfunctions satisfy the orthogonality conditions:

$$\sum_{i=1}^{N} \xi_k \xi_m \begin{cases} = 0, npu\ k \neq 2r(n+1) \pm m, & r = 0,1,2 \ldots \\ \neq 0, npu\ k = 2r(n+1) \pm m, & r = 0,1,2 \ldots \end{cases} \quad (**)$$

In this case, the following designations can be introduced:

$$C'_{km} = \sum_{i=1}^{\infty} sin\frac{k\pi x_i}{a} sin\frac{m\pi x_i}{a} = \begin{cases} -\frac{n+1}{2}, k = 2r(n+1) - m; r = 1,2 \ldots \\ \frac{n+1}{2}, k = 2r(n+1) + m; r = 0,1,2 \ldots \\ 0, k \neq 2r(n+1) \pm m, \quad r = 0,1,2 \ldots \end{cases}$$

$$C'_{km} = \sum_{i=1}^{\infty} cos\frac{k\pi x_i}{a} cos\frac{m\pi x_i}{a} = \begin{cases} -\frac{n+1}{2}, k = 2r(n+1) - m; r = 1,2 \ldots \\ \frac{n+1}{2}, k = 2r(n+1) + m; r = 0,1,2 \ldots \\ 0, k \neq 2r(n+1) \pm m, \quad r = 0,1,2 \ldots \end{cases}$$

$$\sum_{i=1}^{\infty} cos\frac{k\pi x_i}{a} sin\frac{m\pi x_i}{a} = 0,$$

$$\sum_{i=1}^{\infty} sin\frac{k\pi x_i}{a} cos\frac{m\pi x_i}{a} = 0.$$

In this case, the system of equations takes the form:

$$D_{kn}(a_{kn}) - \quad H_{kn}\sum_{m=1}^{\infty}\sum_{p=1}^{\infty}\sum_{k=1}^{\infty}\sum_{n=1}^{\infty} a_{mp}\, C'_{km}C'_{knp} = 0.$$

$$D_{kn}(b_{kn}) - \quad H_{kn}\sum_{m=1}^{\infty}\sum_{p=1}^{\infty}\sum_{k=1}^{\infty}\sum_{n=1}^{\infty} b_{mp}\, C'_{km}C'_{knp} = 0.$$

$$D_{kn}(c_{kn}) - \quad H_{kn}\sum_{m=1}^{\infty}\sum_{p=1}^{\infty}\sum_{k=1}^{\infty}\sum_{n=1}^{\infty} c_{mp}\, C'_{km}C'_{knp} = 0.$$

$$D_{kn}(d_{kn}) - \quad H_{kn}\sum_{m=1}^{\infty}\sum_{p=1}^{\infty}\sum_{k=1}^{\infty}\sum_{n=1}^{\infty} d_{mp}\, C'_{km}C'_{knp} = 0.$$

(10)

We use the notation (*) for a new record of the system (10).

$$D_{kn}\sum_{n=1}^{\infty}\sum_{k=1}^{\infty}\left(\left(\frac{k\pi}{a}\right)^2+\left(\frac{n\pi}{b}\right)^2\right)^2 - \rho h(\omega_{kn})^2)\delta_{km}\delta_{np}$$
$$- 4h\rho_{ij}\sum_{n=1}^{\infty}\sum_{k=1}^{\infty}\omega_{kn}^2\sum_{m=1}^{\infty}\sum_{p=1}^{\infty}C'_{km}C'_{knp})\,\square$$

$$\square e^{i\omega_{kn}t}\begin{pmatrix}a_{mp}\\ b_{mp}\\ c_{mp}\\ d_{mp}\end{pmatrix} = 0\,. \qquad (11)$$

The characteristic equation of the problem has the form

$$D_{kn}\sum_{n=1}^{\infty}\sum_{k=1}^{\infty}\left(\left(\frac{k\pi}{a}\right)^2+\left(\frac{n\pi}{b}\right)^2\right)^2 -$$
$$- \rho h(\omega_{kn})^2)\delta_{km}\delta_{np}$$
$$- 4h\rho_{ij}\sum_{n=1}^{\infty}\sum_{k=1}^{\infty}\omega_{kn}^2\sum_{m=1}^{\infty}\sum_{p=1}^{\infty}C'_{km}C'_{knp})\,\square$$

$$\square e^{i\omega_{kn}t} = 0\,. \qquad (12)$$

From the characteristic equation we find the dependence of ω_{kn} on k and n.

$$(\omega_{kn})^2 == \frac{D_{kn}\left(\left(\frac{k\pi}{a}\right)^2+\left(\frac{n\pi}{b}\right)^2\right)^2}{\rho h\delta_{km}\delta_{np} - 4h\rho_{ij}\sum_{m=1}^{\infty}\sum_{p=1}^{\infty}C'_{km}C'_{np}}$$

$$(\omega_{kn})^2 == \frac{D_{kn}\left(\left(\frac{k\pi}{a}\right)^2+\left(\frac{n\pi}{b}\right)^2\right)^2}{h(\rho\delta_{km}\delta_{np} - 4\rho_{ij}\sum_{m=1}^{\infty}\sum_{p=1}^{\infty}C'_{km}C'_{np})}$$

In the obtained equalities, we use the condition (**):

$$(\omega_{kn})^2 = $$
$$= \frac{D_{kn}\left(\left(\frac{k\pi}{a}\right)^2+\left(\frac{n\pi}{b}\right)^2\right)^2}{h(\rho\delta_{km}\delta_{np} - 4\rho_{ij}\sum_{m=1}^{\infty}\sum_{p=1}^{\infty}\frac{m_1+1}{2}\frac{m_2+1}{2})},$$

where $k = 2r(m_1+1)+m,\ n = 2r(m_2++1)+p.$

$$(\omega_{kn})^2 = \frac{D_{kn}\left(\left(\frac{k\pi}{a}\right)^2+\left(\frac{n\pi}{b}\right)^2\right)^2}{h(\rho + 4\rho_{ij}(m_1+1)(m_2+1))},$$

where $k = 2r(m_1+1)-m,\ n = 2r(m_2++1)+p$ or $k = 2r(m_1+1)+m,\ n = 2r(m_2+1)-p.$

$$(\omega_{kn})^2 = \frac{D_{kn}\left(\left(\frac{k\pi}{a}\right)^2+\left(\frac{n\pi}{b}\right)^2\right)^2}{h\rho}$$

where $k \neq 2r(m_1+1)+m,\ n \neq 2r(m_2++1)+p,$
$k \neq 2r(m_1+1)-m,\ n \neq 2r(m_2+1)+p$ (13)

3 Results

A full spectrum of frequencies (13) of small transverse vibrations of a rectangular plate with lumped masses, located periodically, was obtained, provided that the thickness of the plate remains constant. It should be noted that ρ_{ij} is the density of each of the concentrated masses.

4 Conclusion

Based on the obtained results, the following conclusions were reached:

1. To find the frequencies of natural vibrations of plates loaded with concentrated masses, which are located periodically, the hyperbolic equation is used. As a result, we obtain the values of the natural frequencies of such a rectangular plate. Depending on the type of the given boundary conditions, it is possible to obtain various types of dependence of the natural vibration frequency with taking into account the geometry of the plate and its mechanical properties, for example, dependence on Poisson's ratio, or Young's modulus, etc. It is easy to see that the value of the numerical value of the frequencies strongly depends on the boundary conditions.

2. When solving dynamic problems of this kind, it is possible to use various boundary conditions, including boundary conditions for a supported or embedded edge. This allows us to investigate the differences in the number of obtained natural vibration frequencies of loaded plates.

3. The values of frequencies in problems of natural vibrations of plates loaded with lumped masses increase with decreasing thickness of the considered plate and do not depend on the geometry of lumped masses. It is

obvious that the considered lumped masses can be considered as points having a mass much greater than the mass of the surrounding continuous medium: elastic, viscoelastic, or plastic, at least in some δ-region around this lumped mass.

The use of modern methods of mathematical and functional analysis, the theory of functional series and Fourier series and the theory of functions of a complex variable to solve such problems, allows taking into account the geometric properties of complex structures. The use of methods for solving the equations of mathematical physics in combination with precise formulations of problems in the mechanics of a deformable solid body allows creating new technological solutions for which modern viscoplastic and composite materials can be used. Such materials are necessary for the implementation of new design solutions, since they have low material consumption [6, 7]. It is clear that modern structures created from modern viscoplastic and composite materials should work reliably in the elastic and viscoelastic regions under complex dynamic loads [8, 9].

Acknowledgements

This work was supported by the Ministry of Science and Higher Education of the Russian Federation under project 0707-2020-0034. This work was carried out using equipment provided by the Center of Collective Use of MSUT "STANKIN".

References

1. E.N. Sosenushkin, Vestnik MSTU "STANKIN", **1**, (2010) [in Russian]
2. A.G. Kobelev, Technology layered metal (Metallurgy, 1991) [in Russian]
3. C.B. von der Ohe, R. Johnsen, N. Espallargas, Wear, **288**, 39-53 (2012) https://doi.org/10.1016/j.wear.2012.02.016
4. L. Nikitin, A. Khamraev, E. Yanovskaya, Physics of the Earth and Planetary Interiors, **50**(1), 26-31 (1988) https://doi.org/10.1016/0031-9201(88)90087-8
5. E.A. Yanovskaya, *On the problem of the vibrations of the diaphragm, a lumped* (Sat. "Numerical modeling in mechanical problems", Publishing house of the Moscow University, 1991) [in Russian]
6. V.I. Kolesnikov, A.L Ozyabkin, E.S. Novikov, Friction and Wear., **40**(4), 380-388 (2019)
7. V.M. Musalimov, K.A. Nuzhdin, Journal of Friction and Wear., **40**(1), 51-57 (2019) https://doi.org/10.3103/S1068366619010136
8. K. Nuzhdin, V.M. Musalimov, Procedia Engineering, **199**, 1478-1483 (2017) https://doi.org/10.1016/j.proeng.2017.09.410
9. S.N. Grigoriev, V.D. Gurin, M.A. Volosova, N.Y. Cherkasova, Materialwiss. Werkstofftech., **44**(9), 790-796 (2013) https://doi.org/10.1002/mawe.201300068

Multirotor Heavy Lift Drone

Fausto Freire Carrera[1,2*], *Pablo* Tamayo[2], *Olga* Chadrina[1], and *Sergei* Kharlamov[2]

[1]Faculty of Engineering Sciences and Industries, Mechatronics, UTE University, Rumipamba y Bourgeois, Ecuador
[2]Robotics Laboratory, Central University of Ecuador, EC-170129, Av. América, Ecuador

Abstract. This article begins by analysing the fields of application of cargo drones at an international level, to then determine the most used configurations and with the best performance to lift important loads, from which a coaxial octocopter drone with X8 configuration was designed and built, with a payload capacity of up to 20 kg and a flight range of 45 minutes. The relevance to build in Ecuador of this type of drones was also analysed and thus minimize technological dependence. The results showed that it starts with 31% of national components and that by executing activities aimed at reducing dependency, in a period of 5 years it can reach 63.8% of national components for the construction of drones of these characteristic.

1 Introduction

According to J'son & Partners Consulting [1] the world drone market size in 2017 was 7.3 billion dollars, and the 53% corresponded to the military market. However, according to the market structure, 84% corresponds to consumer drones, 15% to commercial drones and only 0.5% corresponds to military drones, due to the high costs of this type of drones.

It is expected that the cost of the components for the building of a drone will decrease between 7% and 27% by 2020 due to the intervention in the market of new commercial actors. Also, the drone market will grow and reach 9.5 billion dollars.

According to Gartner, 2.1 million drones were built in 2016. It was expected that in 2017 this number will reach 3 million, with an average annual growth of 34%.

In the world, people have been actively working on regularizing the usage of drones in different situations [2,3]. Considering its continuously growing spectrum of use, securing the drones becomes a priority to reduce the risks related with their utilization. Different companies begin to offer insurance services to the drone owners. Nowadays, drones are even displacing satellites, and this trend is expected to continue growing. The emergence of millions of jobs related to the exploitation of drones in various aspects of human life is forecast.

The drones´ development is conditioned by the future usage they will be given. According this, they will be limited in size, speed, etc. For example, for illegal actions such as robberies, violation of human rights (prohibited photographs), transportation of narcotics and illegal drugs a faster and discreet drone is required.

One of the areas where drones are developing more actively is the cargo transport [4]. The drones offer significant advantages over the classic transport such as elimination of traffic jams, lower environmental pollution, minimum delivery times and minimum take-off and landing surfaces.

This article describes the process of planification, analysis, construction, building and economical evaluation of a drone in Ecuador.

2 Materials and methods

2.1 Analysis of energy consumption by number of rotors

The possible combinations and configurations of the rotors used to build a drone and the analysis of the energy consumption is displayed in Figure 1.

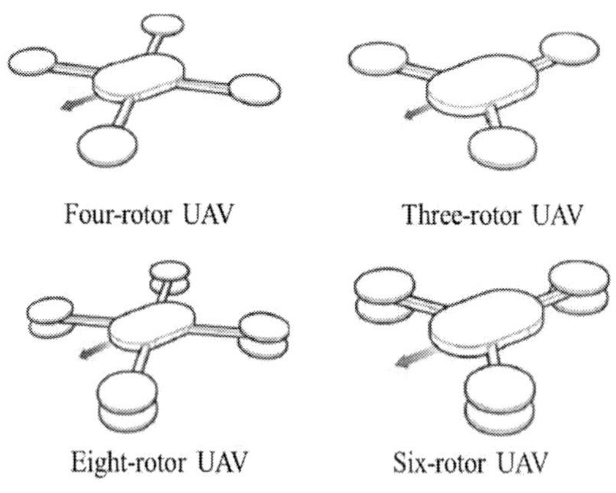

Fig.1. Unmanned Air Vehicle (UAV) Configuration

Dmitri Aleksandrov [5], describes and compares UAVs with different numbers of rotors, obtaining different parameters in terms of energy consumed by each rotor, rotor speed, energy that the rotor must

* Corresponding author: ffreire@ute.edu.ec

https://doi.org/10.1051/epjconf/202124802010

deliver, stationary flight time, etc. Table 1 shows the details of this study.

Table 1. Energy consumption according to the number of drone rotors

Number of rotors	4	3	8	6
Mass (kg)	1.4	1.32	**1.72**	1.56
Force produced by each rotor (or pair) (N)	3.5	4.4	**4.3**	5.2
Thrust force for each rotor (or pair) Fh (N)	0.7	0.88	**0.86**	1.04
Angle α (degrees)	11.31	11.31	**11.31**	11.31
Rotor speed (RPM), in stationary flight	5334	5967	**4656**	5095
Single motor power (W), in stationary flight	32.5	40.3	**23.8**	29.4
Stationary flight time (min)	25.1	27	**17.2**	18.5
Stationary flight time with 1Kg load	35.2	35.7	**29.5**	28.9
Rotor speed (RPM), in horizontal flight	5385	6024	**4700**	5142
Power of a single motor (W), in horizontal flight	33.2	41.2	**24.4**	30
Horizontal flight time (min)	24.5	26.4	**16.7**	18.1
Horizontal flight time with 1 kg load (min)	34.4	34.9	**28.8**	28.2

Table 1 shows that the three-rotor configuration provide the drone with greater efficiency in energy consumption. The difference between a three-rotor and a four-rotor drone configuration is only 1.5%. However, the drones with four rotors offer enhanced maneuverability and stability compared to the three-rotor drones [6].

From the data provided above, it can be concluded that the coaxial octocopter has the best stability characteristics in comparison with other drones, especially regarding the lift of payload. Even though it consumes more energy, the eight-rotor drone is the most appropriate. For this reason, a drone with the "Coaxial Octocopter X8" configuration was selected for this project.

2.2 Selection of the material for the drone

For the project, aluminum, steel and carbon fiber were analyzed. The factors to be considered to select the material were:

- Resistance (fragile)
- Rigidity (flexible)
- Weight (heavy)
- Cost

Aluminum, steel and carbon fiber meet all the criteria mentioned above, but each one is unique in terms of strength, rigidity, weight and cost, as shown in the following Table 2.

Table 2. Comparison of materials

Characteristics	Steel	Aluminum	**Carbon fiber**
Resistance to stress	High	High	Very high
Fatigue resistance	High	High	Very high
Overall thickness	Low	Low	Very low
Own weight	High	Low	Low
Corrosion	Yes	Yes	No
Use	Difficult	Difficult	Easy
Handling	Hard	Easy	Easy
Crossing between elements	Hard	Hard	Easy
Cost of materials	Low	Low	High*
Cost of installation	High	High	Low

Steel has good qualities regarding strength, but its density is high, and it has corrosion problems. Steel´s density is 3 times the density of the aluminum alloys. Aluminum has adequate strength; it is a material widely used in aeronautics. Aluminum has low density and is easy to handle. The strength of carbon fiber is superior to that of steel and aluminum. Carbon fiber offers a lower weight in the same proportion of surface and volume. However, the need to use of several layers of fabric in the areas of union and, above all, the need to homogenize it with resins and adhesives, end up almost equaling in weight with the aluminum.

In any case, although of equal weight, a carbon fiber piece will always be more solid than an aluminum piece. In addition, carbon fiber is resistant to corrosion, fire, chemical inertia and electrical conductivity [7]. Additionally, in the face of temperature variations, carbon fiber retains its shape. With these considerations, we concluded that the material suitable for the project is carbon fiber.

2.3 Design and mechanical simulation of the drone

The design of the drone structure was made using the SolidWorks, for which several designs were assayed to meet the demands of the project [8,9].

During the study it was concluded that the drone should consist of three layers, two to settle the base of the engines, hold in the electronics and power supply and another layer to hold the payload

The studies were carried out applying forces of 50 N in the arms of the structure with different supports and with the three layers of solid material. Then, several assays were performed with different emptying of the layers of the drone. In this article the most important results are shown.

Figure 2 shows the emptying that allowed to minimize the weight of the first lid and obtain good tension, displacement and deformation results by the applied forces. The ends of the cover are carefully analyzed as they are the fastening points for the support tubes of the 8 drone engines.

 https://doi.org/10.1051/epjconf/202124802010

Tension forces (von Mises)

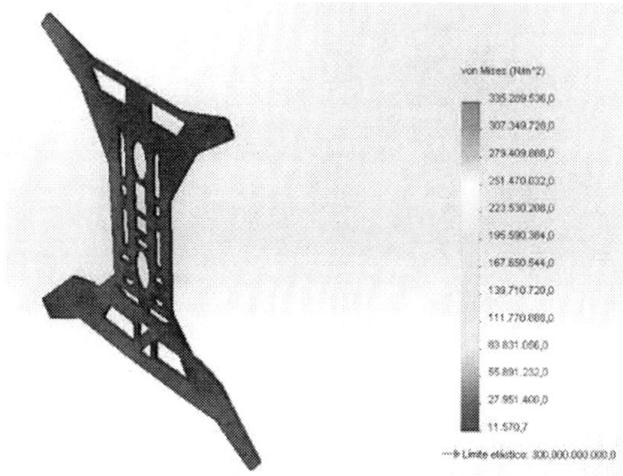

Unitary deformations (ESTRN)

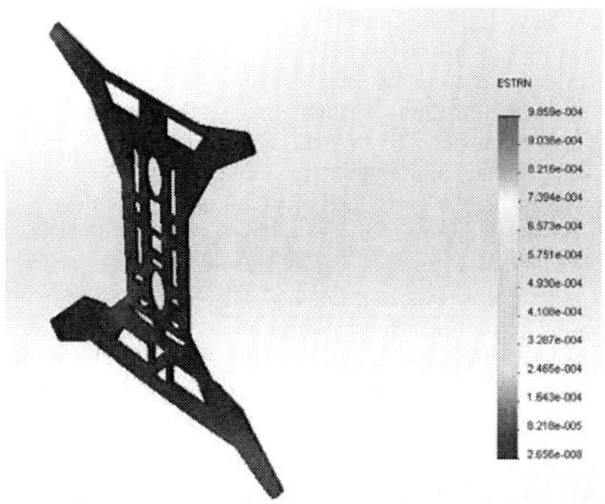

Fig.2. Lid 1 design

Figure 3 shows the design of the second lid, which has a different emptying compared with the Lid 1, as it will serve as a base for the electronic components of control of the drone. The best results were obtained with the forces applied at the ends.

Tension forces (von Mises)

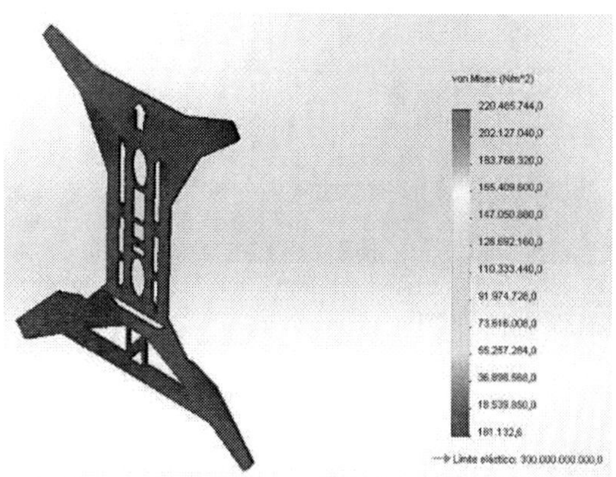

Unitary deformations (ESTRN)

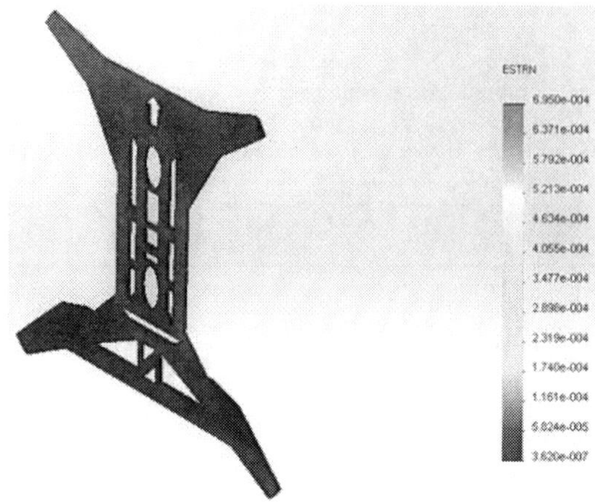

Fig. 3. Lid 2 design

Tension forces (von Mises)

Displacement (URES)

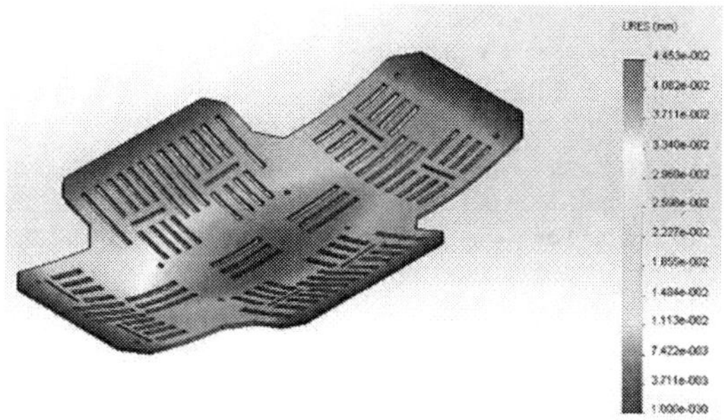

Fig. 4. Lid 3 design

The third lid (Fig. 5) is used to place the useful load of the drone with 4 points of subjection, the applied emptying allows to obtain the best results in terms of tension, displacement and deformation. The mass of the lid is 0.49 kg, made with carbon fiber material, and a reaction force of 31.9 N.

Tension forces (von Mises) Displacement (URES)

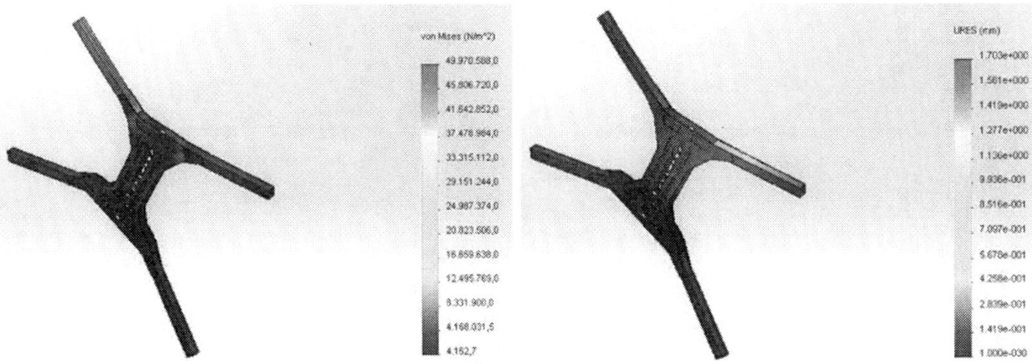

Fig. 5. Assembled drone

After the design stage, the drone was built taking into account all the requirements and the results of the analyzes and simulations of the different drone configurations. Thus, the final prototype met the quality and safety standards necessary for this type of projects.

2.4 Power source selection

For the analysis and selection of the battery type that allows to comply with the requirements of the project, the eCalc simulator was used. This simulator allows the determination of the type of batteries with an accuracy of +/- 15% knowing some characteristics of the drone, such as weight, dimensions, flight height, engine types and number, configuration. Table 3 shows the parameter used for the calculation of the energy needed for the system.

Table 3. Calculation of energy required for the system

Category	Characteristic	Value
General	Engine cooling	Excellent
	Number of rotors	8 (coaxial)
	Model weight	850 g (30 oz)
	Frame size	2000 mm (78.74 inch)
	Fiel elevation	2800 m ASL (9186 ft ASL)
	Air temperature	25 °C (77 °F)
	Pressure (QNH)	1013 hPa (29.91 inHg)
Battery cell	Type (Cont./max. C) – charge state	LiPo 22000 mAh – 15/25C – full
	Configuration	6 S, 2 P
	Cell capacity	22000 mAh (44000 mAh total)
	Maximum discharge	85%
	Voltage	3.7 V
	Weight	476 g (16.8 oz)
Controller	Type	Max 50 A
	Current	50 A cont. (50 A max.)
	Weight	65 g (2.3 oz)
Motor	Manufacturer – Type (Kv)	Tiger motor – U8-16[2] (100)
	KV (w/o torque)	100 rpm/V
	No-load Current	0.6 A @ 22.2 V
	Limit (up to 15s)	750 W
	Weight	240 g (8.5 oz)
Propeller	Type – yoke twist	T-Motor CF – 0°
	Diameter	29 inch (736.6 mm)
	Pitch	9.9 inch (251.5 mm)
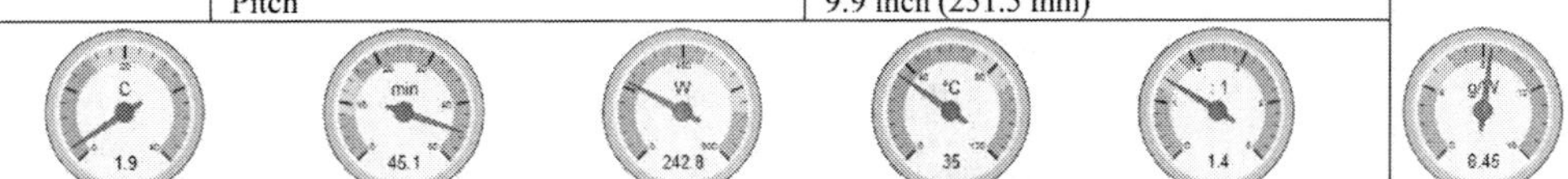		
Load Estimated maximum temperature	Hover flight time Thrust-weight	Electric power Specific thrust

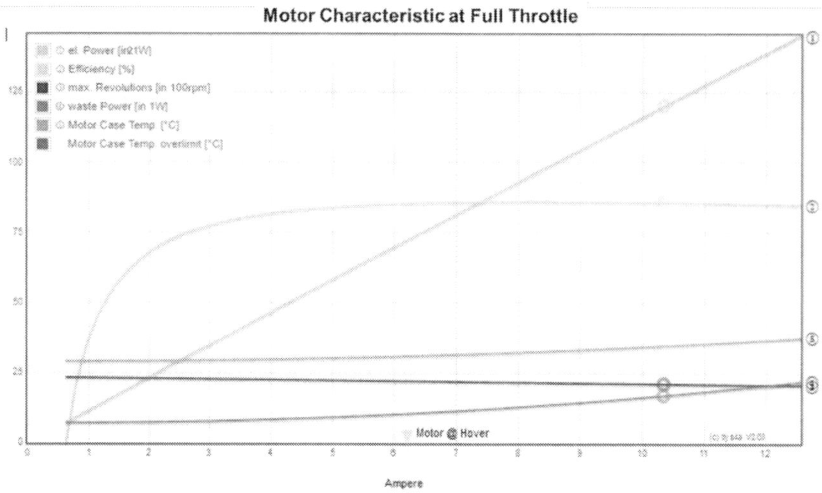

Fig. 6. Drone configuration for source selection

From the analysis of figure 6, we concluded that the best option is to use two LiPo batteries of 22000 mAh 15/25, since these allow to reach the 45 minutes of autonomous flight with load, power and necessary thrusts, to fly at 2800 meters of height.

2.5 Drone control system

The hardware components were specially selected to work with ArduPilot, an Open Source autopilot board that uses an ATMega328 chip, and allows the connection of sensors, GPS and other. It also permits the control of the drone by radio, an automatic control following a previously established trajectory and flight through designated points [10]. ArduPilot supports the sending of telemetric information from the sensors on board to the control station, which can be a computer, telephone or tablet and records data in the memory aboard the drone.

Ardupilot has many advantages over DJI NAZA, Wookoong-M, ZeroUAV (YS-X4, YS-X6), Mikrocopter, Hovefly Pro, and uThree. Ardupilot has a low price, full autopilot, supports the possibility to place 166 flight points, edit the route in flight, permit a remote configuration of the parameters, supports GEO-FENCE, has an open data transmission protocol, supports configurations for tricopters, quadricopters, hexacopters, octacopters, Y6, X8, helicopters, airplanes, cars and ships, supports flight simulation with Mission Planner.

The technical characteristics of the components of the drone to meet the objective are presented in Table 4.

Table 4. Technical specifications of the drone components.

Component	Description		Amount
Engines	Characteristics	U8-10 KV: 170rpm/v no load current : 1.1a/22.2v max watts/time : 750w/180s max amps/time : 24a/180s resistance ohm : 186mΩ case height/length : 86.8c26.5mm weight in grams : 240g	8
Propellers	Characteristics	29x9.5prop 104.9g/sheet Sandwich-like construction	8
Flight controller		45A, 8S	1
Carbon fiber tubes		Rectangles 30x40mm, thickness: 2mm, 380mm long	4
Plates		600x600mm, thickness: 3mm 250x300, thickness:1mm	1 2
Landing gear		double bar, carbon fiber 16x15x500mm base, carbon fiber 10x9x400mm	2
Drone fasteners		Thickness: 3mm, aluminum	8
Three-floor structure		First layer: GPS Second layer: arms of the Drone Third layer: holds landing gear and electronic section	

a)

b)

Fig. 7. View of the drone a) deployed, b) retracted

2.6 Analysis of national added value

The drone is made of 147 elements (Fig. 7), divided into three groups according to the importance of each of them within the project. These categories are: parts/pieces that represent 66% of the drone, materials – 28% and software – 6 % of the drone. An economic analysis was carried out according to the origin of each group, whether Ecuadorian or foreign. Additionally, an economic projection to 5 years was performed, if there are performed investments and research in this area. The results are displayed in Figure 8 and 9.

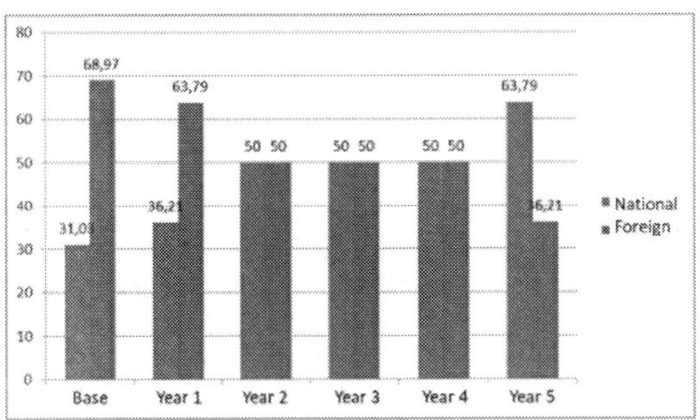

Fig. 8. Distribution by components at 5 years

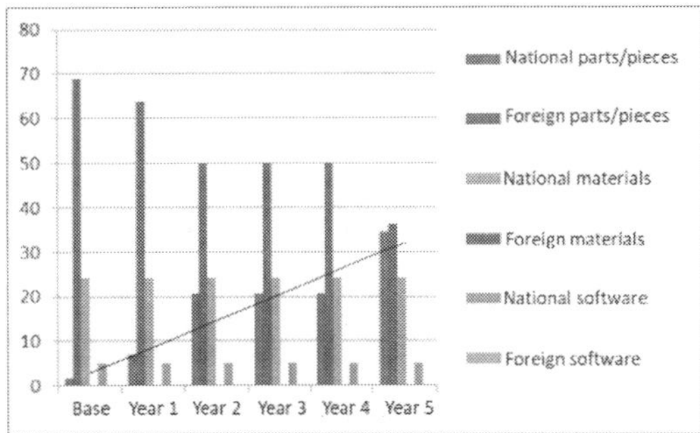

Fig. 9. Distribution by group at 5 years

3 Conclusions

In Figure 9, the year of acquisition of the drone components, the foreign component represents 68.97%, and it decreases to 36.21% in the fifth year. The national (Ecuadorian) component at the beginning represents 31.03% and at the fifth year it represents 63.79%, being the national parts/pieces component that grows more according to the trend. It is shown that this type of project is viable at a medium term and will create jobs and added value to the country. All of this without considering the increase of jobs in the provision of services related to the manufacture and incorporation of this technology in the various fields of production, defense and research in Ecuador.

References

1. J'son & Partners Consulting, Drone market in Russia and in the world, (2018)
2. C. Zhang, J.M. Kovacs, PA, **13**, 693–712 (2012)
3. N.D. Bublik, D.V. Chuvilin, G.A. Shafikov, ESJ, **2**(10), (2018)
4. A.M. Almeshal, M.R. Alenezi, A.K. Alshatti, Information, **11**(9), 442 (2020)
5. D. Aleksandrov, I. Penkov, Parnu (2012)
6. F. Sarghini, A. De Vivo, ChET, **58**, 625-630 (2017)
7. M. Budinger, A. Reysset, A. Ochotorena, S. Delbecq, AST, **98**, (2020)
8. H. Yang, Y. Lee, SY. Jeon, et al, ISR10, 79–93 (2017)
9. T. Kilby, .B. Kilby, Make: Getting started with drones, Makermedia, (2016)
10. K. Gabuev, V. Gongalo, N. Kucherenko, A. Shipko, ATIB-P, **10**, (2018)

https://doi.org/10.1051/epjconf/202124802011

A System for Measuring Water Levels in Open-Air Irrigation Canals

Fausto Freire Carrera[1,2*], *Harry Saltos* Sanchez[1], *Yelitza* Garcia-Orellana[3], and *Olga* Chadrina[1]

[1]Facultad de Ciencias de la Ingeniería e Industrias, Mecatrónica, Universidad UTE, Rumipamba y Bourgeois, Ecuador
[2]Laboratorio de Robótica, Universidad Central del Ecuador, EC-170129, Av. América, Ecuador
[3]Decanato de Agronomía, Universidad Centroccidental "Lisandro Alvarado", Barquisimeto, Venezuela

Abstract. The measurement of the water level in a canal allows us to know its flow. This information is essential for the correct administration of the water resource. Manual quantification can be replaced by micro controlled systems with an internet connection. Thus, the measurement and recording of water levels in open canals in real time is achieved. The chosen micro controlled system was based on ultrasonic instrumentation with GPRS communication. It had a wide cellular network to achieve an effective connectivity in rural areas, a power system combining solar energy with pollutant free batteries and the capacity to store data. The obtained percentage and average errors were lower than the permissible error specified in the requirements. Consequently, the evaluated measurement system is reliable for the evaluation of canals in a real-world setting.

1 Introduction

The increased demand for water resources for urban and industrial use has had a detrimental effect on the water available for agricultural production [1, 2]. Therefore, a better management of the water available for irrigation is now essential.

When evapotranspiration exceeds precipitation, it is necessary to implement irrigation systems to guarantee sustainable agricultural production. Despite the building of reservoirs, water distribution systems are not efficient enough. This produces losses that increase costs and decrease the availability of water resources for farmers [3].

Traditionally, water level measurements were made with manual gauging systems [4] that offered imprecise results. Technological innovations have led to the development of more accurate electronic systems or systems based on diffuse controllers. In Ecuador, although the water distribution systems have been upgraded, they do not have a remote monitoring system to provide information on the water level in the canals.

Certain manoeuvrers are performed during the flow of water in the canals. These include tactics designed to divide the flows in one or several lateral branches, and those aimed at keeping certain stationary states constant. For the latter, the canal requires the shortest possible transition state and insignificant oscillations in water levels to avoid extreme variances in the deliveries [5]. This manoeuvre causes an inadequate measurement of the water level, which leads to unnecessary water costs, electricity consumption and longer pump and gate operations [6].

Notwithstanding the technical requirement, there is no adequate monitoring system to determine the protocols for managing floodgates and activating pumps in the irrigation canals. The consequent occurrence of unrecognized leaks and infiltrations leads to inefficient use of the water resource [7].

Therefore, it is important to check the water level in the network of irrigation canals to control the quality of the water distribution. The current mechatronics' knowledge of the measurement systems based on sensors and microprocessors [8, 9] has allowed the optimization of systems used for monitoring, recording and measurement of water in the irrigation canals.

Using appropriate instrumentation, it is possible to detect changes in the water level in a canal, as well as any anomaly along the distribution network. This allows the implementation of corrective or preventive solutions. Therefore, automatic control of the water level in canals plays a key role in the performance of irrigation projects [10].

The modernization of control and measurement systems has gained ground all over the world. This upgrading can generate hydraulic models with proper height monitoring and data recording. Whereas the operation of the irrigation canals was improved mainly in two ways. First, adopting new technologies that enhance the interaction between the operator and the system, and second, using digital sensors to control and supervise the canals [11, 12]. The technification of the irrigation systems through instrumentation, control and remote operation was an effective way to increase the efficiency of water distribution in areas of shortage, and resolve the constant disputes over its use.

* Corresponding author: ffreire@ute.edu.ec

In this context, various irrigation districts in Mexico have sought to increase the percentage amount of water used. As a result, the crop yields were higher due to the improved irrigation efficiency, optimized energy consumption, and reduced environmental and economic impact of water losses. The consequences observed in the yields were also influenced by the decrease in the duration of the manoeuvres and the reduction of the operators' expense. The modernization of the instrumentation for measuring and monitoring water flow through the irrigation canals was achieved using mathematical models and an automated gauging and measurement system based on fuzzy logic [13, 14].

Ecuador is an entirely agricultural country with important irrigation systems for farming, most of which lack automated systems for measuring the volume of water during the distribution process. The objective of this research was to evaluate an automated system for recording, controlling, and monitoring water levels in the irrigation canals. It was based on a mechanism of collection, storage and transmission of data using microprocessors, sensors and a cellular network. These features will allow its use in rural areas of difficult access. The proposed system is based on the use of solar energy combined with heavy metal-free batteries that guarantee its operation with a minimum environmental impact.

2 Materials and methods

In the field, a system capable of measuring the water level in the canal was used for data collection. The instrumentation employed was suitable for the working conditions. In addition, the integrity of the components was ensured having a strong and resistant protection against external environmental conditions.

A communication system with access to the internet was developed to transfer the water level measurement data from the field to the operator. This system was made compatible with the technology used in the later stages: data acquisition and user interface display.

The acquisition stage of the measurement system must be autonomous to ensure its continuous work even in areas without electrical grid.

Two tests were conducted with the monitoring device placed 4 meters above the base of the irrigation canal to evaluate the system's performance and validate the accuracy of the quantifications. Using the sensor, 182 water level measurements were made with a five-minute interval. Of these, 104 quantifications were performed with a 41 cm and 78 with a 5 cm water level. Thus, the behaviour of the system at different water levels was observed. The data obtained was analysed using descriptive statistical parameters.

The water measuring system comprises various components that are described in detail below.

2.1 Sensor

The MB7060 XL-MaxSonar-WR ultrasonic sensor by MaxBotix [15] was selected because of its outstanding characteristics for operating outdoors. It does not require to be in direct contact with the water and is waterproof.

This sensor has an analogue output and one centimetre spatial resolution. The ultrasonic sensor self-calibrates based on the input voltage according to the Equation 1.

This self-calibration capability reduces or renders the need for sensor maintenance, almost nil, compared to other devices.

$$h_s = Vcc/1024 \qquad (1)$$

Where

hs = the distance between the sensor and the water surface of the canal.

Vcc = reference voltage

2.2 Microcontroller

The ATMEGA 328p microcontroller was used because it is compatible with free software and has an operating current below 200mA at 5V. This microcontroller is the most efficient option for information processing at the data acquisition stage due to the optimal working frequency.

It is worth mentioning that the use of a Programmable Logic Controller (PLC) was not considered, despite its superior intrinsic characteristics of robustness compared to microcontrollers. The hardware characteristics of PLC do not necessarily add value to the system design, along with an unwarranted increase in the cost of the device.

2.3 Communication technology

The evaluated irrigation canals were comprised in the Catarama Project, located in the rural area of Los Rios province. In this area, the mobile phone signal is not strong enough to use the Global System for Mobile Communication (GPRS) connectivity.

A data transmission module was selected to access the internet via cell phone. The module was compatible with the chosen microcontroller and the working conditions. This SIMCom [16] cellular communication module had GSM/GPRS SIM800L capacity and was able to work at higher temperatures than the competitors. It has a low power consumption and a high speed for data transfer. In addition, its ability to work with a signal strength as low as -114 dBm is outstanding, and makes it ideal for internet connection.

The SIMCom module was connected to the ATMEGA328P microcontroller through TX-RX serial communication. For this communication the SIM800L has an auto-tuning system with the "autobaud" working frequency.

2.4 Power source

For research purposes, photocells that capture solar energy were used to generate electric power. They were

https://doi.org/10.1051/epjconf/202124802011

chosen given the ease of implementation and energy production.

Data on average annual insolation were considered to choose a suitable solar panel [17]. After the evaluation of the full power characteristics, the monocrystalline panel SYSM10S was selected as it is capable of supplying up to 10W to the system.

2.5 Circuit design

Most of the selected components and modules were powered with 5V, therefore, the DC-DC LM2596s integrated circuit module was used. This integrated circuit module was responsible for regulating the voltage up to the 5V required by the system. In addition, a voltage sensor module was added to monitor the performance of the solar panel.

2.6 Software development

The program for processing sensor data and sending information to the database was developed using the ARDUINO IDE platform, compatible with the AVR ATMEGA328p, following the process shown in Figure1.

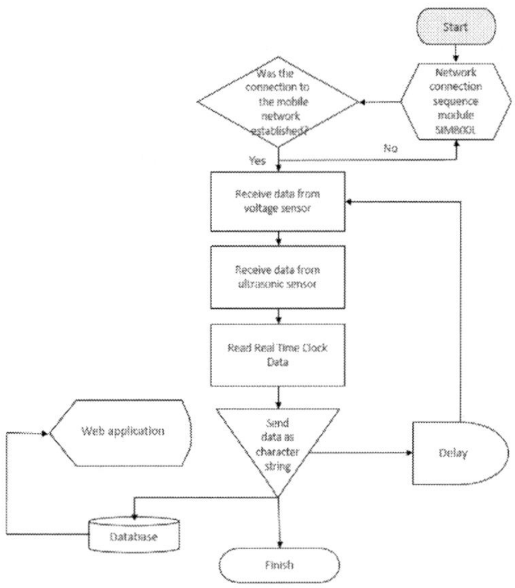

Fig.1. Software design flow chart.

2.7 Database

The database recorded information on the date and time of the measurement, the device identification code and the power source voltage. As a security measure, for the design of the application it was necessary to limit the access to the information and its manipulation. Consequently, a table was created to supervise the access of the users to the platform. The password and user type ere also registered. The database used was of type MariaDB.

2.8 Web Application

The web application was designed to consult the information stored in the database and generate reports on records displayed. A built-in template-based PHP code generation tool was used for application development. Thus, a fast and efficient management of the information and the linking of tables previously created in the were achieved. The linked charts were the CAUDREGISTRO transactional table and the SEGURIDADES security table. Hereby, the CAUDREGISTRO dynamic table was created, which allows the creation of diagrams.

2.9 Flow calculation

The calculation of the water flow in the irrigation canal was done by measuring the water level using ultrasonic technology. This technology was used as it did not require direct contact between the sensor and the water to take measurements. In addition, it can be placed on the canal out of reach of third parties.

Considering the typical cross section of a trapezoidal canal, the placement of the sensor on the canal is showed in (Figure 2).

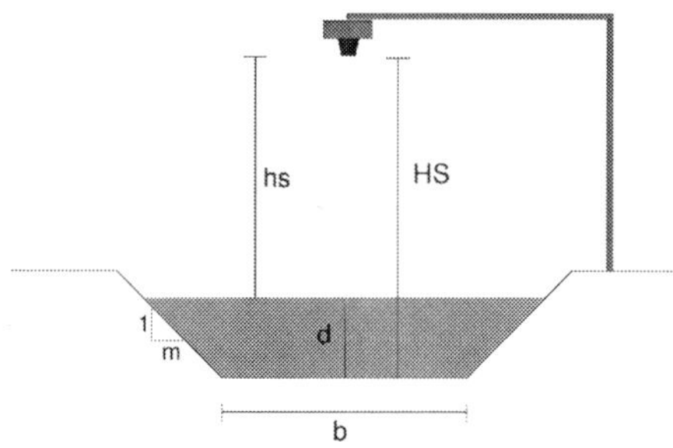

Fig.2. Sensor arrangement on the irrigation canal. d = canal water tie-back; HS = distance between the sensor and the canal base; hs = distance between the sensor and the water mirror of the canal; b = width of the base of the canal; m = inclination of the slope.

Equation 2 showed the difference between the heights involved in the measurement of the canal. Using this equation, the water level (d) can be obtained in terms of the sensor measurement.

$$d = HS - hs \tag{2}$$

Where:
d = canal water tie-back
HS = distance between the sensor and the base of the canal
hs = distance between the sensor and the water mirror of the canal.

Posterior to the computing of the water tie-back (d) and using the dimensions of the canal, Equation 3 [18] was applied. Thus, the average speed of the water in the canal was estimated.

$$v = \frac{1}{n}\left(\frac{bd + md^2}{b + 2\sqrt{m^2+1}}\right) * s^{\frac{1}{2}} \quad (3)$$

Where:
v = average water speed ($m.s^{-1}$)
n = Manning's roughness coefficient, depends on the material of the carcass.
b = width of the base of the canal (m)
d = water pressure (m)
m = inclination of the slope
S = inclination of the canal floor (mm^{-1})

Equation 4 was used to obtain the water flow in the canal at a given time.

$$Q = A * v = (bd + md^2) * v \quad (4)$$

Where
Q = water flow ($m3\ s^{-1}$)
A = Area of the canal (m^2)
v = average water speed ($m\ s^{-1}$)
b = width of the canal base (m)
d = water pressure (m)
m = inclination of the slope

3 Results and discussion

3.1 Choice of instrumentation

The water level measurements were made in an open chanel. Therefore, one of the conditions taken into account for the selection of the instrumentation was its ability to operate outdoors. Ultrasonic technology was selected because it did not require direct contact between the sensor and the water to perform measurements [19]. In addition, the sensor could be located on the canal out of reach of third parties. This type of technology had a number of advantages when compared to hydrostatic, capacitive and mechanical methods. The benefits are summarized in Table 1.

Table 1. Technologies for level measurement

	Hydrostatic	Capacitive	Ultrasonic	Mechanical
Advantages	Easy to operate	No moving parts	No need of direct contact with water. No moving parts. Localization on the canal prevents manipulation by third parties	Easy to use Easy to install
Limitations	Calibration requirements. Need of direct contact with water. Need of submersible equipment or additional means of communication	Need of direct contact with water, which accelerates its deterioration. Need of a probe that limits its measuring capacity	Sensitivity to disturbances like foam or very turbulent flows.	Deterioration of its components by mechanical action. Need of direct contact with water, Increased maintenance requirements

The second criterion for the choice of instrumentation was the ability of the microprocessor to transmit the measured data. The selected microcontroller made it possible to collect the data from aforementioned ultrasonic sensor, transform this data into information and then send it to the web through a communication module. In [20], it is indicated that the cost of microprocessors reduced the cost of sensor installation. However, affordable controllers with high data storage and transmission capacity are currently available on the market.

The choice of the controller (ATMEGA 328p) was based on its compatibility with free software, its operating current that does not exceed 200mA at 5V, and its optimal working frequency. Additionally, ATMEGA 328p had a lower acquisition cost than other microcontrollers (Figure 3), which improves the viability of the project.

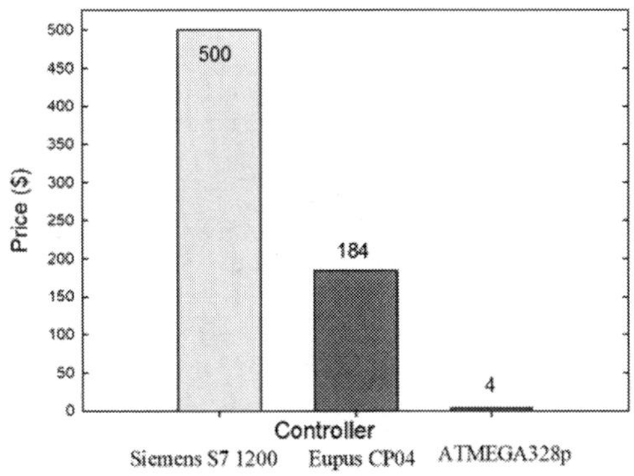

Fig. 3. Cost in US dollars, other PLCs vs ATMEGA328p

3.2 Communication technology

Internet connection was achieved through the mobile network. The cellular network had considerable advantages over other Internet connection technologies. The most relevant advantages were its wireless autonomy for sending data and its cost-benefit ratio due to its installation versatility [21]. The advantages of this network are described in Table 2.

Table 2. Internet connection technologies.

	Wi-Fi	Satellite Connection	Fiber Optics	Cellular
Advantages	Wireless. Installation of repeaters is possible. Data communication at high speeds.	High-speed data transmission. Wireless availability anywhere.	Very high data speed	Entirely wireless. Coverage in all urban and rural areas. Inexpensive installation.
Limitations	Low range of coverage Unavailable in the project's area of influence	High installation and maintenance costs	Unavailable in the project's area of influence. High costs	Low speed compared to other technologies. High cost.

Another advantage that stood out is the wide coverage of the cellular network in urban and rural areas of Ecuador. The project was located in the rural area of Los Rios province, as is the Catarama Project. Mobile network coverage was strong enough to use GPRS connectivity along the canals.

In the South American countries, which are crossed by the Andes mountain range, data transmission becomes complicated. Among the efforts to improve it, [22] the use of a DTT and 4G-LTE cellular transmission system in Colombia stood out. This DTT technology provided benefits that allowed Internet connectivity improvement in rural areas, minimizing connection costs per user. Furthermore, it reduced the need for specialized infrastructure building, since it reused the existing analog and/or digital TV infrastructure. Its applications could be used in data transmission projects for environmental and agricultural purposes as proposed in this research.

3.3 Power source

The measurement system was located in a rural area, so an autonomous power system that ensured constant data acquisition throughout the year had to be chosen. Besides, this energy source must be sustainable and low-pollution. Thus, a mixed system was chosen, combining solar energy with the use of low-polluting batteries. In this sense, [23] stated that the use of solar energy was relevant for irrigation systems in isolated areas, due to the high cost of using another energy sources.

In this project, photocells were used to capture solar energy and generate electrical power, since this is an easy method to obtain energy. Given that the canal water measuring modules required 11.5W of power from source, the photocells did not guarantee sufficient autonomy for continuous data acquisition. Thus, the system had to rely on the use of batteries, which were chosen according to their cost, unit of mass and volume.

When reviewing the advantages and disadvantages of the different types of batteries (Table 3), the lead acid batteries stood out, both for their cost-benefit and for their safety. These batteries had a lower risk of explosion under the given working conditions, even though they had a lower charge density per unit of mass compared to other batteries. Undoubtedly, the low charge density per unit of mass and volume makes the devices less portable. However, this is not a relevant factor for the proper functioning of the device, since portability is not required in this project.

Table 3. Types of batteries.

	Lithium Polymer (Li-Po)	Lithium Ion (Li-Ion)	Lead and Acid	Nickel-Metal Hybride
Advantages	Robust and flexible. High density W/h. Less risk of electrolyte spillage.	Lower costs. High density Wh/kg. No memory effect.	Safer installation. Lowest risk of explosion among all. Inexpensive.	High load density. Inexpensive.
Limitations	Higher costs. Reduced energy delivery time.	Thermal runaway. Special control needed during transport.	Low density Wh/kg	Memory effect. Exceptional protection and monitoring needed.

The selected power source guarantees portability, system autonomy and adequate storage capacity. However, lead acid and nickel batteries were discarded due to the risk of contamination of the river and surrounding areas, since heavy metals are present in their composition [24, 25].

This whole system allowed the calculation of the long-term water flow, that permitted the recognition of the water supply in the canal and the operation of the irrigation system. Thus, guaranteeing an equitable water delivery to the farmers who benefit from the irrigation system of the Catarama River.

3.4 Water level measurement

The results of the statistical analysis for the 41 cm depth are shown in Table 4. This analysis showed an average of 40,81 cm, a coefficient of variation of 2,21 and a percent error of 0,09.

Table 4. Results obtained for the 41 cm water depth.

Parameter	Value
Real size (cm)	41,00
N	104
Mean of the sensor (cm)	40,82
Minimum value (cm)	38,00
Maximum value (cm)	43,00
Standard deviation	0,90
Coefficient of variation	2,21
Percentage of error	0,45
Percent error	0,09

The fact that the values measured by the system were similar to the real values, indicated that the use of the sensor can replace manual gauging systems. The proposed system allowed simultaneous measurements to be carried out continuously during different seasons of the year and environmental conditions. In addition, this measurement system performed the storage and transmission of data, which allowed a systematic evaluation of the operation of the irrigation system. The use of sensors allowed the non-dependence from the operators, which usually increased operating costs [26]. It also facilitated the water level measuring in rural populations that are difficult to access (irrigation systems in the Andean regions of South America [27,13]) or regions where manual measurements are difficult due to social conflicts or insecurity (some irrigation districts in Mexico or Colombia [28, 29]).

The operation of the sensor was tested at two water levels, to evaluate its behaviour when changes in depth occur. Table 5 shows the water height for a water level of 5 cm, observing that the percent error in none of the cases reached 1%.

Table 5. Results obtained for a 5 cm water depth.

Parameter	Valor
Real size (cm)	5,00
N	104
Mean of the sensor (cm)	4,59
Minimum value (cm)	3,00
Maximum value (cm)	6,00
Standard deviation	0,63
Coefficient of variation	13,79
Percentage of error	8,21
Percent error	0,07

The reliability of the measurement system at different water levels allowed its use in the monitoring and control of water levels in open canals. In this type of canals, environmental conditions could affect the measurements, while inadequate measurements lead to prolonged floodgate openings. Excessive opening time leads to water losses and waste of operating time and energy [26]. Furthermore, the structure of the sensor allows it to be placed at different points in the canal, and thus identify leaks and filtrations that reduce the efficiency of the irrigation system [30].

On one hand, the system provides the measurement of water levels with values close to the established standards. Moreover, it provides a method for the acquisition and recording of data that is easily transmitted from remote sites to decision making areas using clean energy.

4 Conclusions

The percentage and average errors were lower than the permissible error. Consequently, the measurement system is reliable for calculating the flow rate in canals under real conditions.

The use of the mobile network as a means of web communication allowed the remote monitoring of water levels. It is important to highlight its ability to work in rural areas or areas of difficult access due to its wireless mode of operation.

The use of solar energy in the devices combined with non-polluting auxiliary energy sources ensures 100% autonomy of the system.

The verification of the correct functioning of the web platform was successful. Its subsequent execution allowed the visualization of new information on the height of water in the canals and the storage of data on the measurements made by the sensor.

References

1. A.L. Arenas, M.R. Ochoa, J.P. Miranda, RLCT, **9**(1), 100-105 (2017)
2. R. Pacheco Vega, A, **27**(74), 219-257 (2014).
3. J.A. Cuchí, P. M. Ramos, J. A. Rausa, IAFJC, **30**, 21-38 (2018)
4. O.M. Chamba, H. Benavides-Muñoz, RCT, **11**(1), 29-36 (2018)
5. E.M. Calderon, R. Sotomayor, IEEE LAT, **14**(2), 471-476 (2016)
6. W.H. Nam, E.M. Hong, J.Y. Choi, IS, **34**(2), 129-143(2016)
7. A. Altamirano-Aguilar, J.B. Valdez-Torres, C. Valdez-Lafarga, J.I. León-Balderrama, M. Betancourt-Lozano, T. Osuna-Enciso, TCA, **8**(4), 79-99 (2017)
8. C.A. Bolu, J. Azeta, F. Alele, E.O. Daranijo, P. Onyeubani, A.A. Abioye, JP, **1378**(3) (2019)
9. B.V. Raju, IJIRSET, **6**(7) (2017)
10. Y. Zhang, Y. Shen, WIR:W, **6**(3) (2019)
11. M. Azúa-Barrón, M.A. Vázquez-Peña, R. Arteaga-Ramírez, R. Hernández-Saucedo, RMCA **8**(1), 1-12 (2017)
12. J. C. Negrete, ASA, **2**(7) (2018)
13. F. Rojas, V. Orsag, RIIARN, **4**(2), 7-19 (2017)
14. G. Ruiz-García, H. Hagras, H. Pomares, I.R. Ruiz, IEEE TFS, **27**(12), 2381-2395 (2019).
15. Maxbotix[Datasheet], https://www.maxbotix.com/documents/XL-MaxSonar-WR_Datasheet.pdf (2020)
16. Simcom[Datasheet],https://simcom.ee/documents/SIM800L/SIM800L%20SPEC170914.pdf
17. CONELEC, Atlas solar del Ecuador, http://energia.org.ec/cie/wp-content/uploads/2017/09/AtlasSolar.pdf
18. R. Manning, ICE, **24**, 179-207 (1895).
19. J. C. Ayala López, R. Albóniga Gil, RCTA, **24**, 91-99 (2015)
20. Y.A. Rivas-Sánchez, M.F. Moreno-Pérez, J. Roldán-Cañas, IA, **23**(1), 19-31(2019)
21. Z.H.A.O. Liang, L.I. Feng, CED, **9**, 2552-2554 (2005)
22. C.D. Radicelli-García, M. Pomboza-Floril, L. Cepeda-Astudillo, DYNA, **85**(204), 319-324 (2005).
23. A.R. Mossande, O. Brown Manrique, A. Mujica Cervantes, C. Mata Rodriguez, I. Osorio León, RCTA, **24**(2), 11-17 (2015)
24. A.O. Afolayan, GCH, **2**(3) (2018)
25. K.A.M. Godoy, P.L.D. Angarita, RI, **39**(86), 131-156 (2015)
26. S. Mohammadi, A.V. Nadaraja, D.J. Roberts, M.H. Zarifi, SA:P, **303**, 111663 (2020)
27. I. Sepúlveda, R. Molina, M. Delgado, J. Guerrero, EA, **51**, 185-206 (2015)
28. V.M. Vázquez, E.C. Poyato, J.A. Díaz, J.L.M. Lugo, M.L. Hernández, RFCA, **49**(2), 149-168 (2017)
29. C. Quiroga Manrique, D. Vallejo Bernal, RCA, **55**(1), 59-89(2019)
30. S. Barkhordari, S.H. Shahadany, S. Taghvaeian, A.R. Firoozfar, J.M. Maestre, CEA, **168**, 105122 (2020)

Acoustic Safety Management at the Locksmith's Workplace Mechanic-Repairman to Replace the Filter Material in the Technological Process of Obtaining of Apatite Concentrate

Ekaterina Chalovskaya[1,*], *Tatiana* Kaverzneva[1,2], *Igor* Skripnik[3], *Vladimir* Senchenko[4] and *Ilya* Klochihin[1]

[1]Peter the Great St.Petersburg Polytechnic University, Higher school of technosphere safety, RU-195251, Saint-Petersburg, Russian Federation
[2]St Petersburg Academic University, RU-194021, Saint-Petersburg, Russian Federation
[3]Saint-Petersburg University of State Fire Service of Emercom of Russia, RU-196105, Saint-Petersburg, Russian Federation
[4]Volgograd branch of Rostelecom PJSC, RU-400048, Volgograd, Russian Federation

Abstract. In this work, measures are being developed to reduce the noise level at the workplace of a mechanic-repairman to replace filter cloth. Based on the results of production control and a thorough assessment of working conditions, the source of noise was determined to be the VDN-17 fan. The analysis of Russian and foreign studies in the field of the negative impact of noise on the human health and methods of dealing with industrial noise has been carried out. As a result, a technical method of reducing negative impact of noise was selected and the acoustic efficiency of the muffler for the VDN-17 fan was calculated.

1 Introduction

According to the Federal State Statistics Service, the percentage of workers employed in harmful and hazardous conditions to the total number of employees in 2018 decreased in relation to 2017 by 2% [1].

The Federal State Statistics Service data for 2018 showed that the most harmful and hazardous activities are: mining - 54.7% and manufacturing - 43.2% [1]. Thus, workers employed in the mining industry are constantly exposed to harmful factors, one of the most dangerous is exposure to occupational noise, which leads to diseases of the cardiovascular system and hearing impairment. According to the results of the study [2], it was revealed that the mining industry ranks first in the prevalence of hearing diseases. This is confirmed by many studies of scientists and the Federal state statistics service, according to data for 2018, the working conditions of employees of the mining and processing industry are characterized by the presence of harmful factors: chemistry - 10.1%, noise - 35.9%, as well as the severity of the labor process - 38,5% [1], [3]. Analysis of the impact of harmful and dangerous factors shows that workers in the mining industry have an average and high risk of developing occupational diseases [4].

According to the data of the state mining supervision for 2018, the total number of workers employed at hazardous production facilities in the mining industry was 226 thousand people, and the volume of rock mass production increased by 5% compared to the previous reporting period, respectively, accidents and injuries to workers occur. As can be seen from Appendix 3, the number of injured workers is decreasing every year, but the number of accidents has been stable over the past 4 years. According to the report of the Federal Service for Environmental, Technological and Nuclear Supervision for 2018, the number of deaths decreased by 1.6 times compared to 2017 [5]. A study based on accident statistics over a ten-year period found that 85% of all injuries are associated with mining equipment, and 90% of fatalities with mobile equipment, including hand tools. Truck operators and repairmen are at the highest risk [6].

This research examines the workplace of a repairman for replacing filter cloth (hereinafter referred to as a repairman in the Polar Division). This profession in various fields of industry is often associated with the negative impact of the following factors: industrial noise, vibration, microclimate parameters, exposure to harmful substances, the severity and intensity of the labor process [3,6, 7]. The constant exposure to occupational noise has a negative effect on the state of the nervous system, such as irritability, headaches, fatigue, sleep disturbance, memory impairment and, accordingly, this leads to occupational diseases [8].

The aim of the study is to develop measures to reduce the noise level at the workplace of the mechanic-repairman to replace the filter cloth. To solve it, following main tasks:

1. Analysis of noise affecting the mechanic-repairman;
2. Analysis of noise sources;
3. Development of measures to reduce noise exposure;

have been accomplished.

* Corresponding author: chalovskaya.ek@edu.spbstu.ru

2 Methods

2.1 Noise reduction techniques

When considering industrial noise as a biological factor, it was found that the noise affecting the employee's body leads to a decrease in performance.

Hearing loss from exposure to noise is the second most common condition in the mining industry. In the study [9], it is proposed to reduce the noise level from two rotating cutting drums by increasing the rigidity of the outer plates, which is the most practical and durable. This method was able to organize noise reduction by 3 dB. The most effective is the fight against noise at its source [10].

The author in the article argues that the effect of increased noise entails an increase in the costs of an employee's efforts to perform work. The use of various noise absorbers is proposed, including their designs. The results of the study [11] claim that every 5-10 years the increase in noise emission is about 5 dBA. According to the results of the analysis, diseases most often occurred in persons with a long work experience. The author proposes to use noise protection structures in order to combat occupational diseases. We are talking about the development of noise-absorbing panels on the basis of the laboratory of the Department of Technosphere Safety of the Federal State Budgetary Educational Institution of Higher Professional Education IzhSTU named after M.T. Kalashnikov ".

According to the Ministry of Labor and Social Protection, sensorineural hearing loss is detected in 30-40% of all recorded occupational diseases. The article [12] discusses active headphones with the ability to adjust the sound volume, that is, to amplify quiet sounds and at the same time to make excessively loud sounds quieter, and they can also have built-in walkie-talkies for the convenience of employees' communication.

Thus, it is possible to implement both technical and organizational measures to combat the negative impact of noise. The most effective technical measure is to tackle the source of the noise, but this is often a complex process. The most common is the use of collective means of protection, such as installing noise-absorbing panels, this is also a very costly process. The most frequent and inconvenient for workers is the use of personal protective equipment, such as headphones or ear plugs [13].

2.2 Analysis of noise sources based on production control fata at the workplace of a mechanic-repairman to replace filter cloth

In industrial shops, many machines and devices are used that generate high levels of noise, which is the main cause of hearing loss.

At the workplace of a repairman, according to the results of production control, an increased noise level was revealed at the marks of disk and belt vacuum filters. This is due to the fact that these control points are located near the noise source - the fans of the VDN-17 drying drums. Centrifugal blowing fans of single inlet type VD are designed to supply air to the furnaces of steam boilers [14]. The overall and connecting dimensions of the smoke exhausters are shown in Appendix 6, and the technical characteristics in table 1.

Table 1. Technical characteristics of VDN-17 draft machine

Types of machines	Power, kWt	Rotation frequency, min^{-1}	Productivity, m^3 / h	Total pressure, Pa	Weight, kg
VDN-17	160	980	73000	282	3100

2.3 Method and calculation of an active silencer made of basalt fiber

One of the methods of collective protection is the use of mufflers, which are used to reduce aerodynamic noise. According to the principle of operation, noise mufflers are: active type (due to the sound-absorbing material), reactive type (due to the reflection of a part of the sound energy back to the source) and combined [15].

The paper proposes to use an active type muffler, which is effective at medium and especially high frequencies. An active muffler in the form of a rectangular channel will be located at the outlet of the VDN-17 fan along the perimeter of the pipeline. the fan is fastened to the unloading chamber in the form of a rectangular pipe with an internal width of -850 mm and a length of 635 mm. To prevent the absorbing material from blowing out, coverings with metal mesh, perforated material, etc. are used.

The muffler is a thin-walled rectangular parallelepiped with a length of 850 mm, a width of 632 mm and a depth of 1100 mm, the inner part of which is sheathed with a heat-resistant mesh with a wire thickness of 1.6 mm to prevent material blowing out.

The calculation of noise reduction (acoustic efficiency) when using an active silencer for a rectangular duct is carried out according to the following formula (1):

$$\Delta L = 1,1\varphi(\alpha)\cdot P\cdot l/S \qquad (1)$$

where $\varphi(\alpha)$ - conditional sound absorption coefficient of the muffler lining, depending on the sound absorption coefficient α of the used adsorbent (table 2);
l and P - respectively the length and diameter of the cross-section of the muffler channel, m
S - channel cross-sectional area, m^2 [16].

Table 2. Values of the conditional sound absorption coefficient of the muffler lining depending on the sound absorption coefficient α of the used absorbent

α	0,1	0,2	0,3	0,4	0,5	0,6	0,7	0,8	0,9	1
$\varphi(\alpha)$	0,1	0,2	0,35	0,5	0,65	0,8	1,2	1,6	2,0	4

Using formula 1, we calculate the acoustic efficiency of an active muffler for VDN-17:

$$\Delta L=1,1\varphi(\alpha)\cdot 2(0,85+0,632)\cdot 1,1/0,85\cdot 0,632 \qquad (2)$$

$$\Delta L=6{,}68\ \varphi(\alpha) \quad (3)$$

When c hoosing a s ound-absorbing m aterial f or an active muffler, i t is necessary to take into acco unt properties s uch a s: flammability, hazardousness t o t he health o f workers, s ervice l ife, et c. T he m aximum operating t emperature of glass wool is 500 ° C , s lag wool - 300 ° C, b asalt fiber - 1000 ° C . B ased o n practical d ata for the use of active mufflers, depending on th e o perating c onditions, a material was c hosen - basalt wool. T hey were selected o n t he b asis t hat gases with a n e levated t emperature of a bout 900 ° C pa ss through VDN-17.

During operation of the VDN-17 blower fan, intense noise i s e mitted i nto t he working space, t he frequency characteristic of which is shown in table 3.

Table 3. Frequency response of noise from VDN-17

Index	Measurement area	Average geometric frequencies of octave bands, Hz						
		125	250	500	1000	2000	4000	8000
Noise from the VDN-17 fan	Injection	112	117	116	113	111	107	101
	Suction	108	113	112	109	107	103	97
	Around the body	103	107	106	103	100	95	89

Since t he muffler is i nstalled a t th e fan o utlet, it i s necessary t o cal culate i n t wo zo nes: d ischarge a nd around the casing.

3 Results

The in itial d ata a nd r esults o f c alculating t he effectiveness of measures for installing a silencer made of a semi-rigid " PP-80" b asalt s lab i n acco rdance with GOST 9573-2012 with a sound absorber thickness of 30 mm in the injection zone are summarized in table 4.

Table 4. The results of calculating the effectiveness of measures to install a silencer made of basalt fiber in the injection zone

Index	Average geometric frequencies of octave bands, Hz						
	125	250	500	1000	2000	4000	8000
Sound Pressure Level from the VDN-17 fan, dB	112	117	116	113	111	107	101
Sound absorption coefficient α	0,08	0,3	0,64	0,89	0,95	0,81	0,73
Conditional sound absorption coefficient $\varphi(\alpha)$	0,08	0,35	0,96	1,96	3	1,64	1,32
Silencer acoustic efficiency, dB	1	3	7	14	22	12	10
Sound Pressure Level of VDN-17 fan after installing a silencer on it, dB	111	114	109	99	89	95	91
Standard values of the Sound Pressure Level, dB	87	82	78	75	73	71	69
Exceeding the norms after installing the muffler, dB	24	32	31	24	16	24	22

First, it is necessary to determine t he s ound absorption coefficient of the semi-rigid basalt slab PP-80 from t he reference d ata. Using table 2, we calculate the value o f t he c onditional s ound a bsorption c oefficient of the muffler l ining de pending on t he s ound a bsorption coefficient α. Then we calculate the acoustic efficiency of t he muffler d epending on t he g eometric mean frequency using formula 3. For example, at 125 Hz, the acoustic efficiency of the muffler will be:

$$\Delta L_{(125)} = 6{,}68\ \varphi(\alpha)_{(125)}=6{,}68\cdot 0{,}8\approx 1\text{(dB)} \quad (4)$$

where $\varphi(\alpha)_{(125)}$ – the c onditional s ound a bsorption coefficient o f the muffler l ining, which depends on the sound a bsorption c oefficient α of t he s emi-rigid b asalt PP-80 slab.

Thus, t he s ound p ressure l evel o f t he V DN-17 f an after in stalling a muffler o n i t i n the d ischarge zo ne at 125 Hz:

$$L_{p2(125)}= L_{p1(125)}- \Delta L_{(125)}=112-1=111\text{(dB)} \quad (5)$$

where $L_{p1(125)}$ – the sound pressure level of the VDN-17 fan b efore i nstalling a muffler o n it in th e d ischarge zone.

Exceeding t he standard v alues of sound pr essure levels after installing a muffler at 125 Hz:

$$\Delta = L_{p2(125)}-L_{stand\ (125)}=111-87=24\text{(dB)} \quad (6)$$

where $L_{stand\ (125)}$ – the standard va lue o f t he sound pressure level at 125 Hz.

Similar cal culations for t he g eometric mean frequencies of the octave bands of 250, 500, 1000, 2000, 4000 and 8000 Hz were carried out in the excel program, the results are shown in Table 4. For the geometric mean frequencies o f the oc tave ba nds o f 31. 5 a nd 63 H z, calculations were not performed, since it was previously said that the active type muffler is effective at medium and high frequencies.

Based o n th e c alculation r esults, we will b uild th e noise characteristic in the injection zone before and after the installation of the muffler (Fig.1).

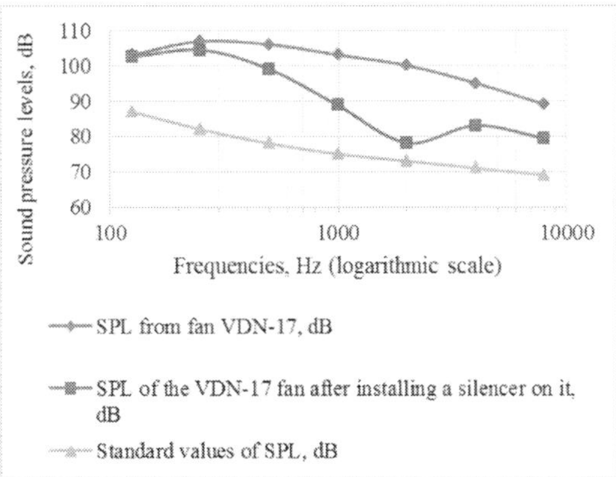

Fig. 1. Noise characteristics in the discharge zone before and after installation of the muffler

Similar calculations were carried out to assess the effectiveness of measures to install a silencer made from a semi-rigid " PP-80" basalt slab in accordance with GOST 9573-2012 with a sound absorber thickness of 30 mm in the measurement zone around the body, the results are shown in table 5.

Table 5. Results of calculating the effectiveness of measures for installing a basalt fiber muffler in the measurement zone around the body

Index	Average geometric frequencies of octave bands, Hz						
	125	250	500	1000	2000	4000	8000
Sound Pressure Level from the VDN-17 fan, dB	103	107	106	103	100	95	89
Sound absorption coefficient α	0,08	0,3	0,64	0,89	0,95	0,81	0,73
Conditional sound absorption coefficient $\varphi(\alpha)$	0,08	0,35	0,96	1,96	3	1,64	1,32
Silencer acoustic efficiency, dB	1	3	7	14	22	12	10
Sound Pressure Level of VDN-17 fan after installing a silencer on it, dB	102	104	99	89	78	83	79
Standard values of the Sound Pressure Level, dB	87	82	78	75	73	71	69
Exceeding the norms after installing the muffler, dB	24	32	31	24	16	24	22

Based on the results of the calculation, we construct the noise characteristic in the measurement zone around the body before and after the installation of the muffler (Fig.2).

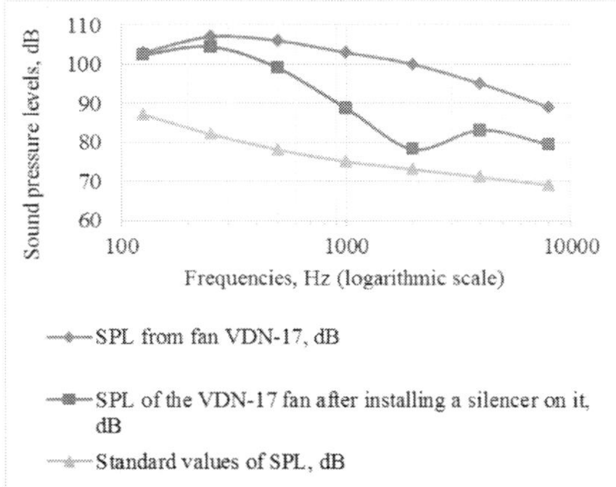

Fig. 2. Noise characteristic in the measuring area around the body before and after installing the muffler

As a result of the calculation, the acoustic efficiency of the muffler made of basalt semi-rigid slab PP-80 was determined and the sound pressure levels in the measurement zones were calculated: injection and around the body. Based on Figures 2 and 3, it can be seen that the muffler is most effective in the high frequency range.

The muffler is made in the form of a rectangular channel and is installed at the outlet of the VDN-17 fan. In Figure 4, the number 8 denotes the VDN-17 fan, and the location of the muffler is indicated in green.

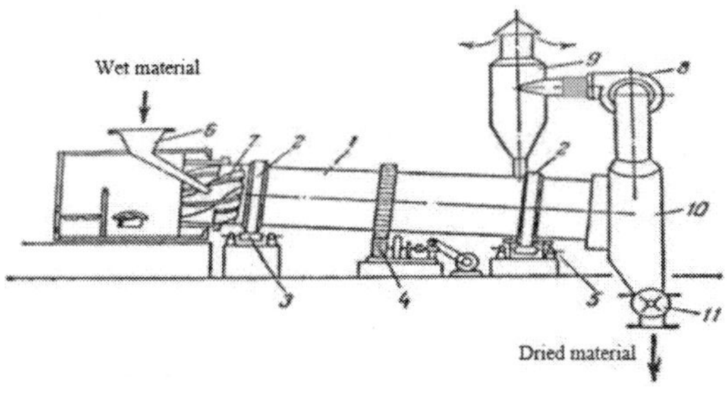

Fig. 3. Principle of operation of the drying drum (1 - drum; 2 - bandages; 3 - support rollers; 4 - transfer; 5 - support rollers; 6 - feeder; 7 - blades; 8 - fan; 9 - cyclone; 10 - unloading chamber; 11 - unloading device)

4 Discussion

The study is dedicated to the development of measures to reduce the negative impact of industrial noise on the fitter-repairman to replace the filter cloth, during which the sources of noise generation were identified and methods of dealing with it were proposed.

During the work, on the basis of statistical data and scientific research, it was revealed that the mining industry is the most harmful and hazardous. World practice shows that workers in this industry belong to groups of increased risk of negative impact of industrial noise, which leads to diseases of the hearing organs, nervous system and others [17, 18].

This work confirms the relevance and necessity of developing measures to reduce the negative impact of increased industrial noise on workers in the mining industry.

References

1. Federal state statistics service. [Electronic source]. https://www.gks.ru/working_conditions (accessed on 01.02.2020)
2. E. Kerns, E.A. Masterson, C.L. Themann, G.M. Calvert, American Journal of Industrial Medicine, **61**(6), 477 -491 (2018) https://doi.org/10.1002/ajim.22833
3. L.K. Karimova, M.K. Gainullina, O.V. Grebeneva, E.R. Shay-Khlislamova, L.M. Mavrin, "O sostoyanii uslovij truda rabotnic gorno-obogatitel'noj fabriki", 22 (2017) [in Russia]
4. T.V. Gorokhova, E.M. Egorova, N.P. Setko, E.B. Beilina. "Osobennosti professional'noj zabolevaemosti rabotnikov gorno-obogatitel'nogo proizvodstva", 98 (2018) [in Russia]
5. Annual report of the Federal service for environmental, technological and nuclear supervision for 2018. [Electronic source]. https://zinref.ru/000_uchebniki/04600_raznie_7/215_otchet_po_ekoloc_atomnome_nadzoru_2018/008.htm (accessed on 01.02.2020).
6. E. Stemn, Safety and Health at Work, **10(2)**, 151 - 165 (2019) https://doi.org/10.1016/j.shaw.2018.09.001
7. A.G. Kochnev, V.Ya. Bershadskij, Molodezh i nauka, **8**, 90 (2018) [in Russia]
8. L.K. Karimova and other, Zdorov'e naseleniya i sreda obitaniya, **3**(288), (2017) [in Russia]
9. E.A. Masterson, J.A. Deddens, C.L. Themann, S. Bertke, G.M. Calvert, American Journal of Industrial Medicine, **58**(4), 392-401 (2015).
10. A.P. Savel'ev, S.V. P'yanzov, A.N. Skvorcov, Al'manah mirovoj nauki, **2-1**, 142-143 (2015) [in Russia]
11. D.V. Parahin, Vystavka innovacij-2015 (osennyaya sessiya), 66-67 (2015) [in Russia]
12. A.S. Pirogov, M.I. Spel'nikova, Ugol', **6**, 54-55 (2015).
13. N.V. Rumyanceva, T.T. Kaverzneva, *Uchebno-metodicheskoe posobie: Opredelenie norm vydachi sredstv individual'noj zashchity metodicheskie rekomendacii po provedeniyu prakticheskih zanyatij po discipline «Ohrana truda»* (Spb.: Izd-vo Politekhn., un-ta, 23, 2019)
14. Tekhnicheskie harakteristiki i opisanie. Tyagodut'evye mashiny VDN I DN (VDN 6,3, DN 6,3, VDN 8, DN 8, VDN 9, DN 9, VDN 10, DN 10, VDN 11,2, DN 11,2, VDN 12 5, DN 12,5, VDN 13, DN 13, VDN 15, DN 15, VDN 17, DN 17, VDN 19, DN 19, VDN 21, DN 21. GK "Novye tekhnologii" : Ventmash. [Electronic source]. https://ventmash.nt-rt.ru/images/manuals/1vdn.pdf (accessed on 25.02.2020)
15. N.A. Evstigneeva, S.V. Karev, MADI (GTU), **44** (2007)
16. *Proizvodstvennaya sanitariya i gigiena truda. Snizhenie aerodinamicheskogo shuma glushitelyami: Metodicheskie ukazaniya po vypolneniyu raschetno-graficheskoj raboty.* (Nacional'nyj mineral'no-syr'evoj universitet «Gornyj», Sost. V.V. Smirnyakova, Spb., 2012).
17. S.K. Karabalin and other, Mezhdunarodnyj zhurnal prikladnyh i fundamental'nyh issledovanij, **12**, 71-75 (2019) [in Russia]
18. N.N. Mazitova and other, Medicina truda i promyshlennaya ekologiya, **2**, 48-53, (2017) [in Russia]

Mathematical Estimation Methods and Models for Industrial Companies

Olga Stikhova*

Moscow State Technological University "STANKIN", RU-127055, Moscow, Russia

Abstract. The collateralized debt obligations and credit default swaps applications are shown in this paper. The industry obligations secondary market risk estimation methods are considered in this work. The new methods taking into account statistically significant parameters for industrial credit derivatives portfolio are offered for single-name investment risks numerical experiments realization. The mathematical estimation of tranche were shown. The single and multiple name default obligations necessary mathematical modeling methods and formulae for the industrial materials manufacturers derivative credit tools market are shown. It is determined that the portfolio of synthetic debt tools is made of the given parameters. The task of a loss derivative tranches mathematical estimation is solved. Late defaults raise the equity tranches payment required sums with high spreads, early defaults reduce. Also the functional characteristics required for an estimation huge debts problem solving are partly considered in this paper. The problem of the default modeling for market tools and numerical simulation of the obligations influence on conditions of current bistability mode are shown here. Some credit derivatives of industrial manufacturers are demonstrated in the modeling process of default as an example. It is found that the model is an additional factor help us to estimate the default opportunity.

1 Inverse Distribution

The present work is concerned with the problem of the default modeling for market tools and numerical simulation of the obligations influence on conditions of current bistability mode. By example some credit derivatives of industrial manufacturers are demonstrated in the modeling process of default [1]. One of the market basic products for hedging credit risk is credit default swap (CDS). CDS forms the basis for more complex structures. In any credit event the credit protection seller carries out a rate CDS payment value for the protection buyer benefit for the default emitter obligation determined in the agreement. Let $(X_n, n \in \square)$ stationary casual sequence given on some filtered stochastic space $(\Omega, \Im, (\Im_n)_{n\geq 0}, P)$. Suppose that X_n behavior dynamics is described as in [2]. Consider that the conditional average μ_n and the volatility σ_n are $\Im_{n-1}$ measurable. Let's take $F_X(x)$ of (X_n) as distribution process function and $F_{X_{n+1}+\ldots+X_{n+k}|\Im_n}(x)$ is distribution conditional function for subsequent k measurements on a $\Im_n$ filtration. We find quantile of such distributions for where N can be large enough and enter also conditional mathematical expectation. To calculate them we should construct the appropriate models of dynamics

1.1 The problem statement

1.1.1 Dynamic models

The distribution conditional function quantile expression $F_{X_{n+1}+\ldots+X_{n+k}|\Im_n}(x)$ is

$$x_q^n(k) = \inf(x \in \square : F_{X_{n+1}+\ldots+X_{n+k}|\Im_n}(x) \geq q). \quad (1)$$

Take also a conditional mathematical expectation

$$S_q^n(k) = E(\sum_{j=1}^{k} X_{n+j} | \sum_{j=1}^{k} X_{n+j} > x_q^n(k), \Im_n) \quad (2)$$

As

$$F_{X_{n+1}|\Im_n}(x) = P(\sigma_{n+1}Z_{n+1} + \mu_{n+1} \leq x | \Im_n) = F_Z((x - \mu_{n+1})/\sigma_{n+1}), \quad (3)$$

then an expression of x_q^n and S_q^n are

$$x_q^n = \mu_{n+1} + \sigma_{n+1} z_q,$$

$$S_q^n = \mu_{n+1} + \sigma_{n+1} E(Z | Z > z_q), \quad (4)$$

* Corresponding author: olgitast2011@mail.ru

where z_q - distribution function quantile $F_Z(z)$. To calculate (4) we should construct the appropriate dynamic models for μ_n and σ_n. We offer to use model *AR(1)*[1] for conditional average μ_n and model *GARCH(1,1)* [2] for volatility σ_n

$$\mu_n = \phi X_{n-1},$$

$$\sigma_n^2 = \alpha_0 + \alpha_1 \varepsilon_{n-1}^2 + \alpha_2 \sigma_{n-1}^2, \quad (5)$$

where $\varepsilon_n = X_n - \mu_n$, $\alpha_0 > 0, \alpha_1 > 0, \alpha_2 > 0,$ $\alpha_2 + \alpha_1 < 1,$ $|\phi| < 1.$

The $\tilde{\theta} = (\tilde{\phi}, \tilde{\alpha}_0, \tilde{\alpha}_1, \tilde{\alpha}_2)$ parameters estimations can be received by a maximum likelihood method. The conditional average $(\tilde{\mu}_1, ..., \tilde{\mu}_n)$ and volatility $(\tilde{\sigma}_1, ..., \tilde{\sigma}_n)$ estimations can be calculated by consecutive substitution in expressions (5) at the appropriate choice of initial meanings. Then the excess meanings and conditional average estimations are

$$(z_1, ..., z_n) = \left(\frac{x_1 - \tilde{\mu}_1}{\tilde{\sigma}_1}, ..., \frac{x_n - \tilde{\mu}_n}{\tilde{\sigma}_n} \right)$$

$$\tilde{\mu}_{n+1} = \tilde{\phi} x_n, \ \tilde{\sigma}_{n+1}^2 = \tilde{\alpha}_0 + \tilde{\alpha}_1 \tilde{\varepsilon}_n^2 + \tilde{\alpha}_2 \tilde{\sigma}_n^2, \quad (6)$$

where $\tilde{\varepsilon}_n = x_n - \tilde{\mu}_n$.

The distribution function considerably differs from normal distribution, assumed in standard GARCH-model as the financial temporary series analysis has shown [3].

1.1.2 Loss industry obligations secondary market

The loss check of the industry obligations of the secondary market and the estimation methods are submitted in this paper, the concept "derivative" of standard financial actives is considered. The collateralized debt obligations (CDO) and credit default swaps (CDS) are appreciated, and also their estimation problem solving functional characteristics are considered.

Necessary modeling methods and mathematical formulae of the single and multiple name credit default swaps are shown for the market of the industrial materials manufacturers. It is determined that the portfolio of synthetic debt tools is made of the given parameters. The CDS spread is fixed so its meaning of a bonus part and the parts of protection were equal:

$$s = \frac{E[\int_0^T (1-R_t)\lambda_t e^{-\int_0^t (r_u+\lambda_u)du} dt]}{E[\sum_{i=1}^M e^{-\int_0^{T_i} (r_t+\lambda_t)dt} \Delta_i + \sum_{i=1}^M \int_{T_{i-1}}^{T_i} (t-T_{i-1})\lambda_t e^{-\int_0^t (r_u+\lambda_u)du} dt]} \quad (7)$$

Default intensity: $\lambda = \frac{1}{\Delta} \ln(\frac{s\Delta}{(1-R)} + 1)$

1.2 Tranches estimation

The single-name default mathematical model is

$$\tau = \inf\{t : N_t = 1\},$$

where τ – default time, $N = (N_t)_{t \geq 0}$ – default process, $\lambda = (\lambda_t)_{t \geq 0}$ - right non-negative continuous and left limited function of default intensity.

The default probability before the time t, in view of the default absence of the emitter and market information up to the moment of time s:

$$P_{def}(t|s) = 1 - E[e^{-\int_s^t \lambda_u du} | F_s] \quad (8)$$

The single-name manufacturer tranches inverse distribution estimation methods are considered. The task of a loss derivative tranches mathematical estimation is solved [3]. Late defaults raise the equity tranches payment required sums with high spreads, early defaults reduce.

Spread CDO:

$$s = \frac{\int_0^T E[e^{-\int_0^s r_u du} dL_s^{(K_1,K_2)}]}{\sum_{i=1}^M E[e^{-\int_0^{T_i} r_u du} \Delta_i \frac{2 - L_{T_{i-1}}^{(K_1,K_2)} - L_{T_i}^{(K_1,K_2)}}{2}} \quad (9)$$

expected losses:

$$EL_{(K_1,K_2)}(t) = \frac{1}{K_2 - K_1} \int_{K_1}^1 (\min(x, K_2) - K_1) dF(t,x) =$$

$$= \frac{1}{K_2 - K_1} \left(\int_{K_1}^1 (x - K_1) dF(t,x) - \int_{K_2}^1 (x - K_2) dF(t,x) \right) \quad (10)$$

distribution of losses :

$$F_\infty(x) = P[X \leq x] = \Psi\left(\frac{\sqrt{(1-\rho)}\Psi^{-1}(x) - K}{\sqrt{\rho}} \right) \quad (11)$$

2 Modelling

In the given work the new mathematical estimation formulae for inverse distribution of tranches are developed. The multiparametrical model based on generalized hyperbolic copula with generalized hyperbolic borders (CGH), taking into account the various mixed distribution and thus, various boundary areas distributions, keeping desirable one-dimensional distribution GH properties is offered.

The GH-copula density $f_{GH_d}(x,\lambda,\chi,\psi,\mu,\Sigma,\gamma)$ depends on some parameters such as K_λ - the modified Bessel function of the third rank with an applicable domain range □ , heavy tail λ, χ– scaling rang, ψ–form, μ– shift, Σ– covariation matrix, γ– skewness.

The parameters $\mu \in$ □ ; $\chi \geq 0, \psi \geq 0$, if $\lambda < 0$; $\chi > 0, \psi > 0$, if $\lambda = 0$; $\chi \geq 0, \psi > 0$, if $\lambda > 0$ and $|\Sigma| := 1$ for recognition. The vector X □ $GH_d(\lambda,\chi,\psi,\mu,AA',\gamma)$ is defined as

$$X \overset{d}{=} \mu + \omega\gamma + \sqrt{\omega} AZ \tag{12}$$

where $Z \overset{d}{=} N_k(0, I_k)$, ω □ $GIG(\lambda,\chi,\psi)$ - positive scalar casual variable, not dependent from matrixes $Z, A \in$ □ $^{d \times k}$ and μ,γ are belong to □ ^{d}.

We assume, that the indefinitely large portfolio with an active profit corresponds to copula model CGH, table 1.

Table 1. Various distributions tranches DJ iTraxx estimation.

%	Market	Gauss	t(4)-t(4) t(3)- t(3)	NIG1 NIG2	CGH
0-3	24.3%	24.3%	24.3%	24.3 %	24.3%
3-6	157.3bp	238.7 bp	167.3bp 148bp	157.3 bp 157.3 bp	157.3bp
6-9	47.1bp	59.01 bp	37.21bp 44.7bp	37.6b p 37.7b p	48.1bp
9-12	21.7bp	18.9b p	28.98bp 27.8bp	30.1b p 30.23 bp	21.7bp
12-22	11.73bp	8.78b p	16.66bp 17.7bp	17.73 bp 17.1b p	11.7bp
absol ute error	24.3%	115,3 bp	31.85bp 25.81bp	21.34 bp 21.75 bp	8.3bp
corre latio n	157.3bp	17.4%	32.71 bp 27.81 bp	16.3 % 15.8 %	23.9%

2.1 Research process

Take $\Omega(x) = K - \upsilon F_{GH}^{-1}(x) - m$ and contingent loss $Y=y$, part of the default papers in portfolio X and $p(y)$ we have:

$$\lim_{N\to\infty} P(X \geq x) = P\big(a(Y)Y \leq \Omega(x), Y \leq \theta\big) + \\ + P\big(a(Y)Y \leq \Omega(x), Y > \theta\big) \tag{13}$$

And then the distribution integrated function is

$$F_\infty(x) = 1 - \lim_{N\to\infty} P(X \geq x) = \\ = 1 - \left[F_{GH}\left(\min\left(\frac{\Omega(x)}{\sqrt{a}}, \theta\right)\right) + \\ + 1_{\frac{\Omega(x)}{\sqrt{b}} > \theta} \left(F_{GH}\left(\frac{\Omega(x)}{\sqrt{b}}\right) - F_{GH}(\theta) \right) \right] \tag{14}$$

During the research of the single-name quotation process it is marked that the distribution F (x) is not trivial and depends on the various emitters defaults in a base portfolio as the occurrence is disproportionate default losses results to heavy tails. In a case of CDO also it is necessary to take into account default time as the payments depend on cost which decreases during the contract period if default occurs [4].

It is offered the multiple names credit default probability model. In a case of n emitters during the time τ^i for each emitter i, i=1..., n the default indicator function $N_t^i = 1_{\{\tau^i \leq t\}}$ i=1, ..., n. The default time is determined as the first moment of time, when the function meaning $\left(e^{-\int_0^t \lambda_s^i ds}\right)_{t \geq 0}$ becomes less then U^i uniform distributed at [0,1]:

$$\tau^i = \inf\{t : e^{-\int_0^t \lambda_s^t ds} \leq U^i\}, \tag{15}$$

where the $U=(U^1,..., U^n)$ dependence fixed by copula C multidimensional function with standard one-dimensional margins.

The probability of non-defaults up to the moment of time t at default absence of the emitters $k \neq j$ up to the moment of time s, default of the emitter j at the moment of time u and the presence of the large number of information parameters is

$$\tilde{p}_{surv}^{i,-j(u)}(t|s) = \frac{\frac{\partial}{\partial x_j} C(\gamma_s^{-i,-j}, \gamma_t^i, \gamma_u^j)}{\frac{\partial}{\partial x_j} C(\gamma_s^{-j}, \gamma_u^j)} \tag{16}$$

2.2 Default probability

With the portfolio losses $x \in [0,1]$ and the default threshold value $K(t) = F_{GH}^{-1}(p(t))$, where $p(t)$ is neutral to risk default probability of each portfolio emitter

$$p_n(y) = F_{GH_i}\left(\frac{K - ay}{\sqrt{1-a^2}}\right) \qquad (17)$$

the portfolio loss distribution is:

$$F_\infty(t,x) = 1 - F_{GH_d}\left(\frac{1}{a}\left(K(t) - \sqrt{(1-a^2)}F_{GH_i}^{-1}(x)\right)\right) \qquad (18)$$

The tranche expected losses [5]:

$$EL_{(K_1,K_2)}(t) = \frac{1}{K_2 - K_1}\int_{K_1}^{K_2}(x - K_1)dF_\infty(t,x) + \left(1 - F_\infty(t,K_2)\right) \qquad (19)$$

In this work single-name default probability model is used to the premium part estimation, the saved payments are taken into account, default time distribution is calibrated under the CDS rates.
For the basic portfolio CDO valuation the multiple-name default probability model is used [5]:

$$V_{prot}(0) = E\left[\int_0^T e^{-\int_0^s r_u du} N_{tr} dL_s^{(K_1,K_2)}\right] \qquad (20)$$

and

$$V_{prem}(0) = \sum_{i=1}^{M} E\left[e^{-\int_0^{T_i} r_u du} s\Delta_i N_{tr} \frac{2 - L_{T_{i-1}}^{(K_1,K_2)} - L_{T_i}^{(K_1,K_2)}}{2}\right] \qquad (21)$$

The value of protection is determined by the size of the expected tranche default losses and the value of tranche premium part is calculated as the current price of all expected payments on spread, where an interval in payments is $\Delta_i = T_i - T_{i-1}$.

2.3 Dependence illustrations

To estimate the price of a base portfolio of CDS the single-name model of probability is used. The default time distribution was calibrated based on CDS. The credit default swap transaction is initiated at the zero moment of time with term of the expiration T., $\forall t \le T$. All bonus payments are accomplished in a time interval $0 < T_1 < \ldots < T_M = T$ at the expected sum of payments N and annual spread rate s and where $\lambda = (\lambda_t)_{t \ge 0}$ is a default intensity function of the appropriate emitter and discount is $E[e^{-\int_t^T r_s ds} | F_t] > 0$.

In case of the default before the term of protection, the seller of protection should carry out compensatory payment $(1 - R_n)N$, $\tau \le T$.

At an estimation of a bonus part the saved payments are taken into account.

At the estimation of the CDO tranche price of a base portfolio the multiple –names default probability model is used [4].

There are no yet credit cases the CDO emitter regularly pays the premium (insurance payment) to the tranche investor.

In a default case the investor (the seller of protection) pays to the CDO emitter (buyer of protection) the rate of the suffered losses sum.

The following premium (insurance payment) is paid in view of a deduction of the sum of the losses.

The tranche CDO protection price covers the losses of a base portfolio between two given threshold meanings K1 and K2.

N_p is an expected income of the appropriate portfolio, annual s tranche spread and $N_{tr} = (K_2 - K_1)N_p$ is an expected tranche income. Dates of payments are $0 < T_1 < \ldots < T_M = T$,

where T is a term of the obligations execution, t0 < t1 is a date of an estimation. The bonus payments are carried out in time tk for the period of payment from tk-1 up to tk.

The premium in time tk is paid at stipulated cost at the given moment of time. The emitter i, $1 \le i \le n$ in a portfolio has a default with intensity λi.

The proportional trance default loss $L_t^{(K_1,K_2)}$ up to the moment of time t (% from tranche nominal).

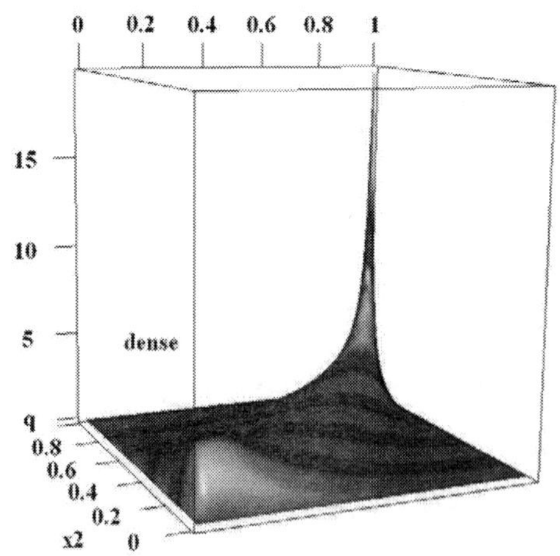

Fig. 1. GH-copula density $c_{GH}(u_1, \ldots, u_d)$.

Thus cost parameters CGH model of credit derivatives depend on the limit independence and tail dependence better than used earlier multidimensional models (fig.3).

For GH-copula the density $f_{GH_d}(x,\lambda,\chi,\psi,\mu,\Sigma,\gamma)$ is:

$$f_{GH_d} = \frac{(\sqrt{\psi\chi})^{-\lambda}\psi^{\lambda}\left(\psi+\gamma'\Sigma^{-1}\gamma\right)^{\frac{d}{2}-\lambda}}{(2\pi)^{\frac{d}{2}}|\Sigma|^{\frac{1}{2}}K_{\lambda}(\sqrt{\psi\chi})}\times \tag{22}$$

$$\times\frac{K_{\lambda-\frac{d}{2}}\left\{\sqrt{\left(\chi+(x-\mu)'\Sigma^{-1}(x-\mu)\right)\left(\psi+\gamma'\Sigma^{-1}\gamma\right)}\right\}e^{(x-\mu)'\Sigma^{-1}\gamma}}{\left\{\sqrt{\left(\chi+(x-\mu)'\Sigma^{-1}(x-\mu)\right)\left(\psi+\gamma'\Sigma^{-1}\gamma\right)}\right\}^{\frac{d}{2}-\lambda}}$$

where

K_{λ} - modified Bessel function of the third kind with the applicable domain □ ,

λ - tail index,

χ – scalability factor,

ψ – form parameter,

μ – shift parameter,

Σ– covariance matrix,

γ – asymmetry parameter.

Parameters $\mu \in$ □ ; $\chi \geq 0, \psi \geq 0$,

if $\lambda < 0; \chi > 0, \psi > 0$, if $\lambda = 0$; $\chi \geq 0, \psi > 0$,

if $\lambda > 0$ and $|\Sigma| := 1$ for an identification.

Vector X □ $GH_d(\lambda,\chi,\psi,\mu,AA',\gamma)$ is $X \overset{d}{=} \mu + \omega\gamma + \sqrt{\omega}AZ$, where $Z \overset{d}{=} N_k(0, I_k)$, ω □ $GIG(\lambda,\chi,\psi)$ - matrix $Z, A \in$ □ $^{d\times k}$ independent positive scalar random variable and μ,γ are parameters at □ ^{d}.

3 Results

The calculation methods of conditional quantile and the mathematical expectations of random stationary sequences for extreme values are considered in this work. For the description of the sequences characteristics dynamics we used parametrical models AR(1) and GARCH (1,1). The generalized Pareto distribution is used as modeling function of conditional distribution. It is shown by experiments, that the offered approach is more effective, than the standard normal distribution [6].

The single and multiple-name default process mathematical modeling task for some industrial production manufacturers is solved. The correlation between the emitter credit derivative default probability in a portfolio and the tranche market cost. The credit derivatives market researches and analysis results are shown that the important empirical property is the correlation dynamics influence on a default time [7-9].

The computing experiments on parameters modeling of the default credit derivatives for industrial manufacturers are carried out both with the generated samples and with the real data. As opposed to earlier multiparameter models these methods and formulae allow to simulate simultaneously tail dependence and various boundary areas distributions. The results verification is done and they have shown high efficiency [10]. The one and several industrial credit derivative debt default process mathematical modeling task having the increase importance for the enterprises, corporations and other structures is solved.

The correlation between credit derivative default probability of the emitter in a portfolio and the grown-up tranche market cost growth fall are established at the equity tranche cost increase. The research results and the credit derivative market analysis have revealed the important empirical property of credit derivative market indices costs - influence of correlation dynamics on the default time.

Based on the developed models the new estimation algorithm of default credit derivative for industrial corporations is offered. It allows to make qualitative market researches of the credit derivatives. Given algorithm can be used on practice for one and several emitters.

References

1. A. Pagan, J. of Empirical Finance, **3**(1), 15-102 (1996) https://doi.org/10.1016/0927-5398(95)00020-8
2. O.V. Stikhova, Vestnik MSTU «STANKIN». Scientific referenced journal, **4**(23), 81 (2012) [in Russia] ISSN 2072-3172.
3. O.V. Stikhova, Fund. Phys. Math. Prob. T-T. Sys. Mod., **16**, 97 (2014) [in Russia] ISBN 978-5-8037-0621-2.
4. R.F. Engle, Econometrica, **50**(4), 987-1007 (1982) https://doi.org/10.2307/1912773
5. O.V. Stikhova, Mnps-2015, 46 (2015) [in Russia] ISBN-978-5-8037-0655-7.
6. D.M. Mason, The Heritage Foundation, **2262** (2009)
7. J. Sidenius, V. Piterbarg, L. Andersen, International Journal of Theoretical and Applied Finance, **11**, 163–197 (2008)
8. A. Kalemanova, R. Werner, *A Short Note on the Efficient Implementation of the Normal Inverse Gaussian Distribution* (Risklab Germany. Nymphenburger Str., 2006)
9. A.J. McNeil, R. Frey, P. Embrechts, *Quantitative Risk Management: Concepts, Techniques and Tools* (Princeton University Press, 2005)
10. O.V. Stikhova, Fund. Phys. Math. Prob. T-T. Sys. Mod., **18**(120) (2017) [in Russia] ISBN 978-5-8037-0710-3.

Designing an Expert System for Recognizing the Emotional State of an Enterprise Employee

A.A. Sekin[1,*], N.A. Bychkova[2,*]

[1]Master's student in the direction 09.04.01, Moscow State Technological University "STANKIN", RU-127055, Moscow, Russia
[2]Ph.D. Associate Professor of Management and Informatics in Technical Systems, Moscow State Technological University "STANKIN", RU-127055, Moscow, Russia

Abstract. The emotional state of an employee of any enterprise influences both the efficiency of work performance and the quality and stability of the final result. Management of production processes, taking into account the monitoring of the emotional state of the employee, is a rather urgent task that allows minimizing the risks of deviations from the specified level of product quality and production safety. However, the quality of the assessment of this influence is currently subjective and is based both on the personal opinion and competences of the expert conducting the monitoring, and on the tools used by him for assessing the emotional state. At the same time, the use of modern intelligent automated methods and tracking systems will reduce the distortion of expert judgment. Creation of an expert system for analyzing the emotional state of employees of an enterprise will make it possible to recognize the emotions of a particular employee with a fairly high degree of accuracy, accumulate a system of knowledge and generate analytical conclusions and predictions of behavior based on it, compile an emotional portfolio of each employee and draw conclusions about the ability to perform a certain type of work and current states. This paper presents the concept of an algorithm of an expert system (hereinafter referred to as ES), which is able, on the basis of the data obtained on the individual methods of non-verbal expression of an employee's emotions, to assess the influence of his emotions on the quality of his work. The article reflects the results obtained in the framework of the implementation of the Agreement on research No. 05.601.21.0019 dated November 29, 2019.

1 Introduction

Today, the issue of using emotional monitoring systems (EM) at enterprises is becoming more and more relevant. Such systems are software tools for security services and employees who manage human capital, such as HR analysts and heads of structural divisions. With EM tools, you can minimize safety risks as well as increase production efficiency. The main focus of the emotional monitoring system is the development of emotional intelligence - the use of employees' emotions for a more effective solution [1-4]. The efficiency of an employee's work depends on the emotions he experiences. Some emotional states force you to focus on completing a task, thereby increasing its effectiveness, while others greatly distract from work, thereby worsening its results. Moreover, the strength of the impact of emotions on a person is not the same for everyone and varies depending on individual characteristics. Consequently, such a factor as emotions experienced by employees significantly affects the efficiency of the enterprise and must be taken into account [5, 6].

Working with the emotional intelligence of an employee is the task of his immediate supervisor, however, assessing the strength of the influence of emotions on the task performed by the employee is subjective and may not be very accurate. An expert system capable of recognizing the emotional state of an employee with a sufficiently high accuracy, generating analytical conclusions and predictions of behavior based on it, can help in solving problems of optimization and quality control of business processes. The developed ES for analyzing the emotional state of employees of the enterprise should accurately recognize the emotions of employees and take into account the individual characteristics of the non-verbal manifestation of emotions of each. The tasks of a system of this type are the need to identify and personify these features with the involvement of an expert, the subsequent analytical processing of the obtained statistical data based on the developed rules of the knowledge base, as well as the accumulation of experience and self-learning of the system.

2 Results and discussion

The projected expert system for analyzing the emotional state of enterprise employees can be represented in the form of four main subsystems:

* e-mail: [1]alex_ligh.t@mail.ru, [2]nabych@mail.ru

- The subsystem for collecting data on the individual characteristics of the manifestation of emotions of the company's employees works in the mode of inputting knowledge of the expert system and consists of two modules:
    - A neural network that registers and interprets changes in the respondent's facial expressions that occurred in response to the question asked;
    - The user interface that the respondent works with. This interface is a window with questions and answers to them.
- Knowledge base of the expert system. This subsystem is based on knowledge, which contains information about all the individual characteristics of the non-verbal expression of emotions of each specific employee. The data in the database are presented in the form of production rules (fact 1: if the corners of this employee's lips have lifted, then fact 2: he is more likely to be angry), where the variable will be the feature of the employee's expressions of emotion, and the value is the emotion most likely experienced in this case. The value affects the corresponding weight of the neural network that interprets facial expressions.
- A subsystem for the interpretation of emotions, which analyzes the experienced emotion from the input video stream, taking into account the features existing in the knowledge base. This subsystem is based on two convolutional neural networks, one of which interprets facial expressions of the muscles of the face, and the second interprets the context of the environment;
- A subsystem for assessing the quality of the enterprise personnel. This part of the program is a solver that, according to some static parameters and the results of the execution of the emotion interpretation module, analyzes the dependence of the employee's work quality on his emotions and generates on its basis behavior forecasts and conclusions about current abilities to perform a certain type of work. Reports on the results obtained are sent to the manager's user interface window.

To obtain data on individual ways of expressing certain emotions, the most effective option would be to conduct one interview with each of the employees. The knowledge acquisition module is responsible for this. During the interview, the respondent receives questions on the user interface window that should provoke an emotional reaction. This reaction will be recorded and interpreted by the convolutional neural network. The data received from it is converted into a form that can be added to the knowledge base of the system. The neural network, which is responsible for the interpretation of the incoming information, is combined and it receives data obtained from several sources at once: the facial expressions of the muscles of the interviewee's face and the context of his environment.

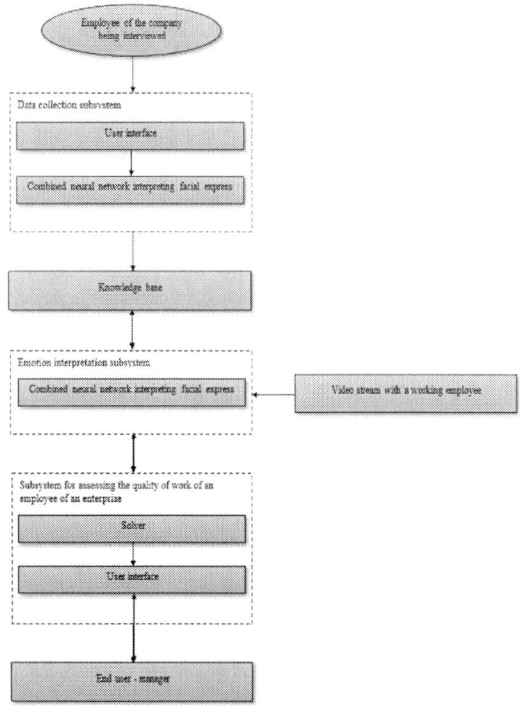

Fig. 1. General structure of the developed expert system

The emotion interpretation subsystem includes a neural network for detecting emotions, which is almost identical to the neural network from the knowledge acquisition module. This combined network analyzes the emotions of a working employee based on data on facial expressions and the environment obtained from the video stream, taking into account the data in the knowledge base of the system. The result of the work of this subsystem is sent to the subsystem for assessing the quality of the work of the enterprise personnel.

Based on the data obtained in the user interface of the system, the manager can draw conclusions about which job to assign this or that employee.

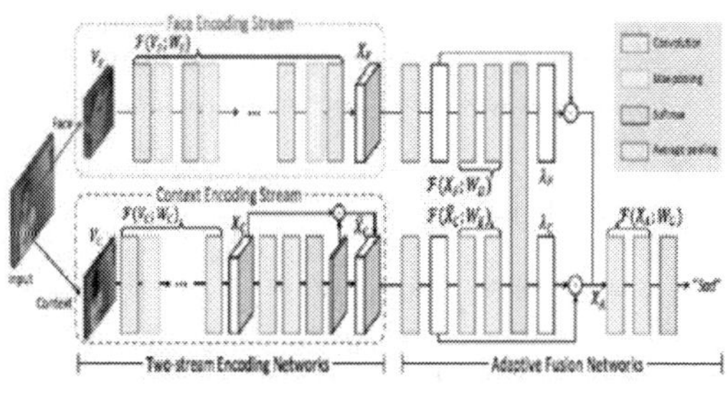

Fig. 2. Configuration of the CAER-Net consisting of dual-stream coding networks and adaptive merge networks

When developing a system whose functions include the recognition of emotions experienced by a person, it is necessary to take into account the high complexity of such a task caused by the variety of ways of expressing emotions in people, as well as a possible attempt on the part of a person to hide their emotions (even in people themselves, the recognition accuracy is relatively poor is large and varies greatly from person to person). For these reasons, you cannot rely on data from just one source (for example, facial expressions). A person has many ways to convey information about his emotional state to others, many of them are not accessible to consciousness, and it is very difficult to control them deliberately, such as narrowing / dilating the pupils. However, it is not necessary to take into account all the manifestations of emotions, a few will suffice. The criterion of fidelity in this case is accuracy. These facts indicate the possibility of correctly choosing the analyzed combinations of the manifestation of emotions and accurately determining a person's feelings.

As a result of the analysis of existing systems, it was decided that a neural network using a combined method of emotion recognition based on two convolutional neural networks will be used as an emotion interpretation module in the system. An example of such a neural network can be the Korean CAER-net, which is able to recognize emotions by facial expressions and the context of a person's environment, with an accuracy of 74.51%. In general, the architecture of this network is a system of two convolutional neural networks, one of which analyzes the face of a person in the picture, and the other, separately from the first, is the context of the environment. On the last layers, the data are combined and the result is given on them, which speaks of the emotions experienced.

To realize the ability to accurately recognize human emotions, in addition to the correct interpretation of facial expressions and analysis of the environment, the system needs to take into account the individual characteristics of the expression of emotions in each specific person. For the possibility of such accounting, it is necessary to collect data personally about each specific person. The best option is to observe the behavior of each individual person for some time, and it is important that there are external stimuli to which he will have to actively react. A suitable environment can be created by interviewing a person. In the process of conducting such an interview, the respondent should be asked questions that will provoke an emotional reaction in him, which will then be analyzed by the expert. Accordingly, the questions will have to evoke various emotional reactions (fear, joy, etc.). An expert is also needed to create a list of the questions themselves.

Due to the difficulty of identifying the most important features of the manifestation of emotions in a particular person, the module for obtaining data on the individual characteristics of the manifestation of emotions of company employees cannot be fully automated [7-10]. The data obtained by the expert must be converted into a form understandable for the system. That is, the expert himself will have to evaluate the strength of the expression of expressed emotions and highlight the key elements of the interviewee's facial expressions that indicate this or that experience, and then enter them into the ES interface.

3 Conclusion

The study allows us to conclude that the ability of the head of an enterprise to accurately recognize the emotions of other people and take them into account when assigning them to perform a particular task is an important factor that can significantly increase production efficiency [11]. However, the subjective nature of such an assessment can greatly affect its accuracy. Tools for analyzing the emotional state of enterprise employees can help in solving this problem. To build such a system, it is necessary that it be able to accurately recognize the emotions experienced by a person. The ES designed in this work takes into account the individual characteristics of the non-verbal manifestation of the internal emotional state of a particular employee, which can significantly increase the accuracy of emotion recognition. Based on the data obtained during the interpretation of the emotions of a particular employee, the expert system draws up a report in which it indicates the current state of the employee and draws conclusions about the ability to perform a certain type of work. This information can help a leader to more effectively manage business processes in the enterprise

The article reflects the results obtained in the framework of the implementation of the Agreement on research for the Ministry of Science and Higher Education of the Russian FederationNo. 05.601.21.0019 dated November 29, 2019

References

1. C.E. Izard, *The psychology of emotions* (New York; London: Plenum press, Cop. 1991) ISBN 0-306-43865-8.
2. P.V. Simonov, *The emotional brain. Physiology, neuroanatomy, psychology of emotions* (Moscow, 1981)
3. D. Evans, *Emotions. A brief introduction to emotions* (Moscow, 2008)
4. Context-Aware Emotion Recognition Networks [electronic resource]: scientific report / Yonsei University-electron. Text messages. - Seoul: Yonsei University, 2019. - access mode: https://caer-dataset.github.io
5. A.V. Shiller, Values and meanings, **4** (2018)
6. E.G Berger., I.Yu Bublik, Colloquium-journal, **10**(62) (2020)
7. E.V. Ryumina, A.A. Karpov, Scientific and technical bulletin of information technologies, mechanics and optics, **2** (2020)
8. L.M. Chervyakov, A.V. Oleynik, N.A. Bychkova, Vestnik of Mechanical Engineering, **11**, 80-82 (2019) [in Russia]
9. D.I. Starkov, N.V. Eliseeva, N.A. Bychkova, Theory and practice of project education, **3**(11), 93-95 (2019) [in Russia] ISSN: 2587-5922

10. A.P. Patrusheva, N.A. Bychkova, Theory and practice of project education, **3**(11), 91-93 (2019) [in Russia]
ISSN: 2587-5922
11. A.A. Rubtsov, N.V. Eliseeva, In the collection: Quality Management, Materials of the XIX International Conference "Quality Management", 224-229 (2020)
ISBN: 978-5-98604-757-7

Neopatrimonial Model of Public Administration and its Transformation in the Context of Global Uncertainty

Viacheslav Maracha [1, 2*], *Sergey* Bespalov [1]

[1]Russian Presidential Academy of National Economy and Public Administration (RANEPA), RU-119571, Moscow, Russian Federation
[2]Financial University under the Government of the Russian Federation, Department of System Analysis in Economics, RU-125993 Moscow, Russian Federation

Abstract. The coordinate system for describing practices of public administration in developed countries is set by three basic models: rational bureaucracy, New Public Management, and New Governance, which are "layered" on top of each other. However, in many developing countries, these models work differently, and instead of an upward movement, a spiral of reforms and counter-reforms is obtained. Moreover, in the face of global uncertainty generated by the transformation of the world order and the intensified COVID-19 pandemic, it is even more difficult to borrow the best practices from highly developed countries. The institutional model of neopatrimonialism, which includes the informal core and the formal periphery of the system of state institutions, has serious potential for understanding the specifics of government in developing countries. The value-neutral interpretation of neopatrimonialism opens up a broader view of such problems as state interventions into the economy, the relationship between power and property, the national model of democracy, etc. For Russia, the authors propose the redistributive neopatrimonialism model. Neopatrimonialism appears to be a phase in the cycle "mobilization system – stagnation – neopatrimonialism – renewed mobilization system". It is hypothesized that the next phase will be built taking into account the principles of open network organization and solidarity.

1 Introduction

The coordinate system for describing the transformation of public administration models in developed countries is set by three main public administration models: the classical Weberian model of "rational bureaucracy", project-target model "the New Public Management" (NPM) and social-network model "the New Governance" [1]. However, attempts to apply these approaches to describe the processes in many countries of "catching up" development, all the more to put them in the basis of transformations of the state apparatus, often lead to very specific results: models in these cases work in a completely different way, and instead of an upward movement, as it were "layering" models on top of each other, we get a spiral of reforms and counter-reforms. Moreover, in the face of global uncertainty generated by the transformation of the world order and the intensified COVID-19 pandemic, the best practices of developed countries no longer seem so immutable.

In this situation, the institutional model of neopatrimonialism possesses a serious potential both for the interpretation of the attempts at transformations that have already taken place and for the design of reforms in public administration [2]. This is especially true for the aspects of reforms related to the adaptation of their institutional design to existing informal institutions, forecasting the consequences of reforms and reducing the risks of counter-reforms.

The institutional model of neopatrimonialism includes the informal core and formal periphery of the system of public institutions. In classical "normative" institutions, the situation is reversed: the core is formal, while the periphery is informal. The research hypothesis, based on the proposed balanced model of institutions, is that the balance of formal and informal components of public administration institutions (or "skew" in one direction or another) is largely determined by the context of the wider systems: the sphere of public life, which includes a given institution, and the institutional order in general. In particular, the authors question the value-laden interpretation of neopatrimonialism, which identifies the latter with "unworthy governance", which must be overcome as soon as possible [3].

The value-neutral interpretation of neopatrimonialism as a control loop within the framework of statehood opens up a broader view of such problems as the role of the state in the economy, the ratio of power and property, the national model of democracy, allowing to adequately highlight and reasonably use the successful experience of the BRICS countries and some APR countries (Japan, South Korea, Singapore, Malaysia, etc.). At the same time, a clear understanding of the system stability mechanism allows predicting the factors of its instability and reducing the risks of

* Corresponding author: maratcha@yandex.ru

imbalance. In connection with the above, the problem of measuring the influence of informal relations on the decision-making processes of state bodies ("the level of neopatrimonialism") becomes urgent.

A significant part of the paper is devoted to the specifics of the Russian model of public administration, which we define as "redistributive neopatrimonialism". To construct the model, we used the theory of institutional matrices by S. Kirdina-Chandler [4] and the distinction between the coordination structures of the "market" and "razdatok" proposed by O. Bessonova [5], which is interpreted from the organization theory point of view (the difference between "vertical" hierarchical, "horizontal" market and "mixed" network structures) [6]. Based on Bessonova's idea about the cyclical nature of Russia's institutional dynamics, the authors propose a hypothesis that redistributive neopatrimonialism is a transitional phase in the cycle "mobilization system – stagnation – neopatrimonialism – renewed mobilization system", and the next phase will be based on the principles of open network organization and solidarity.

2 Global trend in public administration development and peculiarities of Russian practice

As shown in our previous works, the current global trend in public administration is the transition to the third generation of public administration practices (Public Administration 3.0), based on the idea of a "viable" state, combining the qualities of "strong", "effective" and "inclusive" [1]. As noted in the Introduction, these qualities of the state correspond to three basic ideal models of public administration from the practice of developed countries: "rational bureaucracy", NPM and New Governance, combined in "multimodal" public administration. Public Administration 3.0 is built on the principle of multimodality, the essence of which is that the tools of different models are used depending on the nature of the tasks being solved and are combined to solve complex problems for which the tools of any one model are not enough (figure 1).

These changes in the practice of public administration are taking place against the background of larger-scale processes that are engulfing public institutions and institutional governance.

The first large-scale process is the "networkization" of the economy and society, which is expressed in the following package of systemic changes [7]:

1) the economy and society are being transformed on the basis of network interaction with the dominance of horizontal ties;

2) governance systems and institutions based on the domination of hierarchical relations are gradually becoming uncompetitive, and they are being replaced by completely new ways of organization and self-organization: in the field of entrepreneurship, innovative clusters and ecosystems are formed, a "sharing" type of economy appears;

3) the change in the governance paradigm introduced by the network organization also affects the social sphere, public areas, interaction between the authorities and citizens; at the same time, network communities become new actors in the social space – they change the prevailing socio-economic practices, set requirements for infrastructure and urban environment, for the subject orientation and "targeting" of city and regional programs;

4) network interaction becomes a condition for the formation of collaborations and innovative environments in all of the above-mentioned areas – "living institutions" appear that can stimulate the growth of competitiveness in various areas of socio-economic development;

5) the state becomes open and "inclusive", increasing flexibility and developing a culture of interaction with public institutions, boosting the degree of their involvement and public participation in governance.

The second large-scale process is digitalization. In view of the fact that digitalization permeates all spheres of social life, while intensively interacting and intertwining with "networkization" in a complex way, we are talking about the process of "digital transformation" of the world (i.e. the economy, state and other spheres of society), which is expressed in the following social and institutional changes:

1) the state is turning into a public platform for the provision of high-tech services to citizens and entrepreneurs, as well as for the interaction of stakeholders in the name of public interests;

2) the proliferation of network connections and relationships is greatly simplified, which facilitates overcoming inter-level and inter-departmental barriers in state and municipal administration;

3) new channels and opportunities are opening up for horizontal interaction between the state authorities and stakeholders, including for creation communication platforms and transition to a partnership communication model.

The above changes make public administration transform into Public Administration 3.0, intensively developing the tools of all three basic models from the practice of developed countries and using them "multimodally". The latter means an integrated approach, in which a complex problem is divided by systems analysis into simpler ones, while the latter are solved using the most appropriate tools from different models. But this requires a higher level of coordination, since the use of tools from different models is based on the involvement of various competencies, which are either concentrated in specialized government bodies, departments or agencies, or a project team should be created for solving a complex problem, including specialists with the necessary competencies. In this case, the processes of solving simple problems, into which the original complex problem is "decomposed", must be coordinated in such a way that at a certain moment "gather" into a common result without loss of quality. This is exactly what digitalization and "networkization" should contribute to.

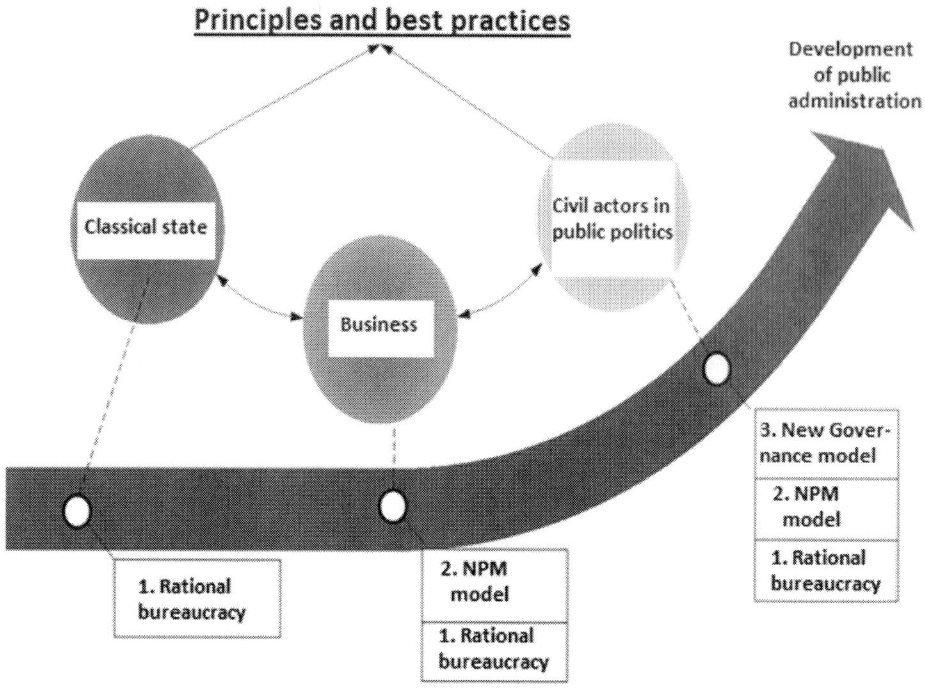

Fig. 1. The ratio of three ideal models of public administration from the practice of developed countries.

What is really happening in the practices of applying public administration basic models in Russia, if we compare them with the best practices of developed and even some successfully developing countries? We have already written about this in our previous works [1; 7], therefore, we restrict ourselves to a concise presentation of the diagnostic results.

1. In Russia, the three basic models work in a completely different way than in developed countries, and instead of an upward movement, "layering" the models on top of each other (Figure 1), we get a "spiral" of reforms and counter-reforms. Moreover, in the face of global uncertainty generated by the transformation of the world order and the intensified COVID-19 pandemic, the best practices of developed countries no longer seem so immutable.

2. The use of the tools of the classical model is burdened by the historical incompleteness of the processes of the modern "rational bureaucracy" formation.

3. The introduction of new models of public administration is taking place in the conditions of unsolved problems of the previous historical stage: the Russian bureaucracy is burdened with "post-Soviet" problems and is struck by the "traditionalism" of informal ties, which is typical for the majority of developing countries, and personal and clan patronage at the lower and middle levels of the government apparatus [8].

4. The use of tools of the New Governance model is associated with political restrictions:

- communication with the population and other stakeholders often has a formal and imitative nature, while the state tries to limit itself to "convenient" dialogue partners, and sometimes even creates them artificially;
- strategic documents and decisions, if they are submitted for public discussion, then usually at the final stage of their development, when it is already difficult to change anything in them (substitution of the discussion on the essence by "popular approval").

5. The market-oriented NPM model is being introduced with the following peculiarities ("three substitutions"):

- substitution of posing problems with "PR", striving for "varnishing reality";
- substitution of meaningful goal setting with the definition of target indicators;
- lack of independent feedback channels, substitution of meaningful analytics with infographics.

6. Project and program-target management is used not so much to achieve the stated goals, but to "slicing" budgets. Development and focus on the priorities for which these methods were created are not provided.

7. Attracting specialists with high industry specialization and unique competencies to the civil service is limited by insufficient motivation. Loyalty is valued, and the meritocratic system is not built.

8. The dominance of the "technocratic" approach to digitalization, while underestimating the role of the human factor and low motivation (point 6) against the background of the problems described in points 1-5,

leads to the conservation of outdated public administration institutions: the "digitization" of ineffective procedures only complicates their improvement.

3 The concept of neopatrimonialism as an attempt to explain "hybrid" forms of government. Measuring neopatrimonialism

The concept of patrimonialism goes back to the theoretical constructions by M. Weber, who viewed it as a method of domination based on the appropriation of public power (both political and economic) for private purposes; in other words, governing the state as a patrimonium – private property [9]. Neopatrimonialism is usually understood as the implementation of such a form of rule (essentially traditional) in modern conditions, and after Weber, this is opposed to rational (legal) domination. It is important to emphasize that the category of neopatrimonialism cannot be reduced only to the nature of the political regime, since, according to N. Robinson [10], V. Gel'man [11] and others, it also covers the mechanisms of governance – political and economic ruling of the state. In systems of the neopatrimonial type, the ratio of the formal and informal components of institutions, hierarchies, and networks, as well as closed and open networks, is extremely significant. Neopatrimonialism can be considered as one of the variants of the limited-access order in the terminology of D. North and his co-authors [12].

The concept of neopatrimonialism arises as a response of specialists in developing countries to classical ideas about public administration within the framework of the general Weberian paradigm of patrimonial and rational bureaucracy analysis. Instead of the dominance of a rational bureaucracy, which Max Weber feared might swallow up market and political decision-making centers around the world, during the twentieth century in most developing countries, after a series of anti-colonial revolutions, peculiar hybrid forms of administration developed. In the early 1970s, S. Eisenstadt proposed one of the first interpretations that distinguish traditional patrimonialism and modern neopatrimonialism [2]. The further intellectual trajectory of the neopatrimonialism concept is associated with the comprehension of this stable alternative to a rational bureaucracy and state institutions in Africa, Latin America, Asia, and other regions [13]. Another important characteristic of the bureaucracy in developing countries is the significant diversity and coexistence of fundamentally different models and mechanisms of administration within the public administration system. A significant part of government bodies can be quite close to Weber's model of rational bureaucracy, interspersed with New Public Management and participatory institutions from the New Governance model, but regional administrations and part of central authorities can be more adequately described in terms of neopatrimonialism.

In this work, we rely on the conceptualization of G. Erdmann and U. Engel [14], which allowed us to synthesize the key approaches of predecessors and at the same time offer a clear differentiating definition of this phenomenon and concept in comparison with other related terms: patrimonialism, clientelism, patronage, patron-client relationships. In general, the most important thing for understanding the specifics of neopatrimonialism seems to be precisely the mixture of formal and rational mechanisms, on the one hand, and interpersonal, patron-client or informal, on the other hand. At the same time, formal legal institutions perform not only their usual functions but acquire secondary hidden functions. Common neopatrimonial practices imply both a fairly widespread possibility of strict adherence to specific rules or powers and a significant amount of arbitrariness or discretion, suggesting the possibility of informal transactions and decisions. In this perspective, neopatrimonialism is associated with the peculiarities of personnel selection and promotion, the payment system, the relationship of employees with external political and economic stakeholders, cultural norms, the quality and style of regulations, rules, and legislative acts, as well as law enforcement.

This structural uncertainty allows actors within the system to engage in complex negotiations or bargaining over whether a formal order will be applied in a particular case or an informal solution can be found. At the same time, the possibility of applying sanctions for non-compliance with formal procedures remains, so that the rule is not only fictitious but also has additional significance. Sanctions for its violation become a control mechanism within the administrative hierarchy, in which violations are quite common. Rational bureaucracy and its subsequent improved versions (NPM and the participation model or New Governance) are generally more effective, neopatrimonial mechanisms are also effective in certain situations. The application of this concept to understanding the models of administration in Russia is due to the idea of their mosaic nature.

So for transition countries it becomes relevant the problem of measuring the influence of informal relations on the decision-making processes performing by government bodies ("the level of neopatrimonialism"). For example, Rachel Sigman and Staffan Lindberg from the Varieties of Democracy Institute (V-Dem – https://www.v-dem.net/en) use empirical tools to assess the levels of neopatrimonialism in African political regimes [15]. Further we consider the general method of measuring neopatrimonialism.

V-Dem project is one of the largest-ever social science research-oriented data collection programs which co-ordinates the network of almost 3000 country experts. The V-Dem indicators (approximately 350 in total) are based on assessments by experts who are situated both within and outside each country. The latest version of the dataset, V10, covers 202 countries from 1789-2019 with annual updates to follow. It is co-hosted by the University of Gothenburg and University of Notre Dame. The V-Dem dataset includes measures of both de jure and de facto aspects of political regimes [16]. Neopatrimonialism includes elements of patrimonial (mostly informal) and rational-bureaucratic rule which co-exist and are sometimes interwoven. It implies de facto practices guided by norms or informal rules that differ from those typically associated with the corresponding formal institutions, so the large selection of de facto indicators is very helpful for measuring neopatrimonialism.

For conceptualising neopatrimonialism Sigman and Lindberg apply the model by Bratton and Van de Walle [17] who define a neopatrimonial regime as one that includes clientelism, presidentialism and the use of public resources for private/political benefit (a particular case of the third is corruption). Following this conceptual model, Sigman and Lindberg use indicators from the V-Dem dataset to construct an index of neopatrimonial rule. In addition to an overall index measuring the latent concept of neopatrimonial rule predicted by indicators across these three dimensions, they also produce sub-indices for each of the three dimensions. The sub-indices permit finer-grained analysis to better understand patterns of variation across the three dimensions and how the different dimensions relate to one another. Each of the four indices are derived using a Bayesian factor analysis (BFA) [18] which allows to estimate the latent concepts from the means of the posterior distributions of each of the V-Dem indicators included in each dimension. The overall Neopatrimonialism index uses the BFA technique on the pool of 16 indicators employed in the three sub-indices [16].

These measurements show that the degree of neopatrimonialism in African countries varies vastly within the continent. There is a clear visible difference between countries scoring lowest and highest on the Neopatrimonialism index. The dimensions of neopatrimonialism are also configured variously. For instance, "clientelistic" regimes tend to also score high on corruption, although there are some exceptions, where moderate levels of political patronage are combined with high corruption. Furthermore, less "presidentialistic" regimes vary considerably in the extent to which their politics are "clientelistic" and plagued by corruption.

Sigman and Lindberg adapted the conceptual model of neopatrimonialism by Bratton and Van de Walle for the particular conditions and development context of African countries. We can made a plausible assumption that other groups of countries (Latin American, East Asian, Arab, Post-Soviet, East European, etc.) will require other conceptual models with different dimensions of the Neopatrimonialism index that will be projected otherwise onto the V-Dem dataset. Analysis of Sigman and Lindberg's work allows us to define the generalised method for constructing Neopatrimonialism index, consisting of the following four steps [19]:

1) based on conceptual considerations, a model is constructed with a few key indicators corresponding to the essential characteristics ("dimensions") of the concept;

2) indicators corresponding to these "dimensions" are disaggregated into a larger number of sub-indices that have a specific empirical relationship within the V-Dem dataset or some other commensurate database;

3) the secondary analysis of the data is carried out using suitable mathematical methods, which makes it possible to statistically verify the empirical consistency of the conceptual model;

4) if the consistency of the conceptual model is verified, then we can use its key indicators to measure (quantify) the essential characteristics of neopatrimonialism in the corresponding group of countries.

4 Redistributive neopatrimonialism as an institutional model of public administration: key characteristics

The diagnosis of the public administration peculiarities in Russia presented in the 2nd part of this paper allows us to propose the following research hypothesis (in three theses):

1. With regard to the conditions of Russia, we can talk about a specific kind of neopatrimonial model – redistributive neopatrimonialism, which:

- includes the interacting informal core and the formal periphery of the public administration system – i.e. corresponds to the general concept of neopatrimonialism, as defined by S. Eisenstadt [2] (in the classical "normative" institutions of public administration, the situation is reversed: the core is formal, and the periphery is informal);
- corresponds to the redistributive nature of the social coordination basic institutions in Russia (primarily in the economic, political and ideological spheres) [4];
- corresponds to the socio-cultural characteristics of the Russian "institutional matrix", which determine the possibilities and limitations of public administration institutional transformations, rebuilding "for themselves" the models borrowed from developed countries;
- undergoes reinterpretation in the context of Russia's institutional development [5] according to the formula "mobilization system – stagnation – neopatrimonialism – renewed mobilization system", becoming a value-neutral model – and not one that is viewed as a model of "unworthy governance" (V. Gel'man [3]) which is to be overcome.

2. The introduction of "redistributive neopatrimonialism" as a public administration model reflecting Russian specifics makes it possible to improve the "coordinate system" defined by the three basic models of public administration mentioned above [20]:

- this "coordinate system" receives a second dimension to describe spiral movements (reforms and counter-reforms);
- this second dimension of the "coordinate system" should be socio-cultural, since the political-economic interpretation of neopatrimonialism,

based on the distinction between extractive and inclusive institutions (according to D. Acemoglu and J.A. Robinson [21]), turns out to be clearly simplified, bringing together all the motives of the participants an informal core to rent extraction and prestigious consumption [22; 23; 11];

- we are moving away from the idea of "catch-up modernization" as a repetition of the developed countries' path.

The proposed alternative is "redistributive neopatrimonialism," which:

- is a systemic model based on a system-institutional approach that takes into account all the main aspects of statehood institutions (economic, political, legal, socio-cultural, etc.);
- examines the reproduction, development, and stability of the Russian statehood in the context of the encompassing system "economic reproduction – political regime – public administration";
- is a value-neutral model and allows considering such characteristics of Russian statehood as sovereignty and "imperialness", the ability to launch mechanisms of political-economic mobilization;
- is open to the experience of different countries and the launch of effective mechanisms for the transfer of foreign experience, taking into account the socio-cultural characteristics of Russia.

3. What new does the hypothesis about the model of "redistributive neopatrimonialism" allow to bring into the characterization of the problem situation? And what are the opportunities it opens up?

First, this hypothesis sets up the analyst to "look at the root" of the problems that were listed above. It is becoming obvious that superficial explanations like gaps in legislation or lack of money are clearly insufficient. And the popular thesis about "bad institutions" gets new depth by pointing to the role of the informal core.

Secondly, the risks of digitalization acquire an additional dimension within the framework of neopatrimonialism. After all, the typical reaction to the problems of public administration under political constraints is "technocratic". But if earlier the "strictness of Russian laws" was at least "compensated by the non-binding nature of their implementation", then as a result of digitalization we risk facing the fact that this saving informal component of Russian institutions will be "cut off" and we will find ourselves in a situation of "digital idiocy". A similar situation was brilliantly played in the famous film by Georgy Danelia: "The aunt was removed – the counting machine was installed." The relevance of this risk for Russia is associated with the neopatrimonial nature of the public administration current model, in which the informal core of governance institutions is significant.

Third, in the context of neopatrimonialism, digitalization not only increases risks but also creates additional opportunities. This happens in the logic of "turning disadvantages into advantages":

- creation of flexible digital platforms using the potential of neopatrimonial institutions self-correction based on their informal component, improvement of inter-level and inter-departmental feedback system;
- a new system of strategic planning and political-economic mobilization based on advanced simulation and scenario modelling (while the informal core makes it possible to move away from rigid Soviet-style directive planning schemes and act flexibly);
- creation of "digital communities", improving the quality of public services and public participation, using the potential of network interactions based on the informal mechanisms of neopatrimonial institutions.

Fourth, the research hypothesis, based on the proposed balanced model of institutions, is that the balance of the formal and informal components of public administration institutions (or "skew" in one direction or another) is largely determined by the context of the encompassing systems: the sphere of public life, which includes this institution, and the institutional order as a whole.

The value-neutral interpretation of neopatrimonialism as a control loop within the framework of statehood opens up a broader view of such problems as the role of the state in the economy, the ratio of power and property, the national model of democracy, etc. allowing to adequately highlight and reasonably use the successful experience of the BRICS countries and some APR countries (Japan, South Korea, Singapore, Malaysia, etc.). At the same time, a clear understanding of the system stability mechanism allows predicting the factors of its instability and reducing the risks of imbalance [20].

5 The place of neopatrimonialism in the cycle of Russia's institutional development

One of the key challenges that neopatrimonialism poses to the functioning of the modern Russian state is the situations it engenders when one is declared in the Constitution, another is written in the laws, and law enforcement is something else. To some extent, this is the usual legal-legislative-governance-social reality (M. Ratz and his colleagues introduced in this connection a special neologism "govern-power" [24, p. 43]), especially if you look at this phenomenon in terms of the difference between theory and practice. On the other hand, these situations point to another important feature of legal reality: the inevitable abstractness of laws and the gap between their formulas and real situations. In this regard, the same authors introduce the idea of a special type of thinking-activity – "norm control", designed to solve the problems arising at these junctions [25, pp. 180-184, 190-192]. But this issue has a third – political – aspect, which opens up in the discussion of special, so-called "hybrid" regimes (E. Shul'man [26]), in which the difference in constitutions, laws, and law enforcement acts as a defining feature of the "policy of power".

In our opinion, the "rut effect" and the problem of the "Russian rut" is best described in terms of institutionalism. They also describe the situation when one is declared in the constitutions, another is written in the laws, and the law enforcement, in this case, is some kind of third. But in countries like Russia, this "usual legal-legislative-governance-social reality" has special characteristics. These characteristics are best captured by the institutional concept of "neopatrimonialism," implying that the institutions of power include a formal shell and an informal core.

Institutions of power (like any other institution) in Western countries also have formal and informal components. If we talk about laws, then the informal component is objectively necessary because the law, concerning a specific situation, due to its normative nature is either too abstract (that is, it allows arbitrary interpretation "at the discretion" of the law enforcement officer), or too casuistic (and then the situation needs to be informal "fit" under the law, and if it turns out to be outlawed, then, again, it will be settled informally "at the discretion" of the law enforcement officer).

But in Western countries, the informal component acts as a "spirit of the law" that does not directly conflict with its "letter", while in countries like Russia, it often turns the law into a "drawbar" that can be turned how the decision-maker wants. As a result, a double standard arises, which is friendly to our close circle and judges others according to the law. And the next logical step required to control the power over "outsiders" is "tightening the screws" when the law becomes such that it is impossible to enforce it literally. Hence the famous aphorism: "The severity of Russian laws is compensated by the non-binding nature of their implementation." But it is compensated only for the time being since everyone is "hooked" in power – and when someone disagrees with the informal "rules of the game", they are immediately treated "according to the law".

Of course, the situation where law implementation is perverted for the sake of "rent-building" and "prestigious consumption" is an important special case of neopartrimonial power institutions. But in the general case, we should talk about the splitting of "normal" institutions into law enforcement megamachines and "secondary" institutions of informal governance, which set goals for law enforcement megamachines. These goals can be economic, but they also can be another. So, it can be a "political order", transmitted through "telephone law", or the interest of the bureaucracy self-preservation. Or there may be quite noble goals of modernization or defense of the country from the aggressor. In other words, the real palette turns out to be much richer than Gel'man's.

How does the decision-making system work in neopatrimonial power institutions? This is an extremely difficult question in the study of public administration. It is extremely difficult to get close to the "decision-making system" at the state level in Russia, especially if we want to avoid psychologism. To do this, it is necessary to grasp the decision-making process as an element of the governance activities system, presenting the latter as a repetitive "governance cycle" having 4 phases [1]:

1) analysis and planning ("preparation" of the decision);

2) decision making;

3) implementation of the adopted decision;

4) control, evaluation, and correction.

Accordingly, even if we do not know anything about how the decision is made, we can consider this process as a black box in the "frame" of the other three elements of the cycle. At the same time, in terms of actors (participants), it is necessary to distinguish between three types of problem situations (or "problem contexts") [27]:

1) unitary (when it is possible to determine a single "decision-maker", and for a controlled system, the goal "descends" from above);

2) pluralist (when several free and equal participants are forced to coordinate goals among themselves);

3) coercive (when there are several participants, but instead of freedom, we have the ability of one of them to force the others to make favourable decisions at the expense of some prevailing resources, that is, the "rules of the game" are unequal, questioned or absent altogether).

Accordingly, we have three types of the ruling (Control & Administration, Management, Governance), in which the decision-making processes are fundamentally different [28].

As noted above (2nd part of this paper), this typology of the ruling is projected onto public administration, giving rise to models of "rational bureaucracy", New Public Management (project-target) and New Governance (social-network), corresponding to the ideas of the "strong", "effective" and "inclusive" state. In the West, these three models (and three ideas), historically appearing in this sequence, do not cancel, but "layered" on top of each other. In Russia and similar medium-developed "non-Western" countries, the implementation of each model and their combination has its own characteristics, which we consider in the framework of the concept of "redistributive neopatrimonialism."

The meaning of the term "redistributive" is to indicate the peculiarity of the powers separation system, which is that the three classical "branches" of power are not independent (as in the framework of the "checks and balances system" characterized Western countries) but receive their real powers from the supreme power [29]. Their formal powers prescribed in the Constitution may differ from their real ones – this is the manifestation of the supreme power role as the informal core of the public administration system. The supreme power itself in such cases follows not so much the "letter" as the "spirit" of the Constitution and laws, acting "from higher considerations," that is, on the basis of the value idea [20] as the informal ideal core of the neopatrimonial state.

Within the framework of the "redistributive neopatrimonialism" concept, the term "redistributive" characterizes not only the state administration and the political system, but also the national economy, which, according to its basic principle of organization and coordination, is not market, but distributive. Therefore, strictly speaking, it cannot even be called "economy" (if "economy" is understood as a market type of organization). Svetlana Kirdina uses the term "redistributive economy" [4], and Olga Bessonova uses the term "razdatok economy", which is understood as an economic system where non-market mechanisms play a dominant role, and market mechanisms play an auxiliary, compensatory role [30].

In 2019, Bessonova proposed a new "integral-institutional" paradigm of the socio-economic development of mankind, within which the market and razdatok are considered as two universal and mutually necessary ways of coordination [5, p. 8].

Bessonova also introduces the concept of the Russian civilization matrix. Its reproduction is based on a razdatok institutional core (dominant institutions), while the market played a subordinate role of compensatory institutions. This allows us to substantiate the hypothesis that the current crisis of the Russian economy and political system is associated with the exhaustion of the quasi-market model that the Russian civilization matrix uses at the transformational phases of its historical evolution.

Bessonova's concept, for all her claims to complexity, is still focused on economics. If we focus on public administration, the "quasi-market" phase can be correlated with neopatrimonialism, which is characterized by the actions of representatives of the informal core, not in public, but in private and group interests, implemented within the framework of state capitalism. But, taking into account the previous trajectory of historical development, it can be assumed that at other phases (Bessonova proposes a cyclic model "perinatal phase – structured phase – phase of institutional exhaustion – transformational phase" [5, p. 17]), representatives of the informal core act primarily in public, not private or group interests.

It is no accident that Bessonova categorically objects to the use of the term "power-property" proposed by the orientalist L. Vasiliev [31] and insists that the basic institution of the distribution economy is public-official property. But at the same time, we observe how at the transformational phase, when the dominant and compensatory institutions of the distribution economy temporarily change places ("quasi-market"), what "in theory" should be public-official property is used in private and group interests, there is a merger of power and property, giving rise to "institutional corruption". In this sense, the term "supreme conditional ownership" used by Svetlana Kirdina in the conceptual framework of institutional matrices and redistribution economics [4] seems to be more accurate (the term "redistribution" is also more preferable for us than "razdatok").

The term "supreme conditional ownership" applies to all four the historical evolution cycle phases of the institutional/civilization matrix since it indicates the main feature: the dominant type of property in the X-matrix states is property, which on behalf of the society is controlled by the supreme power and those persons or organizations that this supreme power "distributed" the corresponding competences (this is indicated by the word "conditional"). At the quasi-market phase, such property can be "distributed", for example, to "oligarchs" – but, as historical experience shows, those "oligarchs" who forget that property has been transferred to them conditionally and violate the terms of its use are ostracized and deprived of this property. It is important to note that formally (by law) this ownership is private, while this ownership is supreme conditional according to informal rules supported by the supreme power, that is, the real core of the neopatrimanal institution system.

Also, control over the most significant property is gradually returning to the state, which paves the way for the transition to the next phase of the cycle, in which only the property that is not needed by the state to serve society remains private. In other words, private property becomes one of the compensatory institutions, and the overall ratio of the dominant and compensatory institutions returns from the "inverted" position to the "normal" one. And if the cycle of Russia's institutional development "according to Bessonova" is reinterpreted in the context of focusing not on the economy, but the state, then the formula will be "mobilization system – stagnation – neopatrimonialism – renewed mobilization system." The mobilization system will correspond to the "perinatal" and "structured" phases of Bessonova's cycle, stagnation – to the phase of "institutional exhaustion", neopatrimonialism – to the "transformational" phase (when redistributive and market institutions "change places") and, finally, the renewed mobilization system will correspond to the transition to a new "round" of the historical cycle of socio-economic evolution.

6 Instead of a conclusion: the institutional dilemma of our time

The foundations of the dichotomous differences between the X- and Y- institutional matrices, as well as the market and razdatok methods of coordination, are clarified by turning to the Marxist theory of social reproduction. K. Marx represented the reproduction cycle of economic activity as a dynamic system of four elements: production, distribution, exchange, consumption: "Production, distribution, exchange and consumption thus form a proper syllogism; production represents the general, distribution and exchange the particular, and consumption the individual case which sums up the whole" [32]. Production and consumption are opposites that can be combined into a system only thanks to the mediating elements, which are system-forming. At the same time, both distribution and exchange are needed to ensure full-fledged economic activity.

In the Asiatic mode of production (Asiatische Produktionsweise) [33], distribution dominates, in the capitalist socio-economic formation – exchange, in the slave-owning and feudal ones they are combined. There is an exchange in the Asiatic mode of production, but it is strictly regulated by the distribution of trade rights and quotas. Under capitalism, exchange develops into a general form of market relations, but the distribution continues to exist within firms and corporations, as well as within families (households). But firms, corporations, and households are "inscribed" in the market system, and exchange relations are system-forming – just as in the Asiatic mode of production these are distribution relations. In this sense, distribution and exchange – like Yin and Yang – are in a dialectical relationship: one element "protrudes" and dominates, but the second does not disappear, it is present, as Hegel would say, "in sublated (aufgehoben) form."

If following O. Bessonova, we consider (re)distribution and exchange as systemic principles of coordination, then it is reasonable to look at them from the point of view of the organization theory, within which it is customary to distinguish three types of relationships and corresponding structures:

1) "vertical" (the principle of hierarchy, or subordination/order);

2) "horizontal" (the principle of cooperation, or an equal contract);

3) "network" (this is the most universal principle of coordination since in the mathematical sense a network is a graph of an arbitrary shape, including both "vertical" and "horizontal" connections).

In economic life, networking can connect large companies that take advantage of corporate hierarchy and economies of scale, and small independent firms that build flexible contractual value chains. In social practice, network relations can connect authorities, acting according to the principles of hierarchy and subordination, with structures of civil society, acting on the principles of self-organization and freedom of association.

In the world, not only the redistributive (hierarchical, order-based, founded on "vertical" ties), but also the market (contractual, based on "horizontal" ties) organization of the economy is gradually being replaced by the network organization. Examples are the "sharing economy" based on the sharing of property (from cars to real estate in so-called "co-living"), as well as innovative clusters and ecosystems built on the principle of co-opetition (combining the words cooperation and competition, which resembles the modern version of socialist competition).

Thus, a completely new level of what in Marxism was called "socialization of production" arises. Only this happens not for the routine labour of industrial society, based on which Marx's Capital was written, but for creative labour in a modern innovative economy (or "knowledge economy"), in which "human capital" becomes the main productive force. In this case, the alienation associated with capitalist ownership of private property is removed (knowledge and know-how are at the real disposal of workers, not capitalists). Accordingly, preconditions arise for the labour to cease to be sold.

To what has been said in this part of the paper, it should be added that the described socialization of production at a new technological level of productive forces in combination with the socialization of consumption processes in the "sharing economy" creates conditions for overcoming the main contradiction of capitalism – between the social form of production and the private form of appropriation. In a situation where the consumption process is socialized, it is logical to socialize the means of production. O. Bessonova calls it "solidarism" [5, p. 24-25]. This was possible before in the so-called "people's enterprises", but in the industrial era, such cases were rather the exception, and within the network organization there are prerequisites for this to become the rule.

Digitalization and the development of modern information technologies (in particular, blockchain technologies) create opportunities for building economic relations without intermediaries and speculators and political relations without corruption and with a minimum of bureaucracy. True, a number of very complex problems of organization and self-organization arise here, associated with the rules for regulating such relations, guaranteeing these rules, and resolving conflicts. In other words, the questions about how the previous functions of the legislative, executive, and judicial authorities will be implemented in the new society, that is, questions about the model of the system of institutions, are being actualized.

The scenario for the future development of neopatrimonialism depends on the path chosen by each particular country to solve the institutional dilemma of our time. Formally, there are three options:

1) a radical transformation of the economic, political, and ideological spheres for the implementation of the developed countries basic models, described in the 2nd part (it actually implies a change in the institutional matrix);

2) consolidation of the neopatrimonial system as a stable form of organization of the economy, politics, ideology, and public administration with the transformation of neopatrimonialism into a "new normality";

3) transition to an integral form of organization of the economy, politics, ideology, and public administration within the framework of the renewed mobilization system.

This paper was prepared as a part of the research work within the framework of the Government order for the RANEPA.

References

1. S.V. Bespalov, V.G. Maracha, Public Administration, **19**, **4** (108), 25–31 (2017)
2. S.N. Eisenstadt, *Traditional Patrimonialism and Modern Neopatrimonialism* (Sage, Beverly Hills, 1973)
3. V.Ya. Gel'man, *Political Foundations of 'Unworthy Governance' in Post-Soviet Eurasia: Outline for a Research Agenda* (Preprint **M-49/16**) (Europ. Univ. in St.Petersburg Pub., 2016) [in Russia]
4. S.G. Kirdina, *Institutional Matrices and the Development of Russia: an Introduction to X-Y theory* (Nestor-History, St. Petersburg, 2014) [in Russia]

5. O.E. Bessonova, World of Russia, **28**(1), 7-31 (2019) [in Russia]
6. M. Bevir, *Governance: a Very Short Introduction* (Oxford Univ. Press, 2012)
7. V.G. Maracha, T.S. Krasnikova, Regional economy. South of Russia, **7**(1), 66-78 (2019) [in Russia]
8. V.G. Maracha, Modern information technologies and IT education, **4**(4), 1003-1010 (2018) [in Russia]
9. M. Weber, *The Theory of Social and Economic Organization* (The Free Press, New York, 1947)
10. N. Robinson, *The Political Origins of Russia's 'Culture Wars'* (Department of Politics and Public Administration University of Limerick, 2014)
11. V.Ya. Gel'man, *Modernization, Institutions and the 'Vicious Circle' of Post-Soviet Neopatrimonialism* (Preprint **M-41/15)** (Europ. Univ. in St. Petersburg Pub., 2015) [in Russia]
12. D.C. North, J.J. Wallis, B.R. Weingast, *Violence and Social Orders. A Conceptual Framework for Interpreting Recorded Human History* (Cambridge Univ. Press, Cambridge, 2009)
13. J.-F. Bayart, S. Ellis, B. Hibou, *The Criminalization of the State in Africa* (James Currey, Oxford, 1999)
14. G. Erdmann, U. Engel, *Neopatrimonialism Revisited – Beyond a Catch-All Concept* (Working Paper **16)** (GIGA Working Papers, Hamburg, 2006)
15. R. Sigman, S.I. Lindberg, *Neopatrimonialism and Democracy: an Empirical Investigation of Africa's Political Regimes* (Working Paper **56**) (V–Dem Institute, Gothenburg, 2017)
16. M. Coppedge et al., *V-Dem Comparisons and Contrasts with Other Measurement Projects*, (Working Paper **45**) (V–Dem Institute, Gothenburg, 2017)
17. M. Bratton, N. Van de Walle, *Democratic Experiments in Africa: Regime Transitions in Comparative Perspective* (Cambridge University Press, Cambridge, 1997)
18. A.D. Martin et al., Package MCMCpack (2020) https://mran.microsoft.com/snapshot/2017-08-1/web/packages/MCMCpack/MCMCpack.pdf
19. V.G. Maracha, S.V. Bespalov, E3S Web of Conferences, **217**, 07029 (2020)
20. V.G. Maracha, S.V. Bespalov, Public Administration: Russia in Global Politics. Proceedings of the VII International Conference, 439-447 (2019) [in Russia]
21. D. Acemoglu, J.A. Robinson, *Why Nations Fail: The Origins of Power, Prosperity and Poverty* (Crown, New York, 2012)
22. V.Ya. Gel'man, Social Sciences and Contemporary World, **6**, 34-44 (2015) [in Russia]
23. V.Ya. Gel'man, Social Sciences and Contemporary World, **1**, 104-116 (2016) [in Russia]
24. M.V. Ratz, S.I. Kotel'nikov, Political Conceptology, **3**, 29-68 (2015) [in Russia]
25. M.V. Ratz, S.I. Kotel'nikov, Political Conceptology, **1**, 156-209 (2014) [in Russia]
26. E.M. Shul'man, *Practical Political Science. Guide to Contact with Reality* (AST, Moscow, 2018) [in Russia]
27. M. Jackson, *Systems Thinking: Creative Holism for Managers* (John Wiley & Sons, Chichester, UK, 2003)
28. V.G. Maracha, *Proceedings of the XIII All-Russian Meeting on Control Problems (VSPU-2019), 1558-1562* (Institute of Control Sciences RAS, Moscow, 2019) [in Russia]
29. N.A. Zaharov, *The System of Russian State Power* (Litera, Moscow, 2002) [in Russia]
30. O.E. Bessonova, *Market and Distribution in the Russian Matrix: from Confrontation to Integration* (ROSSPEN, Moscow, 2015) [in Russia]
31. L.S. Vasil'ev, In: *Types of Social Relations in the East in the Middle Ages*, 60-99 (Nauka, Moscow, 1982) [in Russia]
32. K. Marx, *A Contribution to the Critique of Political Economy* (Lawrence & Wishart, 1971)
33. L. Krader, *The Asiatic Mode of Production: Sources, Development and Critique in the Writings of Karl Marx* (Van Gorcum, Assen, 1975)

Regional Investment Attractiveness's Model Analysis at the Heart of Russian Investment Policy

Evgenii Mitrofanov[1,*], *Alevtina* Kulagina[2], and *Liya* Panferova[2]

[1]Moscow State University of Humanities and Economics, RU-107150, Moscow, Russia
[2]I.N. Ulyanov Chuvash State University, RU-428015, Cheboksary, Russia

Abstract. The article analyzes the investment attractiveness of regions of the Volga Federal District in the Russian Federation using the principal component method. Indicators are selected characterizing investment attractiveness of Russian regions of the Volga Federal District. Functional dependence of the selected indicators are defined, based on which the estimation of level of investment appeal of regions of the Volga Federal District and their clustering. Recommendations for improving the investment policy of Russian regions are given.

1 Introduction

In modern economic conditions, one of the key points of Russia's investment policy is to attract investment in the competition in the global financial market and increase the level of investment attractiveness of the country's regions [1-5]. Determining the level of investment attractiveness of a country and its regions is widely used by potential investors in their activities, which helps to identify weaknesses in investment policy and investment prospects both in an individual subject and in the country as a whole, and also helps to identify factors that affect the country's investment policy and investment attractiveness of subjects, and the need to develop new and improve existing methods for assessing the investment attractiveness of regions of the country [3,5].

Many researchers agree that the investment attractiveness of a region is an integral characteristic of the investment object, which reflects the subjective relationship between the investor and the investment object, i.e. how well the investment object meets the investor's goals. Investment attractiveness includes investment potential and investment risk. In turn, they are divided into a number of indicators. Thus, investment attractiveness is determined by many factors that can be combined in meaning [1,2,4].

2 Results and discussion

Many existing methods of analyzing investment attractiveness are usually related to factor analysis. Therefore, in this study, we will use the principal component method to assess the level of investment attractiveness of regions.

The principal component method is used to decrease the number of variables and construct generalized factors, i.e. principal components that are uncorrelated with each other.

At the first stage of statistical analysis, it is necessary to determine a set of initial indicators that best characterize the objects under study in terms of their investment attractiveness. We selected 15 most significant indicators (table 1).

Table 1. Grouping of indicators used to analyze region's investment attractiveness

Composite indicator	Group indicators
Production group of indicators	x_1 -GRP per capita, RUB
	x_2 - Industrial production index, % of the previous year
Financial and economic group of indicators	x_3 - The balanced financial result of activity of organizations, mln. RUB.
	x_4 - Share of unprofitable organizations, in % of the total number of organizations
	x_5 - Investment in fixed assets per capita, RUB
The employment group of indicators	x_6 - Employment, %
	x_7 - The level of labor productivity in the region (the ratio of GRP to the number of employees), %
Innovative (informational) group of indicators	x_8 -Innovative activity of organizations, %
	x_9 - Percentage of the working population engaged in research and development, people
Infrastructure group of indicators	x_{10} - Retail trade turnover per capita, RUB
	x_{11} - Volume of paid services per capita, RUB

*Corresponding author: mep79@list.ru

Composite indicator	Group indicators
Social group of indicators	x_{12} - The share of the population with monetary incomes below the subsistence minimum, in % of the total population
	x_{13} - The ratio of per capita income to the subsistence minimum, %
Consumer group of indicators	x_{14} - Number of own passenger cars per 1000 people. population, PCs.
	x_{15} - Total area of residential premises per inhabitant on average, sq. m.

Factor analysis is performed using the Statistica application software package. To identify the presence of hidden relationships between indicators, we will find a matrix of paired correlations (table 2). The presence of values exceeding 0,7 in absolute value indicates that there is a relationship between the initial indicators.

Table 2. Correlation coefficient

	x_1	x_2	x_3	x_4	x_5	x_6
x_1	1					
x_2	-0,15	1				
x_3	0,87	0,06	1			
x_4	0,19	-0,18	-0,08	1		
x_5	0,99	-0,18	0,83	0,09	1	
x_6	0,97	-0,01	0,90	0,06	0,96	1
x_7	0,93	0,21	0,87	0,13	0,91	0,96
x_8	-0,71	0,05	-0,73	-0,47	-0,63	-0,70
x_9	-0,91	0,25	-0,73	-0,13	-0,90	-0,83
x_{10}	0,99	-0,19	0,80	0,20	0,99	0,95
x_{11}	0,99	-0,20	0,81	0,23	0,98	0,93
x_{12}	-0,74	0,14	-0,62	0,35	-0,81	-0,73
x_{13}	0,35	-0,10	0,20	-0,26	0,45	0,40
x_{14}	0,99	-0,19	0,81	0,24	0,98	0,95
x_{15}	0,99	-0,17	0,85	0,26	0,96	0,94

Table 2. Correlation coefficient (continuation)

	x_7	x_8	x_9	x_{10}	x_{11}
x_1					
x_2					
x_3					
x_4					
x_5					
x_6					
x_7	1				
x_8	-0,72	1			
x_9	-0,76	0,49	1		
x_{10}	0,90	-0,65	-0,93	1	
x_{11}	0,89	-0,66	-0,94	1,00	1
x_{12}	-0,64	0,13	0,81	-0,78	-0,77
x_{13}	0,29	0,21	-0,34	0,42	0,39
x_{14}	0,90	-0,71	-0,92	0,99	0,99
x_{15}	0,90	-0,73	-0,93	0,98	0,99

Table 2. Correlation coefficient (continuation)

	x_{12}	x_{13}	x_{14}	x_{15}
x_1				
x_2				
x_3				
x_4				
x_5				
x_6				
x_7				
x_8				
x_9				
x_{10}				
x_{11}				
x_{12}	1			
x_{13}	-0,66	1		
x_{14}	-0,73	0,35	1	
x_{15}	-0,71	0,31	0,99	1

We will check the validity of the principal component method and the significance of the correlation matrix using Wilkes test. Let's formulate hypotheses:

H_0 – the correlation matrix is insignificant;

H_1 – the correlation matrix is significant.

Calculate the observed value of the criterion using the formula:

$$\mathrm{X}^2_{\mathrm{obs}} = -\left(\mathrm{n} - \frac{1}{6}(2\mathrm{m}+5)\right)\ln(\det \mathrm{R}) = -\left(11 - \frac{1}{6}(2\cdot 15+5)\right)\ln(1,73\cdot 10^{-8}) = 92,33 \quad (1)$$

The critical value is found in the Pearson distribution table $X^2_{obs} = X^2_{0,05;105} = 82,35$, with significance level 0,05 and the value of the degree of freedom equal to $\frac{1}{2}\mathrm{m}(\mathrm{m}-1) = \frac{1}{2}\cdot 15\cdot(15-1) = 105$.

Thus $X^2_{\mathrm{obs}} = X^2_{\mathrm{crit}}$, therefore the main hypothesis is rejected in favor of the competing one H_1. The found matrix of paired correlations is significant at the significance level of 0,05, i.e. there are hidden connections between the initial features and the use of the principal component method is justified.

The next stage of analysis involves decreasing the dimension of factors. We need to understand how many main components we need to allocate. This can be determined using the Kaiser and scree criteria. The Kaiser criterion selects factors with eigenvalues equal to or greater than 1. According to this criterion, in our case, it is necessary to select 3 factors. According to scree criterion, after the fourth main component, the eigenvalues slow down, so it is worth selecting 4 main components.

On the basis of these criteria, it was decided to limit ourselves to three main components, which account for 92.1% of the total variance (table 3).

Table 3. Eigenvalues and relative contribution to the total variance

№	Eigenvalue	% of the total variance	Cumulative eigenvalue	Total %
1	10,5623	70,4153	10,5623	70,4153
2	1,9426	12,9505	12,5049	83,3658
3	1,3160	8,7732	13,8209	92,1391

№	Eigenvalue	% of the total variance	Cumulative eigenvalue	Total %
4	0,6496	4,3307	14,4705	96,4697
5	0,3444	2,2957	14,8148	98,7654
6	0,1007	0,6710	14,9155	99,4364
7	0,0492	0,3281	14,9647	99,7644
8	0,0259	0,1724	14,9905	99,9368
9	0,0050	0,0331	14,9955	99,9699
10	0,0045	0,0301	15,0000	100,0000

The first main component took the load of 12 indicators. Among them are GRP per capita; the level of employment of the population aged 15-72 years; the volume of paid services per capita; the number of own cars per 1000 people population; the total area of residential premises, on average per inhabitant and others. The first main component can be interpreted as «Coefficients that characterize the standard of living in the region». The second component is defined by two indicators. This is the share of unprofitable organizations and the ratio of average per capita income to the subsistence minimum. We interpret it as «Coefficients of financial capabilities of the region». The third main component contains only one indicator - the industrial production index. Let's call it the «Regional industry coefficient».

Next we calculate the integral assessment of investment attractiveness of the Volga Federal District of Russian Federation according to the weight coefficients:

$$IA = \beta_1 F_1 + \beta_2 F_2 + \beta_3 F_3 \quad (2)$$

$$\mathbf{F_1 = 0,095x_1 + 0,053x_2 + 0,107x_3 + 0,026x_4 + 0,086x_5 + 0,102x_6 + 0,115x_7 - 0,106x_8 - 0,069x_9 + 0,086x_{10} + 0,086x_{11} - 0,04x_{12} - 0,011x_{13} + 0,09x_{14} + 0,094x_{15}} \quad (3)$$

$$\mathbf{F_2 = -0,017x_1 + 0,08x_2 - 0,008x_3 - 0,404x_4 + 0,048x_5 + 0,02x_6 - 0,032x_7 + 0,321x_8 - 0,047x_9 + 0,013x_{10} - 0,004x_{11} - 0,309x_{12} + 0,374x_{13} - 0,027x_{14} - 0,047x_{15}} \quad (4)$$

$$\mathbf{F_3 = 0,018x_1 - 0,691x_2 - 0,235x_3 + 0,273x_4 + 0,044x_5 - 0,11x_6 - 0,24x_7 + 0,104x_8 - 0,146x_9 + 0,079x_{10} + 0,089x_{11} - 0,051x_{12} + 0,134x_{13} + 0,071x_{14} + 0,047x_{15}} \quad (5)$$

β_i – share of the total variance of the i-th factor.

The final assessment of investment attractiveness of the Volga Federal District regions based on the indicators of 2018 is shown in table.4. The subjects are arranged in order of decreasing values of the level of investment attractiveness.

Table 4. Integrated assessment of investment attractiveness of Russian Volga Federal District regions

	F_1	F_2	F_3	IA
Republic of Tatarstan	0,69065	0,61295	-0,58509	**0,55826**
Samara region	0,51758	0,01487	-0,31870	**0,36729**
Nizhny Novgorod region	0,45323	0,24763	-0,47211	**0,33622**
Bashkortostan	0,46803	0,07059	-0,37830	**0,33158**
Perm region	0,41439	-0,05557	-0,29482	**0,28081**
Orenburg region	0,37997	-0,24316	0,05532	**0,26147**
Republic of Udmurtia	0,33549	-0,06923	-0,75793	**0,17450**
Penza region	0,24123	-0,19816	-0,22530	**0,13505**
Saratov region	0,27225	-0,25538	-0,41258	**0,13288**
Ulyanovsk region	0,26850	-0,50392	-0,08130	**0,12663**
Kirov region	0,20308	-0,12959	-0,34929	**0,10372**
Republic of Mordovia	0,18072	-0,23440	-0,24222	**0,08210**
Chuvash Republic	0,07956	-0,03722	-0,54337	**0,00383**
Republic of Mari El	0,08835	-0,49745	0,04453	**0,00184**

In order to identify common features and similarities of regions in terms of investment attractiveness, we will conduct a cluster analysis. The purpose of cluster analysis is to group objects with similar characteristics into clusters.

Since in our case there is no information that any factor is more important for classification than others, we want to take into account the differences in each feature equally, so we will choose the usual Euclidean distance, formula (1), and Ward's method, which often gives fairly compact and well-separated clusters as a distance measure:

$$d_{ij} = \sqrt{\sum_{k=1}^{m} \left(x_{ik} - x_{jk}\right)^2} \quad (6)$$

The final composition of clusters is shown in table 5.

Table 5. Cluster composition

Cluster number	Number of objects in the cluster	Cluster composition

Cluster number	Number of objects in the cluster	Cluster composition
1	1	Republic of Tatarstan
2	4	Samara region, Nizhny Novgorod region, Bashkortostan, Perm region
3	3	Orenburg region, Ulyanovsk region, Republic of Mari El
4	2	Republic of Udmurtia, Chuvash Republic
5	4	Penza region, Saratov region, Kirov region, Republic of Mordovia

To interpret the clusters, we plot the average values of the factors (Fig. 1).

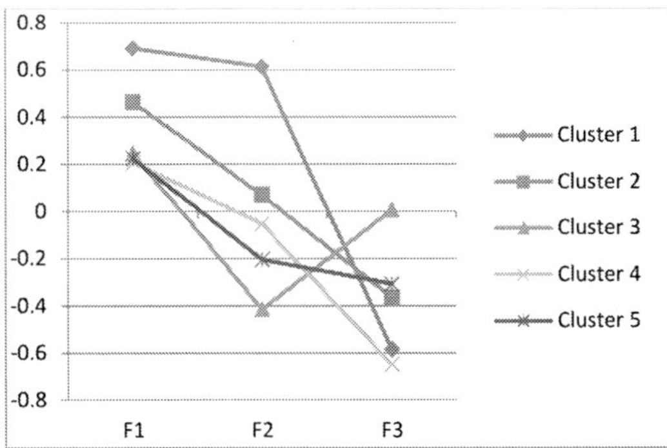

Fig. 1. Graph of average values of factors

3 Conclusion

The leader in terms of living standards and financial opportunities in the region is the first cluster, which includes the Republic of Tatarstan.

The second cluster has a fairly high standard of living in the regions, but the third factor has a negative value, which shows a low index of industrial production in the four regions included in the cluster.

The third cluster was the leader in the industrial production index among other clusters. But at the same time, it has the worst position on the second factor.

The fourth cluster is characterized by the standard of living in the region, but has the last position in almost all factors. Two of the three factor values are negative.

The fifth cluster does not stand out from the others. This cluster has average positions for all factors.

Hence, we can conclude that the regions included in clusters 1, 2 need to improve the level of industrial development in the region in order to increase investment attractiveness.

Regions included in 3, 4, 5 clusters have a large share of unprofitable organizations, as well as a low standard of living. It is possible that decreasing unprofitable organizations and improving the standard of living will lead to an increase in investment attractiveness in these regions.

References

1. A. Danilova, M. Monoyios, A. NG Mathematics and financial economics, **1**, 13-38 (2009).
2. A.G. Kulagina, E.P. Mitrofanov, D.S. Fedyaeva Bulletin of Economics, law and sociology, **1**, 26-30 (2014)
3. Ganju Jitendra, Lucas James American Statistician, **1**, 5-11 (2000).
4. Kim Seong, Sun Dongchu Lifetime data analysis, **3**, 251-269 (2000).
5. V.A. Arkhipova, A.G. Kulagina Economics and entrepreneurship, **2**(103), 1186-1189 (2019)

The Interdependence of Humanistic Values and the Level of Existential Fulfillment of an Intellectual Leader

Marianna V. Mikhaylova[1,*] and *Viktoria A.* Dmitrieva[2]

[1] Moscow State Technological University "STANKIN", Department of philosophy end and social management, RU-127055, Moscow, Russia
[2]Russian State Social University, Department of social ontopsychology, RU-129226, Moscow, Russia

Abstract. Value orientations of the young intellectual leaders cause absolute interest because, by realizing their ambitions, projects and making creative breakthroughs, they will eventually change history, create new directions for the development of science and art. Not only the realization of the talented youth's own potential depends on what they think about the meaning of a person, on the worldview positions, but also whether the future society will be humanistic, whether talent will be cultivated in it. The article presents the results of a statistical sociological study of ideas about humanism, qualities of a leader, value orientations and the level of existential fulfillment of young intellectual leaders - graduate students of a technical university. The level of existential fulfillment of the sample is interconnected with a high vitality and self-awareness of health, ideas about humanism as the flourishing of sciences and art, ideas about freedom as a responsible choice, and the morality of universal justice. However, graduate students deliberately downplay the values "stimulation" and "hedonism" (striving for novelty and deep experience to maintain an optimal level of activity). Although the combination of these values with the value "independence" is significant for creative activity.

1 Introduction

The realization of humanistic values has been in the focus of thinkers in different eras. The main ideas of classical humanism began to be formulated since Socrates, they can be designated as follows: a person has his own internal apriority order, which serves as his criterion of rationality; a person is able to do good and is responsible for his own creative potential and the development of mankind; freedom and human dignity are fundamental values of society [1]. It must be said that the values of humanism are acquiring special significance right now in the period of digital technologies and the increasingly active use of artificial intelligence. The human mind created artificial intelligence, but the idea of a person from the standpoint of existential meaning, longing for one's own being, raises many more questions among talented young people than an ultra-modern "smart" computer. More and more young people admit that social networks, Instagram creates the illusion of pseudo-realization, a race for false values, lead to a decrease in their own self-esteem and an increase in an inferiority complex when it is necessary to compare oneself with a “picture” that is a partial, selective or embellished reflection of reality [2]. This forces the most creative youth, those who can become intellectual leaders, to clarify their own values, cleared of myths and advertising labels in real life. A person capable of developing their potential to the level of functionality for society needs special pedagogy and other life values. “Leaders are generated by vital spontaneity and individual, voluntary, existential-historical self-improvement. These people become great and, perhaps, even the best for the sole purpose - the highest service to humanity." [3] Along with the problems of environmental disasters, the search for new types of energy, there is an acute problem of personality, namely the inability of a person to understand his potential, to achieve authenticity (the way of thinking and actions corresponding to his inner uniqueness and innate talent) and consciously apply his mind and abilities in a specific environment with a creative vector. "The human mind is the first wealth, and one of the highest satisfactions is to become the value quotient, the producer of the greatest being, which, thanks to its own realization, also develops the still necessary humanistic civilization." [4] Regardless of the activity field: mathematics or poetry, an intellectual leader also becomes a conductor of social and humanistic values for his environment and those to whom the results of his work are intended. [5] Humanistic ideas awaken a person's awareness of his central place and value in the universe and history [6]. For the first time, the concept of existential fulfillment, as a description of a person's life quality, began to be used by Viktor Frankl. Reflections on the meaning of human life undoubtedly led to the concept of happiness, which is close to hedonistic in philosophy today in the ordinary sense today. Hedonistic

* e-mail: mari.mikhaylova@list.ru

happiness (Epicurus) is the absence of suffering, well-being, fullness of desires, the taste of life. Existential fulfillment is closer to the concept of eudemonic happiness acquired through dignity according to Aristotle [9]. It means the meaningfulness of one's own life, the ability to contribute to life; the value of morality is assessed in direct proportion to happiness; to wish others' happiness is not so much a moral duty as the happiness of others is a condition of one's own happiness.

For a young intellectual leader, his potential determines the way to achieve existential fulfillment. The inability or frivolous attitude towards the realization of one's own natural potential gives a rise to the feeling of existential melancholy, helplessness, guilt, fear, doubt.

2 Analytical part

In order to study the interdependence of ideas about the humanistic values of an individual and the level of existential fulfillment of a young intellectual leader, a study was carried out in a sample of graduate students of the Russian higher technical school. The number of participants is 58. The average age is 25 years. 72% are male graduate students, 28% are female graduate students. The author's questionnaire, the questionnaire of moral and value choices, the value questionnaire of Schwartz S. [7], the scale of existence of A. Langle and K. Orgler [8] were used as methods. Frequency, correlation and content analysis were used to study the answers, and an ontopsychological approach was used to interpret the results.

The author's questionnaire made it possible to get an idea of how the position of humanism is perceived by graduate students. What do graduate students mean by the concept of humanism. What qualities are considered the most important for the realization of human potential.

2.1 Humanism, humanistic society

The concept of humanism for graduate students is: "spiritual health" and "decent social support for vulnerable groups of the population" (55.17%); "The flourishing of arts and science" (27.59%), "medicine will work so well that there will be no sick people left" (20.69%), "there will be no illiterate people" (see Fig. 1).

Correlation analysis showed that the opinion "humanism is spiritual health is interconnected (p-0.01) with a high self-esteem of vitality. The opinion "humanism is the flourishing of sciences and art" is interconnected (p-0.05) with the ability to feel the existential significance of an action, to orient itself in actions towards meaning, and not just towards the goal (ST, Längle).

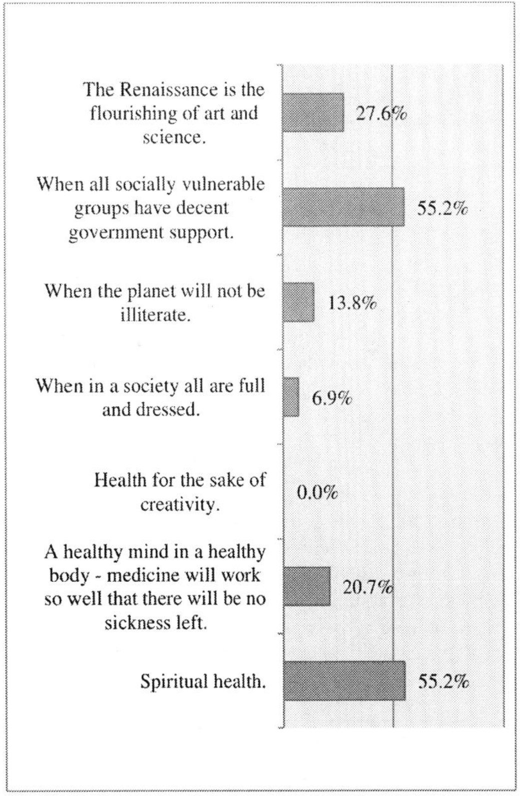

Fig. 1. Associations of graduate students with the concept of humanism.

The question "if we imagine the possibility of the existence of a society consisting of self-realized individuals and geniuses, then what characteristics it would differ" allowed us to determine how much the respondents accept the idea of a society in which everyone has the right and obligation to reveal their talent (see Fig. 3).

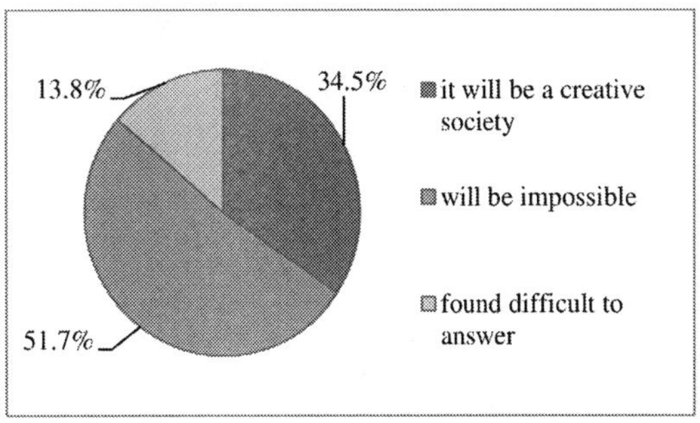

Fig. 2. The opinion of the respondents about the possibility of the existence of a humanistic society.

Almost half of the respondents 51.72% believe such an idea is impossible, 13.79% found it difficult to answer and only 34.48% answered that this is possible, agreeing that such a society would be characterized by creativity, flourishing of culture and various fields of activity.

2.2 Self-realization goals

Ideas about the goals of a person's self-realization reveal the personal orientation of graduate students. Most of 41.38% consider the goal of self-realization to be "achieving individual happiness"; 37.93% "disclosure of one's own abilities, potential" is a focus on self-realization of uniqueness in the classical humanistic sense; 10.34% "leave a trace in history" - an orientation toward power; none of the respondents chose to "plant a tree, build a house, have children" - a generally accepted stereotype.

Correlation analysis revealed that the goal of self-realization "disclosing one's own abilities and potential" is interconnected with the leader's quality "love for one's own business" (p-0.01). This relationship is important for the success of scientific activities. There is also a relationship with the opinion that in "a humanistic society, a person is sure that he is the main and responsible actor capable of self-realization," which means a willingness to go his own non-standard way. The goal of "achieving individual happiness" is significantly (p-0.01) interconnected with the idea that a person is a grain of sand in time. The goal "to leave a mark on history" is significantly interconnected (p-0.01) with the value of power (Schwartz) in practice.

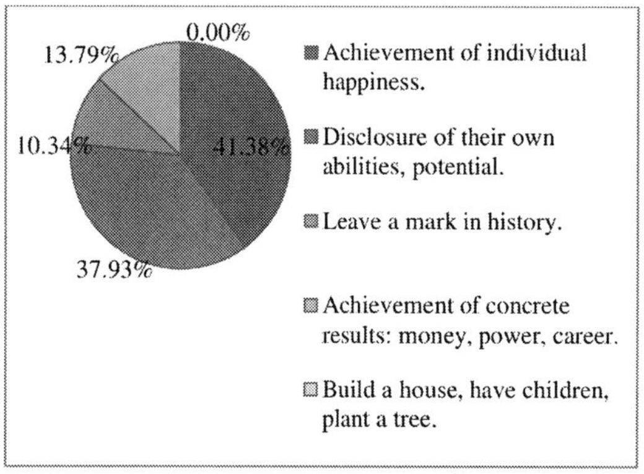

Fig. 3. The goal of human self-realization through the eyes of graduate students.

2.3 Qualities that cause respect for a Human

When asked which person evokes the greatest inner respect, the majority (62.07%) of the respondents answered that this is "a person who fights, acts, is able to transform any problem into victory". About a quarter of the respondents (24.14%) believed that the one who is happy and does not depend on anyone else deserves respect. 10.34% believe that respect should be given to a person who sacrificed his own happiness for the happiness of others, and 3.45% choose the criterion of respect - achieving high results.

One of the open questions concerned the search for a quality that inspires pride in all of humanity (see Fig. 4). The content analysis of the answers allowed all the answers to be divided into seven blocks: kindness-mercy (28.57%), purposefulness-responsibility (17.86%), creativity (14.29%), responsiveness (7.69%), intelligence (10.71%), self-sacrifice (7.14%), communication (7.14%).

The quality of creativity highlighted by the respondents (see Fig. 5) turned out to be significantly (p-0.01) correlated with high self-esteem of vitality and energy, with the concept of "humanism is spiritual health" and morality "internal duty - responsibility to oneself" - in other words: respondents who consider creativity to be the highest achievement of a person in a society of spiritually healthy people make decisions based on the inner truth of things, an inner ethical principle.

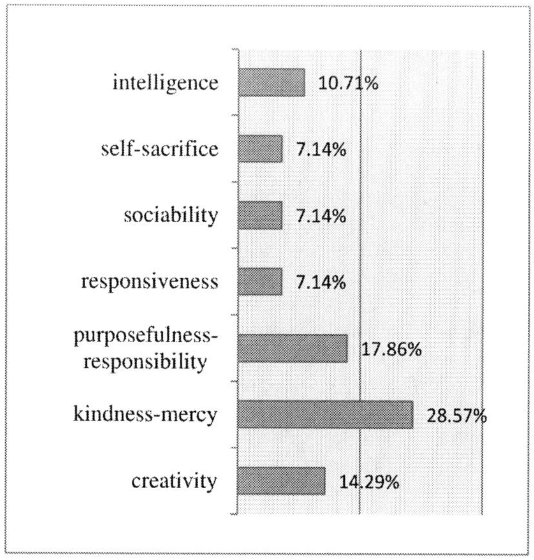

Fig. 4. According to the respondents, the most valuable human quality that inspires pride in humanity.

The opinion that pride in all of humanity evokes human intelligence is interconnected with the idea that in a humanistic society a person is sure that he is the main protagonist (p-0.01), capable of self-realization in a historical context, values of independence and achievement (Schwartz) and is inversely interconnected with the 6th factor of moral choice "role behavior (class ethics in sociology) - in other words, they rely on their intelligence and autonomy to achieve what they want. The opinion that pride in all of humanity causes kindness-mercy is inversely correlated with the 15th factor of moral choice "duty to oneself", with the values of self-achievement (Schwartz) and free emotionality (ST, Längle) - negative correlations seem to indicate ambiguous connotations the concept of "kindness" of the respondents.

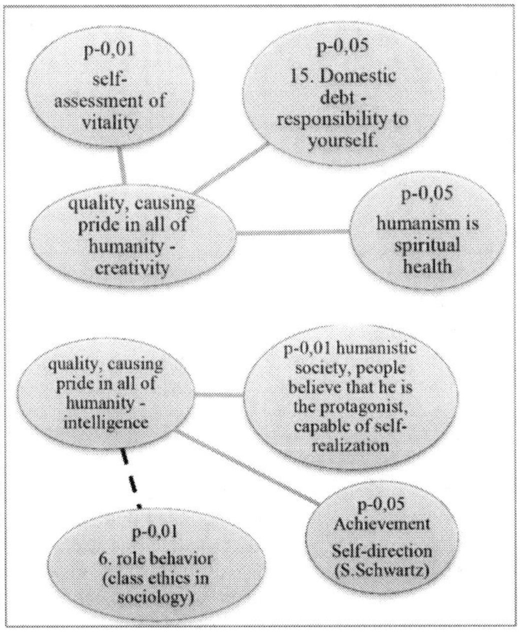

Fig. 5. Correlations with the opinion "creativity causes pride in humanity".

As part of the study of the humanistic orientation of young intellectual leaders, it was interesting to find out what leadership qualities graduate students consider a necessary condition for success. For the scale, a classification of the ontopsychological school was proposed, consisting of seven items, and six more items from the general classification of Stogdill Ralph were added, in total it was proposed to select up to 7 qualities (see Fig. 6). The most important qualities that distinguish an intellectual leader, graduate students consider: "ambition - a strong-willed desire for action", "the highest level of professional knowledge" and "love of one's own business."

The least important characteristics of a leader are recognized as follows: a leader is the expectations of followers (10.34%), a leader is the result of a special atmosphere, respect for family traditions and technical and rational development consistent with one's own identity. The latter quality, apparently, seems insignificant to graduate students, since the level of their education in the direction chosen more than 6-10 years ago seems to be no longer discussed. However, data analysis suggests a significant relationship between the quality of "intuitive rationality" with indicators of existential fulfillment (Längle). Intuition and rational analysis of intuition is one of the least studied areas of psychology, even though intuition always accompanies creative activity, the creation of new fundamental knowledge and invention. But for our sample, this is also a sensitive quality because engineering is an activity focused on accuracy, since it is associated with the safety of technology and special requirements are imposed on it, and intuition is an irrational way of knowing.

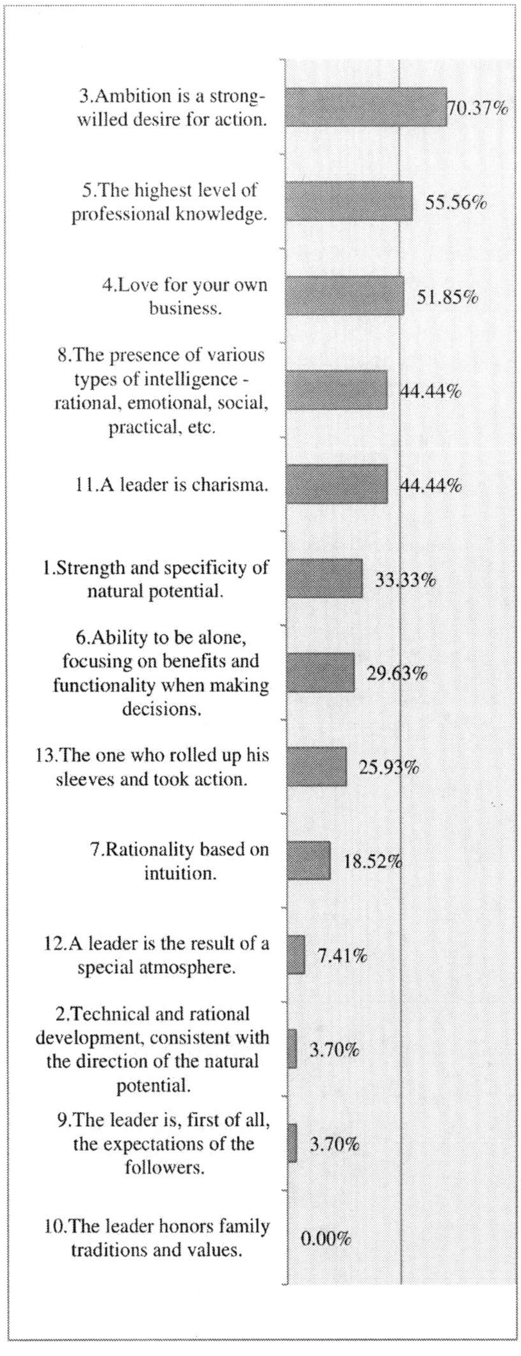

Fig. 6. The qualities necessary for an intellectual leader according to the respondents.

2.4 Freedom

"Freedom of choice or free will": 44.83% of graduate students tend to view freedom as an opportunity of creative development, this opinion is a conscious motivation for self-realization (see Fig. 7). However, this opinion is highly significantly interconnected (p-0.001) with the opinion that in a truly humanistic society a person will always receive support and is significantly interconnected (p-0.01) with the factor of moral choice - religiosity. 41.38% of graduate students see freedom as the ability to take responsibility. The opinion that freedom is a responsibility is inversely related (p-0.01) with the value of achieving socially desirable results in life and the value of power (Schwartz) in the personality profile (see Fig. 8). There is a relationship with ST (realism of perception) (p-0.05) and a significant relationship with P-openness of perception of the world and oneself (Längle). That is, graduate students who believe that to be or not to be free is everyone's personal choice, are satisfied with the quality of their lives and are not inclined to achieve socially desirable values. 24.14% of respondents answered that freedom is a utopia, these graduate students highly appreciate openness to new experience at the level of individual priorities.

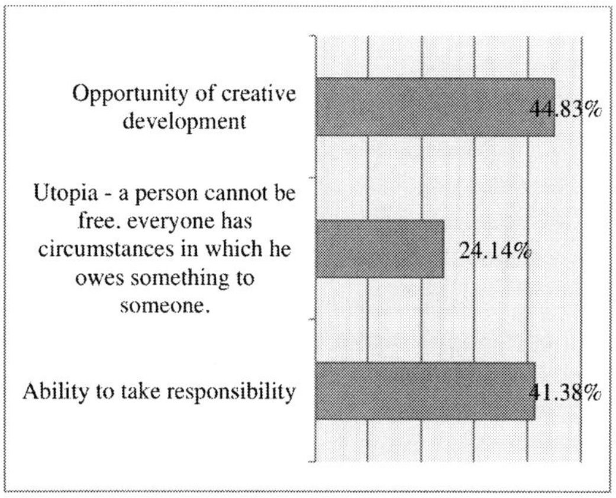

Fig. 7. The respondents' perceptions of freedom.

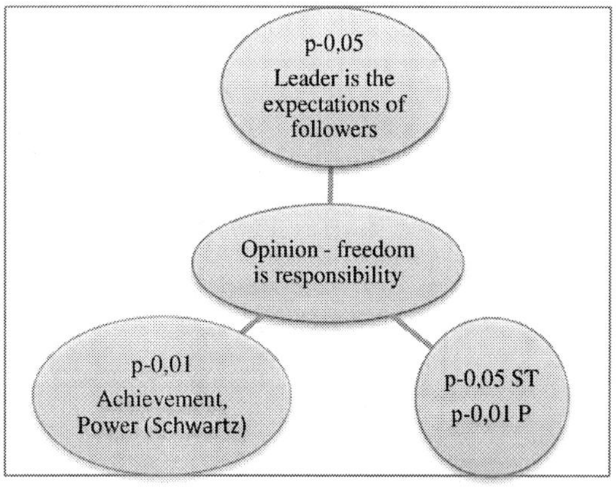

Fig. 8. Interconnection of respondents' ideas about freedom.

2.5 Analysis of the moral and value choices of graduate students.

The 16-factor structure of Dubov [10] is a description of the everyday moral consciousness (at least, the moral consciousness of Russians). And even though the principles representing these factors exist in the minds of each individual in the form of completely individual hierarchies, the named 16 meanings are, in their totality, the main system-forming vectors of public morality characteristic of Russian culture. Relying on them, people think about morality, according to the criteria they have designated, they evaluate their own and other people's actions. For a sample of graduate students of a technical university, the following factors of the methodology used are of the greatest importance: 8th factor - "global survival" (this principle can be interpreted as social Darwinism on a global scale) (4.03); 7th factor "hedonism" (it is one of the variants of hedonism in naturalistic schools, for example, in the history of ethics - Democritus, Socrates, it is based on the principle of pleasure that does not depend on the path of its achievement, the desire for this pleasure may contradict accepted moral concepts) (4.00); The second factor "aversion to immorality" is also called "the intrinsic value of morality" (the main motive of this intrinsic value also includes an aesthetic assessment of the situation, while it should be noted that "aversion to immorality" is expressed in absolute terms stronger than the feeling that good is beautiful) (3.93); 6th factor - "role behavior" (class ethics of sociological schools (3.90); 15th - as "internal duty" or "duty for moral behavior to oneself" (3.86) (see Fig. 9).

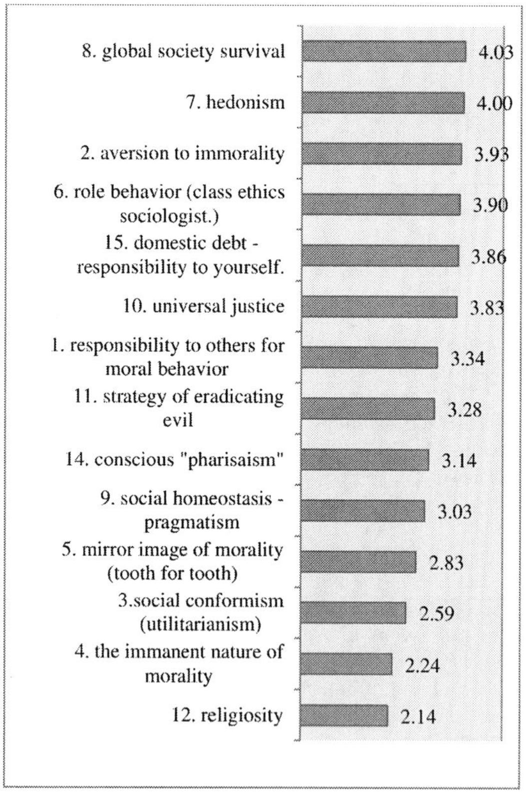

Fig. 9. Results according to the methodology of moral value choices.

According to the research of the author of the methodology (Dubov), the strengthening of 7,8,15 factors is characteristic of the social feelings of optimists (whose attitude to life can be defined as "everything is going well, and when not very well, I successfully overcome difficulties"). The moral motives of the group of intellectually successful young people we are studying can be designated as heterogeneous, multiple: the interests of society, personal pleasure, striving for moral aesthetics and duty to oneself. Moreover, the 2nd factor "the intrinsic value of morality" is the most important difference between our group and the results described by the author of the methodology in other groups.

2.6 Analysis and interpretation of the value questionnaire (Schwartz S.)

The test is usually used to study the dynamics of the transformation of values both in social groups (cultures) and for an individual in connection with his life problems. Values in this questionnaire mean "learned" needs that depend on the culture, environment, mentality of a particular society. The level of normative ideals - reflect the conscious beliefs about how to behave. The level of individual priorities diagnoses the values that the subject is guided by in specific situations of his life. The distribution of the respondents' answers showed that the values of universalism (understanding, tolerance and protection of the well-being of all people and nature), independence (independence of thought and choice of methods of action, in creativity and research activity) and security (safety and stability of society, relations and itself yourself). The lowest scores are for hedonism and stimulation (striving for novelty and deep experiences to maintain an optimal level of activity). To construct Schwartz's circle of values, the obtained mean values on the scales were converted into ranks (see Fig. 10). It should be noted that the pattern of normative values is practically repeated by the pattern of values of individual priorities - the ranking of the values of universalism, independence, security, conformity and kindness coincides. However, it can also be noted that the value of tradition (respect, acceptance of customs and ideas) is more proclaimed than an individual priority, and the opposite situation occurs with the values of hedonism and stimulation - apparently, they are subject to a certain censorship at the level of normative values, then there are deliberately considered unimportant, but at the same time they are used as a practice at the level of individual priorities.

From the point of view of the author of the methodology (Schwartz), the block of values - independence, stimulation and hedonism are an indicator of the individual's creative activity.

Correlation analysis showed that independence is interconnected with the 15th factor of moral choices "internal duty, responsibility to oneself" (p-0.01) and inversely interconnected with the 6th factor "role behavior - class ethics" (p-0.01). Stimulation (openness to novelty) is inversely related to the 10th factor "universal justice - one must reckon with others." Hedonism is inversely related to the 6th factor "role behavior - class ethics".

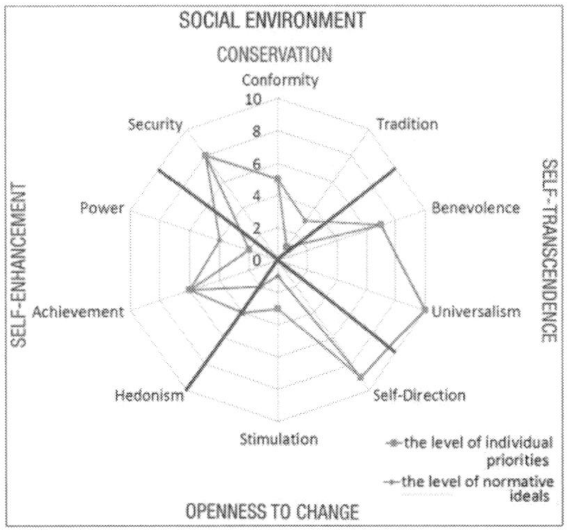

Fig. 10. Schwartz's circle of values for a sample of graduate students.

2.7 Analysis and interpretation of the Längle's questionnaire for a sample of graduate students

The methodology determines the quality of life of an individual associated with the feeling of life being filled with meaning. Existential fulfillment is interpreted as a feeling of accomplishment of the essential meaning, "real" life, the inclusion of one's potential in the dynamics of life.

The average value of the existential fulfillment scales (G-190.32) of the sample is in the middle zone below normal (see Fig. 11). The SD self-distancing subscale measures the stable ability for objective perception, the ability to free oneself from affects and prejudices. The mean SD (31.14) of the sample is below the norm according to the method (35.5) between selective perception and free attention for the perception of the world. The ST self-transcendence subscale measures the ability to feel values, to be guided not only by the goal, but also by the meaning of what is happening. The Freedom F subscale measures the ability to resolve. Responsibility subscale V measures the ability to act based on one's own values.

The analysis of indicators of existential fulfillment of the sample shows an active tendency towards an increase in internal openness, which gives a person the opportunity to go to the requests and offers of the outside world, sensitively correlating them with their own values. There is strength and emotional involvement, which is rather hampered by the constraint in the implementation of one's life. Inadequate norms and goals in the personality of an intellectual leader block openness and inclusion in life. Suppression of personality growth, leaving one's meaning for later leads to a reduction in freedom and responsibility, and can also lead to depression.

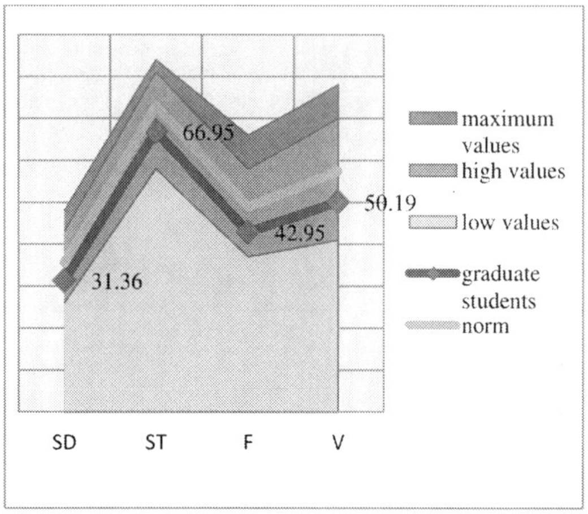

Fig. 11. Langle's Existence Scale for a sample of graduate students.

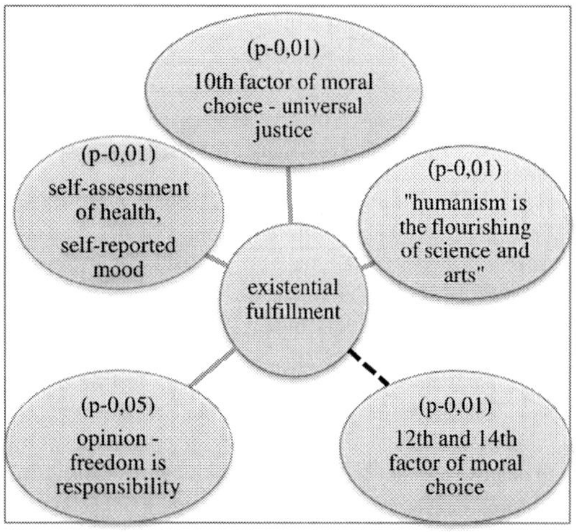

Fig. 12. Correlations with indicators of the scale of existential fulfillment (G)

The most significant interconnections of the parameters of existential fulfillment can be observed with self-esteem of health (see Fig. 12), with self-esteem of mood and vitality, the opinion that "freedom is first responsibility" and "humanism is the flourishing of science and arts". The parameters of existential fulfillment are inversely related to the factors of moral choice: pharisaism and religiosity (Dubov). And they have a direct relationship with the factor of moral choice - universal justice (it is necessary to consider the interests and rights of other people). The moral of universal justice in this interpretation can mean that an intellectual leader is a vector of development not only for himself, but also for others. Respect for others is a natural part of a social leader's personality.

3 Conclusion

In conclusion, we would like to note several features. The comprehension of the proposed questions by graduate students made it possible to highlight important interrelationships of personal positions. If a person's goal is the disclosure of his own abilities, potential, then it is realized through love for his own work, because in a humanistic society he is the main and responsible actor capable of self-realization. If creativity is considered the highest achievement of a person, then because it is a duty to oneself and humanism is then interpreted as the spiritual health of people. Those graduate students who view freedom as an opportunity for creative action expect society to support the initiative, which is certainly true from the point of view of building a humanistic society, but non-functional as a behavioral strategy.

An analysis of graduate students from the point of view of the values of a creative personality (independence, stimulation and hedonism) showed that the independence of thought and action is interconnected with the morality of internal duty, it is responsibility to oneself. The more independence and desire to enjoy life (hedonism) is manifested by a person, the less he or she is guided by class traditional ethical principles. I would like to note that pleasure in this case does not mean generally accepted, but precisely what satisfies the aesthetics of the leader. Openness to novelty and hedonism are reduced by postgraduate students' ideas about their own insignificance, lack of faith in their uniqueness, and a desire to meet the expectations of other people.

References

1. A. Meneghetti, *From historical to eternal humanism*, (Moscow: Antonio Meneghetti Scientific Foundation, 2014)
2. V.A. Dmitrieva, International symposium "Forming leadership for future development", UN, Geneva (Switzerland) (2019)
3. A. Menegetti. "An intellectual leader: what he can and should do" January 5, 2001 in New York (USA), in hall 1 of the UN Crystal Palace
4. A. Meneghetti, *Youth and the Ethics of Life*, (M .: Antonio Meneghetti Scientific Foundation, 2013)
5. M.V. Mikhaylova, M.A. Revina, EPJ Web of Conferences, **224**, 06003 (2019) https://doi.org/10.1051/epjconf/201922406003
6. W.F. Ryan, Ultimate Reality and Meaning, **36**(1-2), 47-63 (2019) https://doi.org/10.3138/uram.36.1-2.47
7. Sh. Schwartz et al, Psychology. Journal of the Higher School of Economics, **9**(2), 43-70 (2012)
8. I.N. Mainina, A.Yu. Vasanov , Psychological Journal, **31**(1), 87-99 (2010)
9. R. Kleszcz, Problems of Science, **53**(1), 29-48 (2017)
10. I.G. Dubov, A.A. Khvostov, Questions of psychology, **5**, 87-99 (2000)

Study of the Distribution and Modification of Writing in Proto-Chinese Language Communities: Correct Identification of Hieroglyphic Signs in the Jiaguwen Inscriptions

Maia Egorova[*], *Aleksandr* Egorov

Peoples' Friendship University of Russia (RUDN University), RU-117198, Moscow, Russia

Abstract. The Chinese language uses the ideographic writing. The outer form of Chinese characters in 3rd century BC and later underwent strong changes, which were caused mainly by change in the technique of writing. In the first century AD the writing style that has survived to the present time. We recall some of main results of the spread and change of linguistic information in certain Chinese community obtained with the help of a dynamic nonlinear model. Undoubtedly, Oracle script writings Jiaguwen (甲骨文) played an important role in this process. We consider it quite legitimate to discuss rather about the written language of the Chinese language, as main link between ancient Chinese, Middle Chinese, and modern Chinese languages Putonghua. In the study of the oldest examples of art, culture and science, for example, such as Jiaguwen, there is a need to study ancient hieroglyphs depicted on the surface of a tortoise shell, copper vessel or stone. New method for correct solving of an inverse problem for pattern recognition is presented. Our method is based on photometry of the studied surface. The obtained data are digitally processed in order to determine the characteristics and parameters characterizing the properties of the investigated surface. The possibilities and limitations of the method, as well as the mathematical model of the groove in the optical range, are given. The main attention is paid to the development of methods for more accurate calculation of the number of solitary grooves and structural elements such as hieroglyphs or ancient Chinese radicals (部首 / bùshǒu). This task is part of the complex inverse problem of pattern recognition, namely: identification of various structural elements, for example, ancient hieroglyphs depicted on the surface of a tortoise shell. To solve this important problem, two main methods were proposed, which are based on the study of both the initial photometric data and the study of their Fourier spectra. The first method allows you to visually identify the location of the characters along the photometric line and determine their approximate number. The second method allows us to get a stable and unambiguous solution to this problem: to determine accurately the number of characters on the photometric line.

1 Introduction

The modern Chinese language i.e. Put onghua (Pǔtōnghuà / 普通话) was created artificially in the middle of the 20th century [1-4]. It is based on the vocabulary and grammar of Mandarin (Mandarin Chinese or North Chinese) and the Beijing dialect for pronunciation, i.e., the phonetics and vocabulary of Putonghua is based on the pronunciation standard of the Beijing dialect, which belongs to the northern group of dialects of the Chinese language.

We consider it quite legitimate to discuss rather about the written language of the Chinese language, as the main connecting link between the ancient Chinese, Middle Chinese, and modern Chinese languages [4]. Three periods are usually distinguished in the development of the Chinese language: Ancient Chinese, Middle Chinese and Modern Chinese. In the history of the living spoken language, the ancient Chinese period ends, most likely, around the sixth century AD, ancient Chinese language from the V-th century BC and to the II-nd century AD can be called classical, the language of earlier monuments is pre-classical, and the language of the third and fourth centuries AD is late ancient Chinese.

The Chinese language uses the ideographic writing; the signs of Chinese writing are called hieroglyphs. In ideographic writing, in contrast to phonetic writing, each sign does not correspond to a sound unit, that is, a sound or syllable, but a significant, a word or morpheme, which are written as a whole, and not divided into their constituent sounds. Different words are spelled differently, even if they sound the same. From this it follows that the same word has been spelled the same way throughout the history of the Chinese language, no matter how its pronunciation changes. Chinese writing does not give us any direct evidence of phonetic changes that have occurred in the language, or information about the phonetic differences between Chinese dialects; the

* Corresponding author: Mey1@list.ru

reverse is true, a text written in hieroglyphs can be read aloud in any of the modern Chinese dialects, as well as in Japanese, Korean, Vietnamese and in any language using Chinese writing.

The outer form of Chinese characters in 3rd century BC and later underwent strong changes, which were caused mainly by change in the technique of writing. In the first century AD the writing style that has survived to the present time first appeared (the model or charter letter kǎishū / 楷书). It differs from previously existing ones in that all the characters in this style are built from a small number of the same basic graphic elements (horizontal line, vertical line, hook, dot, etc.), which completely lose all resemblance to the drawings from which they came.

Oracle bone script writings Jiaguwen (甲骨文, XIV-XI centuries BC) on turtle shells, oracle bones and bronze vessels are among the most ancient Chinese art and culture samples [1-8]. Jiaguwen writings on turtle shells are hieroglyphic inscriptions that record the results of fortune telling or predictions. They are unique in their own way. These objects are often poorly preserved, but are objects of great historical, cultural and scientific value. Basically, their research is devoted to the study of hieroglyphic inscriptions applied to them with the aim of interpreting their content, as well as identifying ancient keys (Chinese radicals) [5-8]. We should note that the structure of the inscriptions practically did not change throughout all periods; they included the date, the name of the fortuneteller, the question, answer and mark of execution; however, the calligraphic style underwent a significant change, from large rough characters of an early period to the smallest ones that are hardly distinguishable by eye. The number of different characters on shells and bones is approximately 5000 characters, of which approximately 1500 are identified with modern characters [8]. In this regard, the most important task arises: to develop a method for recognizing a certain number of important characters depicted on the oracle bones that can be find in the specified list [8].

The present work is devoted to the development of a new method for solving some inverse problem for pattern recognition. The main goal of the work is to demonstrate the possibility of correct identification of groove-type elements that are an integral part of various complex structural elements, according to the obtained digital surface profiles of the sample under study. In particular, in technical applications, these can be various (rectangular, triangular, trapezoidal) solitary grooves, which are the constituent elements of such widely used structures as diffraction gratings (see e.g. [9-11]). In the study of the oldest examples of art, culture and science, for example, such as Jiaguwen, there is a need to study ancient hieroglyphs (keys) depicted on the surface of a tortoise shell, copper vessel or stone. As our study has shown, structural elements such as grooves are part of ancient hieroglyphs, therefore, their more detailed further study, including using rigorous mathematical methods, will provide new results in the study of Jiaguwen (Jiǎgǔwén) [12].

In this article, a new research method is used [12] as a basis, which can be a good addition to the currently used traditional methods (mainly visual) of research [1, 4-8]. This method allows, using digital data of surface photometry, to determine various characteristics and parameters that characterize the properties of the surface under study, including statistical ones [12]. The main advantages of our research method are its contactlessness, informational content, and potentially sufficiently high resolution, for example, by statistical parameters of the surface profile, which allows us to speak about its promising potential in the study, for example, of such ancient examples of art and culture as Jiaguwen.

The results obtained are based on the correct physical and mathematical approach, use rigorous mathematical models and, as a consequence, can be useful when developing other mathematical models, allowing to adequately describe structural elements such as grooves, which are included, for example, in the composition of the most ancient Chinese radicals (部首 / bùshǒu), as well as the hieroglyphs themselves. The developed method can be used in interdisciplinary research, which utilizes methods of mathematical linguistics, lexicostatistics and sociolinguistics, especially in the statistical analysis of hieroglyphic inscriptions.

2 Formulation of the problem. Numerical results and their analysis

2.1. Dynamic nonlinear model of distribution and changes of linguistic information

In our previous works [1, 4] we used a dynamic model of the spread and change of linguistic information in a certain Chinese community. Let us recall some of the main results we obtained earlier. This dynamic model can be described by a nonlinear equation:

$$I_{m+1} = \left[a_1 I_m (M - I_m) + a_2 (M - I_m)^2\right]\lambda , \qquad (1)$$

where I is the value of the analyzed linguistic information, m = 12, ... (m = 1 corresponds to the first "measurement", i.e., I_1 is the initial value, for example, at some initial moment in time t_1); a_1 is a coefficient that characterizes the distribution of linguistic information in the contacts of "unknowing" with "knowing" this information I_n ; a_2 is coefficient that characterizes the impact only on "unknowing"; M is the maximum value of linguistic information; and λ is a task parameter (control parameter). The nonlinear equation (1) allows one to explore the process of changing the spreading linguistic information depending on time and other parameters.

After replacing the variables, equation (1) can be written as follows:

$$y = \lambda_1 x(1-x) + \lambda_2 x(1-x)^2 , \qquad (2)$$

where $y = I_{m+1}/M$, $x = I_m/M$; and $\lambda_1 = a_1 M\lambda$, $\lambda_2 = a_2 M\lambda$ are new control parameters of the system and $0 \le x \le 1$.

We use the data obtained in [1, 4] in a numerical study of the distribution and change of linguistic information in the proto-Indo-European language community to analyze the distribution of language information in the proto-Chinese language community, given that there are no riders in it. As a result, in computer modeling, the ratio of the coefficients λ_1 and λ_2 was selected based on the ratio of the average speeds v_1 and v_2 (in km/h) of different types of pedestrians: $\lambda_1 : \lambda_2 \propto v_1 : v_2 \approx 3:3 \approx 1$. We consider the language policy of China as another important condition for the task that is aimed at linguistic unification, i.e., the creation of one national language in the future [1, 4].

We have conducted an analysis of the spread of language information in the proto-Chinese linguistic community (see e.g. [1, 4]). The main results of a computer simulation were presented in our papers [1, 4]. A brief analysis is given of a trend that originated in antiquity and developed with varying intensity depending on historical eras, to suppress dialects and reduce oral linguistic diversity to a single unified norm, which resulted in the creation in the middle of the 20th century of the national language of Putonghua (Pǔtōnghuà / 普通话). It is noted that final linguistic unification can be achieved no earlier than the year 3500. An estimate of the number of possible elementary "symbols" ("signs", "radicals" or "base units") of a hypothetical proto-Chinese script was obtained of 500–1000 characters for the first time. By basic units we mean the number of primary radicals. To verify our calculations, we examined modern typical SMS messages and made an original comparison of typical hieroglyphs and radicals from these messages with hieroglyphs and radicals on tortoise shells. As a result, matching hieroglyphs and radicals were discovered, which indicates the validity of our approach and the validity of the calculations [4].

We should note that mathematical model of the process of wave propagation and change of linguistics information in this system is described by a system of integral-differential equations and it discussed in sufficient detail in our previous papers (see e.g. [1]), therefore we do not consider it here.

2.2. Photometric analysis of the Jiaguwen example

At this stage of the study, there was an urgent need to develop a method for correct identification of groove-type elements that are an integral part of various complex structural surface elements, according to the obtained digital surface profiles of the sample under study. This problem was identified in our previous research when working with a number of ancient samples. It is especially important in light of the fact that the studied samples were the most important element in the distribution of proto-Chinese script, proto-Chinese and then Ancient Chinese and Middle Chinese languages [1, 4, 12].

Structural elements such as grooves, which are part of the oldest hieroglyphs depicted on the surface of the tortoise shell (XIV-XI centuries BC) are the object under study. The subject of research is the profile of a dielectric surface with hieroglyphs depicted on it. The main task is to determine the number of elementary structural elements and the number of hieroglyphs on the scanning line according to the photometry data of the surface of the sample under study.

In this work, the main attention is paid to the development of methods for a more accurate calculation of the number of grooves and structural elements such as hieroglyphs (ancient keys or bùshǒu). This task is part of a complex inverse problem for pattern recognition, namely: identification of various structural elements, for example, ancient hieroglyphs applied to the surface of a turtle's shell. The solution of this important problem will make it possible to avoid errors at the initial stage, for example, when approximating (fitting) autocorrelation functions (ACF) are found from the obtained digital profiles of the surface of the sample under study [9-12]. As a result, the correct data will be analyzed at all subsequent stages. The importance and urgency of solving this problem is due to the fact that this problem arises in the study of many structural elements of Jiaguwen. In the case of automation of the process of statistical processing of ancient hieroglyphic inscriptions, the solution of this problem will undoubtedly lead to a significant reduction in the time spent on conducting the entire complex of studies of ancient Jiaguwen samples.

Recall that our method is based on photometry of surface profiles, in which a sampling of brightness levels in a discrete set of points is carried out, followed by conversion into digital form [12, 13]. The obtained data (optical response) are digitally processed in order to determine the characteristics and parameters (including statistical ones) characterizing the properties of the investigated surface [12-15]. As a result, from the digitized distribution of the intensity of the light scattered by the surface, one can find, for example, the form of the approximating ACFs, and also determine the corresponding statistical parameters of the roughness of the profiles of the studied surface: the standard deviation and the correlation radius (see details in [9-13]).

In the present work, using the numerical analysis of the digitized photometric data of the profile of the dielectric surface, we got the results that made it possible to obtain a more accurate value of the number of keys (ancient hieroglyphs, bùshǒu). In the future, a previously developed statistical research method can also be applied, which makes it possible to find the statistical characteristics and parameters of the profiles of the studied sample. The shell of a turtle (XIV-XI centuries BC), on the surface of which hieroglyphic inscriptions were written, is a typical example of Jiaguwen (see photo in Fig. 1).

In the photograph given in Fig. 1 a horizontal line (along the axis x) along which the surface was photometric is shown. On the horizontal line, there are 10 groove-type structural elements (numbers from 1 to 10), most of which are undoubtedly part of the hieroglyphs.

Photometric data of the surface profile along the horizontal line shown in Fig. 1 are shown in Fig. 2 and Fig. 3. The signal-to-noise ratio S/N was greater than 7 in average.

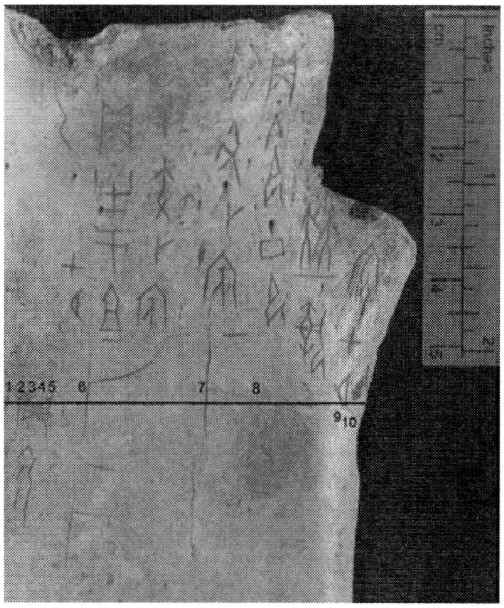

Fig. 1. Jiaguwen example (甲骨文 – "writing on turtle shells"): hieroglyphic inscriptions fixing the results of fortune telling or predictions [7].

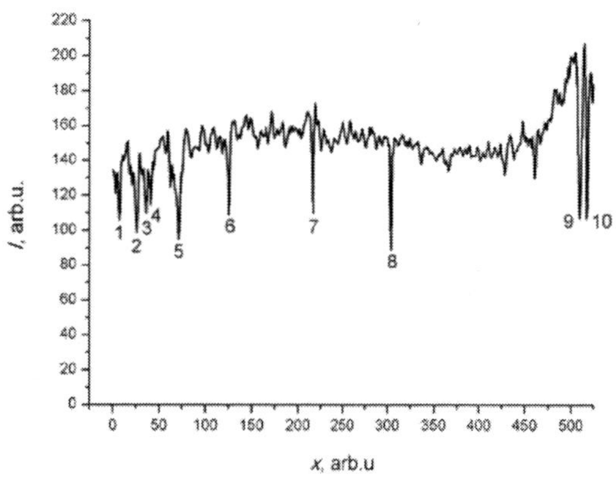

Fig. 2. Photometric data of the surface profile along the horizontal line shown in Fig. 1. 1 arb.u. (along the x axis) ≈ 400 μm.

In Fig. 2 the numbers from "1" to "10" mark the dips in brightness I corresponding to 10 structural elements of the groove type (see Fig. 1), which are part of the oldest hieroglyphs written in the investigated surface.

The level of the ratio S/N could change from realization to realization, but on average it was at least 7.

2.3. The mathematical model of the groove. Scalar scattering

The mathematical model of the groove in the optical range of wavelengths can be described by the following expression (see e.g. [14]):

$$f(x) = \exp[-i2kh(x)], \tag{3}$$

where $k = 2\pi/\lambda$ is wave number, λ is wavelength of radiation incident on an object ($\lambda \approx 0.6$ μm), $h(x) = -h$, at $|x| \leq \Delta x/2$ and $h(x) = 0$ at $|x| > \Delta x/2$, Δx is a groove width. In fact, the quantity $f(x)$ in (3) is a function of the groove shape or the so-called optical profile.

It is important to note that the relationship in the general case of the nonlocal reflection coefficient r and the scattering matrix is determined on the basis of the scattering theory. However, under certain lighting conditions and recording the response, it becomes possible to describe the object by the local reflection coefficient. In particular, if the illumination wave is plane and the projection of the modulus of the wave vector on the plane of the object $q = 0$, then the reflected wave is determined by one column of the scattering matrix. Then the object can be characterized by the local reflection coefficient r, which depends only on the object shape function, which is, in fact, the Fourier transform of the scattering amplitude [14]. It is this method of the signal registration that was used in this paper.

When illuminated by a focused wave, an object can also be described by a local reflection coefficient if such a registration scheme is chosen, in which only one (for example, normally reflected with a projection $q' = 0$) plane wave is recorded. Then the object is described by one row of the scattering matrix, and the corresponding local reflection coefficient is determined by a similar dependence (for more details, see, for example, in [14]).

Such a relatively simple mathematical model of the groove can be used in the case of a scalar consideration of the problem of forming an optical response. Indeed, in the analytical and numerical solutions we use only the distribution of the real function $I(x)$, obtained from photometric data of the profile of the investigated dielectric surface illuminated by a plane wave.

This mathematical model is also valid from a physical point of view, since structural elements of the groove type were investigated, which were homogeneous along the vertical axis y and represented an (almost) rectangular groove in cross-section with a width Δx and depth $h(x)$ much greater λ [14].

If the profile of any groove contains parts that slightly deform a rectangular profile, then parts with dimensions smaller than the wavelength will be smoothed or lost with this approach. Note that with appropriate processing of the response, including in the presence of noise, some of these details can be restored when solving the inverse ill-posed problem. Moreover, in the absence of noise, the image of a localized object (of finite length) can be reconstructed exactly and with superresolution. A detailed discussion of this problem is beyond the scope of this article (see e.g. [10, 14]).

In the scalar approximation, the following expression can be written, connecting the average value of the height $h(x)$ of the profile irregularities and the intensity of light scattered by the surface [10, 15]:

$$h(x) \approx \left(\frac{\lambda}{2\pi}\right) B \ln[I_i / I_r(x)], \quad (4)$$

where B is dimensionless constant that is close to 1 in many practically important cases; $I_{i,r}$ are light intensity incident on a surface and reflected (scattered) from it. Comparison of expressions (3) and (4) shows that the real part $f(x) \propto |\ln[I_i / I_r(x)]|$. In the case of a solitary groove-type object, the profiles $h(x)$ in expressions (3) and (4) are almost identical. In the case of studying an extended surface profile, where there is a set of similar grooves, expression (4) will describe the profile of some groove averaged for a given section of the profile.

Note that the profile $h(x)$ can be found more accurately using expression (3) than using expression (4). However, in the first case, to solve the inverse problem of reconstructing the groove profile, it is necessary to use complex methods of recording the real and imaginary parts of the nonlinear response. For example, register the amplitude and phase of a complex response or a correlation signal that requires the development of complex algorithms for its processing (see, for example, [14]). As a rule, this problem arises when solving various problems of super-resolution, i.e. exceeding the Abbe-Rayleigh diffraction limit [9, 10, 14]. In our case, there is no such need, since the width and depth of the investigated grooves are many times greater than the radiation wavelength (see below). And, therefore, it is sufficient to use the scalar approximation and register only the intensity distribution when photometrying the profile of the investigated dielectric surface $I(x)$. From expression (4) it follows that to determine the surface profile, it is sufficient to know the value λ and the ratio of I_i / I_r.

In Fig. 3 shows the inverted normalized brightness distribution I, shown in Fig. 2, for one structural element of the groove type (marked as "8" in Fig. 2). Using the graph shown in Fig. 4, found the groove width to be reliably determined from the response at 0.5 of the maximum: $\Delta x \approx 0.75$ mm (>> λ). Estimation of its depth Δh: no more than 2–4 mm (>> λ).

If the width Δx (and/or depth Δh) of the groove is comparable to or less than the radiation wavelength λ, then it will be necessary to use additional experimental and numerical possibilities to increase the resolution of the research method described in this work.

As you can see in the photo (see Fig. 1), the width of the vertical line of the cruciform element inside the hieroglyph located above this groove is about 2 times wider: ≈ 1.5 mm. Groove "8" was chosen as an example because it is the least visible visually, especially at the intersection with the horizontal line along which the surface profile was photometric.

Similar data were obtained for grooves "6" and "7", which are better visually distinguishable. As a result, the data obtained allowed not only to demonstrate the resolution of the method, but also to estimate the approximate width of the cutter used by the scribe - most likely it was in the range from 1 to 1.5 mm. In this case, the depth of the structural elements of the ancient hieroglyphs was probably approximately at first 0.5–2 mm. Note that over time, the size of the structural elements of hieroglyphs could probably increase slightly due to natural reasons.

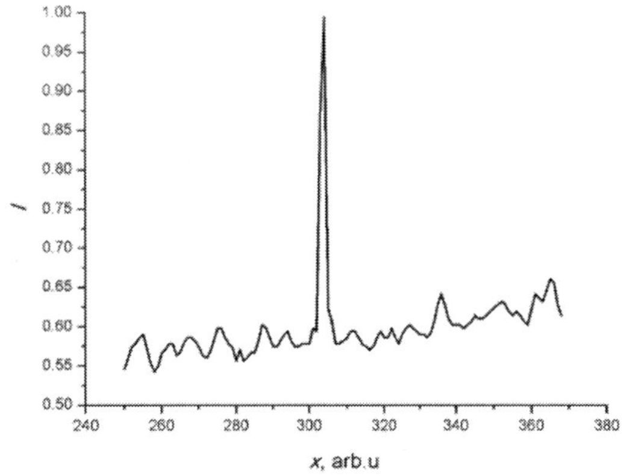

Fig. 3. Normalized inverted profile photometric data of one structural element of the groove type (1 arb.u. ≈ 400 μm).

Analysis of the numerical data of photometry of the surface profile shown in Fig. 2 and Fig. 3, allowed receiving results which will be useful in the following subsection of work for more exact definition of number of Chinese radicals (部首 / bùshǒu). To solve this problem, we will use a statistical approach, in which the useful signal (data containing information about the grooves) is separated from the total signal, which also contains the noise component. To do this, digital data are pre-processed (smoothed) to reduce the impact of different interference.

3 Estimation of the number of ancient Chinese hieroglyphs (radicals / bùshǒu / 部首)

3.1. Estimation of the number of structural elements of hieroglyphic inscriptions such as grooves by the integral method

As an initial step, we define the number of grooves on the scanning line (see Fig. 2). First, indeed, as can be seen from the photometric data in Fig. 2, grooves are the most important structural elements of hieroglyphic inscriptions that record the results of divination or predictions of Jiaguwen (甲骨文). Therefore, the study of grooves is an important stage in solving the complex inverse problem of pattern recognition, namely, the correct identification of ancient hieroglyphs applied to the surface of the tortoise shell. Secondly, it will allow us to demonstrate more widely the possibilities and limitations of our research method. Indeed, as the tortoise shell itself [4-7], and selected for analysis the grooves too can be considered as some test samples.

To achieve this goal, when calculating the number of grooves, an integral method is applied, the essence of which is reduced to: 1) finding the area S_0 under the entire smoothed brightness curve (see Fig. 2); 2) finding the area S_1 under the smoothed brightness curve for one structural element of the groove type (see Fig. 3); 3) determining the number of grooves by dividing S_0 by S_1. After performing the integration procedure, we got: $S_0 \approx 78690$, $S_1 \approx 8112$. As a result, we got: $S_0/S_1 \approx 9.7$. Thus, it was determined that the number of grooves is 10, with an error 3.1%. Note that to find the corresponding areas, we used numerical integration (in arbitrary units). To increase the accuracy, you can use several of the same type of photometry profiles with subsequent averaging over the sample and additional smoothing. Undoubtedly, this will require additional costs of computing resources and time.

3.2. Estimating the number of ancient hieroglyphs using the Fourier transform

In Fig. 4 shows a photograph of a fragment of the sample under study (see the photograph in Fig. 1 in the middle part), which shows a horizontal line (along the axis x) along which the surface photometry was made.

On the horizontal line there are 11 structural elements of the groove type (numbers from "1" to "11"), which undoubtedly are part of 7 hieroglyphs.

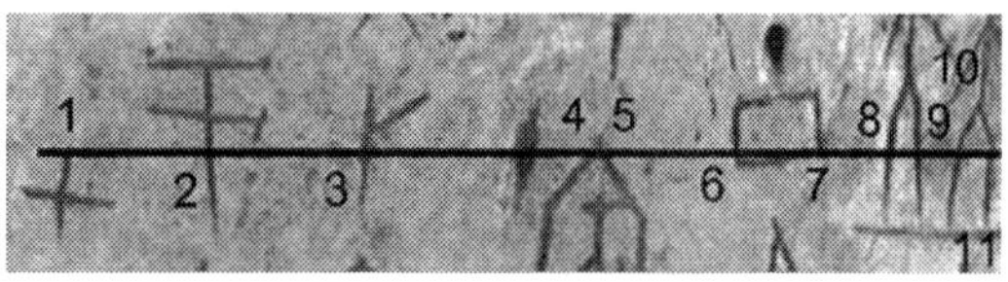

Fig. 4. Photo of the test sample, which shows 11 structural elements of the groove type (numbers from "1" to "11").

Let us show that to estimate the number of Chinese radicals $\Im$ (bùshǒu), in contrast to the previously used method [4, 12] and the integral method described above, one can also use the Fourier transform. As a result, it becomes possible to visually identify the location of hieroglyphs along the scanning line and the ability to determine their approximate number. This method can be used in addition to our previously proposed methods for estimating the number of ancient keys in order to more accurately calculate the value $\Im$.

Let us apply the numerical Fourier transform to the smoothed data of photometry of the surface profile. Fig. 5 shows one of the obtained normalized Fourier spectra $F(q)$ of photometric data of the surface profile (see photo in Fig. 4). The coordinate q characterizes the position of spectral components of the Fourier spectrum $F(q)$ in the Fourier plane [10, 11, 14], q is measured in inverse units of spatial coordinates x and y.

A relatively simple method for determining the number of keys $\Im$ (Chinese radicals) in this case is as follows. The numerically obtained Fourier spectrum $F(q)$ is normalized, by dividing all values of $F(q)$ by the main maximum at $q \approx 364$ arb.u. Then one can find the maximum dips for the function $F(q)$. Then, to count the keys in the function $F(q)$ characteristic dips are found, from which less deep adjacent dips are excluded. In Fig. 5 these dips are in the vicinity of deeper dips, marked with numbers "1–7".

The accuracy of this method can be improved, for example, by setting a certain threshold in terms of the value level of $F(q)$ (see Fig. 5). In this case, one can more accurately determine the number of hieroglyphic characters (keys). For example, setting the threshold at 0.5, we get exactly 7 characters i.e. Chinese radicals (side lobes of lower intensity are neglected). With this approach, obviously, there is no error in the determination, but the question remains of the exact identification of hieroglyphs location.

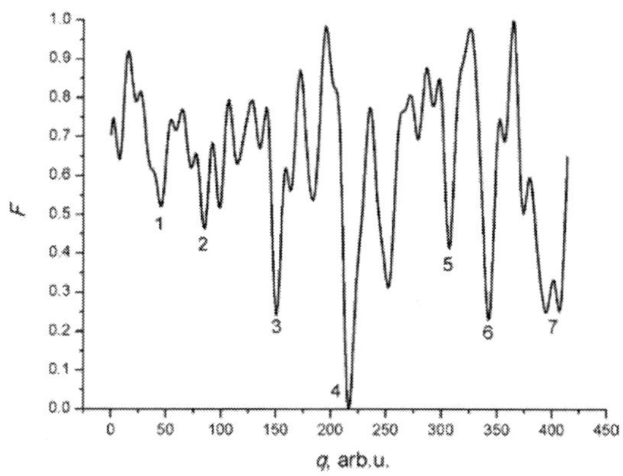

Fig. 5. Normalized Fourier spectrum $F(q)$. $S/N \geq 10$ (1 arb.u. $\approx$ 0.05 μm^{-1}).

It should be noted that the above (integral) method for determining the number of hieroglyphs $\Im$, despite certain simplicity is less accurate than the method given in this section due to some understandable subjectivity.

4 Discussion

By the beginning of the XXI century in the course of the implementation by the PRC government of the program for the dissemination of the official Chinese language Putonghua throughout China, the Chinese language is gradually being introduced into those areas where it was historically not widespread [1, 2, 16]. Definitely, Oracle script writings Jiaguwen played an important role in this process, especially during this period: V-th century BC -

II-nd century AD. This is why the research of Oracle bone script writings Jiaguwen (甲骨文) is so essential.

It is important to emphasize that when solving the problem of finding the number of hieroglyphs $\Im$ using the numerical Fourier transform, we have shown the possibility of not only a stable, but also an unambiguous solution to the problem of determining the number of hieroglyphic signs (Chinese radicals / bùshǒu /部首), as well as their rather precise location.

For a more accurate determination of the number of keys $\Im$ (ancient hieroglyphs), one can perform an effective numerical smoothing of Fourier spectrum. We should note that the smoothing procedure, on the one hand, suppresses the influence of noise, and on the other hand, it can lead to the loss of useful information.

Method for determining the number of ancient keys (bùshǒu) $\Im$ and their precise location requires some improvement and enhancement, at least to exclude false triggering of the numerical Fourier transform when processing the original data with noise. For this purpose, it is possible to improve, for example, the method of counting the number of grooves by the integral method. This will at least improve the accuracy of determining the location of the hieroglyphs.

5 Conclusion

In this paper we primarily recall some of the main results of the spread and change of linguistic information in Chinese social-linguistic community obtained with the help of a dynamic nonlinear model.

A new method of researching writing on turtle shells (Jiaguwen / Jiǎgǔwén / 甲骨文), which dates back to the XIV-XI centuries BC, has been developed. The method is based on photometry of the investigated surface of the sample and subsequent digital processing in order to determine characteristics and parameters characterizing the properties of the investigated surface.

The possibilities and limitations of the method are described. We demonstrated the possibility of accurate counting of the number of solitary grooves and structural elements such as hieroglyphs (ancient keys / Chinese radicals / bùshǒu). The main advantages of the method are its contactlessness and high resolution in terms of the number of hieroglyphs. The method allows, in principle, to automate the process of statistical processing of ancient texts of the Jiaguwen type.

Therefore, the described method is undoubtedly promising in the study of the oldest examples of art, culture and science, like Jiaguwen, especially in cases where visual identification of hieroglyphic inscriptions is difficult, and the elements of the inscriptions are poorly visible or poorly distinguishable. A certain simplicity and clarity of implementation make it possible to apply this method in interdisciplinary research, which involve specialists from different subject areas of knowledge, for example, humanitarian (cultural studies, linguistics, history, etc.).

Considering that the oldest samples of Jiaguwen under study often have poor preservation, as well as high historical, cultural and scientific value, our investigate method can be a good and promising addition to the traditional (mainly visual) methods used to research these antique objects.

Acknowledgements

The publication was supported in part by the "RUDN University Program 5-100". Authors are grateful to the colleagues for kindly advices, which contributed to the best completion of this paper.

References

1. M. Egorova, A. Egorov, T. Solovieva, EPJ WoC, **224**, 06009 (2019)
2. *The Peopling of East Asia: Putting Together Archaeology, Linguistics and Genetics* / Eds. Blench R., Sagart L., Sanchez-Mazas A. (Routledge Curzon, London, 2005) ISBN: 978-0415322423
3. L.S. Vasil'ev, Historical questions, **12**, 86 (1974)
4. M.A. Egorova, A.A. Egorov, T.V. Solovieva, Automat. Doc. and Math. Ling., **54**(2), 92 (2020)
5. D.N. Keightley, *Sources of Shang history: the oracle-bone inscriptions of Bronze Age China* (London, Berkeley, 1985) ISBN: 978-0520054554
6. Oracle bone. URL: https://en.wikipedia.org/wiki/Oracle_bone
7. High-resolution digital images of oracle bones, Cambridge Digital Library. URL: http://cudl.lib.cam.ac.uk/view/MS-CUL-00001-00155/1
8. Request for comment on encoding Oracle Bone Script, L2/15-280. Working Group Document, ISO/IEC JTC1/SC2/WG2 and UTC. 2015-10-21. Retrieved 2016-01-23. URL: https://www.unicode.org/L2/L2015/15280-n4687-oracle-bone.pdf
9. A. Papoulis, *Systems and transforms with applications in optics* (NY, McGraw-Hill, 1981) ISBN: 0898743583
10. A.A. Egorov, Proc. of the 2nd Sci.-Tech. Conf. "Electronics, micro- and nanoelectronics", 57 (2000) ISBN: 5-7262-0344-5
11. J.W. Goodman, *Introduction to Fourier optics* (NY, McGraw-Hill, 2005) ISBN: 9780974707723
12. A.A. Egorov, M.A. Egorova, T.V. Demidova, T.G. Orlova, Scientif. and Tech. Inform. Ser. 2, **8**, 24 (2020)
13. W.K. Pratt, *Digital image processing* (NY, Wiley, 2006) ISBN: 9780471767770
14. D.V. Baranov, A.A. Yegorov, E.M. Zolotov, Las. Phys., **13**(11), 1381 (2003)

15. A.S. Toporets, *Optics of the rough surface* (Moscow, Mashinostroenie, 1988) ISBN: 5-217-00015-5
16. B. Comrie, S. Matthews, M. Polinsky, *The atlas of languages: The origin and development of languages throughout the world* (NY, Facts on File, 2003). ISBN: 978-0816051236

Analytical Processing of Infological Models of Applied Tasks at Design of Information-Active Systems

Olga Novoselova*, and *Anton* Ruzheynikov

Department of information technologies and computing systems, Institute of information technology, Moscow State University of Technology "STANKIN", RU-127055 Moscow, Russia

Abstract. The article describes the concept of the methodology of intellectual labor automation, a formal description of the infological model (ILM) at the object level, and also a description of the analytical processing of ILM. The infological model is a project of an information-active system for a computing environment, independent of software and hardware tools and implementation environments. The infological model describes the static, dynamic and functional components and their linkage. The elements of the static component (the infological structure) are structural unitssuch as "information module", "information entity"and "information attribute". The dynamic component reflects the system of subject accesses to the infological structure. The functional component reflects the system of subject manipulations, which describes the algorithm for performing a class of applied tasks for the computing environment. The main stage of the analytical processing of infological models is the restructuring of their functional components. The method and technique of descriptions restructuring of the system of subject manipulations at the level of structural units of the "information attribute-5" type were developed. The method allows providing reduction of time and space expenses in the computing environment for the performance of the algorithm for solving a class of applied tasks at the design stage of information-active systems.

1 Introduction

The process of modeling information-active, automated, informational and other systems involves the formation of representations of various structures such as those of information and data, structure of the algorithm and their interactions. Currently, many methods and methodologies are known that allow the formation of task models and system designs. One of them is the methodology of intellectual labor automation (MAIL). [1]

The methodology of intellectual labor automation, developed at the Moscow State University of Technology "STANKIN", offers an industrial method of automated systems creation and determines the necessary and sufficient set of models that allows stage-by-stage imaging both the process of solving the problem and the project of an information-active system (IAS). When designing an information-active (automated) system, initial, conceptual, infological and datalogical models are formed. The initial model reflects the process of solving a subject problem by a expert. The conceptual model shows the structure of a domain knowledge system. The infological model is a project or "structrure" of an information-active (automated) system, which is invariant to the program and technical environment and implementation tools. A datalogical model is a project or "technology model" that takes into account the implementation platform. Schematically, the methodology of intellectual labor automation (MAIL) is presented in Figure 1. [1-3]

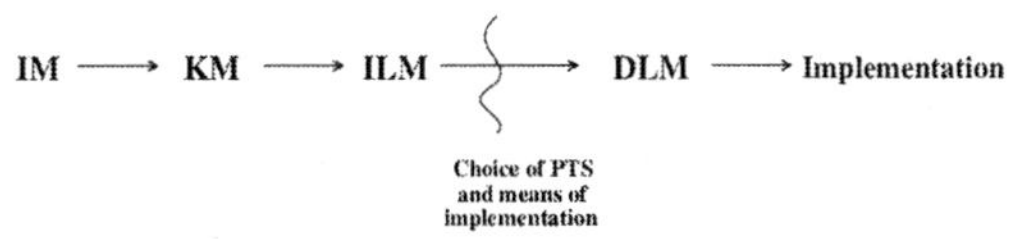

Fig. 1. Methodology of automation of intellectual work, where IM is the initial model, CM - conceptual model, ILM - infological model, DLM - datalogical model, PTS - software and hardware environment

Each model includes the description of informational or static, functional or/and dynamic components, a model in general. Informational (or static) component describes structure of static elements of the model. A functional component is an algorithm of the operations performed within a specific task. A dynamic component describes the structure of restrictions for static structure and accesses to it. The model in general defines coordination of functional or/and dynamic and informational or static components. At the same time, it includes the following components: for the initial model – functional and informational, for the conceptual model – dynamic and static, and for the infological model – functional, dynamic and static.

Each model includes the description of informational or static, functional or/and dynamic components, a model in general. The description of each component is given in the form of special diagrams and specifications.

* Corresponding author: olga.n10@yandex.ru

2 Infological model (ILM)

The purpose of the infological model is to organize and present information and data and the logic of their processing in a computing environment when designing information-active (automated) systems. The model is invariant to the program and technical environment and implementation tools, which allows you to implement a once developed project on various platforms without spending on the initial stages of problem modeling. [1]

The static structure of infological model (infological structure) describes the information structure of a subject task, which is organized as a multilevel structure. Each level is revealed by a set of named information modules (the module is an analogue of a database), consisting of a set of named information entities (an entity is an analogue of a relation). Each of the named informational entities includes a set of named informational attributes. A static structure consists of elements, binary and ternary relations established between elements, derived structures - schemes, binary and ternary relations established between schemes.

The dynamic structure (system of subject accesses) ensures the execution of information actions for writing and reading data necessary for the implementation of functional procedures on the formed information structure. It includes descriptions of subject accesses systems of the 1st and 2nd kind.

The functional structure (system of subject manipulations) is a set of interconnected functional procedures (computational, search, dialogue) for data processing. The subject manipulations system of the 1st kind determines the restrictions for a functional operation - a set of structural units of the "information module-5" type. The subject manipulations system of the second kind defines restrictions on the same structural units of the "information attribute-5" type, taking into account their "family tree" in the infological structure - the structure of location in the schemes.

The description of functional, dynamic and static structures in the formation of the infological model is performed in graphical form (diagrams) and in tabular form (specifications).

Visual notation for the subject manipulations system includes a set of elements (blocks and arrows); structures of these elements: sequence, iteration, alternative, cycle, switch; rules for building diagrams and models from elements and structures. The composition and structure of the diagrams is fixed in the specifications. The 1st kind of subject manipulations system means the use of information only from the level of structural units of the "information attribute-5" type to performing actions (subject manipulations) when executing an algorithm.

The infological model of the object level for the n-th applied task can be formally described as a set of structures and their linkages:

$$ILP2(n) = \\ < L2(n), DE2(n), ZJ2(n), QI2(n), R_2^{IP}(n) > \quad (1)$$

Structural units are divided into types: information module, information entity, information attribute.

$$L2(n) \equiv C(n), C(n) = \bigcup_p C^p(n),$$

$$C^p(n) = \bigcup_i C_i^p(n), C_i^p(n) \cap C_j^p(n) = \emptyset, \quad (2)$$

$$C_i^p(n) \subset C(n), C_i^p(n) = \{c_{il}^p\},$$

where $C_i^p(n) \equiv C_i^1(n)$ - $\boldsymbol{i}$ - a subset of structural units of the "information module" type,
$C_j^q(n) \equiv C_j^2(n)$ - $\boldsymbol{j}$ - a subset of structural units of the "information entity" type,
$C_k^r(n) \equiv C_k^3(n)$ - $\boldsymbol{k}$ - a subset of structural units of the "information attribute" type.

The set of static relations $DE2(n)$ can be represented:

$$DE2(n) = \\ < D2(n), E2(n), \overline{E2}(n), De2(n) >, \quad (3)$$

where $D2(n)$ – multitude of binary relations,
$E2(n)$ – multitude of ternary relations on structural units,
$\overline{E2}(n)$ – multitude of schemes of structural units,
$De2(n)$ – multitude of binary relations on schemes of structural units.

$$D2(n) \subset C(n) \times C(n). \quad (4)$$

$$E2(n) \subset C(n) \times C(n) \times C(n). \quad (5)$$

Multitude $E2(n)$ can be represented as a collection of subsets of a certain type:

$$E2(n) = \{EC_i^p(n)\, C_j^q(n) C_k^r(n)\}. \quad (6)$$

Under the scheme of structural units for subject manipulations we will understand the multitude of elementary schemes (elementary scheme - e_{iljmkn}^2 is determined by the ternary relation of structural units), defined by:

- common structural units of «module» and «entity» type;
- non-repetitive structural units «attribute» type.

Formally, this is represented in the form

$$\overline{e^2{}_{iljm}} = \{e_{iljmkn}^2, \dots, e_{iljmkq}^2\}. \quad (7)$$

Then the scheme of structural units or derived structures is defined as follows:

$$\overline{e_{iljm}^2} = \{\overline{e_{iljmkn}^2}\}. \quad (8)$$

Then the multitude of schemes of structural units for a subject task is defined as follows:

$$\overline{E2}(n) = \bigcup_{il} \overline{E2_{il}}(n), \quad (9)$$

$$\overline{E2_{il}}(n) \subset \overline{E2}(n), \overline{E2_{il}}(n) = \{\overline{e^2_{iljm}}\}.$$

Binary relations can be established between the schemes of structural units:

$$De2(n) \subset \overline{E2}(n) \times \overline{E2}(n). \quad (10)$$

Multitude $De2(n)$ can be represented as a set of subsets of a certain kind:

$$De2(n) = \{ L\bar{E}2_{il}(n), L\bar{E}2_{il}(n)\bar{E}2_{iq}(n), L\bar{E}2_{iq}(n)/\bar{E}2_{il}(n), L\bar{E}2_{il}(n)\bar{E}2_{kn}(n), L\bar{E}2_{kn}(n)L\bar{E}2_{kr}(n)/L\bar{E}2_{il}(n)\}. \quad (11)$$

Binary relations on the schemes of structural units at the object level reflect the procedure of "deployment" (scaling) of sign representations.

The functional relations $ZJ2(n)$ include two types of restrictions - systems of object manipulations of the first $X2(n)$ and second $J2(n)$ kind:

$$ZJ2(n) = < Z2(n), J2(n) > \quad (12)$$

$$Z2(n) = (X2(n), QX2(n)), \quad (13)$$

where $X2(n)$ - the multitude of subject manipulations of the 1st kind and $QX2(n)$ - multitude of binary relations between subject manipulations of the 1st kind.

$$X2(n) = \bigcup_p X2_p(n), X2_p(n) = \{x_{ps}\}. \quad (14)$$

The multitude of subject manipulations is divided into subsets of different levels decomposition:

$$QX2(n) \subset X2(n) \times X2(n), \quad (15)$$

$$QX2(n) = \{LX2_p(n), LX2_p(n)X2_t(n), LX2_t(n)/X2_p(n)\}. \quad (16)$$

A submultitude $LX2_p(n)$ define binary relations of the "ordering" type between subject manipulations of the 1st kind of one subset.

A submultitude $LX2_p(n)X2_t(n)$ define binary relations as type "composition" between subject manipulations of the 1st kind of different subsets.

A submultitude $LX2_t(n)/X2_p(n)$ define the binary relations as type "arrangement" between the subject manipulations of the 1st kind of the one subset.

Since subject manipulations are considered at the object level, any manipulation of the 1st kind is a variant of a certain type always. Since the type of manipulation is fixed on a certain multitude of structural units always, subject manipulation of the 1st kind, as an option, is fixed on a subset of this set of structural units:

$$x_{ps} = x_{ps}(p, t, u_{ps}). \quad (17)$$

The relations describing interrelations of the basic static and functional components of applied task model of object level can be represented in the following form:

$$R_2^{IP-1}(n) \subset C(n) \times C(n) \ldots \times C(n) \times X2(n), \quad (18)$$

$$R_2^{IP-1}(n) = \{r_l^{IP2-1} | r_l^{IP2-1}: (c_{n1}^r, \ldots, c_{nl}^r) \to x_{ps}. \quad (19)$$

3 Analytical processing of the infological model

Analytical processing of infological models includes the following complex of tasks: determination and check of information connectivity between subject manipulations; analysis and restructuring of subject manipulations system; fragmentation of information. [2-4]

Checking information connectivity allows you to determine the possibility of restructuring the system of subject manipulations. To do this, it is necessary to obtain an information description for all subject manipulations (PM) - first elementary, then complicated, taking into account the type of algorithmic structure. As a result, a complete informational description of the algorithm as a whole should be obtained (complicated subject manipulation of the upper level), which is based on the informational description of all elementary (child) manipulations included in it.

Restructuring involves transforming graphical descriptions, which is difficult to do on diagrams without preparation. Therefore, first it is necessary to form a description of the process of performing subject manipulations in a different form. For this, a "conveyor - time step" approach was proposed and the apparatus of block matrices was chosen, which made it possible to present the manipulations system in the form of a structural description. The structural description takes into account the spatio-temporal characteristics of the algorithm execution process - the axis of the conveyors (K) is located vertically and the axis of the time steps (T) is located horizontally. [4, 5]

Let's present a structural description x_{pt}- complex subject dependence in the form of a block matrix H_{pt}:

$$H_{pt} = \begin{bmatrix} H_{pt}^1 & H_{pt}^2 \\ H_{pt}^4 & H_{pt}^3 \end{bmatrix}, \quad (20)$$

where H_{pt}^1 – matrix-column defines the external (input) information connection x_{pt} complex subject manipulation by input structural units of the "information attribute-5" type;

H_{pt}^2 - square matrix that defines the internal informational relationship according to structural units of the "information attribute-5" type between subject manipulations *(p + 1)* of the decomposition level. These subject manipulations are part of the complex subject manipulation x_{pt};

H_{pt}^{3} – matrix-string defines an external (output) information connection x_{pt} complex subject manipulation by output structural units of the "information attribute-5" type;

H_{pt}^{4} – matrix-element. It shows what type of algorithmic construction the description belongs to: 1-sequence, n-iteration, 0-alternative.

The dimension of the matrix H_{pt} is composed of sizes its constituent block matrices H_{pt}^{1}, H_{pt}^{2}, H_{pt}^{3}, H_{pt}^{4} and dimension matrix is $m_{pt} = (m+1) \times (m+1)$.

For each basic algorithmic construction, a type of structural description is defined. For an algorithmic construction of the "iteration" type, the structural description is shown in figure 2, of the "alternative" type, the structural description is shown in figure 3.

$$H_{pt} = \left[\begin{array}{c|c} 1 & x_{(p+1)h} \\ \hline n & 1 \end{array}\right]$$

Fig. 2. Structural description of the algorithmic construction of the "iteration" type

$$H_{pt} = \left[\begin{array}{c|c:c:c:c:c} 1 & x_{(p+1)g} & 0 & \dots & 0 & 0 \\ \hdashline 1 & 0 & \dots & \dots & 0 & 0 \\ \hdashline 1 & 0 & 0 & x_{(p+1)r} & 0 & 0 \\ \hdashline \dots & \dots & \dots & \dots & \dots & \dots \\ \hdashline 1 & 0 & 0 & \dots & 0 & x_{(p+1)m} \\ \hline 0 & 1 & 1 & \dots & 1 & 1 \end{array}\right]$$

Fig. 3. Structural description of the algorithmic construction of the "alternative" type

To form a complete structural description for a task, structural descriptions are created for all complex subject manipulations. Further, the structural descriptions are included in the matrix describing the algorithm as a whole - a complex subject manipulation of the top level. When an element of a matrix (subject manipulation) is disclosed, its structural description specifies the elements that show the informational connectivity of subject manipulations at this level of decomposition. At each step of the iterative formation of a complete structural description, the dimension of the resulting matrix H_{pt}^{*} includes the dimensions of the base matrix H_{pt} and the embedded matrix $H_{(p+1)h}$. As a result, a matrix of the complete structural description of the task is obtained, where the horizontal shows the time spent on the task - time steps, and the vertical spatial chains - conveyors. [4, 6, 7]

Next, an extended structural description $\overline{H_{pt}^{2*}}$ is formed. This matrix is obtained by transposing the elements of the matrix H_{pt}^{2*} relative to the main diagonal.

$$\overline{H_{pt}^{2*}} = H_{pt}^{2*} \cup (H_{pt}^{2*})^{T} \tag{21}$$

Based on the extended structural description $\overline{H_{pt}^{2*}}$, you can transform the description of the system of subject manipulation. To simplify the formal description, the block matrix $\overline{H_{pt}^{*}}$ is copied to the block matrix G_{pt} :

$$\overline{\mathrm{H}_{\mathrm{pt}}^{*}} \equiv \mathrm{G}_{\mathrm{pt}}. \tag{22}$$

$$\mathrm{G}_{\mathrm{pt}} = \begin{bmatrix} \mathrm{G}_{\mathrm{pt}}^{1} & \mathrm{G}_{\mathrm{pt}}^{2} \\ \mathrm{G}_{\mathrm{pt}}^{4} & \mathrm{G}_{\mathrm{pt}}^{3} \end{bmatrix}. \tag{23}$$

Studying the elements showing informational connectivity, one can draw conclusions about the possibility of restructuring the system of subject manipulations. Restructuring is the process of optimizing the algorithm for solving a task in terms of time and space expenses. Determination of the time characteristic (number of time steps) consists in finding the maximum time step number in the optimized structural description of the algorithm. Determination of the spatial characteristic (number of conveyors) consists in finding the maximum conveyor number in the optimized structural description of the algorithm. [8]

To obtain an optimal structural description by time steps, a formal description of the movement of the element of the main diagonal of the matrix-block G_{pt}^{2} - the subject manipulation and all elements associated with it, one time step earlier, was performed. The movement of elements is performed only if there is no informational connectivity between subject manipulations. The absence or presence of informational connectivity between subject manipulations shows the value "0" or "1" in the lower triangular matrix under the main diagonal.

In the G_{pt}^{3}block, the elements move according to the movement of the elements of the main diagonal. Blocks G_{pt}^{1}, G_{pt}^{4} of the block matrix G_{pt} remain unchanged.

As a result of restructuring the system of subject manipulations of the first kind, an optimal structural description in terms of time characteristic is obtained in the form of a block matrix G_{pt}, which contains a horizontally compressed structural description of the matrix $\overline{H_{pt}^{*}}$. [8]

To simplify the description of restructuring process the system of subject manipulations of the first kind by spatial characteristic, the block matrix G_{pt} is copied into the block matrix Q_{pt} : $G_{pt} \equiv Q_{pt}$.

To form a structural description that is optimal along the conveyors, the lack of informational connectivity between the elements describing the subject manipulations is first determined. If the condition is met and two adjacent elements are independent, then the element of the main diagonal (the subject manipulation) is moved to another conveyor located above. Together with this element (the subject manipulation), all elements from this conveyor are moved. The move process is iterative.

Move rules: determine the last element on the main diagonal on the last conveyor (in the last row) of the Qpt matrix; define the element that precedes it - subject manipulation. The presence (1) / absence (0) of informational connectivity is checked between them. If there is no information connectivity, then move. The rules for moving elements of row are similar to the rules for moving when restructuring a structural description by time steps, only they move from the underlying row to the row located above.

When performing the restructuring process through conveyors, the main conveyor of the maximum length and additional conveyors to the main conveyor are formed. These conveyors, as a rule, contain subject manipulations of the "alternative" type as part of a given information-related manipulations chain. The blocks Q^3_{pt}, Q^4_{pt} of the block matrix Q_{pt} remain unchanged.

As a result of performing these actions, a block matrix Q_{pt} is obtained containing a vertically compressed structural description of the matrix G_{pt}.

Fragmentation of information shows the distribution of subject manipulations and their structural units of the "information attribute-5" type on performance time steps.

4 Conclusion

Restructuring is the main task of the analytical processing of system of the subject manipulation. The "conveyor - time step" approach, the rules for the formation and restructuring (optimization) of the structural description of subject manipulations were proposed, which allow the sequential process of the algorithm to be converted into a serial-parallel one based on the analysis of information connectivity between subject manipulations. The restructuring of the system of subject manipulations is very important because it significantly can reduce the algorithm execution time.

Acknowledgments

This article was prepared in the framework of the project No. 17-29-07057 "Development of methods of restructuring and integration for semantic and syntactic representations when creating automation systems for design and management processes" under the RFBR grant. This work was carried out using equipment provided by the Center of Collective Use of MSUT "STANKIN".

References

1. G.D.Volkova, *Methodology of intellectual labor automation. Monograph* (M.: Janus-K, 2013)
2. O.V. Novoselova, G.D. Volkova, Y.M. Solomentsev, Nonlinearity: Problems, Solutions and Application, **2**, 223-238 (NY: Nova Sci. Publ. Inc., 2017)
3. O. Novoselova, A. Ruzheynikov, A. Gavrilov. Analytical Processing of Applied Tasks Conceptual Models at Design of Information-Active Systems, EPJ. Web Conf., **224**, 06008 https://doi.org/10.1051/epjconf/201922406008
4. G.D.Volkova, S.M. Kuryshev, Automation and control in mechanical engineering, **7** (1999)
5. F.R. Gantmakher, *Theory of matrices* (Moscow: Fizmatlit, 2004)
6. G.D. Volkova, N.M. Efromeev, Computer science and information technology: proceedings of the international scientific conference, 80-84 (2012)
7. O.V. Novoselova, Effective tools of modern Sciences – 2014: Materials of the X international scientific-practical conference, 23-30 (2014)
8. O.V. Novoselova, A.A. Ruzheynikov, Science today: challenges and prospects: materials of the international scientific and practical conference, 72-74 (2020)

Representation of Social Systems as Nonlinear, Complex, Self-Organizing and Closely Related to Biosystems and Mathematical Model of Their Evolution

Ekaterina Romanova[1], *Ludmila* Uvarova[2,*]

[1]FGBUN ICTS RAS, RU-127055 Moscow, Russia
[2]Moscow State Technological University "STANKIN", RU-127055, Moscow, Russia

Abstract. On the basis of the systemic and synergistic approaches, the features and regularity of the self-organizing systems, it is concluded that the social systems are nonlinear, open, complex, self-organizing systems. The scheme-model of the general structure of the social system is considered. Much attentions is given to the interactions of the social systems elements with the biosystems. The graphic model-tree, reflecting the association between the social systems and the biosystems. In connection with that the most significant technologies and the scientific trends, connecting with the biosystems: the gens, the genetic engineering, the artificial intellect, the bio- and nanotechnologies, the biomedicine, microbes, viruses are considered. The futures of the dynamics of the social systems are noted. It is shown, that, with a high probability, the spiral evolutionary processes can occur in the social systems. The mathematical model is presented, that describes the spiral structures, based on the Swift-Hohenberg equation.

1 Introduction

For the research, modelling, forecasting of the processes, occurring in the social systems, it is necessary to clearly represent, what type of the systems the social systems belong to. The systemic and synergistic approaches make it possible to believe that social systems are nonlinear, open, complex, informational, self-organizing systems and, therefore, they have the same features, regularity and properties of the self-organizing systems [1 – 9]. Based on the currently available information about the social systems, one can classify them as the self-organizing systems. The specific properties of the social systems, taking into account the peculiarities and regularity of the self-organizing systems, can be illustrated by the scheme shown in the Figs.1 and 2.

Naturally, the social systems can not be considered as isolated. In reality, the social systems are part of the more complex self-organizing system.The relations and the interactions with the technosphere, ecology-economic subsystems, the educational and cognitive spheres, the digital sphere and the biosystems are determined by valuable information [9]. In self-organizing systems the coherent processes occur due to the exchange of the information in the form of the signals both between the parts of the subsystems and the structural levels [4]. Information determines the environment, in which the subsystems are located and the joint actions of which support it.

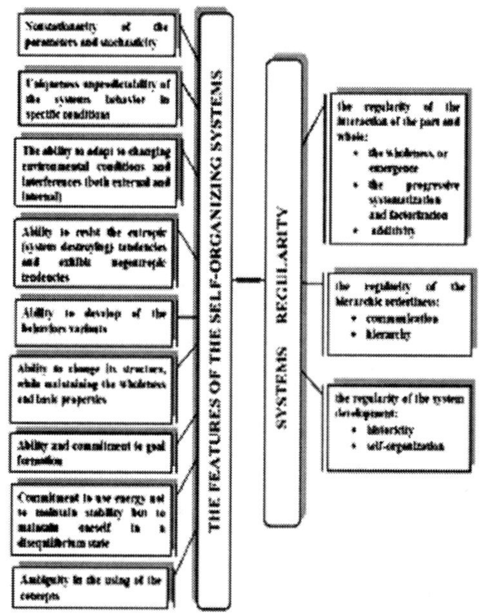

Fig.1. The features and the regularity of the self-organizing systems.

* Corresponding author: uvar11@yandex.ru

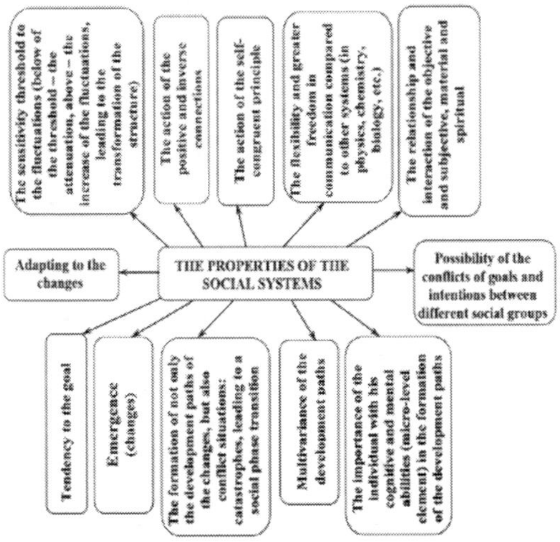

Fig.2. The properties of the social systems.

The information determines the order parameters, that characterize the levels of the system. As a result of the information exchange both the system and the subsystems can pass from one stable state to another, that, is, from one attraction to an other.

Nonlinearity means the multivariance of the system development paths. The social systems are multi-levels. Macro level, on the one hand, in many situations determines the behaviour and the structure of the parts of the system at other levels (meso, micro, nano), and on the other hand, individuals, groups, subgroups, subsystems can influence on the system as a whole. Thus, the structure of a social system can be represented by the following general the scheme-model:

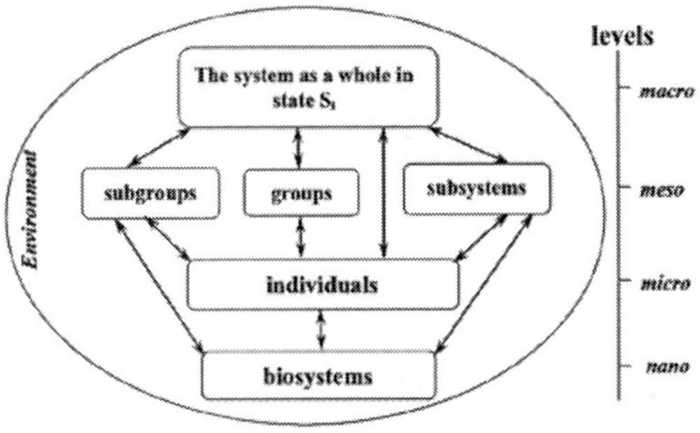

Fig.3. The scheme-model of the social system structure.

In modern conditions, the society has to solve complex problems in connection with the necessary transition to a new level of the development, which is determined by the environment, that is, by all components, the development of new technologies, the managerial problems solving, humanity's entry into the digital sphere, associated with artificial intelligence.

On the one hand, an individual (person) is a single element of the society. That is a microparticle of the complex self-organizing macrosystem. The place of the individual in the surrounding world. In the universe has been discussed for a long time in various directions. The individual (person) is by nature dualistic: it is necessary to distinguish between biological and social components. In the beginning, the biological evolution of man and the evolution of the social relations proceeded in parallel. The present time is characterized by a slowdown in the biological evolution and an intensification of the evolution of the social relations, the collective intelligence of humanity and an increase in the role of the social valuable information. The rate of the development of modern technologies, the technosphere is ahead of the slow development of the individuals, psychics, the social institutions and the technology of the managerial decisions assumption [10]. A person is connected to the society through the culture, economy, ecology, technology and religion. The complication of the technology leads to the complication of the social relations. In such situation in order to conserve a person the society in the process of the self organization must develop new approaches, that allowed a person to orient in the changing reality, which is possible with the help of the science.

2 The graphical model-tree, demonstrating the social signifience of the biosystems

Presently, the influence of the biosystems on the social systems and humans is of particular interest. The connection of almost all components of social system with the biosystems is shown in fig.4

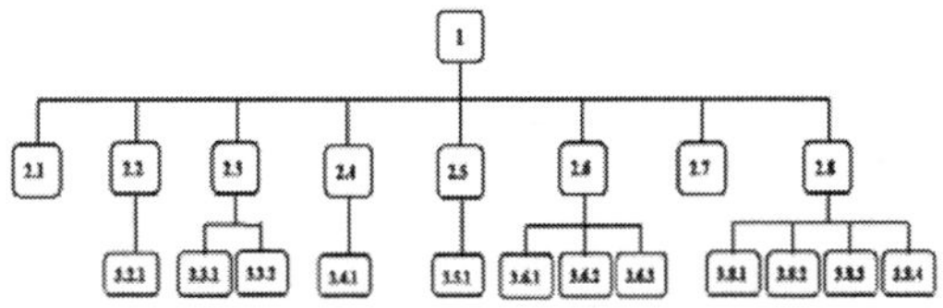

Fig.4. The graphical model-tree illustrating the social significance of the biosystems [5]. Where: 1. Biosystems, 2.1. Artificial intelligence. 2.2. Gen and genetic, engineering. 2.3. Nanobiotechnology. 2.4. Microbes and viruses. 2.5. Individual (person). 2.6. Biomedicine. 2.7. Environment (biosphere): the state of the atmosphere, water basins, soils. 2.8. Biotechnology. 3.2.1. Biological weapons. 3.3.1. Biorobots. 3.3.2. Molecular devices. 3.4.1. Health of the society. 3.5.1. Society. 3.6.1. Public health. 3.6.2. Intellectual biomaterials. 3.6.3. Pharmacology. 3.8.1. Bioelectronics. 3.8.2. Bioindustry of enzymes. 3.8.3. Environmental biotechnologies. 3.8.4. Cell industry.

Microbes and viruses can be concerned to the nanobiosystems, which, as the experience and history of the development human society have shown, affect society (epidemics, new diseases, the fight against which leads to the development of new technologies in medicine). Microbes and viruses are the self-organizing systems, and have the property of the self-defense, manifested in the fact that they mutate under the influence of the man-made preparations to fight them. This is the danger to society on a global scale from the point of view of the nation health. In addition, microbes and viruses may be used to create bacteriological weapons. Currently, the history of the coronavirus demonstrates, how viruses can effect on politics, the economy, the medicine, the market, culture, education and the social-relations, medicine.

Genetic and the genetic engineering are the scientific direction, appeared at the convergence of the sciences. By now, the nanotechnological results, achieved in these areas the significant and promising results for the future of the humanity: the study of genes and gens modifications, and DNA structure. The manipulation with genes, with their subsequent introduction into the organisms cells make it possible to fight which incurable genetic diseases, to create new generations of drugs, to receive the genetically modified products, widely discussed in the scientific world about, their usefulness and danger. The negative results of the noted above developments are the following: 1) the development of various kinds of biological weapons, which are a threat to all of the humanity; 2) works in some laboratories in the world to create chimeras which are unknown how they will end for society.

The biotechnologies [11] on the one hand, also contribute to the development of the biological weapons types, and, on the other hand, solve the agricultural problems, develop alternative ways of obtaining energy, are used to combat environmental pollution, in the food industry. The cellular engineering, associated with the cultivation the cells and the tissues in vitro, plays an important role in the study of the interactions of cells in the tissues, the mechanisms of the emergence and the medical treatment of the cancers, in the breeding work and the creation of new drugs.

The bioelectronics allows the development of the unique, high-precision, control and measuring methods in scientific research, medicine and industry. Biotechnology, in practice, straightens, the seemingly insurmountable differences between production, person, biomicrocosm and allow considering the following chain: biomicrocosm → person → society → production → nature.

In due time, the computer, information, technological revolutions were widely discussed. The present time is characterized by the convergence of the sciences, technologies, digital revolution, which are actually based on the results of the previous revolutions. The formation of the digital sphere is associated with the developments in the field of the artificial intelligence the results of the brain structure studies, the mechanisms of the brain functioning, the achievements of the neurophysiology, the mechanisms of self-assembly of the molecular structures, information transfer in the biosystems, the automatic assembly of viruses, etc [10, 14].

The digital sphere, digital technologies will open the digital world to the humanity and the individuals, change the relationships between the individuals and society. In this connection, it should be noted, that the digital technologies contain, dangers for humans, since the society development can follow the path of consciousness manipulating. The future will show, how the society, the economy, the politics and individual psychology will be transformed.

The biomedicine is a direction, which is successfully developing in the health care and pharmaceuticals. Mutating microbes and viruses pose ever more complex problems for the biotechnologists. The development of the intelligent biomaterials is of great interest in the biomedicine. Such materials are of the natural origin (for example, collagen, keratin) and are used to obtain implantable products, that take into account individual characteristics and are used in the surgery. With the help of such biomaterials, important problems of the interactions of cells with the surfaces of implants and the biological environment are solved, ensuring compatibility with bone, blood, and antibacterial properties. By this means the intellectual materials help to improve the health of both the individual and society as a whole [12].

In the field of the nanotechnology, in increasing frequency more attention is given to the structures and the properties of the biosystem (molecular complexes, enzymes, DNA and RNA molecules and the like), as a result, the direction of bionanotechnology was formed [12 – 15]. So, for example, the natural biosystems copies are reproduced in the form of the combinations of the sensors, detectors, indicators, actuators; the rotaxane molecules are used to create molecular memory. For creating the nanodevices, molecular structural blocks have been developed: bisaminoacids, from which bispeptides of a given shape can be constructed. It is planned to create the nanodevices, capable of the assembling, according to a given program, nanomachines similar to the ribosomes, capable of synthesizing protein molecules from amino acids. The combination of the bionanotechnology and artificial intelligence is used to create of biorobots, that have already been found the application in the socially useful activities.

3 Some development in research on social systems and the possible mathematical model of their evolution

Under the internal or the external influences, the relations between the elements of the social systems can change while maintaining their main qualities. Under the influence of small fluctuations, the social system will adapt to such changes and maintain the structure as a whole. If structural changes exceed a certain threshold of "the sensitivity", the social system can go into a chaotic unstable state, when new internal relationships between elements are formed and the control parameters change. The dynamic social systems in their development pass through the bifurcations points, at which the changes in the stability and the transitions to the new equilibrium states take place. If after passing the bifurcation point there is one variant of a stable state, then the evolution of the social system is predictable, but if a multivariate situation develops, then the evolution uncertainty appears.

In practice, in the dynamics of the multilevel nonlinear social systems, auto-oscillations can form. Irreversible processes can be superimposed on the cyclic processes, and the spiral evolutionary processes are likely to arise [8]. The interaction between the elements of the social systems and their interactions with the systems as a whole is carried out through the information exchange. In the social systems, the conditional and unconditional informations are generated. Valuable information is of great importance for the evolutionary processes [9].

According to H. Haken [16], the public opinion plays the role of the order parameter – the governing parameter which subjugates, the individuals, spreads throughout the system, determing the interaction between the ordered structure and the individual elements of the system. Based on the above – mentioned it is possible to propose for discussion the mathematical model describing the spiral structures based on the Swift–Hohenberg equation:

$$\frac{\partial u}{\partial t} = -\left(\Delta + k^2\right)u + f(u),\ f(u) = \beta u + \varepsilon u^2 - u^3 \quad (1),$$

In the general case, this equation describes the behavior of the system near the bifurcation points, the phase transitions, where β – is a parameter, that determines the proximity of the system to the instability threshold, $\varepsilon < 1$ is a parameter, depending on the environment properties. A special case of the equation (1) in polar coordinates for a homogeneous solution $u = 0$ leads for the vector $r = r(\varphi)$ to the equation of a logarithmic spiral, reflecting the evolution of the social system:

$$r = r_0\, exp\,(\varphi \cdot tg\ \theta) \quad (2),$$

where θ – the angle between the radius-vector of a point on the curve and the tangent to the curve at this point.

4 Conclusion

The systemic, synergistic and informational approaches used in this paper allow drawing the following conclusions:

(1) the social systems are complex, open, multilevel, nonlinear, self-organizing systems that are in an equilibrium stable state with the coherent interactions of all their parts due to the exchanges of the information signals flows;

(2) for the self-organization processes the conditional, unconditional and valuable information play both functional and controlling roles.

The transitions of the social systems from one state to another can occur under the external and internal influences such as oscilationss, which can be of two types:

(1) small oscilations that change the structures and the behaviour of the subsystems but preserve the system as a whole;

(2) oscillations exceeding the system sensitivity threshold and causing the destruction of the structure of the system as a whole. In this connection the dynamic of the social system development can be represented as follows: with small fluctuations the system is capable of the self-healing, and when the system structure is destroyed, a chaotic state is formed, in which it gets to the bifurcation point and randomly chooses a further development path (without external interferences).

In this manner, the prediction of the social system behaviour is difficult, and when modeling the self-organizing, complex social systems, the following factors should be taken into account: (1) all processes occurring in the social systems are self-consisting probabilistic; (2) when modeling such social systems, it is practical to use a probabilistic approach (for example, the mathematical formalism of Markov random processes), modern theories of oscillations and waves, game theory, theory of the decision making [5, 17 – 20].

When studying and modeling all levels of the social systems, one should take into account their connections with to biosystems. Currently, there is a global coronavirus attack an all levels of the social systems. It is clear that viruses and epidemics can be used in the management of the social systems. Many fields related to the biosystems (genetics, biotechnologies, bionanotechnologies, etc.) can influence the dynamics of the entire social systems, individuals and control of their evolution.

The investigation is supported by the Ministry of Science and Higher Education of Russia (Project № 1.7706.2017/8.9). This work was supported by the Ministry of Science and Higher Education of the Russian Federation under project 0707-2020-0034. This work was carried out using equipment provided by the Center of Collective Use of MSUT "STANKIN".

References

1. E. Romanova, L. Uvarova, L. Vasil'eva, Articles. The Fundamental Physico-Mathematical Problems and Modeling of Technical and Technological systems, **18**, 110 (2017)
2. L. Vasil'eva, L. Uvarova, E. Romanova, Articles. The Fundamental Physico-Mathematical Problems and Modeling of Technical and Technological systems, **17**, 64 (2016)
3. *The System Theory and the Systems Analysis Applied to the Organizations Management: Handbook* (Finance and Statistics, Moscow, 2009)
4. H. Haken, *Information and Self-Organization. A Macroscopic Approach to Complex Systems* (KomKniga, Moscow, 2005)
5. E. Romanova, *Systems Analysis and Modeling of the Complex Macro and Micro systems* (Yanus-K, Moscow, 2019)
6. E. Romanova, *Book of Abstracts. The Fourth International Scientific Conference: The Modeling of Nonlinear Processes and Systems* (Yanus-K, Moscow, 2019)
7. E. Knyazeva, S. Kurdyumov, *Foundations of Synergetic Worldview* (KomKniga, Moscow, 2005)
8. V. Weidlich, *Sociodynamics: A Systems Approach to Mathematical Modeling in the Social Sciences* (Book House "LIBROKOM", Moscow, 2010)

9. D. Chernavsky, *Synergetics and Information Theory* (Book House "LIBROKOM", Moscow, 2013)
10. J. Avory, *Information theory and evolution* (SIC. Regular a nd C haotic Dynamics. The I nstitute o f computer studies, Moscow-Izhevsk, 2006)
11. T. Egorova, *The Biotechnology Bases: Training tutorial* (Publishing c enter «A cademy», Moscow, 2003)
12. K. Worden, *The New Intellectual Materials and Constructions. The Properties and Application* (Technosphere, Moscow, 2006)
13. Ch. Pul, Ph. Ouens, *Nanotechnology* (Technosphere, Moscow, 2005)
14. Kr. Shaphmerster, In the World of the Science, **9**, 55 (2007)
15. L. Vasil'eva, L. Uvarova, E. Romanova, *Some Nanotechnology Problems. The Investigation and the Modeling of the Nanosystems: Nanocomplexes, Clusters and Nanoparticles* (Yanus-K, Moscow, 2015)
16. H. Haken, *Secrets of Nature. Synergetics: the study of interaction* (Institute f or Computer R esearch, Moscow-Izhevsk, 2003)
17. N. Karlov, N . Kirichenko, *Oscillations, waves, structures* (FIZMATLIT, Moscow, 2003)
18. E. Wentzel, *Operations, waves, structure* (FIZMATLIT, Moscow, 2003)
19. A. G uts, L. P autova, Y u. F rolova, *Mathematical Methods in Sociology* (Book House "LIBROCOM", Moscow, 2014)
20. E. Romanova, Bulletin of BSTU, **5**, 51 (2018)

Mathematical Model for Collecting and Evaluating Complex Security Factors and Ensuring the Prevention of Threats to Individuals and Society in Cyberspace

Denis V. Poltoranov[1,*] and *Tatiana V.* Karlova[2,**]

[1]Institute For Design-Technological Informatics RAS, RU-127055, Moscow, Russia
[2]Moscow State Technological University "STANKIN", RU-127055, Moscow, Russia

Abstract. This article is devoted to the development of a model and elaboration of the structure of the mathematical formalism for the assessment of the state of comprehensive security in order to prevent threats to the individual and society in cyberspace. The types and objects of information threats are examined, the main properties, which are indicators and parameters we should possess, are revealed and the key stages of status assessment of comprehensive security in state entities, municipalities and organizations are identified.
Keywords. information security, comprehensive security, status assessment, indicators, information threats.

1. Introduction

The system of comprehensive security of each state is a set of bodies and institutions that function in the areas of regulation, prevention and elimination of the consequences of various threats (political, economic, biological, chemical, environmental, etc.). In Russia there is a system of ensuring national security, consisting of state and non-state subsystems. The state system is formed by legislative, executive, and judicial state institutions that take part in the process of making decisions and implementing political, legal, organizational, economic, military, and other measures aimed at ensuring the security of the individual, society, and the state. The non-state system consists of public associations, which include the media and individuals who can influence the formation and implementation of national security policies. The concept of security is a multi-vector and multidimensional concept, which includes a whole set of structural components. [1,2]

The integrated security system of each State consists of the bodies and institutions that operate to regulate, prevent and respond to various threats (political, economic, biological, chemical, ecological and others). The Russian Federation has a national security system consisting of State and non-State subsystems. The State system shall form legislative, executive and judicial State institutions that take part in the process of formulating decisions and implementing measures of a political, legal, organizational, economic, military or other nature; aimed at ensuring the security of individuals, society and the State. The non-State system consists of voluntary associations, which include the media and private individuals who can influence the formulation and implementation of national security policies. The concept of security is a multidimensional and multifaceted one that incorporates a variety of structural components.

Assessing the state of comprehensive security to prevent threats to the person and society in cyberspace is one of the most important tasks that the state solves. Currently, there are no approved methods or recommendations for assessing the state of comprehensive security to prevent threats to the person and society in cyberspace, and the existing international recommendations do not take into account the specifics of the Russian Federation. The types and objects of threats in cyberspace are shown in table 1. [3,4]

Table 1. Objects and types of cyber threats

Source of Threat	Types of Threat
External	
Activities of foreign intelligence and intelligence services, criminal groups and formations	Obtaining unauthorized access to information and monitoring the functioning of information systems; interception and leakage of information through technical channels; (not)controlled distribution of computer viruses and other malicious programs
Illegal activities of individuals	
Internal	
Violation of established regulations on the collection, processing and transmission of information	Leakage/disclosure of information constituting a state, official, or other type of secret; possession of personal data (including financial data) of the company/person; failure, failures, malfunctions of
Intentional actions and unintentional errors of information systems personnel	

e-mail: *depo768@rambler.ru, **karlova-t@yandex.ru

 https://doi.org/10.1051/epjconf/202124803009

Use of unlicensed software, involvement of persons who do not have a license for this type of activity in the creation of information systems of the organization	technical means, information security programs; failures of technical means and software failures in information and telecommunications systems
Transboundary	
Development of cybercrime (financial, piracy, distribution of pornographic and extremist content)	Cardering, Internet acquiring crimes, fraud, illegal business on the Internet, acquisition of intellectual property, recruiting members of international terrorist and extremist groups
Functioning of unlicensed financial institutions and online trading platforms	Conclusion of counterfeit agreements, money laundering, illegal seizure of other people's property

2. Theoretical basis

The theoretical basis of this work was the national scientific works of Russian and foreign scientists on development information management and its subjects, as well as methods, principles and models of historical, genetic and systemic approach the legal and regulatory framework of the Russian Federation and individual States on issues of ensuring state and public security, including in cyberspace; scientific works of T.V. Vladimirova, R.I. Vylkov, A.A. Kalinkin, A.I. Kovaleva, M.A. Petlin, I.Yu.Sayapina and many other domestic and foreign researchers.

3. Discussion

In the methodology presented below, the assessment of the state of comprehensive security to prevent threats to the individual and society in cyberspace (hereinafter referred to as the comprehensive security) is expressed by an integral dimensionless indicator, determined in the range from 0 to 1 and showing the probability with which the integrated security of the object of assessment is ready to perform tasks on destination. In this article, the object of assessment means a federal executive body, a constituent entity of the Russian Federation, a municipality or an organization.

The considered integral indicator of the state of complex security is the upper level of the hierarchy shown in Figure 1, and is determined based on the values of indicators of the state of complex security.

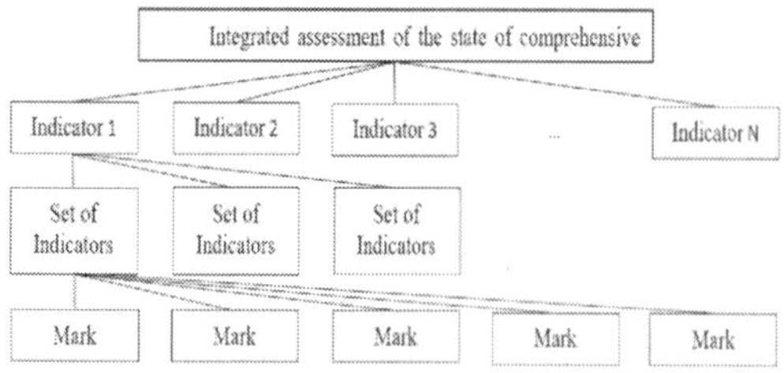

Fig. 1. Hierarchy used to assess the state of comprehensive security.

These indicators should have the following properties:

- reliability - the indicator should characterize progress in achieving a goal or solving a problem and cover the maximum number of aspects of achieving the goal or solving complex security problems;
- objectivity - it is necessary to use indicators that reflect the real state of affairs;
- unambiguity - the definition of the indicator should provide the same understanding of the essence of the measured characteristic, exclude the possibility of double interpretation or understanding;
- cost-effectiveness - reporting data should be obtained at the lowest possible cost, and the indicators used should be based as much as possible on existing procedures for collecting information in the field of comprehensive security;
- reliability - the method of collecting and processing the initial information should allow the possibility of verifying the accuracy of the data obtained;
- timeliness and regularity - to be used for monitoring purposes, reporting data should be updated at least once a year. [5]

Taking into account the above properties, the following indicators of the state of integrated safety are established (depending on the object of assessment, indicators can be supplemented or reduced):

- protection of the individual (society, state);
- readiness of the management system of security agencies;
- the readiness of the security forces;
- stability of functioning of economic objects.

The values of the groups of indicators are calculated by aggregating the corresponding marks of the state of integrated security, which are collected within the framework of information exchange.

Mathematically, such a hierarchy (Figure 1) can be written as:

$$I = f(I_1 \dots I_n);$$
$$I_n = f(M_1^n \dots M_k^n); \; M_k^n = f(m_1 \dots m_i) \,, \qquad (1)$$

where: n – is the number of indicators;
k – number of indicator groups corresponding to the I_n indicator;
i – is the number of marks corresponding to the M_k^n indicator group.

Among the basic principles underlying the theory of management, the central place belongs to the hierarchical subordination in the activities of various systems at various levels. [6,7]

If, as an integral indicator of assessing the state of comprehensive security, the probability of developing instability in society is taken, then various scales can be used to assess the state of comprehensive security: “unstable” - from 0 to 0.7; "Satisfactory" - from 0.7 to 0.8; "Stable" - from 0.8 to 1.

In the following methodological approach to assessing the state of comprehensive security, the Harrington desirability scale is used: “ Very bad” 0.00 - 0.20; “ Bad” 0.20 - 0.37; “ Satisfactory” 0.37 - 0.63; “Good” 0.63 - 0.80; “Very good” 0.80 - 1.00. The choice of marks 0.37 and 0.63 on the desirability scale is explained by the convenience of calculations, since the desirability function is exponential:

$$0{,}37 = \frac{1}{e} \ ; \ 0{,}63 = 1 - \frac{1}{e} \quad . \tag{2}$$

Comprehensive security assessment is carried out in four stages:

Stage 1: "Determination of marks of the state of comprehensive security".

Stage 2: "Calculation of the groups of indicators".

Stage 3: "The calculation of indicators."

Stage 4: "Calculation of the integrated assessment of the state of comprehensive security".

At the 1 st stage, the marks of the state of comprehensive security of the object of assessment are collected. This stage can be considered as a preparatory stage, the reliability of the assessment directly depends on the reliability of the information collected at this stage.

Then, at stage 2, the values of groups of indicators are calculated. Due to the fact that only those that meet the requirements presented above are selected as marks of the state of comprehensive security required for its assessment, the values of the groups of indicators express the share of security (need – actual availability) or the share in the total amount. Accordingly, the values of the indicator groups are in the range from 0 to 1.

At the 3rd stage, the values of the indicators of comprehensive security status are calculated. One of the methods of the theory of experiment planning has been adapted as a methodological basis for this calculation. The value of the generalized Harrington desirability function is taken as the indicator value. This approach, based on the assessment of processes on the Harrington scale of importance, is widely used in quality management systems of organizations to assess their effectiveness. [8]

The calculation sequence of the indicator is as follows:

3.1. Coding the values of indicator groups.

3.2. Calculation of particular values of the desirability of groups of indicators.

3.3. Calculation of the indicator value.

On sub-stage 3.1, the resulting values of indicator groups are coded or, in other words, scaled. This allows to set lower and upper permissible limits of integrated security readiness measurements for different objects of assessment.

Let's give an example: the readiness of comprehensive security in a border state entity of 0,6 can be estimated as "not ready", while the same readiness value for another entity can correspond to the assessment "limited ready".

Thus, the translation of the values of the groups of indicators M into the conditional scale M ' is carried out in order to evaluate various objects of assessment on the same scale. Dependencies for coding:

for state entities, municipalities that are not assigned to civil defense groups, and organizations that are not assigned to civil defense categories by (3):

$$M' = 9{,}73 \cdot M - 5{,}13 \ , \tag{3}$$

for state entities, municipalities assigned to civil defense groups, and organizations assigned to civil defense categories by (4):

$$M' = 12{,}25 \cdot M - 7{,}65 \ . \tag{4}$$

An explanation of the form of equations (3) and (4) and the values of the corresponding coefficients is given below.

On sub-stage 3.2 by (5), particular values of desirability of groups of indicators d_M are calculated by (5):

$$d_M = \exp(-exp(-M')) \ , \tag{5}$$

It is assumed that the relationship between the indicator value and its encoded value is linear, so:

$$M' = a_1 M + a_0 \ . \tag{6}$$

The coded value of M' is expressed from (5) by double logarithm:

$$\ln\ln\left(\frac{1}{d}\right) = -M' \ . \tag{7}$$

Substituting the value of M' in (6), we obtain:

$$a_1 M + a_0 = \ln\left(\frac{1}{\ln\left(\frac{1}{d}\right)}\right) \ . \tag{8}$$

For the minimum and maximum desirable value of the group of indicators M_{min} and M_{max}, respectively, it is possible to create a system of equations (8)

$$\begin{cases} a_1 M_{min} + a_0 = \ln\left(\dfrac{1}{\ln\left(\frac{1}{d_{min}}\right)}\right) \\ a_1 M_{max} + a_0 = \ln\left(\dfrac{1}{\ln\left(\frac{1}{d_{max}}\right)}\right) \end{cases} . \tag{9}$$

The coefficients a_1 and a_0 used in (3) and (4) are expressed for (9).

On sub-stage 3.3 (10), the value of the indicator I_n is calculated as the geometric mean of the particular values of the desirability of the groups of indicators d_M, taking into account their significance.

$$I_n = \sqrt[k]{\prod^{k} d_M^{\beta}} \qquad (10)$$

where: k – is the number of indicator groups that make up the I_n indicator;

β – coefficient of significance of a group of indicators.

The value of the coefficient of significance of a group of indicators is determined by (11) based on its rank assigned according to the results of a survey of experts (1 is the most significant, then in decreasing order of importance) [9]:

$$\beta = \frac{u}{2^{u-1}}, \qquad (11)$$

where: u – the rank of the group of indicators.

The β values for ranks 1 to 9 are shown in Table 2.

Table 2. Significance coefficient Values for ranks 1-9

Significance coefficient β	Rank u
1	1
1	2
0.75	3
0.5	4
0.3125	5
0.1875	6
0.11	7
0.063	8
0.035	9

After determining the values of all groups of indicators, an integral assessment of the state of comprehensive safety is determined at stage 4.

The calculation sequence for its definition is the same as stage 3 and can be presented in three sub-stages:

4.1. Coding of indicator values as per (3) or (4).

4.2. Calculation of the particular values of the desirability of the indicators according to (5).

4.3. Calculation of the value of the integral estimate according to (10).

4. Results and Discussion

In today's world of multiple threats, the concept of security is multidimensional and encompasses many facets of everyday life. Today, many individual spheres of life must be properly regulated and monitored by the State in order to sustain human life. Depending on the type of existing threats, there are also different subspecies of security: national, economic, energy, environmental, etc. According to the number of subjects, the security of the individual, society, and the state is distinguished. In fact, the state's activities are aimed at ensuring the security of the individual and society. Ensuring such security is regulated through the prism of legal regulation and the functioning of the entire system of state and municipal authorities. Depending on existing threats, each State develops its own strategies for the protection of its citizens and its territorial integrity. In the Russian Federation, this strategy has been adopted at the federal level and supplemented by acts of the constituent entities. The concept of comprehensive state security is aimed at creating a system of tools (authorities, services, etc.) which, taken together, are capable of preventing and resolving all kinds of threats to the rights and freedoms of the individual, society and the State. [9, 10]

In conclusion, it should be noted that the advantage of using the described methodological approach is the possibility of using different ranges when assessing the state of integrated safety of various objects of assessment without making changes to the mathematical formalism. In addition, the desirability function has such useful properties as continuity, monotony and smoothness, is not sensitive at extreme values, and reflects well the estimate in the middle zone.

The proposed mathematical model for collecting and evaluating complex security factors and ensuring the prevention of threats to individuals and society in cyberspace may be of interest to specialists in the field of information threats to qualitatively assess the problems of the state of an object.

References

1. Z. Bangwei, Interdisciplinary Description of Complex Systems, **17**(3), 520-545 (2019) https://doi.org/10.7906/indecs.17.3.13

2. G.A. Makeev, Computer Science and Information Technologies (CSIT'2014) Proceedings of the 16 th International Workshop, 101-104 (2014)

3. D.V. Poltoranov, QUALITY. INNOVATION. EDUCATION., **4**(162), 60-66 (2019) [in Russia]

4. T. Karlova, A. Bekmeshov, N. Kuznetsova, 2019 International Conference «Quality Management, Transport and Information Security, Information Technologies» (IT&QM&IS) (2019) doi:10.1109/itqmis.2019.8928412

5. P.A. Knyazev, D.V. Poltoranov, Civil Security Technologies, **13**(2), 92-97 (2016) [in Russia]

6. V.K. Fedyukin *Quality Management of Technological Processes.* (M.: KNORUS, 2013)

7. V.K arlova, E. A. Kirillova, A.Y. Bekmeshov, A.N. Zapolskaya, I.A. Mikhaylov, 2019 International Conference "Quality Management, Transport and Information Security, Information Technologies" (IT&QM&IS) (2019) doi:10.1109/itqmis.2019.8928381

8. R.V. Butkevich, Yu.S. Klochkov, T.S. Yanitskaya, S.A. Yarygin Bulletin of the Samara Scientific Center of the Russian Academy of Sciences, **7**(2), 456-463 (2005) [in Russia]

 https://doi.org/10.1051/epjconf/202124803009

9. T. K arlova, M . Mikhaylova, N . K uznetsova, A. Bekmeshov, D. Poltoranov, E. Obukhova, EPJ Web of Conferences **224**, 0600 4 (2019) https://doi.org/10.1051/epjconf/201922406004
10. A.F. Nevostrueva, Global Media Journal, **1** (2016)

Generation of Electron and Fast Atom Beams by a Grid Immersed in Plasma

Alexander Metel[1,*], *Enver* Mustafaev[1], *Yury* Melnik[1], and *Khaled* Hamdy[2]

[1]Moscow State Technological University "STANKIN", RU-127055, Moscow, Russia
[2]Production Engineering and Mechanical Design Department, Faculty of Engineering, Minia University, EG-61519, Minia, Egypt

Abstract. We present a new method of product processing with beams of accelerated electrons and fast neutral atoms, which are generated by an immersed in plasma grid under a high negative voltage of 5 kV. The electrons appear due to secondary emission from the grid surface provoked by its bombardment with ions accelerated from the plasma. At the gas pressure not exceeding 0.1 Pa the ions with energy of 5 keV reach the grid without collisions in the space charge sheaths near its surface and their current in the grid circuit is by 2-3 times lower than the electron current. At higher pressures accelerated ions due to charge exchange collisions in the sheaths turn into fast neutral atoms leaving the sheaths and forming the beams. With the pressure increasing, the electron beam current diminishes and the current of fast atom beam grows.

1 Introduction

The ion-plasma processing of engineering products allows a modification of their surface properties. Using the plasma immersion ion implantation [1-3], materials can be modified and alloying additives can be introduced into the surface layer. Using sources of metal plasma [4] or metal vapor [5], it allows synthesis of wear-resistant coatings on the products. Ion-plasma treatment reduces the coefficient of friction, increases the useful life of the products, significantly increases their wear resistance and corrosion resistance. It also allows for combined treatment, for example, ion nitriding of products [6, 7] with subsequent deposition of wear-resistant coatings.

In the latter case, a hardened nitride surface layer with a thickness of 50–100 μm prevents brittle fracture of hard coatings with a thickness not exceeding 5 μm. This improves the adhesion of the coating and its resistance to cracking. Hence, the product combines wear resistance with high load bearing capacity and good fatigue strength. In the plasma-immersion ion implantation, the product is immersed in a plasma containing necessary ions, and negative high-voltage pulses are applied to it [8-10]. The product surface bombardment by ions accelerated from the plasma appreciably improves the coating properties [11, 12].

It is impossible to apply high-voltage pulses to dielectric products and for this reason, broad beams of accelerated ions [13, 14] or fast neutral atoms [15, 16] are used to bombard the product surface. The beam sources are also applied to bombard the coating during its deposition and for etching the product surface [17]. All kinds of processing are carried out in discharge plasma at a gas pressure of 0.05–1 Pa [18].

It is possible to dispense with beam sources using an immersed in plasma grids under negative bias potential. Accelerated from the plasma ions will pass through the grid holes and return back on the other side of the grid. At an appropriate gas pressure, they can turn into fast neutral atoms due to charge exchange collisions in the space charge sheaths of the grid. Only fast atoms can leave the sheaths and form neutral beams. It is obvious that the presumable method for generation of fast atom beams should be cheaper in realisation as compared to application of beam sources.

In any case, it is necessary first to develop a model of the beam generation allowing for prediction of the beam parameters dependence on the gas pressure and the discharge current.

2 Beams produced by a flat grid

Let us consider a flat titanium grid *1* (Figure 1) installed in a vacuum chamber filled with plasma *2* and 3 of the glow discharge with electrostatic electron confinement [18] at an argon pressure of 0.1 Pa. When a negative voltage U is applied to the grid, sheaths *4* and *5* of the positive space charge of ions *6* and *7* accelerated from the plasma are formed between the grid and the plasma the sheath width amounting to d.

The current density j_i of these ions on the surface of grid *1* is determined by the emission capacity of the plasma and is expressed by the Bohm formula [19]:

$$j_i = (2e)^{-1/2} en\,(2kT_e/M)^{1/2} = 0.94\times10^{-15}\, n\,[T_e(\text{eV})/M_A]^{1/2}, \quad (1)$$

where e is the Naperian base, n is the plasma density, T_e is the temperature of plasma electrons, e is the

* Corresponding author: a.metel@stankin.ru

elementary charge, k is the Boltzmann constant, M is the ion mass in kilograms, and M_A is the ion mass in atomic mass units $A = 1.66\times10^{-27}$ kg. For a plasma with a density of $n = 2\times10^{16}$ m^{-3} and an electron temperature $T_e = 1$ eV, the ion current density on the grid is equal to $j_i = 3.2$ A/m^2.

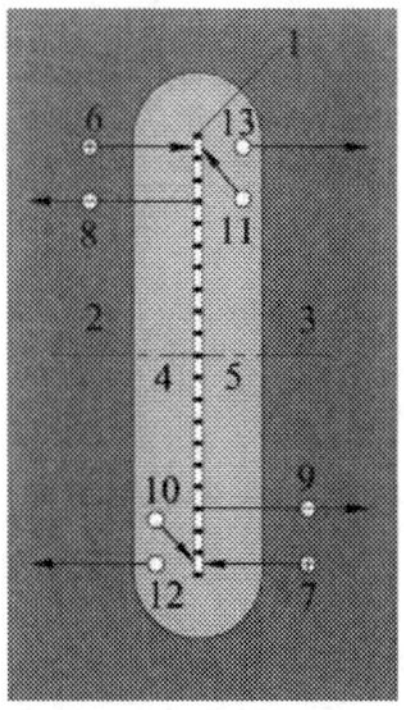

Fig. 1. Schematic of broad electron and fast atom beams generation using a flat grid immersed in plasma under negative voltage.

If we take into account the coefficient of secondary electron emission $\gamma_i = 1.3$ for argon ions with an energy of 4 keV, then the current density of electrons 8 and 9 emitted by the grid is equal to $j_e = \gamma_i j_i = 1.3\times3.2$ A/m$^2 \approx 4.2$ A/m^2, and the total current density of ions and electrons on the surface of the grid amounts to $j = j_i(1+\gamma_i) = 7.4$ A/m^2. The width d of the space charge sheaths is determined by the Child-Langmuir expression [20]:

$$j_i=(4/9)\varepsilon_0(2e/M)^{1/2}U^{3/2}/d^2, \quad (2)$$

If the ion mass $M = AM_A$ is expressed in atomic mass units $A = 2.7\cdot10^{-11}$ kg, then the ion current density

$$j_i=5.46\times10^{-8}1/M_A)^{1/2}U^{3/2}/d^2 \quad (3)$$

With a grid voltage $U = 4000$ V, $M_A = 40$ and $j_i = 3.2$ A/m^2 we get $d^2 = 6.82\times10^{-4}$ m^2 and $d = 2.6\times10^{-2}$ m. At a pressure of 0.1 Pa, the sheath width $d = 2.6$ cm is less than the mean free path of gas molecules $\lambda \approx 7$ cm and less than the ion charge exchange length $\lambda_c \sim 20$ cm. Therefore, accelerated from plasma 2 and 3 ions 6 and 7 fly to the grid surface without collisions.

Let us consider the reciprocation of ions in the sheaths. To determine the average path of N ions in the space charge sheaths prior to absorption by the grid, we divide them into groups of ions with equal path lengths l_k (k = 1,2,3…). At the transparency η of the grid $N(1-\eta)$ of ions are absorbed by the grid, and $N\eta$ of ions fly through its holes, slowdown and turn back to fly in the opposite direction. Hence, the path of the first group of ions in the amount of $N(1-\eta)$ is equal to $l_1 = d$.

When $N\eta$ of ions move backward through the grid, $N\eta(1-\eta)$ of ions are absorbed, and $N\eta^2$ of ions fly through its holes, slowdown and turn back for a flight in the opposite direction. Therefore, the path length of the second group of ions in the amount of $N\eta(1-\eta)$ is equal to $l_2 = 3d$.

When $N\eta^{k-1}$ of ions move backward through the grid, $N\eta^{k-1}(1-\eta)$ of them are absorbed by the grid, while $N\eta^k$ of ions fly through its holes, slowdown and turn around to fly in the opposite direction. Therefore, the path length of the k-th group in the amount of $N\eta^{k-1}(1-\eta)$ is equal to $l_k = (2k-1)d$.

To obtain the average path length of ions in the sheaths, it is necessary to average the path lengths l_k of all groups according to the number of ions in the groups. Therefore, the average path length of ions reciprocating in the sheaths is equal to

$$L = [l_1N(1-\eta) + l_2N\eta(1-\eta) +\ldots + l_kN\eta^{k-1}(1-\eta) +\ldots]/N =$$

$$= (1-\eta)[d+3d\eta+5d\eta^2+ \ldots + (2k-1)d\eta^{k-1} +\ldots] =$$

$$= d(1-\eta)[2\textstyle\sum_{k=1}^{\infty} k\,\eta^{k-1} - \sum_{k=1}^{\infty} \eta^{k-1}]. \quad (4)$$

Taking into account that $\sum_{k=1}^{\infty}\eta^{k-1} = (1-\eta)^{-1}$, and $\sum_{k=1}^{\infty} k\,\eta^{k-1} = (1-\eta)^{-2}$, the average path of ions in the space charge sheaths amounts to

$$L = d(1+\eta)(1-\eta)^{-1} \quad (5)$$

In our case, at $d = 2.6$ cm and $\eta = 0.8$, the average path of accelerated ions in the sheaths amounts to $L = 23.4$ cm. At argon pressure of 0.01 Pa the ion charge exchange length $\lambda_c \sim 2$ m by an order of magnitude exceeds L. Therefore, no charge exchange collisions occur in the sheaths, and no fast atoms are produced. Only two beams of secondary electrons with energy $eU = 4$ keV and current density $j_e = 4.2$ A/m^2 are formed, which propagate in opposite directions. When the pressure grows to 0.1 Pa, λ_c diminishes to ~ 20 cm and at $\lambda_c < L$ fast atoms appear.

The energy $e\varphi$ of a fast neutral atom corresponds to the potential φ of the point of its production in the sheath with a width of d at a distance x from the grid surface. The potential distribution in the sheath is described by the expression [19]:

$$\varphi = U(1-x/d)^{4/3} \quad (6)$$

The maximum energy ~ eU is possessed by fast atoms formed directly in the grid holes, i.e. at $x \approx 0$ and φ equal to the potential difference U between the plasma and the grid (Figure 1). With increasing distance x from the grid, the energy of atoms decreases to zero at $x = d$ at the boundary of the discharge plasma.

With a further increase in pressure, the ions accelerated from the plasma turn into fast atoms even before crossing the plane of the grid. On average, at a distance from the plasma boundary equal to the charge exchange length λ_c, the accelerated ion becomes a fast atom and a slow ion appears. The latter is accelerated in the electric field, travels the same distance ~ λ_c and turns into the second fast atom. As a result, one ion with the energy $eU\lambda_c/d$ and (d/λ_c-1) fast atoms with the same energy come to the grid. For this reason, the power density of ions on the grid and proportional to the power density current density j_e of electron beams decrease with increasing pressure, while the power transported by fast atoms increases. This means that with increasing pressure, the equivalent current of fast argon atoms increases, while the current of electron beams and the intensity of the grid sputtering decrease.

When replacing a flat grid with a concave one (Figure 2), beams are formed in the same way. However, the trajectories of ions 2, accelerated from plasma 3 in sheath 4 at the convex surface of grid 1, as well as fast atoms 6 produced in collisions with gas molecules 5 converge at its focal point.

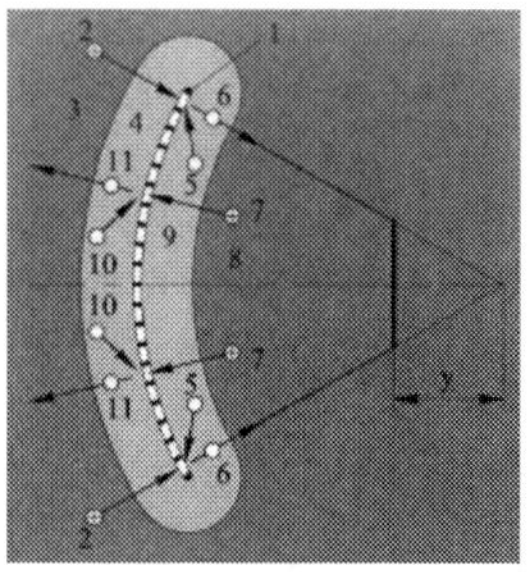

Fig. 2. Generation of compressed beam of fast atoms using a grid with concave surface.

With an increase in the distance from the center of the concave surface of the grid 1, the flow density of fast atoms increases. This makes it possible to multiply the rate of the substrate sputtering in comparison with a flat grid and to increase the heating temperature of the substrate.

3 Experimental study of the fast atom beam formation by a concave grid

The verification of the theoretically obtained regularities of beam formation was carried out using a system shown in Figure 3. A round 20-cm-diameter grid with a 20-cm-curvature radius of its concave surface was fixed to a high-voltage feedthrough inside the system chamber. On the grid surface, holes with a diameter of 7 mm are evenly distributed at a distance of 8 mm between their centers. In the focus of the grid at a distance of 6 cm from the chamber wall, a holder for the samples to be processed by the beam is installed.

At an argon pressure in the chamber of $p \sim 0.5$ Pa, an increase in the voltage between the anode and the chamber to several hundred volts leads to breakdown and the establishing of a glow discharge with a current in the anode circuit I_a up to 4 A and a voltage of $U_d = 400$–500 V. The chamber turns filled with a brightly glowing homogeneous plasma, separated from its walls by a cathode sheath, and from the grid - by a grid sheath of theion space charge. When the camera and the grid are equipotential, both space charge sheaths are of the same width.

The voltage between the chamber and the anode is independent of the argon pressure p at a constant current I_a in the anode circuit and the pressure ranging from 0.2 to 1 Pa. However, at $p < 0.2$ Pa, a further decrease in pressure causes an increase in the discharge voltage U_d,

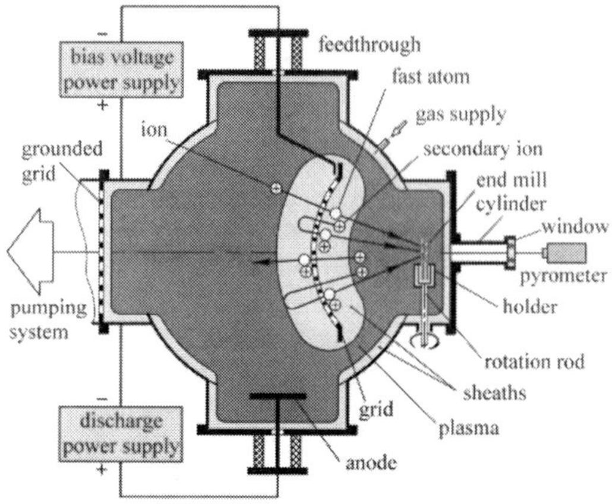

and at p = 0.02 Pa it reaches a value of $U_d \sim 1$ kV.

Fig. 3. Schematic of the beam formation by a grid.

With an increase in the voltage U between the chamber and the grid from zero to 5 kV, the current I in the grid circuit at a constant current I_a in the anode circuit approximately doubles, the width d of the sheath between the plasma and the grid increases to 5-10 cm, and the discharge voltage U_d decreases approximately two times.

The density of gas atoms at room temperature and pressure $p = 0.02$ Pa is equal to $n = 0.5\times10^{19}$ m^{-3} [21] and the mean path length of argon ions between collisions with charge exchange, called the charge exchange length, is $\lambda = 1/n\sigma = 1$ m. Here we took into account that the cross section of charge exchange collisions for argon ions with an energy of 5 keV is equal to $\sigma = 2\times10^{-19}$ m^2 [22-23].

The width of the grid sheath d ~ 0.05 m is much less than the recharging length λc, and therefore the formation of fast atoms in the sheath does not occur. All ions drawn from the plasma bombard the grid and cause secondary electron emission from its surface.

At a pressure of $p = 0.02$ Pa, the grid emits only electron beams with an energy of $e(U+U_d)$, propagating from the accelerating grid in opposite directions. Their total current $I_e = I\gamma_i/(1+\gamma_i)$ is approximately equal to $0.63I$, where I is the current in the grid circuit, and $\gamma_i = 1.7$ is the coefficient of ion-electron emission. The surface of the plasma emitting ions onto the concave surface of the grid is less than the surface of the plasma emitting ions onto its convex surface (Fig. 3). Therefore the current of the electron beam diverging to the chamber walls is higher than the current of the electron beam converging on the sample.

With an increase in pressure to 0.2 Pa, accelerated ions turn in the sheath into fast atoms escaping from the sheath. The energy of each of them, eφ, corresponds to the potential φ of the point in the sheath in which it appeared.

In this regard, the energy of fast atoms varies from zero to $e(U+U_d)$. The secondary ions formed as a result of charge exchange collisions are accelerated in the sheath and also turn into fast neutral atoms that form a beam.

With an increase in pressure p from 0.2 to 2 Pa, the number of fast atoms increases tenfold, their energy decreases from 5 keV to 500 eV, and the currents of neutral beams increase.

As for the accelerated electrons, their current decreases with increasing pressure, their energy is equal to 5 keV and practically does not change over the entire range from 0.02 to 2 Pa. Measurement of the voltage between the anode and the holder showed that its potential does not exceed 20 V.

With increasing pressure p, the current in the grid circuit increases to I = 0.36 A at p = 0.4 and to I = 0.48 A at p = 1.5 Pa. There is also a decrease in the discharge voltage to U_d = 170 V at p = 0.4 and to U_d = 55 V at p = 1.5 Pa.

When a 1-mm-thick flat titanium target is placed in the grid focus, a 2-cm-diameter brightly glowing circle is observed on its surface heated by the fast atom beam. The temperature grows with increasing the gas pressure p and changes with the distance from the left edge of the target, reaching the maximum value of 700°C at p = 0.4 and 900°C at p = 1.5 Pa in the center of the glowing circle.

An increase in the heating intensity with increasing pressure confirms the theory's conclusion about an increase in the flux of fast neutral atoms with a decrease in the charge exchange length λ of ions below the width d of the space charge sheath near the grid.

Fast atoms propagating from the concave surface of the 200-mm-diameter grid bombard the target within a circle with a diameter not exceeding 10 mm. However, due to thermal conductivity, the width of the target temperature distribution must exceed the width of the neutral atoms flux distribution.

For a more accurate estimate of the beam diameter, we measured the distribution of the etching rate by fast atoms of a polished titanium target with a mask on its surface. The straight boundary between them was located horizontally (Fig. 4).

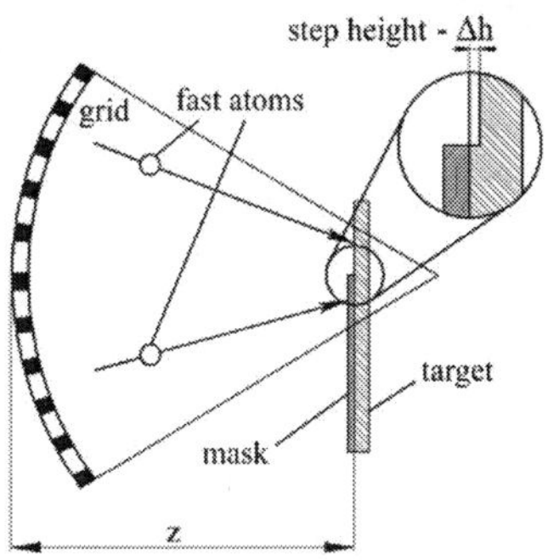

Fig. 4. The formation of a step on the surface of the target installed at a distance z from the grid, due to bombardment with a beam of fast argon atoms.

The target was bombarded with fast atoms at a current in the anode circuit I_a = 1 A and an argon pressure p = 1.5 Pa. Then using a mechanical profile meter Dektak XT the height difference between the covered with the mask and the open substrate surface was measured along the border between them.

The sputtering rate of the target surface was determined by dividing the height of the step by the time of its bombardment with fast argon atoms. The measurement results presented in Figure 5 made it possible to determine the effective beam diameter of fast neutral argon atoms.

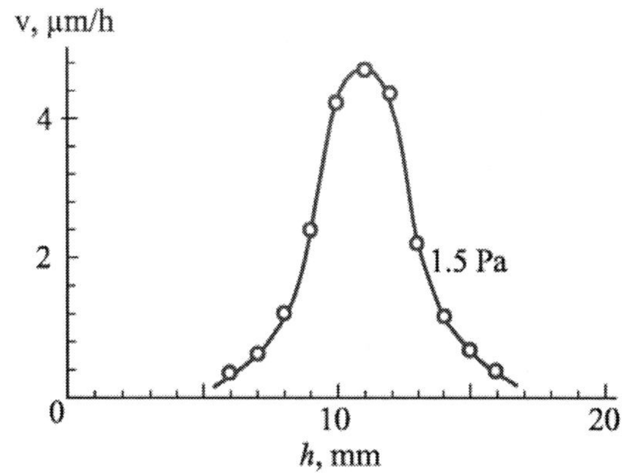

Fig. 5. The distribution of the target etching rate over itssurfaces at a the grid voltage of U = 5 kV, the argon pressure of 1.0 Pa and the anode current of I_a = 1 A.

Since the sputtering rate v is proportional to the flux density of fast atoms, the width of the distribution shown in Figure 5 gives an effective diameter D of the fast atom beam equal to 5 mm in the focal plane of the grid at a distance of 20 cm from its center.

Figure 6 shows the dependence of the effective beam diameter D on the distance Z between the target and the grid. The diameter values were determined in a similar way from the distribution of the sputtering rate of the flat polished substrates.

As the distance Z between the target and the grid increases, the diameter D decreases from 45 mm at Z = 17 cm to 6 mm at Z = 20 cm, and as the distance increases from Z = 21 cm to Z = 24 cm, the diameter D increases from 7 mm up to 52 mm.

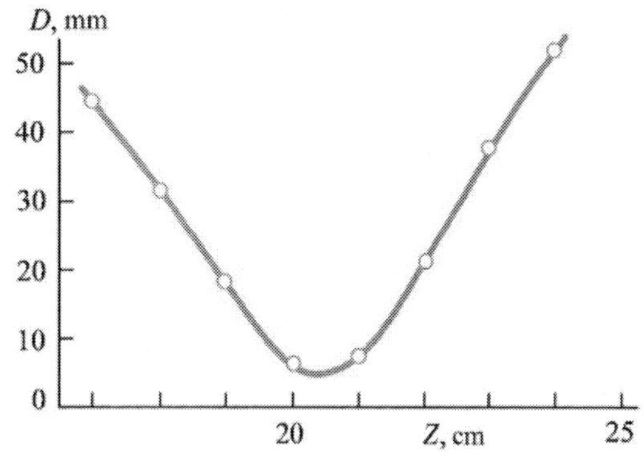

Fig. 6. The dependence of the beam diameter D on the distance Z between the grid and the target.

This means that by changing the distance between the grid and a product to be processed, it is possible to achieve that the beam diameter exceeds the overall dimensions of the product rotating in the plasma, and that fast atoms bombard its entire surface.

The experimental results show that the processing with a grid immersed in plasma is universal. At the gas pressure not exceeding 0.1 Pa, it allows heating the products with electron beams without etching their surfaces. At the gas pressure exceeding 1 Pa, it allows a uniform etching of the product surface with fast atoms without blunting its sharp edges. The surface etching rate depends on the product material and fast atoms energy [24]. The new method for surface processing with fast atoms can replace the methods for plasma electrolytic polishing [25], laser-plasma polishing of metals [26], surface polishing with pulsed electron beams [27] and also can be used for micro- and nanoelectronics [28].

Acknowledgments

This work was supported by the Russian Science Foundation under grant 20-19-00620. This work was carried out using equipment provided by the Center of Collective Use of MSUT "STANKIN".

References

1. *Handbook of Plasma Immersion Ion Implantation and Deposition, A. Anders (Ed.)* (JohnWiley & Sons, Inc., New York, 2000)
2. N.N. Koval, A.I. Ryabchikov, D.O. Sivin, I.V. Lopatin, O.V. Krysina, Yu.H. Akhmadeev, D.Yu. Ignatov, Surf. Coat. Technol., **340**, 152 (2018)
3. D. Vempaire, J. Pelletier, A. Lacoste, S. Béchu, J. Sirou, S. Miraglia, D. Fruchart, Plasma Phys. and Contr. Fusion., **47**, A153 (2005)
4. A. Anders, *Cathodic Arcs: From Fractal Spots to Energetic Condensation* (Springer, New York, 2008)
5. A. Anders, J. Appl. Phys., **121**, 171101 (2017)
6. J.C. Avelar-Batista, E. Spain, J. Housden, A. Matthews, Surf. Coat. Technol., **200**, 1954 (2005)
7. S. Grigoriev, A. Metel, NATO Science Series, Series II: Mathematics, Physics And Chemistry, **155**, 147 (2004)
8. C. Ruset, E. Grigore, Surf. Coat. Technol., **156**, 159 (2002)
9. C. Ruset, E. Grigore, G.A. Collins, K.T. Short, F. Rossi, N. Gibson, H. Dong, T. Bell, Surf. Coat. Technol., **174-175**, 698 (2003)
10. E. Grigore, C. Ruset, K. Short, D. Hoeft, H. Dong, X.Y. Li, T. Bell, Surf. Coat. Technol., **200**, 744 (2005)
11. A.S. Metel, V.P. Bolbukov, M.A. Volosova, S.N. Grigoriev, Yu.A. Melnik, Instrum. Exp. Tech., **57**, 345 (2014)
12. M. Volosova, S. Grigoriev, A. Metel, A. Shein, Coatings, **8**, 287 (2018)
13. A.V. Hayes, V. Kanarov, B. Vidinsky, Rev. Sci. Instrum. **67**, 1638 (1996)
14. E.M. Oks, A.V. Vizir, G.Yu. Yushkov, Rev. Sci. Instrum. **69**, 853 (1998)
15. A.S. Metel, S.N. Grigoriev, Yu.A. Melnik, V.P. Bolbukov, Instrum. Exp. Tech., **55**, 288 (2012)
16. S.N. Grigoriev, Yu.A. Melnik, A.S. Metel, V.V. Panin, Instrum. Exp. Tech., **52**, 602 (2009)
17. A. Metel, S. Grigoriev, Yu. Melnik, V. Panin, V. Prudnikov, Japan. J. Appl. Phys., **50**, 08JG04 (2011)
18. A.S. Metel, S.N. Grigoriev, Yu.A. Melnik, V.V. Panin, Plasma Phys. Rep., **35**, 1058 (2009)
19. M.A. Lieberman, A.J. Lichtenberg, *Principles of Plasma Discharges and Materials Processing* (Hoboken: John Wiley & Sons, 2005)
20. I. Langmuir, Phys. Rev., **33**, 6 (1929)
21. E.W. McDaniel, *Collision Phenomena in Ionized Gases* (Willey, New York, 1964)
22. A.V. Phelps, J. Phys. Chem. Ref. Data., **20**, 557 (1991)
23. A.V. Phelps, C.H. Greene, J.P. Burke, J. Phys. B: At. Mol. Opt. Phys., **33**, 2965 (2000)
24. M. Kaminsky, *Atomic and ionic impact phenomena on metal surfaces* (Springer-Verlag, Berlin, 1965)
25. I. Danilov, M. Hackert-Oschätzchen, M. Zinecker,G. Meichsner, J. Edelmann, A. Schubert, Micromachines, **10**, 214 (2019)
26. A.M. Chirkov, A.P. Rybalko, Ju.I. Rogalskij, et al, Patent of the Russian Federation No. 2381094 of 10.02.2010
27. N.N. Koval, A.D. Teresov, Y.F. Ivanov, E.A. Petrikova, Patent of the Russian Federation No. 2619543 of 16.05.2017
28. V.P. Kudrya, Yu.P. Maishev. Mikroelektronika, **47**, 51 (2018)

Collisional Relaxation of Fast Electrons Energy in Gas

Alexander Metel[1,*], *Enver* Mustafaev[1], *Yury* Melnik[1], and *Khaled* Hamdy[2]

[1]Moscow State Technological University "STANKIN", RU-127055, Moscow, Russia
[2]Production Engineering and Mechanical Design Department, Faculty of Engineering, Minia University, EG-61519, Minia, Egypt

Abstract. We present results of theoretical and experimental study of collisional relaxation of fast electrons energy in gas. The dependence on the gas pressure p and electron energy ε of the mean pass Λ of fast electrons injected into a gas being sufficient to spend on ionization all their initial energy ε has been calculated. It was found that Λ is directly proportional to ε^2 and inversely proportional to the gas pressure. To sustain glow discharge with electrostatic confinement of fast electrons, Λ should be less than the mean way to the anode of emitted by the cathode electrons.

1 Introduction

To transport accelerated particles and atoms of deposited materials from the particles sources to the surface of the workpiece in the working vacuum chamber, the length of their free path must exceed the size of the workpiece ~ 10 cm, and therefore these particles are obtained using a high-frequency [1], vacuum arc [2] and magnetron [3] discharges at a gas pressure below ~ 0.1 Pa. To fill a large working volume with a homogeneous plasma, a discharge with thermionic cathodes and a peripheral magnetic field on the surface of the chamber, is also used [4]. In this case, a cylindrical hollow cathode with an inner diameter of several millimeters made of a refractory metal is often used as a thermionic cathode [5]. As for the glow discharge with a cold hollow cathode [6], the range of its applications for surface treatment was limited by the relatively high working gas pressure. There were no data in the literature on a cold cathode glow discharge at a gas pressure below 1 Pa [7-14]. There were no well-established ideas about the factors that determine the value of its lower working pressure [14-16] and the mechanism of the discharge. There was also no general opinion about the physical processes causing the effect of the hollow cathode, which reflects back electrons emitted from the plasma enclosed inside it. At the same time, the results of research and practical application of the discharge indicated that a glow discharge with a hollow cathode is capable of forming ion and electron beams [17-20], as well as a dense and homogeneous plasma throughout the volume of the cathode hollow.

Most researchers explained that the extinction of the hollow cathode glow discharge is caused by the loss of electrons emitted by the cathode with a decrease in the gas pressure. either as a result of their absorption by the cathode surface or their escape from the cathode [21-23]. In the second case, the lower pressure limit of the discharge should depend on the ratio between the average path length L of electrons inside the cathode before escaping from the cathode and the energy relaxation length Λ hereinafter referred to as the average path length of a fast electron, which it has to pass in order to spend all its energy on the gas ionization. When Λ exceeds L, the electrons spend only part of the energy on the gas ionization inside the cathode. Therefore, with increasing Λ, the ionization intensity in the cathode decreases and the discharge extinguishes.

2 Beams produced by a flat grid

In discharges with electrostatic traps at low gas pressure, electrons emitted by the cathode pass through the cathode sheath without collisions and enter the plasma with an energy of $\varepsilon_0 = eU$ from hundreds to thousands of eV, corresponding to the cathode fall of potential U. Each of them travels a path exceeding the size of the discharge space by hundreds and thousands times and forms tens of new free electrons inside the trap. Here it is necessary to trace the entire path of a fast electron and determine how many new electrons and with what energy it forms both in equipotential plasma and in electric fields of space charges.

To determine Λ we use a well-known experimental fact: the number N of free electrons produced by a fast electron injected into a gas is directly proportional to its initial energy. The ionization cost W of a new electron is constant for any initial energy of the fast electron and is equal to 46 eV for helium, 37 eV for neon, 26 eV for argon, 24 eV for krypton, 37 eV for nitrogen, 36 eV for hydrogen, and 33 eV for oxygen [24]. This allows us to abstract from the details of real events occurring with the electron and restrict ourselves to formalism, which considers only its ionization collisions, in which on formation of a new electron is spent not the ionization energy E_i, but the ionization cost W. To determine the electron mean free path between such collisions, we will

* Corresponding author: a.metel@stankin.ru

https://doi.org/10.1051/epjconf/202124804002

use the experimental values of the ionization cross section σ given in the monograph by Massey and Barhop [25].

As the distance dx travels, the number N_0 of electrons with the initial energy ε decreases as a result of ionizing collisions:

$$dN_0 = -N_0 dx/\lambda_0, \quad (1)$$

where $\lambda_0 = 1/n\sigma(\varepsilon)$, n is the density of gas molecules, and $\sigma(\varepsilon)$ is the dependence of the ionization cross section on the electron energy ε. After the formation of a new free electron, the initial energy ε of a fast electron decreases on average by W. Electrons with energy $(\varepsilon - W)$ appear, and the number of these electrons N_1, as they travel the distance dx, changes by

$$dN_1 = -N_1 dx/\lambda_1 + N_0 dx/\lambda_0, \quad (2)$$

where $\lambda_1 = 1/n\sigma(\varepsilon - W)$. After j collisions, the electron energy decreases to $\varepsilon - jW$, and the number N_j of electrons with indicated energy changes with an increase by dx in the path by

$$dN_j = -N_j dx/\lambda_j + N_{j-1} dx/\lambda_{j-1}, \quad (3)$$

where $\lambda_j = 1/n\sigma(\varepsilon - jW)$ and j = 1,2,.... , (N-1). After N collisions, the electron energy becomes less than W, and it can no longer ionize the gas.

Let us consider ionization of argon by electrons with an energy of 4 keV. After the formation of the first 20 free electrons the energy of each electron decreases on average by $20W \approx 520$ eV, i.e., from 5 to 3.48 keV. The ionization cross section with the indicated decrease in energy varies within the range $\sigma = (3.5 \pm 0.2) \times 10^{-17}$ cm^2. Therefore, on the path length x corresponding to a given change in energy, the value $\lambda = 1/n\sigma$ can be assumed to be approximately constant. In this case, the solution of the equations (1) and (3) gives the following expressions for the dependence of the number N_j of electrons with an average energy $(\varepsilon - jW)$ on the relative length $y = x/\lambda$ of the traversed path

$$N_j(y) = N_0(y^j/j!)\exp(-y). \quad (4)$$

The total number of electrons with energy $\varepsilon \geq \varepsilon - kW$ after their path $x = y\lambda$ is equal to $N_0 M_k(y)$, where

$$M_k(y) \equiv \sum_{j=0}^{k} \frac{y^j}{j!} exp(-y) \quad (5)$$

Figure 1 shows the dependences of the relative number $N_j(y)/N_0$ of electrons with different energies from ε-20W to ε on the length of the traversed path y, and Figure 2 - fractions M_k(y) of electrons with energy $\varepsilon \geq \varepsilon$-k$W$.

Figure 3 shows the energy distributions of electrons after they have traveled a path of 5 λ, 10 λ and 20 λ.

The envelopes of discrete spectra in Fig. 3 present the energy distribution of electrons, which traversed a path of the same length. After passing the distance kλ, the electron energy can have different values in the range from ε-(3k/2)W to ε-(k/2)W, but the energy ε-kW is most likely. This means that passing kλ, it produces k new free electrons.

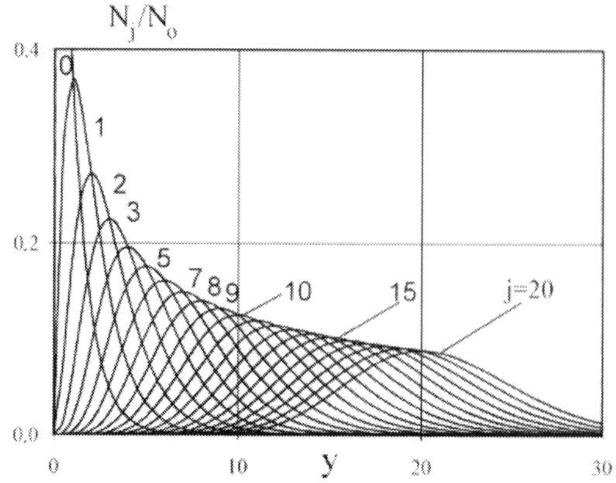

Fig. 1. Relative numbers $N_j(y)/N_0$ of electrons versus the length of the traversed path y.

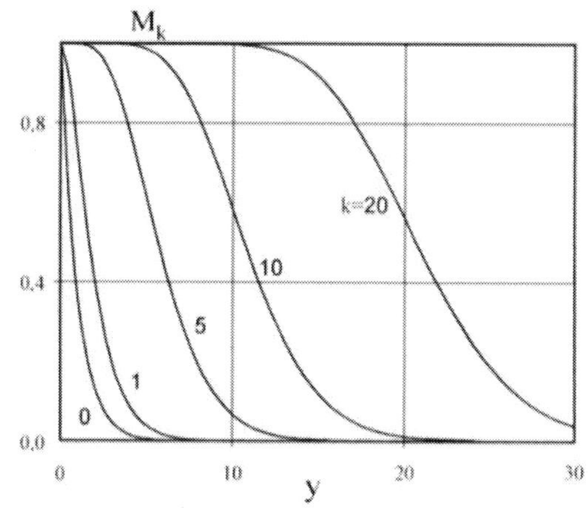

Fig. 2. Fractions M_k (y) of electrons with energy $\varepsilon \geq \varepsilon$-k$W$.

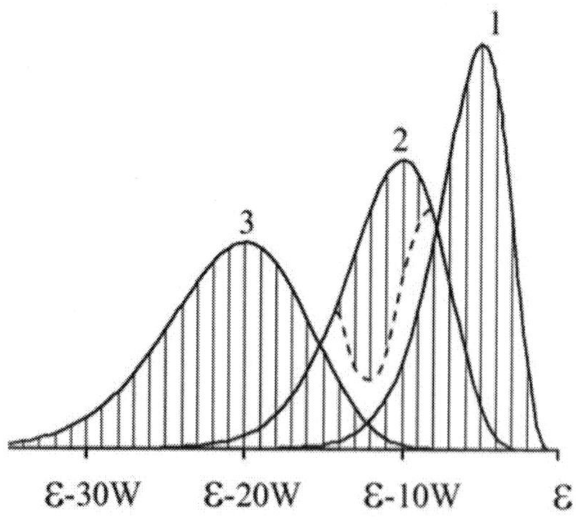

Fig. 3. Energy distributions of electrons traveled a path of 5λ (1), 10λ (2) and 20λ (3).

This conclusion is valid only under the condition that the electron energy does not go beyond the range in which the approximation of λ constancy is permissible.

After passing the distance kλ/2, the energy of all electrons does not really go out of the range from ε-k*W* to ε (Fig.2), however, after the path kλ the number of such electrons diminishes by about two times. Therefore, for k = 20 we can quite reasonably represent only the initial section of the path with a length of 10λ as a sequence of 10 segments of the path, each with a length of $\lambda = 1/n_o\sigma$ and suppose that at the end of each segment a new free electron appears and the energy of the fast electron decreases by *W*.

To reduce the error with a further increase in the length of its path, we must abandon the constancy of λ and assume that the first electron appears at the end of a segment with the length of $\lambda_0 = 1/n_o\sigma(\varepsilon)$, the second electron appears at the end of a segment with the length of $\lambda_1 = 1/n_o\sigma(\varepsilon\text{-W})$, at which the energy of the fast electron is equal to (ε-W), the third electron appears at the end of a segment with the length of $\lambda_2 = 1/n_o\sigma(\varepsilon\text{-}2\text{W})$, at which the energy of the fast electron is equal to (ε-2W), at which the fast electron energy is equal to (ε-2W), etc. As a result, the electron passes a sequence of *N* segments and at the end of each of them forms a new free electron and loses energy of *W*. In total, the electron forms $N = \varepsilon/W$ of new free electrons and passes the path

$$\Lambda\,(p,\varepsilon) = N\lambda_N(p,\varepsilon), \qquad (6)$$

where $\lambda_N(p,\varepsilon)$ is the path λ_j between two successive ionizing collisions, averaged over the entire range of the electron energy from ε to *W*

$$\lambda_N(p,\varepsilon) = \\ = \frac{1}{n_o(p)(\varepsilon - W)} \int_W^{\varepsilon} \frac{d\varepsilon}{\sigma(\varepsilon)} \frac{1}{n_o(p)(N-1)} \int_1^N \frac{dz}{\sigma(W_z)} \qquad (7)$$

The dashed lines in Figure 4 show the dependence of the argon ionization cross-section on the electron energy [25], the dashed-dotted lines show the mean free path λ_0 of an electron with energy ε between ionizing collisions, and the solid lines shows the dependence on the initial energy $\varepsilon = WN$ of the electron path between ionizing collisions λ_N calculated by formula (7) at room temperature 293 K and argon pressure 1.0 Pa ($n_o = 2.5\times10^{14}$ cm^{-3}).

At the electron energy ε ranging from 40 to 400 eV the average electron path between ionizing collisions λ_N can be supposed independent of ε and the energy relaxation length $\Lambda\,(p,\varepsilon) = (\varepsilon/W)\lambda_N(p,\varepsilon)$ is directly proportional to the energy ε.

At the electron energy ε exceeding 1 keV the average electron path between ionizing collisions λ_N can be supposed proportional to ε and the energy relaxation length $\Lambda\,(p,\varepsilon) = (\varepsilon/W)\lambda_N(p,\varepsilon)$ is proportional to ε^2.

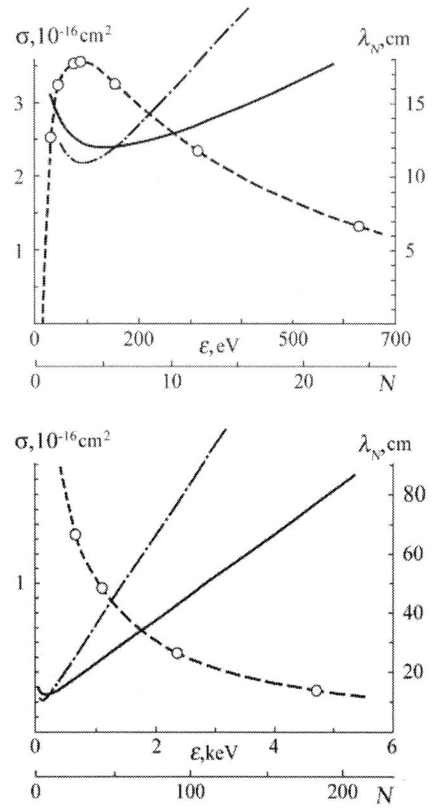

Fig. 4. Dependence on the electron energy ε of the argon ionization cross-section (dashed lines), the mean free path λ_0 of an electron with energy ε between ionizing collisions (dashed-dotted lines) and the average electron path between ionizing collisions λ_N at room temperature 293 K and argon pressure 1.0 Pa (solid lines).

3 Electrostatic traps

Figure 5 presents a schematic of the most wide-spread electrostatic traps based on hollow cathodes featuring trajectory in discharge plasma 1 of electron 2 emitted by the cathode 3 bombarded by ion 4. This electron is hundreds times reflected in a space charge sheath 5 by the voltage of power supply 6 from all parts of the cathode surface enveloping the trap. For this reason distribution of the electron velocity in all the traps is quite isotropic. The distance traveled by electrons before escaping from the trap volume *V* through aperture of losses S_o is equal to

$$L = 4V/S_o \qquad (8)$$

and can by hundreds times exceed the trap width.

Numerous reflections of emitted by the cathode electrons in the cathode sheath ensure an adequate distance traveled by these electrons and, hence, sustaining of the discharge at low pressure.

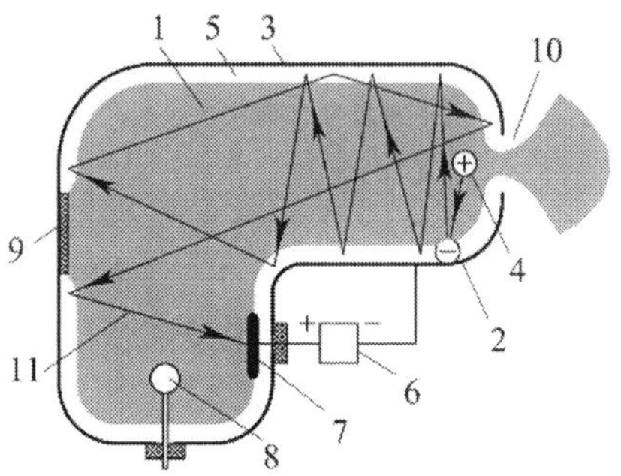

Fig. 5. Electrostatic trap shaped as a hollow cathode.

On their long way to the anode emitted by the cathode electrons visit all parts of the trap and for this reason they are distributed quite homogeneously. Probability of the gas ionization by those electrons and the plasma density are also quite uniform. The lower pressure limit of the discharge should depend on the ratio between the average path length L of electrons inside the cathode before escaping from it and the energy relaxation length Λ. Hereinafter the latter is referred to as the average path length of a fast electron, during the passage of which its energy decreases to the ionization threshold E_i. When the energy relaxation length Λ exceeds L, fast electrons spend only part of their energy on ionizing the gas inside the cathode. To compensate the electron losses it is needed to produce fast electrons in the cathode sheath. At an unchanged discharge current for an increase in the number of additional fast electrons from the sheath, it is needed to increase its width. In accordance with the Child-Langmuir law the sheath width grows with the cathode fall of potential. This is the reason for the discharge voltage increase with the pressure decreasing.

Another type of electrostatic trap is shown in Figure 6.

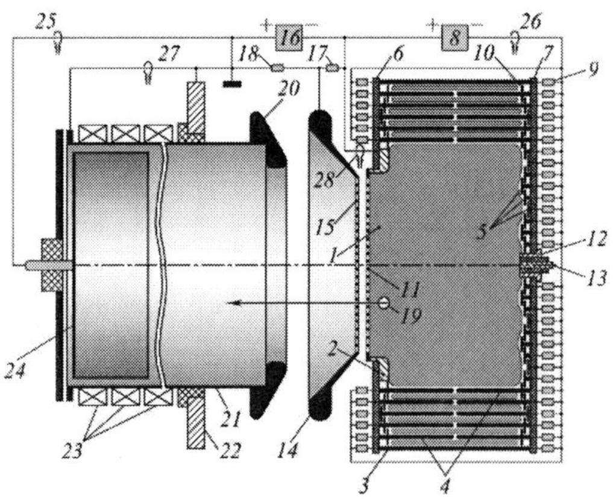

Fig. 6. Schematic of an electron gun with plasma emitter generated by the glow discharge with a multirod cathode.

The plasma emitter 1 of electrons is produced by the glow discharge between anode 2 and a multirod cold cathode. The cathode comprises an outer hexagonal envelope composed of 204 parallel 5-mm-diameter and 200-mm-long cathode rods 3, which are spaced by 1.5 mm, and located inside the envelope hexagonal layers of 5-mm-diameter and 98-mm-long cathode rods 4 the distance between their axes amounting to 15 mm, as well as 23-mm-diameter and 1-mm-thick cathode discs 5.

All the cathode rods and discs are fastened on two dielectric flanges 6 and 7, they are isolated from each other and connected to the negative pole of power supply 8 through resistors 9 with 430-Ω resistance. To prevent from a contact of the dielectric flanges 6 and 7 with the plasma filling the gaps between rods 4, each rod has a 23-mm-diameter and 1-mm-thick metal disk 10. Disks 10 and discs 5 are placed in three layers with a distance of 1 mm between them.

Eight hexagonal layers are formed with 1104 inner rods, and diameter of emissive grid 11 fastened on the anode 2 is equal to 110 mm. A photograph of this anode and inner rods with discs mounted on dielectric flange 6 is presented in Figure 7. In the center of opposite flange 7 is positioned an igniting arrangement with coaxial electrodes 12 and 13. After between those electrodes a high-voltage pulse is applied, the central part of the trap, which is free of rods, is filled with pulsed arc plasma.

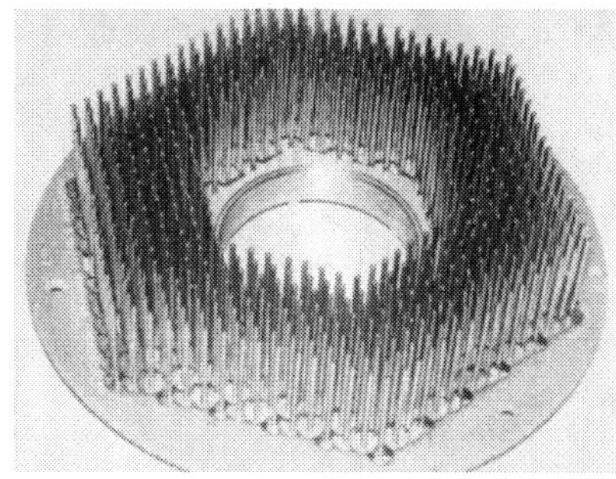

Fig. 7. Photograph of a multirod electrostatic trap.

Dielectric flanges are fixed to each other with three steel spacers, which are also used to fasten resistors *9*. The distance between emissive grid *11* and accelerating grid *15*, which is connected to focusing electrode *14*, amounts to 13 mm. Both grids are 1-mm-thick and have 85 pairs of aligned 6-mm-diameter apertures, distance between aperture centers amounts to 10 mm.

As an accelerating voltage power supply is used a high-voltage square pulse generator *16* based on pulse forming networks. A time delay between switching on the networks allows a decrease of the pulse top ripple down to 1%. Generator *16* with 20-Ω impedance induces 200-μs-wide high-voltage pulses with amplitude up to U = 500 kV across an active load composed of connected in series resistors *17* with 2-Ω resistance and *18* with 18-Ω resistance. Electrons *19* are accelerated in the gap between grids *11* and *15* by the voltage drop across resistor *17* up to energy of $0.1eU$ and then they are accelerated in the gap between focusing electrode *14* and grounded electrode *20* by the voltage drop across resistor *18* up to energy of eU.

Electrode *20* is fastened on 0.8-m-long electron beam track *21*, which is isolated from lateral flange *22* of the vacuum chamber. Displacement of track *21* in horizontal direction relative to flange *22* makes it possible to vary the distance between electrode *20* and grid *15* from 70 to 120 mm. To move in vertical direction electrode *14* together with the whole emissive assembly, which is fastened to a segmented condenser-type high-voltage feedthrough, a special positioner is used, which is mounted inside the central segment of the feedthrough. The positioner also enables canting angle variation of the assembly axis and rotation of the assembly around the vertical axis of the vacuum chamber.

Outside the chamber on track *21* (Fig. 6) are mounted solenoids *23* producing magnetic field with induction up to 0.02 T. At the track end is mounted collector *24* intended to measure using Rogowski coil *25* the current of electron beam, which passed along the track. Current pulses in the circuits of multirod cathode, track *21* and anode *2* are measured, respectively, with Rogowski coils *26*, *27* and *28*.

The plasma emitter of electrons produced using glow discharge with a multirod electrostatic trap made it possible to produce pulsed beams with homogeneous distribution of 700-A current of 300-keV electrons over 170-mm-diameter cross-section of the beam at the pulse width of ~ 200 μs. Connection of about one thousand cathode rods to the discharge power supply through individual resistors prevents the glow discharge with up to 1 kA current from transition to arc and thus ensures homogeneity of the plasma emitter.

At one and the same discharge current I in the cathode circuit the beam current I_b in the circuit of collector is rising with increasing the accelerating voltage and stops rising at $I_b = I$, when all electrons produced in the discharge enter the accelerating gap. At the maximal beam current 700 A of 300-keV electrons the energy capacity of the beam at the pulse width of 200 μs amounts to 40 kJ and at the beam cross-section area of 0.025 m^2 the surface density of energy absorbed by the collector bottom exceeds 10^6 J/m^2. The results obtained overwhelm those previously presented anywhere.

The above broad pulsed beams based on collisional relaxation of fast electrons energy in the plasma emitters of beam and plasma sources were successfully applied for the surface modification of the machinery products. There are many other technological systems based on collisional relaxation of fast electrons energy, for instance [26-29].

Acknowledgments

This work was supported by the Russian Science Foundation under grant 20-19-00620. This work was carried out using equipment provided by the Center of Collective Use of MSUT "STANKIN".

References

1. K. Suzuki, S. Okudaira, N. Sakudo, Jap. J. Appl. Phys., **16**, 1979 (1977)
2. *Handbook of vacuum arc science and technology: fundamentals and applications, Eds by R. L. Boxman, Ph. J. Martin, D. Sanders* (Park Ridge, New Jersey, U.S.A.: Noyes Publications, 1995)
3. A. Anders, Surf. Coat. Technol., **205**, 51 (2011)
4. M. Benhassine, Revue Phys. Appl., **19**, 545 (1984)
5. D.J. Willins, R.L.F. Boyd, J. Phys. D: Appl. Phys., **6**, 1447 (1973)
6. B.I. Moskalev, *Discharge with a hollow cathode.* (Moscow: Energy, 1969) [in Russian]
7. A. Güntherschulze, Zeitschr. für techn. Phys., **11**, 49 (1930).
8. T. Musha, J. Phys. Soc. Japan, **17**, 1440 (1962)
9. C. Popovici, M. Somesan, Appl. Phys. Letters, **8**, 103 (1966).
10. M. Yamashita, M. Kimura, Jap. J. Appl. Phys., **19**, L449 (1980)
11. K. Danzmann, M. Kock, J. Phys. B: At. Mol. Phys., **14**, 2989 (1981)
12. I. Iova, I. Chera, M. Ganciu-Petcu, Rev. Roum. Phys., **29**, 519 (1984).
13. E.M. Veldhuizen F.J. Hoog, J. Phys. D: Appl. Phys., **17**, 953 (1984)
14. W. Stephenson, Phil. Magazine., **2**, 556 (1926)
15. H. Helm, Zeitschrift Naturforsch., **27a**, 1812 (1972)
16. Yu.M. Kagan, J. Phys. D: Appl. Phys., **18**, 1113 (1985)
17. H.L.L. Paassen, E.C. Muly, R.J. Allen, Proc. of the National Electronics Conference., **18**, 590 (1962)
18. H.L.L. Paassen, E.C. Muly, R.J. Allen, Proc. of the National Electronics Conference., **18**, 597 (1962)
19. D. Bloess, I. Kamber, H. Riege, Nucl. Instrum. and Meth., **205**, 173 (1983)
20. J. Mizeraczyk, M. Neiger, Appl. Phys., **B33**, 17 (1984)
21. A. Metel, S. Grigoriev, Y. Melnik, M. Volosova, E. Mustafaev, Machines, **8**, 42 (2020)
22. A.S. Metel, S.N. Grigoriev, Yu.A. Melnik, V.P. Bolbukov, Instrum. Exp. Tech., **55**, 288 (2012)
23. A. Metel, V. Bolbukov, M. Volosova, S. Grigoriev, Yu. Melnik, Instrum. Exp. Tech., **57**(3), 345 (2014)
24. A. Dalgarno, *Atomic and Molecular Processes, ed. D.R. Bates* (Academic Press, New York, 1962)
25. H.S.W. Massey, E.H.S. Burhop, *Electronic and Ionic Impact Phenomena* (Oxford: Clarendon press, 1952)
26. M. Volosova, S. Grigoriev, A. Metel, A. Shein, Coatings, **8**, 287 (2018)
27. S.N. Grigoriev, Yu.A. Melnik, A.S. Metel, V.V. Panin, Instrum. Exp. Tech., **52**, 602 (2009)
28. S. Grigoriev, A. Metel, NATO Science Series, Series II: Mathematics, Physics and Chemistry, **155**, 147 (2004)

https://doi.org/10.1051/epjconf/202124804002

29. A. Metel, S. Grigoriev, Yu. Melnik, V. Panin, V. Prudnikov, Jap. J. Appl. Phys., **50**, 08JG04 (2011)

Mathematical Models of the Geometry of Micro Milling Cutters

Alexander Isaev[1] , *Ramil* Khamzin[1], *Artem* Ershov[1] and *Marco* Leonesio[2]

[1]Moscow State University of Technology "STANKIN", RU-127055, Moscow, Russia
[2]Institute of Intelligent Industrial Technologies and Systems for Advanced Manufacturing of the National Research Council, STIIMA-CNR, IT-20133, Milano, Italy

Abstract. Micromachining is an up-to-date technology widely used in different advanced areas like electronics, aerospace and medical industries. For manufacturing components with highest precision and lowest surface roughness, small-sized end mills with working diameter of less than 1 mm are often used. In this paper, in order to determine the functional relationships between structural strength, cutting properties and geometry of small-sized cutting tools, the mathematical models of working part of micro milling cutters were derived.
Keywords. Micro milling cutter, micro machining, mathematical modelling.

1 Introduction

Micro cutting tools allows producing complex-shaped parts with high quality meeting the strict requirements for the components of electronics, aerospace, medical industry and others. The efficiency and effectiveness of micromachining depends on many variables including workpiece material properties, cutting tool parameters [1-8], process parameters, machine tool characteristics and many others.

When micro cutting the advanced materials such as titanium and its alloys, it is of highly important to take into account their physical and mechanical properties causing certain difficulties in their machining such as relatively low modulus of elasticity produces vibrations during cutting, low ductility causes negative chip shrinkage, low thermal conductivity and high heat resistance intensively generate heat which affects the heat balance during machining with heat transfer through cutting tool resulting in increased tool wear, relatively low modulus of elasticity producing vibrations and chattering during cutting, and chemical interactions with gases at elevated temperatures appearing during cutting that increase the hardness of workpiece superficial layer [2,3,4,7]. To increase tool life, the cutting geometry of micro milling cutters should be thoroughly analysed at the design stage. For dynamically unstable milling process, the actual values of kinematic clearance angles depending on the process parameters should be calculated.

2 Methods

2.1 Influence of rounding radius of the cutting edge on cutting data parameters

The efficiency of chip formation during milling is largely determined by the ratios f_z/ρ and a_e/ρ, where f_z is the feed per tooth, a_e is the width of cut, ρ is the cutting edge rounding radius (Fig.1). These conditions are especially important when machining with small-sized milling cutters. When feed per tooth and cutting depth are lower than the rounding radius, i. e. $f_z < \rho$ and $a_e < \rho$, the cutting capacity of the milling tooth get worse. In this case, the actual rake angle becomes negative, consequently instead of cutting the intensive wear of workpiece material is observed leading to the increase of cutting forces. The compression deformations in the contact layer of the workpiece material produce wear hardening [5]. This worsens the cutting conditions for the following cutting tooth. The same effects present in up-milling where at the tooth entrance the thickness of the cut layer is zero.

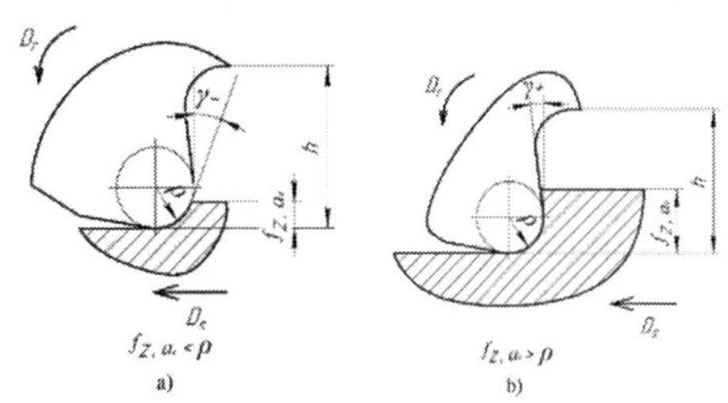

Fig. 1. Cutting edge rounding radius related to the cutting data parameters: a) conditions for inefficient chip formation, b) conditions for effective chip formation

During manufacturing of the cutting tool, the value of ρ is affected by the parameters as follows [9-22]:

- cutting material grade and its grain and pore size;
- wear resistant coating deposition parameters, process parameters, i. e. grinding wheel parameters, grinding speed, feed rate and cutting depth;
- roughness of rake and flank surfaces;
- wear resistant coating deposition parameters;
- process parameters, i. e. grinding wheel parameters, grinding speed, feed rate and cutting depth;

To determine the minimum allowable values of the process parameters required for stable chip formation, the geometry of a micro milling cutter of the GF500T series manufactured by Guhring was analyzed. According to the manufacturer's catalog, the end mill has the following geometric parameters: outer diameter d=0.5 mm, helical angle ω=30°, peripheral rake angle γ=8°, and the corner radius R=0.01 mm. The recommended cutting data for processing titanium alloys are: feed per tooth f_z = 0.007 mm/tooth, width of cut a_e=0.1 mm, axial depth of cut a_p=0.05 mm.

The general geometrical parameters of the micro end mill were measured using measuring machine Walter Helicheck Plus. The cutting edge rounding radii were measured with GFM MikroCAD premium+ Microscope. It was found that the rake angle measured at a distance of 0.6 mm from the tool top is negative and is equal to 2pprox.. –8°. The cutting edge rounding radius is irregular and takes values in the range of 7...9 μm in different cross-sections normal to the helical cutting edge (Table 1).

2.2 Mathematic models for calculating a cutting speed angles and working kinematic clearance angles for down and up milling

During the cutting, both the angle of cutting speed η and the working kinematic clearance angle α_p measured between the flank surface and the direction of the resultant cutting speed change. The behavior of the angle η is affected by the milling direction. When down milling, at the point of engagement into the material, the angle η has the maximum value and gradually decreases with moving to the point of disengagement, increasing the value of the angle α_p. In up-milling, the angle η gradually increases as the cutter engages into the material thereby reducing the working clearance angle α_p (Fig. 2).

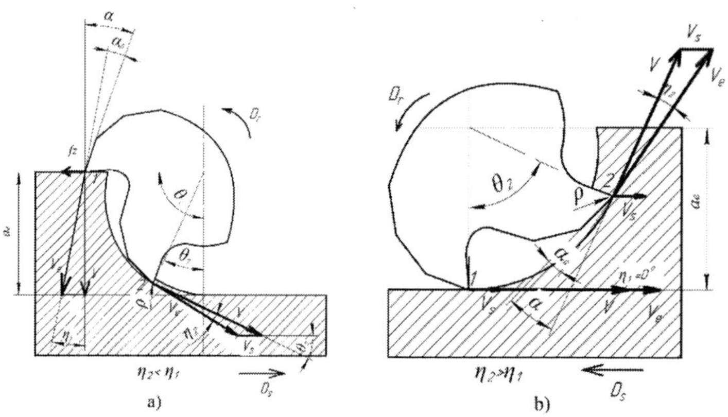

Fig. 2. Change of the angle η n the working plane at different moments of the working stroke of the mill: a – in down-milling, $\eta_2 < \eta_1$, b – in up-milling, $\eta_2 > \eta_1$; 1 – the point at the moment when the cutting edge engages into the workpiece, 2 – the point engaged into the workpiece; V – the velocity of the main cutting movement at the considered cutting point; V_e – the velocity of the resulting cutting speed; V_s – the velocity of the feed; η – the cutting speed angle; θ_1, θ_2 – the angles of contact of cutting tooth with the workpiece surface at the engagement and at some instance, respectively; a_e – the radial depth of cut; α_p – the kinematic clearance angle in the working plane; α – the clearance angle, ρ – the cutting edge rounding radius

With the decrease in the angle α_p, a greater amount of heat is absorbed by cutter thus reducing the tool life and deteriorating the surface quality due to decreasing clearance between the flank and the workpiece surfaces, especially when machining using micro end mills.

To create mathematical models for calculating the α_p angle basing on the cutting data, the schemes of milling process in the working plane (Fig. 3) are used. The kinematic clearance angle is equal to:

$$\alpha_p = \alpha - \eta \quad (1)$$

Table 1. Designed geometry (from catalogue) compared with actual geometrical parameters of micro milling cutters

	Outer diameter ***d*, mm**	**Helical angle ω, °**	**Peripheral rake angle γ, °**	**Corner radius R, mm**	**Cutting edge rounding radius ρ, mm**
Catalog data	0,5	30	8,0	0,01	N/A
Measured data (ID 1)	0,486	30,05	-8,4	0,01	7…9
Measured data (ID 2)	0,478	28,4	4,03	0,01	7…8

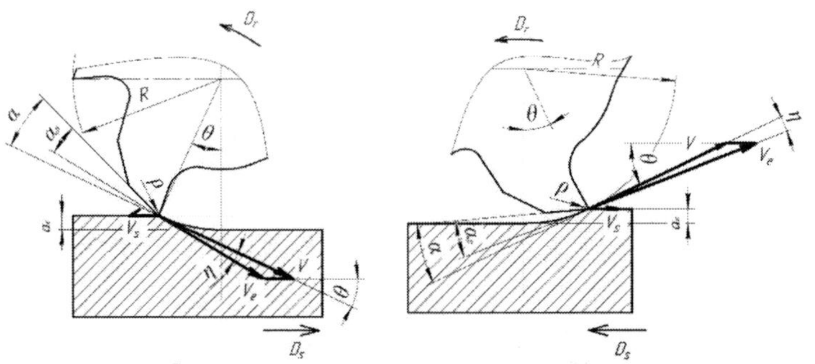

a) b)

Fig. 3. Schemes of milling for determination the angles η and α_p, a – in down milling, b – in up-milling, *V* – the velocity of the main cutting movement at the considered cutting point; V_e – the velocity of the resulting cutting speed; V_s – the velocity of the feed; η – the cutting speed angle; θ_1, θ_2 – the angles of contact of cutting tooth with the workpiece surface at the engagement and at some instance, respectively; a_e – the radial depth of cut; α_p – the kinematic clearance angle in the working plane; α – the clearance angle, ρ – the cutting edge rounding radius, D_s – the feed velocity, D_r – the main cutting movement velocity

To determine the speed of the resulting velocity $\mathbf{V_e}$ we should first calculate the contact angle θ:

$$\cos\theta = 1 - \frac{t}{R} \qquad (2)$$

According to the diagram in Fig. 3, we can express the resulting cutting speed
- for down-milling:

$$\boldsymbol{V_e^2 = V^2 + V_s{}^2 - 2VV_s\cos\theta} \qquad (3)$$

- for up-milling:

$$\mathbf{V_e^2 = V^2 + V_s{}^2 - 2VV_s}\cos(180-\theta) \qquad (4)$$

- for up- and down-milling:

$$\mathbf{V_s{}^2 = V^2 + V_e^2 - 2VV_e}\cos\eta \qquad (5)$$

Taking into account (1), the equations (2–4) take the form:

$$V_e^2 = (2\pi R)^2 + f_z{}^2 - 4\pi R S_z\left(1-\frac{a_e}{R}\right) \qquad (6)$$

$$V_e^2 = (2\pi R)^2 + f_z{}^2 - 4\pi R S_z\left(\frac{a_e}{R}-1\right) \qquad (7)$$

$$f_z{}^2 = (2\pi R)^2 + V_e^2 - 4\pi R V_e\cos\eta \qquad (8)$$

From equation (7) we obtain:

$$\cos\eta = \frac{(2\pi R)^2 + \mathbf{V_e^2} - f_z{}^2}{4\pi R\mathbf{V_e}} \qquad (9)$$

Substituting equations (6) and (7) into equation (9) and simplifying it, we obtain the mathematical models for calculating:
- the angle of the cutting speed during down-milling:

$$\eta = \arccos\left(\frac{2\pi R^2 - f_z R + f_z a_e}{R\sqrt{(2\pi R - f_z)^2 + 4\pi f_z a_e}}\right) \qquad (10)$$

- the angle of the cutting speed during up-milling:

$$\eta = \arccos\left(\frac{2\pi R^2 - f_z a_e + f_z R}{R\sqrt{(2\pi R + f_z)^2 - 4\pi f_z a_e}}\right) \qquad (11)$$

Finally, we obtain the mathematical models for calculating the α_p angle
- for down-milling:

$$\alpha_p = \alpha - \arccos\left(\frac{2\pi R^2 - f_z R + f_z a_e}{R\sqrt{(2\pi R - S_z)^2 + 4\pi f_z a_e}}\right) \qquad (12)$$

- for up-milling:

$$\alpha_p = \alpha - \arccos\left(\frac{2\pi R^2 - f_z a_e + f_z R}{R\sqrt{(2\pi R + f_z)^2 - 4\pi f_z a_e}}\right) \qquad (13)$$

The obtained mathematical models allow us to determine in the working plane the dependence of the α_p

angle on the milling width a_e and the feed f_z for end mills of different diameters. The mathematical models determine the maximum value of the α_p angle: for down-milling at the engagement of the milling tooth into the workpiece, and for up-milling at the disengagement of the milling tooth from the workpiece.

For micro milling cutters with helical teeth, the geometric parameters in the working plane relate to the geometric parameters in the normal cutting plane as follows:

$$\tan \alpha_n = \frac{\tan \alpha_p}{\cos\omega}, \quad (14)$$

where α_n is the clearance angle in the normal secant plane to the main cutting edge, α_p is the clearance angle in the working plane, ω is the helix angle. Therefore, $\alpha_n < \alpha_p$.

2.3 Analysis of kinematic clearance angles of micro end mill

The obtained mathematical models allow determining the dependence of the working kinematic clearance angle α_p on 3 variables, i. e. the working radius of the micro end mill, the radial depth of cut a_e and the feed per tooth f_z measured in different cutting directions. To determine what parameter is the most influential on changing the kinematic clearance angle, it is necessary to define their influence separately. To do this, the dependency graphs of the kinematic clearance angle for micro end mills with the diameters d=1 mm and d=0.5 mm and clearance angle 17° were plotted according to the obtained mathematic models with the following parameters:

- radial depth of cut a_e with the feed per tooth f_z=0,02d (Fig. 4a) and f_z= 0.01, 0.03 and 0.05 mm/tooth (Fig. 5);
- feed per tooth f_z with the radial depth of cut a_e=0,05d (Fig. 5b) and a_e= 0,01, 0,03, 0,05 mm (Fig. 6).

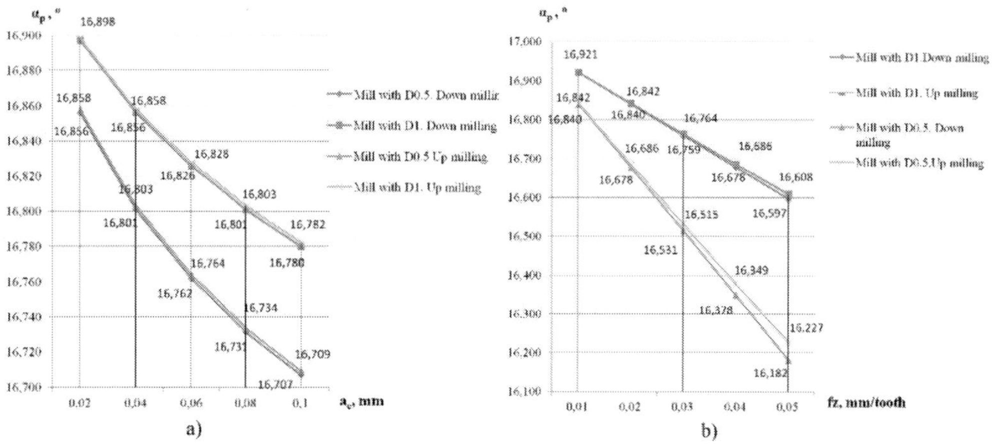

Fig. 4. Dependencies of the kinematic clearance angle α_p for micro end mills with d=1mm and d=0.5mm both having clearance angle α=17° on: a) radial depth of cut a_e with feed per tooth f_z=0.05d; b) feed per tooth f_z with radial depth of cut a_e=0.05d

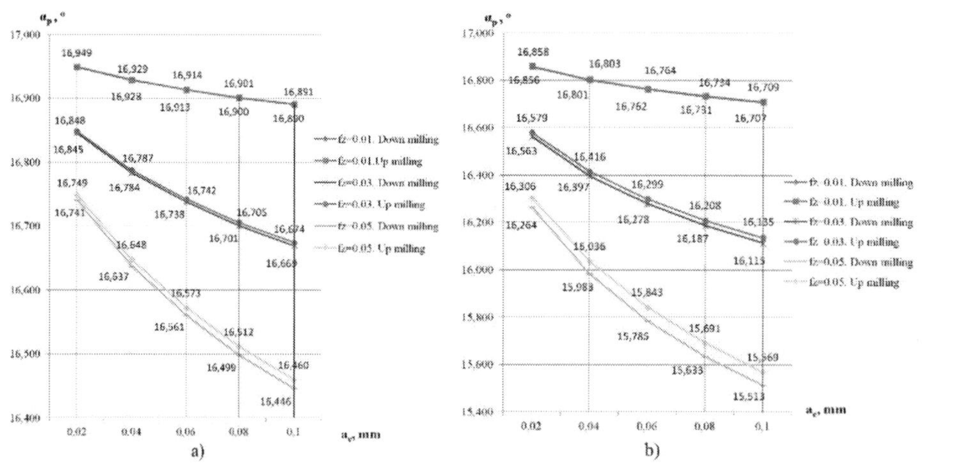

Fig. 5. Dependencies of the kinematic clearance angle α_p on radial depth of cut a_e with feeds per tooth of 0.01, 0.03 and 0.05 mm for micro end mills: a) d=1 mm, b) d=0,5m

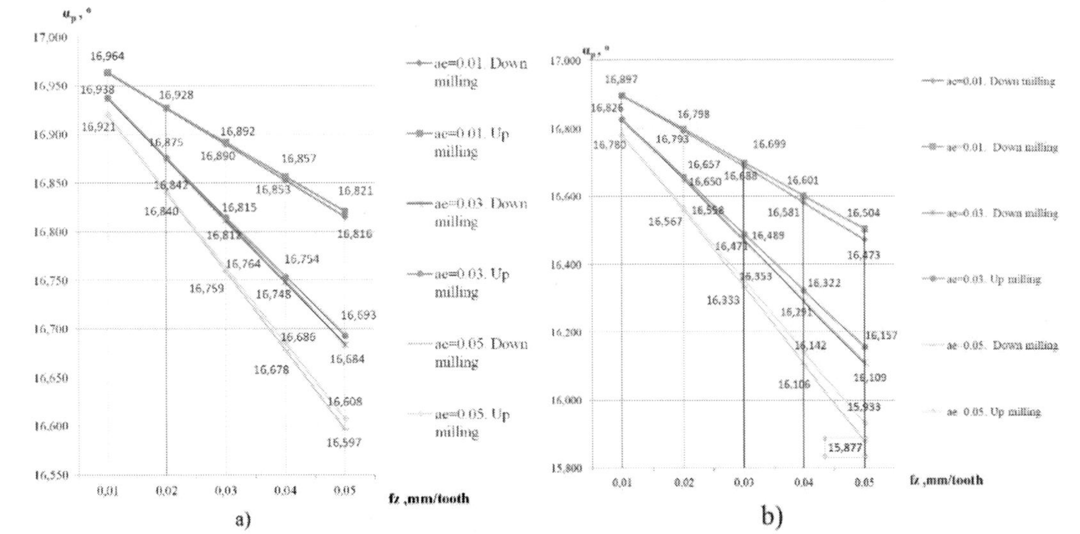

Fig. 6. Dependencies of the kinematic clearance angle on feed per tooth f_z with radial depth of cut a_e=0.01, 0.03 and 0.05 mm for micro end mills: a) d=1 mm, b) d=0,5mm

According to the graph of dependence $\alpha_p(f_z)$ shown in Fig. 4b and 6, a hypothesis for a linear dependence between these parameters was introduced. To confirm this hypothesis and to define a shape of the function $\alpha_p(f_z)$, we have used the approximation tools of the Microsoft Excel. Using this software, the linear function was defined as the most accurate approximation of dependence between the kinematic clearance angle and the feed per tooth since its coefficient of determination has the maximum possible value, R^2=1. The most relevant approximation type for the dependence of kinematic clearance angle α_p on the radial with of cut a_e is a polynomial function of second order. Therefore, the relationship between α_p and f_z has the following form:

$$\alpha_p(f_z) = -Af_z + B \quad (15)$$

The functional relationship between α_p and a_e is as follows:

$$\alpha_p(a_e) = Aa_e^2 - Ba_e + C, \quad (16)$$

where A, B and C are the function coefficients.

Using the cutting conditions recommended for machining the titanium alloys with micro end mills GF500T by Guhring, the graph plots (Fig. 7) were constructed. The graph plot for the $\alpha_p(a_e)$ at such cutting conditions is approximated by the linear function. Basing on this statement, the calculation of coefficients was performed and the separate influence of parameters a_e and f_z on the value of the angle α_p was identified.

To calculate the coefficients A and B in (15) and (16), the least squares method was used as follows:

$$A = \frac{n\sum f_{zi}\alpha_{p_i} - \sum f_{zi}\sum \alpha_{p_i}}{n\sum f_{zi}^2 - \sum f_{zi}\sum f_{zi}} \quad (17)$$

$$B = \frac{\sum \alpha_{p_i}\sum f_{zi}^2 - \sum f_{zi}\sum f_{zi}\alpha_{p_i}}{n\sum f_{zi}^2 - \sum f_{zi}\sum f_{zi}}, \quad (18)$$

where f_{zi} is the i-th value of feed per tooth; α_{pi} is the i-th value of kinematic clearance angle with i-th value of feed per tooth; n is the total number of values.

Taking into account the calculated values of A and B, the equation for the kinematic clearance angle depending on the feed per tooth for up-milling (line 2 in Fig. 7) takes the following form:

$$\alpha_p(f_z) = -21{,}230f_z + 18{,}054 \quad (19)$$

For down milling (Fig. 7, line 1):

$$\alpha_p(f_z) = -22{,}568f_z + 18{,}136 \quad (20)$$

The values of A and B in (17) and (18) are also calculated using the same method of least squares:

$$A = \frac{n\sum a_{ei}\alpha_{p_i} - \sum a_{ei}\sum \alpha_{p_i}}{n\sum a_{ei}^2 - \sum a_{ei}\sum a_e} \quad (21)$$

$$B = \frac{\sum \alpha_{p_i}\sum a_{ei}^2 - \sum a_{ei}\sum a_{ei}\alpha_{p_i}}{n\sum a_{ei}^2 - \sum a_{ei}\sum a_{ei}}, \quad (22)$$

where a_{ei} is the i-th value of the radial depth of cut; α_{pi} is the i-th value of a kinematic clearance angle at i-th value of radial depth of cut; n is the total number of reference values. Thus, the equation of the kinematic clearance angle depending on the radial depth of cut for up-milling (Fig. 7, line 4) is as follows:

$$\alpha_p(a_e) = -0{,}956a_e + 16{,}891 \quad (23)$$

For down milling (Fig. 7, graph 3) the equation of dependence looks:

$$\alpha_p(a_e) = -0{,}956a_e + 16{,}890 \qquad (24)$$

In the derived pairs of equations (19), (20) and (23), (24), the difference between coefficients *A* is more pronounced than the difference between coefficients *B*. Moreover, the value of *A* in equations (23) and (24) is greater than one in formulas (19) and (20). Thus, the feed per tooth at the recommended cutting conditions is more influential on the kinematic clearance angle α_p than the radial depth of cut.

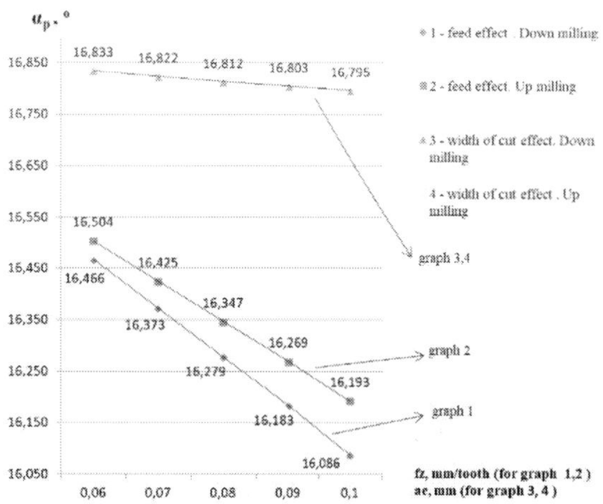

Fig. 7. Graph plots and equations of the kinematic clearance angle α_p depending on depth per tooth and radial width of cut with the recommended cutting data a_e=0,1 mm and f_z=0,007 mm/tooth for micro end mills with GF500T by Guhring with *d*=0.5mm

3 Results and discussion

The created mathematical models allow designing micro end mills with the geometrical parameters optimal in strength and determining the range of optimal feed values and width of cut a_e which must meet the conditions $\rho < f_z < h$ and $\rho < a_e < h$.

According to the measurement protocol, the cutting edge radius is irregular along the helical cutting edge having the values in the range of 7…9 µm. For small-sized milling cutters, the cutting edge radius variation of 25% is very significant since at the recommended feed per tooth f_z = 0.007 mm the chip formation is inefficient in all cross-sections.

The actual measured value of the rake angle on the peripheral part of helical cutting edge turned out to be negative instead of the positive value showed in the catalog. This may result in significant variability in cutting process parameters when milling. At the same time, the negative geometry is a feasible solution for micro milling cutters made of advanced materials like reinforced cutting ceramics. These kinds of cutting tools may be suitable for use in machining medical components made of nickel and cobalt alloys because the negative design makes the cutting edge stiffer and stronger.

4 Conclusions

The derived mathematical models allow designing micro end mills with the geometry optimal in strength. These mathematical models can be used to determine the reasonable range of optimal cutting data for micro milling process.

The negative values of rake angle acquired from measuring machine do not correspond with the catalog of the micro milling cutter manufacturer. Alternatively, the negative cutting geometry can be used for micro end mills made of cutting ceramics because this design feature makes the cutting edge stiffer and stronger.

With decreasing diameter of micro end mill, the influence of feed per tooth f_z and a_e on the kinematic clearance angle α_p increases.

The difference between the values of the kinematic clearance angle α_p for the down and up milling increases with growing a_e and f_z and with decreasing diameter of the micro end milling cutter.

The feed per tooth f_z has a greater effect on the change of the kinematic clearance angle α_p than the radial depth of cut a_e. The dependence function of $\alpha_p(a_e)$ has a polynomial form, and, in this case, second-order polynomial is the most suitable approximation.

Acknowledgments

This research was funded by Ministry of Science and Higher Education of the Russian Federation, Grant No. 0707-2020-0031. This work was carried out using equipment provided by the Center of Collective Use of MSUT "STANKIN".

References

1. V.V. Kuzin, S.N. Grigoriev, M.Yu. Fedorov, Part 2: Microlevel, J. Frict. Wear, **36**(1), 40-44 (2015)
2. M.A. Volosova, S.N. Grigor'ev, V.V Kuzin, Part 4. Action of Heat Flow. Refract Ind Ceram, **56**, 91–96 (2015)
3. A.A. Vereschaka, S.N. Grigoriev, N.N. Sitnikov, A. Batako, Surface and Coatings Technology, **332**, 198-213 (2017)
4. V.V. Kuzin, S.N. Grigor'ev, M.A. Volosova, Refract. Ind. Ceram., **54**, 376–380 (2014)
5. S.N. Grigoriev, O.V. Sobol, V.M. Beresnev, I.V. Serdyuk, A.D. Pogrebnyak, D.A. Kolesnikov, U.S. Nemchenko, J. Frict. Wear, **35**(5) 359-364 (2014)
6. A.S. Metel, S.N. Grigoriev, Yu.A. Melnik, V.P. Bolbukov, Instrum. Exp. Tech., **55**(1), 122-130 (2012)
7. A. Metel, V. Bolbukov, M. Volosova, S. Grigoriev, Yu. Melnik, Instruments and Experimental Techniques, **57**(3), 345–351 (2014)
8. O.V. Sobol', A.A. Andreev, S.N. Grigoriev, V.A. Stolbovoy, Problems of Atomic Science and Technology, **4**, 174-177 (2011)

9. A. Metel, V. Bolbukov, M. Volosova, S. Grigoriev, Y. Melnik, Surf. Coat. Technol., **225**, 34-39 (2013)
10. D.S. Rechenko, A.G. Kol'tsov, Russ. Eng. Res., **32**(2), 179-181 (2012)
11. S. Grigoriev, Y. Melnik, A. Metel, Surf. Coat. Technol. **156**(1–3), 44–49 (2002)
12. V. Grechishnikov, S. Grigoriev, P. Pivkin, M. Volosova, A. Isaev, D. Nikitin, I. Minin, EPJ Web of Conferences, **224**, 05001 (2019)
13. X. Kong, Z. Ding, L. Xu, L. Zhu, J. Zhang, C. Wu, A. Isaev, EPJ Web of Conferences, **224**, 05009 (2019)
14. V.A. Grechishnikov, Y.E. Petukhov, P.M. Pivkin, et al. Meas Tech., **58**, 848–853 (2015) https://doi.org/10.1007/s11018-015-0806-z
15. I.A. Kovalev, P.A. Nikishechkin, A.S. Grigoriev. 2017 International Conference on Industrial Engineering, Applications and Manufacturing (ICIEAM), St. Petersburg, 1-4 (2017)
16. S.N. Grigoriev, P.M. Pivkin, V.A. Grechishnikov, Y.E. Petukhov, M.A. Volosova, A.B. Nadykto, Proc. SPIE (International Society for Optics and Photonic, **11540**, 115401E, 2020)
17. S.N. Grigoriev, M.A. Volosova, A.A. Okunkova, S.V. Fedorov, K. Hamdy, P.A. Podrabinnik, ... , A.N. Porvatov, J. Manuf. Mater. Process., **4**(3), 96 (2020) https://doi.org/10.3390/jmmp4030096
18. P.M. Pivkin, A.B. Nadykto, V.A. Grechishnikov, M.A. Volosova, I.V. Minin, S.N. Grigoriev, Proc. SPIE (International Society for Optics and Photonic, **11540**, 115401G, 2020)
19. I.A. Kovalev, M.S. Babin, P.A. Nikishechkin., MATEC Web of Conferences, **298**, 00110. (2019)
20. S. Grigoriev, M. Volosova, V. Grechishniko, P. Pivkin, P. Peretyagin, A. Zelensky, Proc CIRP, **95**, 337-342 (2020)
21. D.S. Rechenko, A.V. Popov, K.V. Averkov, V.A. Sergeev, Russ. Eng. Res., **32**(5-6), 511-512 (2012)
22. P.A. Nikishechkin, N.S. Grigoriev, N.Yu. Chervonnova, MATEC Web of Conferences, **298**, 00064 (2019)

Solution of the Rodriguez Equation in Modeling Volumetric Geometric Accuracy of Multi-Coordinate Systems

M.M. Stebulyanin[*], *Ya.I.* Pimushkin

Moscow State Technological University "STANKIN", RU-127055, Moscow, Russia

Abstract. The article describes the solution to the Rodrigues equation for determining the volumetric accuracy of multi-axis CNC-controlled systems. An algorithm for calculating the position of the axis of a rotary kinematic pair in problems of volumetric accuracy of mechanical motion of a portal-type system with an additional pair of rotation. The algorithm is based on the analytical solution of the Rodrigues equation in the inverse problem of finding the vector of the final rotation of the known modulus from the known initial and final values of the characteristic vector of the rotated rigid body. In contrast to the well-known direct problem, where based on a finite rotation vector known in direction and magnitude, and the initial value of the characteristic vector of a body, its final value is found, the inverse problem of the Rodrigues equation is not that common due to the nonlinearity and need to solve a nonlinear coupled system of second order equations. The results of this work make it possible to expand the dimension of the space of generalized coordinates of the system analyzed for the volumetric accuracy from three to four. This is expected contribute to the development of ultra-precise systems of controlled mechanical movement. The analytical results of this study were verified by comparing with numerical solutions of the inverse problem in Maple.

1 Introduction

Multi-axis systems with software-controlled movement of units and mechanisms are a an important part of modern mechanical engineering, in terms of both the technological equipment (machine tools, industrial robots) and measuring technologies (coordinate measuring machines). Highly efficient use of such systems at the moment is no longer possible without a systemic increase in the volumetric accuracy (VA) of the actuator equipment, that reflects the ability to accurately reproduce complex three-dimensional surfaces with a measuring tip or cutting edge of a tool [1-3].

For the first time, the concept of volumetric accuracy was introduced in the work of P.A. McKeown, J. Loxham [4], and, in the last decade, the problem of increasing VA has become widespread [5, 6]. In Russia, this concept was introduced in the work of Serkov [7]. Quantitatively, the volumetric accuracy is characterized by the volumetric error (VE), which for each point of the machining workspace is a vector directed from the specified (nominal) position of a certain pole of the tool to its actual position. To study the volumetric accuracy, many methods and approaches such as the Denavit-Hartenberg method [8], the method for describing the kinematics of a rigid body (rigid body kinematics) [9], the kinematics of a set of rigid bodies (Multiple-body kinematics) [10] and methods of matrix simulation [11] are used:. In ref. [12-13] articles, not only systematic but also random characteristics of equipment accuracy are considered

The ultimate goal of the study of volumetric accuracy is the construction of algorithms for correcting the real motion of a mechanical system in a generalized space relative to another, considered under the condition of an "ideal" arrangement of the axes of the kinematic pairs of the system. At the same time, it is known that, for example, with gantry three-coordinate kinematics, the axes of the telescopic pairs of the machine tool do not form an ideal orthogonal trihedron, and this discrepancy increases with the wear of the equipment.

2 Results and discussion

To correct the errors of the three-dimensional portal motion caused by this discrepancy, it is necessary to know the true position of the physical axes of translational displacements. The task becomes more complicated if the fourth rotational axis is included in the generalized space. In this case, it is not possible to correct motion errors by using only translational motion along the axes of the portal kinematic scheme. To solve the problem of motion correction, here it is also necessary to know the true position in space of the axis of rotation of the fourth kinematic pair.

Let explain what has been said in Figure 1, which conventionally shows a mechanism with three translational motions in a stationary Cartesian coordinate system and a fourth link, which is a rotor in a kinematic pair of a rotational type, with a stator connected to a coordinate system co-directed with the system. Translational linear movements are carried out along the

[*] e-mail: mmsteb@rambler.ru

physical axes of a generally non-Cartesian system with a common origin at point O, which, due to design inaccuracies, does not coincide with the origin of the system. The 3x1 vector of generalized coordinates of linear displacements will be denoted by.

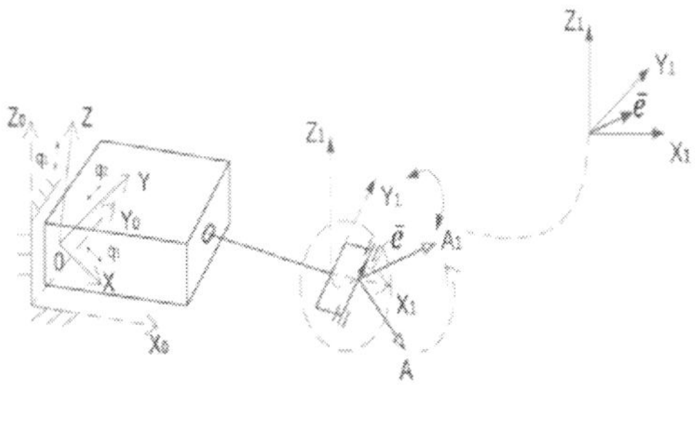

Fig. 1. Deviation of the physical axis of rotation from the ideal position

When turning about the real axis of the pair, the direction of which is set by the unit vector, the effector link moves to a position that can be obtained in the system, for example, on the basis of the theory of finite turns. Why it is necessary to obtain the radius vector of the effector pole in a stationary system, you can use the equation:

$$\bar{r}_A = \bar{r}_1 + \boldsymbol{C}\bar{q} + \bar{r}_a,$$

where C - constant 3x3 is the calibration matrix obtained on the basis of the calibration procedure of translational (linear) movements in the physical axes of telescopic pairs;

$\bar{r}_1$ - radius vector of the origin of the Cartesian system in the Cartesian system effector

$\bar{r}_a$ - radius vector in coordinate system

In the calibration procedure, which is known in principle and therefore is not considered in this work, the values and are obtained. There remains the problem of calculating a vector that completely depends on the true direction of the unit vector of the axis of the rotational kinematic pair.

The true direction of this axis, although it differs from the nominal one by an unknown value, remains unchanged and does not depend on the movement in telescopic kinematic pairs, which makes it possible to determine its unit vector $\bar{e}$.

3 Rodrigues' rotation formula

To find the real position of the axis of a rotational pair, as will be shown below, it is possible to use the Rodrigues' rotation formula:

$$\bar{\rho}' = \bar{\rho} + \frac{1}{1+\frac{1}{4}|\bar{\theta}|^2}\bar{\theta} \times \left(\bar{\rho} + \frac{1}{2}\bar{\theta} \times \bar{\rho}\right)$$

$\bar{\rho}$ - • radius vector of the initial position of the effector pole;

$\bar{\rho}'$ – • radius vector of its final position;

$\bar{\theta} = 2\mathrm{tg}\frac{\chi}{2}\bar{e}$, где χ - угол поворота вектора $\bar{\rho}$.

When applying this formula, it is possible to solve two problems:

- the direct problem of orientation of a rigid body, when a change in the position of the body in space is found for a known rotation vector;
- an inverse problem, in which the vector of the final rotation is determined at the known value of its modulus for the known initial and final positions of the body.

While the direct problem is often solved in analytical mechanics, solving the inverse problem, when the Rodrigues formula is considered as an equation with respect to $\bar{\theta}$,commonly used codes do not provide any solution. For example, Mathcad does not find an analytical solution in symbolic form. Maxima was unable to find an analytical solution in four and a half hours, and eventually stopped itself. After a day of analytical calculations, Maple receives a solution on forty pages, and from the point of view of mathematics, such a solution is too long to be used in practice as shown in Table 1.

Table 1. Results of solving the Rodrigues equation by software packages

	Software Application package	Solution or problems
	Mathcad	Analytical solution in symbolic form not found
	Maxima	4.5 hours of searching for an analytical solution – solution not found
	Maple	25 hours - analytical solution in the form of 40 pages of text (the solution is boundless)

Therefore, the authors have proposed their own analytical solution, which is considered below.

4 Solution of the Rodrigues equation

Let's write down the Rodrigues formula:

$$\bar{\rho}' = \bar{\rho} + \frac{1}{1+\frac{1}{4}|\bar{\theta}|^2}\bar{\theta} \times \left(\bar{\rho} + \frac{1}{2}\bar{\theta} \times \bar{\rho}\right) \quad (1)$$

Let us assume that the vectors $\bar{\rho}'$ and $\bar{\rho}$ are known (for example, given by the results of an experiment).

It is required to find the rotation vector $\bar{\theta}$ with its known modulus$|\bar{\theta}|$.

In Fig. 2. shows an illustration of the vectors involved in the Rodrigues formula.

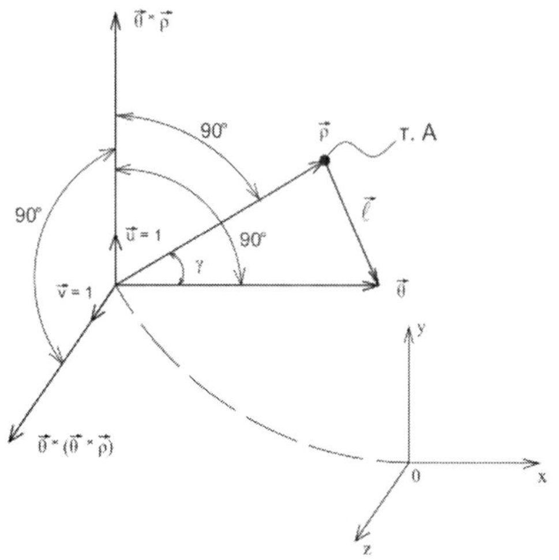

Fig. 2. Rodrigues' rotation formula illustration.

Let be

$$\alpha = 1 + \frac{1}{4}|\bar{\theta}|^2$$
$$x_A = \rho_1, y_A = \rho_2, z_A = \rho_3$$

Let's rewrite formula (1) as:

$$2\alpha\overline{\Delta\rho} = 2\bar{\theta} \times \bar{\rho} + \bar{\theta} \times (\bar{\theta} \times \bar{\rho}),$$

where

$$\overline{\Delta\rho} = \bar{\rho}' - \bar{\rho} = \bar{r} \quad (2)$$

and further we will consider it as an equation with $\bar{\theta}$. In this case, the projections of the vectors on the axes of the coordinate system $Oxyz$ will be accompanied by indices 1, 2, 3, and the unit vectors of the axes Ox, Oy, Oz will be denoted $\bar{\imath}, \bar{\jmath}, \bar{k}$, respectively.

Multiplying scalarly (2) by, we obtain equation (4):

$$2\alpha(\bar{r}, \bar{\theta}) = 0 + 0 = 0 \Rightarrow r_1\theta_1 + r_2\theta_2 + r_3\theta_3 = 0 \quad (3)$$

Then we notice the fulfillment of the following equalities:

$$2\alpha(\bar{r}, \bar{\rho}) = (\bar{\theta} \times (\bar{\theta} \times \bar{\rho}), \bar{\rho}) \quad (4)$$

Next, we notice that the following equalities are fulfilled:

$$\bar{\theta} \times \bar{\rho} = |\bar{\theta}||\bar{\rho}| \sin\gamma \cdot \bar{u}$$

$$\bar{\theta} \times (\bar{\theta} \times \bar{\rho}) = |\bar{\theta}|^2 |\bar{\rho}| \cdot \sin\gamma \cdot \bar{v}$$

$$(\bar{\theta} \times (\bar{\theta} \times \bar{\rho}), \bar{\rho}) = |\bar{\theta}|^2 |\bar{\rho}|^2 \sin\gamma \cdot \cos(90^o + \gamma) = -|\bar{\theta}|^2 |\rho|^2 \sin^2\gamma$$

From here we get:

$$\sin^2\gamma = -2\alpha\frac{(\bar{r}, \bar{\rho})}{|\bar{\theta}|^2|\bar{\rho}|^2}, \quad \cos^2\gamma = 1 + 2\alpha\frac{(\bar{r}, \bar{\rho})}{|\theta|^2|\rho|^2}$$

and, therefore:

$$\cos\gamma = \sqrt{\frac{|\bar{\theta}|^2|\bar{\rho}|^2 + 2\alpha(\bar{r}, \bar{\rho})}{|\bar{\theta}|^2|\bar{\rho}|^2}} \quad (5)$$

Now we will find an analytical expression for equation (4), sequentially calculating vector result $\bar{\theta} \times \bar{\rho}$ multiplication and further$\bar{\theta} \times (\bar{\theta} \times \bar{\rho})$.. Performed:

$$\bar{\theta} \times \bar{\rho} = \bar{\imath}(\theta_2\rho_3 - \rho_2\theta_3) + \bar{\jmath}(-\theta_1\rho_3 + \rho_1\theta_3) + \bar{k}(\theta_1\rho_2 - \rho_1\theta_2)$$

Let's introduce a change of variables:

$$\theta_1' = \theta_2\rho_3 - \rho_2\theta_3$$
$$\theta_2' = -\theta_1\rho_3 + \rho_1\theta_3$$
$$\theta_3' = \theta_1\rho_2 - \rho_1\theta_2$$

Using the new variables, let's write an expression for: $\bar{\theta} \times (\bar{\theta} \times \bar{\rho})$:

$$\bar{\theta} \times (\bar{\theta} \times \bar{\rho}) = \begin{vmatrix} \bar{\imath} & \bar{\jmath} & \bar{k} \\ \theta_1 & \theta_2 & \theta_3 \\ \theta_1' & \theta_2' & \theta_3' \end{vmatrix} = \bar{\imath}(\theta_2\theta_3' - \theta_2'\theta_3) + \bar{\jmath}(-\theta_1\theta_3' + \theta_1'\theta_3) + \bar{k}(\theta_1\theta_2' - \theta_1'\theta_2) = \bar{\imath}[\theta_2(\theta_1\rho_2 - \rho_1\theta_2) - \theta_3(\rho_1\theta_3 - \rho_3\theta_1)] + \bar{\jmath}[\theta_3(\theta_2\rho_3 - \rho_2\theta_3) - \theta_1(\theta_1\rho_2 - \rho_1\theta_2)] + \bar{k}[\theta_1(\rho_1\theta_3 - \rho_3\theta_1) - \theta_2(\theta_2\rho_3 - \rho_2\theta_3)] = \bar{\imath}(\rho_2\theta_1\theta_2 - \rho_1\theta_2^2 - \rho_1\theta_3^2 + \rho_3\theta_1\theta_2) + \bar{\jmath}(\rho_3\theta_2\theta_3 - \rho_2\theta_3^2 - \rho_2\theta_1^2 + \rho_1\theta_1\theta_2) + \bar{k}(\rho_1\theta_1\theta_3 - \rho_3\theta_1^2 - \rho_3\theta_2^2 + \rho_2\theta_2\theta_3)$$

From here we find

$$(\bar{\theta} \times (\bar{\theta} \times \bar{\rho}), \bar{\rho}) = \rho_1\rho_2\theta_1\theta_2 - \rho_1^2\theta_2^2 - \rho_1^2\theta_3^2 + \rho_1\rho_3\theta_1\theta_3 + +\rho_2\rho_3\theta_2\theta_3 - \rho_2^2\theta_3^2 - \rho_2^2\theta_1^2 + \rho_1\rho_2\theta_1\theta_2 + \rho_1\rho_3\theta_1\theta_3 - \rho_3^2\theta_1^2 - \rho_3^2\theta_2^2 + \rho_3\rho_2\theta_2\theta_3 == -(\rho_1\theta_2 - \rho_2\theta_1)^2 - (\rho_1\theta_3 - \rho_3\theta_1)^2 - (\rho_2\theta_3 - \rho_3\theta_2)^2$$

Therefore, equation (4) is obtained in the form (6):

$$(\rho_1\theta_2 - \rho_2\theta_1)^2 + (\rho_1\theta_3 - \rho_3\theta_1)^2 + (\rho_2\theta_3 - \rho_3\theta_2)^2 = -2\alpha(\bar{r}, \bar{\rho}) \quad (6)$$

In this case, the following equalities are fulfilled:

$$\begin{cases} \theta_1^2 + \theta_2^2 + \theta_3^2 = |\bar{\theta}|^2 \\ (\theta_1 - \rho_1)^2 + (\theta_2 - \rho_2)^2 + (\theta_3 - \rho_3)^2 = \ell^2 \end{cases}$$

where $|\bar{\theta}|^2$ – a known value equal to $4tg^2\frac{\chi}{2}$,
where is known χ - angle of rotation of the vector $\bar{\rho}$

Expanding the brackets, we get:

$$\theta_1^2 - 2(\theta_1\rho_1 + \theta_2\rho_2 + \theta_3\rho_3) + \theta_2^2 + \theta_3^2 + \rho_1^2 + \rho_2^2 + \rho_3^2 = \ell^2$$

or

$$\theta_1\rho_1 + \theta_2\rho_2 + \theta_3\rho_3 = \frac{|\bar{\rho}|^2 + |\bar{\theta}|^2 - \ell^2}{2} = c \quad (7)$$

Since $\ell^2 = |\bar{\rho}|^2 + |\bar{\theta}|^2 - 2|\bar{\rho}||\bar{\theta}|\cos\gamma$, by virtue of formula (5) we arrive at the expression:

$$c = |\bar{\rho}||\bar{\theta}|\cos\gamma = \sqrt{|\bar{\theta}|^2|\bar{\rho}|^2 + 2\alpha(\bar{r},\bar{\rho})}$$

Equation (7) is combined with equation (3):

$$\begin{cases}\theta_1\rho_1 + \theta_2\rho_2 + \theta_3\rho_3 = c \\ r_1\theta_1 + r_2\theta_2 + r_3\theta_3 = 0\end{cases} \quad (8)$$

Now, from (8), we express in terms θ_3 of the quantities θ_1, θ_2:

$$\theta_1 = \frac{r_2 c}{r_2\rho_1 - r_1\rho_2} + \frac{r_3\rho_2 - r_2\rho_3}{r_2\rho_1 - r_1\rho_2}\theta_3 = a_1 + b_1\theta_3$$
$$\theta_2 = \frac{r_1 c}{r_1\rho_2 - r_2\rho_1} + \frac{r_3\rho_1 - r_1\rho_3}{r_1\rho_2 - r_2\rho_1}\theta_3 = a_2 + b_2\theta_3 \quad (9)$$

Substituting these values into equation (6), we get:

$$\left(\rho_1(a_2 + b_2\theta_3) - \rho_2(a_1 + b_1\theta_3)\right)^2 + \left(\rho_1\theta_3 - \rho_3(a_1 + b_1\theta_3)\right)^2 +$$
$$+\left(\rho_2\theta_3 - \rho_3(a_2 + b_2\theta_3)\right)^2 = -2\alpha(\bar{r},\bar{\rho})^2 \quad (10)$$

Let us introduce the notation.

$$\rho_{13} = \rho_1^2 + \rho_3^2; \rho_{23} = \rho_2^2 + \rho_3^2; \rho_{12} = \rho_1^2 + \rho_2^2$$

After reducing the homogeneous terms in (10), we arrive at the equation:

$$(\rho_{13}b_2^2 + \rho_{23}b_1^2 + \rho_{12} - 2\rho_1\rho_2b_1b_2 - 2\rho_1\rho_3b_1 - 2\rho_2\rho_3b_2)\theta_3^2 +$$
$$+(2\rho_{13}a_2b_2 + 2\rho_{23}a_1b_1 - 2\rho_1\rho_2(b_2a_1 + b_1a_2) - 2\rho_1\rho_3a_1 - 2\rho_2\rho_3a_2)\theta_3 +$$
$$+\rho_{13}a_2^2 + \rho_{23}a_1^2 - 2\rho_1\rho_2a_2a_1 = -2\alpha(\bar{r},\bar{\rho})$$

Notate:

$$\rho_{13}b_2^2 + \rho_{23}b_1^2 + \rho_{12} - 2\rho_1\rho_2b_1b_2 - 2\rho_1\rho_3b_1 - 2\rho_2\rho_3b_2 = L_2$$

$$2\rho_{13}a_2b_2 + 2\rho_{23}a_1b_1 - 2\rho_1\rho_2(b_2a_1 + b_1a_2) - 2\rho_1\rho_3a_1 - 2\rho_2\rho_3a_2 = L_1$$

$$\rho_{13}a_2^2 + \rho_{23}a_1^2 - 2\rho_1\rho_2a_2a_1 = L_o$$

the solution with respect to θ_3, assuming the angle γ to be acute, is obtained in the form:

$$\theta_3 = \frac{-L_1 + \sqrt{L_1^2 - 4L_2\left(L_o + 2\alpha(\bar{r},\bar{\rho})\right)}}{2L_2}$$

After that, using formulas (9), we find the values of θ_1 and θ_2. Vector θ found. Dividing by $\bar{\theta}$ $|\bar{\theta}|$, we find the unit vector of the rotation axis $\bar{e}$.

5 Validation of the solution in the Maple software package

To validate the solution, the authors used the inverse method implemented in Maple. For the a priori given components of the vectors θ^- and ρ, the algorithm described above was checked for validity. According to the results of repeated modeling, the components of the vector θ^- certainly agreed with those specified with an accuracy of up to the fourth decimal place, with which the calculations themselves were carried out. In this case, pairs of roots were obtained, which was due to the solution of this equation, depending on the value of the angle γ, acute or obtuse.

6 Conclusion

In this paper, the Rodrigues equation for determining the volumetric accuracy of multi-axis CNC-controlled systems has been solved. An algorithm for calculating the position of the axis of a rotary kinematic pair in problems of volumetric accuracy of mechanical motion of a portal-type system with an additional pair of rotation was proposed based on the analytical solution of the the inverse problem of finding the vector of the final rotation of the known modulus from the known initial and final values of the characteristic vector of the rotated rigid body. The obtained results make it possible to expand the dimension of the space of generalized coordinates of the system analyzed for the volumetric accuracy from three to four that can be used as the basis for achieving the dimension of the universal generalized space of a controlled mechanical system, while reducing amount of probabilistic estimates related to the problem of the volumetric accuracy. This is expected to make an important contribution to the development of ultra-precise systems of controlled mechanical movement . The analytical results obtained in this work were verified by comparing with the numerical solutions of the inverse problem in Maple.

References

1. S.N. Grigoriev, D.A. Masterenko, V.I. Teleshevskii., P.N. Emelyanov, Measurement Techniques, **55**(11), 1311-1315 (2013)
2. V.I. Teleshevskii, V.A. Sokolov, Y.I. Pimushkin, Journal of Physics: Conference Series, **998**(1), 012034 (2018)
3. V.I. Teleshevskii, V.A. Sokolov Y.I. Pimushkin, Journal of Physics: Conference Series, **1260**(3), 032038 (2019)
4. P.A. McKeown, J. Loxham, Ann. CIRP, **22**(1) (1973)
5. H. Schwenke et. al., Ann. CIRP, **57,** 660–675 (2008)
6. H. Schwenke, M. Franke, J. Hannaford, Ann. CIRP, **54**(1), 475–478 (2005)
7. N. A. Serkov, Vestnik of Scientific and Technical Development, **3**(55), 26-46 (2012)

8. L.N. Lamikiz, López de Lacalle, O. Ocerin, D. Díez, E. Maidagan, International Journal of Advanced Manufacturing Technology, **37**, C. 122–139
9. M. Rahman, J. Heikkala, K. Lappalainen, International Journal of Machine Tools & Manufacture, **40**, 1535-1546 (2000)
10. A.C. Okafor, Y.M. Ertekin, International Journal of Machine Tools & Manufacture, **40**(8), 1199–1213 (2000)
11. Y. Lin, Y. Shen, International Journal of Advanced Manufacturing Technology, **21**, 243–248 (2003)
12. K.Gee. Ahn, D.W. Cho, International Journal of Machine Tools & Manufacture, **40**, 2235–2248 (2000)
13. Q. Cheng, Q. Feng, Z. Liu, P. Gu, L. Cai, Proceedings of the Institution of Mechanical Engineers, Part C: Journal of Mechanical Engineering Science, **229**, 2534-2550 (2015) 10.1177/0954406214562633.

Investigation of the Stress-Strain State of the Cavity Profiles for Osteosynthesis Screw Heads

Daria Balova[1,*], *Denis* Rechenko[1]

[1]Omsk State Technical University, RU-644050, Omsk, Russia

Abstract. In the modern world, taking into account the active lifestyle of people, the dynamics of growth of complex fractures is observed. Their treatment is associated with rigid fixation of fragments with plates and screws, that is, with the help of osteosynthesis. In the surgical operating room during osteosynthesis is not possible to use auxiliary tools, which can be used to fix the screw with the required screw-in force. As a result, the screw breaks either during surgery or during further operation. This can be avoided by examining the stress-strain state of the screw head for osteosynthesis. By simulating the screw turning process, it is possible to determine the maximum stresses appearing in the screw heads with different profiles, and to choose or develop the profile of the recess under the key, which will be optimal in terms of stress-strain state and manufacturability.

1 Introduction

The number of fractures of different degrees of complexity around the world is increasing every year. Meltona [1] showed that in Russia the number of patients with hip fractures alone will grow, according to World Health Organization forecasts, to 6 million by 2050, or by 300% compared to 1.5 millions in year 1990. This explains a growing demand on implants and other elements such as screws.

All existing developments and products in the field of fastening elements of screws for osteosynthesis are of foreign origin, and are made by "Zimmer" (international company), "Koenigsee" (Germany), "Double Medical" (China) and others. There are practically no modern domestic analogues of these products, and taking into account the fact that the share of Russian production in this market is about 5 % of the world market ("Osteosynthesis", "Osteomed", "TitanImplant"), this area, despite being quite promising, is yet to be developed.

Osteosynthesis is the prompt insertion of a retainer (plate) and its fastening with screws or screws for osteosynthesis [2, 3]. The main difficulty in a surgical operating room for this type of manipulation is the impossibility to use a special instrument, such as torque wrenches. This leads to a number of problems related to the fixation of osteosynthesis plates. If the osteosynthesis screw is not tightened enough during operation, it may unscrew and break, damaging the soft tissues in the fracture area. In case of exceeding the allowable tightening force - the screw may be damaged directly in the process of screwing, it leads to the fact that during the surgical intervention the surgeon will have to pull it out, install a new screw (if possible), thereby increasing the duration of the operation and the amount of anesthesia. In each of these two cases, the risks of additional harm to the patient health increase [4, 5].

One of the most common cases of osteosynthesis screw breakage is the buckling of the recess and breaking of the screw head. This is due to the stresses that occur in the screw head during the tightening process, which primarily depend on the keyhole profile. Fig. 1 shows standard profiles used in the manufacture of osteosynthesis screws.

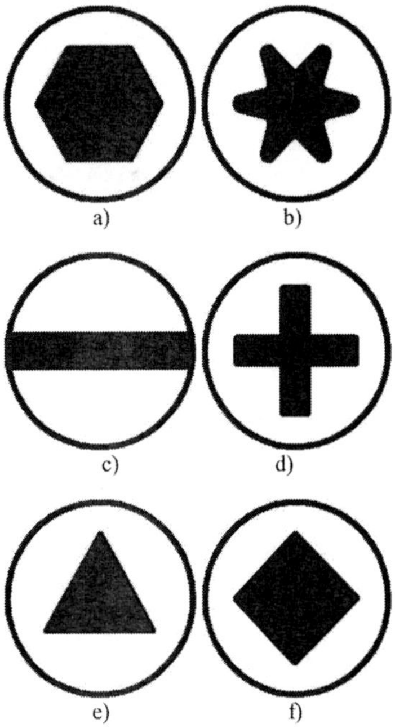

---
* Corresponding author: rvbd@mail.ru

g)

Fig. 1. Standard screw head wrench recess profiles. a) hexagon; b) torx; c) flat slot; d) cruciform slot; e) triangle; f) square; g) pentagon

Despite such a variety of profiles, the problem of their buckling and breaking of the screw head still remains unsolved due to significant stresses and deformations.

The main goal of the study is to develop a turnkey depression profile that will reduce the number of screw breakages for osteosynthesis. To solve this problem, it is necessary to investigate the stress distribution inside the screw head in the process of screw turning, which can be obtained using three-dimensional modeling [6, 7].

In this study, the stress-strain state of the screw heads for osteosynthesis was simulated using standard types of turnkey recess profiles. The heads were modeled using SolidWorks software according to GOST R 50582-93 (ISO 5835-91) [8]. Since osteosynthesis screws with a nominal thread diameter of 3.5 mm are most often used, the heads with these parameters were modeled [9, 10] (Fig. 2).

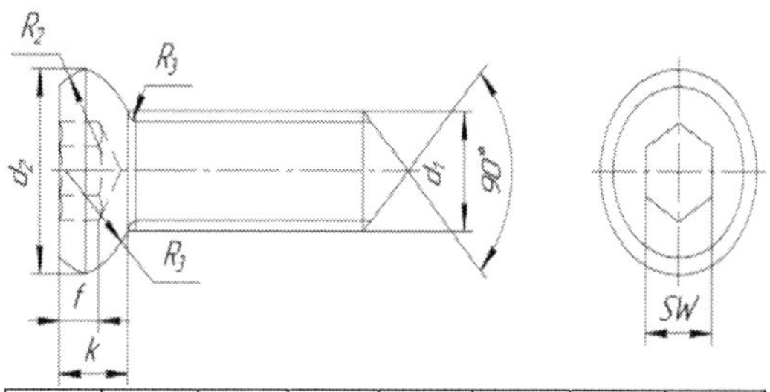

d_1, mm	d_2, mm	f, mm	k, mm	R_1, mm	R_2, mm	R_3, mm	SW, mm
3,5	6	1,5	2,6	3	2,5	0,3	2,5

Fig. 2. Screw drawing for osteosynthesis GOST R 50582-93 (ISO 5835-91).

The diameter of the circle SW entered in the figure is selected as the main parameter that determines the size of the profile for the key screw head for osteosynthesis. The choice of material was determined taking into account the requirements to metal materials intended for use in production of surgical implants according to GOST R ISO 5 832-2-2014 [11]. Chemically pure titanium Ti was used as a material for modeling the stress-strain state of screw heads for osteosynthesis [12, 13].

2 Experiments

Cutting speeds of up to 400-500 m/s can be achieved by Stress on the keyhole profile of the head screw for osteosynthesis was taken for the reason that the average person when tightening the screw develops a torque of 3-5 N/m. In operation, stresses, deformations and displacements were calculated taking into account the applied 4 N/m torque. In this case, the load was applied evenly distributed on the contact surfaces of the profile of the recess under the turnkey screw head. As a result of the applied load in the body of the screw head for osteosynthesis there is a stress-strain state that determines the level of buckling and deformation of the recess profile. Simulation of the screw-in process for osteosynthesis allows to determine a rational profile in terms of stresses, deformations and movement. In fig. 3 shows the result of simulation using different standard profiles.

As a result of modeling for each screw head turn-key indentation profile the maximum values of stresses, displacements and deformations arising during the screw-in process are obtained (Table 1).

From the results obtained, we can conclude that the smallest stresses, displacements and deformations occur on pentahedral, hexagonal and square profiles and have the following values: maximum stress within σ = 1270.59-1296.16 MPa, maximum displacement e = 12.9-13.4 μm, maximum strain Δ = 0.0083-0.0091.

Table 1. Maximum values of stresses, displacements and deformations using different standard profiles

Profile form	Maximum stress, σ MPa	Maximum movement, e μm	Maximum deformation, Δ
Pentagon	1270,6	13,0	0,0083
Hexagon	1274,6	12,9	0,0091
Square	1296,2	13,4	0,0086
Cruciform slot	1314,5	14,5	0,0093
Flat slot	1411,3	24,9	0,0099
Triangle	1967,8	19,3	0,0099
Torx	1970,8	15,2	0,0106

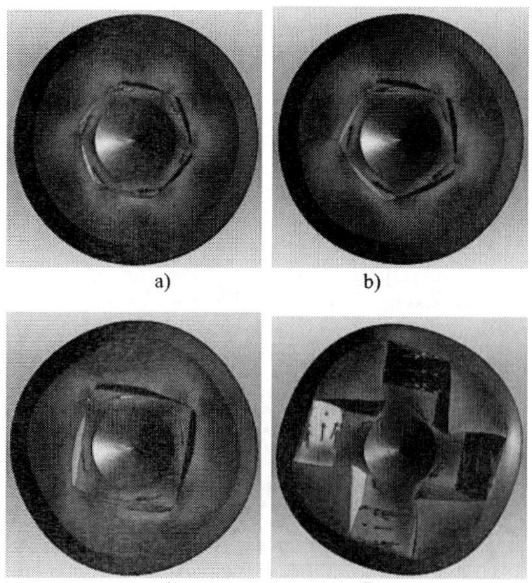

e)

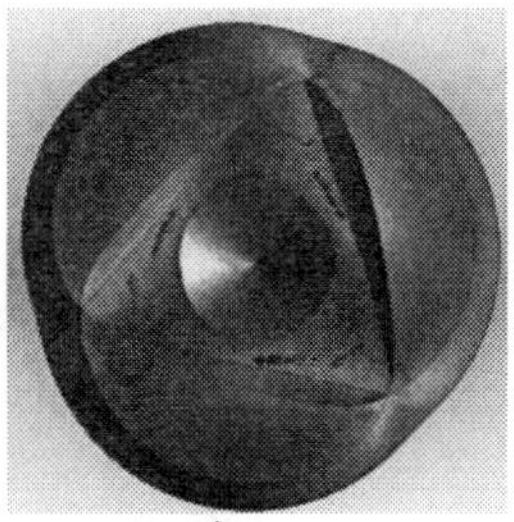
f)

g)

Fig. 3. The result of simulation using different standard profiles; a) hexagon; b) pentagon; c) square; d) cruciform slot; e) flat slot; f) triangle; g) Torx

The practical implementation of these profiles shows unstable results in the installation of screws for osteosynthesis, i.e. there is a risk of breakage of the head of the screw for osteosynthesis as a result of buckling during their insertion and operation. In this case, in the manufacture of standard profiles of the recessed turnkey there are certain difficulties due to the fact that the existing process of manufacturing screws involves obtaining a turnkey profile by means of stamping operations, which causes the profile corners to be rounded. At the same time, the corresponding keys do not have such roundings, as a result of which this difference in the shape of the key and the key recess profile leads to an uneven distribution of stresses in the screw head. Also the shape of the keyhole profile of the head of the screw for osteosynthesis determines the contact area of the key with the profile. There is always a certain gap between the key and the key recess of the screw head. Because of this, the key is free running and the contact area between the key and the recess at the moment of applying the force required for screwing in the section is actually reduced to a point or a line. Therefore, if the torque created by turning the screw will be higher than allowed, the corners of the key will crush the profile of the recess in the screw head, causing it to be inoperable.

This is also a defining criterion in the distribution of voltages. That's why there was a need to create and study non-standard shapes of recesses with no sharp angles.

The analysis of existing shapes showed that of interest are such keyed indentations profiles for osteosynthesis screw heads as the Ryolo's triangle (the result of intersection of three circles, the figure of constant width) and the ellipse. Also in the work was built and calculated such a profile as a rounded triangle, which is also a figure of constant width.

The proposed profiles are distinguished by the use of radius curves, with the Ryolo's triangle, as well as standard profiles have angular shapes.

Three-dimensional modeling made it possible to determine stresses, deformations and movements in the proposed profiles. Modeling results are shown in Fig. 4 and in Table 2.

Table 2. Maximum values of stresses, displacements and deformations using various special profiles

Profile form	Maximum stress, σ MPa	Maximum movement, e μm	Maximum deformation, Δ
Rounded triangle	1309,46	13,67	0,0087
Ryolo's triangle	1249,64	13,47	0,0084
Ellipse	2185,40	32,65	0,0127

The use of turnkey profiles without sharp angles makes it possible to facilitate the process of manufacturing screws and improve their quality. The larger contact area between the key and the profile, compared to standard profiles, ensures an even distribution of stresses inside the screw head for osteosynthesis, thus reducing the number of breakages.

As a result of modeling of standard profiles it has been established that such profiles as square, pentagon and hexahedron are rational in terms of stresses, deformations and displacements. The values of maximum stress are within σ = 1270.59-1296.16 MPa.

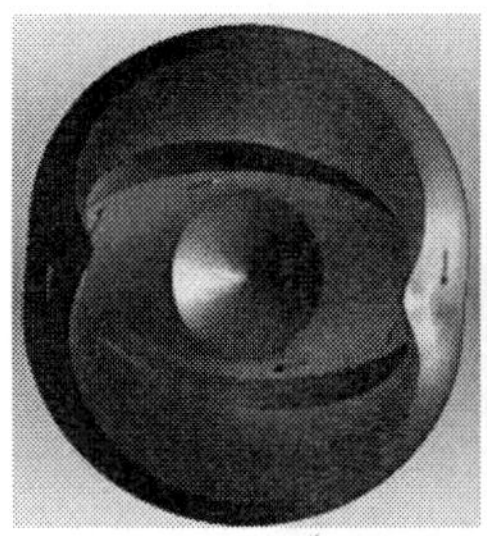
a)

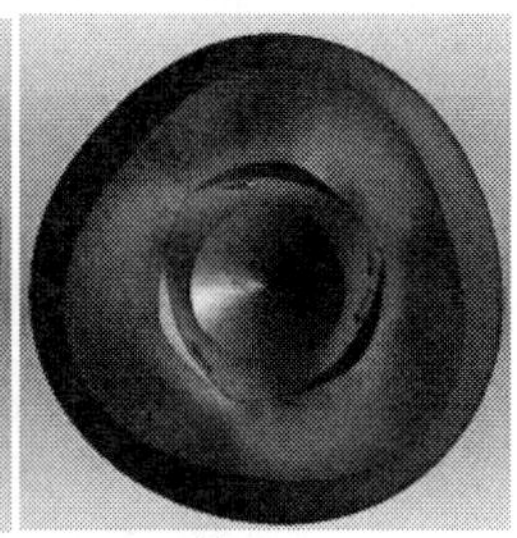
b)

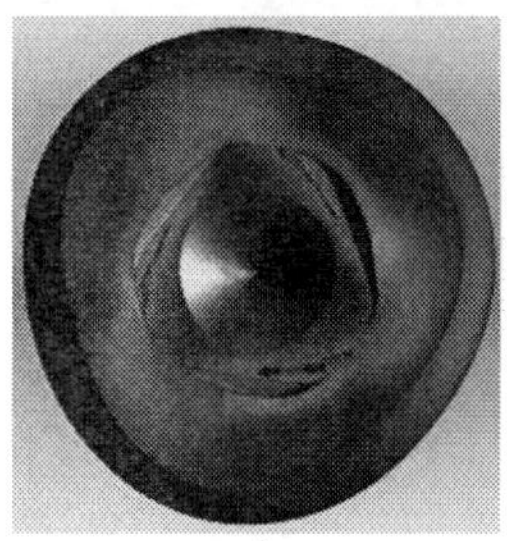
c)

Fig. 4. The result of modeling with the use of various special profiles. a) ellipse; b) Ryolo's triangle; c) rounded triangle

As a result of the simulation of special profiles, rational were such profiles as the Ryolo's triangle and the rounded triangle. The maximum stress values for these profiles are within σ = 1249.64-1309.46 MPa.

The ex periment car ried o ut with t he h elp o f t hree-dimensional modeling i s i deal an d t he act ual s tresses created i n r eal co nditions i n t he screw head will b e different from t he cal culated o nes. T his i s b ecause a person c annot i nsert t he screw into t he b one with t he same force e very time, it is also q uite common for t he key to the cavity to be installed with a distortion, resulting i n a ch ange o f co ntact ar ea a nd l oad distribution.

In o rder t o make the cal culations a s cl ose t o r eal as possible, it is necessary t o co nduct a multifactorial experiment. T he number o f n ecessary e xperiments i s calculated b y formula n = 2 k, where k is the number o f factors. . T he m ain f actors i nfluencing t he t ensions created in the screw head during the screwing-in process are t he co ntact ar ea b etween t he key an d t he r ecess profile S (mm) a nd t he t orque g enerated b y t he M кр (H·m). Consequently, the number o f factors k = 2 , an d the number of experiments n = 4.

The torque was selected in the range M кр = 3… 10, the co ntact ar ea f or each r ecess p rofile was cal culated individually.

3 Conclusion

Simulations has s hown th at t he r ational p rofiles o f t he recessed head o f the screw in terms o f the stress-strain state are square (σ_{max} = 1296,2 MPa) , pentagon (σ_{max} = 1270,6 M Pa), he xagon (σ_{max} = 127 4,6 M Pa), R yolo's triangle (σ_{max} = 1249, 64 MPa) and a rounded t riangle (σ_{max} = 1309,46 MPa). F rom the technological point o f view, with accounting for t he c omplexity of production and the contact area between the key and the profile, the rational f orm o f t he p rofile o f t he r ecess f or osteosynthesis screw head is a rounded triangle.

Acknowledgments

The w ork was c arried o ut with t he f inancial support of the G rant C ouncil o f the President o f t he R ussian Federation for s tate support o f young Russian s cientists and for s tate support of leading scientific s chools o f t he Russian Federation, grant number MD-345.2020.8.

References

1. L.J. M elton, B one, **14**, 1 -8 (2002) https://doi.org/10.1016/8756-3282(93)90341-7
2. K.K. R omanenko, A .I. B elostotsky, D.V. Prozorovsky, et al., Orthopedics, traumatology and prosthetics, **1**(578), 68-75 (2010.) [in Russian]
3. R. Lakatos, M. Herbenick, eMedicine (2007)
4. K.A. Berdyugin, A.K. Chertkov, D.I. Stadler, et al., Spine surgery, **4**, 19-24 (2010) [in Russian]
5. G.G. Golka, A.I. Belostotsky, I.A. Saturday, et al., Orthopedics, traumatology, and prosthetics, **4**(585), 72-80 (2011)
6. A.A. Alamovsky, *SolidWorks/COSMOSWorks. Engineering Analysis by Finite Element Method* (Moscow: DMK Press, 2004)
7. A.S. Nekhlopochin, S .N. N ekhlopochin, G.V. Syrovoy, Russian Journal of Biomechanics, **21**(1), 88-101 (2017)
8. *GOST R 50582-93 (ISO 5835-91) Implants for Surgery. Metal bone screws with special thread, spherical head and hexagon socket on a turnkey basis. Dimensions (adopted as an interstate standard GOST 30400-95 (ISO 5835-91)). 1995–01–01* (Moscow: Standards Publishing House, 1993)
9. F. B aumgart, J . C ordey, K. Mo rikawa, e t a l., International Jornal of the Care of the Injury, **24**, 1–17 (1993) https://doi.org/10.1016/0020-1383(93)90312-T
10. A.M. Aronov, V .F. P ichugin, S .I. T verdokhlebov, *Methodological bases of development and organization of manufacture of medical devices* (Tomsk : Wind, 2007)
11. *GOST R ISO 5832-2-2014 Implants for Surgery. Metal materials. Metal materials for surgery. Part 2. Unalloyed titanium. 2016–01–01* (Moscow: Standardinform, 2015)
12. J. S chatzker, I nt O rthop., **20**(4), 244 –252 (1996) https://doi.org/10.1007/s002640050073
13. A.V. P opkov, T he g enius of or thopedics. **3**, 9 4-99 (2014) [in Russian]

Synergistic Effects Under Ion-Beam Modification of Metals

Anatoly Borisov[1,2,*], *Boris* Krit[1,2], *Igor* Suminov[2], *Mikhail* Ovchinnikov[3] and *Sergey* Tikhonov[1]

[1]Moscow Aviation Institute (National Research University), RU-125993 Moscow, Russia
[2]Moscow State Technological University "STANKIN", RU-127994 Moscow, Russia
[3]Skobeltsyn Institute of Nuclear Physics, Moscow State University, RU-119991 Moscow, Russia

Abstract. The combined effect of ion and laser beams on physical and mechanical properties of metal and alloy surfaces has been studied. The technique of determining the main parameters of polyenergetic ion implantation using a vacuum-arc ion source is proposed and evaluated. It is found that treatment with titanium ions and the subsequent laser thermal hardening increase microhardness of steel 45 and U8 up to 6 times.

1 Introduction

The material processing with concentrated energy flows allows changing physical and chemical surface properties in a wide range. The ion and laser beams are extensively used to modify the surface of materials. In the context of ion processing, in particular ion implantation, surface properties of materials are modified as a result of the implantation of high-energy ions causing changes in composition of elements (doping) and structure-phase states of subsurface layers. It is possible to change directly such properties of materials as microhardness, wear resistance, corrosion resistance, heat resistance as well as friction, optical, magnetic, emission properties, etc. by using ion implantation.

The progress of laser techniques has caused wide application of lasers for the materials processing. One of the perspective directions of laser technology from the point of view of materials properties modification is surface laser hardening in which the structure-phase state of subsurface layers is changed by pulse heat treatment. The prospects of such technology are determined by the possibility to create various new surface structures with improved physical, chemical and mechanical properties in sufficiently thick layers using relatively simple equipment.

The analysis of different technological processes showed that the efficiency of laser processing methods could be significantly increased by creating the laser combined methods based on the combination of laser sources with other sources of synergistic effect on materials.

For the ion processing of structural materials, a significant change in properties occurs when the concentration of dopant reaches a density comparable to the density of a solid material. For this reason, the required irradiation fluences are about 10^{16}-10^{18} cm^{-2} (high-dose range). For the high-dose ion implantation (HDII) the ions with the energy, in which their stopping range in the material exceeds the thickness of the sputtered layer, are used. The depth of implantation for the ions with energies ~10^4-10^6 eV are 0,1-1 μm [1,2].

HDII can be considered as complex chemical, thermal and mechanical impact on materials. HDII has several advantages over other methods of material processing (e.g., vacuum deposition).

Nowadays, works on modification of properties of structural materials by ion-plasma flows which allow to change physico-mechanical, electrophysical, electrochemical properties of a surface layer of materials and work-pieces from them are intensively developing [2]. In these works the requirements to monoenergeticity and purity of the ion beam are reduced through the necessity to increase the productivity (processing dose 10^{16}-10^{18} cm^{-2}) and the area of the processed work-pieces (respectively, ion beam cross section) up to 10^2-10^3 cm^2.

The non-separation industrial ion implanters, with the vacuum-arc pulsed metal ion sources of MEVVA type (metal vapor vacuum arc) have high productivity [2,3]. It is based on a complex composition of the ion beam both for ion charge and energy. The energy of single-charged ions is determined by the accelerating voltage, the energy of multi-charged ions is a multiple of the energy of single-charged ions, therefore, this ion implantation is polyenergetic.

The present work is aimed to study the complex (synergistic) effect on the microhardness of the surface of structural and instrumental steels of polyenergetic metal ion beams of, produced using a vacuum-arc source, and laser thermal hardening.

2 Experimental techniques

The non-separation ion implanter with pulse vacuum-arc ion source used in the work is illustrated by the scheme shown in Fig. 1.

* Corresponding author: anatoly_borisov@mail.ru

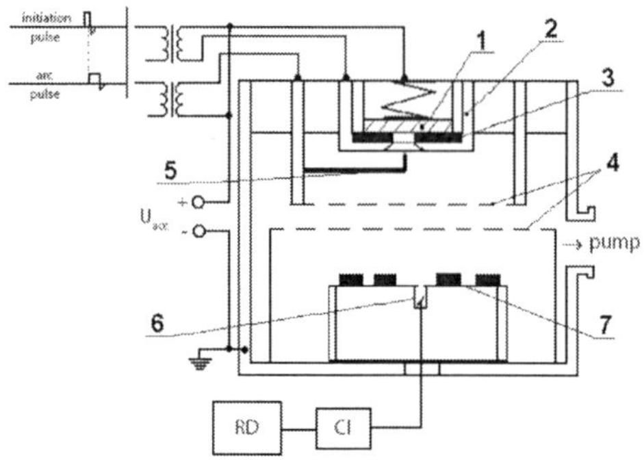

Fig. 1. Scheme of the implantation installation with pulse vacuum-arc ion source [4].

Arc ignition is initiated by electric breakdown of ceramic ring insulator - *3* between replaceable water-cooled cathode - *1* with diameter 10 mm and ignition electrode - *2*. The arc discharge is supported by a pulse power supply providing about 300 A current of the arc during 300 μs. The acceleration of ions of cathode material, generated by vacuum arc, takes place in the electrostatic field between tungsten meshes of ion-optical system - *4*. Application of rod anode insert - *5* allows more than 3 times increase in current of extractable ions due to change of indicator of ions output from cathode. The dosimetry of ion irradiation of samples, installed on the object table - *7*, is performed using the Faraday cylinder - *6* and current integrator (CI). The current integrator converts the charge, which passed through the Faraday cylinder, into the frequency of pulse repetition, which is then summed up by recalculating device (RD). The implanter allows obtaining wide aperture round cross-section (diameter of output aperture 30 cm) metal ion beams with current 0.1-1 A, pulse duration ~300 μs, pulse repetition frequency up to 50 Hz. Maximum value of accelerating voltage U_{acc}=50 kV.

The conditions of irradiation of steel samples (St. 45, 40X13, U8) and graphite selected as a witness - sample of polyenergetic ion implantation are presented in Table 1.

Table 1. Ion implantation parameters and Rutherford spectrometry results for graphite ion implantation

Cathode material	Ion charge concentrations in the beam q=1$^+$, 2$^+$, 3$^+$, 4$^+$, 5$^+$, %	q_{mean}	U_{acc}, kV	F, 10^{17} cm^{-2}	Rutherford backscattering spectroscopy and simulation results		
					D_{Ruth}, 10^{17} cm^{-2}	S	D_{sim}, 10^{17} cm^2
Ti	11, 75, 14	2,1	35	6,0	1,7	1,55	3,0
Al	38, 51, 11	1,7	35	0,4	0,73	0,67	0,4
Nb	1, 24, 51, 22, 2	3,0	35	3,0	0,39	3,24	1,34
Cu	16, 63, 20,1	2,0	35	0,6	0,6	0,99	0,59
Cr	10, 68, 21, 1	2,1	35	4,2	2,5	1,75	3,0
Pb	36, 64	1,6	20	0,3	0,2	3,07	0,24

The element composition of the implanted layer was analyzed using the Rutherford backscattering spectroscopy (RBS) of a 2 MeV helium ion beam obtained at the electrostatic ion accelerator of the SINP MSU using the same method as in [4, 5].

Laser processing was carried out on the laser installation "Kvant-18" with pulse duration 8 ms, pulse overlapping factor 0.7 and laser beam power density 2 W/cm^2, which preserved the surface microgeometry and did not melt it.

Microhardness was measured according to Vickers on the multipurpose measuring complex "METALPLAN" by LEITZ at the indenter loads from 50 to 100 g.

3 Results and discussion

The experimental data obtained in the present work on polyenergetic high-fluence implantation (Table 1) were considered in the framework of the analytical model of high-fluence ion implantation, which takes into account one dominant factor - sputtering of the target by ion bombardment [5].

Under polyenergetic implantation, when the energy of ions is determined by their charge q=1,2,3,..., the partial concentration cross sections of ions $N_q(x)$ in the target can be described in a form:

$$N_q(x) = \frac{N_0 \cdot n_q}{2S}\left[erf\left(\frac{x + \frac{F}{N_0}S - R_q}{\sqrt{2}\cdot\Delta R_q}\right) - erf\left(\frac{x - R_q}{\sqrt{2}\cdot\Delta R_q}\right)\right], \quad (1)$$

where, N_0 - concentration of target atoms; n_q - relative concentration of ions with charge q in the beam; R_q and ΔR_q - projected range and mean square straggling of projected range of ions with charge q, respectively; F - fluence of implantation. The sputtering yield for polyenergetic ion irradiation:

$$S = \sum_{q=1}^{n} S_q n_q, \quad (2)$$

where S_q is the sputtering yield of the target by ions with charge q. Values of S_q were calculated using SRIM software.

Total concentration cross section of implanted atoms:

$$N(x) = \sum_{q=1}^{n} N_q(x). \quad (3)$$

Integrated dose of implantation:

$$D(x) = \int N(x)dx \quad (4)$$

An example of calculations of partial and total concentration cross sections is shown in Figs. 2, 3. The total concentration cross section is significantly determined by the charge composition of the ion beam (Fig. 2).

The analysis of RBS spectra in case of irradiation with fluence above 10^{17} cm^{-2} revealed the presence of impurities implanted in the samples. The detected impurities (not more than 5% of the implanted element) such as tungsten, molybdenum, aluminum are the elements of the ion source. In the case of titanium ion implantation, oxygen was also found, the atomic concentration of which is comparable to that of implanted titanium, which indicates the possible formation of titanium oxides such as TiO and Ti_2O_3 on the sample surface.

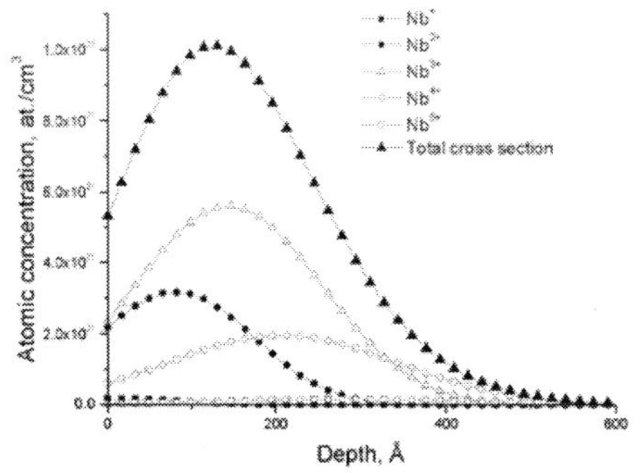

Fig. 2. Partial and total concentration cross sections of niobium implanted in 40X13 steel under irradiation fluence $F=3\cdot10^{17}cm^{-2}$.

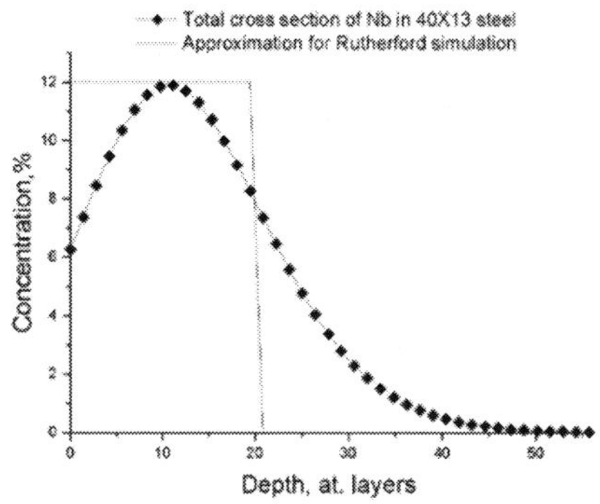

Fig. 3. Total cross section of Nb in steel 40X13 and its approximation for RBS simulation.

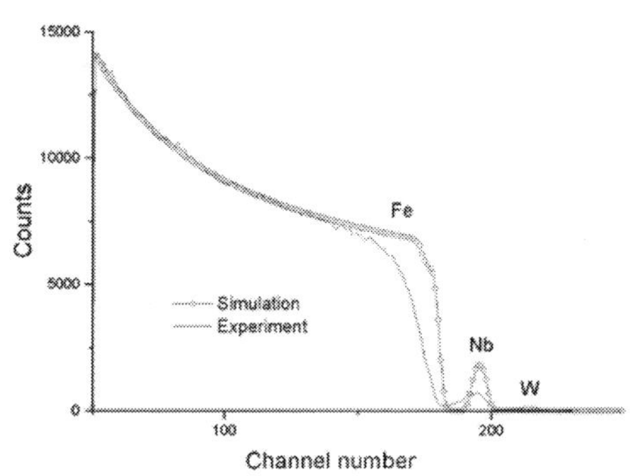

Fig. 4. Comparison of calculated and experimental RBS spectrum for Nb implanted in steel 40X13.

Some differences in the peak shape for Nb implanted in 40X13 steel (Fig. 4) may also be related to the selected approximation of its simulated cross section (Fig. 3).

Since the concentration profile of the implanted element is significantly influenced by the ion charge, which determines the ion stopping range in the material and sputtering of the surface, as well as their partial composition in the beam, on which the overall profile depends, the results of the difference between the simulation of polyenergetic ion implantation and the experiment (Table 1) can be associated with the following factors:

1. The difference of the used data in [3] by the charge composition of n_q ion beams produced by MEVVA type vacuum-arc ion sources (Table 1) with the corresponding composition for the implanter ion source.
2. The concentration of the implanted element depends on the sputtering yield S, which is a function of ion charge q. The value of projected run, on the other hand, depends on the energy of ions, which is a product of charge q and accelerating voltage U_{acc}. Thus, because of this incompatibility of charge data, it is unknown what effect on the difference between the calculated and experimental RBS spectra was greater: sputtering yield or projected range.
3. In this model, the values of projected ranges, stragglings and sputtering yield calculated with the SRIM program were used. This program simulates sputtering in linear cascade mode. In our case, heavy ions are implanted and, consequently, a thermal peak mode is possible, which is not considered in the SRIM program.
4. The presence of impurities may also affect the concentration cross section of the implanted dopant, since the presence of impurities is not taken into account in the simulation of the implantation process.

Table 2 shows the results of microhardness measurement at loading on the indenter 50 g and parameters of ion implantation for steels 40X13, st. 45, U8.

Table 2. Ion implantation parameters and microhardness of steels.

#	Ion	U_{acc}, kV	F, 10^{17} cm^{-2}	Irradiation time, min	Microhardness		
					40X13	st. 45	U8
1	Without irradiation				301±4	105±4	104±3
2	Cu	35	$5.8\cdot10^{16}$	110	380±5	126±3	151±3
3	Pb	20	$3\cdot10^{16}$	52	393±6	113±3	154±6
4	Al	35	$4\cdot10^{16}$	70	389±5	123±5	129±4
5	Ti	35	$2\cdot10^{17}$	135	537±10	159±6	183±5
6	Nb	35	$3\cdot10^{17}$	162	546±9	158±5	175±6

The depth of ion stopping range, calculated using the SRIM program, in steel does not exceed 0.1 micrometers, and the indenter bulge depth is one micrometer. The effect of increasing microhardness in such cases is usually associated with the long-range effects of ion implantation [6].

One of the mechanisms of such effects is the radiation stimulated diffusion of point defects and formation of dislocation loops, which cause the lattice distortion and hardening of the material [1, 6]. The flow of interstitial atoms in depth of sample can cause dislocation creep. Thus, radiation defects and elastic stresses can penetrate to much greater depths than the stopping range of ions, and have a significant impact on the properties of the surface layers.

It is found that significant increase in microhardness of the surface is observed at ion bombardment with titanium and niobium ions.

After ion implantation, laser thermal hardening of samples from structural and instrumental steels was carried out at "Kvant-18" installation. Results of microhardness measurement are shown in fig. 5.

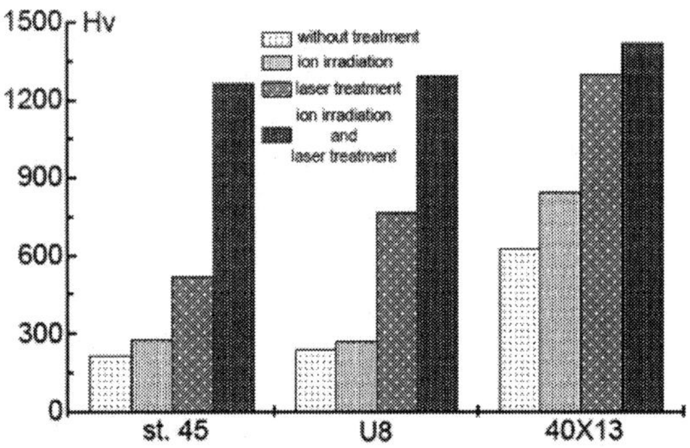

Fig. 5. Influence of titanium ion implantation and laser treatment on microhardness of 45, U8, 40X13 steels

A dramatic increase in microhardness of steels under combined ion-laser treatment can be explained by the influence of radiation defects formed by ions and the implanted impurities on the subsequent formation of the hardening structure during laser thermal hardening, changes in the microstructure of the surface and, in particular, an increase in the dispersion of martensite.

4 Conclusions

Possibilities of analytical methods of determination of the main parameters of polyenergetic ion implantation with the use of vacuum-arc ion source have been analyzed using Rutherford backscattering spectroscopy of implanted samples.

The impact of ion implantation on surface hardening of structural and instrumental steels was studied. It is found out that ion implantation increases microhardness of the steel surface by 20-80% depending on the type of ions. The best results of microhardness increase of steels U8, 45 and 40X13 in 1,5-1,8 times were obtained by treatment with titanium and niobium ions at the irradiation fluences $\sim 10^{17}$ cm^{-2}.

The effects of laser heat treatment on microhardness of steels have been studied. The obtained results show an increase in microhardness of U8, 45 and 20X13 steels surface by 1.3-2.5 times.

The results show that combined ion-laser treatment of steels is promising. It was found that treatment with titanium ions and subsequent laser thermal hardening increase microhardness of 45 and U8 steel up to 6 times, and thermally pre-hardened steel 40X13 - more than 2 times.

Acknowledgments

The study was carried out with the financial support of the Russian Foundation for Basic Research within the framework of scientific project No. 21-79-30058.

References

1. K.J. Kirkby, R.P. Webb, Encyclopedia of Nanoscience and Technology, **4**, 1-11 (2004)
2. J.M. Poate, G. Foti, D.C. Jacobson, *Surface Modification and Alloying by Laser, Ion, and Electron Beams* (Springer Science Business Media, New York, 1983)
3. Ian G. Brown, *The Physics and Technology of Ion Sources* (John Wiley & Sons, University of California. Berkeley, California, 1989)
4. A.M. Borisov, B.L. Krit, S.A. Tikhonov, I.V. Suminov, V.S. Kulikauskas, V.G. Sukharev, Bull. Russ. Acad. Sci. Phys., **64**, 4 (2000)
5. A.M. Borisov, N.V. Borodulina, B.L. Krit, S.A. Tikhonov, J. Appl. Phys., **1**, 89-93 (2004)
6. R. Smith, M. Jakas, D. Ashworth, R. Oven, M. Bowyer, I. Chakarov, R.P. Webb, *Atomic & Ion Collisions in Solids and at Surfaces* (Cambridge University Press, Cambridge, 1997)

https://doi.org/10.1051/epjconf/202124804007

Integrated Probabilistic Surface Roughness Assessment for Various Processing Methods

Igor Bobrovskij[1,2,3,*], *Petr* Zibrov[2], *Aleksey* Luk'yanov[3], *Olesja* Levitskih[3]

[1]Almetyevsk State Oil Institute, RU-423450, Tatarstan, Almetyevsk, Russian Federation
[2]Togliatti State University, RU-445020 Togliatti, Russian Federation
[3]Samara Scientific Center of Russian Academy of Science, RU-443001, Samara, Russian Federation

Abstract The surface texture characteristics are specified by international standards, which include dozens of parameters, basically geometrical. The probability of filling the rough surface layer with a material from the reference line and through the layer thickness is proposed as a quantitative roughness assessment. It takes account of geometrical parameters, shape, height and frequency of surface irregularities. An example is given of studying a surface with processing defects, and a comparative analysis of surface treatment methods is presented. The calculation results are summarized in a table of probabilistic characteristics of surfaces for various processing methods.

1 Introduction

The probabilistic parameters of presence of a material within a section of the surface layer and through the entire layer thickness are evaluated using the computational and graphical method for studying the rough surface profilograms obtained from various processing methods.

The probability of filling density $P(S_{\alpha i})$ in i-th section relative to the surface reference line is calculated as [1-3]:

$$P(S_{\alpha i}) = \frac{M_i}{l},$$

where $M_i = \sum_{j=1}^{m_i} M_{ij}$, l – profilogram base length.
M_{ij} – width of the j-th peak in the i-th section;
m_i – the number of peaks in the specified section;
l and M_{ij} are measured in the same linear units shown on the profilogram (Fig.1);
$0 \le P(S_{\alpha i}) \le 1$ – the probability of filling density in the i-th section, a dimensionless value.

It should be noted that in the Abbott curve definition there is an assumption about its correspondence to the probability density, which is not correct, since the probability density characterizes the maximum of the probability change rate at the point of the interval. In the present case the amount of the material in the section changes and the formula above corresponds to the graphical probability definition of the considered event.

$P(S_\alpha)$ curves constructed for real profilograms are close to parabola form. Their equations, which best fit the obtained statistical data, can be written as:

$$s = (1-\alpha)(1-\beta\alpha).$$

The dimensionless coefficient β is obtained from the statistical data P(Sαi) and αi using the least squares method:

$$\beta = -\frac{\sum_{i=1}^{n}[P(S_{\alpha i}) - (1-\alpha_i)](1-\alpha_i)\alpha_i}{\sum_{i=1}^{n}(1-\alpha_i)^2\alpha_i^2} = -\frac{\xi}{\eta}$$

Here $\xi = \sum_{i=1}^{n} \xi_i$, $\eta = \sum_{i=1}^{n} \eta_i$, $\xi_i = [P(S_{\alpha i}) - (1-\alpha_i)](1-\alpha_i)\alpha_i$, $\eta_i = (1-\alpha_i)^2\alpha_i^2$, β is taken with its own sign.

The probability of a material within a section of the surface layer and through the entire layer thickness is found from the following formula [4-8]:

$$P(V_{\alpha i}) = 1 + \frac{1+\beta}{2}\alpha_i + \frac{\beta}{3}\alpha_i^2$$

For a smooth curve $P(S_i)$ along with the indicated relation, the graphical integration by the chord method can be used:

$$P(V_{\alpha i}) = \frac{1}{k}\left[\frac{P(S_0)+P(S_k)}{2} + \sum_{j=1}^{k-1} P(S_j)\right], \; i=k=1,2,\ldots,n,$$

where n – the number of sections of the rough layer along the OZ axis.

2 Results and discussion

In accordance with the formula $P(V_{\alpha=1})$, which gives value of the probability of the through-thickness filling the rough layer with a material, the Simpson's formula is suitable, adapted for the present case.

$$P(V_{\alpha=1})_{Sim} = \frac{1}{6m}\left[1 + 2\sum_{i=1}^{m-1} P(S_{\alpha 2i}) + 4\sum_{i=1}^{m} P(S_{\alpha 2i-1})\right].$$

[*] e-mail: bobri@yandex.ru

Here m is integer and $n = 2m$, n – number of sections across the rough layer thickness.

The relative difference in results found by Simpson's formula and the approximating parabola satisfying the least squares method will be:

$$\delta = \left|1 - \frac{P(V_{\alpha-1})_{Sim}}{P(V_{\alpha-1})}\right| \cdot 100\%$$

It quantitatively characterizes the difference between the found characteristics.

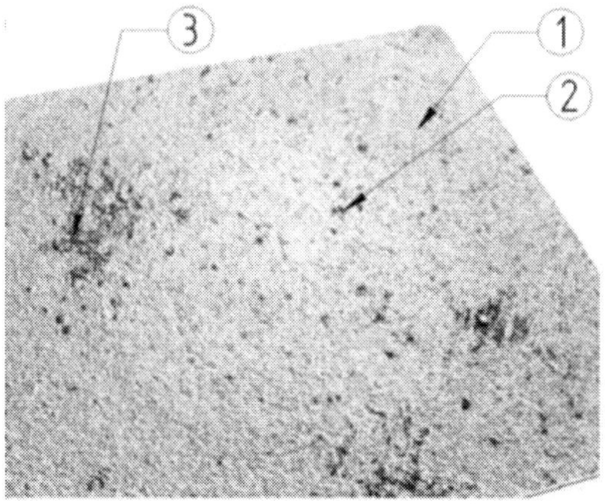

Fig.1. Photo of a processed surface with defects.

Length = 319.375µm Pt = 146.675nm Scale = 200.000 nm

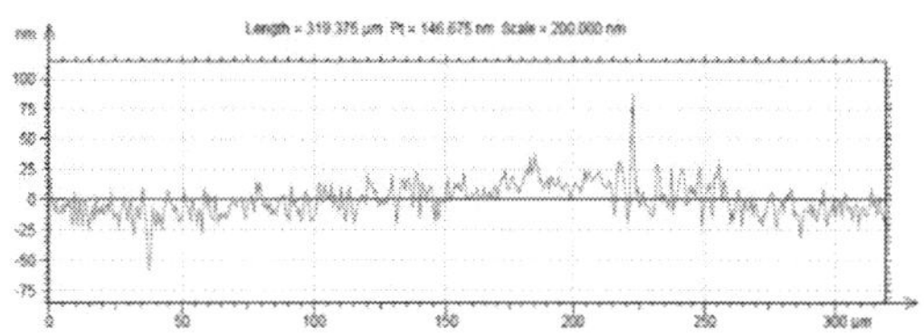

Fig. 2. Example of an imperfect surface profilogram.

As an example, the studies of the surface shown in Figure 1 [9-11] are presented; its profilogram is given in Figure 2. The defects in areas 2 and 3 are noted due to: surface pores in area 2 and material adhered to the ball during rolling is redistributed to the area 3.

The original profilogram is divided into three sections. A computational and graphic study of the probability of the material presence in the surface irregularities is conducted using complex parameters according to the dependences presented above, and obtained results are compared.

The use of dimensionless characteristics makes it possible to choose the base length and the profilogram width for various sizes of a rough layer, compare them in relative units and identify the features of each section considering the trends in microroughness.

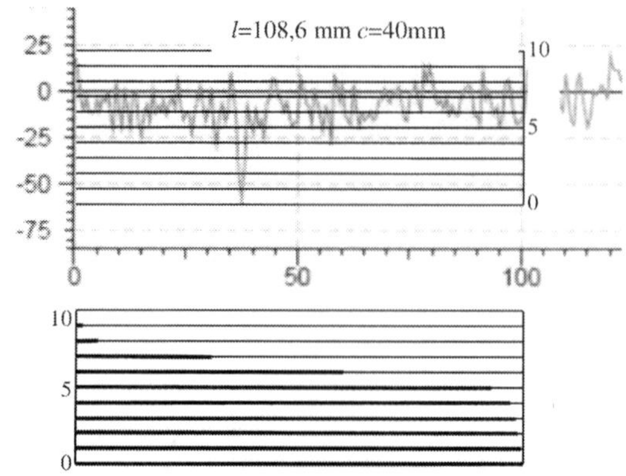

Fig. 3. Profilogram and diagram of filling the rough layer with a material along the 0-100 section of the profilogram depicted on fig. 2.

Table 1. Results of probabilistic parameters calculation along the length 0 -100 mm of the profilogram.

α_i	0	0,1	0,2	0,3	0,4	0,5
$P(S_{\alpha i})$	1	0,9917	0,9880	0,9816	0,9742	0,9227
ξ_i	0	0,0083	0,0301	0,0591	0,0898	0,1057
$P(V_{\alpha i})_{тр}$	1	0,9959	0,9929	0,9902	0,9871	0,9794
$P(V_{\alpha i})$	1	0,9934	0,9806	0,9616	0,9364	0,9050
s_i	1	0,9837	0,9488	0,8953	0,8232	0,7325

Table 1. Continuation

0,6	0,7	0,8	0,9	1,0
0,6261	0,2753	0,0497	0,0076	0
0,0545	-0,0052	-0,0240	-0,0083	0
0,9453	0,8747	0,7857	0,7016	0,6314
0,8674	0,8236	0,7736	0,7174	0,6551
0,6232	0,4953	0,3488	0,1837	0

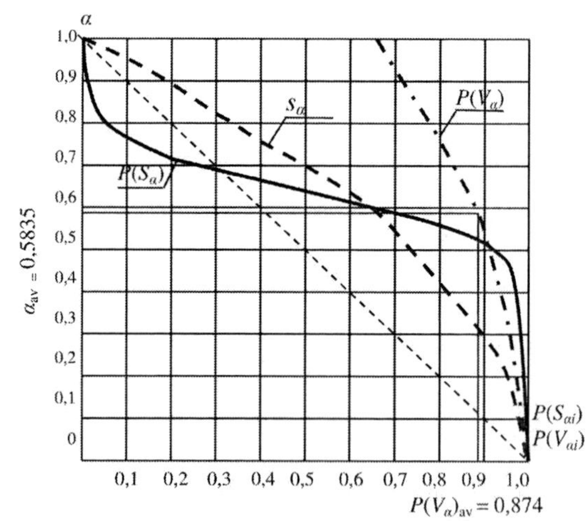

Fig. 4. P(Sα), P(Vα) and s(α) curves constructed for the profilogram and diagram depicted on fig. 3 (first section of the profilogram on fig. 2).

Quantitative values in the calculation formulas for the first section on fig. 2.

$$\xi=0{,}31,\ \eta=0{,}3333,\ \beta=-0{,}9301;$$

$$P(V_{\alpha i}) = 1-(0{,}03495+0{,}3100\alpha_i)\alpha_i;$$

$$s_i=(1-\alpha_i)(1+0{,}9301\alpha_i);$$

$$\delta = 1-\frac{0{,}6314}{0{,}6551}=0{,}036 \text{ or } 3{,}6\%;$$

$$P(V_\alpha)_{av}=0{,}8740;\ \alpha_{av}=0{,}5835;\ P(V_{\text{тр}})_{av}=0{,}8986.$$

l = 106,5 mm c = 30 mm

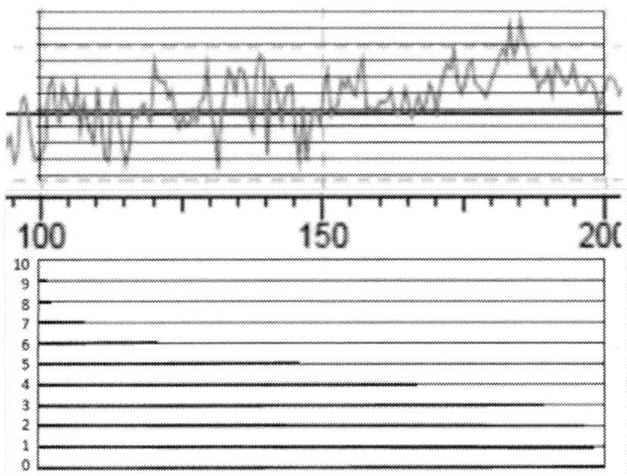

Fig. 5. Profilogram and diagram of filling the rough layer with a material along the 100-200 section of the profilogram depicted on fig. 2.

Table 2. Results of probabilistic parameters calculation along the length 100-200 mm of the profilogram.

α_i	0	0,1	0,2	0,3	0,4
$P(S_{\alpha i})$	1	0,9793	0,9606	0,8714	0,6488
ξ_i	0	0,0071	0,0257	0,0360	0,0117
$P(V_{\alpha i})_{\text{тр}}$	1	0,9897	0,9798	0,9585	0,9089
$P(V_{\alpha i})$	1	0,9407	0,8827	0,8260	0,7707
s_i	1	0,8820	0,7680	0,6580	0,5520

Table 2. Continuation

0,5	0,6	0,7	0,8	0,9	1,0
0,4493	0,2006	0,0676	0,0150	0,0089	0
-0,0127	-0,0479	-0,0488	-0,0296	-0,0082	0
0,8370	0,7516	0,6634	0,5856	0,5219	0,4702
0,7167	0,6640	0,6127	0,5627	0,5140	0,4667
0,4500	0,3520	0,2580	0,1680	0,0820	0

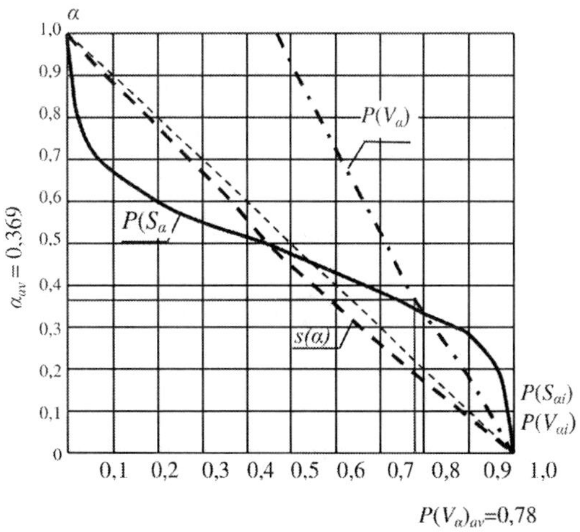

Fig. 6. P(Sα), P(Vα) and s(α) curves constructed for the second section of the profilogram depicted on fig. 2.

Quantitative values in the calculation formulas for the second section on fig. 2.

$$\xi=-0{,}0667;\ \eta=0{,}3333;\ \beta=0{,}2001;$$

$$P(V_{\alpha i}) = 1-(0{,}6-0{,}667\alpha_i)\alpha_i;$$

$$s_i=(1-\alpha_i)(1-0{,}2001\alpha_i);$$

$$\delta = 1-\frac{0{,}4667}{0{,}4702}=0{,}0074 \text{ or } 0{,}74\%;$$

$$P(V_\alpha)_{av}=0{,}7878;\ \alpha_{av}=0{,}369;\ P(V_{\text{тр}})_{av}=0{,}7879.$$

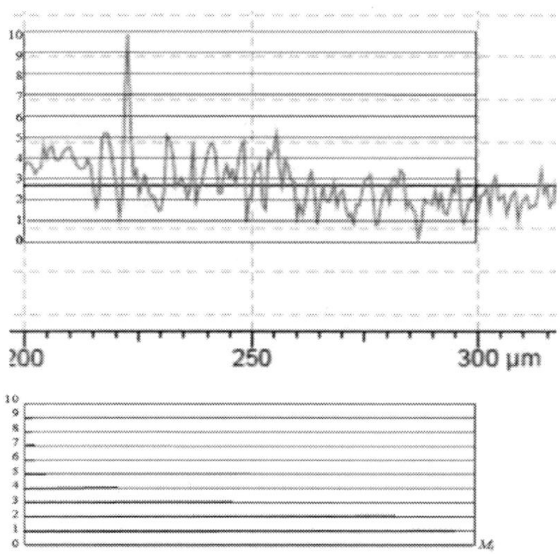

Fig. 7. Profilogram and diagram of filling the rough layer with a material along the 200-300 section of the profilogram depicted on fig. 2.

https://doi.org/10.1051/epjconf/202124804007

Table 3. Results of probabilistic parameters calculation along the third section (200-300 mm).

α_i	0	0,1	0,2	0,3	0,4	0,5	0,6	0,7	0,8	0,9	1,0
$P(S_{\alpha i})$	1	0,9713	0,7568	0,4446	0,1676	0,0395	0,0164	0,0160	0,0096	0,0085	0
ξ_i	0	0,0064	-0,0069	-0,0536	-0,1038	-0,1151	-0,0921	-0,0596	-0,0305	-0,0082	0
$P(V_{\alpha i})_{\text{тр}}$	1	0,9857	0,9249	0,8168	0,6891	0,5720	0,4813	0,4149	0,3645	0,3251	0,2930
$P(V_{\alpha i})$	1	0,8851	0,7795	0,6831	0,5961	0,5183	0,4497	0,3904	0,3404	0,2997	0,2682
s_i	1	0,7749	0,5776	0,4080	0,2663	0,1524	0,0066	0,0008	-0,0224	-0,0251	0

Table 3. Continuation

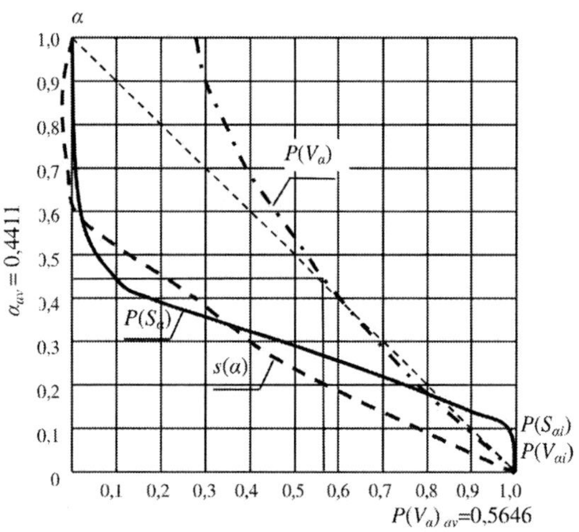

Fig. 8. P(Sα), P(Vα) and s(α) curves constructed for the third section of the profilogram depicted on fig. 2).

Quantitative values in the calculation formulas for the third section on fig. 2.

$$\xi = -0{,}4634,\ \eta = 0{,}3333,\ \beta = 1{,}3903.$$

$$P(V_{\alpha i}) = 1 - (1{,}1952 - 0{,}4634\alpha_i)\alpha_i;$$

$$s_i = (1 - \alpha_i)(1 - 1{,}3903\alpha_i);$$

$$P(V_a)_{av} = 0{,}5646,\ \alpha_{\text{cp}} = 0{,}4411,\ P(V_{\text{тр}})_{av} = 0{,}6244.$$

$$\delta_{1,2} = 0{,}2553 \approx 25{,}5\%,$$

$$\delta_{1,2} = 0{,}536 \approx 53{,}6\%,$$

$$\delta_{1,2} = 0{,}6231 \approx 62{,}3\%.$$

The results of the computational and graphic study show that the probabilities of filling the rough layer with a material have significant difference at each of the three sections, that is from 25.5% to 62.3%. Such variation indicates the inhomogeneity of the rough layer.

Similar characteristics were obtained for various methods of surface treatment. They are consolidated in the table 4.

Table 4. Results of computational and graphic studies of surface texture probabilistic parameters for different processing types.

№	Type of treatment	$P(V_{\alpha=1})$	Relative difference, δ %	$P(V)_{av}$	α_{av}
1	Boring, rough honing	0,31096	$\delta^a_{1,2}$=33,6	0,5951	0,445
2a	Rough honing	0,4681	$\delta_{a,b}$=33,97	0,7215	0,47
2b	Fine honing	0,7089	$\delta^b_{1,2}$=56,13	0,9265	0,68
3a	Grinding - neck, Salnik	0,5084	$\delta_{a,b}$=5,11	0,7865	0,50
3b	Grinding - neck, Flange	0,5358	$\delta_{b,c}$=1,51	0,7969	0,52
3c	Grinding - neck, Main 1	0,5440	$\delta_{a,c}$=6,54	0,7700	0,77
4a	Grinding - neck, Pin 1	0,5172	$\delta_{a,b}$=4,93	0,7638	0,50
4b	Grinding - neck, Main 1 (according to ISO 13505)	0,7677	$\delta^b_{3,4}$=30,21	0,870	0,77
5a	Wide burnishing, Salnik	0,5146	$\delta_{a,b}$=10,29	0,7688	0,52
5b	Wide burnishing, Flange	0,5736	$\delta^a_{4,5}$=0,34	0,8117	0,55
6a	Polishing, Main 4	0,6889	$\delta_{a,b}$=4,53	0,9205	0,67
6b	Polishing, Pin 2	0,7216	$\delta^b_{5,6}$=20,51	0,9250	0,68

Thus, the probabilistic characteristics of filling a rough layer with a material give integral assessment of the part surface texture, taking account of the geometry, shape, frequency and height of the layer peaks and valleys represented by the profilogram.

The results of computational and graphic studies of the influence of the part surface treatment methods on the probabilistic characteristics of the rough layer filling shows that their relative difference varies from 0.34% to 56.13%. This range makes it possible to quantitatively evaluate the compliance of the surface texture with quality standards with a sufficiently high accuracy using proposed probabilistic characteristics.

3 Conclusions

This work leads us to the following conclusions:

1. The probability of filling the rough surface layer with a material derived from profilogram analisys is proposed to assess the surface texture.
2. Analytical dependences are obtained for calculating the probability of filling the rough layer with a material along layer sections, and the probability of the through-thickness filling.
3. An example is introduced of a computational-graphic analysis of process-induced surface defects.
4. Probabilistic characteristics of surfaces obtained from various processing types are presented.

Acknowledgements

This research was funded by the grant of Russian Science Foundation (project NO 19-19-00171).

References

1. Geometrical Product Specifications, ISO 4287, (1997)
2. Geometrical Product Specifications, ISO 4288 (1996)
3. R. Deltomber, K.J. Kubiak, M. Bigerelle, Scanning, **36**(1), 150-160 (2014) http://dx.doi.org/10.1002/sca.21113
4. P.F. Zibrov, O.G. Zibrova, Bulletin of the Volga University named after V.N. Tatishchev (Translated from russian), **2**(3), 32-36 (2015) [in Russia]
5. S.N. Grigoriev, G.M. Martinov, Procedia CIRP, **46**, 525-528 (2016) DOI: 10.1016/J.PROCIR.2016.04.036
6. S.N. Grigoriev, D.A. Masterenko, V.I. Teleshevskii, P.N. Emelyanov, Measurement Techniques, **55**(11), 1311-1315 (2013)
7. S.N. Grigoriev, G.M. Martinov, Procedia CIRP, **1**(1), 238-243 (2012)
8. M.A. Volosova, S.N. Grigor'ev, V.V. Kuzin, Refract Ind Ceram., **56**, 91–96 (2015)
9. V.V. Kuzin, S.N. Grigor'ev, M.A. Volosova, Refract. Ind. Ceram., **54**, 376–380 (2014)
10. M.A. Volosova, S.N. Grigor'ev, V.V. Kuzin, Refract. Ind. Ceram., **55**, 565 (2015)
11. A.S. Metel, V.P. Bolbukov, M.A. Volosova, S.N. Grigoriev, Yu.A. Melnik, Instrum, Exp. Tech., **57**(3), 345 (2014)

Development and Power Calculation of a Grinding Wheel Design for Ultra-High-Speed Grinding

Denis Rechenko[1,*], *Renat Kamenov*[1]

[1]Omsk State Technical University, RU-644050, Omsk, Russia

Abstract. Increasing the structural reliability at ultra-high-speed grinding is an important problem because it defines safety of metalworking. The purpose of work is to achieve the cutting speed up to 400-500 m/s by developing a new design of the grinding wheel with the possibility of using different grinding materials. Grinding wheel design for ultra-high-speed grinding was developed based on calculations carried out using the tear design method, which allows determining the maximum possible circumferential speed, and finite element method used to determine the equivalent stresses, plastic deformations, circumferential and radial movements . The possibility of development of a grinding wheel design for ultra-high-speed grinding has been tested experimentally. The developed designs of grinding wheels show may poddibly be used on super high-speed technological equipment for processing of parts made of hard-to-machine materials. Calculations and comprehensive experimental tests of the design of grinding wheels for ultra-high-speed grinding carried pit in this work allow determining the serviceability and the maximum possible cutting speed and provide useful information for the further development of grinding tools.

1 Introduction

An analysis of the existing designs of grinding wheels for high-speed grinding shows that the maximum circumferential machining speed of up to 160 m/s can be achieved (Table 1). Standard grinding wheels for high-speed grinding have a strength that allows processing at speeds up to 60-160 m/s [1-17]. After refining the existing designs, it is possible to reach cutting speeds of about 250 m/s in laboratory conditions without any possibility of further industrial applications of the refined design. In the case, when the cutting speed limit is exceeded, the individual components are torn off or the grinding wheel housing is torn apart. There exist the following criteria for the grinding wheel design:

- the grinding wheel can run at the highest possible cutting speed without risk of rupture;
- the ability to use different grinding materials;
- simplicity of design with a minimum number of stress concentrators;
- reliability of the clamping design of grinding segment.

Table 1. Grinding wheel designs.

№	Sketch	Strength	Description of how to increase the strength of a grinding wheel structure
1	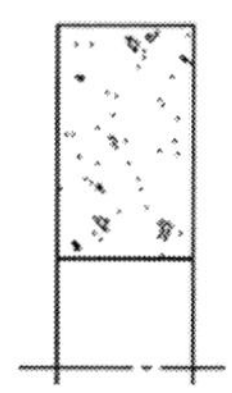	+	Strengthening of the non-working central part of the circle due to the application of fine-grained mixtures forming a dense, strong composition with binder or impregnation of the central part of the circle on the ceramic binder with strong thermoplastic compositions. Working speed - up to 60 m/s.
2	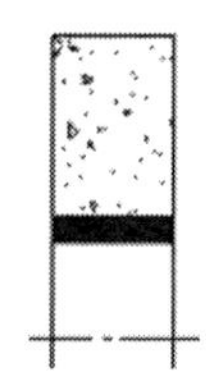	++	Reinforcement of the central part of the circles by pressing in a bushing made of particularly strong materials. Working speed up to 80 m/s.

* Corresponding author: rechenko-denis@mail.ru

3	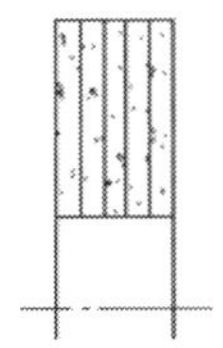	++	Strengthens circles on the bakelite bond by installing GRP mesh gaskets inside the circle. Working speed - up to 100 m/s.
4	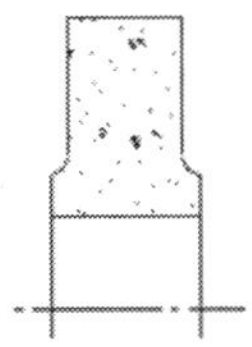	++	Manufacture of circles of variable cross-section with a thickening of the central part, smoothly descending to the working part. Working speed - up to 100 m/s.
5	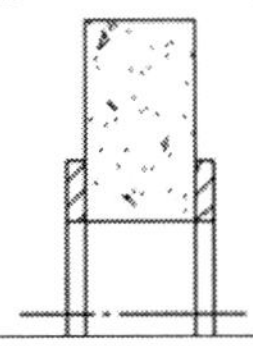	+++	Hardening due to metal discs glued to the ends of the grinding wheel. Working speed - up to 120 m/s.
6	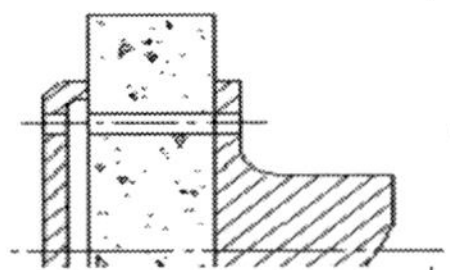	++++	Sharpening by using a one-piece grinding wheel design with small holes to allow attachment to the plate. Working speed - up to 120 m/s.
7	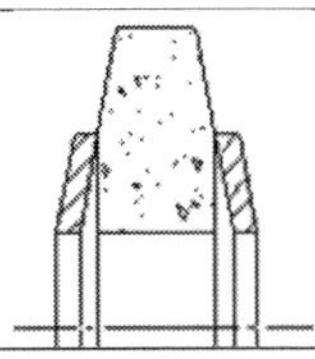	++++	Reinforcing by using special spring flanges. Working speed - up to 120 m/s.
8	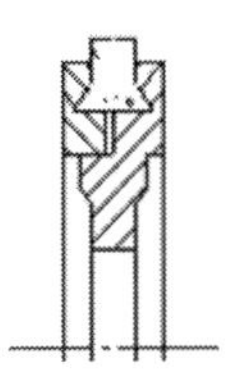	+++++	Peening by the use of prefabricated circular structures consisting of segments attached to the plate. Working speed - up to 160 m/s.

Creating a new design that allows for safe grinding at the highest possible circumferential speed is a complex task that requires a careful consideration of the strength, maximum possible operating speeds, the system of attachment of the grinding elements and fixing the body in the spindle shaft.

Some designs of grinding wheels include the use of metal bandages, shroud, etc., which significantly reduces the possibility of their rupture. The analysis of the existing designs has been shown that a grinding wheel for ultra-high-speed processing needs an all-metal case with the smallest number of parts and that the safety margin factor should be 1.4 (GOST R 52857.1-2007).

2 Problem statement

Cutting speeds of up to 400-500 m/s can be achieved by designing a new grinding wheel design with the possibility of using different grinding materials. Grinding wheel design for ultra-high-speed grinding was calculated using the tear design method, which allows to determine the maximum possible circumferential speed and finite element method to determine the equivalent stresses, plastic deformations, circumferential and radial movements in the process.

3 Analysis and development of bindings of grinding segment

Development of replaceable grinding segments, namely the circuit and the mechanism of their fixation is an important task, because the performance of the entire process unit depends on it. Different grinding segment characteristics can only be implemented with a one-piece grinding wheel housing with mounting sockets for replacement segments. Studies of aerodynamic work have shown that the most appropriate use of cylindrical grinding segments. From possible schemes of fastening of grinding segments, the author offers and considers the most technological and having a simple design (Fig. 1) [2-7].

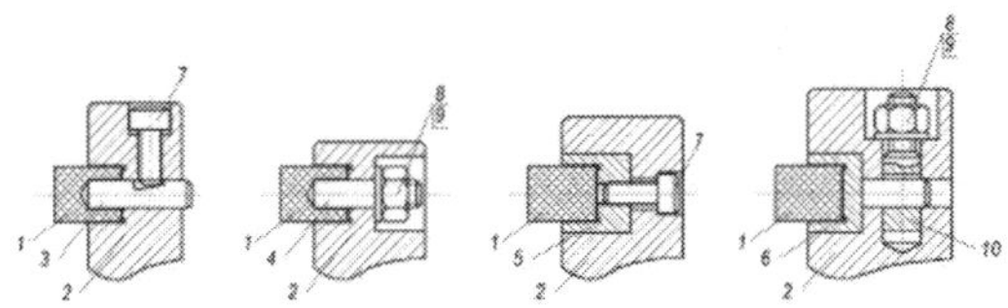

a) b) c) d)

1 - grinding head; 2 - body of grinding wheel; 3 - oblique groove pin; 4 - threaded pin; 5 - beaker; 6 - beaker on pin; 7 - fastening bolt; 8 - fastening nut; 9 - washer; 10 - trunnion block

Fig. 1. Fastening structures for replacement grinding segments: a) fastening with a radial bolt; b) fastening with an axial nut; c) fastening with an axial bolt; d) fastening with a radial nut.

The presented designs of fastening of grinding segments provide reliable fastening, thus fastening with a radial bolt (Fig. 1, a) provides simplicity of realization of this design on the basis of standard grinding heads with a pin. The use of the attachment structure with an axial nut (Fig. 1, b) is quite dangerous, as the attachment force also works on detaching the diamond head from the pin. The use of axial bolt fixing (Fig. 1.c) seems to be the most rational, due to the simplicity of its construction and the clamping force being provided exclusively on the bucket in which the grinding segment is located. The design of the attachment with a radial nut (Fig. 1.d) is reliable, but complex in execution and operation.

The results of strength calculations of various abrasive and binder materials are shown in Figure 2, the shear stress dependence on the grinding wheel circumferential speed is presented.

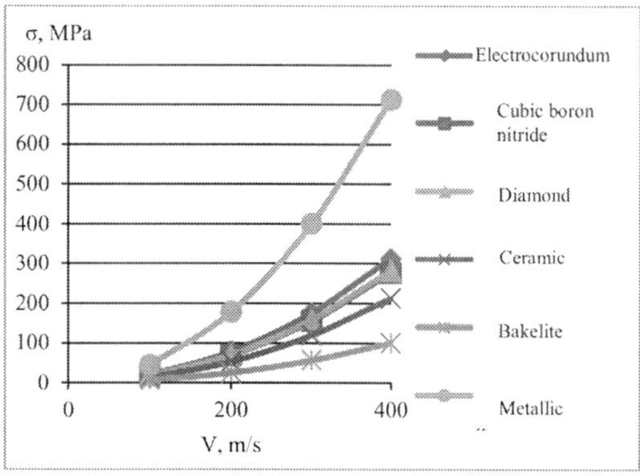

Fig. 2. Cut voltage dependence on grinding wheel circumferential speed.

It can be seen from the dependence that the bakelite ligament has shear stresses at a speed of 400 m/s about 100 MPa. Standard grinding inserts have a strength limit of about 30-40 MPa, so it is necessary to develop grinding inserts on the bakelite binder, providing the ability to work at speeds up to 400 m / s, while ensuring the appropriate shear strength.

According to the conducted tests, the bending strength limit for the given developed formulations is 120.2...153.8 MPa, which provides the coefficient of safety margin in the range of 1.2-1.5.

The number of grinding segments is determined by the heat dissipation condition during grinding and the effective power condition. When calculating the different number of heads involved in grinding, the working area must be determined. The working surface area of the diamond heads is 4 pcs., is 8% – S_{heads} =1257 mm^2 and S_{work} =15708 mm^2. Accordingly, at 16 pcs. – S_{work} = 62832 mm^2.

4 Determining the maximum permissible grinding wheel speed for high-speed machining

Grinding wheel calculation comes down to determining the maximum permissible cutting speed at which a grinding wheel is capable of machining. The maximum circumferential speed of the grinding wheel is influenced by the body material, dimensions and shape. The calculation is carried out for three proposed seating options: nozzle, striking face and end face (Fig. 3), based on the allowable tension on the tension of the material of the body of the circle [8-15].

Grinding wheels for rupture are calculated according to the allowable stress and the stress arising from centrifugal forces from the condition:

$$\sigma \leq [\sigma], \quad (1)$$

where σ – tensile stress arising during operation due to centrifugal forces, MPa; $[\sigma]$ – permissible tensile stress, MPa.

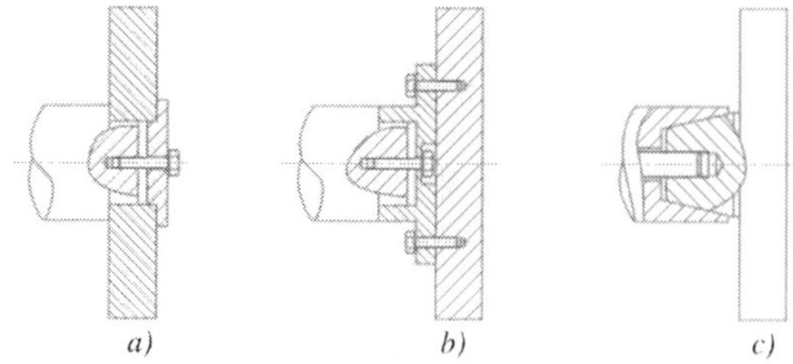

Fig. 3. Fitting options for an ultra-fast grinding wheel: a) nozzle mount; b) striking mount; c) end mount.

The calculation of tensile stresses is based on formulas:

$$\text{– for a disk} -\sigma = \frac{\rho \cdot V_r^2}{2}; \quad (2)$$

$$\text{–for a ring–}\ \sigma = \frac{R \cdot \rho \cdot V_r^2}{R-r}, \quad (3)$$

where ρ – ringing wheel body material density, kg/m^3; V_r – linear grinding wheel speed at the calculated radius, m/s; R – calculated grinding wheel radius, m; r – landing radius, m.

Based on the expressions (1, 2 and 3), the maximum allowed circumferential speed is calculated by the formulas:

$$\text{– for a disk} - V_r = \sqrt{\frac{2 \cdot [\sigma]}{\rho}};$$

$$\text{– for a ring} - V_r = \sqrt{\frac{(R-r) \cdot [\sigma]}{\rho \cdot R}}.$$

Figure 4 and Table 2 show the results of ho using calculations for different materials. Overall dimensions R = 125 mm; r = 38 mm.

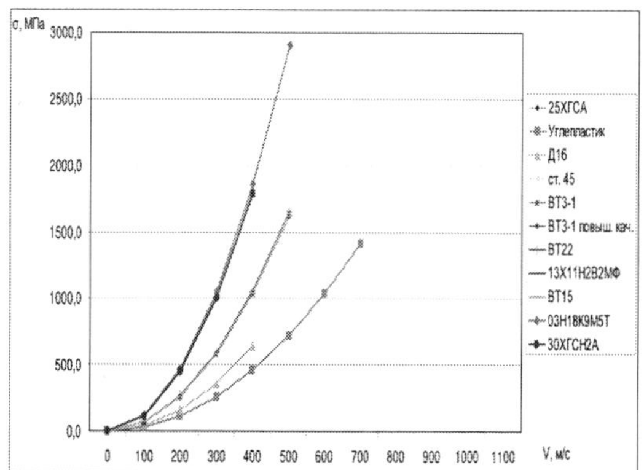

Fig. 4. Dependence of tensile stress on the circumferential speed of the grinding wheel in different materials.

Application of 30KhGSN2A or VT3-1 of the raised quality as the case of a grinding wheel allows to receive break rates to 667,8 and 720,2 m/s accordingly. Taking into account the safety factor of 1.4 we will get the maximum cutting speed of 484.2 and 543.0 m/s respectively.

 https://doi.org/10.1051/epjconf/202124804008

Table 2. Grinding wheel body calculation results from different materials.

№	Grinding wheel body material	Permissible tensile stress, [σ], MPa	Density, ρ, kg/m^3	Maximum allowable linear break rate, V_r, m/s		Maximum permissible operating speed with safety factor in mind, V_r, m/s	
				a	*b c*	*a*	*b c*
1.	**AlCu4Mg1**	460	2770	576,6	340,0	411,6	242,8
2.	**C45**	600	7800	392,2	231,4	280,2	165,3
3.	**13Kh11N2 VMF-Sh**	950	7800	493,5	291,2	352,5	208,0
4.	**25KhGSA**	1080	7850	524,6	309,4	374,7	221,0
5.	**VT3-1**	1180	4500	724,2	379,3	517,3	270,9
6.	**VT22**	1280	4620	746,0	440,1	532,9	314,3
7.	**VT15**	1500	4500	816,5	481,7	583,2	344,0
8.	**30KhGSN 2A**	1620	7770	645,7	380,9	461,2	272,1
9.	**03H18K9 M5T**	1920	8050	688,5	406,2	491,8	290,1

Titanium alloy VT3-1 is of higher quality in comparison with other materials and existing structures due to its lower density and high tensile strength. However, due to its density and correspondingly the final weight of the grinding wheel, the use of this material is limited by its vibration resistance, i. e. the design does not have sufficient vibration resistance. Therefore, the material 30KhGSN2A capable of providing the required cutting speeds at vibration disturbances was used [16-17].

5 Experimental results and discussion. Grinding wheel load and stress analysis

Calculation and distribution of stresses in a grinding wheel was made by a method of finite elements with application of calculation and analytical program ANSYS. The work presents the calculations of strength, static strength, elastic and elastic plastic calculations. The author considers two variants of arrangement of machining elements: horizontally and at an angle of 45 degrees and offers designs (figure 5 and 6). The arbor and spindle are also calculated.

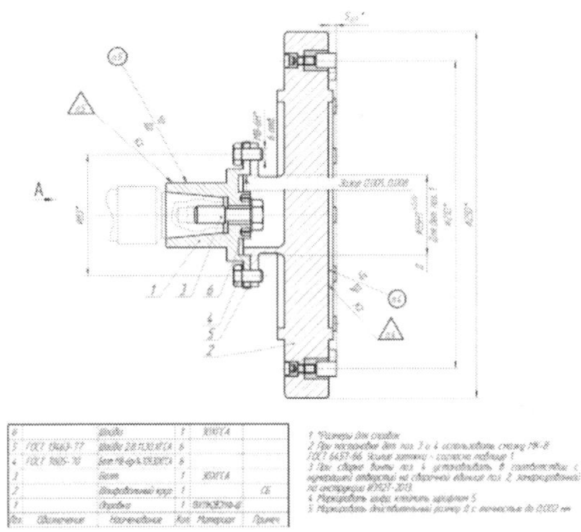

Fig. 5. Grinding wheel for ultra-fast grinding with horizontal machining units

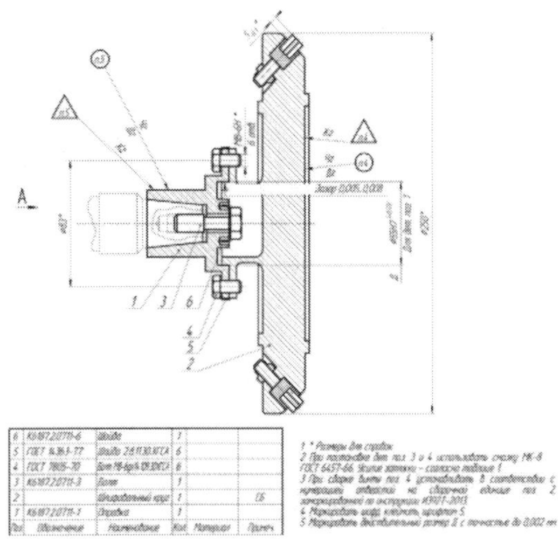

Fig. 6. Grinding wheel for ultra-fast grinding with 45 degree grinding angles

The developed high-speed grinding and grinding wheel has the following main parameters:

1) maximum circle diameter up to 250 mm;
2) maximum width up to 40 mm;
3) maximum working speed up to 400 m/s;
4) number of machining elements - 16 pieces;
5) two arrangement options for the machining elements.

The grinding wheel consists of the body and the machining elements. Body material - 13Kh11N2V2MF-Sh ($\sigma_{0,2}$ = 950 MPa). Properties of the material: E = 196200 MPa (20000 kgf/mm^2) – modulus of elasticity; μ = 0,3 – Poisson's ratio; σ_v = 1100 MPa – tensile strength; $\sigma_{0,2}$ = 950 MPa – yield strength; δ = 1,5% – percentage elongation; H = 1360 MPa – hardening module; ρ = 7,8 g/sm^3 – density.

The weight calculation of the grinding heads is shown in Table 3. With horizontal arrangement M = 21,718 g, with 45-degree machining position M = 29,663 g.

Table 3. Grinding head weight calculation.

Location head	Details	Material	Density ρ, g/sm³	D, mm	Dimensions d, mm	L, mm	Volume V, mm³	Mass M, g
Horizontal	Body	Abrasive	4,15	10	3	13	929,13	3,856
	Bucket	C45	7,8	14	10	8	603,19	4,705
			7,8	14	6	5,9	741,42	5,783
			7,8	9	0	2,1	133,6	1,042
			7,8	9	5	2,1	94,5	0,737
	Screw	13Kh11 N2V2M F-Sh 1	7,8	6	0	8	226,19	1,764
			7,8	4,5	0	4	63,617	0,496
			7,8	10	0	9	706,86	5,513
			7,8	8,4	0	3	-166,3	-1,297
			7,8	6	0	4	-113,1	-0,882
								21,718
At an angle 45 degrees	Body	Abrasive	4,15	10	3	13	929,13	3,856
	Bucket	C45	7,8	14	10	8	603,19	4,705
			7,8	14	0	6	923,63	7,204
			7,8	9	0	2	127,23	0,992
			7,8	9	5	2	90	0,702
			7,8	8	0	20	1005,3	7,841
	Nut		7,8	14	8	4,5	466,53	3,639
			7,8	9,5	8	4,5	92,775	0,724
								29,663

Centrifugal force of grinding heads with horizontal arrangement C = 30156 N (3074 kgf), with an angle of 45 degrees C = 32108 N (3273 kgf).

Housing with horizontal grinding heads

The volume solid model of the grinding wheel body is built. A sector (1/6 part) is cut out of the body. This sector is taken as a calculation model. The cyclo-symmetric model is fixed along the axis of rotation and in the circumferential direction[18-20].

Elements in the form of a tetrahedron are used to generate a finite element grid. The sector is divided into NEL = 1796237 finite elements and contains KN = 316998 nodes. Calculation of the stress-strain state (SSS) of the wheel is made from the action of centrifugal forces occurring in the body, and from the centrifugal forces of grinding heads C = 30156 N (3074 kgf). Operating speed n = 34725 min^{-1} (ω = 3636 rad/s). Figures 7-9 show the results of the case SSS under working conditions. The greatest stresses are in the grinding head holes and are local in nature. Equivalent voltages in them σ_e^{max} = 942 MPa. Equivalent plastic deformation in the holes $\varepsilon_p = 0,035\%$. Voltages in the astragal molding of the carrier pin fabric 875 MPa. Voltages in the hub 803 MPa.

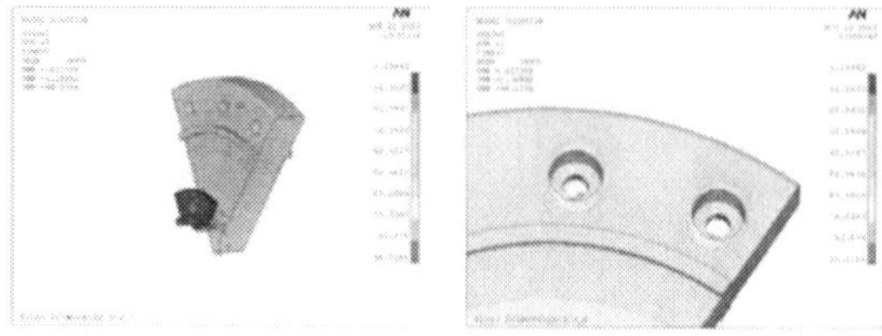

a) b)

Fig. 7. Equivalent stresses in sector (a) and grinding head holes (b).

Radial movements:
flange holes (83 mm diameter) UR = 0.0163 mm;
holes for grinding heads (105 mm diameter) UR = 0.184 mm.

Axial movements holes for grinding heads on the abrasive side (105 mm diameter) UZ = 0.065 mm. The housing is evenly stretched out in radial direction and narrowed in axial direction.

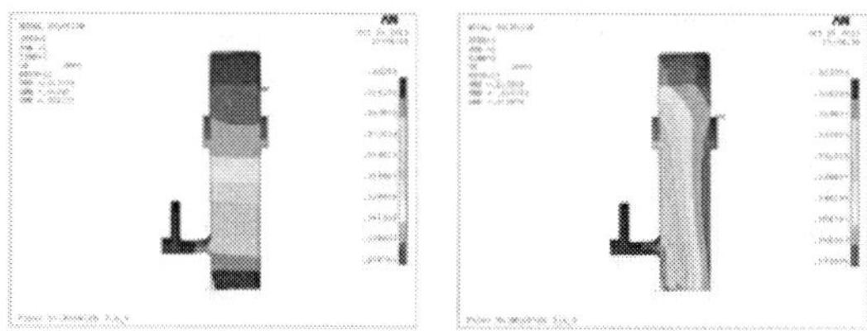

a) b)

Fig. 8. Radial movements (a) and axial movements (b).

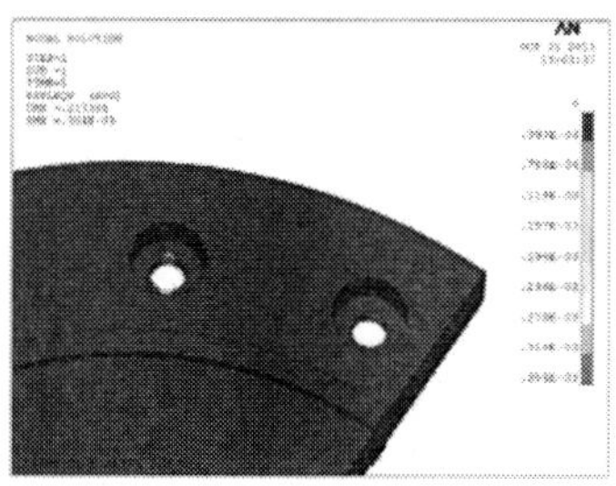

Fig. 9. Equivalent plastic deformation.

When calculated according to the two-dimensional scheme of the stress state with stress averaging on the thickness of the disk, the safety margin Km = 1.651 in the center of the disk. Allowance for destructive rotational speed Kv = 1.425. Stocks are sufficient.

Enclosure with angled grinding heads

The volume solid model of the grinding wheel body is built. A sector (1/16 part) is cut out of the body. This sector is taken as a calculation model. The cyclo-symmetric model is fixed along the axis of rotation and in the circumferential direction.

Elements in the form of a tetrahedron are used to generate a finite element grid. The sector is divided into NEL = 898532 finite elements and contains KN = 160671 node. Calculation of the stress-strain state (SSS) of the wheel is made from the action of centrifugal forces arising in the body and from the centrifugal forces of grinding heads C = 32108 N (3273 kgf). Operating speed n = 30558 min^{-1} (ω = 3200 rad/s). Figures 10-12 show the SSS on the hull under working conditions

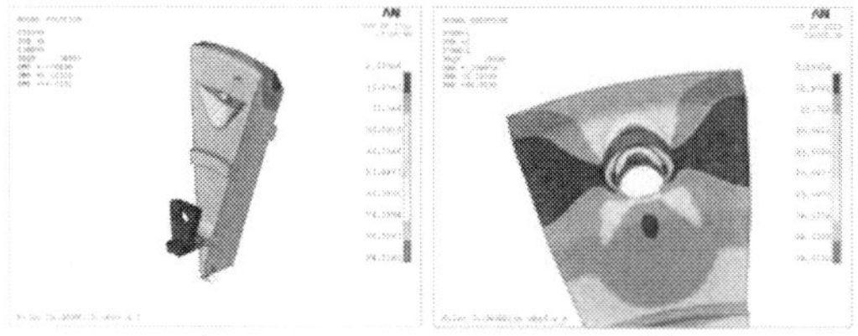

a) b)

Fig. 10. Equivalent tensions in sector (a) and into the grinding head opening (b).

a) b)

Fig. 11. Radial movements (a) and axial movements (b).

The greatest stresses are in the grinding head holes and are local in nature. Equivalent voltages in them σ_{ekq}^{max} = 948 MPa. Equivalent plastic deformation in the holes ε_p = 0,3%. Voltages in the astragal molding of the carrier pin fabric 722 MPa. Voltages in the hub 604 MPa.

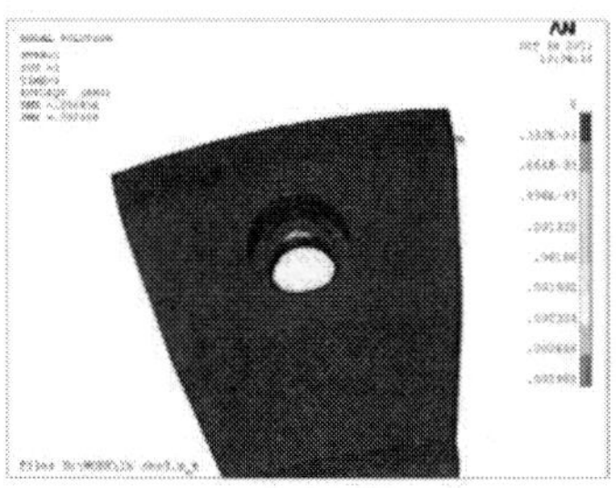

Fig. 12. Equivalent plastic deformation

Radial movements:
flange bores (83 mm diameter) U_R = 0,0137 mm;
grinding head bores on the abrasive side (105 mm diameter) U_R = 0,127 mm.
Axial movements:
grinding head holes on the abrasive side (105 mm diameter) U_Z = 0,07 mm.

The body in the central part is evenly stretched out radially and narrowed axially. The periphery of the casing bends to the abrasive side. When calculated according to the two-dimensional scheme of the stress state with stress averaging on the thickness of the disk, the safety margin Km = 2.132 in the center of the disk. Inventory at destructive speed Kv = 1.619.

Calculating mandrel and spindle

In order to assess the failure to open the joint between the spindle and mandrel, calculations of these elements from the action of centrifugal forces were made. Figures 13 and 14 show the equivalent stresses and radial movements. At the points of contact between the spindle and mandrel, the mandrel displacement is 3 μm, while the spindle displacement is 0.5 μ m. This results in a conicity of 1 : 5 to the axial displacement of the mandrel by δ = (3 - 0.5) ·5·2 = 25 μm.

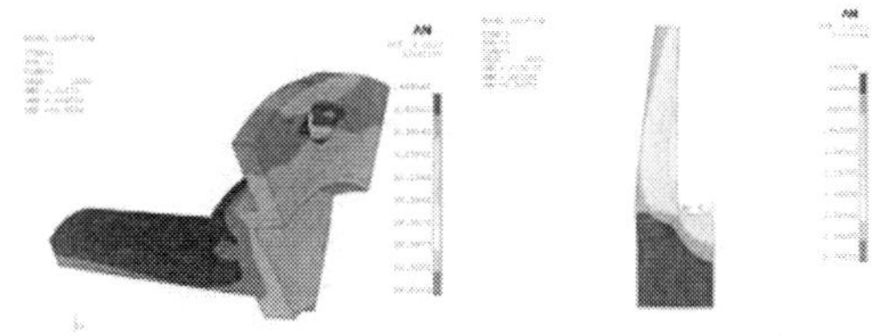

a) b)

Fig. 13. Equivalent stresses in mandrel (a) and spindle (b).

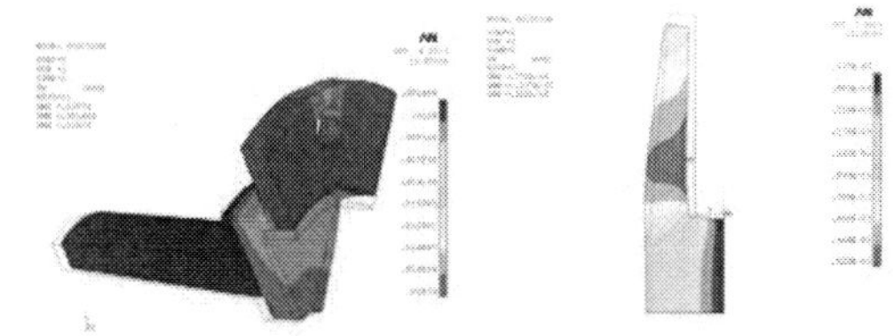

a) b)

Fig. 14. Radial movements in mandrel (a) and spindle (b).

This movement causes the tightening to be loosened. For this reason, a special flow screw has been designed to compensate for axial movement.

Radial displacement of flange bores (83 mm diameter) U_R = 0.017 mm, almost identical to the movements of the housing flange.

6 Conclusion

This work leads us to the following conclusions:

1. The strength of high-speed grinding wheels has been calculated. Two variants of arrangement of machining elements are considered: horizontally and at an angle of 45 degrees. The horizontal version of the grinding heads is preferable. The casing deforms uniformly without bending. Manufacturing this casing is less time-consuming.
2. The mandrel and spindle are calculated. According to calculations, a special screw with a flow was designed to compensate the axial movement of the mandrel relative to the spindle.
3. Radial movements in the area of the flange holes (83 mm in diameter) at the housing and mandrel are nearly the same.

Acknowledgments

The work was carried out with the financial support of the Grant Council of the President of the Russian Federation for state support of young Russian scientists and for state support of leading scientific schools of the Russian Federation, grant number MD-345.2020.8.

References

1. M.M. Paley, *Grinding and grinding technology of the cutting tool / M.M. Paley, L.G. Dibner, M.D. Flyd* (M.: Mashinostroenie, 1988)
2. Pat. 101666 Russian Federation, MPK B 24 D 7/06. Grinding w heel f or high-speed m achining / D.S. Rechenko, K.V. Averkov, A.Yu. Popov; applicant an d o wner o f O mGTU - № 2010139548/02; de clared. 24. 09.2010; o p. 27.01.2011, Bulletin No. 3. – 2 c.: silt.
3. Pat. 1 46505 R ussian F ederation, I PC B 24D 3/32. Grinding to ol / D.S. Rechenko, B.P. Kudryashov, A.Yu. Popov; applicant and owner of OmGTU - № 2014123264/02; de clared. 06. 06.2014; o p. 10.10.2014, Bulletin No. 28. – 1 c.: silt.
4. Pat. 2440229 Russian Federation, IPC B24B 1/00, B24B 9/ 16, B 28D 5/ 02. Method of s uperhard materials p rocessing / D.S. Rechenko, A.Yu. Popov. 08.02.2010; op. 20.01.2012, Bulletin No. 2. – 1 c.
5. Pat. 2547980 Russian Federation, IPC B24B 3/36. Method of sharpening the metal-cutting tool blade with a gr inding wheel / D.S. Rechenko, A.Yu. Popov; applicant and owner of OmGTU - № 2013142129/02; de clared. 13. 09.2013; op. 10.04.2015, Bulletin No. 10. – 1 c.
6. Pat. 55665 Russian Federation, MPK B24D 17/00. Grinding wheel f or di amond a nd a brasive machining / D.S. Rechenko, Yu.R. Nurtdinov, A.Yu. Popov; applicant and owner of OmGTU - № 2006111080/22; de clared. 05.04.2006; o p. c it. 27.08.2006, Bulletin No. 24. - 2 p.: silt.
7. D.S. Rechenko, *Improvement of quality of the high-speed sharpening of the carbide tools by the diamond circles with the interrupted surface*: (Dissertation ... Cand. of Sciences, Tomsk, 2009)
8. V.A. Rybitskiy, S ynthetic diamonds., **3**, 32–35 (1978)
9. L.N. Filimonov, *High-speed grinding / L.N. Filimonov. - L.N* (Filimonov: M ashinostroenie, Leningr. departing, 1979)
10. L.N. Filimonov, *Grinding wheel resistance / L.N. Filimonov. - L.N* (Filimonov: M ashinostroenie, 1973)
11. D.S. R echenko, A .G. K ol'tsov, Russian Engineering Research, **32**(2), 179 – 181 (2012)
12. D.S. Rechenko, A.Y. Popov, K.V. Averkov, V.A. Sergeev, Russian Engineering Research, **32**(5-6), 511 – 512 (2012)
13. L. T ian, Y . F ub, J . X ub, H . Li , W . D ing, International J ournal o f Machine T ools & Manufacture, **89**, 192–201 (2015)
14. C. W ang, Q. F ang, J . C hen, Y . Li u, T . J in, International J ournal o f Advanced M anufacturing Technology, **83**, 937–948 (2016)
15. L. Y ang, Y . F u, J . X u, Y . Liu, M aterials a nd Design, **88**, 827–836 (2015)
16. Z. Zhao, Z. Fu, J. Xu, Z. Zhang, The International Journal o f Advanced M anufacturing T echnology, **87**, 3545–3555 (2016)
17. Y. Zhu, W. Lu, Y. Sun, D. Zuo, The International Journal o f Advanced M anufacturing T echnology, **89**, 1269–1277 (2017)
18. S.N. Grigoriev, V.A. Grechishnikov, M.A. Volosova, X . Jiang, P .M. Pivkin, In IOP Conference S eries: M aterials S cience a nd Engineering, **971**(2), 022065 (2020) doi: 10.1088/1757-899X/971/2/022065
19. S.N. Grigoriev, P.M. Pivkin, V .A. Grechishnikov, Y.E. Petukhov, M .A. Volosova, A.B. Nadykto, Emerging I maging a nd S ensing T echnologies for Security a nd D efence V; an d A dvanced Manufacturing Technologies for Micro-and Nanosystems i n S ecurity a nd D efence I II, **11540**, 115401E (2020) https://doi.org/10.1117/12.2574389
20. S.N. Grigoriev, M.A. Volosova, A. A. Okunkova, S.V. Fedorov, K . Hamdy, P .A. Podrabinnik, ... , A.N. Porvatov, J. Manuf. Mater. Process., **4**(3), 96 (2020) https://doi.org/10.3390/jmmp4030096

Model of the Ultimate Stress State of the Coated Tool Cutter Material

Aleksei Osinnii[1,*], *Jury* Bublikov[2], *Anatoli* Chigarev[3], *Anna* Okun'kova[1], *Nataliya* Kapustina[1]

[1]Moscow State Technological University "STANKIN", RU-127055, Moscow, Russia.
[2]IDTI RAS, Moscow, Russia.
[3]Belarusian State University, BL-220030, Minsk, Belarus.

Abstract. The article proposes a model of the ultimate stress state of the material of the coated tool cutter. It is found that with an increase in the fracture toughness of a tool in connection with the material ductility the machining accuracy deteriorates due to arising elastic-plastic vibrations of the tool cutter. In case when no ultimate stress state is reached, that is, a tool operates in the elastic region, then an alternating stress distribution diagram is realized for the tool cutter at the beam approximation. Therefore, in addition to the frictional vibrations, arising from the interaction between the tool cutter and a workpiece, the elastic vibrations can arise, which affects the machining accuracy and the service life of the coated tool cutter. The use of coatings makes it possible not only to increase the wear resistance of cutting tools, but also to transform the stress distribution diagrams of the normal σ_N and tangential τ_γ contact stresses acting on the rake face of the cutting tool. In particular, it is possible to control the length of the total contact area between the chips and the tool rake face.

1 Introduction

The study of the ultimate stress state of a tool cutter is an important task aimed at predicting the service life of a coated tool cutter and its fracture pattern [1]. The deposition of wear-resistant coatings is one of the prevailing methods for improving the performance properties of cutting tools [2, 3]. Accordingly, the predictive modeling of the cutting properties of coated tools is an important scientific and practical challenge. Various approaches are being used to find a solution to the above challenge. In particular, the method of finite element model (FEM) is being increasingly used [4-12], as well as the shear-lag model [13] and the discrete element method (DEM) [14]. The number of other methods includes the Bower and Fleck's equation and the Hamilton's equation [15] and the microcrack formation theory [16]. In [17,18], the studies were focused on the application of the methods based on the linear elastic fracture mechanics (LEFM). The issues concerning the influence of various coating defects (in particular, microdroplets) on the general working efficiency of a cutting tool were considered separately [19-24]. The probabilistic methods that take into account the stochastic nature of the deposition process are often used to simulate the process of deposition of coatings on metal-cutting tools [25-31]. Moreover, the cutting process is also a stochastic process that requires an appropriate approach during the modeling. In recent years, the wear-resistant coatings with new compositions and architectures are being actively developed. In particular, the multilayered coatings are being widely used due to their contribution to the additional significant improvement in the working efficiency of the metal-cutting tools [32-37]. At the same time, the methods of coating deposition are being improved to control the properties of the coatings by creating the systems of complex composition and architecture [38-47]. The wear-resistant coatings of new generation are able to change significantly the cutting conditions, and this fact should be taken into account during the modeling of the state of a tool cutter [48-70]. The aim of this study is to create a model of the ultimate stress state of the material of the coated tool cutter, taking into account the stochastic nature of loading during the cutting process and the effect of the multilayered wear-resistant coatings of new generation.

2 Developing a model for the ultimate stress state of the material of coated tool cutter

Let us consider the stress state of the tool cutter in the plane of the rake face (Fig. 1): the plane of *BXY*, which is affected by the average contact specific loads that create the distributed pressure *P*, demonstrates a diagram shown in Fig. 1 b.

Then, in the plane of *BXY*, a plane stress state is realized in the quadrangle of *BACE*. Here, the region of *ARB* means an absolutely rigid body, that is, a region without plasticity. In fact, this region has linear

* Corresponding author: dr.a.veres@yandex.ru

elasticity, and in order to simplify the model, it will be assumed to be the *ARB*. In case when the ultimate stress state is reached, the stresses of σ_x, σ_y, τ_{xy} will satisfy the equations of quasi-static equilibrium.

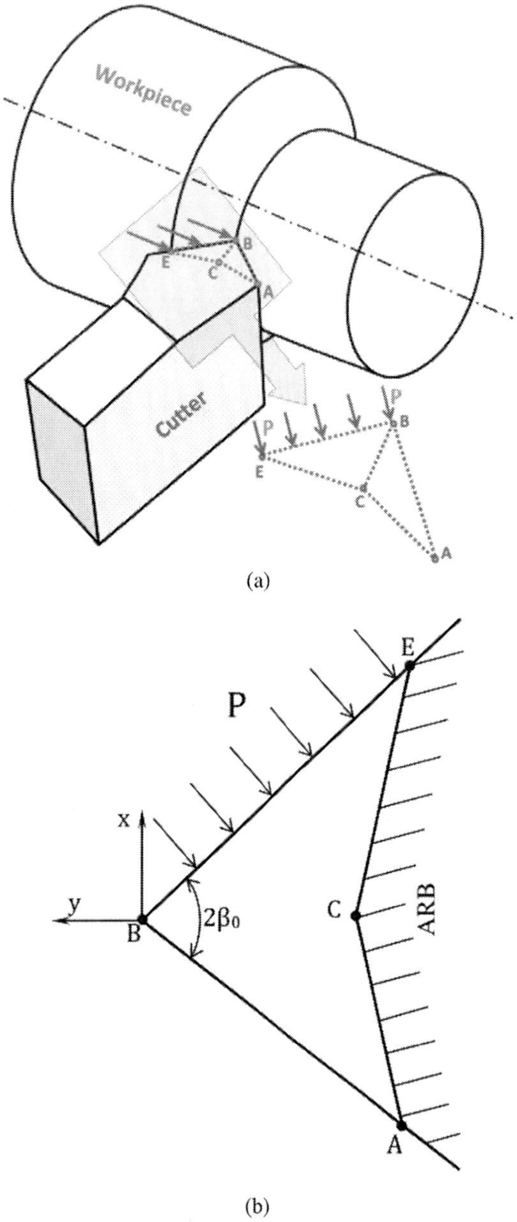

Fig. 1. Location of the region under consideration: (a) based on the cutting pattern and (b) in the plane of *BXY*, in which the pressure *P* is distributed. *ARB* is the absolutely rigid body, the angle $2\beta_0 = \varepsilon$ (apex angle).

$$\frac{\partial\sigma_x}{\partial x} + \frac{\partial\tau_{xy}}{\partial y} = 0$$
$$\frac{\partial\tau_{xy}}{\partial x} + \frac{\partial\sigma_y}{\partial y} = 0 \quad (1)$$

and the condition of ductility:

$$\left(\sigma_x - \sigma_y\right)^2 + 4\tau_{xy}^2 = 4k^2 \quad (2)$$

The rates of plastic deformations determined by the relationships are as follows:

$$\varepsilon_x = \frac{\partial v_x}{\partial x}, \varepsilon_y = \frac{\partial v_y}{\partial y},$$
$$\varepsilon_{xy} = \frac{\partial v_x}{\partial y} + \frac{\partial v_y}{\partial x}, \quad (3)$$
$$v_x = \frac{\partial u}{\partial t}, v_y = \frac{\partial v}{\partial t}$$

where *u*, υ are the displacements along the *x*, *y* axis.

The conditions for isotropy of the plastic state and incompressibility of the material are as follows, respectively:

$$\frac{\varepsilon_x - \varepsilon_y}{\sigma_x - \sigma_y} = \frac{\varepsilon_{xy}}{\tau_{xy}} = \lambda \geq 0 \quad (4)$$

$$\varepsilon_x + \varepsilon_y = 0 \quad (5)$$

The solution to the problem in Cartesian coordinates *x*, *y* can be simplified by taking the coordinates α, β, when the coordinate lines are directed along the lines of the maximum displacements, that is, along the slip curves. As is known, a slip curve is a line, each point of which touches the area of the maximum tangential stress. In the coordinates α, β, the relationships for the rates of deformation are simplified (Fig. 2).

$$\varepsilon_\alpha = \varepsilon_\beta = 0, \varepsilon_{\alpha\beta} \geq 0 \quad (6)$$

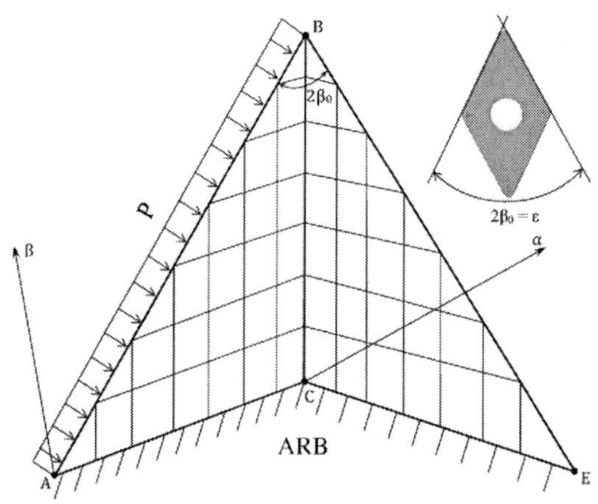

Fig. 2. Field of material slip curves in the region of plastic deformation of the cutting wedge of the tool, when $\varepsilon = 2\beta_0 \leq \pi/2$.

Figure 2 exhibits the field of material slip curves in the region of plastic deformation of the tool, for the case when $\beta_0 \leq \pi/4$, when the stress fields in the area of *ABC* with the loaded boundary of *AB* and in the area of *CBE* with the free boundary of *BE* are uniform, but different, and, therefore, the stresses on the line of *BC* are discontinuous. In case when $\beta_0 \leq \pi/4$, the grid of slip curves has the shape depicted in Fig. 3.

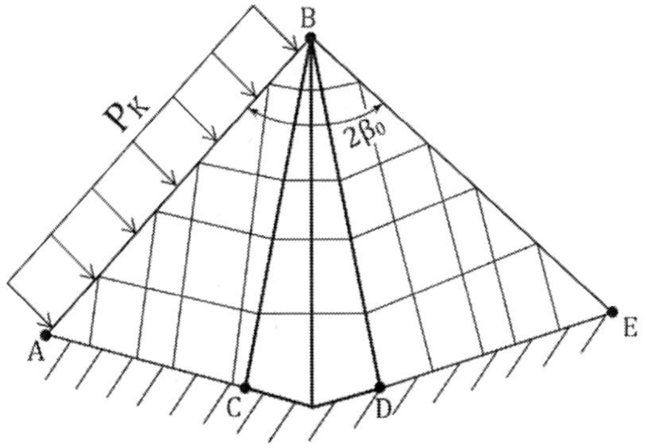

Fig. 3. Grid of slip curves, when $\beta_0 \leq \pi/2$.

The ultimate pressure, corresponding to the field of slip curves and the continuous stress field in Fig. 3, can be calculated as follows:

$$P_K = 2k\left[1 + \left(2\beta_0 - \frac{\pi}{2}\right)\right] \quad (7)$$

When the area of the angle, where the slip curves form a fan-shaped pattern, tends to zero, we get the slip curve geometry in Fig. 3, which realization requires the ultimate pressure equal to:

$$P_K^{(1)} = 2k\left[1 + sin\left(2\beta_0 - \frac{\pi}{2}\right)\right] \quad (8)$$

It should be noted that this solution also satisfies all the conditions for the stresses when $\beta_0 \geq \pi/4$, if the grid of slip curves looks like depicted in Fig. 4.

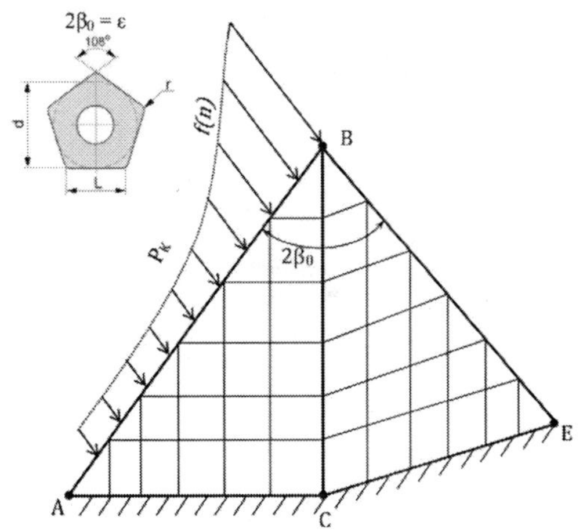

Fig. 4. Grid of slip lines, when $\varepsilon = 2\beta_0 \geq \pi/2$.

Let us now consider fields of the plastic flow velocities in accordance with the field of slip curves depicted in Fig. 4. For the coordinates α, β, the boundary of AB is set by the equation of $\alpha = \beta = \eta$. Then the plastic deformation rate for the boundary of AB in a normal plane is $f(\eta)$. Let us denote the velocity components in the directions of α_1, β as u, v, respectively, then at the discontinuity line of BC the velocities are equal to zero. The conditions (6) are as follows:

$$\frac{\partial u}{\partial \alpha} = \frac{\partial v}{\partial \beta} = 0 \quad (9)$$

The solutions (9) will be written as:

$$u = u(\beta), \quad v = v(\alpha) \quad (10)$$

Taking into account (1), it is possible to build the field of the plastic flow velocities for the area of ABC. For simplicity, let us assume that $AC = 1$ (Fig. 5).

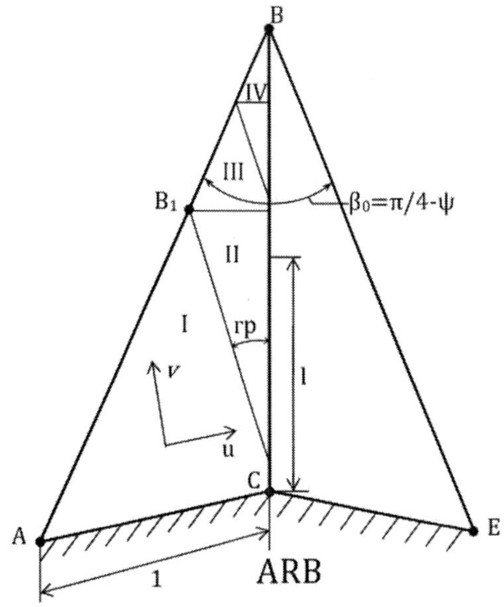

Fig. 5. Field of plastic flow velocities for the area of ABC at $AC = 1$.

At the boundary of AC of the plastic zone at the ARB, $v = 0$ and thus $v = 0$ in the area I ACB_1. The solution can be written as

$$v_I = 0 \quad (11)$$

The directions of α, β are inclined at the angle π/η to AB, and the boundary velocities at AB are expressed analytically in the following form:

$$f(\eta) = u(\eta) - v(\eta) \quad (12)$$

While constructing the solutions for velocities, the equations (12) are usually required to determine $u(\beta)$ in the region, in which v is already known, for example, in the area I under consideration. The equation (12) can be written as follows:

$$u(\beta) = \sqrt{2}f(\beta) + v(\beta) \quad (13)$$

While considering the section of the AB border in the area I from the equation (13), let us define $u(\beta)$ along this border to be used on the boundary of the areas I and II, and we get:

$$u_I = u_{II} = \sqrt{2}f(\beta) \tag{14}$$

Based on the condition at the discontinuity boundary of BC, it can be found that the velocity components in the direction of CB will be equal to zero, which can be written as follows:

$$u\, sin\, \psi + v\, cos\, \psi = 0 \tag{15}$$

where ψ is the angle between the direction of CB and the increasing direction of β.

The equation of the BC line is:

$$\beta = \frac{\alpha-1}{\tan\psi} \tag{16}$$

Taking into account (16), the condition (15) is as follows:

$$v(\alpha) = -tu\left(\frac{\alpha-1}{t}\right), t = tan\, \psi \tag{17}$$

By applying the equation (17) to the section of BC in the area II, we get:

$$v_{II} = v_{III} = -t\sqrt{2}f\left(\frac{\alpha-1}{t}\right) \tag{18}$$

When continuing this process using the equations (13) and (17), we get solutions in the form of the following series:

$$\begin{gathered} u = \sqrt{2}f(\beta) - t\left[\sqrt{2}f\left(\frac{\beta-1}{t}\right) - t\left\{\sqrt{2}f\frac{\beta-1/t-1}{t} - \cdots\right\}\right] \\ v = -t\left[\sqrt{2}f\left(\frac{\alpha-1}{t}\right) - t\left\{\sqrt{2}f\frac{\alpha-1/t-1}{t} - \cdots\right\}\right] \end{gathered} \tag{19}$$

At the point B, it follows from the equation (19) that:

$$\begin{gathered} u = \sqrt{2}f_B\sum_{i=0}^{\infty}(-1)^i t^i = \frac{\sqrt{2}f_B}{1+t} \\ v = -\sqrt{2}f_B\sum_{i=1}^{\infty}(-1)^{i+1}t^i = -\frac{\sqrt{2}f_B t}{1+t} \end{gathered} \tag{20}$$

By introducing $t = \tan\psi$ into (20) and given that $\frac{\sqrt{2}}{(1+t)\cos\psi} = \frac{1}{\cos\left(\frac{\pi}{n}-\psi\right)} = \frac{1}{\cos\beta}$ we get that the limit of the velocity component at the point B will be equal to zero for the velocity component, parallel to BC and $f_B/\cos\beta_0$ for the velocity component, not perpendicular to B. Therefore, the plastic flow velocity in the plastic region defines the velocity of the elastic line of BC in the point B, which is determined uniquely from the condition for the plastic flow velocity on BC and finds the component of the plastic flow velocity at the point B, normal to AB.

Due to the symmetry, it is clear that the plastic flow velocity field, obtained for the area of ABC, can be also applied to the area of CDE, thus finding the distribution to EB, similar to the distribution of normal velocity to AB.

Before the solution may be distributed, it is necessary to check that the constraints on $f(\eta)$ under the condition of $\varepsilon_{\alpha\beta} \geq 0$ satisfy the equation (6) so that the plastic flow velocity field corresponds to the stress field.

The mechanical interpretation concerning the discontinuity line of BC can be obtained from the concept of the elastic-plastic beam deformation. In the given case, the area of ABC can be considered as a beam of variable cross-section, embedded along ACE and loaded along AB by a uniformly distributed load. This model is well suited for small values of β_0.

For simplicity, let us assume that the area of ABC is rectangular, then the stress distribution diagram in the elastic state looks like ***a*** (Fig. 6), ***b*** – in elastic-plastic state, or ***c*** – when some of the areas are in the plastic state.

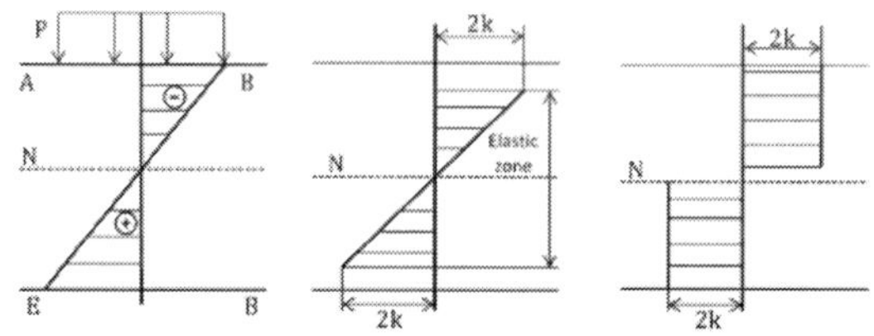

Fig. 6. Stress distribution diagrams in the elastic state.

The neutral line is a line of discontinuity in the plastic stresses. The relationship between the velocity $f(\eta)$ and η looks as follows:

$$\begin{gathered} f(\eta) = \frac{1}{\sqrt{2}}\varphi\left(\frac{\eta}{cos\,\psi}\right)cos\,\psi + \varphi\left(\frac{\eta-1}{sin\,\psi}\right)sin\,\psi\, H(\eta-1) \\ u = \varphi\left(\frac{\beta}{\cos\psi}\right)\cos\psi, v = -\varphi\left(\frac{\alpha-1}{\sin\psi}\right)\sin\varphi \\ \varepsilon_{\alpha\beta} = \varphi'\left(\frac{\beta}{\cos\psi}\right) - \varphi'\left(\frac{\alpha-1}{\sin\psi}\right) \end{gathered} \tag{21}$$

$$\varphi(v) = 0, \varphi'(\ell) \geq 0$$

In case of a tool with the angle of the $\frac{\pi}{2}$ line, the field of plastic deformations and deformation rates increases towards the cutting edge. The stresses change the sign when crossing the neutral line.

3 Conclusions

Thus, with an increase in the fracture toughness of a tool through the material ductility the machining accuracy deteriorates due to the arising elastic-plastic vibrations of the tool cutter. When no ultimate stress state is reached, that is, a cutting tool operates in the elastic region, then, according to Fig. 6, an alternating stress distribution diagram is realized for the tool cutter at the beam approximation. Therefore, in addition to the frictional vibrations, arising from the interaction between the tool cutter and a workpiece, the elastic vibrations may occur, which affects the machining accuracy and the service life of the tool cutter. Due to the application of a multilayered structure of the coatings, the above phenomena can be leveled. The use of coatings makes it possible not only to increase the wear resistance of cutting tools, but also to transform the stress distribution diagrams of the normal σ_N and tangential τ_γ contact stresses acting on the rake face of the cutting tool. In particular, it is possible to control the length of the total contact area between the chips and the tool rake face. Due to the above, it is possible to achieve a decrease in the σ_N and τ_γ contact stresses, improve the heat removal from the cutting zone, increase the thermal conductivity of the surface layers of the tool (the length of the plastic contact $C_{\gamma P}$ increases), which, in turn, reduces the specific thermomechanical loads on the cutting edge.

Acknowledgments

This research was funded by Ministry of Science and Higher Education of the Russian Federation, Grant No. 0707-2020-0025. This work was carried out using equipment provided by the Center of Collective Use of MSUT "STANKIN".

References

1. V. Astakhov, J. Davim, *Tools (Geometry and Material) and Tool Wear* (Machining. Springer, 2008)
2. A.A. Vereschaka, S.N. Grigoriev, N.N. Sitnikov, A.D. Batako, Wear, **390-391**, 209, (2017)
3. A. Vereschaka, V. Tabakov, S. Grigoriev, N. Sitnikov, G. Oganyan, N. Andreev, F. Milovich, Wear, **420–421**, 17–37, (2019)
4. P. Bansal, P.H. Shipway, S.B. Leen, Surf.Coat. Technol., **200**(18-19), 5318, (2006)
5. M. Bäker, Comput. Mater, Sci., **64**, 79, (2012)
6. Z. Xia, W.A. Curtin, B.W. Sheldon, Acta Mater, **52**, 3507, (2004)
7. J.H. Lee, Y.F. Gao, K.E. Johanns, G.M. Pharr, Acta Mater, **60**, 5448, (2012)
8. H.C. Hyun, F. Rickhey, J.H. Lee, M. Kim, H. Lee, Eng. Fract. Mech., **134**, 304, 2015)
9. K. Holmberg, A. Laukkanen, H. Ronkainen, K. TribolInt, **38**, 1035, (2005)
10. T.L. Anderson, *Fracture Mechanics: Fundamentals and Applications* (CRC Press, Boca Raton, 1994)
11. K. Holmberg, A. Laukkanen, E. Turunen, T. Laitinen, Surf.Coat. Technol., **247**, 1-13, (2014)
12. A.L. Mohd Tobi, P.H. Shipway, S.B. Leen, Tribol Int, **58**, 29, (2013)
13. K. Wang, F. Zhang, R.K. Bordia, Materials, **11**(4), 497, (2018)
14. N. Ferguen, Y. Mebdoua-Lahmar, H. Lahmar, W. Leclerc, M. Guessasma. Surf. Coat. Technol., **371**, 287, (2019)
15. M. Białas, Surf. Coat. Technol., **202(24)**, 6002, (2008)
16. Y. Gu, K. Chen, R. Liu, M.X. Yao, R. Collier, Surf. Coat. Technol., **309**, 536, (2017)
17. X.-F. Wu, R.A. Jenson, Int. J. Eng. Sci., **49**(3), 279, (2011)
18. X.-F. Wu, R.A. Jenson, Y. Zhao, Mech. Mater., **69**(1), 195, (2014)
19. E. Harry, M. Ignat, Y. Pauleau, A. Rouzaud, P. Juliet. Surf. Coat. Technol., **125**, 185, (2000)
20. L. Aihua, D. Jianxin, C. Haibing, C. Yangyang, Z. Jun. Int. J. Refract. Hard. Met., **31**, 82, (2012)
21. A. Vereschaka, V. Tabakov, S. Grigoriev, N. Sitnikov, F. Milovich, N. Andreev, J. Bublikov, Wear, **438**, 203069, (2019)
22. A. Vereschaka, S. Grigoriev, N. Sitnikov, F. Milovich, A. Aksenenko, N. Andreev, Int. J. Adv. Manuf. Technol., **102**(9-12), 2953, (2019)
23. V. Alexey, Key. Eng. Mater., **581**, 62, (2014)
24. M. Antonov, I. Hussainova, F. Sergejev, P. Kulu, A. Gregor, Wear, **267**, 898, (2009)
25. M. Grujicic, S.G. Lai, J. Mater. Sci., **36**(12), 2937, (2001)
26. C.K. Birdsall, PIC-MCC., IEEE Trans. Plasma Sci., **19**(2), 65, (1991)
27. D.T.K. Kwok, C. Cornet, IEEE Trans. Plasma Sci., **34**, 2434, (2006)
28. E. Bultinck, A. Bogaerts, New J. Phys., **11**, 103010, (2009)
29. M. Siemers, A. Pflug, T. Melzig, K. Gehrke, A. Weimar, B. Szyszka, Surf. Coat. Technol., **241**, 50, (2014)
30. H. Cansizoglu, M. Yurukcu, M.F. Cansizoglu, T. Karabacak, Thin Solid Films, **583**(1), 122, (2015)
31. S. Kumar, W.A. Curtin, Materials Today, **10**(9), 34, (2007)
32. G. Skordaris, K.-D. Bouzakis, P. Charalampous, Surf. Coat. Technol., **265**, 53, (2015)
33. B.D. Beake, L. Ning, Ch. Gey, S.C. Veldhuis, A. Komarov, A. Weaver, M. Khanna, G.S. Fox-Rabinovich, Surf. Coat. Technol., **279**, 118, (2015)
34. L. Ning, S.C. Veldhuis, K. Yamamoto, Int. J. Mach. Tools Manuf., **48**, 656, (2008)
35. A. Vereschaka, V. Tabakov, S. Grigoriev, A. Aksenenko, N. Sitnikov, G. Oganyan, A. Seleznev, S. Shevchenko, Surf. Coat. Technol., **357**, 218, 2019)
36. A. Vereschaka, V. Tabakov, S. Grigoriev, N. Sitnikov, N. Andreev, F. Milovich, Wear, **416**, 772, (2018)
37. B.D. Beake, G.S. Fox-Rabinovich, Surf. Coat. Technol., **255**, 102, (2014)
38. S.N. Grigoriev, M.A. Volosova, A.A. Vereschaka, N.N. Sitnikov, F. Milovich, J.I. Bublikov, S.V. Fyodorov, A.E. Seleznev, Ceram. Int., **46**(11), 1824, (2020)
39. W. Kalss, A. Reiter, V. Derflinger, C. Gey, J.L. Endrino, Int. J. Refractory Met. Hard Mater., **24**(5), 399, (2006)
40. K. Bobzin, CIRP J. Manuf. Sci. Technol., **18**, 1, (2017)
41. A.A. Vereshchaka, A.S. Vereshchaka, O. Mgaloblishvili, M.N. Morgan, A.D. Batako, Int. J. Adv. Manuf. Tech., **72**(1-4), 303, (2014)

42. S. Grigoriev, Y. Melnik, A. Metel, Surf. Coat. Technol., **156**(1–3), 44, (2002)
43. O.V. Sobol', A.A. Andreev, S.N. Grigoriev, V.A. Stolbovoy, Probl. At. Sci. Technol., **4**, 174, (2011)
44. V.V. Kuzin, S.N. Grigor'ev, M.A. Volosova, Refract. Ind. Ceram., **54**, 376, (2014)
45. V.V. Kuzin, S.N. Grigoriev, M.Yu. Fedorov, J. Frict Wear, **36**(1), 40, (2015)
46. A.S. Metel, S.N. Grigoriev, Yu.A. Melnik, V.P. Bolbukov, Instrum. Exp. Tech., **55**(1), 122, (2012)
47. A.S. Vereschaka, S.N. Grigoriev, V.P. Tabakov, E.S. Sotova, A.A. Vereschaka, M.Y. Kulikov, Key Eng. Mater., **581**, 68, (2014)
48. S.N. Grigoriev, O.V. Sobol, V.M. Beresnev, I.V. Serdyuk, A.D. Pogrebnyak, D.A. Kolesnikov, U.S. Nemchenko, J. Frict. Wear, **35**(5), 359, (2014)
49. A. Metel, V. Bolbukov, M. Volosova, S. Grigoriev, Y. Melnik, Surf. Coat. Technol., **225**, 34, (2013)
50. A.S. Vereschaka, S.N. Grigoriev, E.S. Sotova, A.A. Vereschaka, Adv Mat Res, **712-715**, 391, (2013)
51. A.A. Vereschaka, A.S. Vereschaka, S.N. Grigoriev, A.K. Kirillov, O. Khaustova, Procedia CIRP, **7**, 311, (2013)
52. A.A. Vereschaka, S.N. Grigoriev, N.N. Sitnikov, G.V. Oganyan, A. Batako, Surf.Coat. Technol., **332**, 198, (2017)
53. M.A. Volosova, S.N. Grigor'ev, V.V. Kuzin, Refract Ind Ceram, **56**, 91, (2015)
54. A. Vereschaka, V. Tabakov, S. Grigoriev, N. Sitnikov, F. Milovich, N. Andreev, C. Sotova, N. Kutina, Surf. Coat. Technol., **385**, 125402, (2020)
55. 55. A. Vereschaka, A. Aksenenko, N. Sitnikov, M. Migranov, S. Shevchenko, C. Sotova, A. Batako, N. Andreev, Tribol. Int. **128,** 313 (2018)
56. 56. V. Alexey, Adv Mat Res, **712-715**, 347, (2013)
57. 57. J. Gradišek, I. Grabec, S. Siegert, R. Friedrich, Mech Syst Signal Process, **16**(5), 831, (2002)
58. 58. A. Vereschaka, M. Volosova, A. Chigarev, N. Sitnikov, A. Ashmarin, C. Sotova, J. Bublikov, D. Lytkin, Coatings, **10**(1), 63, (2020)
59. 59. S. Grigoriev, A. Vereschaka, F. Milovich, V. Tabakov, N. Sitnikov, N. Andreev, T. Sviridova, J. Bublikov, Surf. Coat. Technol., **401**, 126258, (2020)
60. M.A. Volosova, S.N. Grigor'ev, V.V. Kuzin, Refract. Ind. Ceram., **55**, 565, (2015)
61. M. Volosova, S. Grigoriev, A.Metel, A. Shein, *Coatings, 8,* 287, (2018)
62. M.A. Volosova, S.N. Grigor'ev, V.V. Kuzin, Refract. Ind. Ceram., **56**, 197, (2015)
63. M.A. Volosova, S.N. Grigor'ev, V.V. Kuzin, Refract. Ind. Ceram., **55**, 487, (2015)
64. V.V. Kuzin, S.N. Grigoriev, M.A. Volosova, J. Frict. Wear, **35**, 505, (2014)
65. M.A. Volosova, S.N. Grigoriev, E.A. Ostrikov, Mech. Ind., **17**, 720, (2016)
66. Kunarev, E.A., Makarova, T.A., Lyubomudrov, S.A., Russ. Eng. Res., **41(2)**, 175, (2021)
67. Meurer, M., Augspurger, T., Tekkaya, B., Lima, A.P., Bergs, T. Procedia CIRP, **87**, 444 (2020)
68. Ma, W., Shuang, F. Int. J. Adv. Manuf. Technol., **105**(5-6), 1933 (2019)
69. Komvopoulos, K., Erpenbeck, S.A. Journal of engineering for industry, **113**(3), 253 (1991)
70. Xie, L.-J., Schmidt, J., Schmidt, C., Biesinger, F. Wear, **258**(10), 1479 (2005)

Reliability Calculation for a Multilayered Coating Based on a Model with Independently (Sequentially) Degrading Layers

Aleksei Osinnii[1,*], *Anatoli* Chigarev[2], *Jury* Bublikov[3], *Mikhail* Mosyanov[1], *Anton* Kekalo[1]

[1]Moscow State Technological University "STANKIN", RU-127055, Moscow, Russia
[2]Belarusian State University, BL-220030, Minsk, Belarus.
[3]IDTI RAS, Moscow, Russia.

Abstract. The paper proposes methods to predict the reliability of a multilayered coating based on a model with independently (sequentially) degrading layers. For the mathematical description of the process of crack penetration into the depth of a layered solid body from the free surface with certain roughness, the propagation of a crack tip into a layer adjacent to the free boundary of the coating is considered. This stochastic process is described through the specification of all multipoint probability distributions (densities), and it represents a Markov process. A mathematical model, described by equations of the parabolic type, is proposed to describe the above-mentioned process. Based on the above model, graphs are constructed to describe the change in the time of the crack propagation through a layer depending on the surface roughness and the ultimate strength (ductility) of the coating. The crack growth rate will be in the range from 10^{-6} mm/s (when the surface roughness is low) to 10^{-4} mm/s (when the surface roughness is high).

1 Introduction

Brittle fracture occurring as a result of the active cracking is a significant factor in the wear and failure of metal-cutting tools [1]. In most cases, modern metal-cutting tools are manufactured with wear- resistant coatings [2,3]. The challenge of predicting the reliability of the coatings of various compositions and architecture and, in particular, of modeling the coating resistance to brittle fracture is crucial and urgent. The brittle fracture of coatings can be modeled using various approaches. The traditional methods treat the crack growth as a deterministic process. In this case, a finite element model (FEM) is being widely used [4-12]. In particular, the influence of such factors as specific features of the coating material, the number of loading cycles, and the surface microgeometry was studied [4,5]. Several papers developed models to simulate the coating fracture during the scratch testing [6-9]. The effect of the substrate roughness on the fracture pattern on the coating was investigated [10]. The influence of defects in a coating on its fracture pattern was considered [11-18]. To predict the crack resistance of a coating, a shear-lag model was used which made it possible to take into account the through thickness inhomogeneity of the coating [19]. The discrete element method (DEM) was also used [20]. To predict the brittle fracture resistance of the coatings, the Bower/Fleck's and Hamilton's equations were applied [21]. The fracture toughness model was also considered, in which crack tip opening displacement was used to predict the crack propagation velocity [22]. In [23,24], linear elastic fracture mechanics (LEFM) was used to model the process of cracking in a coating depending on the substrate microgeometry, the properties of the coating material, and the applied loading.

However, in fact, the process of crack growth is a stochastic process, during which only the average crack growth rate can be determined, with the fallibility of n(t), which average value is $\langle n(t)\rangle=0$. To implement the above approach, the probabilistic methods are used to take into account the stochastic nature of both the process of the coating formation and the process of the coating fracture [25-31]. It should also be considered that most of the modern coatings have a multilayered architecture, with alternating layers with different mechanical properties [32-37]. The application of the modern technologies and methods of the coating deposition [38-47] has significant influence on the properties of the coatings, which should be taken into account when developing the modern methods of modeling. The specific features of cracking and brittle fracture of the coatings of various compositions and architecture were considered in several papers [48-7]. The objective of this study is to develop methods to run the reliability calculation of a multilayered coating based on a model with independently (sequentially) degrading layers.

* Corresponding author: dr.a.veres@yandex.ru

2 Model of crack propagation in a coating layer

A real surface is characterized by a certain level of roughness, that is, a set of irregularities occurring with relatively small steps along the base length. The roughness refers to the microgeometry of solid objects and determines their most important performance properties. Surface microdepressions are stress concentrators and potential areas for the start of crack formation. A crack can develop on the coating surface by forming a grid of surface cracks (cracking) or growing deep into the coating material. A crack directed deep into the coating material consists of two surfaces intersecting along the crack front line. During the process of its development, a crack can bend, branch, and behave in a complex manner; however, in general, the fracture mechanics considers the propagation of cracks in a plane, along a rectilinear path. Thus, in order to describe mathematically the process of crack propagation into the depth of a layered solid object from the free surface with a certain roughness, the consideration will be focused on the motion of a crack tip (a point at the crack front) in a layer, adjacent to the free interface from an initial microcrack, a depression on the surface. Let one of the coordinate axes be directed along a rectilinear inplane trajectory, while the other axis is directed along the roughness midline (Fig. 1).

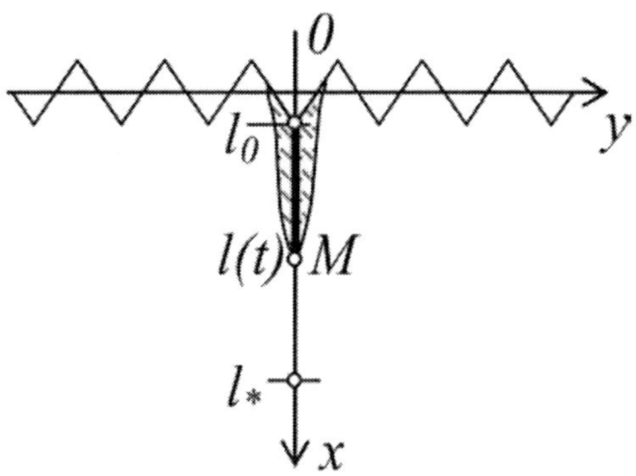

Fig. 1. Coordinate system and a model of crack propagation deep into the layered coating structure.

Figure 3 depicts l_0 as the depth of the depression relative to the roughness midline, which can be considered as the initial point of crack propagation, from which the crack begins to grow deep into the coating material, along the axis x, l_* is the thickness of the upper layer of the coating, $l = l(t)$ is the length of the crack at the time point t, a random function. During the process of crack growth, the function $l(t)$ can either grow or remain constant at some time point, and therefore, a model of random walk along a straight line is chosen. As is known, a random process is described by specifying all multipoint probability distributions (densities) or moments. For example, a two-point distribution is known to be written as:

$$P(l_1, t_1, l_2, t_2) = \Pi(l_2, t_2 | l_1, t_1)\, P(l_1, t_1) \tag{1}$$

Where $\Pi(l_2, t_2 | l_1, t_1)$ describes the probability that at the time point t_2, the crack has the length $l(t_2) = l_2$ provided that at the time point t_1, it has the length $l(t_1) = l_1$.

Then, if assumed that the information about the crack length l_{21}at the time point $t_2$depends only on the information about the crack length $l_1$at the time point t_1, then all multipoint probability distributions are described in terms of the conditional (transition probabilities, probability densities) $\Pi(l_2, t_2 | l_1, t_1)$. This is a Markov process, and it is possible to develop a mathematical model for it described by parabolic equations. In general, such equations cannot be solved analytically, but a solution can be reached using the finite difference method (FDM) (1D) or finite element method (FEM) (2D, 3D). The solution depends on the final and boundary conditions, thickness of layers and their physical and mechanical properties (elastic modulus, fracture strength, and stress intensity factors). This means it is necessary to specify $A(l), B(l)$ in the equation (1). $A(l)$ describes the forward motion of the crack and depends on the values listed above. $B(l)$ describes a random error in the estimation of the crack propagation velocity dl/dt due to its random behaviour. It is assumed that the crack propagation will be quite smooth, without any sudden accelerations and stops. Based on the above, let $B(l) = \frac{N_0}{2}$ *- const.* A linear approximation is assumed for $A(l)$, with thin layers, so that

$$A(l) = A(l_0 + \Delta l) \approx A(l_0) + A'(l_0)\Delta l \tag{2}$$

where $\Delta l = l - l_0, A' = \frac{dA}{dl}$

Then the linear model of crack development can be written in the following form:

$$\frac{dl}{dt} = (A_0 + A_1 l) + \frac{N_0}{2}\,\delta(t) \tag{3}$$

It can also be assumed that $\Pi(l_2, t_2 | l_1, t_1)$ depends not on the time points t_1, t_2, but on their difference $t_2 - t_1 = \tau$, i.e. there is a steady state of the crack development process. This assumption is quite admissible because of the progressive crack growth mechanism, for example, under cyclic loads, depending not on the specific exposure time pointst_1, t_2, but on the exposure duration:

$$\Pi(l_2, t_2 | l_1, t_1) = \Pi_t(l_2 | l_1) = \Pi(l_2 | l_1, t_2 - t_1) = \Pi(l_2 | l_1, \tau) \tag{4}$$

Then for $\Pi_\tau(l_2 | l_1)$, two types of equations will be obtained:
the direct equation:

$$\frac{\partial \Pi_\tau}{\partial \tau} = \frac{\partial}{\partial l_2}(l_2 \Pi_\tau) + \frac{\partial^2 \Pi_\tau}{\partial l_2^2} \tag{5}$$

the inverse equation:

$$\frac{\partial \Pi_\tau}{\partial \tau} = -l_1 \frac{\partial \Pi_\tau}{\partial l_1} + \frac{\partial^2 \Pi_\tau}{\partial l_1^2} \tag{6}$$

The solution of $\Pi_\tau(l_2|l_1)$ describes the steady-state Gaussian Markov process:

$$\Pi_\tau(l_2|l_1) = \frac{1}{\sqrt{2\Pi(1-l^{-2\tau})}} exp\left\{-\frac{(l_2-l_1e^{-2\tau})^2}{2(1-e^{-2\tau})}\right\} \quad (7)$$

For $P_1(l)$:

$$P_1(l_1) = \frac{1}{\sqrt{2\Pi}} e^{-1/2 l_1^2} \quad (8)$$

The condition of normalization for $\Pi_\tau(l_2|l_1)$ should be satisfied:

$$\int \Pi_\tau(l_2|l_1)dl_2 = 1 \quad (9)$$

The probability that the crack length $l(t)$ reaches the value of $l = l_1$ is calculated by the formula:

$$P_1(l_1) = \int \Pi_\tau(l_2|l_1)P_1(l_2)dl_2 \quad (10)$$

As is known, the Gaussian approximation for a random variable is determined by two parameters: the mean (expected value) and the dispersion.

$$m_l(l) = \langle l(t)\rangle, D_l(t) = \langle (l(t) - \langle l(t)\rangle)^2\rangle \quad (11)$$

And it can generally be written as

$$P(l,t) = \frac{1}{\sqrt{2\Pi D_l(t)}} exp\left\{-\frac{(l(t)-m_l(t))^2}{2D_l(t)}\right\} \quad (12)$$

Here, the equations for $m_l(t)$ and $D_l(t)$ are found from the equation (3) by the method of averaging:

$$\frac{dm_l(t)}{dt} = (A_0 + A_1)m_l(t) \quad (13)$$

$$\frac{dD_l(t)}{dt} = 2D_l + Bm_l(t) \quad (14)$$

The solutions of (13) and (14) are expressed as

$$m_l(t) = m_l(0)exp[(A_0 + A_1)t]$$
$$D_l(t) = \frac{N_0}{2} \quad (15)$$

Let us take a generalize model of the crack growth in the form:

$$\frac{dl}{dt} = -\gamma l + \frac{N}{4l} + n(t) \quad (16)$$

Where $N = \omega l_* \sqrt{\frac{(1-{}^{l_0}/_{l_*})G^2}{\sigma_0 E}}, \gamma = \omega \frac{(1-\nu)E}{(1-{}^{l_0}/_{l_*})G}$

For the problem of the crack growth, of the main interest is the time during which a crack cuts through a layer. It is clear that on a set of identical layers, the time period T during which a crack cuts through a layer is a continuous random variable which can be described using probabilistic moments, dispersion, etc.

Let us present the mean time $T_1 = \langle T\rangle = T\left(0, \frac{l_0}{\sigma}, \frac{l_*}{\sigma}\right)$ in a form convenient for numerical calculations:

$$T_1\left(0, \frac{l_0}{\sigma}, \frac{l_*}{\sigma}\right) = \frac{1}{2\gamma}\sum_{K=1}^{\infty}\frac{1}{K!K}\left[\left(\frac{l_*^2}{2\sigma^2}\right)^K - \left(\frac{l_0^2}{2\sigma^2}\right)^K\right] \quad (17)$$

An error in the calculation of T_l, the average operating time of a layer, is determined by the dispersion D, which can be written in a form convenient for calculations as follows:

$$D\left(0, \frac{l_0}{\sigma}, \frac{l_*}{\sigma}\right) = \frac{1}{2\gamma}\sum_{n=1}^{\infty}\frac{1}{2n^2}\sum_{K=0}^{2n-1}\frac{(-1)^{K+1}}{K!}$$
$$\left[l^{l_*^2/2\sigma^2}\left(\frac{l_*^2}{2\sigma^2}\right)^K - l^{l_0^2/2\sigma^2}\left(\frac{l_0^2}{2\sigma^2}\right)^K\right] \quad (18)$$

If no roughness is taken into account, i.e. $l_0 = 0$, then (18) is simplified to the following form:

$$D\left(0, \frac{l_0}{\sigma}, \frac{l_*}{\sigma}\right) = \frac{1}{2\gamma^2}\sum_{n=1}^{\infty}\frac{1}{2n^2}$$
$$\left\{1 - l^{l_*^2/2\sigma^2}\left[1 + \sum_{K=1}^{2n-1}\frac{(-1)^K}{K!}\left(\frac{l_*^2}{2\sigma^2}\right)^K\right]\right\} \quad (19)$$

The model (16) considers two terms that behave differently and take into account softening and resistance to crack growth. It is known from the fracture mechanics that with an increase in the crack length, the forces required for its growth decrease, which is taken into account by the second term in the formula (16). The coefficient N_0 depends on the physical and mechanical properties (Yung's modulus, yield strength), while the coefficient γ takes into account the opposite tendency in fracture. Figure 2 depicts in the dimensionless axes the curves describing the relationships between γT_1 and l_*/σ at the given values of the roughness and layer thickness, built based on the formula (10).

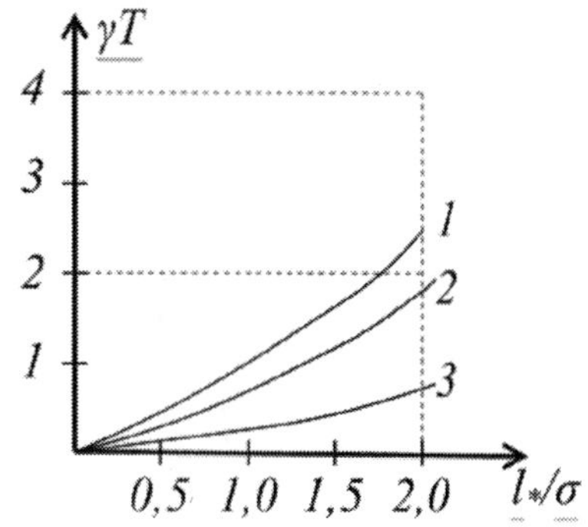

Fig. 2. Curves describing the relationships between γT and l_*/σ at the given roughness and layer thickness.

1. $l_0/l_* = 0, N = 10^{-6} mm/sec$ is low roughness
2. $l_0/l_* = \frac{1}{4}, N = 10^{-5} mm/sec$ is mean roughness
3. $l_0/l_* = \frac{3}{4}, N = 10^{-4} mm/sec$ is high roughness

Figure 3 exhibits the curves estimating an error (dispersion), also built based on the formula (10).

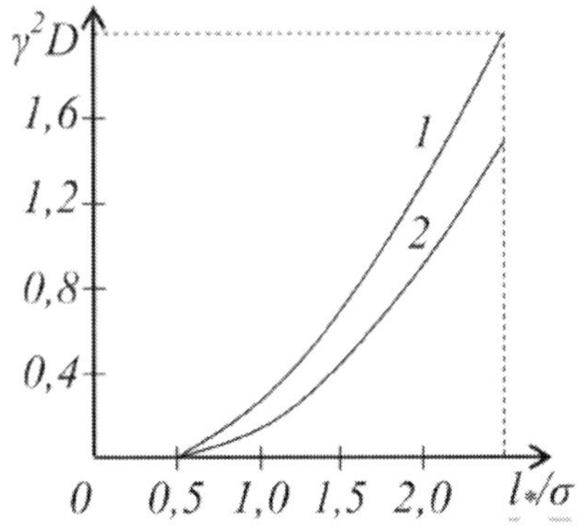

Fig. 3. Curves for estimating an error (dispersion) $\gamma^2 D$ on l_*/σ.

1. Under the conditions:

$$l_0/l_* = 0, N = 10^{-6}\, {mm}/{s},\ \gamma = {10^{-5}}/{s}$$

2. Under the conditions:

$$l_0/l_* = \frac{3}{4}, N = 10^{-5}\, {mm}/{s},\ \gamma = {10^{-4}}/{s}$$

The connection between N_0 and γ is described based on the formula of $\sigma^2 = \frac{N_0}{4\gamma}$.

3 Conclusions

The paper proposes the methods to predict the reliability of a multilayered coating based on a model with independently (sequentially) degrading layers. A crack propagates along the x-axis depending on the time t or the number of loading cycles and is generally a random function. For the mathematical description of the process of crack penetration into the depth of a layered solid body from the free surface with certain roughness, the propagation of a crack tip into a layer adjacent to the free boundary of the coating starting from an irregularity (depression) on the surface is considered. This stochastic process is described through the specification of all multipoint probability distributions (densities), and it represents a Markov process. The mathematical model, described by equations of the parabolic type, is proposed to describe the above-mentioned process. Based on the above model, the graphs are constructed to describe the change in the time of the crack propagation through a layer depending on the surface roughness and the ultimate strength (ductility) of the coating. The crack growth rate T will be in the range from 10^{-6} mm/s (when the surface roughness is low) to 10^{-4} mm/s (when the surface roughness is high). The dispersion D also increases with an increase in roughness.

Acknowledgments

This research was funded by Ministry of Science and Higher Education of the Russian Federation, Grant No. 0707-2020-0025. This work was carried out using equipment provided by the Center of Collective Use of MSUT "STANKIN".

References

1. V. Astakhov, J. Davim, *Tools (Geometry and Material) and Tool Wear* (Machining. Springer, 2008)
2. A.A. Vereschaka, S.N. Grigoriev, N.N. Sitnikov, A.D. Batako, Wear, **390-391**, 209, (2017)
3. A. Vereschaka, V. Tabakov, S. Grigoriev, N. Sitnikov, G. Oganyan, N. Andreev, F. Milovich, Wear, **420–421**, 17–37, (2019)
4. P. Bansal, P.H. Shipway, S.B. Leen, Surf.Coat. Technol., **200**(18-19), 5318, (2006)
5. M. Bäker, Comput. Mater, Sci., **64**, 79, (2012)
6. Z. Xia, W.A. Curtin, B.W. Sheldon, Acta Mater, **52**, 3507, (2004)
7. J.H. Lee, Y.F. Gao, K.E. Johanns, G.M. Pharr, Acta Mater, **60**, 5448, (2012)
8. H.C. Hyun, F. Rickhey, J.H. Lee, M. Kim, H. Lee, Eng. Fract. Mech., **134**, 304, 2015)
9. K. Holmberg, A. Laukkanen, H. Ronkainen, K. TribolInt, **38**, 1035, (2005)
10. T.L. Anderson, *Fracture Mechanics: Fundamentals and Applications* (CRC Press, Boca Raton, 1994)
11. K. Holmberg, A. Laukkanen, E. Turunen, T. Laitinen, Surf.Coat. Technol., **247**, 1-13, (2014)
12. A.L. Mohd Tobi, P.H. Shipway, S.B. Leen, Tribol Int, **58**, 29, (2013)
13. E. Harry, M. Ignat, Y. Pauleau, A. Rouzaud, P. Juliet. Surf. Coat. Technol., **125**, 185, (2000)
14. L. Aihua, D. Jianxin, C. Haibing, C. Yangyang, Z. Jun. Int. J. Refract. Hard. Met., **31**, 82, (2012)
15. A. Vereschaka, V. Tabakov, S. Grigoriev, N. Sitnikov, F. Milovich, N. Andreev, J. Bublikov, Wear, **438**, 203069, (2019)
16. A. Vereschaka, S. Grigoriev, N. Sitnikov, F. Milovich, A. Aksenenko, N. Andreev, Int. J. Adv. Manuf. Technol., **102**(9-12), 2953, (2019)
17. V. Alexey, Key. Eng. Mater., **581**, 62, (2014)
18. M. Antonov, I. Hussainova, F. Sergejev, P. Kulu, A. Gregor, Wear, **267**, 898, (2009)
19. K. Wang, F. Zhang, R.K. Bordia, Materials, **11**(4), 497, (2018)
20. N. Ferguen, Y. Mebdoua-Lahmar, H. Lahmar, W. Leclerc, M. Guessasma. Surf. Coat. Technol., **371**, 287, (2019)
21. M. Białas, Surf. Coat. Technol., **202**(24), 6002, (2008)
22. Y. Gu, K. Chen, R. Liu, M.X. Yao, R. Collier, Surf. Coat. Technol., **309**, 536, (2017)
23. X.-F. Wu, R.A. Jenson, Int. J. Eng. Sci., **49**(3), 279, (2011)
24. X.-F. Wu, R.A. Jenson, Y. Zhao, Mech. Mater., **69**(1), 195, (2014)
25. M. Grujicic, S.G. Lai, J. Mater. Sci., **36**(12), 2937, (2001)

26. C.K. Birdsall, PIC-MCC., IEEE Trans. Plasma Sci., **19**(2), 65, (1991)
27. D.T.K. Kwok, C. Cornet, IEEE Trans. Plasma Sci., **34**, 2434, (2006)
28. E. Bultinck, A. Bogaerts, New J. Phys., **11**, 103010, (2009)
29. M. Siemers, A. Pflug, T. Melzig, K. Gehrke, A. Weimar, B. Szyszka, Surf. Coat. Technol., **241**, 50, (2014)
30. H. Cansizoglu, M. Yurukcu, M.F. Cansizoglu, T. Karabacak, Thin Solid Films, **583**(1), 122, (2015)
31. S. Kumar, W.A. Curtin, Materials Today, **10**(9), 34, (2007)
32. G. Skordaris, K.-D. Bouzakis, P. Charalampous, Surf. Coat. Technol., **265**, 53, (2015)
33. B.D. Beake, L. Ning, Ch. Gey, S.C. Veldhuis, A. Komarov, A. Weaver, M. Khanna, G.S. Fox-Rabinovich, Surf. Coat. Technol., **279**, 118, (2015)
34. L. Ning, S.C. Veldhuis, K. Yamamoto, Int. J. Mach. Tools Manuf., **48**, 656, (2008)
35. A. Vereschaka, V. Tabakov, S. Grigoriev, A. Aksenenko, N. Sitnikov, G. Oganyan, A. Seleznev, S. Shevchenko, Surf. Coat. Technol., **357**, 218, 2019)
36. A. Vereschaka, V. Tabakov, S. Grigoriev, N. Sitnikov, N. Andreev, F. Milovich, Wear, **416**, 772, (2018)
37. B.D. Beake, G.S. Fox-Rabinovich, Surf. Coat. Technol., **255**, 102, (2014)
38. S.N. Grigoriev, M.A. Volosova, A.A. Vereschaka, N.N. Sitnikov, F. Milovich, J.I. Bublikov, S.V. Fyodorov, A.E. Seleznev, Ceram. Int., **46**(11), 1824, (2020)
39. W. Kalss, A. Reiter, V. Derflinger, C. Gey, J.L. Endrino, Int. J. Refractory Met. Hard Mater., **24**(5), 399, (2006)
40. K. Bobzin, CIRP J. Manuf. Sci. Technol., **18**, 1, (2017)
41. M.A. Volosova, S.N. Grigor'ev, V.V. Kuzin, Refract. Ind. Ceram., **55**, 565, (2015)
42. S. Grigoriev, Y. Melnik, A. Metel, Surf. Coat. Technol., **156**(1–3), 44, (2002)
43. M.A. Volosova, S.N. Grigor'ev, V.V. Kuzin, Refract. Ind. Ceram., **55**, 487, (2015)
44. V.V. Kuzin, S.N. Grigor'ev, M.A. Volosova, Refract. Ind. Ceram., **54**, 376, (2014)
45. V.V. Kuzin, S.N. Grigoriev, M.Yu. Fedorov, J. Frict Wear, **36**(1), 40, (2015)
46. A.S. Metel, S.N. Grigoriev, Yu.A. Melnik, V.P. Bolbukov, Instrum. Exp. Tech., **55**(1), 122, (2012)
47. M.A. Volosova, S.N. Grigor'ev, V.V. Kuzin, Refract Ind Ceram, **56**, 91, (2015)
48. A.S. Vereschaka, S.N. Grigoriev, V.P. Tabakov, E.S. Sotova, A.A. Vereschaka, M.Y. Kulikov, Key Eng. Mater., **581**, 68, (2014)
49. S.N. Grigoriev, O.V. Sobol, V.M. Beresnev, I.V. Serdyuk, A.D. Pogrebnyak, D.A. Kolesnikov, U.S. Nemchenko, J. Frict. Wear, **35**(5), 359, (2014)
50. A. Metel, V. Bolbukov, M. Volosova, S. Grigoriev, Y. Melnik, Surf. Coat. Technol., **225**, 34, (2013)
51. A.S. Vereschaka, S.N. Grigoriev, E.S. Sotova, A.A. Vereschaka, Adv Mat Res, **712-715**, 391, (2013)
52. A.A. Vereschaka, A.S. Vereschaka, S.N. Grigoriev, A.K. Kirillov, O. Khaustova, Procedia CIRP, **7**, 311, (2013)
53. A.A. Vereschaka, S.N. Grigoriev, N.N. Sitnikov, G.V. Oganyan, A. Batako, Surf.Coat. Technol., **332**, 198, (2017)
54. A. Vereschaka, V. Tabakov, S. Grigoriev, N. Sitnikov, F. Milovich, N. Andreev, C. Sotova, N. Kutina, Surf. Coat. Technol., **385**, 125402, (2020)
55. A. Vereschaka, A. Aksenenko, N. Sitnikov, M. Migranov, S. Shevchenko, C. Sotova, A. Batako, N. Andreev, Tribol. Int. **128,** 313 (2018)
56. V. Alexey, Adv Mat Res, **712-715**, 347, (2013)
57. J. Gradišek, I. Grabec, S. Siegert, R. Friedrich, Mech Syst Signal Process, **16**(5), 831, (2002)
58. A. Vereschaka, M. Volosova, A. Chigarev, N. Sitnikov, A. Ashmarin, C. Sotova, J. Bublikov, D. Lytkin, Coatings, **10**(1), 63, (2020)
59. S. Grigoriev, A. Vereschaka, F. Milovich, V. Tabakov, N. Sitnikov, N. Andreev, T. Sviridova, J. Bublikov, Surf. Coat. Technol., **401**, 126258, (2020)
60. A.A. Vereshchaka, A.S. Vereshchaka, O. Mgaloblishvili, M.N. Morgan, A.D. Batako, Int. J. Adv. Manuf. Tech., **72**(1-4), 303, (2014)
61. M. Volosova, S. Grigoriev, A.Metel, A. Shein, Coatings, *8*, 287, (2018)
62. M.A. Volosova, S.N. Grigor'ev, V.V. Kuzin, Refract. Ind. Ceram., **56**, 197, (2015)
63. O.V. Sobol', A.A. Andreev, S.N. Grigoriev, V.A. Stolbovoy, Probl. At. Sci. Technol., **4**, 174, (2011)
64. V.V. Kuzin, S.N. Grigoriev, M.A. Volosova, J. Frict. Wear, **35**, 505, (2014)
65. M.A. Volosova, S.N. Grigoriev, E.A. Ostrikov, Mech. Ind., **17**, 720, (2016)
66. Moës, N., Dolbow, J., Belytschko, T. Int J Numer Methods Eng., **46**(1), 131, (1999)
67. Kumar, K.S., Van Swygenhoven, H., Suresh, S. Acta Materialia, **51**(19), 5743, (2003)
68. Moës, N., Belytschko, T. Engineering Fracture Mechanics, **69**(7), 813, (2002)
69. Rabczuk, T., Belytschko, T. Int J Numer Methods Eng., **61**(13), 2316, (2004)
70. Miehe, C., Welschinger, F., Hofacker, M. Int J Numer Methods Eng., **83**(10), 1273, (2010)

https://doi.org/10.1051/epjconf/202124804011

Determination of Rational Design and Geometric Parameters of a High-Performance Drill Based on a Mathematical Model of the Cutting Part

Petr Pivkin[1,2,3,4], *Vladimir* Grechishnikov[2], *Artem* Ershov[1,2,3,4] *Vladimir* Kuptsov[3] and *Xiaohui* Jiang[5]

[1]Dept. of High-Efficiency Processing Technologies, Moscow State University of Technology "STANKIN", RU-127055, Moscow, Russia
[2]Dept. of Cutting Tools and Shaping Technologies, Moscow State University of Technology "STANKIN", RU-127055, Moscow, Russia
[3]Lab. of Micromachining Technologies, Moscow State University of Technology "STANKIN", RU-127055, Moscow; Russia
[4]Center of Cognitive Technologies and Machine Vision, Moscow State University of Technology "STANKIN", RU-127055, Moscow; Russia
[5]Department of Mechanical Design, Manufacturing and Automation, University of Shanghai for Science and Technology, CH-200093 Shanghai, China

Abstract. Processing of high-precision holes in one technological operation is an urgent problem of advanced manufacturing. Processing of precise holes in parts for aerospace and machine-building industries with a diameter of up to 30 mm is performed during countersinking, deployment or grinding operations. These operations are applied only if there already exists a pre-treated hole. Monolithic three-fluted drills have been becoming common for processing high-precision holes of 7-8 quality over the last few years. The processing of various types of materials such as stainless steels, cast iron and heat-resistant steels requires rational geometric and structural parameters of the cutting tool. The nature of the load distribution between all the teeth during drilling plays a huge role in the processing efficiency. Even load distribution between the three teeth and a positive geometry improves self-centering and reduces the deviation from the specified axis of the hole. The drill sharpening provides positive geometry along the entire main cutting edge. The influence of the geometric parameters of the cutting edge of the screw groove on the shape of the drill bit is equally important. Existing approaches to the design of the thinning do not account for the influence of the geometric parameters of the cutting edge on the section of the screw groove. Analytical approaches to modelling of the main cutting edges are typically married with difficulties associated with achieving a smooth change in the angle of inclination to the tangent of the cutting edge. The complexity of the aforementioned task is largely due to the presence of critical points at the interface of the spiral groove and thinning. Determining the rational shape of two sections of the main cutting edge performed in this study is a complicated task that includes several steps needed to find the number of nodal points. Achieving a positive rake angle in the normal section to the cutting edge at the gash area that was formed via a special sharpening is one of the most important results of this paper. The rational shape of the cutting edge and the front surface provides an increase in the strength of the cutting part by 1.3 times.

1 Introduction

Three-fluted designs of monolithic drills have better self-centering due to the convergence of three cutting edges at one point at the top and a more uniform load distribution due to three cutting teeth [1], which together increases the productivity of hole processing with high precision requirements. Since the advent of three-pronged designs, various designs have been developed that can improve accuracy, durability or processing performance. The working surfaces of such structures have a complex profile geometry. They provide a more rational distribution of geometric parameters. Thus, the back surface of the cutting part of the drill as a screw is sharpening, allows not only smoother distribution of the values of the back angle along the main cutting edge, but also lower values on the transverse front cutting edge due to the stabilization of the cutting process. With the radial shape of the cutting part [1-2] reducing stress concentrations at the edge corners, the shape of the back surface is formed along a curve lying on a spherical surface, thereby forming a complex profile surface. The curvature on the undercut also forms the front surface of a complex shape.

The complexity of designing such drills lies in the formation of these surfaces with taking into account the interaction of the totality of all shapes and geometric parameters. Thus, the design of the gash depends on the shape and location of the edge on the screw groove section, which in turn depends on the shape and geometric parameters of the chip groove and the back surface.

Existing approaches to the design of such types of drills do not functionally link the shape of the cutting part and the groove profile, in order to achieve the desirable shape of the main cutting edge that affects the stability of the cutting process. The straight shape of the cutting edge is optimal for the best cutting stability due to a uniform distribution of geometric parameters along the edge. An equally important design element is sharpening of the main cutting edge to reduce the length of the section with a negative front angle, which prevents the cutting process during drilling. Existing approaches to the design of the gash do not take into account the shape of the edge. There are various types and shapes of gash, with the most common being straight and radius gashes. Straight-line sharpening is performed at a certain angle from the main cutting edge to the top of the drill, forming a sharp transition at the junction of the two sections. The radius sharpening is also performed at some angle to the vertex, but in a radius shape. In this case, a sharp transition can be avoided at the expense of the optimal curvature of the undercut. In the first case, the sharp transition forms a zone of stress concentrations, which under load during cutting can lead to breakage of the drill. In the second case, the uneven distribution of the angle of inclination along the entire section of the cutting edge contributes to an uneven load, which leads to a loss of productivity. The optimal shape of the undercut should take into account the shape of the main cutting edge on the chip groove section [3-47].

2 Results and Discussion

As discussed earlier, the section of the main cutting edge determines the starting point and shape of the gash. It is worth noting that with a different shape of the main cutting edge, the curvature and position of the gash will also change. The formation of a straight cutting edge shape requires the construction of a mathematical model of the surfaces forming the cutting edge. Such surfaces are the screw surface of the chip groove and the conical surface. It is their intersection that forms the main cutting edge in the undercut area. The profile of the chip groove in the radial section is determined by the formula (1).

The slope of the cutting edge in the chip groove section is determined by the angle "ι" of the blade relative to the line on the conical surface connecting the beginning of the cutting edge at the corner and the tip of the drill.

$$Xk(X) = X \cdot \cos\left[\frac{2\pi \cdot \left[R - \sqrt{\left(R + X \cdot tan\left(\frac{\pi}{2} - \iota\right)\right)^2 - X^2}\right]}{T \cdot \tan(\varphi)}\right] + \sin\left[\frac{2\pi \cdot \left[R - \sqrt{\left(R + X \cdot tan\left(\frac{\pi}{2} - \iota\right)\right)^2 + X^2}\right]}{T \cdot \tan(\varphi)}\right] \cdot \left(R + X \cdot \tan\left(\frac{\pi}{2} - \iota\right)\right) \quad (1)$$

$$Yk(X) = cos\left[\frac{2\pi \cdot \left[R - \sqrt{\left(R + X \cdot tan\left(\frac{\pi}{2} - \iota\right)\right)^2 - X^2}\right]}{T \cdot tan(\varphi)}\right] \cdot \left(R + X \cdot tan\left(\frac{\pi}{2} - \iota\right)\right) - X \cdot sin\left[\frac{2\pi \cdot \left[R - \sqrt{\left(R + X \cdot tan\left(\frac{\pi}{2} - \iota\right)\right)^2 + X^2}\right]}{T \cdot tan(\varphi)}\right] + sin\left[\frac{2\pi \cdot \left[R - \sqrt{\left(R + X \cdot tan\left(\frac{\pi}{2} - \iota\right)\right)^2 + X^2}\right]}{T \cdot tan(\varphi)}\right] \cdot \left(R + X * tan\left(\frac{\pi}{2} - \iota\right)\right) \quad (2)$$

The profile of the chip groove consists of two circles tangent to each other, as well as tangent to the inner diameter of the drill. The circle rk1 also touches the front surface located at an angle γk.

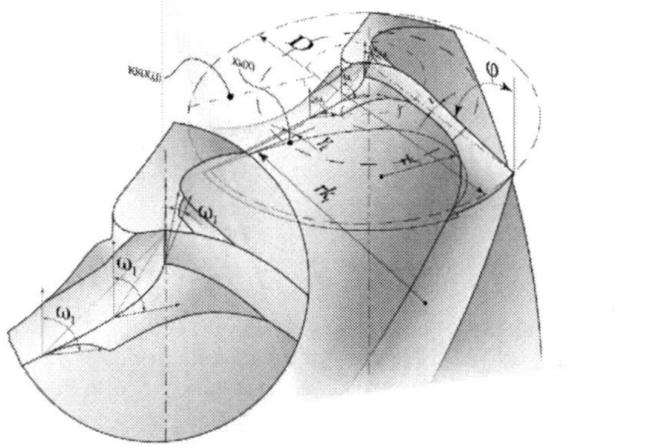

Fig. 1. Chart of the three-fluted drill design

The profile of the chip groove consists of two circles tangent to each other, as well as tangent to the inner diameter of the drill. The circle rk1 also touches the front surface located at an angle γk.

The next step, after modeling the main working surfaces, based on a mathematical model, is to build a section of the gash. In the polar coordinate system, on a plane perpendicular to the tool axis, a spline tangent to the cutting edge is defined with a different number of reference points. The resulting spline is projected onto a conical surface. The description of the conical surface is determined by the formula (3):

$$KK(X,Y) = \begin{bmatrix} X\cdot\cos(\varphi) - \dfrac{\sin(\varphi)\cdot\left(R-\sqrt{X^2+Y^2}\right)}{tan(\varphi)} \\ \sin(\theta)\cdot\left[X\cdot\sin(\varphi) + \dfrac{\cos(\varphi)\cdot\left(R-\sqrt{X^2+Y^2}\right)}{\tan(\varphi)}\right]\Big) \\ +Y(\phi,\theta)\cdot co\,s(\theta) \end{bmatrix} \quad (3)$$

The spatial curve on the surface is the cutting edge on the gash area.

Thus, changing the number of nodal points of the sharpening edge from 3 to 10, we get 7 different designs of the cutting part of the drill. It should be noted that on a design with 4 reference points, the shape of the cutting edge is slightly distorted. An edge with 3 nodal points loses its original shape, acquiring a more rectilinear curvature. Changing the shape of the cutting edge introduces a limitation in the formation of optimal curvature, which entails a narrower range of processed materials. The structures with the 9th, 7th and 5th reference points are shown in Figure 2.

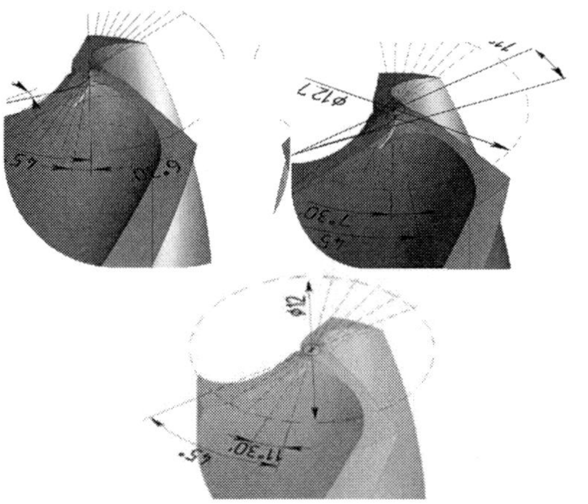

Fig. 2. Areas of the gash with the 9th, 7th and 5th reference points

The next step is to study the curvature of the gash section, by measuring the angle of inclination of the cutting edge ω1 (Fig. 1), along the entire section on each structure. – the angle between the tangent to the cutting edge and the axis of the drill in a plane perpendicular to the line passing through the test point and the tip of the drill.

Measurements of each structure were carried out at 11 equidistant points. The graph will allow you to determine the design with the most uniform distribution of the angle of inclination of the cutting edge.

The graph shows that the smoothest distribution of the angle on structures with 3 and 4 reference points, as a result of a more rectilinear shape of the edge. These structures, as mentioned above, have distortions of the edge shape, so they cannot be selected as optimal. The design with 5 reference points is devoid of this disadvantage, as a result of which it is defined as an optimal design.

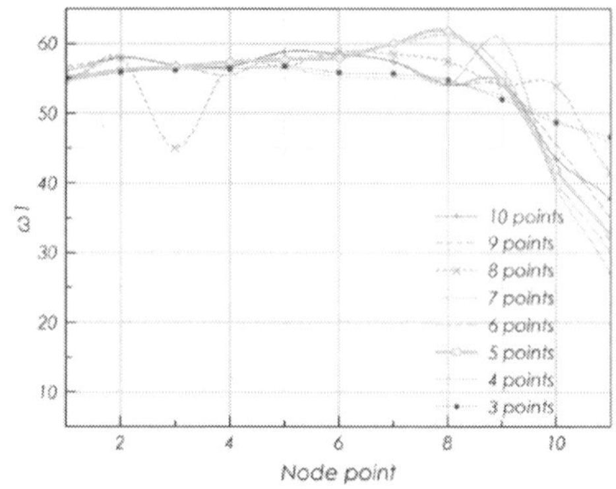

Fig. 4. Graph of the change of the angle ω1 along the cutting edge in the section of the gash

To confirm the increase in the performance of this design compared to similar ones formed with a large number of nodal points, the final stage of the study is the analysis of structures in the ANSYS engineering analysis system.

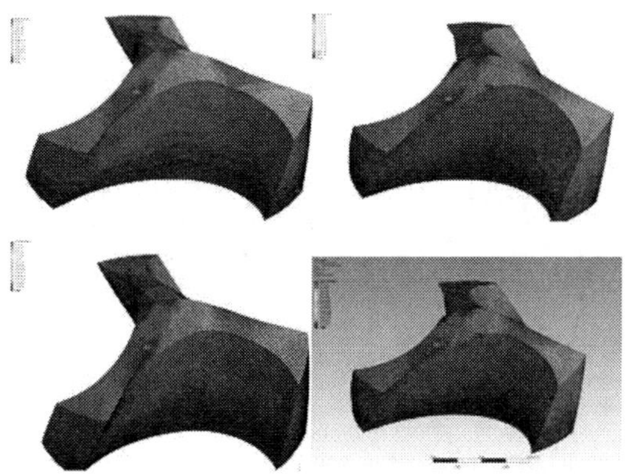

Fig. 3. The results of engineering analysis of structures with 8,7,6 and 5 control points

Table 1. The results of the internal stresses of each structure

Number of edge reference points on the thinning	Maximum internal stress values, Mpa
5	307
6	346
7	398
8	422

It follows from the table that the lowest values of internal stresses are shown by a design with 5 nodal points on the cutting edge of the studied area.

3 Conclusion

The design with the lowest value of internal stresses and a uniform distribution of the values of the angle of inclination of the cutting edge, allows you to increase productivity by more than 27% compared to analagic ones. Which allows us to draw the following conclusions:

1. A new three-fluted drills design method based on cutting theory and analytical geometry allows developing various designs with variable cutting edge geometry.

2. New functional connections allow getting rid of a sharp drop in the values of geometric parameters along the cutting edge, which increase the strength of the cutting part.

3. The developed model can be used as a theoretical basis for studying the tangent angle to the cutting edge. The new method makes it possible to develop designs of three-fluted drills with a straight cutting edge on the section of the chip groove of a rational shape, which increases the efficiency of the drill design during processing.

Acknowledgments

This research was funded by Ministry of Science and Higher Education of the Russian Federation, Grant No. 0707-2020-0031. This work was carried out using equipment provided by the Center of Collective Use of MSUT "STANKIN".

References

1. N. Baroiu, D. Boazu, C. Vasilache, T. Virgil, Appl. Mech. Mater, **371**, 509-513 (2013)
2. H. Jung-Fa, D. Psang, Int. J. Mach. Tools Manuf. **42**, 1181-1193 (2002)
3. N. Baroiu, V. Teodor, S. Berbinishi, N. Oancea, Proc. IN-TECH, (2011)
4. H. Jung-Fa, Int. J. Mach. Tools Manuf., **45**(7–8), (2005)
5. L. Xiong, H. Shi, Y. Chen, ISAAT **9**, 493-499 (2006)
6. P. Pivkin, A. Ershov, A. Nadykto, *IEEE (FarEastCon*, Vladivostok, Russia, 2020)
7. L. Xiong, H. Shi, Y. Chen, Chin. Mech. Eng., **18**, 1165-1167 (2007)
8. N. Baroiu, N. Berbinschi, Fascicola suplimentara. **58**(72), 31-34 (2013)
9. P. Pivkin, V. Grechishnikov, A. Ershov, A. Nadykto, Proc. SPIE 11540 (Advanced Manufacturing Technologies for Micro- and Nanosystems in Security and Defence III, 2020)
10. J. Hsieh, Int. J. Adv. Manuf. Technol., **45**, 967–977 (2005)
11. S.N. Grigoriev, V.D. Gurin, M.A. Volosova, N.Y. Cherkasova, Materwiss. Werksttech, **44**, 790-796 (2013)
12. B. Nicusor, T. Virgil, B. Silviu, O. Nicolae, Adv. Mat. Res., **1036**, 259-264 (2012).
13. H. Shi, [A Study on Curved Edge Drills. In: Metal Cutting Theory], Springer, Cham, (2018)
14. Y.E. Petukhov, N.V. Kolesov, S.Y. Yurasov, Russ. Eng. Res., **34**, 374–380 (2014).
15. S.N. Grigoriev, V.I. Teleshevskii, Meas. Tech., **54**, 744–749 (2011)
16. S.N. Grigoriev, M.P. Kozochkin, P.M. Pivkin, IEEE *(FarEastCon, Vladivostok, Russia, 2020)*
17. S.N. Grigoriev, O.V. Sobol, V M. Beresnev, et al., J. Frict. Wear, **35**, 359–364 (2014).
18. S.P. Radzevich, Florida: CRC Press., **738** (2014).
19. T.-R. Lin, J. Mater. Process. Technol., **127**(1), 493-499 (2002)
20. F. Tang, J. Bai, X. H. Wang, Int. J. Adv. Manuf. Technol., **73**(5-8), 659-667 (2014).
21. D. Livia, P. Dan, N. Brndas, M. Calin, R. Johannes, Int. J. Adv. Manuf. Technol., **83**(9-12), 2111–2127 (2015)
22. S.N. Grigoriev, V.A. Grechishnikov, M.A. Volosova, X. Jiang, P.M. Pivkin, Mater. Sci. Eng. C, **971**, 2 (2020)
23. B. Nicuşor, B. Silviu, T. Virgil, O. Nicoale, Tech. Mach. Build, 2012, 17-26 (2012)
24. C. Fetecau, N. Oancea, F. Stan, J. Mater. Process.Technol. **209**(7), 3460-3468 (2009)
25. S.N. Grigoriev, V.A. Sinopalnikov, M.V. Tereshin, V.D. Gurin, Meas. Tech., **55**, (2012)
26. S. Barnes, P. Bhudwannachai, A. Dahnel, Proc. ASME, **2B**, (2013).
27. V.V. Kuzin, S.N. Grigoriev, M.A. Volosova, J. Frict. Wear, **36**(1), 40, (2015)
28. A.V. Gusarov, S.N. Grigoriev, M.A. Volosova, Y.A. Melnik, A. Laskin, D.V. Kotoban, A.A. Okunkova, J. Mater. Process. Technol., **261**, 213 (2018)
29. R.S. Khmyrov, S.N. Grigoriev, A.A. Okunkova, A.V. Gusarov, Phys. Procedia, **56**, 345 (2014)
30. S.N. Grigoriev, G.M. Martinov, Procedia CIRP, **46**, 525 (2016)
31. S.N. Grigoriev, D.A. Masterenko, V.I. Teleshevskii, P.N. Emelyanov, Meas. Tech, **55**(11), 1311 (2013)
32. V.A. Grechishnikov, Y.E. Petukhov, P.M. Pivkin, Parts. Meas. Tech., **58**, 848–853 (2015)
33. S.N. Grigoriev, M.A. Volosova, A.A. Okunkova, S.V. Fedorov, K. Hamdy, P.A. Podrabinnik, P.M. Pivkin, M.P. Kozochkin, A.N. Porvatov, J. Manuf. Mater. Process., **4**, 96 (2020)
34. S.N. Grigoriev, M.A. Volosova, A.A. Okunkova, S.V. Fedorov, K. Hamdy, P.A. Podrabinnik, P.M. Pivkin, M.P. Kozochkin, A.N. Porvatov, Technologies., **8**, 49 (2020)
35. S.N. Grigoriev, G.M. Martinov, Procedia CIRP, **1**(1), 238 (2012)
36. M. Volosova, S. Grigoriev, A.Metel, A. Shein, Coatings, **8**, 287, (2018)
37. A.S. Metel, V.P. Bolbukov, M.A. Volosova, S.N. Grigoriev, Yu.A. Melnik, Instrum. Exp. Tech., **57**(3), 345, (2014)

38. M.A. Volosova, S.N. Grigor'ev, V.V. Kuzin, Refract. Ind. Ceram., **55**, 565, (2015)
39. M.A. Volosova, S.N. Grigor'ev, V.V. Kuzin, Refract. Ind. Ceram., **56**, 197, (2015)
40. M.A. Volosova, S.N. Grigor'ev, V.V. Kuzin, Refract. Ind. Ceram., **55**, 487, (2015)
41. V.V. Kuzin, S.N. Grigoriev, M.A. Volosova, J. Frict. Wear, **35**(6), 505, (2014)
42. V.V. Kuzin, S.N. Grigor'ev, M.A. Volosova, Refract. Ind. Ceram., **54**, 376, (2014)
43. S. Berbinschi, N. Baroiu, T. Virgil, N. Oancea, New technologies in manufacturing, (2011)
44. S.N. Grigoriev, M.A. Volosova, A.A. Okunkova, S.V. Fedorov, K. Hamdy, P.A. Podrabinnik, ... , A.N. Porvatov, J. Manuf. Mater. Process., **4**(3), 96 (2020) https://doi.org/10.3390/jmmp4030096
45. P.M. Pivkin, A.B. Nadykto, V.A. Grechishnikov, M.A. Volosova, I.V. Minin, S.N. Grigoriev, Emerging Imaging and Sensing Technologies for Security and Defence V; and Advanced Manufacturing Technologies for Micro-and Nanosystems in Security and Defence III, **11540**, 115401G (2020) https://doi.org/10.1117/12.2574391
46. S.N. Grigoriev, V.A. Grechishnikov, M.A. Volosova, X. Jiang, P.M. Pivkin, IOP Conference Series: Materials Science and Engineering, **971**(2), 022065 (2020) doi: 10.1088/1757-899X/971/2/022065
47. P.M. Pivkin, V.A. Grechishnikov, A.A. Ershov, A.B. Nadykto, Emerging Imaging and Sensing Technologies for Security and Defence V; and Advanced Manufacturing Technologies for Micro- and Nanosystems in Security and Defence III, **11540**, 1154014 (2020) https://doi.org/10.1117/12.2574392

Effect of Geometrical Parameter Cutting Edge on the Toroidal Forming Surface of a Solid End Mill

Petr Pivkin*1,2,3,4, Ilya Minin[1,2,3,4]

[1]Dept. of Cutting Tools and Shaping Technologies, Moscow State University of Technology "STANKIN", RU-127055, Moscow, Russia;
[2]Dept. of High-Efficiency Processing Technologies, Moscow State University of Technology "STANKIN", RU-127055, Moscow, Russia,
[3]Lab. of Micromachining Technologies, Moscow State University of Technology "STANKIN", RU-127055, Moscow; Russia;
[4]Center of Cognitive Technologies and Machine Vision, Moscow State University of Technology "STANKIN", RU-127055, Moscow; Russia;

Abstract. Solid end mill with a toroidal forming section are the most productive solutions in the field of processing difficult-to-process materials. The process of forming the cutting geometry of such cutters is performed on grinding machines with CNC with the use of abrasive grinding wheels. The formation of the cutting edge on the toroidal section is carried out by discrete movements of the working bodies of the machine in accordance with the control program. The disadvantage of the stepper drive used in CNC machines is the limited discreteness of the described trajectory resulted in the loss of accuracy when forming radius, arc, spherical and toroidal sections. Loss of accuracy in the formation of elements of the working geometry of a solid end mill can negatively affect the tool life, and the structural strength of the joints of the transition of working surfaces. The study is aimed at identifying the necessary and sufficient number of points that describe the spatial shape of the cutting edge with acceptable loss of accuracy.

1 Inroduction

The formation of surfaces of parts made of hard-to-work materials, such as heat-resistant nickel steels and stainless alloys, is accompanied by a high power load cause by hardening. In order to structurally increase the strength of the cutting tool, the tops of the teeth are shaped in the form of radius rounding that allows to improve the distribution of power loads [1-4]. In the process of cutting, the radius surface forms a toroidal shaping section, which provides the possibility of machining end, cylindrical and shaped surfaces of various configurations.

Designing milling cutters with a toroidal end mill includes the solution of forming problems consisting in providing the required angular parameters of the cutting geometry. The cutting edge on the toroidal cutting part is formed at the intersection of the front surface of the screw chip groove, the end front surface and the back surface specified by the rear angles α_1 and α_2. The shape of the edge depends on the size of the radius forming the toroidal surface, the anterior angle γ, the section of the constant geometry S_c and the transition section to the back of the tooth V_d. Figure 1 shows the key parameters of the chip groove profile.

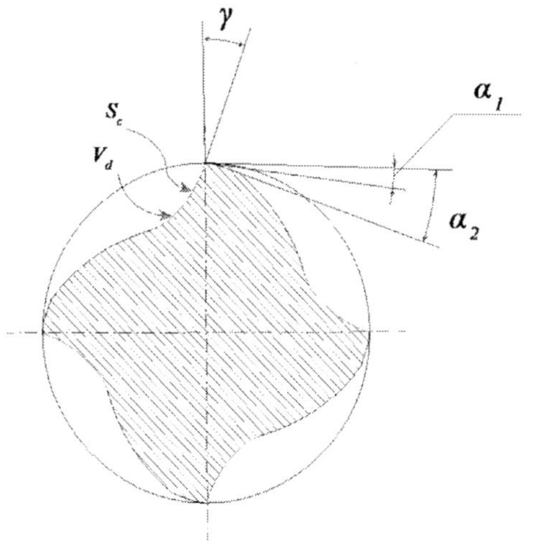

Fig. 1. Parameters of the chip groove profile.

Milling cutters with a toroidal forming surface are produced mainly on CNC grinding machines. Shaping of the front and back surfaces of the tool is performed by grinding abrasive wheels of various configurations suitable for processing specific surfaces. The choice of the most suitable trajectory of the grinding wheel is

*PMPivkin@gmail.com; phone +74999729520

 https://doi.org/10.1051/epjconf/202124804012

determined by ensuring that the shape of the formed areas and the sharpness of the geometry of the cutting wedge being processed correspond each other.

Stepper motors on the modern grinding machines are used as drive elements that carry out the main movements of the tool and the workpiece. One of the disadvantages of such a system is the restriction on the permissible number of individual movements per unit of time. Exceeding the limit can lead to an interruption of the control program execution or to the skipping of the trajectory stages, which can lead to a violation of the accuracy requirements for the shape being formed. Radius, spherical, arc-shaped and other sections of the trajectory that have curvature are approximated in accordance with the accepted limit on the throughput of the CNC system. As a result of approximating, a loss of accuracy in shaping the working surfaces of the tool appears, which can lead to a distortion of the geometry of the cutting edge that adversly affects the accuracy and performance of the entire cutting tool [5-50].

2 Results and Discussion

Defining the necessary and sufficient number of points that gives shape to the spatial line of the cutting edge it is possible to reduce the load on stepper motor tool grinding machine, determining the location of these points can reduce the overall form error of the cutting edge.

At the first stage of the study, a three-dimensional model of a solid end mill with a toroidal forming surface was developed. The model was based on the working surface of the torus, defined using parameterization elements, expressed in the established functional dependencies between the parameters of the end mill and the dimensions of the forming and guiding toroidal surface. The equation of the toroidal generating surface in parametric form is presented as equation 1.

$$T(\varphi,\psi) = \begin{cases} x(\varphi,\psi) = \left(\left(\frac{D}{2} - r\right) + r \cdot \cos(\psi)\right)\cos(\varphi) \\ y(\varphi,\psi) = \left(\left(\frac{D}{2} - r\right) + r \cdot \cos(\psi)\right)\sin(\varphi) \\ z(\varphi,\psi) = r \cdot \sin(\psi) \end{cases}$$

$$\varphi \in [0, 2\pi), \psi \in [-\pi, \pi],$$

where D is the diameter of the cutting part, r is the radius at the top of the tooth.

In [10.1051 / epjconf/201922405001], the process of mathematical modeling of the working part of a solid end ceramic cutter with a toroidal forming surface is described in detail. The shape of the cutting edge is functionally related to the sections of the front surface, rear surfaces and the section of the undercut formed at the intersection of the front end surface and the screw chip groove. The sharpening ensures that the front angle on the cutting part is constant, which best affects the distribution of the power load during the cutting process.

Nodal points were formed on the toroidal surface, defining the shape and location of the cutting edge on the studied area. Each nodal point $N_i \in (T)$ was located with a uniform step in the first quarter of the generating circle with radius r. Figure 2 shows a schematic isometric image of a toroidal surface with nodal points located on it.

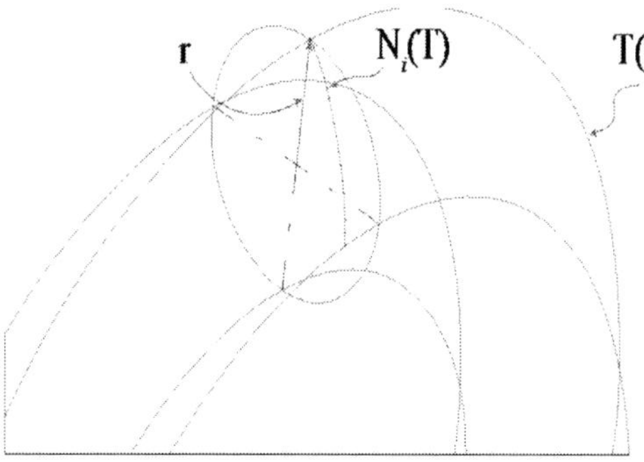

Fig. 2. Toroidal surface with nodal points located on it.

In order to determine the necessary and sufficient number of nodal points, it was necessary to exclude them from the general series with different frequency. By measuring the degree of deviation of the new cutting edge formed between the remaining nodal points, you can determine the degree of influence of both a specific point and a section of the arc formed by a circle on the shape of the cutting edge. Table 1 shows the results of the study.

Table 1. result of measuring the deviation of the cutting edges from the nominal one.

	Edge 2	Edge 3	Edge 4	Edge 5	Edge 6	Edge 7	Edge 8	Edge 9	Edge 10
Point	Error	Error	Error	Error	Error	Error	Error	Error	Error
0	0	0	0	0	0	0	0	0	0
1	0,04802	0,0991	0,13505	0,18739	0,21695	0,30316	0,30731	0,33388	0,37371
2	0	0,04687	0,09069	0,15345	0,19337	0,0378	0,30932	0,3413	0,40362
3	0,01031	0	0,03864	0,09854	0,14283	0,2655	0,27245	0,30606	0,38422
4	0	0,01967	0	0,04335	0,08761	0,2183	0,22138	0,25627	0,34226
5	0,0035	0,0142	0,01442	0	0,0334	0,15725	0,16798	0,20113	0,28873
6	0	0	0,00571	0,02545	0	0,10429	0,11706	0,14681	0,23048
7	0,03358	0,05322	0,04326	0,09274	0,07911	0	0,01285	0,03721	0,10978
8	0	0,02926	0	0,06624	0,05923	0,01312	0	0,01397	0,06937
9	0,00476	0	0,03314	0,03254	0,02469	0,01386	0,00139	0	0,03257
10	0	0,03843	0,05813	0	0,01832	0,00374	0,009	0,0046	0
11	0,04802	0,0239	0,02081	0,07498	0,03772	0,08444	0,07106	0,10009	0,12711

	Edge 2	Edge 3	Edge 4	Edge 5	Edge 6	Edge 7	Edge 8	Edge 9	Edge 10
Point	Error	Error	Error	Error	Error	Error	Error	Error	Error
12	0	0	0	0,05563	0	0,05834	0,04318	0,08825	0,14616
13	0,00082	0,00273	0,02323	0,03867	0,02716	0,02899	0,00914	0,0695	0,15613
14	0	0,00279	0,04854	0,02102	0,04591	0	0,02834	0,04464	0,15539
15	0,04948	0	0,07458	0	0,05928	0,02579	0,06617	0,01452	0,14219
16	0	0,0785	0	0,07335	0,0298	0,04965	0	0,07929	0,21396
17	0,00399	0,04295	0,0239	0,04033	0,0188	0,0233	0,02564	0,04035	0,16859
18	0	0	0,04746	0,00063	0	0,00797	0,04581	0	0,10398
19	0,0339	0,04561	0,07228	0,04604	0,03433	0,04765	0,06706	0,03912	0,01554
20	0	0	0	0	0	0	0	0	0
Summ Error	0,23638	0,49723	0,72984	1,05035	1,10851	1,44315	1,77514	2,11699	3,46379
Mid Error	0,023638	0,038248	0,048656	0,065647	0,069282	0,084891	0,10442	0,124529	0,192433
Max Error	0,04948	0,0991	0,13505	0,18739	0,21695	0,30316	0,30932	0,3413	0,40362
Crit. Point	15	1	1	1	1	1	2	2	2
Critical Point Number	10	13	15	16	16	17	17	17	18

After analyzing the results, it can be concluded that the transition to the cutting edge at the end section has the greatest impact. The degree of error increases with a decrease in the number of nodal points, but for results 3 – 5 deviations have a periodic form, which can be compensated by shifting the nodal points from a given position. In charts 1 to 3 presents the deviations of the results of 3 to 5, respectively.

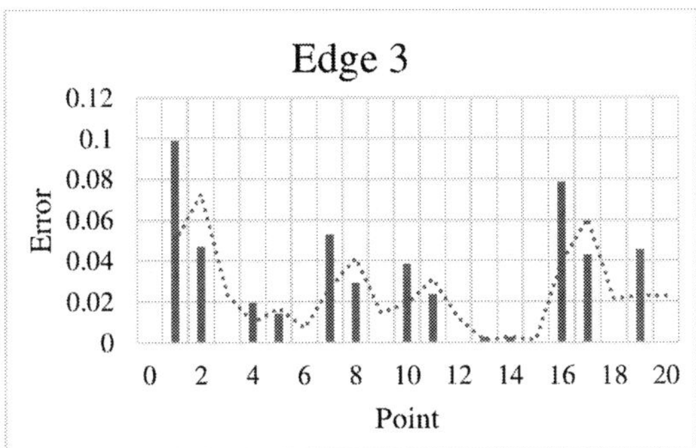

Graph 1. Deviations of grid points from the nominal for a result of 3.

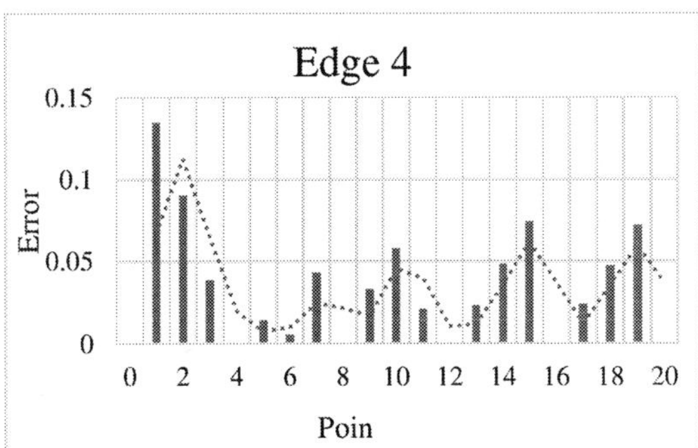

Graph 2. Deviations of the nodal points from the nominal ones for result 4.

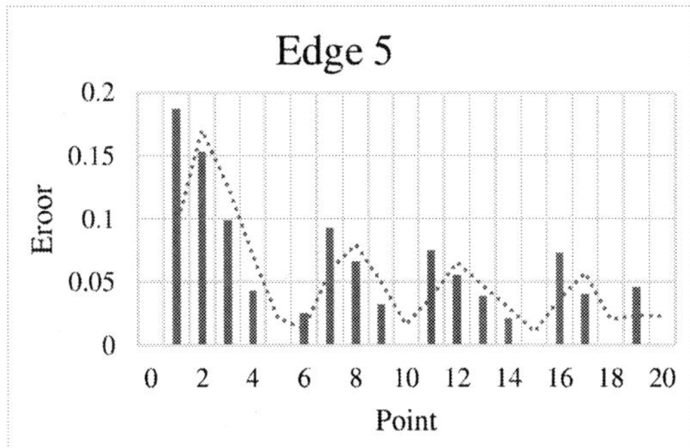

Graph 3. Deviations of nodal points from nominal ones for result 5.

By making adjustments at the most critical nodal points, you can improve the accuracy of the cutting edge shape. Thus, it is possible to reduce the required number of nodal points for setting the spatial line of the cutting edge, which helps to reduce the necessary approximations on CNC grinding and grinding machines when preparing the control program.

3 Conclusion

A model of the end mill with toroidal section and the cutting edge lying on it was developed. By sifting out, with different frequency, the nodal points that define the shape of the spatial line of the cutting edge, it was found that there is a periodicity in the resulting deviations that can be compensated and, thus, the loss of accuracy of the cutting edge shape from nominal as well as the number of approximations on CNC machines can be significantly reduced. Thus, it is possible to significantly reduce the load on the stepper motor and, at the same time, allow for only a small reduction in the accuracy of the edge being formed.

Acknowledgments

This research was funded by Ministry of Science and Higher Education of the Russian Federation, Grant No. 0707-2020-0031. This work was carried out using equipment provided by the Center of Collective Use of MSUT "STANKIN".

References

1. T. Chen, X. Liu, C. Wang et al., Int J Adv Manuf Technol **80,** 567–579 (2015).
2. S. Kim, Key Eng. Mater. **96**, 33-80 (1994).
3. B. Bitterlich, S. Bitsch, K. Friederich, J. Eur. Ceram. Soc. **28**(5), 989-994 (2008).
4. T. Zlámal, J. Hajnyš, J. Petrů, I. Mrkvica, Adv. Sci. Technol. Res. J. **12**(1), 236-243 (2018).
5. S.N. Grigoriev, V.A. Grechishnikov, M.A. Volosova, et al., Russ. Eng. Res. **40**, 79–82 (2020).
6. Ji Wei, L. Xianli, W. Lihui, M. Yue, Wu Xuefeng, J. Manuf. Process **19**, 205-211 (2015).
7. G.M. Martinov, P.A. Nikishechkin, A.S. Grigoriev, N.Yu. Chervonnova, Autom. Remote. Control. **80**(3), 584–591 (2019).
8. S.N. Grigoriev, G.M. Martinov, Proc. CIRP **14**, 517-522 (2014).
9. S.N. Grigoriev, G.M. Martinov, Proc. CIRP **41**, 858-863 (2016).
10. F. Molaiekiya, M. Aramesh, S.C. Veldhuis, Wear. **203191**, 446–447 (2020).
11. L. Tan, C. Yao, J. Ren, D. Zhang, Int. J. Adv. Manuf. Technol. **88**, 2589-2602 (2016).
12. J. Hsieh, Y. Tsai, Int. J. Adv. Manuf. Technol. **29,** 912–921 (2006).
13. V.V. Kuzin, S.N. Grigoriev, M.Yu. Fedorov, J. Frict. Wear. **36**(1), 40-44 (2015).
14. R. Alexander, K. Iyas, K. Andreas, J. Eur. Ceram. Soc. **36**(3), 705-717 (2016).
15. V.V. Kuzin, S.N. Grigor'ev, M.A. Volosova, Refract. Ind. Ceram. **54**, 376–380 (2014).
16. L. Xing, D. Zhaoqian, Jianxin, *Tran Tech Publications*, (1995).
17. Z. Wang, Liu, Int. J. Refract. Hard. Met. **55**, 24-32 (2016).
18. N.V. Kolesov, Y.E. Petukhov, Russ. Eng. Res. **27**, 812–814 (2007).
19. A. Metel, V. Bolbukov, M. Volosova, S. Grigoriev, Yu. Melnik, Surf. Coat. Technol. **225**, 34-39 (2013).
20. V.V. Kuzin, S. Grigoriev, Appl. Mech. Mater. **486**, 32–35 (2013).
21. L. Xiong, B. Li, Int. J. Adv. Manuf. Technol. **82**, 889–900 (2016).
22. A.S. Metel, S.N. Grigoriev, Y.A. Melnik, et al., Instrum. Exp. Tech. **55**, 288–293 (2012).
23. V.A. Grechishnikov, S.N. Grigoriev, P.M. Pivkin, A.V. Isaev, M.A. Volosova, P.Yu. Peretyagin, Russ. Eng. Res. **39**, 1084–1087 (2019).
24. V. Grechishnikov, S. Grigoriev, P. Pivkin, M. Volosova, A. Isaev, D. Nikitin, I. Minin, EPJ Web Conf. **224**(05001) (2019).
25. Z. Wang, Liu, Int. J. Refract. Metals Hard Mater. **55**, 24–32 (2016).
26. Y.E. Petukhov, N.V. Kolesov, S.Y. Yurasov, Russ. Eng. Res. **34**, 374–380 (2014).
27. Y.E. Petukhov, A.A. Vodovozov, Russ. Eng. Res. **35**, 670–673 (2014).
28. S.C. Von, M. Stoetzer, M. Ruecker, M. Wagner, P. Schumann, N.C. Gellrich, Int. J. Oral. Maxillofac. Implants. **29**(4), 942-948 (2014).
29. R.S. Khmyrov, S.N. Grigoriev, A.A. Okunkova, A.V. Gusarov, Phys. Procedia **56**, 345 (2014)
30. S.N. Grigoriev, D.A. Masterenko, V.I. Teleshevskii, P.N. Emelyanov, Meas. Tech, **55**(11), 1311 (2013)
31. V. Krishnaraj, S. Samsudeensadham, R. Sindhumathi, P. Kuppan, Proc. Eng. **97**, 251–257 (2014).
32. L. Berglind, E. Ozturk, Springer, 57-93 (2019).
33. B.A. Oluwamayokun, S. Wei-Ming, H.W. Pihua, Mach. Sci. Technol. **16**(1), 111-130 (2012).
34. T. Matsumura, T. Furuki, E. Usui, LEM21 2003, 783-788 (2003).
35. S.P. Radzevich, M&C Publishers, California, San Rafael (2008).
36. S.P. Radzevich, CRC Press, Boca Raton Florida, 738 (2014).
37. S.N. Grigoriev, M.A. Volosova, A.A. Okunkova, S.V. Fedorov, K. Hamdy, P. A. Podrabinnik, A.N. Porvatov, J. Manuf. Mater. Process., **4**(3), 96 (2020).
38. P.M. Pivkin, A.B. Nadykto, V.A. Grechishnikov, M.A. Volosova, I.V. Minin, S.N. Grigoriev, Proc. SPIE **11540**, 115401G (2020).
39. S.N. Grigoriev, V.A. Grechishnikov, M.A. Volosova, X. Jiang, P.M. Pivkin, IOP Conf. Ser.: Mater. Sci. Eng. **971,** 022065 (2020).
40. P.M. Pivkin, V.A. Grechishnikov, A.A. Ershov, A.B. Nadykto, Proc. SPIE 11540, 1154014 (2020).
41. Wei-Fang, C. Wei-Ya, J. Mater. Process. Technol. **121**(2–3), 217-225 (2002).
42. D.S. Rechenko, A.Y. Popov, K.V. Averkov, V.A. Sergeev, Russian Engineering Research, **32**(5-6), 511 – 512 (2012)
43. Q. Guo, Bo Zhao, Yan Jiang, Wu. Zhao, Precis. Eng. **51**, 659-681 (2018).
44. A.V. Gusarov, S.N. Grigoriev, M.A. Volosova, Y.A. Melnik, A. Laskin, D.V. Kotoban, A.A. Okunkova, J. Mater. Process. Technol. **261**, 213 (2018).

45. A. Kovalev, P.A. Nikishechkin and A.S. Grigoriev, ICIEAM - 2017, 1-4 (2017).
46. K. Vinod, A. Eakambaram, A. Arivazhagan, Proc. Eng. **97**, 1237-1246 (2014).
47. B. Bergmann, B. Denkena, T. Grove, et al., Int. J. Precis. Eng. Manuf. **20**, 37–44 (2019).
48. M. Volosova, S. Grigoriev, A. Metel, A. Shein, Coatings, 8, 287, (2018)
49. M. Tiffe, R. Aßmuth, J. Saelzer, et al., Prod. Eng. Res. Devel. **13**, 459–467 (2019).
50. B. Denkena, D. Biermann, CIRP Annals **63**(2), 631-653 (2014).

Analysis of Positioning Adjustment Approaches for Cutting Inserts of Thread Cutters

Vladimir Grechishnikov[1], *Alexander* Isaev[1], *Mikhail* Kozochkin[1], *Pyae Phyo Aung*[1]

[1] Moscow State University of Technology "STANKIN", RU-127055, Moscow, Russia

Abstract. The inclination angle of cutting inserts of thread cutters should be adjustable to provide the required geometry when machining threads. In serial production, this adjustment is realized by changing the carbide shims. In the present paper, the analysis of relationships between geometric parameters and positioning accuracy errors for thread cutters with inserts is presented.

1 Introduction

In Russian standard GOST 24705-81 "Metric thread. Main dimensions" the application of the threads with different pitches is described and regulated. This allows to take into account the purpose and requirements of the thread when designing new products, but increases the range of threading tools due to a change in the thread helix angle. Using the universal cutters with the cutting edge inclination angle $\lambda = 0°$ with a variety of the thread helix angles $\tau \neq 0$ does not guarantee optimal cutting conditions and significantly degrades the quality of production.

2 Methods and Results

To reduce the product range and to create optimal cutting conditions for machining threads using lathe turning machines, various designs of cutters were created with the ability to tilt the cutting element at the thread helix angle related to the workpiece axis along the pitch diameter in order to optimize the geometric parameters of the cutter [1-3]. This is confirmed by the diagram shown in Fig. 1, which shows two options for the position of the cutting insert. In Fig. 1a the thread helix angle is $\tau>0$ and the cutting-edge inclination angle $\lambda=0°$, i.e. the cutting angles on the right and left edges of the cutting part are different. In Fig. 1b, due to the basing shim with the angle $\tau=\lambda$, the cutting conditions for the cutting edges become symmetric.

The values of the clearance angles at the side cutting edges (Fig. 1a) at the angle of inclination of the cutting element $\lambda = 0°$ are determined by the formulas [4-5]:

$$\gamma 1= +\tau; \quad (1)$$

$$\gamma 2= -\tau; \quad (2)$$

$$\alpha 1=\alpha st - \tau; \quad (3)$$

$$\alpha 2=\alpha st + \tau; \quad (4)$$

$$tg\tau=p/(\pi dc), \quad (5)$$

where γ_1,γ_2 are the rake angles on the left and right cutting edge, respectively; α_1,α_2 are the clearance angles on the left and right cutting edge; τ is the thread helix angle on the pitch diameter; α_{st} is the clearance angle of cutting edges in static coordinate plane; P is the thread pitch; d_c is the pitch diameter, mm;

For this case, the basing shim is manufactured with the inclination angle $\lambda= 0$.

When fitting the rake face of the cutter by d_c in the plane normal to the turns of thread (Figure 1, b) at $\lambda_{insert}=\lambda=\tau$, the angles on the left and right cutting edges are determined by the formulas:

$$\gamma 1=\gamma 2=0; \quad (6)$$

$$\alpha 1=\alpha 2=\alpha st \quad (7)$$

The cutting conditions become equal for the right and left sides of the cutting edge [6-28].

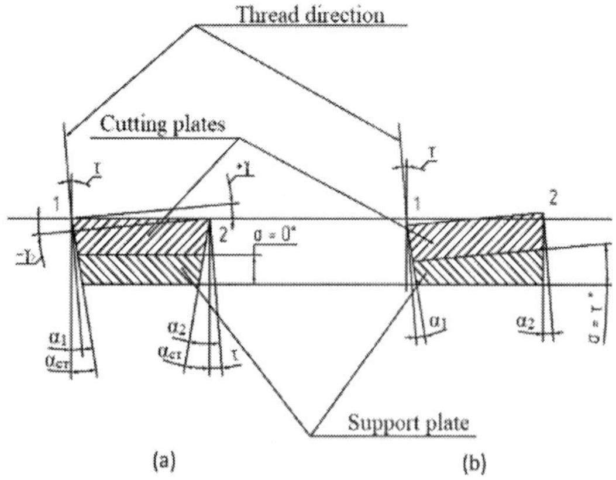

Fig. 1. Scheme of fitting the thread inserts: a) $\lambda=\lambda_{insert}=0°$, b) $\lambda=\lambda_{insert}=\tau$

* Corresponding author: avisz@yandex.ru

Selection of the dimensions range for basing shims using the values of the angle λ is performed by various analytic approaches. For a large dimension range, the number of shims decreases, but it becomes more difficult to choose a suitable angle of the shim. When the dimension range is small, the available quantity of shims increases. Thus, on average, the cutting tools manufacturers have from 7 to 8 shims providing a range of angles from -2.0 to 4.5 degrees.

There are technical solutions for the design of thread cutters with the adjustable angle λ which allows to shorten a dimension range for shims.

The thread cutting tool with a rotary holder having the adjustable cutting-edge inclination angle (with λ > 0) is shown in Figure 2(Patent 60414).

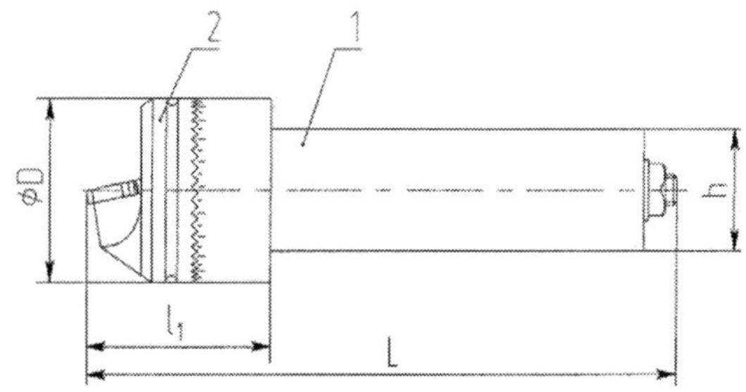

Fig. 2. Illustration of patent No. 60414 (RU): 1 – thread cutter's body; 2 – adjustable head

Threading tool with a rotary tool holder consists of body (1) and holder (2) having a carbide insert fastened with a screw. On the left end of the cutter body, the serrations are made, which in contact with the holder fix a position of the thread insert relative to the cut thread taking into account the change in the angle τ.

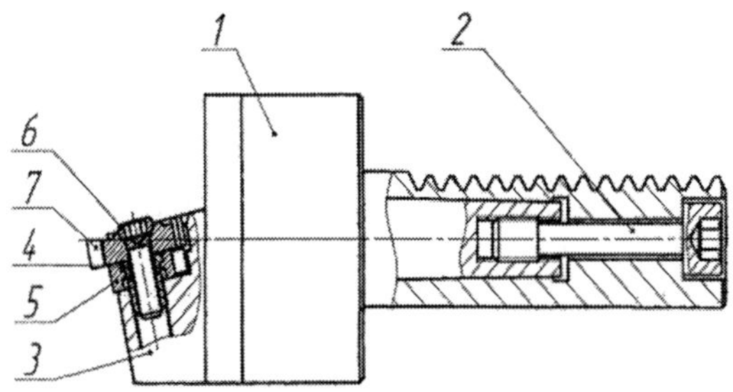

Fig. 3. Illustration of patent No. 60414 (RU): 1 - body; 2 - clamping screw; 3 - rotary holder; 4 - fixing spring of shim; 5 - shim; 6 - fixing screw; 7 - cutting insert

Another variant of the thread cutting tool with a rotary tool holder consists of body (1), scale holder (2), cutting replaceable insert (3), fixed screw (4) and basing shim (5). The body has a form of a stepped cylinder with a flattened surface and corrugations on it in the tail part and a vernier scale in the head part. The axis of rotation of the holder intersects the axis of the machine spindle and coincides with the symmetry axis of the cutting-edge tip. This axis is located in a plane passing through the cutter's tip parallel to the main plane. The basing shim is located between the cutting insert and the holder.

These designs are not widely used due to the complexity of their manufacture and large overall dimensions. And, as a consequence, the base shims remain the main option for adjusting the angle λ.

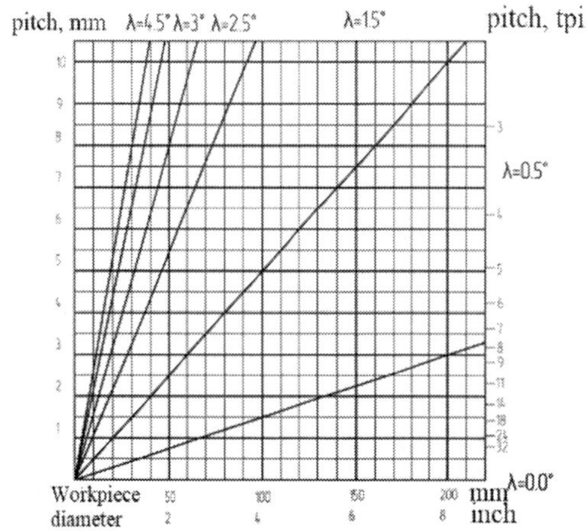

Fig. 4. Diagram of the selection of basing shims from the Vargus catalog

In the today's market of metalworking tools, there are two solutions for the choice of the dimension range of the basing shims. Different cutting edge inclination angles λ allow providing different cutting conditions on the side cutting edges and reduce the number of carbide shims, the cost of which is comparable with the cost of the cutting insert.

The selection diagram for the shims is shown in Figure 4 (for products by Vargus) and Figure 5 (for products by Sandvik Coromant). According to the selection diagram and depending on the diameter of the workpiece and the thread pitch, one can determine the required basing shim. Table 1 summarizes both solutions in a form convenient for analysis.

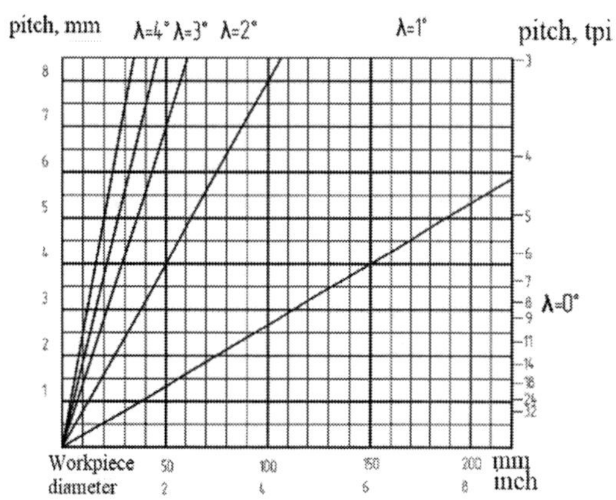

Fig. 5. Diagram of the selection of basing shims from the Sandvik Coromant catalog

Table 1. Recommendations by manufacturers of cutting tools for choosing the interval of dimensions

Cutting edge inclination angle λ (Vargus)	Thread helix angle τ		Cutting edge inclination angle λ (Sandvik Coromant)	Thread helix angle τ	
	min	max		min	max
-1,50	-2,00	-1,00	-2,00	-2,50	-1,50
-0,50	-1,00	-0,25	-1,00	-1,50	0,50
0,00	-0,25	0,25	0,00	-0,50	0,50
0,50	0,25	1,00	1,00	0,50	1,50
1,50	1,00	2,00	2,00	1,50	2,50
2,50	2,00	3,00	3,00	2,50	3,50
3,50	3,00	4,00	4,00	3,50	4,50

For convenience and comparison of options, tabular values are represented in the form of graphs (Figure 6 and 7). The charts show the difference in recommendations of the cutting tools producers.

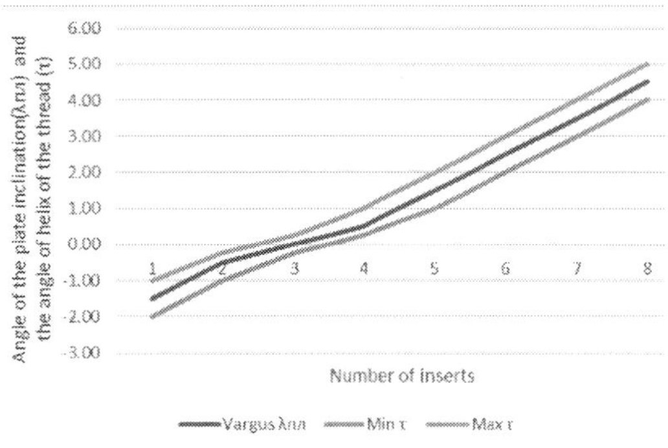

Fig. 6. Graphs for selection of λ angle by Vargus

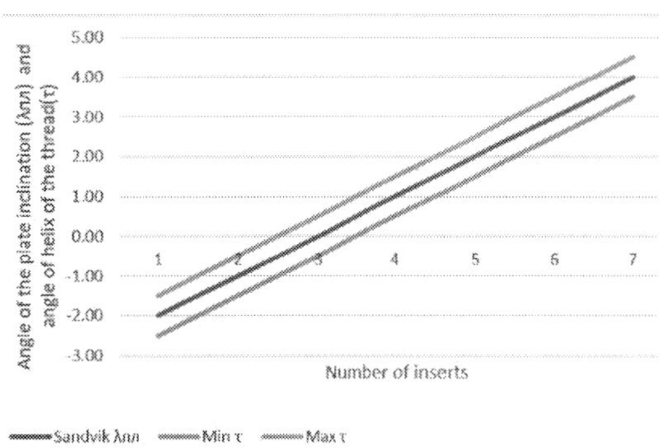

Fig. 7. Graphs for selection of λ angle by Sandvik Coromant

To evaluate the positioning accuracy of the basing shims, GOST 24705-81 was used. This standard has a fairly wide range of threads with different diameters and pitches. Figure 8 shows a diagram of the values of the thread helix angles depending on the diameter and pitch.

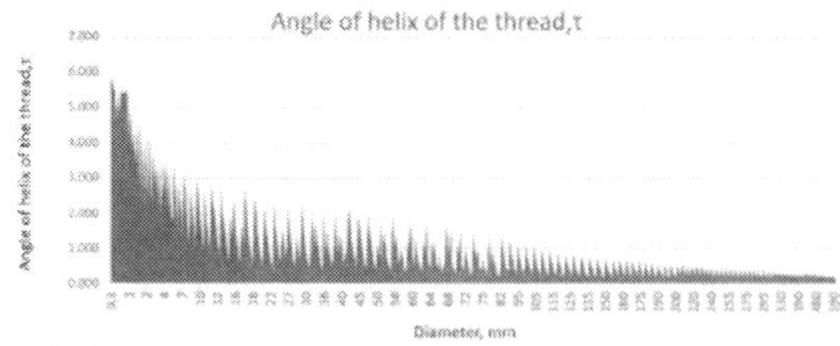

Fig. 8. Diagram of the thread helix angles depending on the diameter and pitch according to GOST 24705-81.

After combining the graph of the choice of λ and the sorted diagram of the thread helix angles τ, it can be concluded that the basing shims by Vargus are closer to the calculated values of the thread helix angles the thread on the d_c (Figure 9).

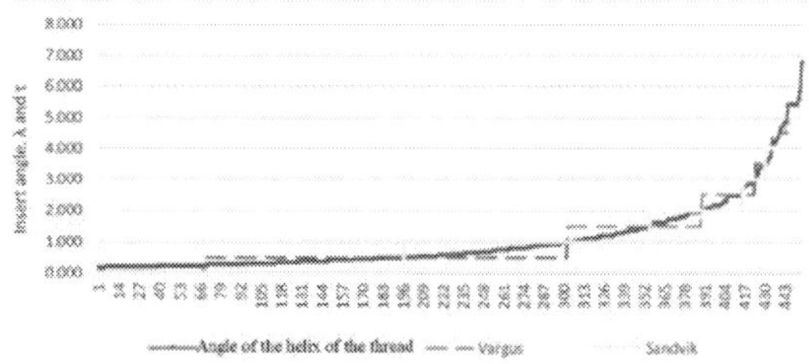

Fig. 9. Change of the thread helix angles and the choice of angle τ according to GOST 24705-81 and the choice of λ according to the manufacturers of the basing shims

3 Conclusions

The basing shims by Vargus are more preferable for machining threads according to GOST 24705-81 compared to Sandvik Coromant shims as they provide optimal cutting angles for the right and left cutting edges.

Increasing the range of basing shims reduces the installation errors of cutting inserts calculated using the values of the thread helix angle τ.

Acknowledgments

This research was funded by Ministry of Science and Higher Education of the Russian Federation, Grant No. 0707-2020-0031. This work was carried out using equipment provided by the Center of Collective Use of MSUT "STANKIN".

References

1. V.A. Grechishnikov, Y.E. Petukhov, V.B. Romanov, A.V. Isaev, P.V. Domnin, P.M. Pivkin, Rus. Eng. Res., **38**, 1018-1021 (2018)
2. V.A. Grechisnikov, Rus. Eng. Res. **40**, 51-54 (2020)

3. V.A. Grechishnikov, Y.E. Petukhov, P.M. Pivkin, A.A. Timofeev, A.V. Isaev, V.B. Romanov, I.V. Minin, L.E. Shvartsburg, S.V. Yurasov, Rus. Eng. Res., **40**, 768-771 (2020)
4. S.N. Grigoriev, G.M. Martinov, Procedia CIRP, **1**(1), 238 (2012)
5. V.A. Grechishnikov, S.N. Grigoriev, P.M. Pivkin, M.A. Volosova, A.V. Isaev, D.V. Nikitin, I.V. Minin, *EPJ Web of Conferences*, **224**, 05001 (2019)
6. V.V. Kuzin, S.N. Grigoriev, M.A. Volosova, Refract. Ind. Ceram., **54**, 376, (2014)
7. V.V. Kuzin, S.N. Grigoriev, M.A. Volosova, J. Frict. Wear, **36**(1), 40, (2015)
8. V.A. Grechishnikov, Y.E. Petukhov, P.M. Pivkin, A.V. Isaev, S.V. Bushuev, V.B. Romanov, Meas. Tech., **58**, 848–853 (2015)
9. A.V. Gusarov, S.N. Grigoriev, M.A. Volosova, Y.A. Melnik, A. Laskin, D.V. Kotoban, A.A. Okunkova, J. Mater. Process. Technol., **261**, 213 (2018)
10. A.S. Metel, V.P. Bolbukov, M.A. Volosova, S.N. Grigoriev, Y.A. Melnik, Instrum. Exp. Tech., **57**(3), 345, (2014)
11. R.S. Khmyrov, S.N. Grigoriev, A.A. Okunkova, A.V. Gusarov, Phys. Procedia, **56**, 345 (2014)
12. X. Kong, Z. Ding, L. Xu, L. Zhu, J. Zhang, C. Wu, A. Isaev, EPJ Web of Conferences, **224**, 05009 (2019)
13. A.V. Isaev, V.A. Grechishnikov. Rus. Eng. Res., **30**, 413-417 (2010)
14. S.N. Grigoriev, D.A. Masterenko, V.I. Teleshevskii, P.N. Emelyanov, Meas. Tech, **55**(11), 1311 (2013)
15. A.V. Isaev, M.P. Kozochkin, Meas. Techniq., **56**, 1155-1161 (2014)
16. M.A. Volosova, S.N. Grigoriev, A.S. Metel, A.A. Shein, Coatings, **8**, 287, (2018)
17. R.M. Khusainov, R.M. Khisamutdinov, S.Y. Yurasov, S.A. Belov, O.V. Goryacheva, V.A. Grechishnikov, A.V. Isaev, V.B. Romanov, Rus. Eng. Res., **37**, 344-347 (2017)
18. M.A. Volosova, S.N. Grigoriev, V.V. Kuzin, Refract. Ind .Ceram., **56**, 197, (2015)
19. S.N. Grigoriev, V.A. Grechishnikov, M.A. Volosova, X. Jiang, P.M. Pivkin, IOP Conference Series: Materials Science and Engineering, **971**, 022065 (2020)
20. S.N. Grigoriev, M.P. Kozochkin, P.M. Pivkin, 2020 International Multi-Conference on Industrial Engineering and Modern Technologies (FarEastCon) (2020)
21. P.M. Pivkin, A.A. Ershov, A.B. Nadykto, 2020 International Multi-Conference on Industrial Engineering and Modern Technologies (2020)
22. P.M. Pivkin, V.A. Grechishnikov, A.A. Ershov, A.B. Nadykto, Emerging Imaging and Sensing Technologies for Security and Defence V; and Advanced Manufacturing Technologies for Micro- and Nanosystems in Security and Defence III, **11540**, 1154014 (2020)
23. S.N. Grigoriev, P.M. Pivkin, V.A. Grechishnikov, Y.E. Petukhov, M.A. Volosova, A.B. Nadykto, Emerging Imaging and Sensing Technologies for Security and Defence V; and Advanced Manufacturing Technologies for Micro-and Nanosystems in Security and Defence III, **11540**, 115401E (2020)
24. P.M. Pivkin, A.B. Nadykto, V.A. Grechishnikov, M.A. Volosova, I.V. Minin, S.N. Grigoriev, Emerging Imaging and Sensing Technologies for Security and Defence V; and Advanced Manufacturing Technologies for Micro-and Nanosystems in Security and Defence III, **11540**, 115401G (2020)
25. S.N. Grigoriev, M.A. Volosova, A.A. Okunkova, S.V. Fedorov, K. Hamdy, P.A. Podrabinnik, P.M. Pivkin, M.P. Kozochkin, A.N. Porvatov, Technol., **8**(3), 49 (2020)
26. S.N. Grigoriev, M.A. Volosova, A.A. Okunkova, S.V. Fedorov, K. Hamdy, P.A. Podrabinnik, P.M. Pivkin, M.P. Kozochkin, A.N. Porvatov, J. Manuf. Mater. Process., **4**(3), 96 (2020)
27. S.N. Grigoriev, M.A. Volosova, V.A. Grechishnikov, P.M. Pivkin, P.Y. Peretyagin, A.A. Zelensky, Procedia CIRP, **95**, 337-342 (2020)
28. A.V. Isaev, R.R. Khamzin, I.V. Minin, P.M. Pivkin, M. Leonesio, MATEC Web of Conferences, **329**, 03073 (2020)

Mathematical Modeling of Production Processes of Discrete Machine-Building Enterprises Based on the Interaction of Simulation Systems and Operational Planning Systems

Vitalii A. Dolgov[1], *Petr A.* Nikishechkin[1,2,*], *Aleksandr A.* Leonov[2], *Sergey S.* Ivashin[1], *Nikita V.* Dolgov[1]

[1]LLC "Digital Systems Factory", Moscow, Russia
[2] Moscow State Technological University "STANKIN", RU-127055, Moscow, Russia

Abstract. Analysis of production systems (PS) of discrete multi-nomenclature machine-building enterprises is a complex task, its solution is necessary to support decision-making during technical re-equipment, modernization or technological preparation of production. The paper shows a concept of joint use of operational scheduling systems and simulation modeling systems to improve the efficiency and adequacy of PS analysis. The problem of determining the deviation of the planned state of the PS from the simulated state and evaluating the level of stability and stability of the PS behaviour on its basis is considered. It is revealed that the proposed approach allows us to more adequately determine the timing of the production program, assess the stability of the PS behaviour when using various planning logics and algorithms, and choose the best one for subsequent use in a real PS.

1 Introduction

Discrete multi-nomenclature machine-building enterprises are subject to periodic changes in the product range and production program. It has a significant impact on balancing the workload of workplaces and often requires making appropriate adjustments to the technological and organizational solutions used. A machine-building enterprise is a complex organizational and technical system and such decisions should be made reasonably on the basis of the results of a comprehensive analysis of performance indicators in the implementation of a promising production program. To a greater extent, the tasks of engineering today are relevant for discrete multi-nomenclature production, which is more susceptible to changes in the nomenclature, production program and the influence of other factors [1-3].

Solving the problems of analysis of the production system (PS) in the context of assessing the sufficiency of production capacities for the implementation of the production program of output, justifications for making changes to the PS, justification of investments, implementation of diversification projects, as well as risk assessments of such projects, are impossible without calculating and modeling the implementation of the production program with a given degree of detail, taking into account both the configuration and composition of production capacities, and planning and management algorithms [4].

One of the ways to solve the problem of analyzing and modeling the PS work of is to create its digital twin, as evidenced by a large number of publications in this field [5-11].

2 Modeling and analysis of production systems of discrete machine-building enterprises

From point of view of accurate planning of work at machine-building enterprises, systems of classes APS (Advanced Planning & Scheduling Systems) and MES (Manufacturing Execution Systems) are of interest. Such operational planning systems have a wide range of capabilities: assessment of the implementation of the production plan, drawing up detailed plans for the operation of equipment, performance analysis, dispatching of products. When developing production schedules in systems of this class, computational methods are used: methods of combinatorial optimization, theory of schedules, heuristic methods, which determine the logic of production management. Work schedules can be used to assess changes in the time of work load factors, the overall efficiency of OEE equipment (Overall Equipment Efficiency), production cycles of manufacturing products, etc. [2,7,11].

At the same time, a significant limitation of the calculation methods is the lack of the possibility of accounting for random events (equipment failures, supply failures, the occurrence of defects, etc.), the ability to work with indistinctly specified information in the form of value intervals, etc. Thus, the PS in systems of the MES/APS class is considered as a stable closed system, in which the occurrence of any event affecting the execution of the production program is set by a schedule.

* Corresponding author: npa@digitalfabrika.ru

However, practice shows that the PS of discrete multi-nomenclature machine-building enterprises are characterized by instability of the processes of management and transformation of the material flow. Therefore, the use of systems based only on calculation methods, including systems of the MES/ASP class, for the analysis of PS of multi-nomenclature machine-building enterprises does not provide the required reliability of the results [12,13].

The effects of random events are taken into account by systems based on modeling the behaviour of the PS during the execution of the production program. The simulation system is a tool that allows us to simulate the behaviour of the system when performing specified processes. Simulation systems (SIM) are based on the use of situational logic of decision-making in the management of material flow. Such systems include PlantTwin, Siemens Plant Simulation, DELMIA, AnyLogic, etc. [3].

For mass and in-line production, it is sufficient to specify the basic input data (information about products, processes and resources), as well as basic control algorithms (Fig. 1a) as the initial data for conducting process simulation with an acceptable level of adequacy.

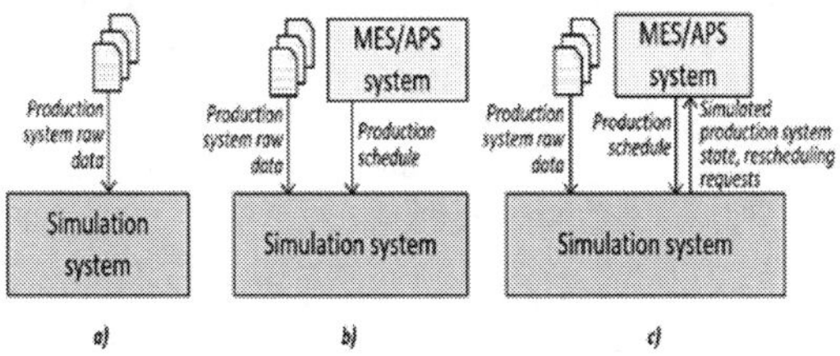

Fig. 1. Different approaches to simulation modeling

Speaking of discrete multi-nomenclature production, to simulate the operation of its production system, the source data must additionally contain the production schedule generated in MES/APS systems. It is important to note that the algorithms used to form the production schedule largely depend on the result of modeling and the implementation of the production program on time. At the same time, in contrast to MES/APS operational scheduling systems, simulation systems allow us to take into account additional factors that affect the behaviour of the production system over time and its functioning. These factors include: non-constant time of operations, failures, equipment breakdowns, non-constant time of transportation of parts, delivery delays, etc. Thus, the implementation of the interaction of MES/APS systems with simulation systems and the transfer of the production schedule to it can give a synergistic effect and significantly increase the adequacy of modeling [3,14].

Figure 1b shows the interaction scheme of the simulation system with the MES/APS system. In this case, the behaviour of the PS is modeled when executing a production schedule obtained from systems of the MES/APS class. At the same time, the above-described probabilistic and situational factors can affect the implementation of the production plan, reduce the stability of the system and lead to a deviation from the specified deadlines for the implementation of the production program.

Figure 2 shows a general graph of changes in the state of the production system over time when planning the execution of the production program, and then modeling its operation using a simulation system [15].

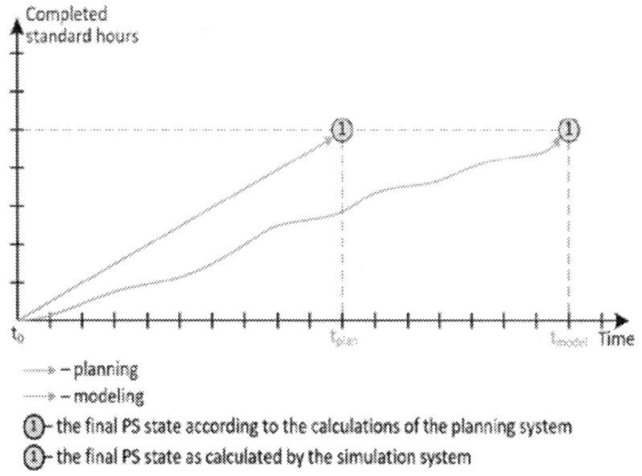

Fig. 2. Graphs of changes in the planned and simulated state of the PS during the execution of the production program

The figure shows that during the modeling process, discrepancies with the plan will invariably accumulate, which is caused by the above-mentioned additional factors, which, unlike planning systems, the simulation system takes into account. Ultimately, this is expressed in the form of a shift in the simulated timing of the production program. It is important to note that the tasks of the simulation system do not include the management and formation of corrective control actions for the return and "catch-up" of the planned dates calculated in the operational-calendar planning systems. The tasks of the simulation system include exclusively modeling the execution of the generated schedule, taking into account a large number of factors [16-19].

Factors affecting the stability of the production system, taken into account in simulation systems, can significantly affect the timing of the production program, as well as lead to the impossibility of its implementation. When a real production system is functioning, in such cases, work is usually rescheduled, i.e. operational scheduling systems form a new up-to-date production schedule based on the current actual state of the PS.

Taking into account the above, the paper proposes a statement of the problem of implementing iterative interaction of operational-calendar planning systems and simulation modeling systems for the possibility of performing rescheduling cycles in the modeling process. Thus, the simulation of the functioning of the PS becomes more adequate and close to the work of the real PS. Such a statement of the problem requires the implementation of feedback from the simulation system to the MES/APS system, to transfer the current simulated state of the PS to them and send a request for rescheduling (Fig. 1C) [3,12,20-22].

Figure 3 shows the change in the state of the PS over time in accordance with the calculations of the planning system, as well as the results of simulation modeling.

https://doi.org/10.1051/epjconf/202124804014

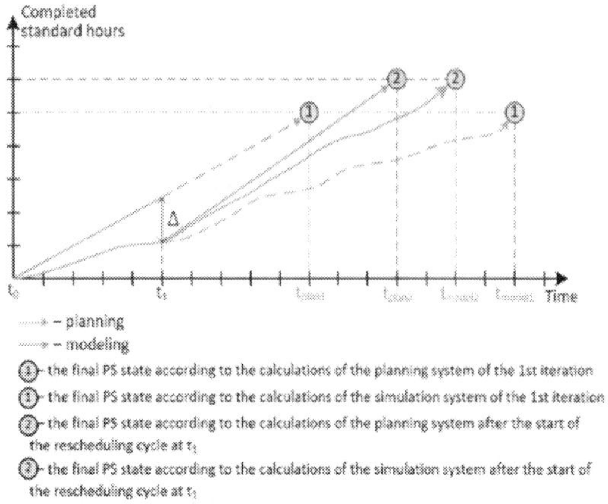

Fig. 3. Graphs of changes in the planned and simulated state of the PS during the implementation of the production program with the implementation of the rescheduling cycle

As can be seen from figure 3, the graph contains one rescheduling cycle at the point t1, at which the critical deviation (value Δ) of the simulated PS state from the planned one, expressed in the form of standard hours worked, was determined. The idea of implementing iterative interaction between the simulation system and operational scheduling systems is that during the simulation process, the planned and simulated work volumes are compared, expressed in standard hours. When the specified level of deviation of the simulated PS state from the planned one is reached (the value Δ), the experiment control module initiates a command to reschedule and form a new schedule in the MES/APS system [23].

The increase in standard hours is mainly caused by the excess of the number of defective products obtained during modeling from the specified norm during the development of the production schedule.

When implementing this approach, the simulation system allows us to determine the deviation of the simulated state of the PS from the planned one in time and acts as a kind of virtual dispatcher, which simulates the state of the PS when executing a given schedule and reports deviations from the specified plan, which initiates a command to reschedule work (Fig. 1c). Thus, the iterative formation of the production schedule and its verification is implemented. In the actual operation of the PS, MDC/MDA systems are used as a tool for evaluating the actual implementation of the plan and the presence of critical deviation factors, which allow real-time monitoring of the PS state [3,24].

The results of simulation of a PS with a cycle of rescheduling can be seen that the time of completion of the production program after the start of the cycle of rescheduling decreased in comparison with if system simulation will continue to run for a given plan, given the backorder from him. We can also see that when rescheduling cycles are implemented, the amount of standard hours worked may increase compared to the original calculated data. This may occur due to the fact that during the modeling process there was an excess of the permissible level of incorrigible defects in a number of nomenclature items, which led to an increase in the labor intensity of the production program and the production cycle [22-25].

An important task for the implementation of this approach to the analysis of PS and the forecast of the timing of the production program is to determine the fact of the occurrence of a critical deviation, which should cause the start of the rescheduling cycle [26-30]. This deviation can be expressed in quantitative terms or represent an event, the occurrence of which can disrupt the stability of the PS and require the start of a rescheduling cycle. The main reasons for the need to restart the development of the production schedule:

- deviation from the deadlines for the implementation of the specified production program. This factor can be expressed in quantitative terms, but it should be taken into account that this deviation may be different for different nomenclature items. In this case, it is necessary to determine the priority, i.e. deviations from the production time of which parts can lead to a significant violation of the stability of the PS, and for which parts this may be permissible;
- equipment failure. If the PS has equipment of the same type as the failed one, or an alternative equipment is specified for performing the necessary operation, then the occurrence of this event may not lead to the need for rescheduling. However, if there is no possibility of using other equipment, or all the equipment units from the group are out of order, this will lead to a significant deviation from the set plan and will require rescheduling;
-an incorrigible defect. If the modeling process reveals an excess of the norm of an incorrigible defect of the part/product, this should lead to the start of the rescheduling cycle, since the implementation of the production plan becomes impossible.

3 Conclusions

The article shows that the analysis of the PS of discrete machine-building enterprises is the basis for decision-making support during technical re-equipment, modernization or technological preparation of production. The proposed scheme of joint use of operational scheduling systems and simulation modeling systems will increase the efficiency and adequacy of the PS analysis. The problem of determining the deviation of the planned state of the PS from the simulated state and evaluating the level of stability and stability of the PS behaviour on its basis is considered. This approach will allow us to more adequately determine the timing of the production program, evaluate the stability of the PS behaviour when using various planning logics and algorithms, and choose the best one for subsequent use in a real PS.

Acknowledgments

Funding: This research was funded by Ministry of Science and Higher Education of the Russian Federation, Grant No. 0707-2020-0025.

References

1. S.N. Grigoriev, V.A. Dolgov, A.A. Leonov, IOP Conference Series: Materials Science and Engineering, **971** (2020)
2. A.I. Borovkov, Yu.A. Ryabov, Proceedings of the scientific and practical conference with foreign participation "Digital transformation of the economy and industry", 234-245 (2019)
3. P.A. Nikishechkin, S.S. Ivashin, V.E. Chernenko, A.A. Malykhanov, N.V. Dolgov, MATEC Web of Conferences, **329**, 03075 (2020)
4. S.N. Grigoriev, V.A. Dolgov, P.I. Umnov, N.N. Fokin, O.S. Proshkina, MATEC Web of Conferences, **329**, 03076 (2020)
5. J. Davis, T. Edgar, J. Porter, J. Bernaden, M. Sarli, Computers and Chemical Engineering, **47**, 145-156 (2012)
6. Y. Lu, C. Liu, K. Wang, H. Huang, X. Xu, Robotics and Computer Integrated Manufacturing, **61** (2020)
7. C.L. Tomovic, L.B. Ncube, A. Walton, M. Grieves, International Journal of Manufacturing Technology Management, **19**, 167-179 (2010)
8. A. Prokhorov, A.I. Borovkov, *Digital twin. Analysis, trends, world experience* (AliancePrint, Moscow, 2020)
9. Y. Cai, B. Starly, P. Cohen, Y. Lee, Procedia Manufacturing, **10**, 1031-1042 (2017)
10. J. Cheng, H. Zhanga, F. Tao, C. Juang, Robotics and Computer Integrated Manufacturing, **62** (2020)
11. V.E. Arkhangelskii, Requirements for production planning systems in the context of the concept "Industry 4.0 "[Electronic resource], VII international forum "Information technologies in the service of the military-industrial complex of Russia". Yalta, (2018, April 24-26), URL: http://итопк.RF/wp-content/uploads/2018/05/Arhangelskij.pdf (2018)
12. A.V. Gusarov, S.N. Grigoriev, M.A. Volosova, Y.A. Melnik, A. Laskin, D.V. Kotoban, A.A. Okunkova, J. Mater, Process. Technol, **261**, 213 (2018)
13. V.E. Chernenko, A.A. Malykhanov, N.V. Dolgov, Conference proceedings Simulation modeling. Theory and practice (IMMOD-2019), 37-46 (2019)
14. S.N. Grigoriev, G.M. Martinov, Procedia CIRP, **46**, 525 (2016)
15. S.N. Grigoriev, V.A. Dolgov, A.A. Leonov, IOP Conference Series: Materials Science and Engineering, **971** (2020)
16. M.A. Volosova, S.N. Grigoriev, E.A. Ostrikov, Mech. Ind., **17**, 720 (2016)
17. R.A. Nezhmetdinov, P.A. Nikishechkin, A.N. Nikich, International Russian Automation Conference (RusAutoCon), 1-4 (2018)
18. A.S. Metel, V.P. Bolbukov, M.A. Volosova, S.N. Grigoriev, Yu.A. Melnik, Instrum. Exp. Tech., **57**(3), 345 (2014)
19. P.A. Nikishechkin, N.Yu. Chervonnova, A.N. Nikich, MATEC Web Conf, **224** (2018)
20. S.N. Grigoriev, D.A. Masterenko, V.I. Teleshevskii, P.N. Emelyanov, Meas. Tech, **55**(11), 1311 (2013)
21. G.M. Martinov, P.A. Nikishechkin, A.S. Grigoriev et al., Automation and Remote Control, **80**(3), 584–591 (2019)
22. S.N. Grigoriev, G.M. Martinov, Procedia CIRP, **1**(1), 238 (2012)
23. P.A. Nikishechkin, N.Y. Chervonnova, A.N. Nikich, IOP Conference Series: Materials Science and Engineering, **709**(4), 044065 (2020)
24. R.S. Khmyrov, S.N. Grigoriev, A.A. Okunkova, A.V. Gusarov, Phys. Procedia, **56**, 345 (2014)
25. V.A. Grechishnikov, Y.E. Petukhov, P.M. Pivkin, Meas Tech, **58**, 848–853 (2015)
26. A.P. Okunev, A.I. Borovkov, A.S. Karev, D.O. Lebedev, V.I. Kubyshkin, S.P. Nikulina, A.A. Kiselev, Russian Engineering Research, **39**(6), 453-458 (2019)
27. G.M. Martinov, R.L. Pushkov, L.I. Martinova, N.V. Kozak, S.V. Evstafieva, MATEC Web of Conferences, **329**, 03026 (2020).
28. S.N. Grigoriev, P.M. Pivkin, V.A. Grechishnikov, Y.E. Petukhov, M A. Volosova, A.B. Nadykto, **11540**, 03076 (2020)
29. S.N. Grigoriev, M.A. Volosova, A.A. Okunkova, S.V. Fedorov, K. Hamdy, P.A. Podrabinnik, A.N. Porvatov, J. Manuf. Mater. Process., **4**(3), 96 (2020)
30. S.N. Grigoriev, V.A. Grechishnikov, M.A. Volosova, X. Jiang, P.M. Pivkin, IOP Conference Series: Materials Science and Engineering, **971**, 022065 (2020)

Models for Managing Production Systems of Machine-Building Enterprises Based on the Development and Using of Their Digital Twins

Vitalii A. Dolgov[1], *Petr.* A. Nikishechkin[1,2,*], *Vladimir* E. Arkhangelskii[3], *Pavel* I. Umnov[4] and *Alexey* A. Podkidyshev[1]

[1]LLC "Digital Systems Factory", Moscow, Russia
[2]Moscow State Technological University "STANKIN", RU-127055, Moscow, Russia
[3]LLC "AAMC", RU-121471, Moscow, Russia
[4]PJSC «DSPP», RU-141700, Moscow region, Dolgoprudny, Russia

Abstract. The paper discusses the goals and objectives of creating digital twins of the production system of a machine-building enterprise. The data structure of the information model of the production system of a machine-building enterprise, which is the basis for building a digital twin, is presented. The paper shows the main approaches to managing a production system based on the construction of its digital twin. It is revealed that along with traditional approaches to PS management by forming recommendations in terms of PS engineering and its operation, the choice of the most rational PS management algorithms that take into account the peculiarities of the production process organization and ensure the formation of production schedules that take into account PS reliability indicators has a great potential. It is proposed to use a specialized language of DPML to describe the information model of the PS through the "product-process-resource" paradigm, which ensures the coordination of the processes of forming recommendations in terms of engineering and operation of the PS, as well as the choice of the most rational algorithm for managing the PS.

1 Introduction

Modern trends in the development of information technologies in the world in recent years are inextricably linked with the emergence of new approaches to the design and organization of production. An important transition period was the emergence of the concept of "Industry 4.0", the main idea of which is to digitalize production, increase the level of automation and change the role of man in production. In the digital economy, new technologies associated with high - tech industry are emerging - digital twins and digital shadows. These concepts have many interpretations and definitions today, but in a generalized form, a digital twin is a virtual copy of an object/process/system with a given level of compliance (adequacy). The main task of building any digital twin is to diagnose and predict the operation of the object/process/system. a digital twin of an object / process / system can be created and used at various stages of the object/process/system life cycle to solve specific tasks. The concept of digital twins is relevant in many industries, such as energy, transport, and construction, but it primarily works in the high-tech industry [1-3].

The creation of digital twin of objects/products is in some way solved with the use of product support and life cycle management technologies: CALS (Continuous Acquisition and Life Cycle Support) and PLM (Product Lifecycle Management) – organizational and technical systems that support all information about the product and related processes from design and production to decommissioning [4].

Speaking about the creation of a digital twin of a complex system, for example, a machine-building enterprise, it can be noted that today there are no real examples, as well as tools that allow creating a digital model of an entire enterprise with a high level of adequacy, taking into account all production processes (technological, resource management, accounting, labor protection, safety, etc.), their regulatory and resource support [5].

Thus, it is advisable to talk about the gradual construction of a digital twin of the enterprise, in accordance with clear aims and objectives that need to be reached through its functioning. This paper discusses the issues and aspects of building a digital twin of a production system (PS) of an enterprise, includes work equipment, products, production processes (technological processes, maintenance and repair processes, transportation and storage processes) and production planning systems and modeling and analysis tools, the joint functioning of which makes it possible to produce products [2,6-9].

* Corresponding author: npa@digitalfabrika.ru

2 Theoretical aspects of building a digital twin of the production system of a machine-building enterprise

Building a digital twin of the PS can be divided into two stages: digital twin of PS prototype and digital twin of PS instance (Fig. 1).

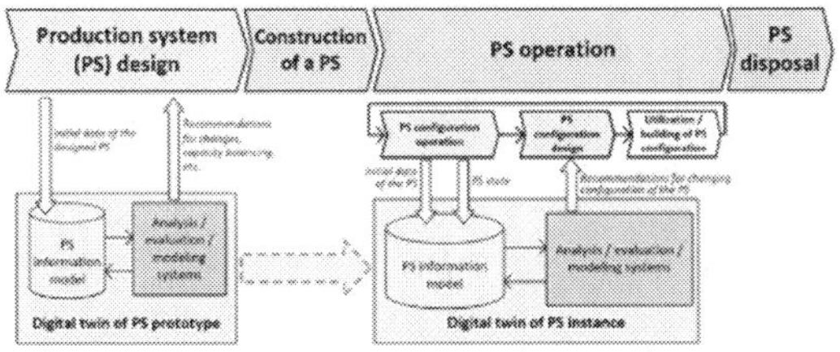

Fig. 1. Building a digital twin of a production system at various stages of its life cycle

At the design stage of the production system, a digital twin of the PS prototype is developed, on the basis of which technological solutions are tested and recommendations for changing design solutions are formed [7-11]. It is possible to distinguish the main characteristics of digital twin of the PS prototype of a machine-building enterprise:
- developed before the production system is available based on design decisions;
- does not include measurement data and reports coming from a real system;
- based on a highly accurate mathematical model;
- allows us to evaluate the behavior of a real system based on its model;
- used to further create the digital twin instance of the PS [5].

At the operational stage, a digital twin of the PS instance is being developed. At this stage, the PS is subject to periodic configuration changes due to changes in the product range and production program. The digital twin of PS instance of the machine-building enterprise is developed on the basis of the digital twin of the PS prototype and has additional characteristics:
- corresponds to a specific PS;
- allows us to diagnose and evaluate the behaviour of the PS based on its model, including statistical data on its work;
- contains information about the model settings, control parameters, sensor data and statistics about the operation of the PS.

As figure 1 shows, both the digital twin of the PS prototype and the digital twin of the PS instance are based on its information model, as well as on analysis, evaluation and modelling systems [10].

Build a digital twin PS of machine - building enterprises are almost impossible without functioning information systems of the enterprise, which is the source of data for constructing the information model of the substation and performing coordinated control of production and business processes at various levels of enterprise management [11].

The information model of the production system should contain machine-oriented data for their subsequent processing in analysis and modeling systems. the data set contained in the information model should be refined for specific purposes and tasks of the production system analysis. Figure 2 shows the data flows for the formation of the information model of the PS and further analysis and modeling in the corresponding systems.

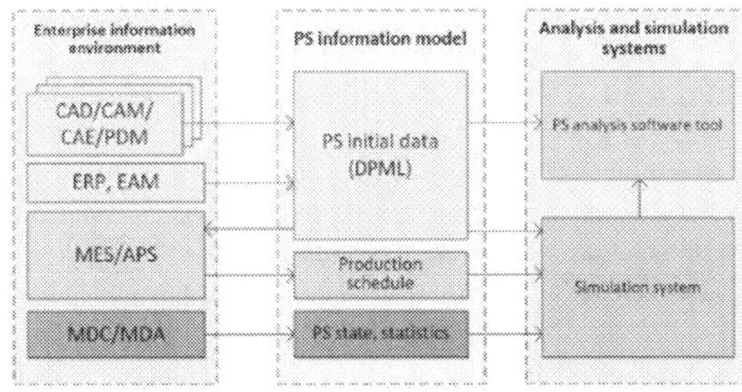

Fig. 2. Architecture for building a digital twin of a PS instance

The information model of the PS can be divided into three blocks: the initial data of the PS, the production schedule formed in accordance with the production plan, as well as a set of data on the current state of the PS and statistics of its work. To describe the initial data on the PS, it is proposed to use the universal language for describing products, processes and resources DPML (Discrete Process Modeling Language), which allows describing the PS in the form of a block diagram in a visual graphical notation. table 1 provides a description of each of the data blocks, the main sources of information, the format of presentation and the degree of variability of data within a single PS cycle.

Table 1. Structure of building an information model of the PS for the development of a digital twin

Data	References	Data description format	Estimate of variability of the data during operation
Nomenclature of manufactured products (products)	Production preparation and product lifecycle management systems (PDM, CAD/CAE, CAM/CAPP)	Universal description language DPML (Discrete Process Modeling Language)	Conditionally constant data
Production processes			Conditionally constant data
Production resources	Enterprise resource and production management systems (ERP)		Conditionally constant data
Production schedule	Production process management and planning systems (MES, APS)	Specialized schedule file	Periodically changing data
PS status, statistics	Monitoring and dispatching systems (MDC, MDA)	Database	Changing data

The main block of the information model is the initial data on the PS described in the "product-process-resource" paradigm, which can be attributed to conditionally unchanged data within a single cycle of PS operation. This block contains information about the range of manufactured products, production processes and production resources of the PS in the form of a DPML model. This model is a data source for the formation of local information models for various consumer information systems, which can include MES/APS systems, as well as PS analysis and modelling systems [10,12].

System MDC/MDA class required to collect information about the machines and monitoring jobs, which allows to collect statistical data on the PS and determine its reliability taken into account in the further analysis of the PS and the simulation of its operation. MDC/MDA systems automatically record the time and duration of operation/downtime/emergency conditions of the machine, as well as allow to collect data on technological modes (feed, speed, load), currently running control program (CP), error codes, causes of downtime, etc [13].

The production schedule formed in systems of the MES/APS class can be considered the main tool for managing the PS, provided that the basic data block - the source data of the PS presented in the form of a DPML model-is relatively unchanged. PS can have different levels of stability in individual areas, and for their accounting, it is necessary to conduct a preliminary analysis of PS, as well as verification of the generated production schedules. To solve these problems, a comprehensive solution is proposed, including analytical (computational) methods, as well as simulation modelling [14-17].

Analysis of the enterprise's PS and decision-making support for its management can be carried out using a specialized software tool, the main functions of which are:

- pre-processing and verification of data required for PS analysis;
- calculation of a set of production and economic indicators of the efficiency of the PS functioning on the basis of initial data (product range, technological processes, production resources, production plan). The tool provides the ability to calculate indicators for different levels of enterprise decomposition (enterprise, shop, site, workplace);
- analysis of simulation results, including calculation of performance indicators of the enterprise's PS;
- generating analytical reports based on the results of calculating performance indicators and modeling results necessary to support decision-making by managers at various levels [5,18].

The software tool for conducting production system analysis allows us to perform a basic calculation of PS performance indicators both on the basis of initial data about the system and on the basis of simulation results, as well as to analyze and compare them with actual indicators obtained from MDC/MDA systems. The production system at this stage is considered as a stable system. The result of the software tool is the formation of analytical reports containing the necessary data in a visually understandable form, providing decision-making support for managers at various levels [8,9,19-22].

The next stage of the analysis is the simulation of the functioning of the entire PS or its individual divisions. simulation modeling in the form of an algorithm reproduces the logic of the PS functioning in time and allows us to take into account the sequence and interaction of production processes, as well as random events. In simulation modeling, the PS is considered as an unstable system and random factors arising in the PS are taken into account. System simulation discrete industries PlantTwin allows to consider, besides the basic information about the production system, the following number of additional factors that affect the functioning of the PS: irregular execution time of technological operations; delays and deviations of the timing of delivery of components; different time transportation of workpieces between production units; limitation of the buffer capacity (warehouses, storage) in various stages of the production cycle; unscheduled repairs of process equipment with regard to crashes; repairs of equipment by operating time, etc. [23-26]. Accounting for the above factors in other operational scheduling systems is usually not carried out, but in the course of production, these factors can have a significant impact on the implementation of the production plan.

Thus, at this stage of the analysis of the production system with the use of simulation modeling, it is possible to solve the following tasks:

- quantitative assessment of the production capacity of the considered production system for the implementation of the production program;
- justification of the necessary composition and quantity of technological equipment (jobs) to run a given production program;
- validation of production plans and schedules using simulation;
- analysis of the behavior of the production system and development of organizational and technological recommendations to improve its efficiency.

Also, systems of this class allow to model different experiments and simulate the system behavior under different scenarios: for example, to assess the impact of organizational and technological solutions in the work of the PS, which allows for detailed and rapid analysis of design decisions [12,27].

The results of the analysis of PS based on building its digital twin, one way to control is to change the logic and algorithms for building a production schedule that does not require modification of the unit source data PS. Since simulation systems take into account a number of additional factors that affect the functioning of the PS, based on the simulation results, it is possible to give a more adequate answer about the feasibility of the production schedule and the stability of the PS with the selected control logic. thus, for effective PS management, it is necessary to perform preliminary verification of production schedules generated using various algorithms and determine the best of them [28-30].

Another way to manage the PS is to form recommendations regarding the engineering of the PS or the operation of the PS, i.e. changing the initial data of the PS. Best practices for engineering PS can be attributed to change of the scheme of cooperation, the composition of jobs and equipment, structures, technological processes, technological equipment and technological conditions, the degree of automation of the technological process (reduction of the human factor), technological procedure of manufacturing products, increasing the effectiveness of the system of technological preparation of production and flexibility of PS (reduction of changeover time). Suggestions for changing the structure of workshops, organization of the maintenance and repair system, processes and algorithms of the logistics system, conditions for the development of production schedules, work schedules of workshops and sites, personnel composition can be considered as recommendations for the operation of PS [31-34].

3 Conclusions

Thus, the solution of the problem of effective PS management using its digital twin is based on the task of creating an adequate and consistent information model that defines the requirements for the data structure necessary to solve the corresponding problems, as well as for the composition and interaction of information systems containing and processing this data. The problems of analysis, modelling, valuation, forecasting and management of the PS is impossible without a coherent, logically related source data of it, as18 well as the complex interactions of systems of operative-calendar planning and systems analysis, evaluation and modelling of the PS.

It is shown that along with traditional approaches to PS management by forming recommendations in terms of PS engineering and its operation, the choice of the most rational PS management algorithms that take into account the peculiarities of the organization of the production process has a great potential. The selected algorithms ensure the formation of production schedules that take into account the reliability indicators of the PS. The production schedules formed in this way provide a higher stability of the PS during their execution, which is confirmed by the results of analysis and modelling using both computational methods and simulation modelling [20, 21, 26].

The article suggests the use of a specialized DPML language to describe the information model of the PS through the "product-process-resource" paradigm, which ensures the coordination of the processes of forming recommendations in terms of engineering and operation of the PS, as well as the choice of the most rational algorithm for managing the PS.

Acknowledgments

Funding: This research was funded by Ministry of Science and Higher Education of the Russian Federation, Grant No. 0707-2020-0025.

References

1. S.N. Grigoriev, V.A. Dolgov, A.A. Leonov, IOP Conference Series: Materials Science and Engineering, **971** (2020)
2. A.I. Borovkov, Yu.A. Ryabov, Proceedings of the scientific and practical conference with foreign participation "Digital transformation of the economy and industry", 234-245 (2019)
3. A.V. Gusarov, S.N. Grigoriev, M.A. Volosova, Y.A. Melnik, A. Laskin, D.V. Kotoban, A.A. Okunkova, J. Mater. Process. Technol., **261**, 213 (2018)
4. V.E. Arkhangelskii, *Requirements for production planning systems in the context of the concept "Industry 4.0* "[Electronic resource], VII international forum "Information technologies in the service of the military-industrial complex of Russia". Yalta, (2018, April 24-26), URL: http://итопк.RF/wp-content/uploads/2018/05/Arhangelskij.pdf (2018)
5. R.A. Nezhmetdinov, P.A. Nikishechkin, A.N. Nikich, International Russian Automation Conference (RusAutoCon), Sochi: IEEE, 1-4 (2018)
6. C.L. Tomovic, L.B. Ncube, A. Walton, M. Grieves, International Journal of Manufacturing Technology Management, **19**, 167-179 (2010)
7. A. Prokhorov, A.I. Borovkov, *Digital twin. Analysis, trends, world experience* (AliancePrint, Moscow, 2020)
8. Y. Cai, B. Starly, P. Cohen, Y. Lee, Procedia Manufacturing, **10**, 1031-1042 2017.
9. J. Cheng, H. Zhanga, F. Tao, C. Juang, Robotics and Computer Integrated Manufacturing, **62** (2020).
10. V.E. Arkhangelskii, *Operational model of production as a standard component of means of operational planning of order-based production* [Electronic resource], VI international forum "Information technologies in the service of the military-industrial complex of Russia", Izhevsk, (2017, June 20-22). URL:http://aamc.ru/wp-content/uploads/2018/06/ITOPK2017-ArkhangelskyVE-WithNotes_v102.pdf (2017)
11. P.A. Nikishechkin, N.Yu. Chervonnova., A.N. Nikich, Lecture Notes in Electrical Engineering 18761100, **641,** 181-188 (2020)
12. G.M. Martinov, R.L. Pushkov, L.I. Martinova, N.V. Kozak, S.V. Evstafieva, MATEC Web of Conferences, **329**, 03026 (2020).
13. S.N. Grigoriev, V.A. Dolgov, P.I. Umnov, N.N. Fokin, O.S. Proshkina, MATEC Web of Conferences, **329**, 03076 (2020)
14. J. Davis, T. Edgar, J. Porter, J. Bernaden, M. Sarli, Computers and Chemical Engineering, **47**, 145-156 (2012)
15. Y. Lu, C. Liu, K. Wang, H. Huang, X. Xu, Robotics and Computer Integrated Manufacturing, **61** (2020).

16. S.N. Grigoriev, V.A. Dolgov, P.A. Nikishechkin, N.V. Dolgov, IOP Conference Series: Materials Science and Engineering, **971** (2020)
17. V.A. Dolgov, A.A. Kabanov, Automation. Modern technologies, **72(4)**, 178-184 (2018)
18. P.A. Nikishechkin, S.S. Ivashin, V.E. Chernenko, A.A. Malykhanov, N.V. Dolgov, MATEC Web of Conferences, **329**, 03075 (2020)
19. V.A. Dolgov, A.A. Podkidyshev, I.V. Datsyuk, A.A. Kabanov, M.A. Vasiltsov, Automation. Modern technologies, **72**(8), 350-354 (2018)
20. M.A. Volosova, S.N. Grigoriev, E.A. Ostrikov, Mech. Ind., **17**, 720 (2016)
21. S.N. Grigoriev, D.A. Masterenko, V.I. Teleshevskii, P.N. Emelyanov, Meas. Tech, **55**(11), 1311 (2013)
22. R.S. Khmyrov, S.N. Grigoriev, A.A. Okunkova, A.V. Gusarov, Procedia **56**, 345 (2014)
23. P.A. Nikishechkin, N.Y. Chervonnova, A.N. Nikich, IOP Conference Series: Materials Science and Engineering, **709**(4), 044065 (2020)
24. S.N. Grigoriev, G.M. Martinov, Procedia CIRP, **46**, 525 (2016)
25. G.M. Martinov, P.A. Nikishechkin, A.Al. Khoury, A. Issa, IOP Conf. Ser.: Mater. Sci. Eng, **919**(3), 032030 (2020)
26. G.M. Martinov, R.L. Pushkov, L.I. Martinova, N.V. Kozak, S.V. Evstafieva, MATEC Web of Conferences, **329**, 03026 (2020).
27. S.N. Grigoriev, G.M. Martinov, Procedia CIRP, **1**(1), 238 (2012)
28. V.A. Grechishnikov, S.N. Grigoriev, P.M. Pivkin, et. al., EPJ Web Conf., **224,** 05001 (2019)
29. A.S. Metel, V.P. Bolbukov, M.A. Volosova, S.N. Grigoriev, Yu.A. Melnik, Instrum. Exp. Tech., **57**(3), 345 (2014)
30. V.A. Grechishnikov, Y.E. Petukhov, P.M. Pivkin, Meas Tech, **58**, 848–853 (2015)
31. A.P. Okunev, A.I. Borovkov, A.S. Karev, D.O. Lebedev, V.I. Kubyshkin, S.P. Nikulina, A.A. Kiselev, Russian Engineering Research, **39**(6), 453-458 (2019)
32. S.N. Grigoriev, P.M. Pivkin, V.A. Grechishnikov, Y.E. Petukhov, M A. Volosova, A.B. Nadykto, **11540**, 03076 (2020)
33. S.N. Grigoriev, M.A. Volosova, A.A. Okunkova, S.V. Fedorov, K. Hamdy, P.A. Podrabinnik, A.N. Porvatov, J. Manuf. Mater. Process., **4**(3), 96 (2020)
34. S.N. Grigoriev, V.A. Grechishnikov, M.A. Volosova, X. Jiang, P.M. Pivkin, IOP Conference Series: Materials Science and Engineering, **971**, 022065 (2020)

Application of Circuit Modeling in the Study of Spark Formation During Electroerosion Treatment

Artur Porvatov[1,4], *Mikhail* Kozochkin[1], *Ilya* Minin[1,2,3,4], *Andrey* Maslov[1], *Vladimir* Kuptsov[3], *Marian* Morozkin[3], *Vladimir* Kuznetsov[5], *Marko* Leonesio[6], *Alexander* Zelensky[1], *Petr* Pivkin[1,2,3,4]

[1]Dept. of High-Efficiency Processing Technologies, Moscow State University of Technology "STANKIN", RU-127055, Moscow, Russia,
[2]Dept. of Cutting Tools and Shaping Technologies, Moscow State University of Technology "STANKIN", RU-127055, Moscow, Russia;
[3]Lab. of Micromachining Technologies, Moscow State University of Technology "STANKIN", RU-127055, Moscow; Russia;
[4]Center of Cognitive Technologies and Machine Vision, Moscow State University of Technology "STANKIN", RU-127055, Moscow; Russia;
[5]Moscow Polytechnic University, RU-107996; Moscow; Russia;
[6]Institute of Industrial Technologies and Automation, Italian National Research Council, IT-00187 Rome, Italy.

Abstract. Electrical discharge machining (EDM) of a workpiece is a complex, fast-flowing process characterized by alternating (intermittent) states: short circuit, idle and spark formation. The discontinuity of the EDM process means that the processing is carried out in single pulses, which are formed by a special pulse generator. The parameters of the generator pulses can be divided into time and electrical. The time period and duration of the pulses, as well as the difference between these two parameters (the "silence" interval), are considered temporary. The electric ones include the amplitude value of the voltage, the maximum permissible current, and the polarity of the pulses. in addition, depending on the device of the generator, the pulses can be composite, for example, have an igniting pulse with a higher voltage and a lower current than the main (working) pulse. In this work, we have developed a interelectrode gap model that allows not only to obtain values of electrical parameters, but also to evaluate and to optimize the electrical parameters of materials being processed with known electrical properties. The key advantage of this model is its modularity, which allows to add new functional blocks, which describe external and internal influences, for example, the concentration of erosion products, uneven electrical conductivity of the workpiece, and others, without changing its structure.

1 Introduction

The EDM process is influenced by various factors such as the electrical conductivity of the workpiece and liquid, the state of the electrode, the parameters of the generator and others. One of them is the state of the interelectrode gap (IEG). The state of the IEG has concentrated most of the parameters that are difficult or impossible to control in the EDM process, so its assessment is carried out indirectly, for example, by current. However, this does not give a complete picture of what is happening with EDM, and even more so does not allow you to estimate in advance certain EDM parameters necessary for setting up the generator [1].

The IEG is the distance between the electrode and the workpiece [2]. With EDM, the field strength in the IEG increases up to the discharge with a decrease in this distance. conditionally, depending on the interelectrode distance, the IEG can be in three states (fig. 1):

1) idling (XX), at h2 < IEG < ∞;
2) operating pulses (ri), at h1< ieg < h2;
3) short circuit (short circuit), at 0 < ieg < h1,

where h1 and h2 are the lower and upper bounds for the occurrence of discharges, respectively.

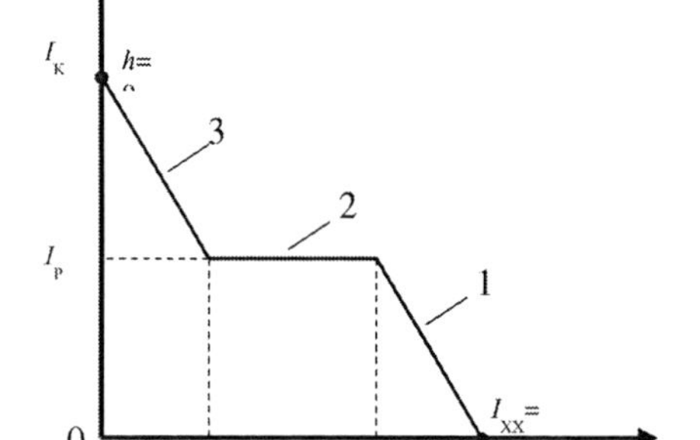

Fig. 1. Dependence of current (I) on the value of IEG (h) for EEP

The IEG model should take into account the transition from one state, since the EEP process itself depends on it. The description of the IEG model can be simplified by introducing assumptions:

1) when xx discharges are absent, there is no material removal and the IEG behaves like a dielectric;

2) when short-circuit discharges are present, but the removal of material does not occur and the IEG behaves as an ideal conductor.

Thus, the main interest is in area 2 (see Fig. 1), in which the material is removed.

In this area, a discharge occurs in the IEG (during the operation of the generator and the formation of current pulses), which takes place in 6 stages:

1) the appearance of an electromagnetic field between the electrode and the workpiece;
2) ionization of the fluid in the IEG;
3) formation of a conduction channel;
4) spark discharge-conversion of electrical energy into thermal energy (formation of a plasma channel) - destruction of the material;
5) crater formation;
6) repeat steps 1-5.

At stage 4, a plasma channel is formed in the IEG and the material of the electrode and part is destroyed, which occurs unevenly for both of them and depends on the polarity of the applied voltage. Thus, it can be concluded that with an increase in voltage or a decrease in distance, the resistance of the IEG before discharge decreases sharply and increases after it. The increase in resistance is due to the ionization of the liquid in the IEG after discharge. the time associated with" de-ionization " (restoration of the IEG resistance after discharge), along with the pulse time, is often entered as a parameter for controlling the eep process, and the pulse repetition period is calculated as their sum. also, after the discharge, erosion products are formed, which reduce the resistance of the IEG and interfere with the passage of pulses between the electrode and the part. These two parameters can be combined in one concept – electrical strength.

2 Results and Discussion

The electrical strength of the working fluid (RJ) strongly depends on the presence of various impurities, including erosion products. Taking as a basis the concept of concentration, which describes the percentage of erosion products in the IEG from 0 to 100%, it is possible to express the breakdown strength of RYE contaminated with erosion products as a function of the distance and concentration of erosion products between the electrode and the part [3–4]:

$$\begin{cases} h_{МЭП} \approx h_{пр} = \dfrac{U_И}{E_{пр}} \\ E_{пр.з.} = E_{пр}\left(1 - \dfrac{K_З}{100\%}\right) \end{cases}$$

where the voltage pulse (In), EPR – electric strength of uncontaminated working fluid (V/m), map – value MEHP (m), p – value breakdown of IEG (m), KZ is the concentration of erosion products in the map (from 0 to 100%), EPR.z. – tension breakout RJ contaminated with products of erosion.

Sometimes researchers use the schematic representation model of the IEG, shown in Fig. 2. Based on the current-voltage characteristic of the IEG [2], the circuit model of the IEG will be characterized by equivalent parameters: inductance Leq, resistance Req and capacitance Ceq (Fig. 2)

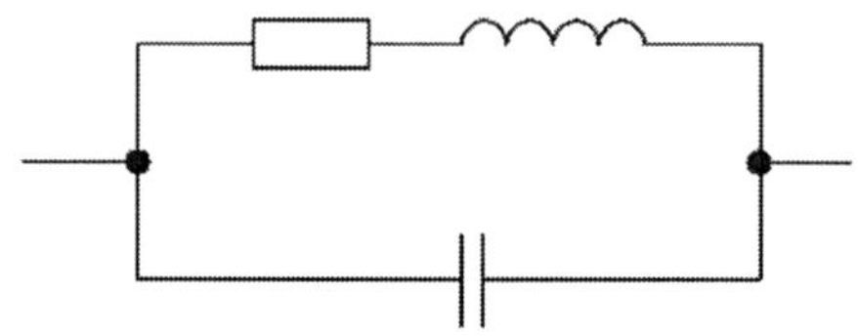

Fig. 2. Schematic model of IEG with equivalent L_{eq}, R_{eq}, C_{eq}

Equivalent parameters (L_{eq}, R_{eq}, C_{eq}) can be determined using the equations:

$$R_{eq} = \frac{d}{\sigma \cdot S} \tag{1}$$

and

$$C_{eq} = \varepsilon_0 \varepsilon_r \frac{S}{d} \tag{2}$$

Where $S = \pi \cdot r_э^2$ is the area of the end of the cylindrical electrode (rэ is the radius of the electrode), d is the distance of the IEG, σ is the specific conductivity of the medium, εr is the permittivity of the medium, εr = 8.854·10-12 F/m is the electric constant,

Knowing the processing parameters (the rate of change of current di/dt and the value of voltage (V_L (t)) in the IEG zone), the equivalent inductance can be estimated by the formula

$$V_L(t) = L(t)\frac{dt}{dt}$$

However, this model does not take into account the fact that these parameters can change, since the geometry and mutual location of the workpiece and the electrode, as well as the concentration of erosion products in the IEG during the EEP process, change. It is also necessary to take into account the properties of the plasma channel formed during the passage of the discharge and the processing parameters set by the generator – pulse time, duty cycle, maximum and average current. In [3], the inductance of the plasma channel that occurs when the current is discharged as estimated as

$$L(t) = 2d\left(\ln\left(\frac{2d}{a(t)}\right) - 0.75\right) \quad (3)$$

where L (t) is the inductance of the plasma (ngn), d is the IEG, and a (t) is the radius of the plasma channel (mcm), depending on the current and properties.

$$a^2(t) = \left(\frac{4}{\pi^2 \rho_0 \zeta \sigma}\right)^{1/3} \int_0^t i^{2/3}(t)dt \quad (4)$$

where i is the channel current (A), p_0 is the initial gas density, σ is the conductivity, and ζ is the temperature coefficient [4].

As can be seen from (4), the current values of current and voltage, which are largely determined by the EEP pulse generator, also affect the equivalent electrical parameters of the MEP. Thus, the MEP model should additionally include a model of a pulse generator, the main parameters of which are:

- Operating current I, A;
- Operating voltage V, V;
- operating frequency f, Hz:

$$f = \frac{1}{T} = \frac{1}{t_{on} + t_{off}}$$

where T is the period equal to the sum of the pulse durations (t_{on}) and its absence (t_{off});

- pulse duty cycle, q, or the inverse of it, D_u (duty cycle):

$$D_U = \frac{1}{q} = \frac{t_{on}}{t_{on} + t_{off}}$$

- pulse fill factor for a discharge of duration ti, which may not coincide with ton:

$$D_I = \frac{t_I}{t_{on} + t_{off}}$$

Modern generators are complex power electronics devices with a large number of feedbacks to analog and digital controllers [5]. However, the principle of operation of most of them is similar and consists in the formation of an output pulse with the specified parameters (see above). In this regard, the generator consists of:

1) voltage source;
2) of control unit;
3) one or more managed keys.

Construction of the model and simulation was carried out in the environment of circuit modeling LTspice XVII (Linear Technology). Figure 3 shows a block diagram of the model for studying the EEE process, as well as a schematic model of the MEP (Block 4) based on formulas (1), (2) and (3).

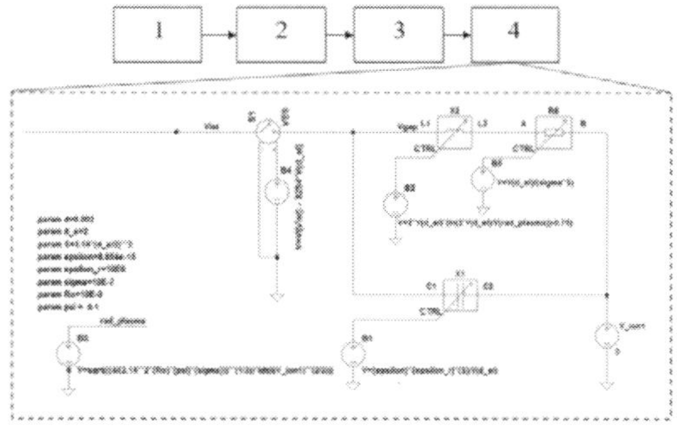

Fig. 3. The block diagram model MEP generator: voltage source (1), the control unit (2), two managed key (3), schematic model of interelectrode gap (4).

Model of inductance, capacitance and resistance in block 4 – MEP (Fig. 3) described in the form of macromodels using the Spice language, the parameters of which vary according to the given laws (see formulas above) using a controlled voltage source.

The generator circuit is based on a half-bridge circuit

As a result of modeling, the time dependences of current and voltage were obtained, as well as the interdependence of the conductivity of the MEP and the current at breakdown, the values of the MEP and the breakdown voltage, and many others. Figure 4 shows an example of the time dependence of current (I (R2)) and voltage (V (vist)) with a linear change in the MEP value from 1000 microns (XX from 0 to 2.5 ms) to 0 (short-circuit for 7ms) and vice versa.

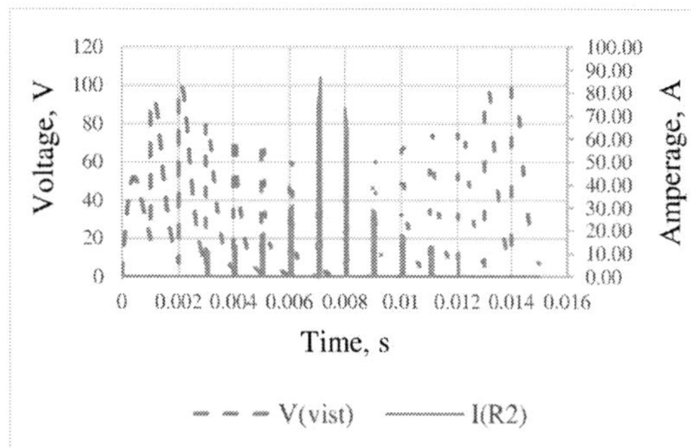

Fig. 4. Simulation results for linear reduction and increase of the interelectrode gap (voltage-dashed line, current – solid).

The data obtained in the simulation are in good agreement with other experimental data [6]. Taking into account the peculiarities of the behavior of current and voltage during EEE, it is possible to evaluate the effectiveness of a particular mode of processing various materials.

The adequacy of the obtained results of circuit modeling was evaluated by comparison with the results of experimental studies on the developed experimental stand [7-34]. Figure 5 shows a joint graph of the results of the experiment (a) and simulation (b) at the operating pulse.

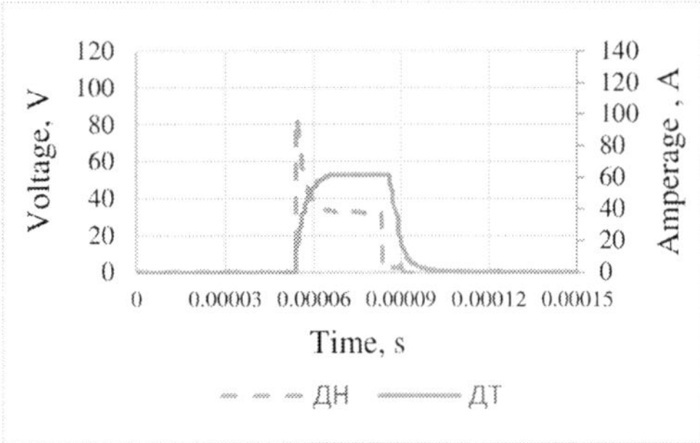

a)

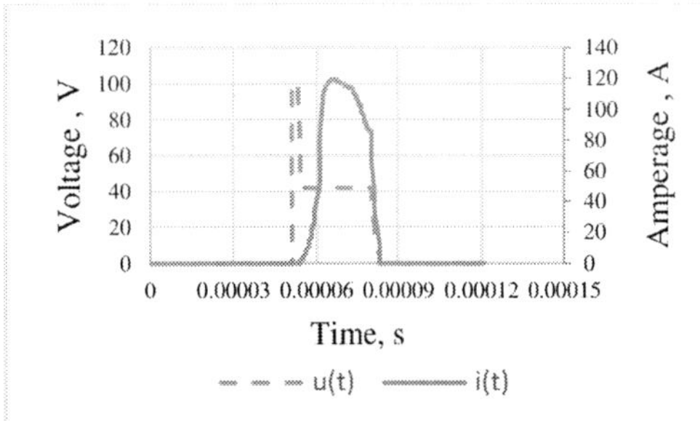

b)

Fig. 5. Comparison of the results of the experiment (a) and simulation (b) on the example of a single pulse at a working pulse.

As can be seen from Fig. 5, the voltage at RI (discontinuous graph) is the same in shape and mean values during simulation and experiment. small deviations are due to the characteristics of the generator of the experimental installation and parasitic inductances of the wires of this generator. The average values of the currents (solid graph) do not match. in this case, there is a limitation of the input range of the measuring system, in particular the analog-to-digital converter, so the current has a significant cut of the pulse tip.

3 Conclusion

The developed model allows not only to obtain values of electrical parameters, but also to evaluate and to optimize the electrical parameters of materials being processed with known electrical properties. The key advantage of this model is its modularity, which allows to add new functional blocks, which describe external and internal influences, for example, the concentration of erosion products, uneven electrical conductivity of the workpiece, and others, without changing its structure.

Acknowledgments

The study was carried out with the financial support of the Russian Foundation for Basic Research within the framework of scientific project No. 21-19-00790. This work was carried out using equipment provided by the Center of Collective Use of MSUT "STANKIN".

References

1. T.R. Raul, A. Saha, H. Majumder, et al., J Braz. Soc. Mech. Sci. Eng., **41**, 300 (2019).
2. S.N. Grigoriev, M.A. Volosova, A.A. Okunkova, S.V. Fedorov, K. Hamdy, P.A. Podrabinnik, P.M. Pivkin, M.P. Kozochkin, A.N. Porvatov, J. Manuf. Mater. Process., **4**(3), 96 (2020).
3. V. Grechishnikov, S. Grigoriev, P. Pivkin, M. Volosova, A. Isaev, D. Nikitin, I. Minin, EPJ Web Conf. **224**, 05001 (2019).
4. B. Wang, Z. Liu, Int. J. Refract, Metals Hard Mater., **55**, 24–32 (2016).
5. Fan Yinsheng, Bai Jicheng, Li Qiang, Li Chaojiang, Cao Yan, Li Zhengkai, Procedia CIRP, **42**, 28 – 33 (2016).
6. M. Hogg, I. Timoshkin, S. MacGregor, M. Given, M. Wilson, T. Wang, IEEE, **2014**, 391-394 (2014).
7. R.S. Khmyrov, S.N. Grigoriev, A.A. Okunkova, A.V. Gusarov, Phys. Procedia, **56**, 345 (2014).
8. A.S. Metel, V.P. Bolbukov, M.A. Volosova, S.N. Grigoriev, Yu.A. Melnik, Instrum. Exp. Tech., **57**(3), 345 (2014).
9. S.N. Grigoriev, G.M. Martinov, Procedia CIRP, **46**, 525 (2016).
10. V.A. Grechishnikov, Y.E. Petukhov, P.M. Pivkin, et al., Meas Tech., **58**, 848–853 (2015).
11. V.V. Kuzin, S.N. Grigoriev, M.A. Volosova, J. Frict. Wear, **35**(6), 505 (2014).
12. S.N. Grigoriev, M.A. Volosova, A.A. Okunkova, S.V. Fedorov, K. Hamdy, P.A. Podrabinnik, P.M. Pivkin, M.P. Kozochkin, A.N. Porvatov, Technologies, **8**(3), 49 (2020).
13. S.N. Grigoriev, D.A. Masterenko, V.I. Teleshevskii, P.N. Emelyanov, Meas. Tech., **55**(11), 1311 (2013).
14. M.A. Volosova, S.N. Grigor'ev, V.V. Kuzin, Refract. Ind. Ceram., **56**, 197, (2015).
15. A.S. Metel, V.P. Bolbukov, M.A. Volosova, S.N. Grigoriev, Yu.A. Melnik, Instrum. Exp. Tech., **57**(3), 345 (2014).
16. M. Gostimirovic, P. Kovac, M. Sekulic, Sadhana Acad. Proc. Eng. Sci., **43**(5), (2018).
17. S.N. Grigoriev, M.P. Kozochkin, A.N. Porvatov, P.S. Zhavoronsky, Jiang Xiaohui, P.M. Pivkin, EPJ Web Conf., **224**, 05002 (2019).
18. V.V. Kuzin, S.N. Grigoriev, M.A. Volosova, J. Frict. Wear **36**(1), 40 (2015).
19. D.S. Rechenko, A.G. Kol'tsov, Russian Engineering Research, **32**(2), 179 – 181 (2012)
20. M. Volosova, S. Grigoriev, A.Metel, A. Shein, Coatings, **8**, 287 (2018).
21. P.A. Nikishechkin, N.S. Grigoriev, N.Yu. Chervonnova. MATEC Web of Conf., **298**, 00064 (2019).
22. X. Chen, Z. Wang, J. Xu, et al., J. Clean. Prod., **188**, 1 – 11 (2018).

23. S.N. Grigoriev, M.P. Kozochkin, P.M. Pivkin, IOP Conf. Ser.: Mater. Sci. Eng., **971**, 022065 (2020)
24. S.N. Grigoriev, M.A. Volosova, A.A. Okunkova, S.V. Fedorov, K. Hamdy, P.A. Podrabinnik, A.N. Porvatov, Technologies, **8**(3), 49 (2020).
25. S.N. Grigoriev, M.A. Volosova, A.A. Okunkova, S.V. Fedorov, K. Hamdy, P.A. Podrabinnik, A.N. Porvatov J. Manuf. Mater. Process., **4**(3), 96 (2020).
26. S.N. Grigoriev, G.M. Martinov, Procedia CIRP, **1**(1), 238 (2012).
27. I. A. Kovalev, P. A. Nikishechkin and A. S. Grigoriev. Approach to programmable controller building by its main modules synthesizing based on requirements specification for industrial automation. In: 2017 International Conference on Industrial Engineering, Applications and Manufacturing (ICIEAM), St. Petersburg, pp.1-4 (2017)
28. C. Zhang, R. Zou, Z. Yu, et al., Int J Adv Manuf Technol, **106,** 5269–5276 (2020).
29. M.A. Volosova, S.N. Grigor'ev, V.V. Kuzin, Refract. Ind. Ceram., **55**, 487 (2015).
30. YC. Lin, JC. Hung, HM. Lee, et al., Int J Adv Manuf Technol, **92,** 2801–2808 (2017).
31. X. Liu et al., IEEE Transactions on Plasma Science, **43**(6), 1959-1968 (2015).
32. M.S. Sohani, V.N. Gaitonde, B. Siddeswarappa, et al., Int J Adv Manuf Technol **45,** 1131 (2009).
33. A. Mohd, S.G. Darius, Md. F. Bahari, Int. J. Mach. Tools Manuf., **47**(7-8), 1214-1228 (2007).
34. X. Wang, Z. Liu, M. Qiu, Z. Hui, Z. Tian, Y. Huang, Mater. Manuf. Process., **29**(11-12), 1367–1373 (2014).

Thermodynamic Model of Wear Intensity of Cutting Tools With Coatings

Sergey Grigoriev[1,*], *Mars* Migranov[1], *Artur* Migranov[1], *Abdumalik* Seitkulov[2]

[1] Moscow State Technological University "STANKIN", RU-127055, Moscow, Russian Federation
[2]Kazakh National Technical University, KZ-050013, Almaty, Republic of Kazakhstan

Abstract. On the basis of thermodynamics of nonequilibrium processes the model is developed, allowing to predict wear resistance and others tribotechnical parametres hardloaded high-temperature interfaces. For the predicted adaptation of friction surfaces and a decrease in the wear rate, it is necessary to use tool materials and wear-resistant coatings containing chemical elements and compounds capable, under certain conditions, of chemically interacting with the material being processed and with the environment, creating non-equilibrium secondary structures with negative entropy production. Contributions of additional impacts on the cutting zone due to the use of coolant, unsteadiness of the cutting mode elements, and other relevant factors should also be considered.

1 Introduction

It is well known [1, 2] that the main phenomena during friction are concentrated in a thin near-surface layer. The thermodynamic aspects of the state of this layer and the relationship of wear with this state are of scientific and practical interest. It was convincingly shown in [3] that any tribosystem should be considered as an open nonequilibrium thermodynamic system. In works [4, 5, 6], the principle of secondary dissipative heterogeneity is formulated, according to which, in the process of friction, phenomena of structural adaptability (adaptation) of contacting materials occur, in which all types of interaction of bodies are localized in a thin-film object - secondary structures (SS).

In accordance with this principle, secondary structures are necessary for the dissipation of energy during its transition from the friction zone to the rubbing bodies, and the dissipation of energy should occur with the lowest rate of entropy growth. Secondary structures perform protective functions, limiting the spread of interaction inside rubbing bodies and reducing the intensity of this interaction, therefore their appearance corresponds to the Le Chatelier principle [3, 5].

2 Theoretical and experimental hypothesis

It is known [7] that the change in entropy of any thermodynamic system is equal to

$$dS = d_eS + d_iS, \quad (1)$$

where d_eS is the change in entropy due to the exchange of matter and energy with the external environment;

d_iS - change in entropy due to "uncompensated transformation", i.e. this entropy is produced by irreversible processes within the system.

In an equilibrium stationary state, the change in entropy over time t is equal to zero, i.e.

$$\frac{dS}{dt} = \frac{d_iS}{dt} + \frac{d_eS}{dt} = 0 \quad (2)$$

and thus $\frac{d_iS}{dt} > 0$, then $\frac{d_eS}{dt} < 0$

In the thermodynamics of irreversible processes, an important role is played by the dissipative component $\frac{d_iS}{dt}$, which describes the production of entropy due to internal sources in the system

$$\frac{d_iS}{dt} = \sum J_k \cdot X_k, \quad (3)$$

where X_k is the generalized force; J_k is the generalized flow caused by it [7, 8].

When developing the principles of rational choice and application of wear-resistant tool materials, let us consider the tribosystem "tool - work material". The following main thermodynamic flows pass through the secondary structures of this tribosystem: heat flow, matter flow, flow of physicochemical transformations and flow of additional influences (use of coolant, unsteadiness of cutting conditions, etc.).

Imagine a system consisting of a rubbing body and energy sources in the friction zone. We believe that energy sources do not have mass, and therefore do not have entropy.

Let us consider the change in the production of entropy of one of the contacting bodies - a cutting tool - when cutting metals. First, we assume that only two

* e-mail: s.grigoriev@stankin.ru

thermodynamic flows pass through the VS of the rubbing surfaces of the tool: the heat flow and the substance flow. Then the change in the production of entropy of the aircraft with time in the system under consideration (without taking into account the relationship between flows) is:

$$\frac{d_i S}{dt} = \frac{dS_q}{dt} + \frac{dS_m}{dt}. \quad (4)$$

Here $\frac{dS_q}{dt}$ - the change in the production of the entropy of the heat flow - is equal to:

$$\frac{dS_q}{dt} = J_q \cdot X_q, \quad (5)$$

where J_q is the heat flux, X_q = (grad T)/T^2 is the thermodynamic force causing the heat flux.

(T - temperature), according to the Biot-Fourier law of heat transfer $J_q = -\lambda \cdot$grad T (λ - thermal conductivity). We assume that only a certain part (X) of the mechanical frictional energy is dissipated by the heat flux, i.e.

$$J_q = f_m \cdot p_r \cdot V \cdot X \ (X < 1) \quad (6)$$

(f_m - coefficient of friction; p_r - specific normal load;
V - cutting-sliding speed).

Then, after the appropriate substitutions and transformations, expression (5) takes the form

$$\frac{dS_q}{dt} = \frac{(f_m \cdot p_r \cdot V \cdot X)^2}{\lambda \cdot T^2}. \quad (7)$$

The change in the entropy of the flow of matter in the BC is equal to:

$$\frac{dS_m}{dt} = J_m \cdot X_m, \quad (8)$$

where Jm is the substance transfer flux; Xm is the thermodynamic force causing the flow of matter;
Xm = (grad C)/T (C is the concentration of the substance). According to Fick's law of transfer of matter; Jm = me – D·grad C (D is the diffusion coefficient). We assume that some part (Y) of the total production of entropy during friction is spent on the formation of the flow of matter, i.e.

$$J_m = f_m \cdot p_r \cdot V \cdot Y \ (Y < 1). \quad (9)$$

Then expression (8) takes the form

$$\frac{dS_m}{dt} = f_m \cdot p_r \cdot V \cdot Y \frac{m_e}{D \cdot T}. \quad (10)$$

After substituting expressions (8), (10) into formula (4), we get:

$$\frac{d_i S}{dt} = \frac{(f_m \cdot p_r \cdot V \cdot X)^2}{\lambda \cdot T^2} + f_m \cdot p_r \cdot V \cdot Y \frac{m_e}{D \cdot T}. \quad (11)$$

According to the theorem of I. Prigogine [7, 8], the change in the production of entropy with time in the system in a stationary state is minimal and stable within certain limits of the change in parameters.

Differentiating expression (11) with respect to cutting speed V (sliding speed) - as an independent variable of the tribological system under consideration - and equating the result to zero, we obtain the condition of stationarity of this system:

$$\frac{2 \cdot f_m^2 \cdot p_r^2 \cdot V \cdot X^2}{\lambda_{cp} \cdot T} - \left| Y \frac{m_e}{D_{cp}} \right| = 0. \quad (12)$$

The sign in front of the second term is taken negative, since from the additivity condition [7] it is proportional to the wear rate of the rubbing body material, and wear particles are removed from the tribosystem with their entropy content.

From expression (12), the flow of matter can be determined:

$$m_e = \frac{2 \cdot f_m^2 \cdot p_r^2 \cdot V \cdot X^2 \cdot D_{cp}}{\lambda_{cp} \cdot T \cdot Y}. \quad (13)$$

In expression (13), the average values of thermal conductivity λ_{cp} and diffusion coefficient D_{cp} in secondary structures are introduced on the assumption that they are independent of the cutting speed.

Obviously, the value of the wear rate of the cutting tool is related to the substance flow m_e, which is determined by the concentration gradient of the substance. And only with wear (removal of saturated layers) is it possible to continue the flow of the substance, since the gradient of its concentration appears again. Then the linear wear rate J_h and the substance flow m_e can be related by the ratio:

$$J_h = \frac{k \cdot m_e}{\rho \cdot V}, \quad (14)$$

where k is the coefficient of proportionality;
ρ - substance density of worn-out particles.

The friction coefficient on the flank surface of cutting tools, as shown in [9], is mainly determined by the adhesion component of the friction coefficient and is equal to

$$f_m = \frac{\tau_{nn}}{p_{rn}}, \quad (15)$$

where τ_{nn} is the shear strength of adhesive bonds;
p_m is the limiting normal stress at the plastic contact.

It is this contact that takes place on the back surface of the cutting tools, i.e. in expression (13) $p_r = p_m$ Then expression (15) can be represented as:

$$J_h = \frac{2 \cdot X^2 \cdot k}{Y} \cdot \frac{\tau_{nn}^2 \cdot D_{cp}}{\lambda_{cp} \cdot T \cdot \rho}. \quad (16)$$

From expression (16) it can be seen that the intensity of tool wear along the flank surface is largely determined by the temperature dependences $\tau_{nn}(T)$.

Formula (16) is an analytical expression for the intensity of wear of cutting tools, which is obtained from the condition of stability of a stationary state with a minimum production of entropy at the frictional contact "tool - workpiece material".

From expression (16), it follows that under conditions when the effect of physicochemical transformations in the friction contact VS is insignificant and can be disregarded:

$$C = \frac{J_h \cdot T}{\tau_{nn}^2} = \frac{2 \cdot X^2 \cdot k}{Y} \cdot \frac{D_{cp}}{\lambda_{cp} \cdot \rho} \cdot \qquad (17)$$

According to (17), the value of C should not change with increasing temperature T.

3 Results and discussion

Tables 1 - 3 present the results of experimental studies and the values of C calculated from these results, depending on the cutting temperature, which changes due to changes in the cutting speed. It can be seen from the tables that for each state of the tribosystem "tool - workpiece material", an increase in temperature T in the investigated range to a certain value, called the characteristic T_{xap}, practically does not change the value of C. This temperature in [10] is called the optimal cutting temperature, i.e. to. at this temperature, the wear rate of the cutting tool is minimal.

Table 1. Turning of KhN70VMTYUB alloy with VK6M cutter; cutting speed V = 15 65 m / min; depth t = 0.5 mm; feed S = 0.08 mm / rev

Temperature cutting T, K	τ_{nn}, MPa	$J_h \cdot 10^7$	$C \cdot 10^{10}$, K / (MPa)^2
873	620	1,1	2,5
923	540	0,9	2,8
973	500	0,8	2,9
1073	380	2	15
1173	220	4	96

Table 2. Milling of steel 30HGSA (HRCэ 35 ... 38) single-tooth end mill T15K6; cutting speed V = 40 – 120 m / min; depth t = 0.5 mm; feed per tooth S_z = 0.1 mm / rev

Temperature cutting T, K	τ_{nn}, MPa	$J_h \cdot 10^5$	$C \cdot 10^8$, K/(MPa)^2
923	600	1,3	3,5
993	350	0,53	4,2
1023	300	0,45	4,8
1053	280	1,0	13
1093	210	1,32	33

Table 3. Turning of KhN70VMTYUB alloy R18F2K8M cutter; cutting speed V = 3 – 10 m / min; depth t = 0.5 mm; feed S = 0.12 mm / rev

Temperature cutting T, K	τ_{nn}, MPa	$J_h \cdot 10^7$	$C \cdot 10^{10}$, K/(MPa)^2
593	780	1,1	1,1
693	740	0,9	1,1
763	640	0,7	1,3
783	620	1,8	3,7
813	600	4,8	10,8

This indicates: firstly, that up to temperatures T_{xap}, physicochemical transformations at a frictional contact cannot play a decisive role in the formation of a material flow and wear of a tool (possibly due to the fact that these transformations are practically absent up to these temperatures) ; secondly, the correspondence of the wear intensity of cutting tools (at least up to temperatures T_{xap}) to expression (16) shows that the state of the BC at the frictional contact "tool - workpiece material" can be considered as stationary with a minimum production of entropy, stable with respect to speed changes (temperature) cutting; thirdly, at temperatures T_{xap} and higher, obviously, new connections are established between the elements of the system (flow of matter and temperature), which can be explained by the appearance of a new structure that promotes ordering of the system. Such results were obtained for substantially different conditions for cutting metals: turning and milling; carbide and high-speed tool materials; heat-resistant alloys and structural steels as workable materials.

It was shown in [11] that if the mechanical energy of friction is the only source of changes in the production of entropy, the thermodynamic system does not lose stability. For self-organization to be possible, more than one independent source of energy dissipation is required. They can be non-spontaneous chemical reactions initiated by friction [12,13]. Consequently, at cutting temperatures Tchar and higher, new sources of energy dissipation appear on the friction contact - non-spontaneous chemical reactions initiated by friction, with excessive production of entropy, some of which can become negative (depending on the type of chemical reactions occurring).

Then the change in the production of entropy of the air force with time in the tribosystem, taking into account the physical and chemical transformations, is equal to:

$$\frac{d_i S}{dt} = \frac{dS_q}{dt} + \frac{dS_m}{dt} + \frac{dS_\phi}{dt} \cdot \qquad (18)$$

Here, the change with time of the entropy of the flow of physicochemical transformations in the air force according to [8] is:

$$\frac{dS_\phi}{dt} = J_x \cdot X_x, \qquad (19)$$

where Jx is the rate of chemical reactions;
Xx - thermodynamic force causing chemical reactions;

Xx = A / T (A is chemical affinity). According to studies [14-17], near thermodynamic equilibrium, the rate of a chemical reaction and chemical affinity are related by the ratio:

$$V_x = L_{11} \cdot R \cdot T \cdot \left(\exp \frac{A}{R \cdot T} - 1 \right), \qquad (20)$$

where L_{11} is the coefficient of conjugate interactions,
R is the gas constant.

Expanding $\left(\exp \frac{A}{R \cdot T} \right)$ in a series and discarding the terms of the second and further degrees of smallness, we get:

$$V_x \approx L_{11} \cdot A \cdot \qquad (21)$$

Let us assume that some part (Z) of the mechanical frictional energy is spent on physicochemical transformations in the VS, i.e.

$$J_x = f_m \cdot p_r \cdot V \cdot Z \; (Z = 1 - X - Y). \qquad (22)$$

Then expression (19) takes the form

$$\frac{dS_{\Phi}}{dt} = f_m \cdot p_r \cdot V \cdot Z \frac{V_x}{L_{11} \cdot T} \cdot \qquad (23)$$

After substituting expressions (11) and (23) into (18), we obtain

$$\frac{d_i S}{dt} = \frac{(f_m \cdot p_r \cdot V \cdot X)^2}{\lambda \cdot T^2} - \cdot$$
$$- \left| f_m \cdot p_r \cdot V \cdot Y \frac{m_e}{D \cdot T} \right| \pm$$
$$\pm \left| f_m \cdot p_r \cdot V \cdot Z \frac{V_x}{L_{11} \cdot T} \right| \qquad (24)$$

If, in the general case, thermodynamic flows of heat, matter, physicochemical transformations and additional influences pass in the secondary structures of the tribosystem "tool - processed material" [18-22], then the change in aircraft production over time will be equal to

$$\frac{d_i S}{dt} = \frac{(f_m \cdot p_r \cdot V \cdot X)^2}{\lambda \cdot T^2} -$$
$$- \left| f_m \cdot p_r \cdot V \cdot Y \frac{m_e}{D \cdot T} \right| \pm \qquad (25)$$
$$\pm \left| f_m \cdot p_r \cdot V \cdot Z \frac{V_x}{L_{11} \cdot T} \right| + J_в \cdot X_в$$

where $J_в$ is the flow of additional impacts;
$X_в$ - thermodynamic force causing the flow of additional influences. Specification of the values of $J_в$ and $X_в$ depends on the type of applied additional influences. In the general case, this component of entropy production does not depend on the characteristics of friction at the frictional contact.

Therefore, for the predicted adaptation of friction surfaces and a decrease in the wear rate, it is necessary to use tool materials and wear-resistant coatings containing chemical elements and compounds capable, under certain conditions, of chemically interacting with the material being processed and with the environment, creating non-equilibrium secondary structures with negative entropy production [23-28]. Additional influences on the cutting zone (the use of coolant, unsteadiness of the cutting mode elements, etc.) should contribute to this.

Thus, the presented experimental results confirmed the passage of self-organization at cutting temperatures Txap and higher and the formation of nonequilibrium dissipative structures on the frictional contact "tool - workpiece material".

Acknowledgments

This research was funded by Ministry of Science and Higher Education of the Russian Federation, Grant No. 0707-2020-0025. This work was carried out using equipment provided by the Center of Collective Use of MSUT "STANKIN".

References

1. I.V. Kragelskyю Mech. Eng., 525, (1977)
2. B.I. Kostetsky. Technics, 395, (1970)
3. L.I. Bershadsky. Knowledge, 253, (1990)
4. B.I. Kostetsky. Technics, 26, (1976)
5. I.S. Gershman. Fric. Wear. **16**, 1, 61-70, (1995)
6. V.S. Ivanova. Fric. Wear. **18**, 1, 74-79, (1997)
7. I. Prigogine. Inostr. lit., 127, (1960)
8. I. Prigogine. Mir, 461, (2002)
9. L.Sh. Shuster. Mashin., 96, (1988)
10. A.D. Makarov. Mech. Eng., 278, (1976)
11. I.S. Gershman. Doct. diss.VNIIZhT
12. G. Heinike. Mir. 582, (1987)
13. S.B Bulgarevich. MGTU im. N.E. Bauman, 60-67, (2005)
14. E.M. Gutman. Metallurgy, 230, (1974)
15. M.A. Volosova, S.N. Grigor'ev, V.V. Kuzin, Refract. Ind. Ceram., **55**, 565, (2015)
16. M. Volosova, S. Grigoriev, A. Metel, A. Shein, Coatings, **8**, 287, (2018)
17. M.A. Volosova, S.N. Grigor'ev, V.V. Kuzin, Refract. Ind .Ceram., **56**, 197, (2015)
18. M.A. Volosova, S.N. Grigor'ev, V.V. Kuzin, Refract. Ind. Ceram., **55**, 487, (2015)
19. V.V. Kuzin, S.N. Grigoriev, M.A. Volosova, J. Frict. Wear, **35**(6), 505, (2014)
20. S.N. Grigoriev, O.V. Sobol, V.M. Beresnev, I.V. Serdyuk, A.D. Pogrebnyak, D.A. Kolesnikov, U.S. Nemchenko, J. Frict. Wear, **35**(5), 359, (2014)
21. O.V. Sobol', A.A. Andreev, S.N. Grigoriev, V.A. Stolbovoy, Probl. At. Sci. Technol., **4**, 174, (2011)
22. A.S. Metel, V.P. Bolbukov, M.A. Volosova, S.N. Grigoriev, Yu.A. Melnik, Instrum. Exp. Tech., **57**(3), 345, (2014)

23. S.N. Grigoriev, M.A. Volosova, A.A. Vereschaka, N.N. Sitnikov, F. Milovich, J.I. Bublikov, S.V. Fyodorov, A.E. Seleznev, Ceram. Int., **46**(11), 18241, (2020)
24. V.V. Kuzin, S.N. Grigor'ev, M.A. Volosova, Refract. Ind. Ceram., **54**, 376, (2014)
25. V.V. Kuzin, S.N. Grigoriev, M.A. Volosova, J. Frict. Wear, **36**(1), 40, (2015)
26. P. Huang, W. Qi, X. Yin, J.O. Choi, X.C. Chen, J.S. Tian, J.X. Xu, H.C. Wu, J.B. Luo, CARBONT, **154**, 203-210 (2019)
27. H.W. Chen, M. McGurr, M. Brickhouse, Proceedings of SPIE, **9472** (2015)
28. J. Zekonyte, T. Polcar, ACS APPLIED MATERIALS & INTERFACES, **7**(38), 21056-21064 (2015)

Modeling of Non-Stationary Processes When Cutting Hard-to-Process Materials

Sergey Grigoriev[1,*], *Mars* Migranov[1], and *Abdumalik* Seitkulov[2]

[1] Moscow State Technological University "STANKIN", RU-127055, Moscow, Russian Federation
[2]Kazakh National Technical University, KZ-050013, Almaty, Republic of Kazakhstan

Abstract. In the conditions of high-speed processing of parts of complex configuration, with a large end and longitudinal length, from hard-to-work steels and alloys, it is difficult to ensure the wear resistance of the cutting tool in the aisles of one technological passage. To ensure the appropriate quality indicators of the surface layer, it is impossible to replace a worn-out cutting tool. In connection with the above, the problem of ensuring the operability (wear resistance) of the cutting tool is acute. The results of theoretical and experimental studies of contact phenomena in blade cutting based on the thermodynamics of non – equilibrium processes and from the standpoint of self-organization of the tribosystem are presented. the developed thermodynamic model of blade processing with variable cutting modes (non-stationary) allows to minimize the wear of the cutting tool and generally increase production efficiency by accelerating the drive of the main movement of the metal-cutting machine.

1 Introduction

A feature of the use and operation of cutting tools on modern metal-cutting equipment equipped with numerical software (CNC) and adaptive control (ADSU) systems is the variability of many parameters of the blade processing process. Operations such as multi-step machining, machining complex contours on a copier, face, vibration and interrupted turning are performed with variable elements of the cutting mode and, as a consequence, with non-stationary temperature-force conditions of loading of the cutting tool. External factors - fluctuations in the stock allowance and their physical and mechanical inhomogeneity, beats and vibrations, variability of technological parameters of the turning process (speed, feed and depth of cut), and internal, associated with changes in forces and temperatures in the cutting zone due to friction and irreversible growth of wear of the cutting tool, is made as significant contribution, to the non-stationarity of blade processing and, in particular, the contact “tool-part” [1-4].

Practical recommendations used by the industry, possible non-stationarity are taken into account by correction factors for a decrease in processing modes or not taken into account at all, which leads to a decrease in productivity and in most cases is not justified [5 – 7].

A wide variety of new materials with increased strength properties, the need to reduce the time of their development and increase the overall efficiency of the cutting tool operation puts forward among the urgent tasks the development of methods for the theoretical study of non-stationary turning in order to determine functional and mathematical models of machinability characteristics suitable for solving optimization and control issues cutting process in various technological conditions based on the relationship of variable elements of the cutting mode with contact phenomena during blade processing.

In the process of cutting, due to tool wear, there is a continuous change in the spatial shape of the contact surfaces. This, even with constant values of the elements of the cutting mode, can lead to a change in the distribution of stresses and temperatures in the actual contact zone, which in turn affect the wear rate [1, 5, 6]. Thus, contact wear and the distribution of specific

loads and temperatures are interdependent processes, which must be taken into account when developing analytical dependencies for calculating the wear rate in non-stationary cutting.

Currently, the generally accepted [1, 4, 8 – 10] is a complex theory of friction and wear, which combines the molecular-kinetic and structural-energy approaches. Its main concept is the need for multiple frictional action to destroy the friction surfaces. In this case, the process of friction is considered as a loss of mechanical energy during the period of relative motion of the contacting areas of materials. Destruction (wear) occurs as a result of the accumulation of internal energy in the deformed volume, spent on the formation of defects in the crystal lattice of rubbing bodies and on increasing the temperature of their surface layers.

* Corresponding author: s.grigoriev@stankin.ru

2 Theoretical and experimental hypothesis

Based on the first law of thermodynamics for the processes of friction and wear, the change in internal energy (ΔW) can be represented in the form

$$\Delta W = \Delta W_{mv} + \Delta W_{ph} + \Delta W_{d} + \Delta W_{f} \quad (1)$$

Where ΔW_{mv} - is the energy spent on separating the wear particle,

ΔW_{ph} - energy spent on structural-phase transformations;

ΔW_{d} - the energy spent on plastic deformation;

ΔW_{f} - energy spent on shaping rubbing surfaces.

Since at the present time there are no analytical dependences, according to which it is possible to estimate with sufficient accuracy all components of dependence (1), we assume in the first approximation that as a result of frictional contact interaction, the internal energy will dissipate due to:
- plastic deformation of a single microroughness as a result of a shift by the average diameter of the contact spot;
- shaping the surface layer of the wear material as a result of the formation of a wear fragment;
- the formation of new free surfaces as a result of the separation of wear particles.

Using the principles of nonequilibrium thermodynamics [5, 7, 8, 10-12], contact processes in unsteady cutting can be most fully described using dissipative functions (DF), which represent the rate of change (dW_i/dt) of the energy spent on a process (for example, plastic deformation of the processed material) per unit of actual contact area (A_r).

$$\overline{\Psi}_i = dW_i / d\tau \cdot 1 / A_r \quad (2)$$

To ensure the irreversibility of the process, it is necessary to have a generalized thermodynamic flow (I_i) and a force ($\Delta\sigma_i$) [4,5], i.e. the supported gradient of the values of the state of the thermodynamic system, which prevents the reverse process and taken with the opposite sign. The DF of such a process is equal to the product of the generalized flow and force

$$\overline{\Psi}_i = I_i \cdot (-\Delta\sigma_i) \quad (3)$$

Taking into account the above assumptions, the change in the internal energy of the frictional dynamic system (ΔW) as a result of the action of external forces can be represented as the sum of the energy spent on dispersing (wear) the instrumental and elastic-plastic deformation of the processed material. In this case, the law of conservation of energy can be represented in the form of a balance of dissipative functions

$$\overline{\Psi}_b = \overline{\Psi}_f + \overline{\Psi}_d \quad (4)$$

where $\overline{\Psi}_b, \overline{\Psi}_f, \overline{\Psi}_d$ - respectively, DF of external forces; the process of plastic deformation of the processed material as a result of shear by the average diameter of the contact spot; dispersion and reshaping of the surface layer of the worn tool material.

Based on the analysis of contact processes during nonstationary turning, the DF of external forces is presented as

$$\overline{\Psi}_b = P_z(\tau) \cdot V(\tau) / (l_1 + h_з) \cdot b \quad (5)$$

where $P_z(\tau)$, $V(\tau)$ - respectively, the current values of the force and cutting speed, N, m / s;

$l_1, h_з$ - respectively, the length of the tool contact on the front and back surfaces, mm;

b - width of the cut layer, mm.

From the analysis of the shape stability of the cutting wedge, the model of the accumulation of damage in the near-contact layers of the tool with variable elements of the cutting mode and the probabilistic nature of the separation of the wear particle [1,3, 13-15],

$$\overline{\Psi}_f = I_h \cdot V(\tau) \cdot \left(\frac{HV_u}{HV_ф}\right)^{\alpha} \cdot erfP \cdot \left[p_r + \frac{12 \cdot (1+\mu)}{E} \cdot \sigma_T^2\right] \quad (6)$$

where I_h is the wear rate of the cutting tool;

$\frac{HV_U}{HV_ф}$ - the ratio of microhardness of the instrumental and processed materials;

α - an indicator that takes into account the effect of cutting temperature;

$erfP$ - probability of separation of wear particles;

p_r - normal specific load in the contact zone N / m;

$\sigma_T = \frac{1}{2\sqrt{3}} HV_u$ - the yield point of the tool material, MPa.

The DF of plastic deformation of the processed material is determined on the basis of the dislocation theory of the process, taking into account the concept of the modified temperature, by the rate of deformation and is determined by the current value of the cutting temperature (deformation ($\Theta(\tau)$) , degree of plastic deformation ($\Delta\gamma$), physical and mechanical properties ofthe processed material ($\sigma_{m.\partial.}, G_{\partial}$)) [6, 8], a See also vibrationcharacteristics of the cutting process (f_r).

$$\overline{\Psi}_в = 10^4 \cdot \Theta(\tau) \cdot \Delta\gamma \cdot \left[f_r \cdot \frac{\sigma_{m.\partial.}}{G_{\partial.}} \cdot \frac{\Theta(\tau)}{\Theta_{отп.}} \cdot exp\frac{\Theta_{отп.}}{\Theta(\tau)}\right]^{\frac{1}{n}} \quad (7)$$

Based on the above and the obtained dependences for the components of the balance of dissipative functions (4 -7), we can write:

$$\frac{p_z(\tau) \cdot V \cdot f}{(l_1 + c) \cdot b} = 0{,}186\theta(\tau) J_\partial J_\theta +$$

$$+ J_h \cdot V \cdot \left(\frac{HV_u}{HV_ф}\right) \cdot erfP(\tau) \cdot \left[P_r + \frac{12 \cdot (1+\mu)}{E} \cdot \sigma_m^2\right], \quad (8)$$

Solving this equation with respect to the tool wear rate J_h, we obtain an expression for calculating the tool wear

$$J_h = \frac{p_z(\tau)\cdot V\cdot f/(l_1+c)\cdot b - 0{,}186\theta(\tau)J_\partial J_\theta}{V\cdot\left(\frac{HV_u}{HV_\phi}\right)^a\cdot erfP(\tau)\cdot\left[P_r+\frac{12(1+\mu)}{E}\sigma_\tau^2\right]} \quad (9)$$

From equation (9) it follows that the main ways to reduce the wear rate of blade cutting tools are:

– reduction of the coefficient of friction f on frictional contact with the processed material (due to the formation of secondary structures and phases on the working surfaces - products of self-organization of the tribosystem, as well as the use of lubricating and cooling technological means and unsteadiness of cutting, contributing to this phenomenon) [16,17];

– increasing the value of the ratio of the hardness of the contacting surfaces of the tool $HV_и$ and the workpiece $HV_ф$ (by reducing the dependence of the physical and mechanical properties of the contacting surfaces on the temperature in the working zone - taking into account the phenomena of self-organization during friction) [18-20];

– optimal change in temperature and power conditions of cutting, using the change in cutting speed over time (acceleration) by the main motion drive (PGD).

3 Results and discussion

To confirm the mathematical model of the wear rate of the cutting tool, a series of experimental studies of the temperature-power parameters with the variability of the elements of the cutting mode were carried out. A comparative analysis of the dependences of the components of the cutting force on the cutting speed, both in stationary and non-stationary turning, showed that they have different levels, i.e. the value of the components of the cutting force when machining with positive accelerations is much lower (Fig. 1).

Based on the results of experimental studies of the cutting force of the "Evrika" programs, mathematical models for the tangential component of the cutting force were obtained. The Evrika application package allows the least squares method to approximate functions, statistical data and is widely used for engineering calculations.

With non-stationary turning (with variable cutting speeds) for one of the investigated pairs of tool and machined materials (XN73MBTU - H10), the dependence of the tangential component of the cutting force on the cutting speed and its acceleration has the form:

$$P_z(\tau) = 219{,}37 + 32{,}632\tau\cdot V_0^{-1{,}4(1\pm a_V)} \quad (10)$$

at S = 0.1 mm / rev; t = 0.5 mm.

With a continuous change in feed in the area of practically applied cut configurations for finishing and semi-finishing turning (S=0,05÷0,15 mm / rev), the components of the cutting forces Px, Py, Pz change proportionally 2: 1, 5: 1 (Fig. 2).

Experiments have shown that the cutting force during non-stationary turning changes non-linearly with a change in feed according to a linear law. When turning with positive rates of change in feed (aS >0), the cutting forces increase monotonically, but decrease monotonically at negative rates of change (aS <0).

The approximate dependence of the tangential component of the cutting force on the feed and the rate of its change (a_s) in time (τ) is as follows:

$$P_z(\tau) = 260{,}66 + 514{,}2\tau\cdot S^{1.3(1+a_s)} \quad (11)$$

at S = 0.5 m / s; t = 0.5 mm.

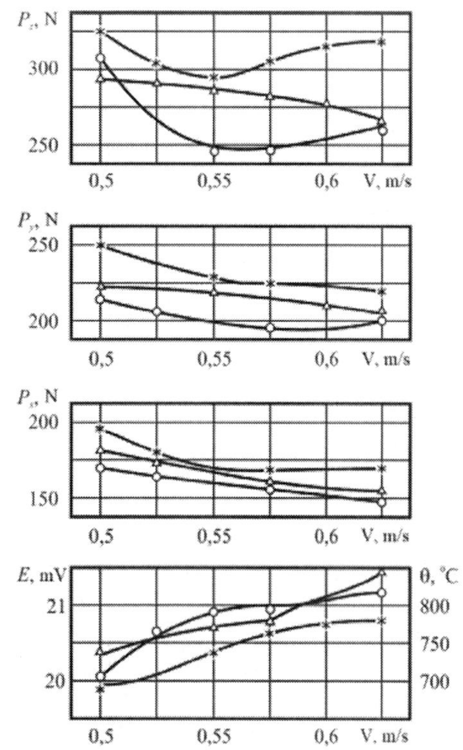

Fig. 1. Dependence of the components of the cutting force, thermo-EMF and temperature on the cutting speed when turning XN73MBTU – H10
(- av> 0; - av <0; - av = 0; s = 0.1 mm / rev; t = 0.5 mm)

Thus, we can say that in non-stationary turning with a change in cutting speed and feed according to a linear law, the cutting force differs significantly in magnitude than in stationary turning (a_v=$a_{s=0}$) and is a function of the magnitude and sign of the rate of their change (a_v, a_s).

The dependences of the cutting temperature on the speed and feed for non-stationary and stationary turning for various combinations of tool and work materials are shown in Fig. 1 and 2.

When analyzing the data obtained, it was found that for the investigated combinations of "cutter - part" the dependences θ(V) have a non-monotonic character from the point of inflection (inversion) when working at optimal cutting speeds, but with a point shift towards high speeds when turning with positive accelerations (a_V>0) in comparison with work with a_V = 0 and towards lower speeds when working with negative accelerations (a_V > 0).

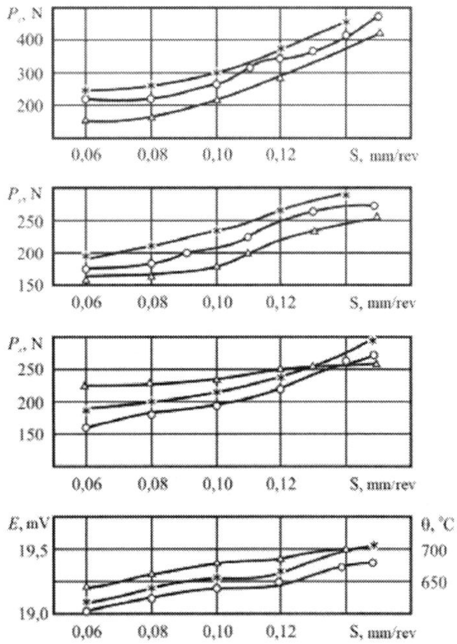

Fig. 2. Dependence of the components of the cutting force, thermo-EMF and temperature on the cutting speed when turning EI698-VK8
(- aS> 0; - aS <0; - aS <0; V = 0.66 m / s; t = 0.5 mm)

Analysis of the dependences of temperature on speed and feed in non-stationary turning confirms that the main influence on the cutting temperature is exerted by the cutting speed and the workpiece material, all other things being equal. During non-stationary turning, as well as during stationary turning of machined materials, tool materials have an ambiguous effect on the cutting temperature, which is associated with the complex nature of their influence, first of all, on the processes of frictional interaction, which largely determine the temperature in the cutting zone.

According to the results of experimental studies, approximation dependences for the cutting temperature were obtained with a continuous change in the cutting speed (V) with acceleration (a_V) in time (τ)

$$\theta(\tau) = 295{,}6 + v_0^{0,8(1 \pm a_v)} \cdot \tau \text{, K} \quad (12)$$

and with a continuous change in feed:

$$\theta(\tau) = 512{,}4 + 33{,}2 S_0^{0,8(1 \pm a_s)} \cdot \tau \text{, K} \quad (13)$$

To concretize the dependence of the components of the balance of dissipative functions (ψ_t, ψ_d, ψ_f) and to ensure the conditions for minimizing the intensity of tool wear (J_h) during non-stationary turning, the dependences of the main output parameters (tangential component of the cutting force P_z, relative linear tool wear $h_{oз}$, cutting speed V, in the form of specific energy $\left(\frac{P_z \cdot V}{F_k}\right)$ - spent on the processes of friction and wear of the cutting tool - from the cutting temperature during non-stationary turning (Fig. 1).

Specific energy expended on the processes of friction and wear, according to [8,9], is expressed as

$$\Delta W = \frac{P_z \cdot V}{F_k}, \frac{H \cdot м}{c \cdot м^2} \quad (14)$$

where F_k is the area of contact surfaces of the tool with the processed material and chips.

Analysis of experimental data (Fig. 3) - dependences of cutting temperature (θ) and specific energy spent on friction and wear processes for all studied pairs during non-stationary turning showed that:
– the graph of dependence has a characteristic break at temperatures corresponding to the optimum temperature of stationary turning;
– at temperatures θ >$\theta_{опт}$, the dependence has a different angle of inclination for different accelerations of PGD;
The specific energy for different accelerations of PGD, both in sign and in magnitude, has a different value (different level) and a minimum during acceleration (a_v>0), as already noted, corresponding to a lower temperature-force tension of the cutting zone.

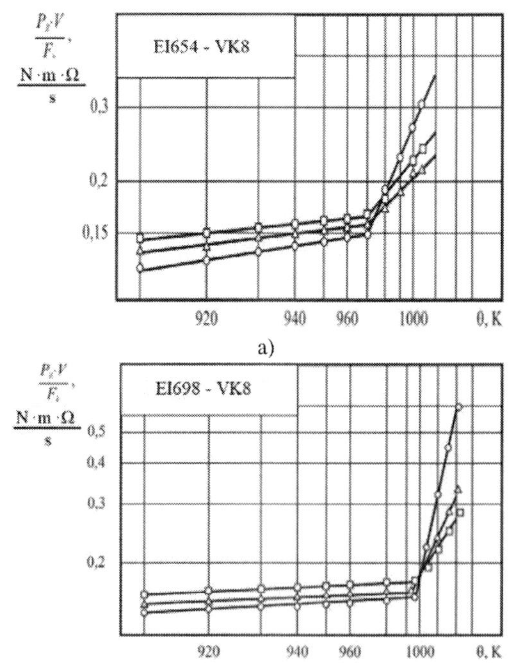

Fig. 3. Dependence of power and electrophysical parameters, speed versus cutting temperature for non-stationary and

discrete turning (S = 0.1 mm / rev; t = 0.5 mm; - av> 0; - av <0; - av = 0)

Experiments carried out with a continuously varying cutting speed with different accelerations of the main motion drive (MD) both in sign and in value ($a_v \neq 0$) showed that for all the pairs under study, wear prevails along the rear main and auxiliary surfaces of the tool. For finishing and semi-finishing (S = 0.05 - 1.5 mm / rev; t = 0.25 - 0.5 mm), the hole on the front surface is located closer to the main cutting edge. As the tool wears out, the holes can join with the wear chamfers, which predetermines the onset of catastrophic wear.

In general, it was found that the dependence of wear on cutting speed is extreme. Moreover, when operating with a positive speed acceleration ($a_v > 0$) , this dependence is flatter compared to $a_v < 0$ or $a_v = 0$ (Fig. 4).

With a change in temperature, as shown by numerous studies [1-5], the wear rate changes non-monotonically. Usually, in the course of a resistance experiment, the temperature in the cutting zone changes continuously as a result of an increase in the wear chamfer along the flank of the cutter. Changing the wear chamfer leads to a redistribution of heat fluxes on the cutting edges of the tool, which in turn changes the intensity of wear. The dependence of the logarithm of the intensity of tool wear on temperature during stationary and non-stationary turning is extreme, with a minimum at a temperature corresponding to the optimum for stationary turning (Fig. 4).

Moreover, this dependence for non-stationary turning has three sections: section I - up to the optimum temperature ($\theta_{опт}$), section II - from the optimal to critical temperatures ($\theta_{кр}$), section III - above the critical temperature. According to studies [2, 3], the appearance of a critical temperature is associated with a change in the dynamic safety factors of the contact or surface strength of the tool and processed materials.

According to the results of wear resistance studies of stationary and non-stationary turning it follows that:

– the dependence of the intensity of tool wear on the cutting speed, both in stationary and non-stationary turning, is extreme. Moreover, with non-stationary turning, the minimum intensity of tool wear is observed at higher cutting speeds;

– the intensity of tool wear during non-stationary turning (a, as) is much less than during stationary turning, which, apparently, as already noted, is associated with a decrease in the temperature-force tension of the cutting zone;

– the minimum of the dependence of the wear rate on the cutting temperature in non-stationary turning is observed at a temperature that is optimal for stationary turning. This confirms the position about the optimal temperature of A.D. Makarov [3] in relation to non-stationary turning;

– the presence of a platform on the dependence of the intensity of tool wear on the cutting temperature, i.e. temperature range $\theta_{кр}$ - opt, determines the area of rational use of any brand of hard alloy when cutting a given workpiece material with a minimum of tool wear intensity and can serve as one of the criteria for the machinability of this material and for determining the boundary conditions for cutting conditions during non-stationary turning.

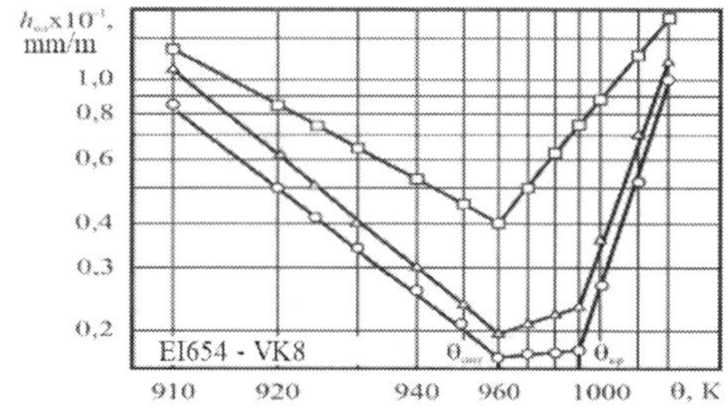

Fig. 4. Relative linear wear versus temperature when turning with different accelerations (S = 0.1 mm / rev; t = 0.5 mm)
Δ - av <0; - av = 0; - av> 0

4 Conclusion

The results of theoretical and experimental studies of contact phenomena in blade cutting based on the thermodynamics of non – equilibrium processes and from the standpoint of self-organization of the tribosystem are presented. Developed thermodynamic model of blade processing with variable cutting modes allows to minimize the wear of the cutting tool by half during high-speed processing of parts of complex configuration with large dimensions from difficult-to-machine steels and alloys while ensuring and maintaining high quality indicators of the surface layer of the processed surface.

Acknowledgments

This research was funded by Ministry of Science and Higher Education of the Russian Federation, Grant No. 0707-2020-0025. This work was carried out using equipment provided by the Center of Collective Use of MSUT "STANKIN".

References

1. S.A. Vasin, A.S. Vereshchaka, V.S. Kushner, *Cutting materials thermomechanical approach to the system of interrelations in cutting* (M.: Publishing house of Bauman Moscow state technical University, 2001)
2. T.N. Loladze, *Strength and wear resistance of the cutting tool* (M.: Mashinostroenie, 1982)
3. A.D. Makarov, *Optimization of cutting processes* (M.: Mashinostroenie, 1976)
4. B.I. Kostetsky, M.G. Nosovsky, L.I. Bershadsky, *Surface strength of materials under friction* (Kiev: Technika, 1976)
5. I.S. Gershman, N.And. Bush, Friction and wear, **16**(1), 61 – 70 (1995)

6. V.S. Ivanova, N.And. Bush, I.S. Gershman, Friction and wear, **18**(1), 74 – 79 (1997)
7. I. Prigogine, *Introduction to the thermodynamics of irreversible processes* (M.: Inostr. lit., 1960)
8. M.Sh. Migranov, Friction and lubrication in machines and mechanisms, **5**, 23-26 (2015)
9. M.Sh. Migranov, *Improving the wear resistance of tools based on the intensification of the processes of adaptation of friction surfaces when cutting metals* (Publishing house "Gilem" of the Academy of Sciences of the Republic of Belarus, 2011)
10. Yu.B. Rumer, M.Sh. Ryvkin, *Thermodynamics, statistical physics and kinetics* (Moscow: Nauka, 1977)
11. Iu.G. Kabaldin, *Selforganization and processes in nanostructured metal cutting* (N.Novgorod, NNSTU n.a. R.E. Alekseev publ., 2014)
12. Y.G. Kabaldin, D.A. Shatagin, M.S. Anosov, A.M. Kuzmishina, Intelligent Control of Technological Systems in Digital Manufacturing, **1**(178), 3-12 (2020) DOI: 10.18698/0536-1044-2020-1-3-12
13. M.A. Volosova, S.N. Grigor'ev, V.V. Kuzin, Refract. Ind. Ceram., **55,** 487, (2015)
14. I.B. Zolotukhin, Sorosovskiy Educational Journal, **1**, 103–106 (1998) [in Russia]
15. S.N. Grigoriev, O.V. Sobol, V.M. Beresnev, I.V. Serdyuk, A.D. Pogrebnyak, D.A. Kolesnikov, U.S. Nemchenko, J. Frict. Wear, **35**(5), 359, (2014)
16. O.V. Sobol', A.A. Andreev, S.N. Grigoriev, V.A. Stolbovoy, Probl. At. Sci. Technol., **4**, 174, (2011)
17. S.N. Grigoriev, M.A. Volosova, A.A. Vereschaka, N.N. Sitnikov, F. Milovich, J.I. Bublikov, S.V. Fyodorov, A.E. Seleznev, Ceram. Int., **46**(11), 18241, (2020)
18. N.K. Krioni, M.Sh. Migranov, S.R. Shechtman, *Innovative technologies of processing of metals in engine* (M. Publishing house "Innovation engineering", 2019)
19. V.V. Kuzin, S.N. Grigor'ev, M.A. Volosova, Refract. Ind. Ceram., **54**, 376, (2014)
20. V.F. Smirnov, *Physics of fractal clusters* (Moscow, Nauka publ., 1998)

Modeling of Thermophysical Phenomena When Cutting with Coated Tools

Marina Volosova[1,*], *Artur* Migranov[1], and *Maksim* Rai[1]

[1]Moscow State Technological University "STANKIN", RU-127055, Moscow, Russian Federation

Abstract. The results of computer simulation of thermophysical phenomena in the contact zone during blade cutting of metals with multi-layer composite wear-resistant coatings that ensure the adaptability of the cutting wedge to friction conditions are presented. On-site experimental studies of the cutting temperature during turning with various coatings, structural-phase analysis of the surface layer of the cutting tool to explain the mechanism of formation of secondary structures with a shielding effect – the effect of self-organization.

1 Introduction

The efficiency of automation of technological processes and the use of modern expensive high-performance machine-tool equipment equipped with mechatronic systems requires a more complete study of the features of the temperature-force conditions of blade metal cutting, where the weakest link in terms of wear resistance is the cutting tool [1-3]. Currently, considerable experience and theoretical and experimental studies have been accumulated to improve the performance of cutting tools by using modern multilayer composite wear-resistant coatings that are capable of adapting to frictional conditions during cutting by forming protective secondary structures.

It is known [4-8] that cutting is a thermodynamic process, and at the same time thermal phenomena in the contact zone have a significant effect, both on the whole on the state of the cutting wedge and on the structural-phase state of the surface layer of the coated tool, due to the activation and passivation of energy ... A change in the internal energy of the tribosystem further leads to the interaction of the activated surfaces with the environment and the formation of protective structures that serve as a kind of shield between the contacting materials. The contact of a tribopair is carried out using secondary structures formed in the friction zone due to self-organization. The importance of using this phenomenon in engineering and technology is undeniable. During friction and wear, the self-organization of the tribosystem materials occurs as a reaction to external conditions caused by the cutting process, which, in many cases, leads to fundamental structural changes in the surface layers of materials. These changes concern many characteristic properties of rubbing surfaces and nearby surface layers (for example, geometric parameters, microstructure, physicochemical and mechanical properties).

Friction always accompanies wear, it is the only "normal" cause of wear and loss of cutting properties by the tool, and the general patterns of friction and wear of an interacting pair of materials of the tool and the workpiece are a practical basis for determining and assigning cutting conditions. Arising directly in the process of cutting and influencing its phenomena, friction itself depends on the elements of the cutting regime, which create specific conditions for the flow of friction, characterized by high pressures (plastic deformation) in the friction zone, a wide range of temperatures and the presence of third bodies (protective secondary structures) on rubbing surfaces, etc. Based on the foregoing, we can assume the existence of an "ideal" in terms of temperature-force conditions of cutting and wear resistance of the cutting tool model of a cutting wedge, which has the form of a puff cake (composite) (Fig. 1), formed by a complex (combined) impact on the surface layer as a result self-organization. In this case, each layer of the "pie": the application of wear-resistant and damping coatings (layers 1 and 2), the implantation of various metals and non-metals (layer 3) and the use of lubricating and cooling technological means (COTS) as a catalyst, is responsible for a certain zone of the wear curve and improves the wear resistance of the tool as a whole.

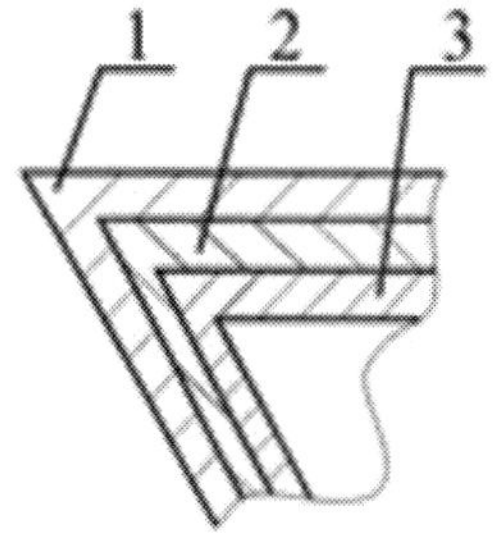

Fig. 1. Cutting wedge model

* Corresponding author: m.volosova@stankin.ru

The aim of the work is to simulate thermophysical processes and phenomena of self-organization during blade cutting with a tool with composite coatings that minimize the intensity of wear of the cutting tool under various temperature and force conditions in the contact zone.

2 Materials and methods

In this work, a series of full-scale experiments in longitudinal turning and milling were carried out to study the tribotechnical characteristics and wear resistance of cutting tools with different coatings in terms of composition and atomic structure of the surface layer of the material. At the same time, the analysis of the literature data showed that in the presence of numerous works [2-5] on the study of contact processes when working with a cutting tool covered with a multilayer composite wear-resistant coating, there are no data on the study of the effect of the use of cooling lubricants and the coatings themselves that have the effect self-organization during friction and the formed protective secondary structures on the contacting surfaces on the distribution of heat fluxes from the cutting zone and control of temperature and power conditions during cutting, by varying the physicomechanical, thermophysical, physicochemical parameters of media and coatings. The above problem can be modeled in the presence of a large number of applied software products for personal computers by using them to significantly reduce the labor intensity, high cost of field experiments and increase the multivariance of solutions.

When considering the contact "tool - part", the heat flux from the cutting zone is distributed between the workpiece, the cutting tool, coming off the chips and the environment [2]. At the same time, with a directed change in the physicomechanical, structural and thermochemical properties of lubricating-cooling technological means and multicomponent composite coatings, it is possible to have a significant effect on the direction and, in general, on the distribution of the heat flow. With an increase in the heat flux transferred to the chips (under conditions of high-speed processing), the thermal loading of the cutting wedge of the tool decreases, followed by a change in adhesive and diffusion wear. Heat removal (recovery) in the cutting zone at low processing speeds provides a lubricating effect due to the formation of rutile from the coatings and leads to a decrease in abrasive wear.

On the basis of well-known works [2, 6] to solve the equation of thermal conductivity and thermophysical parameters in relation to the tool with a coating, systems of differential equations were compiled for each of the sources, followed by the development of a program for their calculations and calculations of temperature fields in the cutting zone were carried out. The calculation results and analysis of the obtained multivariate data showed that, firstly, it is always necessary to take into account the variability of the main components of the equation, secondly, it is impossible to systematize them, and thirdly, it is impossible to give a visual assessment of the temperature field in the cutting zone, which allows regulating the cutting conditions.

Currently, to solve such complex problems and the possibility of assessing the direction and amount of heat flux, the software packages DEFORM, ANSYS and other various versions are increasingly used in engineering and research in thermophysical analysis. This makes it possible to carry out not only modeling, but also to clearly demonstrate the direction of the heat flow at different cutting speeds and to illustrate the contact processes in the cutting zone at different thermophysical properties of multilayer wear-resistant coatings and variability of technological parameters.

In this work, a series of full-scale tests were carried out to determine the tribotechnical characteristics of carbide cutting tools of the VK, TK and TTK groups with various coatings (TiCr)N, (TiAl)N, (AlTi)N, (TiAlCr)N, (AlTiCr)N, as well as by determining the cutting temperature by the natural thermocouple method when turning 40X structural steel with a hardness of 20 HRC on a 16K20 screw-cutting lathe with non-regrowth plates with all of the above coatings. At the same time, preliminary experimental studies and comparative analysis of the data of the mathematical model with the real values of the cutting temperature and wear resistance of the cutting tool applied by various companies (Balzers, Caromant, Carbide, Rimet) were carried out using factory technologies.

The tribological properties were determined using a specially designed adhesion meter [9]. The adhesive meter in Fig. 2, a rotating indenter with investigated coatings 2 was placed between two polished specimens 1 made of the material being processed. To simulate friction conditions similar to the cutting process, the samples and the indenter were heated by an electrocontact method in the temperature range from 150 to 1050 ° C. To create plastic deformation in the contact zone, similar to the cutting process, samples from the processed material 1 were compressed by a force equal to 2400 N. In this case, the adhesive component of the friction coefficient was used to assess the antifriction properties of the layer. This component is important for assessing and predicting the intensity of wear during metal friction. It was defined as the ratio of the shear resistance τ_{nn} caused by the adhesive interaction between the materials of the tool and the workpiece to the normal stress P_{rn} arising at the plastic contact at test temperatures (τ_{nn}/P_{rn}).

In this work, to obtain information about the average contact temperature during cutting with a sufficiently high accuracy, simplicity and reliability, according to the recommendations of [2, 10-13], the natural thermocouple method was used, a schematic diagram of which for turning operations is shown in Fig. 3. To exclude measurement errors from "stray parasitic" thermoelectromotive forces, the part and the cutter were isolated from each other with textolite gaskets [14-16].

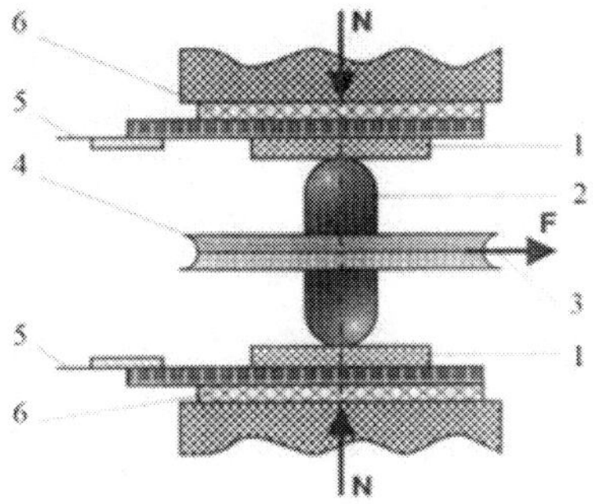

Fig. 2. Model of the setup for the study of tribotechnical properties

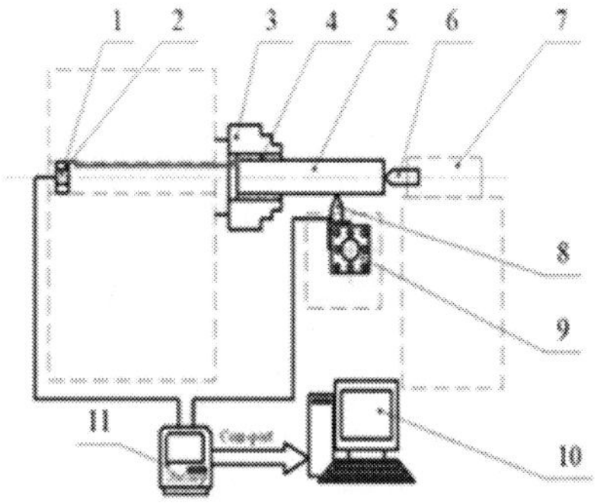

Fig. 2. Scheme of temperature measurement by natural thermocouple method. 1 - mercury current collector; 2, 4 - insulation; 3 - cartridge; 5 - blank; 6 - isolated back center; 7 - tailstock; 8 - tool holder; 9 - cutter with insulation; 10 - computer; 11 - digital voltmeter

The results of temperature experiments when turning with a cutting tool with different coatings are shown in Fig. 3. It was found that the most favorable from the point of view of temperature (thermal loading) of the cutting zone while minimizing the wear rate of the cutting tool is a multilayer composite coating (TiAl)N from Rimet [17-24].

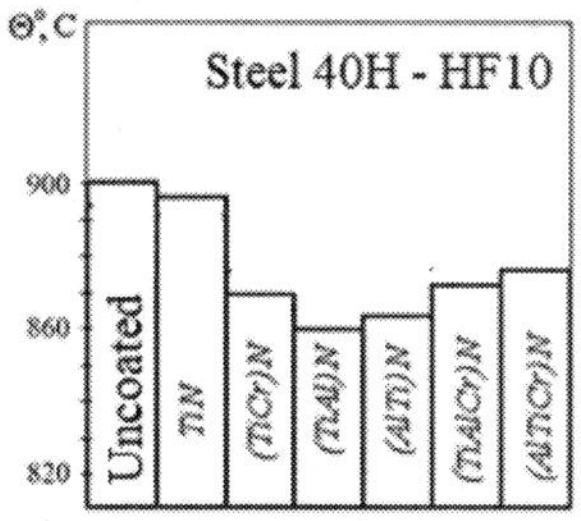

Fig. 3. Effect of coating type on temperature in the cutting zone during longitudinal turning

3 Results of the study

When developing the methodology and modeling, the following assumptions were made: thermal conductivity, specific heat of the tool and processed materials do not depend on temperature: sources of heat generation are distributed in the contact surfaces and in the shear plane; the cleavage plane is perpendicular to the front surface of the cutter; there is no heat exchange with the environment; friction coefficients, cutting forces, chip shrinkage coefficient, length of contact surfaces for cutting tools with different coatings are variable; the speed of movement of the elements of the chip material is the same along its thickness. The values of the intensity of heat release were taken from the works of Professor A.N. Reznikov. To ensure the identity of the experimental data and the simulation results, the constancy of the optimal cutting speed in terms of the intensity of the cutting tool wear was ensured with the variability of the thermal conductivity coefficient and the physicochemical composition of the coatings. In fig. 4 shows the results of calculations in the ANSYS software package for various thermal conductivity coefficients of the material of the part, tool and coating.

Based on the results of field experimental studies and modeling, it was found that the coating material is involved in direct contact with the part, the thermal conductivity of which varies depending on the type of coating. So, in Figure 4a, the thermal conductivity of the coating $\lambda = 0.25$, in Figure 4b $\lambda = 0.15$. In this case, the temperature in the chips increases as the thermal conductivity of the coating changes at the same contact temperature, i.e. most of the heat from the cutting zone is transferred to the chips. Thus, by changing the thermophysical properties of the friction surfaces, the thickness and structural composition of coatings for cutting tools, it is possible to predict and control the thermal regime, thereby minimizing the intensity of wear of the cutting tool/

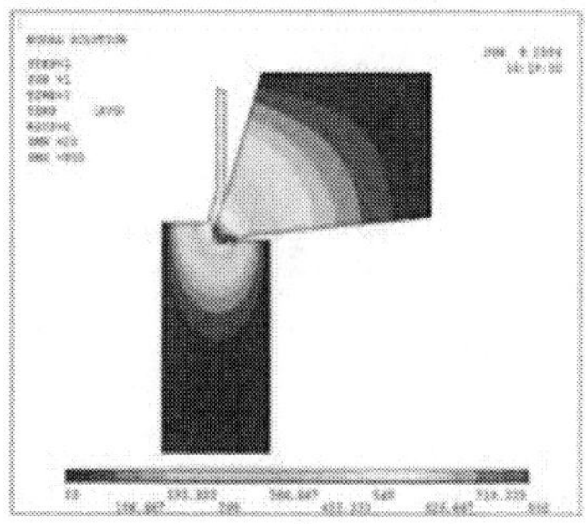

a)

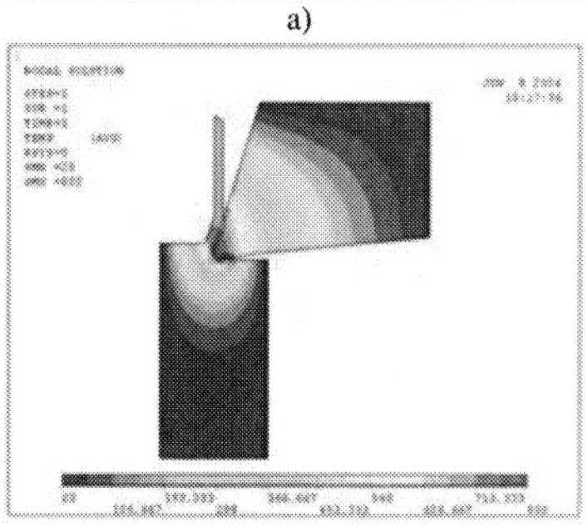

b)

Fig. 4. Calculation of temperature fields in the ANSYS environment: λ = 0.25 (a); λ = 0.15

Interesting results from the point of view of clarity and information content were obtained using the DEFORM software package (Fig. 5-7). In particular, you can see Fig. 5, that at the beginning of cutting with a multilayer coating (step 733), the bulk of the heat is concentrated on the rear surface of the cutting wedge, while due to the high temperature in the cutting zone and adhesion phenomena, the radial wear of the cutting wedge will intensify, which determines the size of the workpiece and the quality indicators of the surface layer. At the same time, after some time in Fig. 7 (step 1294), a redistribution (transition) of the heat flux from the rear to the front surface of the tool occurs, i.e. most of the heat goes into the chips, which is more favorable from the point of view of adhesion and reduction of dimensional wear [2,3]. Modeling in this environment allows, in addition to thermophysical parameters, to take into account technological parameters - cutting speed, the presence of a coating or coolant, as well as the wear of the cutting tool and their effect on the distribution of heat fluxes, which is important when using this technique in real machine-building production.

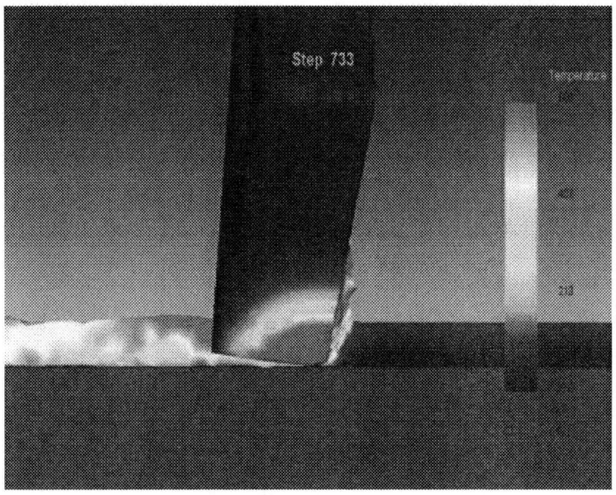

Fig. 5. The picture of thermal phenomena in the contact zone of the cutting coated tool at cutting speed V = 450 m / min

Fig. 6. The picture of thermal phenomena in the contact zone of the cutting uncoated tool at cutting speed $V = 450$ m / min

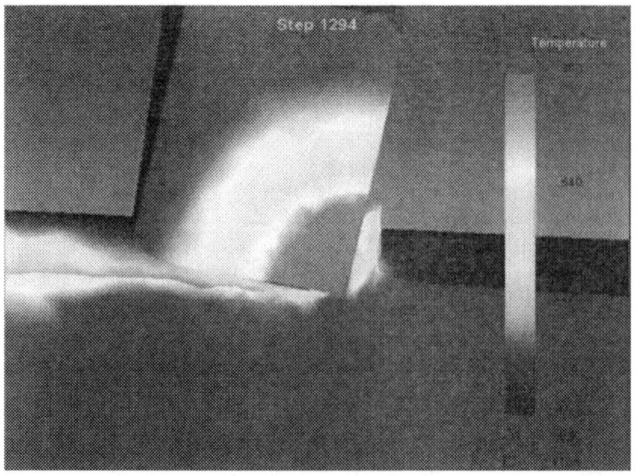

Fig. 7. The picture of thermal phenomena in the contact zone of the cutting coated tool at cutting speed $V = 450$ m / min (after 5 minutes and at h_3 = 0.15 mm)

Comparative analysis of the results of calculating the values of temperatures by mathematical models and data from field experiments showed that the discrepancy is 3 ... 6% and is a fairly objective assessment of the temperature field in the cutting zone [6]. On the basis of theoretical and experimental studies, recommendations on the temperature regime are proposed to ensure the minimization of the intensity of wear of cutting tools with various coatings.

To explain the mechanism of changing the temperature regime and heat flux in the contact zone, additional studies of the surfaces of the cutting tool with coatings after cutting were carried out [7, 8]. To understand the physical mechanism of the phenomena that take place on the surface "cutting tool - workpiece", the chemical and phase compositions of the surface layers were studied using modern techniques and equipment to analyze the composition of the coating surface and the outer surface of the films formed on the edge of the cutting tool, as well as for analyzing the contacting surface of the chips. For these studies, various special samples and oblique sections were made.

The results of spectroscopic studies made it possible to establish that during high-speed cutting, titanium oxides, the so-called rutiles and aluminum oxides with an amorphous structure, are formed on the surface of the coated cutting tool. It is known [1,4,7] that rutiles work as a lubricant, while amorphous-like structures have shielding (for heat flux) and protective (wear-resistant) effects.

4 Conclusion

The results of full-scale experimental studies in turning and computer simulation of thermophysical phenomena when changing the cutting temperature are comparable. In a wide range of changes in the elements of the cutting mode when turning with a cutting tool with wear-resistant coatings for preliminary calculations and the development of promising multilayer composite coatings with the effect of self-organization, one can use the software products DEFORM and ANSYS, and they confirm the adequacy of the model.

High-speed cutting leads to tribooxidation of multilayer composite coatings, which is an important and favorable process for the phenomenon of self-organization and the formation of oxygen-containing compounds on a metal basis, which can act as a shield and protect the tool surface and generally increase wear resistance. In this case, the effect of self-organization makes it possible to create coatings that provide a favorable temperature regime in the contact zone during cutting.

Acknowledgments

The study was carried out with the financial support of the Russian Foundation for Basic Research within the framework of scientific project No. 21-79-30058. This work was carried out using equipment provided by the Center of Collective Use of MSUT "STANKIN".

References

1. G. Polzer, V. Ebeling, Firkowski, Friction and wear, **9**(1), 12 (1988)
2. A.D. Makarov, Mashinostroenie, 278, (1976)
3. S.A. Vasin, A.S. Vereshchaka, V.S. Kushner, MSTU named after N.E. Bauman, 448, (2001)
4. G.S. Fox-Rabinovich, A.I. Kovalev, L.S. Shuster, Y.F. Bokiy, G.K. Dosbayeva, D.L. Wainstein, V.P. Mishina. Wear, **206**(1-2), 214–220 (1997)
5. N.K. Krioni, M.Sh Migranov, Innov. Mash, 377, (2016)
6. M.Sh. Migranov, R.G. Akhmarov, A.M. Migranov, Mekh. fiz. prots. Poverkh, 83–86, (2018)
7. M.Sh. Migranov, R.M. Migranova, Key Engineering Materials, **496**, 75–79, (2012)
8. A.S. Vereschaka, M.Sh. Migranov, A.A. Vereschaka, O. Sharipov. ARPN Jour. of Eng. and App. Scie., **10**(20), 9282–9286, (2015)
9. Patent RF 34249 for utility model. *Pribor dlya issledovaniya adgezionnogo vzaimodeystviya* [Device for the study of adhesive interaction] / Shuster L.Sh., Migranov M.Sh. Declared 06/24/2003. Published 11/27/2003, Bulletin No. 33.
10. T.N. Loladze, *Strength and wear resistance of the cutting tool* (M.: Mashinostroenie, 1982)
11. M. Volosova, S. Grigoriev, A. Metel, A. Shein, Coatings, **8**, 287, (2018)
12. Yu.B. Rumer, M.Sh. Ryvkin, *Thermodynamics, statistical physics and kinetics* (Moscow: Nauka, 1977)
13. I.B. Zolotukhin, Sorosovskiy Educational Journal, 1, 103–106 (1998) [in Russia]
14. Y.G. Kabaldin, D.A. Shatagin, M.S. Anosov, A.M. Kuzmishina Intelligent Control of Technological Systems in Digital Manufacturing, **1**(178), 3-12 (2020) DOI: 10.18698/0536-1044-2020-1-3-12
15. S.N. Grigoriev, O.V. Sobol, V.M. Beresnev, I.V. Serdyuk, A.D. Pogrebnyak, D.A. Kolesnikov, U.S. Nemchenko, J. Frict. Wear, **35**(5), 359, (2014)
16. O.V. Sobol', A.A. Andreev, S.N. Grigoriev, V.A. Stolbovoy, Probl. At. Sci. Technol., **4**, 174, (2011)
17. M.Sh. Migranov, Friction and lubrication in machines and mechanisms, **5**, 23-26 (2015)
18. A.S. Metel, V.P. Bolbukov, M.A. Volosova, S.N. Grigoriev, Yu.A. Melnik, Instrum. Exp. Tech., **57**(3), 345, (2014)
19. V.V. Kuzin, S.N. Grigor'ev, M.A. Volosova, Refract. Ind. Ceram., **54**, 376, (2014)
20. M.Sh. Migranov, Improving the wear resistance of tools based on the intensification of the processes of adaptation of friction surfaces when cutting metals (Publishing house "Gilem" of the Academy of Sciences of the Republic of Belarus, 2011)
21. S.Y. Zhu, J. Cheng, Z.H. Qiao, J. Yang, Tribology International, **133**, 206-223 (2019) DOI: 10.1016/j.triboint.2018.12.037
22. A. Chouhan, T.K. Sarkar, S. Kumari, K.L.N. Sivakumar, H. Sugimura, O.P. Khatri, Applied Surface Science, **538**, 148041 (2021) DOI: 10.1016/j.apsusc.2020.148041
23. S.M. Aouadi, J.J. Gu, D. Berman, Journal Of Vacuum Science & Technology A, **38**(5), 050802 (2020) DOI: 10.1116/6.0000350
24. K. Yang, J.Q. Zeng, H.S. Huang, J.D. Chen, B.A. Cao, Materials, **12**(17), 2744 (2019) DOI: 10.3390/ma12172744

Non-Stationary Temperature Behavior of Spindle Assemblies and Ball Screw Gears for High Precision Metal Cutting Machines

A.P. Kuznetsov[1]

[1]Moscow State Technological University "STANKIN", RU-127055, Moscow, Russia

Abstract. Typical temperature and thermoelastic models of structures of spindle units and ball screw systems of metal-cutting machines were considered. The features of determining the geometric and thermophysical parameters of these models were described in some detail. Both analytical and numerical solutions for estimating time-dependent temperatures are presented. The solutions were obtained for thermoelastic displacements of spindles and ball screw systems for various boundary conditions and limitations of deformations during heating due to the design of these elements of metal-cutting machines.

1 Introduction

Progress in improving the accuracy of metal-cutting machines is inextricably linked and is due to a deeper knowledge of thermal processes and effects occurring in machines needed to control them. Since 1950th till now researchers have performed a number of studies of thermal processes and effects that occur in metal-cutting machines during their operation. The conducted studies of accuracy indicators and machine properties give reason to believe that in the balance of machine accuracy, the share of temperature errors in the period of 1950 – 2020s changed from 20-30% to 70% [1,2]. This follows from the analysis of energy requirements (energy barrier of accuracy), which are necessary and make it possible to achieve a certain level of processing accuracy (Fig.1) [3,4].

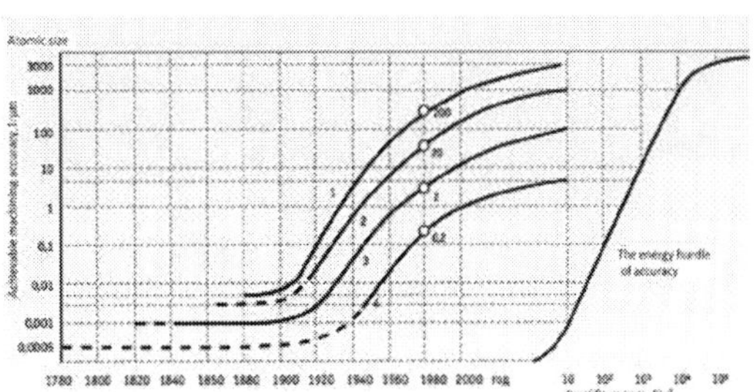

Fig.1. Evolution of the development of achievable machining accuracy of Machine tools:1-traditional, 2 – precision, 3 – high precision, 4-ultra-high precision

In the period of 1980-2020s, types and forms of methods for correcting and compensating the thermal behavior of forming elements of machine tools were developed [5,6], which are based on the use of linear and nonlinear regression methods or correlation between displacements and temperature of machine components and parts. These methods are based on direct methods for measuring temperature errors, indirect methods based on the measured temperature obtained from sensors (or without sensors) [7] located at various points of the machine and the subsequent determination of TCP offsets from the thermal model, indirect methods according to information received from sensors on different parameters (e.g., frequency of rotation, speed, axis, etc.), calculations based on the constructed or selected mathematical model, software-based methods such as thermal or thermo-elastic machine models, predictive models of changes of temperature deformations and machine control based on experimental data on the temperature behavior of the machine [8]. Over the past 30 years, the international organization for standardization (ISO) has developed several standards: ISO 230-3 (Thermal displacements in machine tools); ISO 10791-10 (Thermal displacements in machining centers); ISO 13041-8 (Thermal displacements in lathes), ISO 16907-2015(correction of geometric errors using CNC), which define test procedures for evaluating thermal effects that lead to temperature deformation of the structural and layout structure of the machine and/or its positioning systems. The functional relationship between the measured temperature in the machine units and the amount of displacement can be established experimentally. With good repeatability and fairly stable reproducibility of the results, this functional relationship is expressed by an equation that can be used as the basis for the operation of the control device for compensation of thermal deformations [6,9-11]. Analysis of design and layout solutions of modern metal-cutting machines has shown that in the thermal model of the machine, the spindle assembly [9-11] and ball-screw systems [12,13] of movement of the working bodies of the machine are the main and dominant sources of the thermal instability. Since mathematical methods for describing thermal models [14-22], numerical and experimental stationary thermal conductivity methods, and FEM were most often used, non-stationary processes were either neglected or described using the simplest models of thermal balance without accounting for the full range of boundary conditions in determining temperatures and solving thermoelastic problems in a non-stationary temperature field. As a rule, this is due to the fact that for the accuracy required in the past, it was enough to consider the thermoelastic deformations of the spindle and drive screws as linear elongation of the rod in proportion to the average temperature. This paper presents solutions to non-stationary problems for determining the temperature fields of spindle assemblies and ball screw gears and their temperature deformations for conditions that limit their free movement to that in typical modern metal-cutting machines.

2 Thermal models of spindle units and ball screw units

Analysis of the design of machine components and parts with an indication of the kinematic scheme, configuration of parts, their connections, material, identification of the main sources of heat, their type and location, heat exchange conditions and other factors are necessary for the formation of thermal models of the machine unit or part with specified and reasonable assumptions. The thermal model is the basis for creating a mathematical model written in the form of equations with boundary conditions that reflect the connections of a node or part in terms of their heat exchange. Mounting options used in machine tools, linear motion drive "lead screw-rolling nut", used in metal cutting machines, have four types of mounting supports of the lead screw and the first support always has a mounting of the lead screw and either allows or does not move the lead screw in the axial direction, and the second support is made in four versions: LI - the first and second supports support the lead screw with the possibility of free axial movement; LII - the second support only supports the screw, and the first does not allow you to move in the axial direction; LIII-the second support prevents axial movement of the lead screw; LIY - the second support is missing.

Despite all the variety of design features of building schemes for installing bearings in the front and rear spindle supports, which are determined by the requirements for providing the specified parameters and characteristics of spindle units (radial and axial stiffness, dynamic stiffness, rotation accuracy, spindle runout, speed parameters, required torque, durability, and others), existing spindle unit schemes can be divided into four fundamentally different types in relation to perception (as direct, and reverse) axial loads of the front and rear spindle supports, and, accordingly, the ability to move in the axial direction:

SI-the rear support accepts the axial load in the forward and reverse directions, and the front support moves freely in the axial direction and accepts the radial load;

SII - the front support accepts the axial load in the forward and reverse directions, and the rear support moves freely in the axial direction and accepts the radial load;

SIII-the front and rear supports perceive the axial load in different directions, and one of the supports can move freely in the axial direction under the action of an external axial force;

SIY-the front and rear supports perceive the axial load in different directions, and one of the supports can move freely in the axial direction under the action of an internal axial force.

Thus, the analysis of structures and generalization of basic layout schemes for the construction of both spindle and ball screw units allows us to form a general temperature and thermoelastic models for calculating temperatures and axial displacements, which is due to and determined by the unity of models in the formulation and solution of this problem. Therefore, the thermal model of a part, unit or machine is defined as their geometric image, which adequately reflects the conditions of unambiguity due to the features of thermal processes that are formed by the functional purpose, technological and technical parameters, characteristics and properties under specified operating conditions.

The degree of adequacy of the thermal model is determined by the accepted simplifications, assumptions, and completeness of consideration and accounting in the thermal model of real thermophysical processes that are determined by the goal (s), tasks to be solved, and methods for achieving them.

If the criterion value Bio small $Bi < 0{,}1 \div 0{,}2$, in this case, the dimensionality of the problem decreases as the thermal resistance of the boundary layer is substantially more thermal resistance and the problem in finding the temperature on the surface of the part and any other point of the surface, and the temperature of the part at the Ld almost identical.

So, in table.1. the limits of the values of the smallest size Ld of a part or Assembly of a metal-cutting machine are given, depending on their material (thermal conductivity coefficient λ W/m × 0K) and heat exchange conditions (heat transfer coefficient α W/ m2 × 0K), under which the dimension of the task can be reduced, i.e. the two-dimensional task can be considered as one-dimensional without loss of solution accuracy.For example, for a column (cast iron) or machine bed, the problem can be considered as two-dimensional with a column thickness of less than 0.688 meters, and for a machine spindle (steel 40X), the temperature will be uniform over its diameter, the value of which does not exceed 0.150 m.and the temperature distribution will occur along its length and the problem will be considered as one-dimensional with heat transfer on the surface of the spindle (see Fig.2).

The thermal model of the spindle can be represented geometrically as a solid or hollow cylinder with a constant or variable cross-section, on the outer and inner surfaces of which heat exchange with the environment occurs with a heat transfer coefficient α. In this case, each surface can have its own values of the heat transfer coefficient due to different conditions of heat exchange with the environment. For example, the heat exchange conditions on the surfaces of the protruding part of the spindle may differ greatly from the heat exchange conditions on the inner surface of the spindle, where heat exchange with the environment is significantly difficult. Heat supply to the spindle is surface and is carried out from various heat sources: spindle supports; drive elements; tools or parts that are fixed in the spindle and are heated due to heat generation during cutting.

Table 1. Values of the characteristic size Ld of the machine part (units), to reduce the dimension of the mathematicals task.

λ / α	10	15	20	25	30	35	40	45	50	55	60
2	0.5	0.75	1.0	1.25	1.50	1.75	2.0	2.25	2.50	2.75	3.0
4	0.25	0.38	0.5	0.63	0.75	0.88	1.0	1.13	1.25	1.36	1.5
6	0.17	0.25	0.33	0.42	0.50	0.58	0.67	0.75	0.83	0.92	1.0
8	0.13	0.19	0.25	0.31	0.38	0.44	0.50	0.56	0.63	0.69	0.75
10	0.10	0.15	0.20	0.25	0.30	0.35	0.40	0.45	0.50	0.55	0.6
12	0.08	0.14	0.17	0.21	0.25	0.29	0.33	0.38	0.42	0.46	0.5
14	0.07	0.11	0.14	0.18	0.21	0.25	0.29	0.32	0.36	0.39	0.43
16	0.06	0.09	0.13	0.16	0.19	0.22	0.25	0.28	0.31	0.34	0.38
18	0.06	0.08	0.11	0.14	0.17	0.19	0.22	0.25	0.28	0.31	0.33
20	0.05	0.07	0.10	0.13	0.15	0.18	0.20	0.23	0.25	0.28	0.30

As a thermal model (Fig. 2) of a spindle Assembly and a ball screw Assembly, we take a cylinder of finite length L, on the side surfaces of which heat exchange with the environment occurs with a heat exchange coefficient equal to α, and heat is supplied to the cylinder at the places where the bearings are fixed or only through the end surfaces. The following assumptions are made:

-the problem is considered as one-dimensional and the temperature changes only along the length of the cylinder, and on each section it made constant and does not depend on radius and angle;

-thermal parameters (heat transfer coefficients, thermal conductivity and heat capacity) are functions of temperature, coordinate and time, and the heat transfer part of the end surfaces can be neglected due to the small size of their area in relation to the lateral surface of the cylinder.

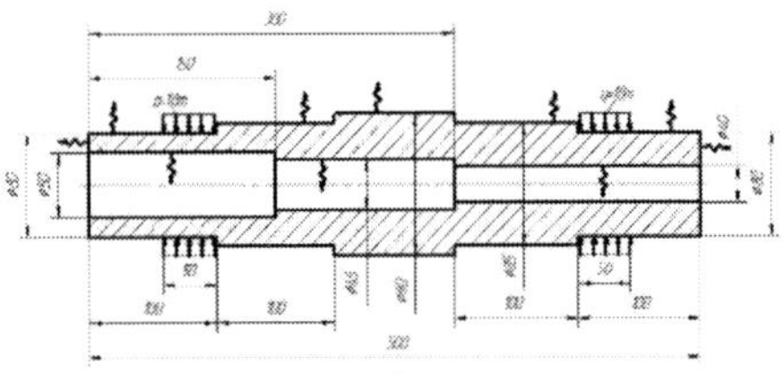

a)

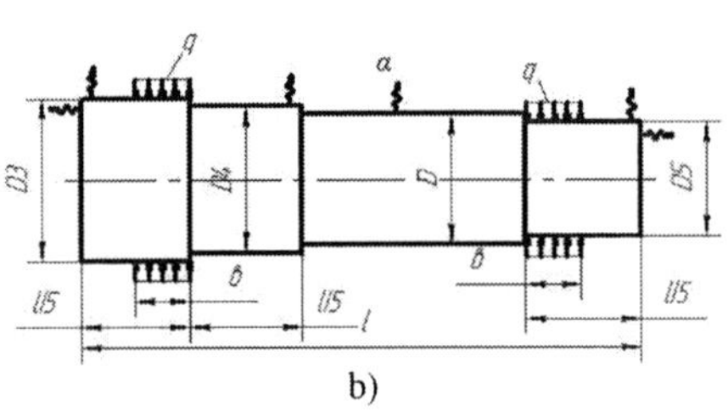

b)

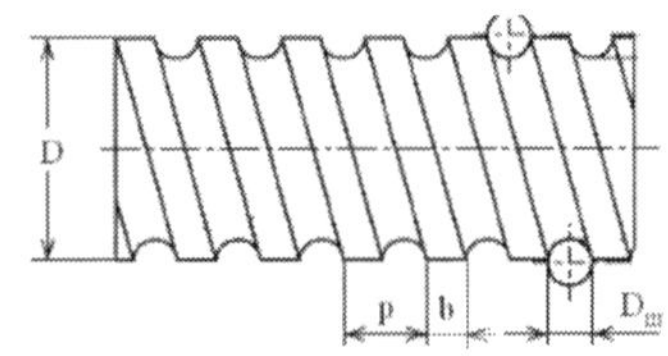

c)

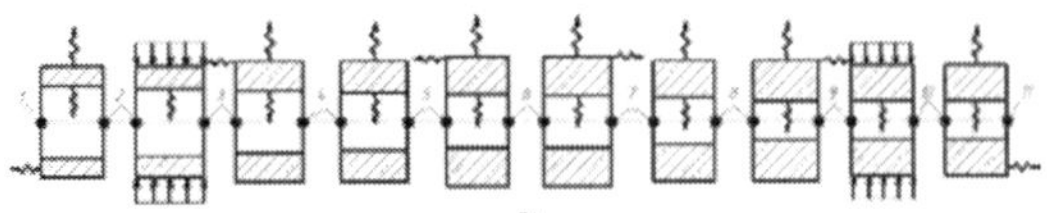

d)

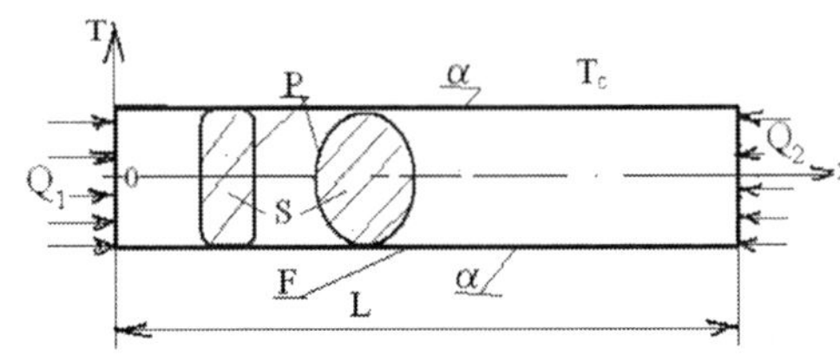

e)

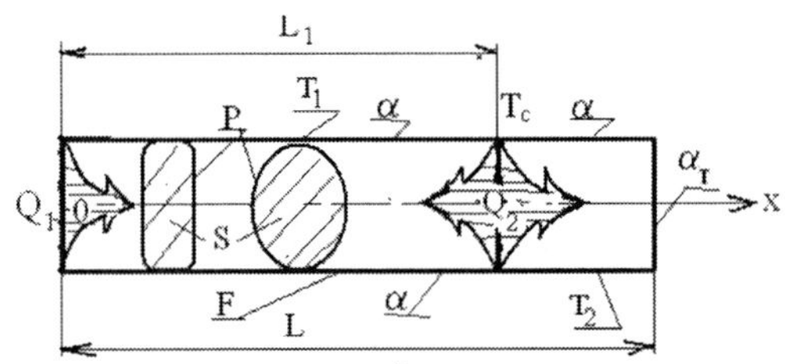

f)

Fig. 2. Thermal image of the spindle – a, b; parameters of the ball screw-c; thermal models for FEM-d; reduced to the cylinder-d, e.

Due to the assumptions made, this thermal model is not completely identical to the actual spindle node. However, in order for the thermal processes to be similar, it is necessary to choose such defining cylinder dimensions that the similarity criteria of Bio and Fourier are equal for both the thermal model and the actual spindle design.

When drawing up a thermal model, almost in most cases it is necessary to change (simplify or reduce to simpler forms) the geometric dimensions of real parts, as shown in Fig.2. In this case, to preserve the equivalence of conditions thermal processes can proceed in two ways.

The first is to reduce the part to a simple geometric shape while maintaining the volume of the real part and the thermal model while maintaining the same heat flow. Then in calculations it is necessary to ensure equality of the areas of the heat exchange surfaces of the thermal model Ftm and the real part Fre , and the heat exchange coefficient should be proportionally changed to the ratio of these areas. The given value, for example, of the heat flow will be equal to Qeq= Qre(Fre,/ Ftm) and accordingly, the heat transfer coefficient of the thermal model will be equal to aeq= are(Fre,/ Ftm). For linear dimensions, their values are determined for the same reasons, for example, Ltm= Vre / Ftm .

Second , if the lengths of the thermal model and the spindle are the same, then the diameter of the cylinder of the thermal model can be found from the equality of the Biot or Fourier criteria. In General, the most accurate value for the cylinder diameter of the thermal model will be determined by the expression:

$$d_{мс} = \alpha_{cp}\left(\Sigma d_i^2 l_i - \Sigma d_j^2 l_j\right) / \left(\Sigma \alpha_i d_i l_i + \Sigma \alpha_j d_j l_j + \Sigma \alpha_{Ti}(d_i^2 - d_j^2)\right) \quad (1)$$

where α_{cp} is the accepted heat transfer coefficient of the thermal model or calculated from the dependence;

$$\alpha_{cp} = [\Sigma(\alpha_i F_i) + \Sigma(\alpha_{Tj} F_{Tj})] / [\Sigma(F_i) + \Sigma(F_{Tj})] \quad (2)$$

where α_i is the heat transfer coefficient of the i-th heat transfer surface Fi(internal and external), α_{Tj} is the heat transfer coefficient of the end surfaces.

In the case when it is necessary to take into account the heat transfer from the end surface instead of the length *L* of the spindle, take its reduced length LPR

$$L_{np} = L + (\alpha_m / \alpha) \cdot (F / P) \quad (3)$$

where P is the perimeter of a cylinder or rod of arbitrary cross-section.

It should be taken into account that the heat transfer coefficient from the screw surface is assumed to be equal to the given value:

$$\alpha_{np} = \alpha_c \cdot \sqrt{[1 + (p / \pi D)^2] \cdot [(\pi r_b + b) / p]} \quad (4)$$

where α_c is the heat transfer coefficient of the cylindrical surface of the screw model , r_b is the radius of the rolling elements, D is the cylinder diameter, p is the screw pitch, and b is the width of the screw profile pad.

For a more accurate determination of the reduced heat transfer coefficient, when considering the screw as a finned surface, the expression for calculating the α_{np}, based on the main provisions of the works, is written:

$$\alpha_{np} = \alpha_c \cdot (1 + E \cdot \varepsilon \cdot \sqrt{[1 + (p / \pi D)^2] \cdot [(\pi r_b + b) / p]}) \quad (5)$$

where E is the coefficient of finning efficiency, and ε is the coefficient of the shape of the edge profile (for a straight edge, it is equal to one).

To determine the temperature axial displacements (Fig. 3), it is necessary to solve the thermoelastic task in a quasi-static formulation, when dynamic effects during heating are not taken into account. In this formulation, the problem allows you to determine the axial deformations of both spindles and lead screws. We will consider this problem with the following assumptions: - the temperature field is axisymmetric;- when heated, the thermophysical and thermomechanical parameters of the material remain constant;- Dugamel's hypothesis is valid, according to which acceleration can be ignored;-the thermoelastic state is uniaxial.

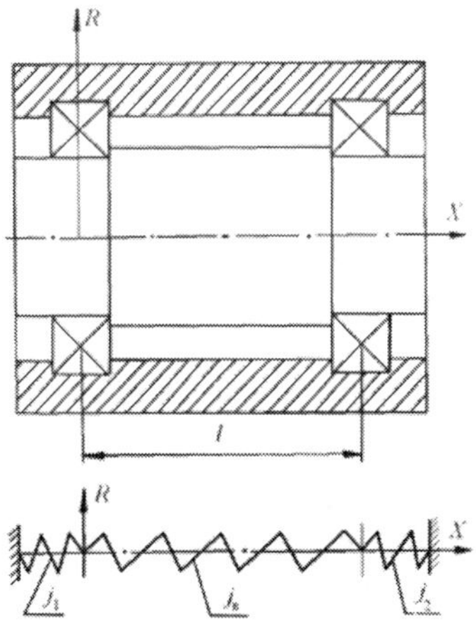

Fig.3. Typical thermoelastic model (the spindle, the lead screw) with elastic links.

Therefore, to determine the value of the thermoelastic displacement of a typical machine element (spindle or lead screw), we present it as a rod of finite length, whose quasi-thermostable connections are located only in the direction of the coordinate axes and have rigidity: the proper axial stiffness of the rod, the axial stiffness of the first and second supports . A typical thermoelastic model for this case is shown in Fig. 3. The non-stationary temperature field of the rod must also be known as the initial data for determining thermoelastic displacements at any time. To determine the non-stationary (i.e. changing in time and space) of the temperature field on the basis of the thermal model Fig.2 it is necessary to solve the heat equation with boundary conditions of the second kind:

$$\partial^2 T / \partial x^2 - m^2 T = a \cdot \partial T / \partial \tau; \quad (6)$$

$$\left.\begin{aligned} \partial T / \partial x \,\big|_{x=0} &= -Q_1 / F \cdot \lambda; \\ \partial T / \partial x \,\big|_{x=l} &= Q_2 / F \cdot \lambda; \\ T(x,\tau) \,\big|_{\tau=0} &= 0 \end{aligned}\right\} \quad (7)$$

where is the cross-sectional area of the cylinder;- values of the heat flow in the rear and front supports;- coefficient inverse to the coefficient of thermal conductivity (- specific heat capacity, - specific weight of the material, - thermal conductivity of the material);- heat transfer coefficient;- thermal conductivity; .

The General solution for determining the non-stationary temperature field has the form [23]:

$$T(x,\tau)=\frac{Q_2\cdot K_{Q_2}}{F\lambda\sqrt{a}2\sqrt{2b}}\left\{\sum_{n=0}^{\infty}\left[e^{-M\sqrt{2b}}erfc\left(\frac{M}{2\sqrt{\tau}}-\sqrt{2b\tau}\right)-e^{M\sqrt{2b}}erfc\left(\frac{M}{2\sqrt{\tau}}+\sqrt{2b\tau}\right)\right]+\right.$$
$$\left.+\sum_{n=0}^{\infty}\left[e^{-M_1\sqrt{2b}}erfc\left(\frac{M_1}{2\sqrt{\tau}}-\sqrt{2b\tau}\right)-e^{M_1\sqrt{2b}}erfc\left(\frac{M_1}{2\sqrt{\tau}}+\sqrt{2b\tau}\right)\right]\right\}+\frac{Q_1\cdot K_{Q_1}}{F\lambda\sqrt{a}2\sqrt{2b}}$$
$$\left\{\sum_{n=0}^{\infty}\left[e^{-U\sqrt{2b}}erfc\left(\frac{U}{2\sqrt{\tau}}-\sqrt{2b\tau}\right)-e^{U\sqrt{2b}}erfc\left(\frac{U}{2\sqrt{\tau}}+\sqrt{2b\tau}\right)\right]+\right.$$
$$\left.+\sum_{n=0}^{\infty}\left[e^{-U_1\sqrt{2b}}erfc\left(\frac{U_1}{2\sqrt{\tau}}-\sqrt{2b\tau}\right)-e^{U_1\sqrt{2b}}erfc\left(\frac{U_1}{2\sqrt{\tau}}+\sqrt{2b\tau}\right)\right]\right\} \quad (8)$$

where indicated:

$$M=\sqrt{a}\left[(2n+1)l-x\right], U=\sqrt{a}\left(2nl-x\right),$$
$$M_1=\sqrt{a}\left[(2n+1)l+x\right],$$
$$U_1=\sqrt{a}\left[(n+1)2l-x\right], 2b=m^2/a, m^2=4\alpha/\lambda d$$

K_{Q_1}, K_{Q_2} - coefficients that take into account the boundary conditions of various heat transfer on the end and other surfaces of the model among them other than the cylindrical one. For example, for different heat exchange on a part of a cylindrical surface and heat exchange on an end surface, the value will be defined as:

$$K_{Q_2}=\frac{D_m}{1+(m_1/m_2)\cdot D_m\cdot th(m_1\cdot L_1)} \quad (9)$$

where indicated:

$$D_m=[K_m\cdot\exp(2(L-L_1)\cdot m_2)+1]/[K_m\cdot\exp(2(L-L_1)\cdot m_2)-1]$$
$$K_m=[m_2+(\alpha_m/\lambda)]/[m_2-(\alpha_m/\lambda)]$$

Table 2. Comparative data for determining temperatures by (8) for different K_{Q_1}, K_{Q_2}.

N п/п	α_1, w/м^2 0K	α_2, w/м^2 0K	α_T, w/м^2 0K	T(0), 0K	T(0.4), 0K
1	25	25	0	15.02	21.87
2	25	25	(Lпр)	14.66	20.54
3	25	25	25	14.66	20.53
4	25	50	0	13.85	17.46
5	25	50	50	13.42	15.85
6	125	25	0	4.51	10.67
7	125	50	0	4.48	9.30
8	25	25	500	13.19	14.99
9	25	50	500	12.41	12.07
10	25	25	1500	12.53	12.51
11	325	50	1500	2.66	5.13

3 Non-stationary temperature deformations

To determine the temperature deformations that are caused by an axisymmetric non-stationary temperature field, it is necessary to solve the General equation in displacements for the thermoelastic model (Fig.3.) with boundary conditions determined by the elastic state of the rod (cylinder) and fixing in the locations of the supports: Equations should be centred and should be numbered with the number on the right-hand side.

$$\frac{\partial^2 U}{\partial x^2}=\beta\cdot\frac{\partial T}{\partial x} \quad (10)$$

$$\varepsilon_{xx}\cdot E-\beta\cdot E\cdot T=\sigma_{xx} \text{ при } x=0$$

$$\varepsilon_{xx}\cdot E-\beta\cdot E\cdot T=-\sigma_{xx} \text{ при } x=L \quad (11)$$

The General expression for determining axial offsets in an arbitrary cross section is [24]:

$$U(x,\tau)=\int_0^x\beta\cdot T(x,\tau)\cdot dx-\frac{x\cdot j_\Sigma}{E\cdot F}\int_0^L\beta\cdot T(x,\tau)\cdot dx-\frac{j_\Sigma}{j_1}\int_0^L\beta\cdot T(x,\tau)\cdot dx=$$
$$=\int_0^x\beta\cdot T(x,\tau)\cdot dx-j_\Sigma\,(j_1^{-1}+\frac{x}{E\cdot F})\cdot\int_0^L\beta\cdot T(x,\tau)\cdot dx \quad (12)$$

where β - coefficient of thermal expansion of the material, E - modulus of elasticity.

For boundary conditions when there is an initial gap δ_0 in the support, and there is a pre-stressed state along the axis, i.e., when the spindle or lead screw is pre-stretched (compressed) by force P_{np}, the generalized expression for estimating the axial displacement at an arbitrary time and in an arbitrary cross section will take the form:

$$U(x,\tau)=\int_0^x\beta T(x,\tau)dx-\frac{x\cdot P_{np}}{EF}-j_\Sigma(j_1^{-1}+\frac{x}{EF})[\int_0^L\beta T(x,\tau)dx-\frac{x\cdot P_{np}}{EF}-\delta_0] \quad (13)$$

The expression (13) allows you to determine the thermal displacements of bodies that can be reduced to the considered thermal model. When considering the thermal displacements of the spindle, we are usually interested in the amount of displacement of its end face . We transform the expression (13) by dividing its left and right parts by the same value equal to the value of the thermal displacement of the spindle free from external loads, i.e., when the spindle supports do not interfere with its displacement, and as a result we get:

$$\frac{U(L,\tau)}{\int_0^L\beta T(x,\tau)dx}=\frac{j_\Sigma}{j_2}-\frac{j_\Sigma}{j_2}\cdot\frac{L\cdot P_{np}}{EF\int_0^L\beta T(x,\tau)dx}+\frac{j_\Sigma\cdot(j_1^{-1}+\frac{L}{EF})\delta_0}{\int_0^L\beta T(x,\tau)dx}=$$
$$=\frac{j_\Sigma}{j_2}-\frac{j_\Sigma}{j_2}\cdot\frac{j_{np}}{j_в}+j_\Sigma\cdot(j_1^{-1}+j_в^{-1})\cdot K_0=K_j\cdot\int_0^L\beta T(x,\tau)dx \quad (14)$$

where indicated:

$$K_j = \frac{j_\Sigma}{j_2} \cdot (1 - K_{np} - K_0) + K_0 = [\frac{j_\Sigma}{j_2} \cdot (1 - K_{np}) + K_0 \cdot (1 - K_{np})]$$

$$K_{np} = j_{np} / j_в, \quad j_в = EF / L,$$

$$j_{np} = P_{np} / \int_0^L \beta T(x,\tau) dx,$$

$$j_\Sigma = [j_1^{-1} + j_2^{-1} + j_в^{-1}]^{-1}, \quad K_0 = \delta_0 / \int_0^L \beta \cdot T(x,\tau) \cdot dx$$

The range of values is from 0 to 1, and the limit values correspond to the following ratios of stiffness values:

$$j_2 = 0 \Rightarrow K_j = 1; \quad j_2 = \infty \Rightarrow K_j = 0$$

$$j_1 = 0 \Rightarrow K_j = 0; \quad j_1 = \infty \Rightarrow K_j = j_в / (j_в + j_2)$$

In order to get an explicit solution for determining non-stationary temperature displacements, it is necessary to substitute the value of the function in the expression (8).

Then, the final solution will be obtained when all the components of the expression (13,14) are sequentially integrated. Subintegral expressions are structurally similar functions. As an example, let's integrate one of the same type of functions:

$$\Omega = \int_0^x e^{-U\sqrt{2b}} erfc\left[\frac{U}{2\sqrt{\tau}} - \sqrt{2b\tau}\right] = \int_0^x \exp\left[-\sqrt{2b}(2nl+\xi)\sqrt{a}\right] erfc\left[\frac{\sqrt{a}(2nl+\xi)}{2\sqrt{\tau}} - \sqrt{b\tau}\right] d\xi \quad (15)$$

Perform the following conversions:

$$e^{-\sqrt{2ba(2nl+\xi)}} = \exp\left[-\xi\sqrt{2ba}\right] \cdot \exp\left[-\sqrt{2ba} \cdot 2nl\right] \quad (16)$$

$$\frac{\sqrt{a}(2nl+\xi)}{2\sqrt{\tau}} - \sqrt{2b\tau} = \frac{\xi\sqrt{a}}{2\sqrt{\tau}} + \frac{\sqrt{a}2nl}{2\sqrt{\tau}} - \sqrt{2b\tau} \quad (17)$$

Substituting in (17), we get:

Denote $2nl\sqrt{a} - \tau\sqrt{8b} = A$ and $A + \xi\sqrt{a} = \mu$ introduce new variables:

$$u = erfc\left(\frac{\mu}{2\sqrt{\tau}}\right)$$

$$du = -\frac{\sqrt{a}}{\sqrt{\pi\tau}} \exp\left(-\frac{\mu^2}{4\tau}\right) d\xi$$

$$v = -\frac{1}{\sqrt{2ba}} \cdot \exp(-\zeta \cdot \sqrt{2ba})$$

$$dv = -\frac{1}{\sqrt{2ba}} d[\exp(-\xi\sqrt{2ba})]$$

Then the expression (15) will look like:

$$\Omega = \exp\left[-\sqrt{2ba} \cdot 2nl\right]\left\{\left(\frac{1}{\sqrt{2ba}}\right)\exp\left(-\xi\sqrt{2ba}\right) \times erfc\left(\frac{\mu}{2\sqrt{\tau}}\right) - \right.$$

$$\left. -\int_0^x \exp\left[-\left(\frac{\mu^2}{4\tau} + \xi\sqrt{2ba}\right)\right] \cdot \left(\frac{\sqrt{a}}{\sqrt{\pi\tau}\sqrt{2ba}}\right) d\xi \right. \quad (18)$$

Transform the newly obtained integrand:

$$\exp\left[-\left(\frac{\mu^2}{4\tau} + \xi\sqrt{2ba}\right)\right] = \exp\left[-\frac{\left(\xi\sqrt{a} + A_1\right)^2}{4\tau}\right] \exp\left(\frac{A1^2 - A^2}{4\tau}\right) \quad (19)$$

where indicated $A_1 = A + 2\tau\sqrt{2b} = 2nl\sqrt{a}$. As a result, the expression is written as follows:

$$\Omega = \exp\left(-2nl\sqrt{2ba}\right)\frac{1}{\sqrt{2ba}}\left\{-\exp\left(-\xi\sqrt{2ba}\right) erfc\left(\frac{\mu}{2\sqrt{\tau}}\right)\Big|_0^x - \right.$$

$$\left. -\frac{\sqrt{a}}{\sqrt{\pi\tau}} \exp\left(\frac{A1^2 - A^2}{4\tau}\right) \int_0^x \exp\left[\frac{\left(\xi\sqrt{a} + A_1\right)^2}{4\tau}\right] d\xi \right\} \quad (20)$$

We introduce new variable integration limits from the given boundary conditions defined by the expressions:
выражениями:

$$\gamma = \frac{\xi\sqrt{a + A_1}}{2\sqrt{\tau}}; \quad \xi = \gamma\frac{2\sqrt{\tau}}{\sqrt{a}} - \frac{A_1}{\sqrt{a}}; \quad d\xi = \frac{2\sqrt{\tau}}{\sqrt{a}} d\gamma.$$

$$\xi = 0 \quad \gamma = \frac{A_1}{2\sqrt{\tau}}; \quad \xi = x \quad \gamma = \frac{x\sqrt{a + A_1}}{2\sqrt{\tau}}.$$

Substituting new variables in the expression and performing integration within the new limits, we get:

$$\Omega = \exp\left(-2nl\sqrt{2ba}\right)\frac{1}{\sqrt{2ba}}\left\{-\exp\left(-\xi\sqrt{2ba}\right) erfc\left(\frac{\mu}{2\sqrt{\tau}}\right)\Big|_0^x - \right.$$

$$-\frac{\sqrt{a}}{\sqrt{\pi\tau}} \exp\left(\frac{A1^2 - A^2}{4\tau}\right) \int_{\frac{A_1}{2\sqrt{\tau}}}^{\frac{z\sqrt{A}+A_1}{2\sqrt{\tau}}} \exp\left(-\gamma^2\right)\frac{2\sqrt{\tau}}{\sqrt{a}} d\gamma \quad (21)$$

Sequentially defining the value of the function x within the integration, we find the expression explicitly:

$$\Omega = \exp\left(-2nl\sqrt{2ba}\right)\frac{1}{\sqrt{2ba}}\left\{\left[-\exp\left(-x\sqrt{2ba}\right)erfc\left(\frac{x\sqrt{a}+A}{2\sqrt{\tau}}\right)+\right.\right.$$
$$\left.+erfc\left(\frac{A}{2\sqrt{\tau}}\right)\right]-\exp\left(\frac{A1^2-A^2}{4\tau}\right)\left[erf\left(\frac{x\sqrt{a}+A1}{2\sqrt{\tau}}\right)-\right.$$
$$\left.\left.-erf\left(\frac{A1}{2\sqrt{\tau}}\right)\right]\right\}=\delta_x\left(-U,A,A_1\right)$$

(22)

Performing similar transformations with all the components of the expression, we find an explicit solution to the problem of determining non-stationary thermoelastic displacements for the thermal model shown in Fig.3.

$$\delta(x,\tau)=\frac{\beta q_1}{F\lambda 2\sqrt{2ba}}\left\{\sum[\delta_x(-U,A,A_1)-\delta_x(U,A_2,A_3)]+\sum[\delta_x(-U_1,A',A'_1)-\delta_x(U_1A'_2,A'_3)]\right\}+$$
$$+\frac{\beta q_2}{F\lambda 2\sqrt{2ba}}\left\{\sum[\delta_x(-M,B',B'_1)-\delta_x(M,B'_2,B'_3)]+\sum[\delta_x(-M_1,B,B_1)-\delta_x(M_1,B_2,B_3)]\right\}-j_2\left(\frac{x}{EF}+\right.$$
$$\left.+\frac{1}{j_1}\right)\frac{\beta q_1}{F\lambda 2\sqrt{2ba}}\left\{\sum[\delta_l(-U,A,A_1)-\delta_l(U,A_2,A_3)]+\sum[\delta_l(-U_1,A',A'_1)-\delta_l(U_1,A'_2,A'_3)]\right\}$$
$$+\frac{\beta q_2}{F\lambda 2\sqrt{2ba}}\left\{\sum[\delta_l(-M,B',B'_1)-\delta_l(M,B'_2,B'_3)]+\sum[\delta_l(-M_1,B,B_1)-\delta_l(M_1,B_2,B_3)]\right\};$$

(23)

where indicated:

$$U=\sqrt{a}(2nl+x);\quad U_1=\sqrt{a}[2l(n+1)-x];$$
$$M_1=\sqrt{a}\,[l(2n+1)+x];\ M=\sqrt{a}[l(2n+1)-x].$$
$$A=2nl\sqrt{a}-2\tau\sqrt{2b};\ A_1=A+2\tau\sqrt{2b};$$
$$A'=2l(n+1)\sqrt{a}-2\tau\sqrt{2b};\ A'_1=A'+2\tau\sqrt{2b};$$
$$B=l(2n+1)\sqrt{a}-2\tau\sqrt{2b};\ B_1=B+2\tau\sqrt{2b};$$
$$B'=l(2n+1)\sqrt{a}-2\tau\sqrt{2b};\ B'_1=B'+2\tau\sqrt{2b};$$
$$A_2=2nl\sqrt{a}+2\tau\sqrt{2b};\ A_3=A_2-2\tau\sqrt{2b};$$
$$A'_2=2l(n+1)\sqrt{a}+2\tau\sqrt{2b};\ A'_3=A'_2-2\tau\sqrt{2ba};$$
$$B_2=l(2n+1)\sqrt{a}+2\tau\sqrt{2b};\ B_3=B_2-2\tau\sqrt{2ba};$$
$$B'_2=l(2n+1)\sqrt{a}+2\tau\sqrt{2b};\ B_3=B'_2-2\tau\sqrt{2b};$$

The expression includes functions δ_x and δ_l that characterize the values obtained after defining the corresponding integral expressions with integration boundaries from 0 to x and from 0 to l, and the parameters on which they depend are indicated in parentheses/Here are several components of the expression (23):

$$\delta_x\left(-U,A,A_1\right)=\frac{1}{m}\exp\left(-2mnl\right)\left\{\left[erfc\left(\frac{A}{2\sqrt{\tau}}\right)-e^{-mx}\cdot erfc\left(\frac{A+x\sqrt{a}}{2\sqrt{\tau}}\right)\right]-\right.$$
$$\left.-\exp\left(\frac{A_1^2-A^2}{4\tau}\right)\left[erf\left(\frac{A_1+x\sqrt{a}}{2\sqrt{\tau}}\right)-erf\left(\frac{A_1}{2\sqrt{\tau}}\right)\right]\right\}$$

(24)

$$\delta_x\left(U_1,A'_2,A'_3\right)=\frac{1}{m}\exp\left[2m(n+1)l\left[\left\{\left[e^{-mx}\cdot erfc\left(\frac{A'_2-x\sqrt{a}}{2\sqrt{\tau}}\right)-erfc\left(\frac{A'_2}{2\sqrt{\tau}}\right)\right]+\right.\right.\right.$$
$$\left.+\exp\left(\frac{A'^2_3-A'^2_2}{4\tau}\right)\left[erf\left(\frac{x\sqrt{a-A'_3}}{2\sqrt{\tau}}\right)+erf\left(\frac{A'_3}{2\sqrt{\tau}}\right)\right]\right\}$$

(25)

$$\delta_l\left(M,B'_2,B'_3\right)=\frac{1}{m}\exp\left[m(2n+1)l\right]\left\{\left[e^{-ml}\cdot erfc\left(\frac{B'_2-l\sqrt{a}}{2\sqrt{\tau}}\right)-erfc\left(\frac{B'_2}{2\sqrt{\tau}}\right)\right]+\right.$$
$$\left.+\exp\left(\frac{B'^2_3-B'^2_2}{4\tau}\right)\left[erf\left(\frac{l\sqrt{a}-B'_3}{2\sqrt{\tau}}\right)+erf\left(\frac{B'_3}{2\sqrt{\tau}}\right)\right]\right\}$$

(26)

A stationary solution to the problem of determining thermoelastic displacements can be found from the expression for the value $\tau\to\infty$. Obviously, you need to find the limits of the functions δ_x,δ_l included in the expression. As an example, we define a number of them:

$$\lim\left[\delta_x(-U,A,A_1\right]=\lim_{\tau\to\infty}\frac{1}{m}\exp(-2mnl)\left\{\left[erfc(-\infty)-\right.\right.$$
$$\left.\left.-\exp(-mx)\,erfc(-\infty)\right]-\exp\left(\frac{A_1^2-A^2}{4\tau}\right)\left[erf(0)-erf(0)\right]\right\}=$$
$$==-\frac{2}{m}\exp(-mnl)\left[1-\exp(-mx)\right]$$

(27)

$$\lim_{\tau\to\infty}\left[\delta_x(U,A_2,A_3)=\lim_{\tau\to\infty}\frac{1}{m}\exp(2mnl)\left\{\exp(mx)\left[erfc(\infty)-\right.\right.\right.$$
$$\left.\left.-erfc(\infty)\right]\right\}=0$$

(28)

Find the limit of the expression function for a multiplier Q_1 :

$$\frac{\beta Q_1}{2F\lambda\sqrt{2ba}}\frac{2}{m}\left\{\sum\left[\exp(-2mnl)\left(1-e^{-mx}\right)\right]+\sum\left[\exp\left[-2m(n+1)l\right]\times\right.\right.$$
$$\left.\left.\times\left(e^{mx}-1\right)\right]\right\}=\frac{\beta Q_1}{F\lambda m^2}\left[\frac{\left(1+cth(ml)\left(1-e^{-mx}\right)\right.}{2}+\frac{\left(cth(ml)-1\left(e^{mx}-1\right)\right.}{2}\right]=$$
$$=\frac{\beta Q_1}{F\lambda m^2}\left[1-\frac{Shm(l-x)}{Sh(ml)}\right]$$

(29)

Similarly, we define the limit of the expression function for a multiplier Q_2 :

$$\frac{\beta Q_2}{F\lambda m^2}\left\{\sum\exp\left[-m(2n+1)l\left(e^{mx}-1\right)+\sum\exp\left[-m(2n+1)l\right]\times\right.\right.$$
$$\left.\times\left(1-e^{-mx}\right)\right\}=\frac{\beta Q_2}{F\lambda m^2}\left[\frac{1}{2}\frac{1}{Sh(ml)}\left(e^{mx}-1\right)+\frac{1}{2}\frac{1}{Sh(ml)}\left(1-e^{-mx}\right)\right]=$$
$$=\frac{\beta Q_2}{F\lambda m^2}\frac{Sh(mx)}{Sh(ml)}$$

(30)

Then it is possible to determine the temperature displacements of an arbitrary cross-section of the lead screw from the following expression, which fully corresponds to the solution in the stationary formulation of the problem [9]:

$$\delta(x) = \frac{\beta}{F \cdot \lambda \cdot m^2}\{Q_2 \frac{Sh(mx)}{Sh(ml)} - Q_1[1 - \frac{Shm(l-x)}{Sh(ml)}]\} - j_{\Sigma} \cdot (j_1^{-1} + \frac{x}{E \cdot F}) \cdot \frac{\beta \cdot (Q_1 + Q_2)}{F \cdot \lambda \cdot m^2} \quad (31)$$

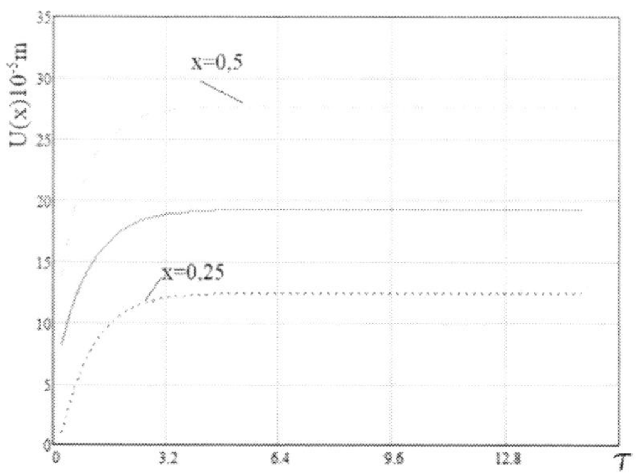

Fig. 4. The calculated values displacement.(23)

4 Conclusion

In this work, the temperature displacements of rod-type heat-active elements were considered for possible variants of their quasi-thermostable bonds, and expressions were obtained for describing and evaluating their values. This type is most typical for both spindle units and running screws.

In many cases of evaluating the thermal behavior of machine parts and assemblies, when their shape differs from the rod type, one can also use rod models, in which these parts are given according to criteria of similarity of processes, geometric similarity, depending on the goals and methods of solving the problem.

References

1. J. Mayr, at al, CIRP Annals - Manufacturing Technology, **61**, 771–791 (2012)
2. Y. Yto, *Thermal deformation in Machine tools.* (McGraw-Hill, 2010)
3. A.P. Kuznetsov, Russian Engineering Research, **37**(3), 171–179 (2017) https://doi.org/10.3103/S1068798X17030157
4. A.P. Kuznetsov, Stankoinstrument,. **3**, 30–41 (2020)
5. S. Mekid, *Introduction to precision machine design and error assessment* (CRC Press, 2009)
6. A.P. Kuznetsov, H.-J. Koriath, The Conference on Thermal Issues in Machine Tools. Proceedings CIRP sponsored conference, Verlag Wissenschaftliche Scripten, Dresden, 47-60 (2018)
7. W. Xian, Hindawi Mathematical Problems in Engineering, **2018** (2018) https://doi.org/10.1155/2018/5752932
8. M. Mares, O. Horejs, L. Havlík, Precision Engineering, **66**, 21–30 (2020) https://doi.org/10.1016/j.precisioneng.2020.06.010
9. E. Abele, Y. Altintas, C. Brecher, CIRP Annals - Manufacturing Technology, **59**(2), 781–802 (2010)
10. Y. Li, et al, Int J Adv Manuf Technol., **72**, 1415–1427 (2014)
11. Y. Li, et al, Int J Mach Tools Manuf., **95**, 20–38 (2015)
12. J. Zapłata, M. Pajor, Precision Engineering, **60**, 160-166 (2019) https://doi.org/10.1016/j.precisioneng.2019.07.011
13. J. Jedrzejewski et al, Journal of Machine Engineering, **19**(4), 27–41 (2019)
14. Z.C. Du, S.Y. Yao, J.G. Yang, Int. J. Precis. Eng. Manuf., **16**, 1571–1581 (2015)
15. H.R. Cao, X.W. Zhang, X.F. Chen, Int J. Mach Tools Manuf., **112**, 21–52 (2016)
16. Li. Sun, et al, Int J Adv Manuf Technol., **88**, 1267–1277 (2017)
17. C. Ma, et al, Int J Adv Manuf Technol., **89**, .3071–3085 (2017)
18. F. Tan, et al, The International Journal of Advanced Manufacturing Technology, **94**, 2861–2874 (2018)
19. Y. Li, et al, The International Journal of Advanced Manufacturing Technology, **99**, 1341–1349 (2018)
20. H. Zhou, et al, Procedia Manufacturing, **26**, 1137–1146 (2018)
21. G. Fu, H. Gong, H. Gao, International Journal of Advanced Manufacturing Technology, **105**, 2039–2055 (2019)
22. Q. Yin, et al,The International Journal of Advanced Manufacturing Technology, **101**, 1699–1713 (2019)
23. A.P. Kuznetsov, *Methods for evaluating thermal deformations of metal-cutting machines and ways to reduce them* (M., Niimash, 1983)
24. A.P. Kuznetsov, *Thermal processes in machine tools* (M.: Technosphere, 2019)

Using the Mathematical Criterion of Gas Pipeline Failure to Determine the Remaining Operation Life

Minsariya Bayburova[1], *Zulfiya* Ismagilova[2]

[1]Candidate of engineering sciences, associate professor, associate professor of the oil and gas transportation and storage department, Almetyevsk State Oil Institute, RU-423450, Tatarstan, Almetyevsk, Russia
[2]Candidate of engineering sciences, associate professor, associate professor of the gas and oil pipe and storage construction and repair department, Ufa State Petroleum Technological University, RU-450064, Ufa, Republic of Bashkortostan, Russia

Abstract: During the years of operation, underground pipelines deteriorate because of aging, wear, damage, and destruction. A comprehensive inspection of the main pipeline revealed various defects on a certain pipeline route section. To identify the remaining operation life of the pipeline, a number of hydraulic tensile strength tests were performed using the cyclic loading method. The authors propose to calculate the minimum allowable wall thickness capable of withstanding the maximum internal pressure, as an alternative to cyclic hydraulic tests. The strength calculations were performed using the well-known mathematical strength criteria proposed by K.V. Zakharov and I.I. Goldenblatt and V.A. Kopnov.
Keywords: oil and gas pipeline, hydraulic testing, pipeline diagnostics, cyclic tests, mathematical strength criterion

1 Introduction

Providing reliable and trouble-free operation of main pipelines transporting crude oil and oil products is an important issue. Pipelines used in oil and gas fields deteriorate over time due to aging, wear, damage, and destruction. Such deterioration can be caused by both corrosive and non-corrosive factors.

The problem of stress corrosion cracking of pipe steels is one of the most urgent in countries with an extensive system of underground gas pipelines.

The development of methods for predicting the state of the linear part of main gas pipelines today is one of the priority areas in the oil and gas industry. The issues of diagnostics and monitoring of stress corrosion cracking are covered in the works of Russian and foreign scientists and other specialists in the field of stress corrosion cracking [1,2,3]. However, methods for assessing the tendency of sections of the linear part of gas pipelines to stress corrosion cracking require further research and development.

Features of operation and reliability of pipelines transporting hydrogen sulfide-containing media are reflected in the works of scientists [4, 5].

Brittle fractures can occur in ductile steel structures at stresses well below the yield strength. The fractures can be caused by defects such as tiny cracks that may occur during steelmaking or appear and develop during the structure operation. Therefore, steel structures require timely diagnostics. Such diagnostics can be performed using modern non-destructive testing methods that help detect cracks and other defects in the course of operation [4].

At the same time, it is important to take into account that the hydraulic tests carried out are quite laborious. To reduce material costs and obtain the desirable result, it can be sufficient to calculate the minimum permissible wall thickness using a phenomenological method for this, considering the pipe material as an averaged solid material – homogeneous anisotropic.

2 Results and discussion

Pipeline transport requires constant engineering maintenance to ensure safe operation and a certain level of quality. The need to keep pipelines in good condition requires researchers to look for new effective methods of pipe inspection in order to detect defects and cracks, as well as corrosion on their surface.

Non-destructive testing as a means of detecting signs of destruction of pipeline material and preventing emergencies is relevant today. Methods for determining the dimensions and orientation of defects are constantly developing, equipment is being improved, research and testing are carried out to identify the characteristics of flaw detector models, and analysis is performed to further improve the operation of flaw detection tools.

A comprehensive inspection of the gas pipeline revealed various defects on a certain pipeline route section. The main defects observed were bulges, blisters, and delaminations in the pipe (Figure 1).

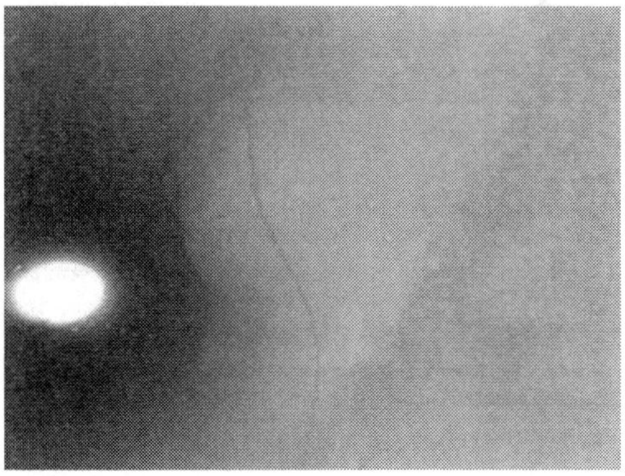

Fig. 1. Detected blister-type defect

The pipes are made of low-carbon low-alloy steel with a 0.18...0.20% carbon content, a tensile strength of at least 520 MPa, yield strength of at least 300 MPa, and impact strength of 500 kJ/m^2 at minus 40^0C. The pipes are made of normalised sheets.

Main design specifications of the section under consideration: design pressure — 5.4 MPa, allowable operating pressure — 3.4 MPa, pipeline diameter — 1,020 mm, and wall thickness —14-16 mm.

To study the behavior of the detected gas pipeline defects during the subsequent operation, a number of tensile strength tests were performed using the cyclic loading method. Pipe samples taken from the linear part of the pipelines with a defect on the internal surface were made in the form of barrels and equipped with plugs (Fig. 2).

Fig. 2. A sample barrel before hydraulic testing

Samples were cut out both along the rolling direction and crosswise. An additional pipeline system with a pressure gauge and a tap to measure cyclic and destruction parameters was installed on the test barrel at a safe distance.

The pressure was initially increased to 6 MPa using a hydraulic pump; after 24 hours it was lowered to 5 MPa; then after another 24 hours — to 4 MPa, etc. At the next stage, the pressure was increased to 7 MPa and the test was repeated. After testing at 8 MPa and 9 MPa, the sample was destroyed by a pressure of 9.5 MPa (Fig. 3).

The tests proved that, despite the identified defects, the pipe's properties remained unchanged, so it could remain operational at the permitted operating pressure of 3.4 MPa;

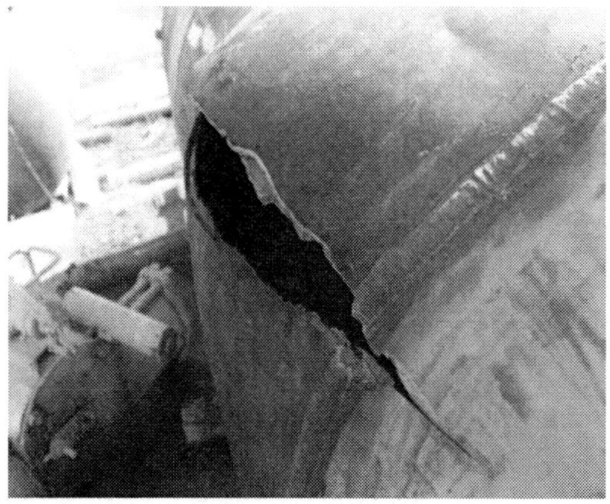

Fig. 3. Destruction of the barrel after hydraulic testing

The tests showed that the tensile strength values lengthwise and crosswise the rolling direction are consistent with the certified pipe material parameters. The same result was obtained by the authors theoretically. It was recognised that there was a metallurgical defect that could be classified as a delamination along the lap formed during the pipe piercing and rolling. Besides, the assumption was made that the damage may have been caused by growing propagation of the metallurgical defect (expanding discontinuities) spread across the entire wall thickness in the course of the gas pipeline operation when exposed to gas pressure and stresses created by low-cycle loads.

Delaminations were observed on the barrel surface after hydraulic testing (Fig. 4). The testing revealed delamination on the surface of the samples cut from used pipes. No delamination was observed in emergency stock pipes.

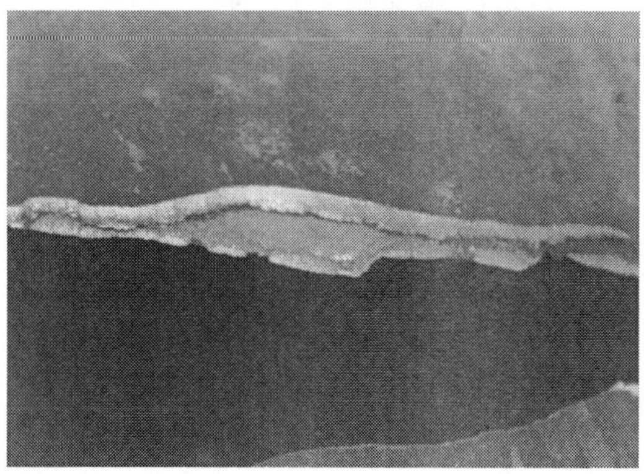

Fig. 4. Delamination of the barrel surface after hydraulic testing

The test results indicate that the cold gas transported by the pipe affects the state of its wall surface and leads to blistering and delamination of the latter. The reason

* Corresponding author:

https://doi.org/10.1051/epjconf/202124804021

for the appearance of such blistering and delamination can be explained by the condition of the gas pipeline operation at the stage when the extracted raw gas is supplied directly to the gas pipeline. A sharp change in gas pressure (from the initial pressure of 12-13 MPa to the operating pressure) leads to a decrease in the gas temperature and a change in the stress state of the pipe material. The temperature jump of the inner and outer surfaces of the pipe reaches 60°C, which in turn leads to the delamination of the pipe steel structure (Fig. 5). For this reason, a delamination-prone pipeline cannot provide the operability envisaged by the technical design.

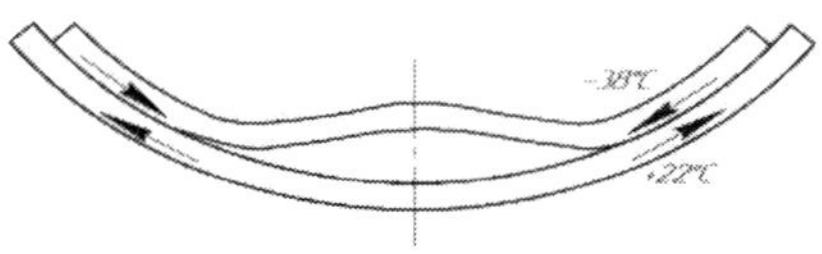

Fig. 5. Pipe wall buckling failure diagram

The hydraulic tests performed are costly and time-consuming. The detected pipe blisters formed as a result of rolling defects or defects that occurred in the course of operation (for example, internal corrosion) could cause a decrease in the wall thickness. Therefore, the allowable wall thickness capable of withstanding the internal operating pressure must be obtained to prevent pipe rupture.

In [6], the anisotropy of mechanical characteristics was revealed as a result of tests performed on flat samples cut out from dismantled pipes (samples were cut along and across the rolling direction).

Considering that metal sheets after rolling or upsetting have orthotropic properties and assuming that pipe steel may be characterised by anisotropic elasticity [6], the minimum allowable wall thickness can be calculated with the use of the strength criteria for orthotropic materials, without performing a tensile test [7].

In order to determine the most effective criterion, it is necessary to analyse the applicability of the criterion for the material used, and it is also necessary to take into account that the mathematical expression of the accepted strength condition is as simple as possible. As an example, we will illustrate the method of selecting the desired criterion. For comparison purposes, we can take two commonly used strength criteria for anisotropic rocks — Zakharov and Goldenblatt-Kopnov criteria [8, 9] — and then use the most suitable one for a particular case.

When assessing the strength of an orthotropic material, its characteristics are determined in the direction of the orthotropic axes and at an angle of 45 degrees to them (Figure 6). The Goldenblatt-Kopnov model takes into account the ultimate shear strength of a material rotated by 45 degrees at positive and negative signs of shear stresses [7].

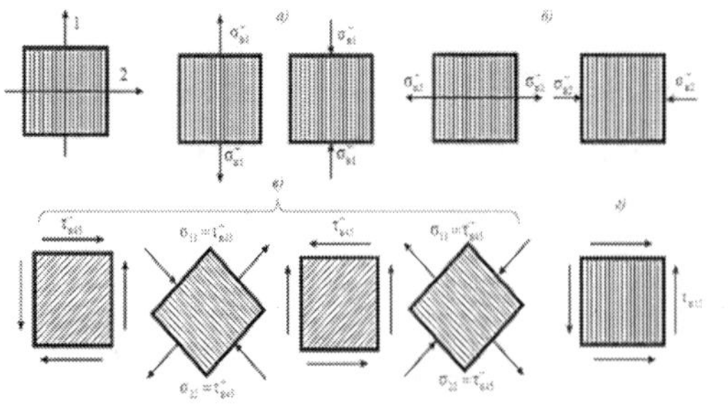

Fig. 6. Diagram of the main experiments to determine the components of the surface tensor for the Goldenblatt-Kopnov strength criterion.

a) $\sigma_{\text{в}_1}^{+}$ *and* $\sigma_{\text{в}_1}^{-}$ are ultimate tensile and compressive strengths in the direction of axis 1;

b) $\sigma_{\text{в}_2}^{+}$ *and* $\sigma_{\text{в}_2}^{-}$ – ultimate tensile and compressive strengths in the direction of axis 2;

c) $\tau_{\text{в}_{45}}^{+}$ и $\tau_{\text{в}_{45}}^{-}$ – ultimate shear strengths along areas inclined to the main ones at an angle of 45 degrees;

d) $\tau_{\text{в}_{1,2}}$ – shear strength at the site with axes 1 and 2.

In developing his criterion, K.V.Zakharov limited himself only to a coordinate system that coincides with the axes of symmetry of the mechanical properties of the material (5). Zakharov writes the strength condition for materials with different tensile and compressive strength as a second-order surface equation:

$$a_1\sigma_x^2 + a_2\sigma_y^2 + a_3\sigma_z^2 + a_4\sigma_x\sigma_y + a_5\sigma_x\sigma_z + a_6\sigma_y\sigma_z + \\ + a_7\sigma_x + a_8\sigma_y + a_9\sigma_z + a_{10}\tau_{xy}^2 + a_{11}\tau_{xz}^2 + a_{12}\tau_{yz}^2 + \\ + a_{13} = 0. \quad (1)$$

In the case of a plane stress state, the strength condition in the principal stresses is as follows:

$$\sigma_1^2 + k_1\sigma_1^2 + k_2\sigma_1\sigma_2 + k_3\sigma_1 + k_4\sigma_2 + k_5 = 0, (2)$$

where $a_1, a_2,\ k_1, k_2 \ldots -$ are constant factors that determine the strength of materials under different loading conditions.

After expressing the factors in terms of the corresponding tensile strength, the criterion (2) looks like [5]

$$\sigma_1^2 + \frac{\sigma_{1p}\sigma_{1c}}{\sigma_{2p}\sigma_{2c}}\sigma_2^2 + \left[1 + \frac{\sigma_{1p}\sigma_{1c}}{\sigma_{2p}\sigma_{2c}} + \frac{\sigma_{1c}-\sigma_{1p}}{\tau_{45}} - \right. \\ \left. - \frac{\sigma_{1p}\sigma_{1c}}{\sigma_{2p}\sigma_{2c}}\left(\frac{\sigma_{2c}-\sigma_{2p}}{\tau_{45}}\right) - \frac{\sigma_{1p}\sigma_{1c}}{\tau_{45}^2}\right] \cdot \sigma_1\sigma_2 + (\sigma_{1c} - \sigma_{1p})\sigma_1 + \\ + \frac{\sigma_{1p}\sigma_{1c}}{\sigma_{2p}\sigma_{2c}}(\sigma_{2c} - \sigma_{2p})\sigma_2 - \sigma_{1p}\sigma_{1c} = 0, \quad (3)$$

where $\sigma_{1p}, \sigma_{2p}, \sigma_{1c}, \sigma_{2c}$ –are the tensile strength values along the main reinforcement directions (p –tension, c – compression), and τ_{45} –the tensile strength at pure shear at $\sigma_1 = \sigma_{1p} = -\sigma_2 = -\sigma_{2c}$.

In this case, the dependence of the ultimate shear strength does not take into account the sign of the shear stress; nonetheless, the Zakharov criterion describes the experimental data satisfactorily.

The simplest representation of the Goldenblatt-Kopnov criterion [8] is a follows:

* Corresponding author:

$$\frac{\sigma_{1c}-\sigma_{1p}}{2\sigma_{1p}\sigma_{1c}}\sigma_1+\frac{\sigma_{2c}-\sigma_{2p}}{2\sigma_{2p}\sigma_{2c}}\sigma_2+$$
$$+\frac{1}{2}\left\{\left(\frac{\sigma_{1p}+\sigma_{1c}}{\sigma_{1p}\sigma_{1c}}\right)\sigma_1^2+\left(\frac{\sigma_{2p}+\sigma_{2c}}{\sigma_{2p}\sigma_{2c}}\right)\sigma_2^2+\right.$$
$$+\left[\left(\frac{\sigma_{1p}+\sigma_{1c}}{\sigma_{1p}\sigma_{1c}}\right)^2+\left(\frac{\sigma_{2p}+\sigma_{2c}}{\sigma_{2p}\sigma_{2c}}\right)^2-\right.$$
$$\left.\left.-\left(\frac{\tau_{45}^{+}+\tau_{45}^{-}}{\tau_{45}^{+}\cdot\tau_{45}^{-}}\right)\right]\right\}^{1/2}=1, \qquad (4)$$

where τ_{45}^{-} and τ_{45}^{+} – pure shear.

To obtain the five strength constants included in the Zakharov criterion (2), five independent uniaxial elongation, uniaxial compression, and pure shear tests must be performed.

The first and second experiment types according to (2) result in

$$\begin{cases}\sigma_{1p}^2+k_3\sigma_{1p}+k_5=0;\\ \sigma_{1c}^2+k_3\sigma_{1c}+k_5=0.\end{cases}$$

Hence we obtain

$$k_3=\sigma_{1c}-\sigma_{1p};\ k_5=-\sigma_{1p}\sigma_{1c}.$$

By performing the third and fourth experiment types, we determine

$$\begin{cases}k_1\sigma_{2p}^2+k_4\sigma_{2p}+k_5=0;\\ k_1\sigma_{2c}^2+k_4\sigma_{2c}+k_5=0.\end{cases}$$

Hence

$$k_1=\frac{\sigma_{1p}\sigma_{1c}}{\sigma_{2p}\sigma_{2c}};\qquad k_4=\frac{\sigma_{2c}-\sigma_{2p}}{\sigma_{2p}\sigma_{2c}}\sigma_{1p}\cdot\sigma_{1c}.$$

The fifth experiment type gives us:

$$k_2=1+\frac{\sigma_{1p}\sigma_{1c}}{\sigma_{2p}\sigma_{2c}}+\frac{\sigma_{1c}-\sigma_{1p}}{\tau_1}+$$
$$+\frac{\sigma_{1p}\sigma_{1c}(\sigma_{2p}-\sigma_{2c})}{\tau_1\sigma_{2p}\sigma_{2c}}-\frac{\sigma_{1c}\sigma_{1p}}{\tau_1^2}.$$

As a result of the sixth experiment type, the same constant factor is determined as

$$k_2=1+\frac{\sigma_{1p}\sigma_{1c}}{\sigma_{2p}\sigma_{2c}}-\frac{\sigma_{1c}-\sigma_{1p}}{\tau_2}-$$
$$-\frac{\sigma_{1p}\sigma_{1c}(\sigma_{2p}-\sigma_{2c})}{\tau_2\sigma_{2p}\sigma_{2c}}-\frac{\sigma_{1c}\sigma_{1p}}{\tau_2^2}.$$

Thus, the Zakharov strength criterion in a plane stress state in a coordinate system that coincides with the material symmetry axes can be represented as

$$\sigma_1^2+\frac{\sigma_{1p}\sigma_{1c}}{\sigma_{2p}\sigma_{2c}}\sigma_2^2+$$
$$+\left[1+\frac{\sigma_{1p}\sigma_{1c}}{\sigma_{2p}\sigma_{2c}}\pm\frac{\sigma_{1c}-\sigma_{1p}}{\tau_{45}}\pm\frac{\sigma_{1p}\sigma_{1c}}{\sigma_{2p}\sigma_{2c}}\left(\frac{\sigma_{2p}\sigma_{2c}}{\tau_{45}}\right)-\right.$$
$$\left.-\frac{\sigma_{1c}\sigma_{1p}}{\tau_{45}^2}\right]\cdot\sigma_1\sigma_2+(\sigma_{1c}-\sigma_{1p})\sigma_1+$$
$$+\frac{\sigma_{1p}\sigma_{1c}}{\sigma_{2p}\sigma_{2c}}(\sigma_{2c}-\sigma_{2p})\sigma_2-\sigma_{1p}\sigma_{1c}=0, \qquad (5)$$

where the upper sign is assumed to be + if $\tau_{45}=$ $=\tau_{45}^{+}=\tau_1$, and the lower sign is assumed to be – at $\tau_{45}=\tau_{45}^{-}=\tau_2$.

Given the Zakharov and Goldenblatt-Kopnov [7] creep-rupture criteria are similar, we can perform necessary strength calculations for a steel barrel using the Zakharov criterion and then compare the results obtained with the results given in [9].

Let us obtain the wall thickness of a closed-end barrel made of low-carbon low-alloy steel with a 0.18...0.20% carbon content that is under a uniform internal pressure for a long time according to the operating conditions.

The Goldenblatt-Kopnov strength criterion is written as [8]:

$$\Pi_{11}\sigma_{11}+\Pi_{22}\sigma_{22}+$$
$$+\sqrt{\Pi_{1111}\sigma_{11}^2+\Pi_{2222}\sigma_{22}^2+2\Pi_{1122}\sigma_{11}\sigma_{22}}=$$
$$=\frac{\alpha+\beta e^{-\lambda t_*^{-k}}}{n}. \qquad (6)$$

where t_* –durability; n – safety factor; α,β,λ,k – factors for a given fiberglass laminate.

Let us transform the strength criterion proposed by K. V. Zakharov (3) as follows:

$$k_1\sigma_1^2+k_2\sigma_2^2+k_3\sigma_1\sigma_2+k_4\sigma_1+k_5\sigma_2=1. \quad (7)$$

Let us put the factor as

$$k_1=\frac{1}{\sigma_{1p}\sigma_{1c}};\quad k_2=\frac{1}{\sigma_{2p}\sigma_{2c}};$$
$$k_3=k_1+k_2+k_1\frac{\sigma_{1c}-\sigma_{1p}}{\tau_{45}}-k_2\frac{\sigma_{2c}-\sigma_{2p}}{\tau_{45}}-\frac{1}{\tau_{45}^2};$$
$$k_4=k_1(\sigma_{1c}-\sigma_{1p})$$
$$k_5=k_2(\sigma_{2c}-\sigma_{2p}).$$

where $\sigma_{1p},\sigma_{2p},\sigma_{1c},\sigma_{2c}$ – tensile strength values in the main directions (p –tension, c – compression).

We put the criterion (7) as follows, taking into account the following values of the main stresses

$$\sigma_1=\frac{pR}{2\delta};\quad \sigma_2=\frac{pR}{\delta}$$

$$k_1\left(\frac{pR}{\delta}\right)^2+\frac{k_2}{4}\left(\frac{pR}{\delta}\right)^2+\frac{k_3}{2}\left(\frac{pR}{\delta}\right)^2+k_4\frac{pR}{\delta}+k_5\frac{pR}{2\delta}=$$
$$=\frac{\alpha+\beta e^{-\lambda t_*^{-k}}}{n}. \qquad (8)$$

* Corresponding author:

where p — pressure; R — radius; and δ – the pipe wall thickness.

The right side of the equation is replaced using the experimental data obtained, proving the relationship between the short- and long-term tensile strength:

$$\sigma_{\text{в}}(t_*) = \sigma_{\text{в}}(t_0)\left(\alpha + \beta e^{-\lambda t_*^{-k}}\right)$$

By putting

$$z = \frac{R}{\delta}; \qquad \frac{\alpha + \beta e^{-\lambda t_*^{-k}}}{n} = \overline{n};$$

we represent (8) as follows:

$$k_1 p^2 z^2 + \frac{k_2}{4} p^2 z^2 + \frac{k_3}{2} p^2 z^2 + k_4 p z + \frac{k_5}{2} p z = \overline{n}$$

After transforming the equation, we obtain

$$z^2 - \alpha z - \text{в} = 0, \qquad (9)$$

where

$$\alpha = \frac{\text{k}_4 + 0{,}5\text{k}_5}{2(\text{k}_1 + 0{,}25\text{k}_2 + 0{,}5\text{k}_3)\text{p}};$$

$$\text{в} = \frac{\overline{\text{n}}^2}{(\text{k}_1 + 0{,}25\text{k}_2 + 0{,}5\text{k}_3)\text{p}^2}.$$

The real root (9) has the following form:

$$z = -\alpha + \sqrt{\alpha^2 + \text{в}}.$$

The wall thickness of a cylindrical barrel that is 1,020 mm in diameter with a safety factor of $n = 1{,}1I$ will be $\delta = 14\,\text{мм}$. Upon comparing the calculation results obtained by the formula (4) with the calculations given in [4], we see that they have only a slight difference.

3 Conclusions

The present study leads us to the following conclusions:

1. Experimental studies have shown that the breaking pressure of pipes with internal delamination is within 9.5 MPa. Considering that this value is 2.76 times higher than the operating pressure in the pipeline equal to 3.4 MPa, it can be argued that further long-term operation of the gas pipeline with such defects is possible.

2. It has been proved that instead of cyclic tests for ultimate pressure to determine the remaining life of a gas pipeline, calculation of the minimum permissible wall thickness capable of withstanding the working pressure in the pipeline is sufficient. The data obtained as a result of mathematical calculations prove the correctness of the proposed theory.

3. The results of calculating the wall thickness of a steel barrel, obtained according to two different criteria, coincide with an insignificant error. This makes it possible to use any of the proposed criteria for the strength calculation.

4. The obtained results expand possibilities of applying the considered strength criteria, enabling justifiable selection of methods to operate pipelines, as well as approaches and modes of performing repair and diagnostics and increasing pipe wall strength.

References

1. V.V. Salyukov, V.V. Kharionovsky, *Main gas pipelines. Diagnostics and technical condition management* (2016) ISBN: 978-5-8365-0451-9
2. S.V. Alimov, Gas Industry. **724**, 10-15 (2015) [in Russia]
3. V. Linton, E. Gamboa, M. Law, Journal of Pipeline Engineering, **4**(1), 207–217 (2007) http://hdl.handle.net/2440/43274
4. A.V. Mitrofanov, N.A. Gafarov, B.V. Kichenko, A.I. Rezvykh, V.A. Polozov, The Science and Research Journal of All-Russia Research Institute of Organisation, Management and Economy in Oil and Gas Industry (NTZh VNIIOENGa), Corrosion prevention and environmental protection, **10**, 2-11 (1996) [in Russia]
5. S.A. Bubnov, I.I. Ovchinnikov, A.A. Bubnov, Vestn. Sam. State Tech. University. **2**(27), 178-182 (2012) [in Russia] DOI: https://doi.org/10.14498/vsgtu960
6. M.M. Aliev, R.R. Kantyukov, F.M. Mustafin, K.F. Ulshina, Gas Industry, **12**, 56-59 (2015) [in Russia]
7. M.M. Bayburova. Abstract of the of Cand. Sci. (Eng.), Thesis, Saratov, 16 (2008)) [in Russia]
8. I.I. Goldenblatt, V.A. Kopnov. Izvestia of the Academy of Sciences of the USSR, Mechanics. **6**, 77-83 (1965) [in Russia]
9. K.V. Zakharov, Plastics, **8**, 61-67 (1961)

* Corresponding author:

Methods of Mathematical Modeling of Automated Machine Systems of Multi-Nomenclature Production

Elena Dumler[1], *Roza* Vakhitova[2], *Diana* Saracheva[2], *Tatyana* Makarova[2]

[1]Ufa State Petroleum Technical University, RU-450064, Ufa, Republic of Bashkortostan, Russia
[2]Almetyevsk State Oil Institute, RU-423450, Tatarstan, Almetyevsk, Russia

Abstract. Automated machine tool systems for multi-product manufacturing were designed for processing parts of a certain nomenclature, which includes separate groups of parts, united by the design, technological, organizational and economic properties. To reduce the number of options in the structure of an automated machine tool system, a target nonlinear design function was applied. Technological conditions were formulated in such a way as to allow synthesizing alternative structures of the designed automated system. At this stage, a matrix of relations was built and a formalized description of technological conditions was given. Specific and generalized technological conditions, which take into account the graph of relations and compatibility condition formulated for any pair of elements, were considered. The article discusses the generalized technological conditions obtained on the basis of the synthesis of compatibility and follow-up properties. The existence of generalized technological conditions for the elements of the automated system of the same name was assessed. The interaction of the elements of an automated system during processing, when the condition of the following and compatibility of technological parameters are met, was studied. By analyzing the functional and technological structure of an automated machine tool system for multi-product manufacturing, the key relationships between its main elements were determined, and the technological environment, a key subject of the system research, was identified.

1 Introduction

The development of integrated computer-aided design systems for automated machine-tool systems (CAD ACS) and specialized computer-aided design systems for individual stages of ACC design [1-3] is of great interest for the production of parts for multi-product manufacturing system.

In order to solve the problems of CAD ACC development, designers were first given the task of integrating automation systems and interconnecting systems, subsystems, elements of these systems at the stages of technological design, including the choice of computer technology [4, 5]. Although the above-mentioned tasks are solved on the basis of system design principles [6, 7], the coordination of the ready-to-use solutions at different (technological, algorithmic, computer, systemic) levels under the real conditions of ACC design is more difficult [8].

In many ways, production efficiency depends on the rational selection of equipment. At present, there does not exist an no methods for separating huge total sets of equipment and machine tools that are most expedient in terms of technical and economic indicators for processing parts of a certain nomenclature.

The design of an automated machine tool system is a complex multidimensional task that includes assessment, modeling, synthesis, analysis and optimization of project proposals. These tasks can be solved using only systematic approach, which accounts for fact that the specificity of complex objects must include both the features of their constituent elements and the nature of various relations and connections between them [9].

Typically, automated machine tool systems for diversified production are designed for processing parts of a certain nomenclature, which includes separate groups of parts, united by design, technological, organizational and economic properties. Therefore, defining a group of parts and nomenclature is an important task [10].Automated machine tool systems for diversified production, as a rule, were designed for processing parts of a certain nomenclature, which included separate groups of parts, united by design, technological, organizational and economic characteristics.

2 Nomenclature and its properties

Nomenclature is a set of parts that have similar design, functional, technological and organizational characteristics [11].

Formal description of a group of parts $D_z, z \in \{\Gamma\}$, where the nomenclature of parts $\{D^H\}$, processing of which is provided by the synthesized automated system:

$$\{D_1\}\cup\{D_2\}\cup\ldots\cup\{D_z\}\cup\ldots\cup\{D_M\}=\{D^H\}.$$

Let us d efine t he p roperties o f t he n omenclature, which we select from the subset of elements of the graph G = (X, O, T, I, D), which determine the structure of the automated machine to ol s ystem, i. e. Let u s d efine t he properties of the nomenclature, which we select from the subset o f e lements o f t he gr aph G = (X, O , T, I , D), which determine the structure of the automated machine tool system, i.e.

Nomenclature define properties th at distinguish subsets o f e lements o f a g raph G = (X, O , T , I , D), determining the structure of the automated machine tool, i.e.
we make the transition from *G = (X, O, T, I)* to *G * = (X ', O', T ', T')*, for which the condition:

$$\{X'\}\subseteq\{X\},\{T'\}\subseteq\{T\},\{O'\}\subseteq\{O\},\{I'\}\subseteq\{I\}.$$

The s et o f tr ansitions *{L}* and t he a ssociated machined surfaces of the set of parts *{Dg}* serve as initial parameters f or s ynthesizing t he s tructure o f an automated machine tool system and for which

$$P_i=\left\lfloor\underset{L}{\wedge}\pi_l^i\right\rfloor. \qquad (1)$$

3 The matrix between the elements of an automated machine tool system

The p redicate m atrix of t he r elationship b etween t he elements $a_x, x\in\{X\}$ and $\pi_l, l\in\{L\}$ the automated machine t ool s ystem s ets t he t echnological capabilities of the equipment used:

$$M=\begin{cases} & \pi_1, & \pi_2, & \ldots, & \pi_l, & \ldots, & \pi_L \\ a_1 & y_{11}, & y_{12}, & \ldots, & y_{1l}, & \ldots, & y_{1L} \\ a_2 & y_{21}, & y_{22}, & \ldots, & y_{2l}, & \ldots, & y_{2L} \\ \ldots & \ldots & \ldots & \ldots & \ldots & \ldots & \ldots \\ a_x & y_{x1}, & y_{x2}, & \ldots, & y_{xl}, & \ldots, & y_{xL} \\ \ldots & \ldots & \ldots & \ldots & \ldots & \ldots & \ldots \\ a_X & y_{X1}, & y_{X2}, & \ldots, & y_{Xl}, & \ldots, & y_{XL} \end{cases} \qquad (2)$$

From formula 1 f or a group of parts $d\in\{D_z\}$, it is obvious that:

$$d=\underset{N'}{\wedge}p_i\left\lfloor\underset{L}{\wedge}\pi_l^i\right\rfloor \qquad (3)$$

for a group of parts $\{D_z\}$

$$\{D_z\}=\underset{M}{\wedge}d_j\left\{\underset{N'}{\wedge}p_i\left\lfloor\underset{L}{\wedge}\pi_l^i\right\rfloor\right\} \qquad (4)$$

for part nomenclature $\{D^H\}$

$$\{D^H\}=\underset{z=1}{\overset{M}{\forall}}D_z\left[\underset{M}{\wedge}d_j\left\{\underset{N'}{\wedge}p_i\left\lfloor\underset{L}{\wedge}\pi_l^i\right\rfloor\right\}\right]. \qquad (5)$$

For e ach tr ansition $\pi_l, l\in\{L\}$ (2), co nstruct a vector $y_l=(y_{1l},y_{2l},\ldots,y_{xl},y_{Xl})$. For each t ransition (2), construct a vector.

Its nonzero el ements determine the eq uipment of the automated machine t ool s ystem $a_x, x\in\{X\}$, o n which this transition can be implemented:

$$\begin{aligned}\pi_1&=(y_{11},y_{21},\ldots,y_{x1},y_{X1})\\ &\ldots\ldots\ldots\ldots\ldots\ldots\ldots\ldots\ldots\ldots.\\ \pi_L&=(y_{1L},y_{2L},\ldots,y_{xL},y_{XL})\end{aligned} \qquad (6)$$

We substituted vectors (6) into (1) for $P_i, i\in\{N'\}$

$$\begin{aligned}p_i=\pi_1^i(y_{11},y_{21},\ldots,y_{x1},\ldots,y_{X1})\wedge\\ \wedge\pi_2^i(y_{12},y_{22},\ldots,y_{x2},\ldots,y_{X2})\wedge\\ \wedge\ldots\ldots\ldots\ldots\ldots\ldots\ldots\ldots\ldots\ldots\\ \wedge\pi_l^i(y_{1l},y_{2l},\ldots,y_{xl},\ldots,y_{Xl})\wedge\\ \wedge\ldots\ldots\ldots\ldots\ldots\ldots\ldots\ldots\ldots\ldots\\ \wedge\pi_L^i(y_{1L},y_{2L},\ldots,y_{xL},\ldots,y_{XL}),\end{aligned}$$

and after the transformations we got:

$$p_i=\wedge\pi_l^i(y_{1l},y_{2l},\ldots,y_{xl},\ldots,y_{Xl})$$

or

$$p_i=\wedge\pi_l^i\left(\underset{X}{\forall}y_{xl}\right),$$

and for the detail

$$d_j=\underset{N'}{\wedge}p_i\left[\underset{L}{\wedge}\pi_l^i\left(\underset{X}{\forall}y_{xl}\right)\right]$$

For a group of parts $\{D_z\}$:

$$\{D_z\}=\underset{M'}{\wedge}d_j\left\{\underset{N'}{\wedge}p_i\left[\underset{L}{\wedge}\pi_l^i\left(\underset{X}{\forall}y_{xl}\right)\right]\right\}, \qquad (7)$$

for part nomenclature $\{D^H\}$:

$$\{D^H\}=\underset{\bar{a}=1}{\overset{M}{\forall}}D_{\bar{a}}\left[\underset{M}{\wedge}d_j\left\{\underset{N'}{\wedge}p_i\left[\underset{L}{\wedge}\pi_l^i\left(\underset{X}{\forall}y_{xl}\right)\right]\right\}\right]. \qquad (8)$$

If t he disjunction o f vectors $y_{xl}=0$ and i f $\underset{X}{\forall}y_{xl}=0$, then the given nomenclature of parts $\{D^H\}$ cannot be manufactured by this automated machine tool system according to (8).

We g ot a p redicate, where $\forall d_j$ – is t he merge quantifier with domain M:

$$\psi\{D^H\} = \underset{M}{\forall} d_j = \left\{\underset{N'}{\wedge} p_i \left[\underset{L}{\wedge} \pi_l^i \left(\underset{X}{\forall} y_{xl}\right)\right]\right\} \quad (9)$$

The r esulting p redicate $\psi\{D^H\} = 1$, i f for each p art $d_j \in \{D^H\} \underset{X}{\forall} y^i{}_{xl} \neq 0$ there ar e many t ransitions $\pi_l, l \in \{L\}$ in the accepted equipment of the automated machine to ol s ystem $a_x, x \in \{X\}$, which i s cap able o f processing any part $d_j, j = \overline{1, M}$.

The analysis of the model described in (9) shows that the num ber of variants of the structure of the au tomated machine tool system is large and therefore, to reduce the number of variants, it is possible to use the target nonlinear function of the MNP ACC design:

$$F \rightarrow \min[\{X\} \wedge \{T\} \wedge \{O\} \wedge \{I\}], \quad (10)$$

if

$$\psi\{D^H\} = 1 \quad (11)$$

The w ork [12] co nfirms t he c orrectness an d v alidity of function (10).

Minimize a lternative structures o f t he a utomated system in order to obtain foreseeable set of options, can be isomorphic mapping relationships between the elements of ACC on the graph G *.

Thus, we g et a d ialectical r elationship b etween th e structures of the nomenclature of parts and an automated machine tool system.

For a n automated machine t ool s ystem, it can b e expressed through the p roperties o f el ements that implement the structure of the system (table 1). Based on the p roperties o f t he el ements o f an automated s ystem, technological co nditions (TC) can b e f ormulated t hat will a llow synthesizing a lternative structures o f t he designed automated system, i.e.:

$$G^* = (X', T', O', I')^{\{TY\}} \Rightarrow G^{**} = (X'', T'', O'', I''),$$

where {TY} is a set of formalized technological conditions, which generates a gr aph o f r elations, o n t he basis of which a matrix of relations is built (Fig. 1) and a formalized description of technological conditions is given.

Let us co nsider t echnological conditions of t wo types: particular and generalized.

Consider t he pr ocess c onditions a re of t wo t ypes: partial and generalized.

Particular te chnological c onditions a re a pplied between el ements o f t he s et o f t he s ame n ame, an d generalized conditions are ap plied between el ements o f unlike sets.

ACC element designation	X'	T'	C'	I'
X'	≡, ↔, ↦, ><	↔, ↦, ><	↔, ↦, ><	↔, ↦, ><
T'		≡, ↔, ↦, ><	↔, ↦, ><	-
O'			≡, ↔	↔, ↦, ><
I'		-		≡, ↔, ><

Fig. 1. Matrix of relations between ACC elements

At the same time, it i s taken into account that dj is a complex part, which, in terms of its structural and technological ch aracteristics and p arameters, r epresents all the details o f the group for which the ACC is designed. F or e xample, the c ompatibility r elationship between t he workpiece an d t he ACC eq uipment i s formally described:

$$\forall d_j \underset{X}{\forall} a_x \underset{S_d}{\exists} S_d \underset{S_x}{\exists} S_x \ni P\{(d_j[S_d^1 = shaft] \equiv a_x[S_x^1 = turning]) \wedge$$
$$\wedge (d_j[S_d^3] \geq a_x[S_x^2]) \wedge (d_j[S_d^4] \geq a_x[S_x^3]) \wedge (d_j[S_d^5] \leq a_x[S_x^4])\} \Rightarrow$$
$$\Rightarrow [d_j \leftrightarrow a_x] \quad (12)$$

Among the set of equipment on the basis of TU (12) for the part $d_j \in \{D^H\}$ under consideration, a s ubset of machines ca n b e d istinguished, acco rding t o t heir parameters, cap able o f e nsuring t he p rocessing o f t his part in accordance with the specified quality parameters.

Thus, by (12), taking into account the graph of relations, we can formulate the compatibility conditions for any pair of elements.

Table 1. Matrix of properties of the main elements of the machine tool system (fragment)

ACC element	Element property	Property name notation	Property value notation
Machin e tool (set *{X}*)	Processing precision	S_x^1	$[S_x^1]$
	Roughness of processing	S_x^2	$[S_x^2]$
		S_x^3	$[S_x^3]$
	Geometric parameters	S_x^4	$[S_x^4]$
	Processing method	S_x^5	$[S_x^5]$
	Power characteristics	S_x^6	$[S_x^6]$

Vehicle (set *{T}*)	A type	S_t^1	$[S_t^1]$
	Positioning accuracy	S_t^2	$[S_t^2]$
	Geometric parameters	S_t^3	$[S_t^3]$
	Power characteristics	S_t^4	$[S_t^4]$
	Lifting capacity	S_t^5	$[S_t^5]$
	Type of transported part or tool	S_t^6	$[S_t^6]$
Rigging (set *{O}*)	A type	S_0^1	$[S_0^1]$
	Processing precision	S_0^2	$[S_0^2]$
	Roughness of processing	S_0^3	$[S_0^3]$
	Geometric parameters	S_0^4	$[S_0^4]$
	Processing method	S_0^5	$[S_0^5]$
	Power characteristics	S_0^6	$[S_0^6]$
	Basing scheme	S_0^7	$[S_0^7]$
Tool (set *{I}*)	A type	S_I^1	$[S_I^1]$
	Processing precision	S_I^2	$[S_I^2]$
	Roughness of processing	S_I^3	$[S_I^3]$
	Geometric parameters	S_I^4	$[S_I^4]$
	Processing method	S_I^5	$[S_I^5]$
	Cutting material	S_I^6	$[S_I^6]$
Surface to be treated (set *{N '}*)	A type	S_P^1	$[S_P^1]$
	Part material	S_P^2	$[S_P^2]$
	Processing precision	S_P^3	$[S_P^3]$
	Roughness of processing	S_P^4	$[S_P^4]$
	Geometric parameters	S_P^5	$[S_P^5]$
	Surface position	S_P^6	$[S_P^6]$
Part (set *{D}*)	A type	S_d^1	$[S_d^1]$
	Part material	S_d^2	$[S_d^2]$
	Processing precision	S_d^3	$[S_d^3]$
	Roughness of processing	S_d^4	$[S_d^4]$
	Geometric parameters	S_d^5	$[S_d^5]$
	Processing method	S_d^6	$[S_d^6]$
	Basing scheme	S_d^7	$[S_d^7]$
	Weight	S_d^8	$[S_d^8]$

The te chnological c onditions o f c ompatibility between the part and the fixture can be represented as:

$$\forall d_j \underset{O}{\forall} pr_o \underset{Sd}{\exists} S_d \underset{S_o}{\exists} S_o \ni P\{(d_j[S_d^7] \equiv pr_o[S_o^7]) \wedge$$
$$\wedge (d_j[S_d^5] - pr_o[S_o^4] \leq k) \wedge (d_j[S_d^6] \equiv pr_o[S_o^5])\} \Rightarrow$$
$$\Rightarrow [d_j \leftrightarrow pr_o] \quad (13)$$

where k i s a co efficient t hat t akes into acco unt t he economically justified conditions for using this device.
where k - factor t hat t akes into acco unt eco nomically viable conditions for the use of this device.

The technological conditions given in (12), (13) have the transitivity property:

$$P[(d_j \leftrightarrow a_x) \wedge (d_j \leftrightarrow pr_o)] \Rightarrow (a_x \leftrightarrow pr_o) \quad (14)$$

4 Simulation of an automated machine tool system

The following relation takes a special place among particular te chnological c onditions, s ince it i s d ecisive not o nly in t he t echnological s equence o f t he arrangement o f t he el ements o f an a utomated machine tool system, but also in the direction of material flows, on which the structural structure of the entire automated system depends.

The r atio of r epetition h as a s pecial p lace am ong private process conditions, as is determined not only in the automated sequencing machine tool a rrangement o f elements, but in the direction of material flow, on which the s tructural c onstruction of th e e ntire a utomated system.

The t echnological sequence of pr ocessing on t he equipment $a_x, a_q, x, q, \in \{X\}$ is d etermined from t he expression:

$$\underset{X}{\forall} a_x \underset{X}{\forall} a_q \underset{S_x}{\exists} S_x \underset{S_q}{\exists} S_q \ni P\{(P(a_x, \vec{U}, a_q) \wedge$$
$$\wedge (a_x[S_x^2] \geq a_q[S_q^2]) \wedge (a_x[S_x^3] \geq a_q[S_q^3])\} \Rightarrow$$
$$\Rightarrow [a_x \leftrightarrow a_q] \quad (15)$$

The s equence i n t erms o f t he d irection o f material flows i n a n a utomated machine tool s ystem c an b e determined from the technological condition connecting the elements of the transport system $b_t t \in \{T\}$ and the main technological equipment $a_x, x \in \{X\}$:

$$\underset{T}{\forall} b_t \underset{X}{\forall} ax \underset{S_t}{\exists} St \underset{S_x}{\exists} S_x \ni P\{(P(b_t, \vec{U}, a_x) \wedge$$
$$\wedge (b_t[S_t^3] \equiv a_x[S_x^4]) \wedge (b_t[S_t^4] \equiv a_x[S_x^6])\} \Rightarrow$$
$$\Rightarrow [b_t \mapsto a_x] \quad (16)$$

The a nalysis s howed that t he generalized technological co nditions were o btained o n t he b asis o f the synthesis of compatibility and succession properties, and since the compatibility condition contains an equivalence condition in an implicit form, then the verification of the existence of generalized technological conditions for the elements of the same name of an automated system is the establishment of the direction of material flows between them, and for dissimilar elements - technological adequacy.

The elements of the automated system interact during processing when the following condition (movement of material flows in the automated system) and the compatibility of technological parameters are met.

For two dissimilar objects, a generalized technological condition will be valid:

$$\underset{T}{\forall} b_t \underset{X}{\forall} ax \underset{S_t}{\exists} St \underset{S_x}{\exists} S_x \ni P\{(P(b_t,\vec{U},a_x)\wedge(b_t[S_t]\equiv a_x[S_x])\}\Rightarrow$$
$$\Rightarrow [b_t >< a_x] \qquad (17)$$

Expression (17), based on the previously defined graph *G(Z,U,P)*, allows us to build a generalized technological condition for several elements of an automated system capable of implementing a certain technological function, the consequence of which is transitivity:

$$[(pr_o >< a_x)\wedge(b_t \overline{><} a_q)] \qquad (18)$$

The truth of expression (18) provides the ability to choose the structure of an automated machine tool system, as shown in Fig. 2.

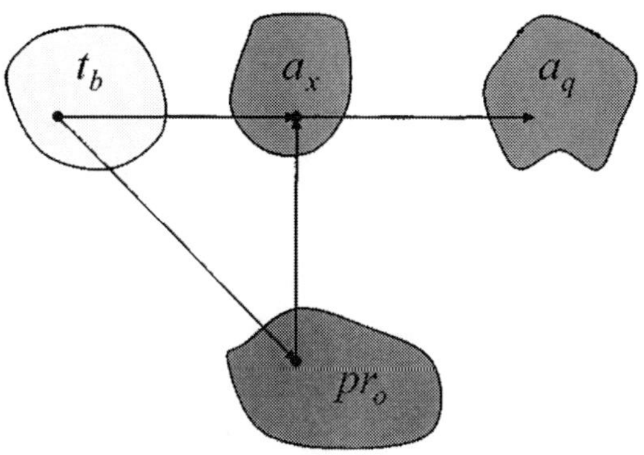

Fig. 2. The choice of elements of an automated system depending on technological conditions.

The resulting structure of the automated system satisfies condition (9), makes it possible to use various target design functions to obtain a quasi-optimal option, without contradicting (10).

By applying the investigated design procedures tailored to the specific optimization objective there is a possibility of a variety of options to get the structure of quasi-optimal automated machining system.

The sequence of mathematical modeling of the process of creating an automated machine tool system is shown in Fig. 3.

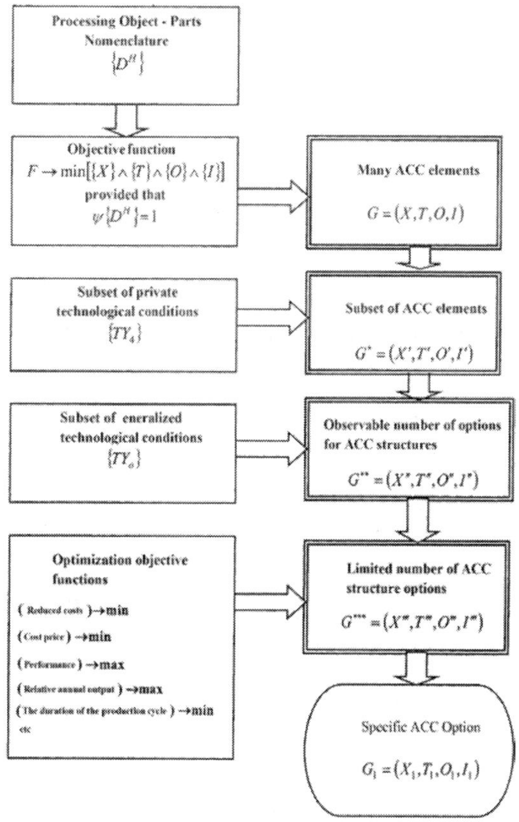

Fig. 3. Mathematical modeling of the process of designing an automated system.

5 Conclusion

The paper discusses the generalized technological conditions obtained on the basis of the synthesis of compatibility and follow-up properties, with the existence of generalized technological conditions for the elements of the automated system of the same name being accounted for simultaneously. The interaction of the elements of an automated system during processing, where the condition of compatibility of technological parameters is met, was investigated.

The analysis of the functional and technological structure of the automated machine tool system of multi-product manufacturing was carried out based on the relationships between its main elements, which describe the technological environment and are a subject of system research.

References

1. R.I. Vakhitova, E.B. Dumler, F.Z., Bulyukova, E.F. Zakharova, *Increasing the efficiency of diversified production based on modeling automated technological equipment: monograph* (Almetyevsk, AGNI, 2019, 92 p) ISBN: 978-5-94454-039-3

2. E.B. Dumler, Voronezh State Technical University Bulletin, **3**(5) (2007) [in Russia] ISSN 1729-6501
3. Yu.M. Solomentsev, V.G. Mitrofanov, A.V. Kapitanov, *Automated machining systems, organization of operation of metalworking tools* (Interdisciplinary information service, 4, 2012).
4. I. Shakhnovich, M. Sheikin, Electronics: Science, Technology, Business, **4**, 135 (2014)
5. M.O. Kalinin, V.I. Kalinina. *Control technology of functional stability of control information systems of mechanical engineering* (In the collection: Prospective Development of Science, Engineering and Technologies, 2014) ISBN: 978-5-9905939-3-0V.V. Dodonov. Mechanical engineering, **12**, (2011) ISSN: 0536-1044
6. V.V. Dodonov. Engineering Journal: Science and Innovation, **2**(38) (2015) DOI: 10.18698/2308-6033-2015-2-1370
7. N.I. Dyatchin *Engineering technology and its development as a science* (In the book: Modern technological systems in mechanical engineering. Abstracts of reports, 2006)
8. A.A. Tarasov, V.N. Krutov, V.A. Treyal, A.A. Smirnov, Instruments and technologies, **29**, (2010) [in Russia]
9. G.B. Burdo, A.N. Bolotov, N.V. Ispiryan, S.R. Ispiryan, *Hybrid integrated automated design and management systems* (In the collection: Development of technical sciences in the modern world. Collection of scientific papers on the basis of the international scientific and practical conference, 2017).
10. S.G. Mitin, P.Yu. Bochkarev, Bulletin of the Saratov, State Technical University, P.2, **1**(75) (2014) [in Russia]
11. V.V. Dodonov, Proceedings of higher educational institutions, Mechanical engineering, **12**, (2012)

Simulation of Automated Machine Tool Systems

Elena Dumler[1], *Roza* Vakhitova[2], *Diana* Saracheva[2], *Tatyana* Makarova[2]

[1]Ufa State Petroleum Technical University, RU-450064, Ufa, Republic of Bashkortostan, Russia
[2]Almetyevsk State Oil Institute, RU-423450, Tatarstan, Almetyevsk, Russia

Abstract. A method of multi-nomenclature production in terms of technical and economic indicators was considered. An integrated design system that is able to assess the real relationship in the system "part-machine-transport device-device-tool" was proposed, and arelationship between the elements of an automated machine tool system was examined. The structure of an automated machine tool system, proposed in the form of a mixed graph, represented by sets described by nonlinear equations, was considered. The definitions of the sets of elements for an automated machine tool system that characterize the properties of technological objects were formulated. The main objects of design in the technological project of an automated machine system of multi-nomenclature production were determined and a graphical interpretation of the automated machine tool system was presented. It was revealed that technological equipment, the main technological machines, cutting and auxiliary tools and vehicles adequately describe manufacturing process for group of parts and are the main objects of designing technological projects for an automated machine system of multi-product manufacturing. The relationships between the elements indicated in the graph allow us to formalize and to establish the relationship between the individual components of an automated machine system, indicating in this way the need to develop a formal methodology for synthesizing the structure of an automated machine system.

1. Introduction

The current state of robotics, machine tools, informatics and other related industries, the features of multi-nomenclature production and the level of automation of metalworking at industrial enterprises have made it possible to formulate and to realize the concept of an automated machine tool system of multi-nomenclature production (ACC MNP) [1, 2].

Recently, the structure of the design stages of a multi-nomenclature production, which is based on the principles of group organization, has been developed [3 – 5]. The basic system-structural approaches to design were also developed. The relationship between the elements of the ACC such as the main technological equipment, auxiliary and cutting tools, vehicles and technological equipment at the level of the technological stratum were investigated; however, this investigation was not deep enough to fulfil all practical needs and, thus, need further refinement.

The production efficiency depends, to a large extent on the rational choice of equipment. As for now, no efficient methods of separating, from a huge total set of equipment, a set of machine tools, which are most expedient in terms of technical and economic indicators for processing parts of a certain nomenclature, have been proposed for [6, 7].

It is clear that a thorough investigation of the relationship between the elements of the ACC is possible only when using an integrated design system capable of assessing the real relationships in the system "part - machine - transport device - device - tool" [8, 9].

2. The structure of the automated machine tool system

The structure of the automated machine tool system can be represented as an arbitrary mixed graph:

$$G = (Z, U, P), \qquad (1)$$

where Z is a plurality of elements of an automated machine tool system;

U - many relationships between the elements of the automatedmachine system,

P- is a triple predicate, called an incident.

The set $\{Z\}$ of elements of the automated machine tool system defines a subset of units of vehicles, main process equipment, devices and warehouses for their maintenance, measuring and cutting tools, technological equipment, instrumentation system, control and measuring devices and functional support aids.

Set $\{U\}$, which is determined by the set of information and material relationships that arise between elements of the automated machine tool system during its operation.

The graph G corresponds to the conditions in which: the incident P is set in the ordered triples of elements$z$, y, u, where $y \in \{Z\}$, a $u \in \{U\}$;

$$P(z',u,y') \rightarrow [(z=z' \wedge y=y') \vee (z=y' \wedge y=z')]$$
$$(\forall u \in U)(\forall z, y \in Z)[P(z,u,y) \wedge (\forall z', y' \in Z)]$$

The properties of an arbitrary mixed graph are such that for any value and $u \in \{U\}$ one of three expressions is true:

$$(\forall z, y \in Z)[P(z,u,y) \rightarrow \overline{P}(z,u,y)], \quad (2)$$

$$(\exists z, y \in Z)[P(z,u,y) \wedge P(y,u,z) \wedge (z \neq y)], \quad (3)$$

$$(\exists z \in Z)[P(z,u,z)], \quad (4)$$

Depending on the fairness of the statement, the set $\{U\}$ is divided in pairs into three disjoint subsets $\{\vec{U}, \tilde{U}, \dot{U}\}$, which are respectively called arcs, links and loops.

Thus, the graph describing the structure of the automated machine tool system *G = (Z, U, P)* between its two vertices contains links and arcs running in the same direction, and has several loops at one vertex, i.e. $\vec{U} \neq 0, \tilde{U} \neq 0, \dot{U} \neq 0$.

As an example, figure 1 shows a graphical interpretation of an automated machine tool system, which includes:

Z_{11} –warehouse of blanks, finished parts;
Z_{12}, Z_{13} – stores;
Z_{21}, Z_{22} – transport robots intended for delivery tools and parts for machines;
metalworking machines:
Z_{31}, Z_{32} –two turning, Z_{33}, Z_{34} –two milling;
Z_5 –the processing center;
Z_{36} –grinder;
Z_{41}, Z_{42}– control and measuring equipment;
Z_{51} –tool storage room;
Z_{61} –units of an overload of details;
Z_{71} –management systems.

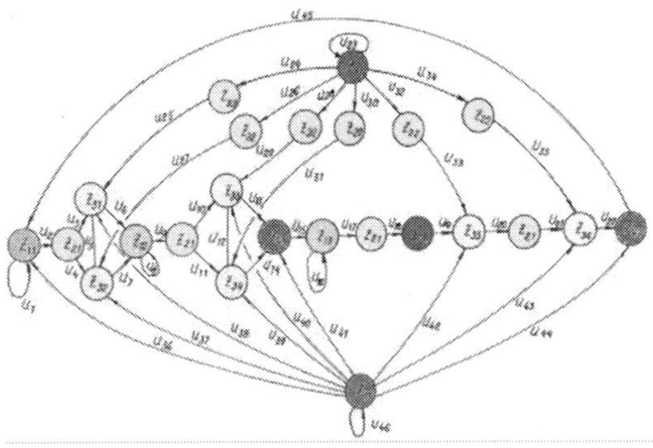

Fig. 1. Graphical interpretation of the structure of the machine tool system

3. Mathematical model of the structure of an automated machine tool system

For the automated machine system under investigation, we will build an incident P and a set of *{Z}, {U}*:

$$Z = \{Z_{11}, Z_{12}, Z_{13}, Z_{21}, Z_{22}, Z_{31}, Z_{32}, Z_{33}, Z_{34}, Z_{35}, Z_{36}, Z_{41}, Z_{42}, Z_{51}, Z_{61}, Z_{71}\}$$

Indicators that satisfy one of the properties for the graph with the incident P described by formulae (2), (3), (4) are true, and the indicators on all other triples of elements are false. In the mixed graph G = (Z, U, P), the existing set of edges U is divided into a subset of arcs:

$$\vec{U} = \{U_2, U_3, U_4, U_6, U_7, U_9, U_{10}, U_{11}, U_{13}, U_{14}, U_{15}, U_{17}, U_{18}, U_{19}, U_{20}, U_{21}, U_{22}, U_{24}, U_{25}, U_{26}, U_{27}, U_{28}, U_{29}, U_{30}, U_{31}, U_{32}, U_{33}, U_{34}, U_{35}, U_{36}, U_{37}, U_{38}, U_{39}, U_{40}, U_{41}, U_{42}, U_{43}, U_{44}, U_{45}\}$$

where $\dot{U} = \{U_1, U_8, U_{16}, U_{23}, U_{46}\}$ is a subset of loops;
$U = \{U_5, U_{12}\}$ is a subset of links.

Subsets $\vec{U}, \tilde{U}, \dot{U}$ determine technological capabilities of warehouses and control system and characteristics of information and material flows of processing equipment in automated machine-tool system.

From the formal model of the automated machine tool system represented by the arbitrary mixed graph *G = (Z, U, P)*, we distinguish the functional medium, which is defined by the expression:

$$G=(X,O,T,I,D_g,I',E_i)i=1,5 \quad (5)$$

where $X = \{a_1, a_2, ..., a_x, a_X\}$ - a set of the main technological equipment units;

$O = \{pr_1, pr_2, ..., pr_o, pr_O\}$ – a plurality of process tooling units (different accessories);

$T = \{b_1, b_2, ..., b_t, b_T\}$ – a plurality of vehicles for tools and parts;

$I = \{in_1, in_2, ..., in_i, in_I\}$ – a plurality of auxiliary and cutting tools;

$D_g = \{d_1, d_2, ..., d_M\}, g \in \{\Gamma\}$ - Multiple Group Processing Technologies parts, as well as design and technological characteristics;

$I' = \{iz_1, iz_2, ..., iz_{i'}, iz_{I'}\}$ – a plurality of monitoring and measuring machines and tools;

E_i, i=1,5receives the following values:

$E_1 = \{e_1^1, e_2^1, ..., e_{E_1}^1\}$ – multiple items of warehouse equipment;

$E_2 = \{e_1^2, e_2^2, ..., e_{E_2}^2\}$ – a plurality of control system elements;

$E_3 = \{e_1^3, e_2^3, ..., e_{E_3}^3\}$ – a many pieces of equipment for the movement of cutting fluids;

$E_4 = \{e_1^4, e_2^4, ..., e_{E_4}^4\}$ – Multiple units of machinery and equipment of the industrial waste management system;

$E_5 = \{e_1^5, e_2^5, ..., e_{E_5}^5\}$ – multiple elements of information flows in the ACC.

The identity of the service purpose, structural and technological elements, manufacturing material, technological process, machinery and equipment, tools and technological equipment used in production are characterized by the groups of parts D_g, $g \in \{\Gamma\}$, and therefore, the expression is valid for them:

$$D_g = \wedge_{j=1}^{M'} d_i = \begin{Bmatrix} d_0 = \{\beta_1, \beta_2, \dots, \beta_\sigma\} \\ p = \{\gamma_1 \gamma_2, \dots, \gamma_\mu\} \\ S_{d_i} = G(N, E) \end{Bmatrix} \quad (6)$$

where M' – is the number of parts in the group;

p_i –characteristics of machined surfaces $i \in \{N'\}$,constituting the group;

S_{di}–the structure of the set of parts dj represented by the mixed graph $G(N,E)$, in which the surfaces correspond to vertices, and the relations and operations - to edges, characterize the mutual relationship of the surfaces described by the expression:

$$E = (R_{p_i}, F_{p_i}), i \in \{N\}$$

d_0–General information about part group Dg, $z \in \{\Gamma\}$including design and process codes, workpiece types, material, and so on.

In different groups Dg, the description parameters $d_0$and $p_i$and the set can be changed, however, all these changes do not affect the conventional design methodology and are quantitative.

According to definition 1 for formula (5), the expression:

$$\{X\}\wedge\{0\}\wedge\{T\}\wedge\{I\}\wedge\{I'\}\wedge\{E_i\}=\{Z\}, i \in 1{,}5, \quad (7)$$

The main process machines and equipment, technological equipment, cutting and auxiliary tools, vehicles, all of the above, ultimately provides the technological process of processing a group of parts and is the main design objects when creating the technological design of an automated machine system of multi-item production.

Thus, in the process design of an automated machine tool system of multi-nomenclature production, the graph described by the expression:

$$G = (X, O, T, I, D), \quad (8)$$

The relationships between the sets $\{X\}$, $\{O\}$, $\{T\}$,$\{I\}$, $\{D\}$, i.e. between the elements of the model presented in formula (8) create analytical relationships and are available for mathematical interpretation, which makes it possible to move to formalized regulation when designing elements of an automated machine tool system of multi-nomenclature production and are the main subject of system research.

Relationships between processing objects represented by processing means and set $\{D\}$ cover sets $\{X\}$, $\{O\}$, $\{T\}$, $\{I\}$ and form relationships arising between model elements (8).

Relationships are based on some properties of process objects of the automated machine tool system, such as equivalence, compatibility, following and interaction.

We formulate definitions characterizing the properties of technological objects.

Equivalence is defined as the equality of technological capabilities or properties of technical objects of an automated machine tool system (definition 3).

This attitude is indicated by a sign$\equiv$.

Let us present this definition as an expression in the form of symbols:

$$\underset{X}{\forall} a_x \underset{X}{\forall} a_q \underset{S}{\exists} S_x \underset{S}{\exists} S_q \ni P\{(a_x[S_x] = a_q\lfloor S_q \rfloor) \wedge (x \neq q)\} \Rightarrow \lfloor a_x = a_q \rfloor \quad (9)$$

where $\{S\}$ - is a set of properties.

Process objects can be shared, provided they are defined by the same properties.

This attitude is indicated by a sign$\leftrightarrow$.

Let us present this definition in the form of characters:

$$\underset{N'}{\forall} p_i \underset{O}{\forall} pr_o \underset{I}{\forall} in_j \underset{S_p}{\exists} S_p \underset{S_o}{\exists} S_o \underset{S_j}{\exists} S_j \ni P\{(p_i\lfloor S_p \rfloor = pr_o[S_o]) \wedge (pr_o[S_o] = in_j\lfloor S_j \rfloor) \wedge$$
$$\wedge (p_i[S_p] = in_j[S_j])\} \Rightarrow [(p_i \leftrightarrow pr_o) \wedge (pr_o \leftrightarrow in_j) \wedge (p_i \leftrightarrow in_j)],$$
$$i \in \{N'\}, o \in \{O\}, j \in \{I\} \quad (10)$$

wherein S_p, S_o, S_j are a plurality of properties of machined surfaces, devices, tools, respectively.

Any element of the automated machine tool system is associated with another compatibility ratio (10).

If the structural, process and organizational-functional properties of a process object are considered in a certain sequence, then these two objects in the production process follow each other.

This attitude is indicated by a sign$\mapsto$.

Let's represent this relationship as symbols:

$$\underset{X}{\exists} a_x \underset{X}{\exists} a_q \underset{T}{\exists} b_t \ni P[P(a_x, \vec{U}, a_q) \wedge \bar{P}(a_q, \vec{U}, a_x) \wedge P(b_t, \vec{U}, a_x) \wedge \bar{P}(b_t, \vec{U}, a_q)] \Rightarrow$$
$$\Rightarrow [(b_t \mapsto a_x) \wedge (a_x \mapsto a_q)] \quad (11)$$

Obviously (11), all elements of the automated machine tool system can be arranged according to the properties (2) to (4) and in the processing sequence.

If process objects work together, have the same properties and implement any functions or actions, then they interact.

The interaction relationship is indicated by the sign $><$.

Let's represent this relationship with characters:

$$\underset{M}{\forall} d_k \underset{N'}{\exists} p_i \in \underset{T}{\exists} b_t \underset{X}{\exists} a_x \underset{O}{\exists} pr_o \underset{I}{\exists} in_j \underset{S_d}{\exists} S_{d_k} \underset{S_p}{\exists} S_p \underset{S_t}{\exists} S_t \underset{S_x}{\exists} S_x \underset{S_o}{\exists} S_o \underset{S_j}{\exists} S_j \ni$$
$$\ni P\{[P(b_t, \vec{U}, a_x) \wedge (d_k[S_d] = b_t[S_t]) \wedge [a_x[S_x]] =$$
$$= p_i[S_p] = pr_0[S_0] = in_j[S_j]\} \Rightarrow$$
$$\Rightarrow [(b_t >< a_x) \wedge (b_t >< d_k)] \wedge [a_x >< p_i >< pr_0 >< in_j], i \in \{1, \overline{N'}\} \quad (12)$$

When several elements of the automated machine tool system are connected to each other by an interaction relationship in pairs, it is possible that elements not included in each pair may interact with each other, i.e.

$$P[(b_t >< a_x) \wedge (a_x >< p_i) \wedge (p_i >< pr_0)] \Rightarrow (b_t >< pr_0)(a_x >< pr_0) \quad (13)$$

Therefore, if the expression (12) is true, then it can be confidently considered that any of the elements of the

ACC can be related by the relation of interaction with any other element thereof.

Figure 2 shows a graph constructed in accordance with the definitions of the relationships between the sets of the process environment of the automated machine system.

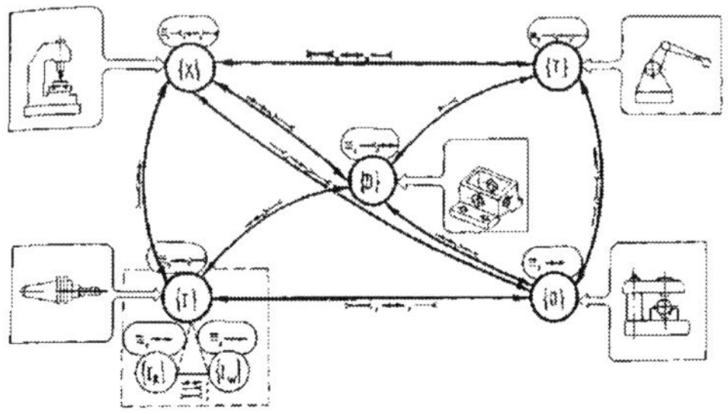

Fig. 2. Graph of relationships of process medium sets of automated system.

On the graph, the relationships between elements of the same or different sets are highlighted.

4. Conclusion

The relationships between the elements indicated on the graph allow formalizing and establishing the relationship between individual components of the automated machine tool system, indicating the need to develop a formal methodology for synthesizing the structure of the automated machine tool system.

The main machines and equipment, cutting and auxiliary tools, vehicles and other relevant components fully describe and provide the functionality of the technological process of manufacturing a group of parts and are the main design objects when developing a technological design of an automated machine system of a multi-nomenclature production.

References

1. S.N. Sharapov, A.V. Zaitsev, Directory, Engineering Journal with Annex, **6**(267) (2019) ISSN: 0203-347X
2. S.G. Mitin, I.A. Razmanov, P.Yu. Bochkaryov, Directory, Engineering Journal with Annex, **5**(242) (2017) ISSN: 0203-347X
3. I.I. Kovtun, G.S. Romanenko, Informatization and Communication, **1** (2020) [in Russia] ISSN: 2078-8320
4. E.G. Krylov, S.V. Shurkhovetskiy, D.T. Arstangaliev, E.V. Reznikova, Magistracy Bulletin, **6**(93) (2019) [in Russia] ISSN 2223-4047
5. S.S. Sergeev, L.G. Bokova, *Promising methods for automating the creation of technological operations in multi-product production* (Collection of articles of the XXV international scientific and practical conference, 2019)
6. R.I. Vakhitova, E.B. Dumler, F.Z., Bulyukova, E.F. Zakharova, *Increasing the efficiency of diversified production based on modeling automated technological equipment: monograph* (Almetyevsk, AGNI, 2019) ISBN: 978-5-94454-039-3
7. E.B. Dumler Voronezh State Technical University Bulletin, **3**(5) (2007) ISSN 1729-6501
8. T.G. Devzheeva, *New design principles in automated systems. Thermophysical and technological aspects of increasing the efficiency of machine-building production* (Works of the IV International Scientific and Technical Conference (Reznikov readings), 2015) ISBN: 978-5-8259-0872-4
9. I.P. Sitdikova, E.B. Dumler, R.I. Vakhitova, N.V. Abdulkina, *Design of automated machine tools for multi-item production* (Achievements, problems and prospects for the development of the oil and gas industry. Proceedings of the IV International Scientific and Practical Conference, 2019). ISBN: 978-5-94454-045-4

Development of a Nonlinear Dependence for Determining the Stress-Strain State of the Pipeline-Composite Bandage System

Robert A. Ramazanov[1], *Igor F.* Kantemirov[1], *Denis A.* Gulin[1], *Indira F.* Makhmudova[1], and *Zulfiya F.* Ismagilova[2]

[1]Ufa State Petroleum Technological University, RU-450062, Ufa, Republic of Bashkortostan, Russian Federation
[2]Almetyevsk state oil institute, RU-423450, Almetyevsk, Republic of Tatarstan, Russian Federation

Abstract. The range of applications of polymeric materials in the oil and gas industries is expanding every year. For example, at present, a number of industrial enterprises are engaged in the production of various tape polymer composite structures that are used to repair pipelines. They are widely used due to the significant advantages of polymeric materials over steel such as low costs, low weight, corrosion resistance and strength. Moreover, most of the operated pipelines are about to reach their standard service life. In the course of technical diagnostics, a significant number of defects of mechanical origin from external influences are constantly revealed. Repair work on pipelines must fully restore the bearing capacity of the site. Therefore, the use of polymer materials for pipeline repair is an important task. Here, we propose to use polymer composite bandages, which will increase the overhaul period. However, a coherent methodology for choosing a material and calculating the required thickness of this structure is not available in the literature, and, therefore, the purpose of this work was to develop such a methodology, as well as to assess the stress-strain state of pipeline-bandage systems. The method presented in this work takes into account the possible anisotropy of the shroud material in the longitudinal and transverse directions and the possibility of plastic deformation of the pipeline material in the annular direction. This work is based on the following assumptions: (a) bilinear dependence of stress on deformation for steel; (b) plastic deformations of the band material are absent; and (3) absolute adhesion between the pipeline and the band. The paper presents the analysis of the market of polymeric materials for pipeline repair, their key properties, and the minimum thickness of the band made of these materials was calculated for various values of the stress concentration coefficient.

1 Introduction

In modern days, new main pipelines are built much less frequently than at the end of the 20th century. As a result, the bulk of the pipelines were produced more than 20 years ago, and, hence, amount of work on repair and reconstruction of main pipeline facilities constantly increases.

Long-term operation of pipelines inevitably leads to aging and, therefore, to defects resulting in a marked decrease in the load bearing capacity of pipeline structures. The defects can be divided into repairable and non-repairable ones [1].

Analysis of the diagnostics data showed that most of the defects are of mechanical origin, with "metal loss" type defects being the most common ones. Stress corrosion of pipe metal is also a harmful process. However, timely repair allows reducing negative impacts of the aforementioned factors on the pipeline system.

In ref. [2], a study of the stress-strain state and the assessment of the strength in the zones of dents on the pipes of technological pipelines were reported. The coefficients of the concentration of stresses and strains have been determined using computer modeling for both elastic and elastoplastic nature of local deformations of materials in the mechanical defect zone. The ultimate deformation and the strength conditions, which ensure the reliability of technological pipelines in the presence of defects of mechanical origin, were estimated.

In work [3], the destruction of a pipeline with a corrosion defect was modeled. For steel 19G, a mathematical model of elastoplastic deformation of metals obtained earlier by the same authors was used and a set of practical recommendations was obtained.

In ref.[4], the stress-strain state of a pipeline section with a defect during repair with a composite coupling was investigated. It was shown that the circumferential stresses along the pipe axis in the repaired section, on average, decreased by half. Calculations were validated using field tests of composite couplings.

To reduce the time for repair work on local defects in the pipe metal, it was proposed to use composite reinforcing bands, which are already used to increase the bearing capacity of a pipeline section, for example, in cramped conditions with a forcedly reduced protective zone [5]. Their use requires calculations with the «installation before operation» method and the design scheme shown in Figure 1.

In this case, the bandage is mounted on a welded and insulated pipe string. The material of the bandage is subjected to tensile loads. The initial data are the strength properties of the bandage material, the value of the tension force and the design value of the working pressure in the pipeline. After the calculation is completed, we obtain the thickness of the band, which will provide the necessary increase in the bearing capacity of the pipeline.

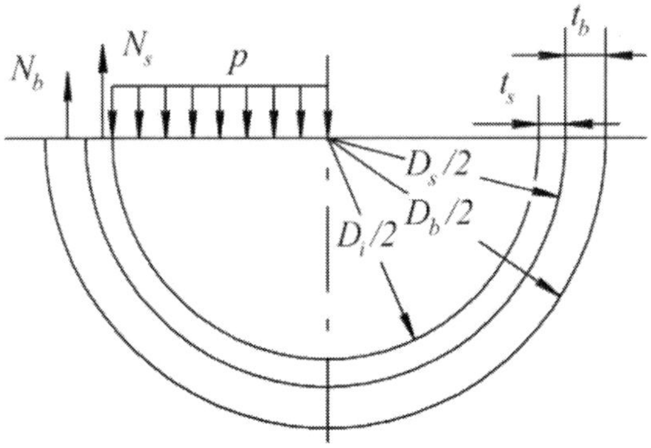

Fig. 1. Equal balance of a half-ring cut from a pipe under the internal pressure (to increase the bearing capacity of the pipeline).

In this case, the bandage is mounted on a welded and insulated pipe string. The material of the bandage is subjected to tensile loads. The initial data are the strength properties of the bandage material, the value of the tension force and the design value of the working pressure in the pipeline. After the calculation is completed, we obtain the thickness of the band, which will provide the necessary increase in the bearing capacity of the pipeline.

Mounting the bandage according to the «on operated pipeline» method provides a completely different operation of the «bandage-pipe» system. The bandage also stretches, but, at the same time, it exerts a compressive force on the pipeline, due to which turning pipeline in an operating mode becomes possible. The design scheme is shown in Figure 2.

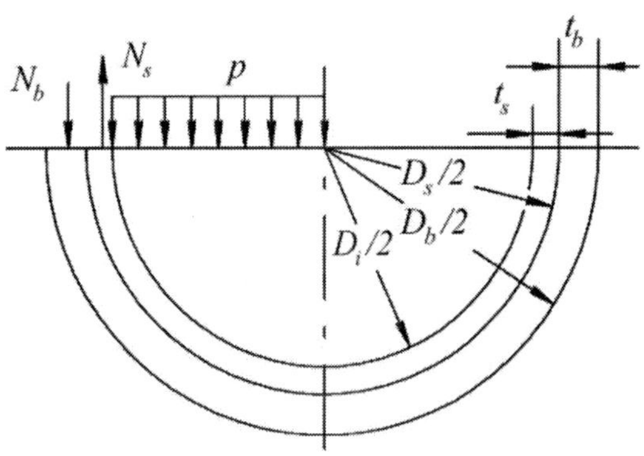

Fig. 2. Equal balance of a half-ring cut from a pipe under the action of internal pressure (during repairing a pipeline section).

The purpose of the calculation is to ensure the safety of the operation of the main pipeline in the case, when a bandage is applied using the second method. The direct result of the strength calculation is a certain value of the minimum thickness of a band made of a specific material with known strength. In addition, the geometrical properties of the pipeline, the type and size of the defect and the minimum pressure that the pipeline must withstand for further operation should be available.

2 Theory

Let us consider the equal balance of the half-ring of the «pipeline-bandage» system under the action of internal pressure (during the repair of a pipeline section) in the annular direction, which will be described by the following expression:

$$\sigma_{2s} \cdot t_s - \sigma_{2b} \cdot t_b = \frac{pD_i}{2}. \quad (1)$$

where σ_{2s} – circular stress in steel, MPa;
t_s – wall thickness, m;
σ_{2b} - circular stress in the material of the band, MPa;
t_b – bandage thickness, m;
p – internal pressure in the pipeline, MPa;
D_i - internal diameter of the pipeline, m.
Circular stresses arising at the site of the defect are proposed to be taken into account using the stress concentrator factor:

$$\sigma_{2s} = \alpha_\sigma \cdot \frac{pD_i}{2}, \quad (2)$$

where α_σ – stress concentration factor.

The stress concentration factor itself is calculated depending on the type of defect (corrosion pits, mechanical marks, scratches, cuts and scuffs, characterized by the length l, width d , depth h and radius of curvature at the vertex ρ for a pipe with wall thickness t s) [6].

It is proposed to calculate the thickness of the band, or rather the number of layers of the wound material, depending on its strength characteristics. The permissible stress value in the band is defined as the elastic limit of the band, taking into account the stresses arising during winding. As a result, the following ratio was obtained for the number of layers of the bandage:

$$n = \frac{pD_i}{2} \cdot \frac{\alpha_\sigma - 1}{\sigma_{u2b} t_{b0} - N}. \quad (3)$$

where n – number of layers;
t_{b0} – thickness of the bandage layer;
σ_{u2b} – elastic strength of the band to the circular direction;
N – tension force of rolled materials.

The resulting value must be rounded up to prevent overvoltage in the cross section of the band material.

To assess the real stress-strain state of the pipeline-bandage system, it is proposed to add the following to equation (1) and, at a known value of the bandage thickness, evaluate the work of the steel and the bandage in the longwise and circular directions.

The material of the band is considered as a non-plastic material, then its properties obey Hooke's law. It is also assumed that the steel in the longwise direction does not achieve plastic deformation. With absolute adhesion under operating conditions, the deformations of the pipe and the band are equal to each other, both in the longwise and circular direction. Let us express the deformation in the circular direction:

$$\varepsilon_2 = \frac{1-\nu_{12b}\nu_{21b}}{E_{2b}}\cdot\sigma_{2b} - \frac{\nu_{12b}}{E_{0s}}\cdot\left(\sigma_{1s} - \nu_{0s}\cdot\sigma_{2s}\right). \quad (4)$$

where E_{2b} – Young's modulus of the anisotropic material of the band in the circular direction, MPa;
ν_{21b} – Poisson's ratio from the action of the circular stress in the longwise direction of the anisotropic material of the band;
ν_{12b} – Poisson's ratio from the action of longwise stress in the circular direction of the anisotropic material of the band.
E_{0s} – Young's modulus of steel in the elastic area, MPa;
ν_{0s} – Poisson's ratio of steel in the elastic area.

The average normal stress of steel in a plane stress state does not take into account the influence of the radial component. Average linear deformation is related to the average normal stress by the ratio of bulk elasticity.

In the framework of the deformation theory of plasticity, the elastoplastic deformation of a pipe can be described by the following equality:

$$2\sigma_{2s} - \sigma_{1s} = 2\cdot\frac{\sigma_{is}}{\varepsilon_{is}}\cdot\left(\varepsilon_2 - \frac{1-2\nu_{0s}}{E_{0s}}\cdot\frac{\sigma_{1s}+\sigma_{2s}}{3}\right). \quad (5)$$

where σ_{is} – intensity of normal stresses in steel, MPa;
ε_{is} – the intensity of linear deformations of steel.

The intensity of normal stresses in steel is described by the following expression:

$$\sigma_{is} = \sqrt{\sigma_{1s}^2 + \sigma_{2s}^2 - \sigma_{1s}\sigma_{2s}}\,. \quad (6)$$

Longwise stress in the pipeline from the action of internal pressure and temperature difference is determined as:

$$\sigma_{1s} = \nu_s\frac{pD_i}{2t_s} - \alpha E_s\Delta t \pm \frac{E_s\left(D_i + 2t_s\right)}{2\rho}, \quad (7)$$

where ν_s – variable Poisson's ratio of steel;
E_s – variable Young's modulus of steel, MPa;
α – coefficient of linear expansion of pipe metal, deg^{-1};
Δt – the calculated temperature difference, taken as positive when heated, °C;
ρ – the minimum radius of elastic bending of the pipeline axis, m.

For steel, the variable Poisson's ratio and the variable Young's modulus are calculated using the well-known formulas [7]:

$$\nu_s = \frac{\frac{1}{2} - \frac{1-2\nu_{0s}}{3E_{0s}}\cdot\frac{\sigma_{is}}{\varepsilon_{is}}}{1 + \frac{1-2\nu_{0s}}{3E_{0s}}\cdot\frac{\sigma_{is}}{\varepsilon_{is}}}, \quad (8)$$

$$E_s = \frac{\sigma_{is}/\varepsilon_{is}}{1 + \frac{1-2\nu_{0s}}{3E_{0s}}\cdot\frac{\sigma_{is}}{\varepsilon_{is}}}, \quad (9)$$

Let us take the «stress-strain» diagram of pipeline steel approximated to bilinear dependence (Figure 3).

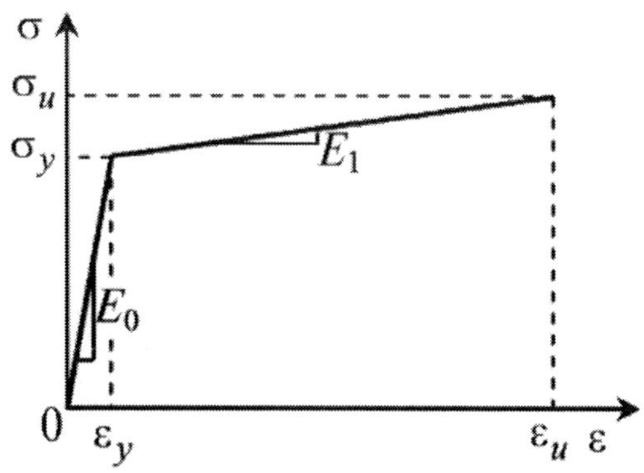

E_0 – modulus of elasticity of steel; E_1 – tangent modulus in the area of elastoplastic deformation; σ_y and ε_y – stress and deformation of the transition point to the section of elastoplastic deformation; σ_u and ε_u – stress and strain corresponding to the ultimate strength of the steel pipe material.

Fig. 3. Diagram of uniaxial tension of steel during deformation in the annular direction.

The problem of determining the value of the flow stress has not been solved. In the literature specific values are given for various steel grades and defect shapes. There is no theoretical dependence, and the magnitude of the flow stress is determined, as a rule, experimentally.

Since the stress intensity is the reduced intensity of shear stresses, expressed in terms of the equivalent normal stress in simple tension, i.e. equivalent tensile stress, then the following system is valid:

$$\begin{cases} \varepsilon_{is} = \frac{2\sigma_{is}}{3E_{0s}}\left(1+2\nu_{0s}\right) \text{ при } \sigma_{is} \le \sigma_y, \\ \varepsilon_{is} = \frac{\sigma_{is}-\sigma_y}{E_{1s}} + \frac{\sigma_y}{E_{0s}} \text{ при } \sigma_{is} > \sigma_y. \end{cases} \quad (10)$$

The tangent modulus of elasticity E1s according to the simplified diagram is numerically equal to the tangent of the angle of inclination of the tangent drawn from the point of yield to the point of break, and is calculated by the following formula [8]:

$$E_{1s} = \frac{\sigma_u\left(1+\frac{\delta}{100}\right)-\sigma_y}{\ln\left(1+\frac{\delta}{100}\right)-\varepsilon_y}, \qquad (11)$$

where δ – the relative elongation.

The anisotropic material of the band presumes different properties in different directions. According to the rule of mixtures, the values of Young's modulus are found along and across the fibers [9]:

$$\begin{cases} E_L = E_f V_f + E_m\left(1-V_f\right), \\ E_T = \dfrac{E_f E_m}{E_m V_f + E_f\left(1-V_f\right)}, \end{cases} \qquad (12, 13)$$

where E_L – Young's modulus of a conventionally homogeneous anisotropic composite material along the fibers, MPa;
E_T – Young's modulus of a conventionally homogeneous anisotropic composite material across the fibers, MPa;
E_f – Young's modulus of fiber filler, MPa;
E_m – Young's modulus of matrix, MPa;
V_f – volumetric content of fiber filler, MPa;
V_m – volumetric content of matrix, MPa;

Poisson's ratio can also be calculated according to the rule of mixtures in the direction along the fiber, and in the transverse direction, provided that the material is orthotropic according to the ratio between Young's moduli and Poisson's ratios [9], [10]:

$$\begin{cases} \nu_{TL} = \nu_f V_f + \nu_m V_m, \\ \nu_{TL} E_L = \nu_{LT} E_T, \end{cases} \qquad (14, 15)$$

where ν_{LT} – Poisson's ratio for direction along the fiber;
ν_{TL} – Poisson's ratio for direction across the fiber;
ν_f – Poisson's ratio of fiber filler;
ν_m – Poisson's ratio of matrix.

Ultimate pressure that the pipeline can experience in the presence of a defect, MPa [11]:

$$P_u = \frac{\sigma_m t_s}{R}\cdot\frac{A_0 - A}{A_0 - A\cdot M^{-1}}, \qquad (16)$$

where σ_m – the yield point, with a bilinear stress-strain relationship, we take $\sigma_m = \sigma_y$, MPa;
R – pipe radius, m;
A – area of metal loss on the projection of the defective area in the longwise section of the pipe wall, mm^2;
A_0 – initial (without corrosion) area of the longwise section of the pipe wall along the length of the defective section, mm^2;
M – Folias factor.

To assess the stress-strain state of the pipeline-bandage system, it is also necessary to determine the factory test pressure (the corresponding formula is given in [7]). The smallest of the defined pressures (factory test or limit for a pipe with a defect) is taken for calculation.

The calculation is done using numerical methods.

Let us evaluate the adequacy of the presented calculation method for various materials of the band. The mechanical mark on the outer surface of the pipeline metal was chosen as a defect. Such a defect is classified as "short".

The rest of the initial data for the calculation are shown in Table 1.

Table 1. Initial data.

Pipeline diameter D, mm	1220
Wall thickness t_s, mm	20
Tension force of roll materials N, kN/m	1.472
Internal pressure in the pipeline P, MPa	7.5
Tensile strength of steel σ_{us}, MPa	550
Yield strength of steel σ_{ys}, MPa	410

To put that into perspective, we present the calculation results in the form of a graph, Figure 4, showing the dependence of the thickness of the band on the selected material at different values of the stress concentration factor:
- tensile strength of the band from 100 to 1600 MPa;
- stress concentration factor from 1.5 to 3.5.

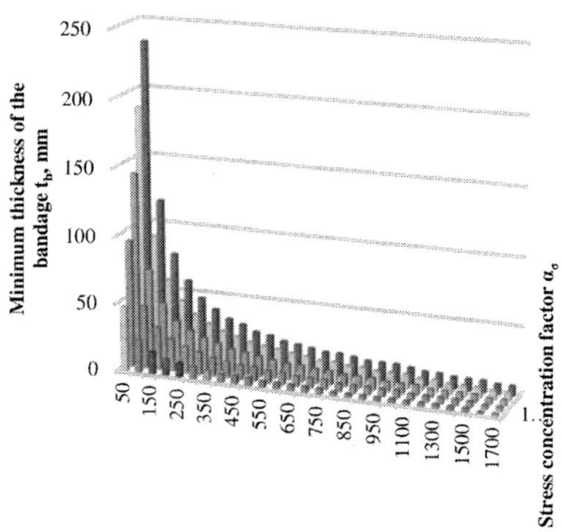

Fig. 4. Dependence of the thickness of the band on the selected material at different values of the stress concentration factor.

Based on the Figure 4, it can be argued that it is preferable to use materials with a tensile strength of 200 MPa or more.

The obtained values of the minimum thickness of the band can find practical application for the known values of the ultimate strength of the band. For this purposes the polymer materials on the market used for pipeline repairs were analyzed. The characteristics of the materials are presented in Table 2.

Table 2. Specifications of materials.

Material	**Tensile strength, MPa**	**Young modulus**	**Poison's ratio**
Carbon fiber (tape) of JSC «Kompozit»	900	150	0.265
Organoplastic of JSC «Kompozit»	1600	80	0.34
Fiberglass of JSC «Kompozit»	600	50	0.28
Fiberglass (epoxy polymer 5-211B and E glass)	200	24.26	0.15
CHT	350÷684	21.6÷28.0	0.3
EARF	950÷1390	24.8÷52.0	0.3
EARF-2	600	38	0.3
ERF	900	52	0.3
COT-301-1	1370	140	0.3
«Clock Spring» Cuffs	900	38	0.3
Fiberglass VPS-53/120	460	21.7	0.28
Tape «Polyterm»	12	0.2	0.4
«WrapMaster» cuff material	620.5	34.4375	0.4

3 Conclusion

The present study leads us to the following conclusions:

1. Out of a number of materials being studied here, only «Polyterm» tape was unable to provide the normal operation of the pipeline system with a reasonable thickness of the bandage layer. The bandage material should be selected based on the rational choice of the minimum wall thickness and the material properties, with tensile strength being the main one;

2. The advantages of the repair using the banding method include the absence of restrictions on the diameter, length and operating pressure of reinforced pipelines;

3. The calculations of the stress-strain state of the bandage should be done using properly chosen the «bandage-pipe» design scheme

4. For the full use of the material, it can be recommended to consider the stress state of the band with taking into account the elastic-plastic behavior of its material;

5. The possibility of using of a bandage for repairing defects should be determined based on capabilities of machines being used and the analysis of the prior use of the bandage for repairing defects with known stress concentration factors.

References

1. RD 153-39.4-067-04, *Methods for repairing defective working oil trunk pipelines* (JSC AK Transneft, 2004)
2. I.N. Budilov, G.V. Kulyasov, Yu.V. Lukashchuk, V.V. Shevela, Bulletin of USATU, **1**(54), 57–62 (2013)
3. S.A. Korneev, I.V. Krupnikov, S.N. Polyakov, V.V. Shalai, Omsk Scientific Bulletin, **1**(87), 48–51 (2010)
4. A.A. Bauer. V.M. Kushnarenko, A.E. Pyataev, Yu.A. Chirkov, D.N. Shchepinov, OSU, (2015)
5. E.A. Lobova, D.A. Gulin, R.A. Fazletdinov, R.R. Musalyamov, Transport and Storage of Oil Products and Hydrocarbons, **1** (2018)
6. S.K. Rafikov, G.S. Sharnina, M.M. Fattakhova, *Calculation of the residual resource of oil pipelines by low-cycle durability of steel pipes* (USPTU, 2011)
7. SP 36.13330.2012. Trunk pipelines (Moscow, 2013)
8. V.A. Vasiliev, A.V. Samoshin, Measurement. Monitoring. Management. Control, **3**(21), 108-114 (2017) DOI: 10.21685/2307-5538-2017-3-15
9. A.S. Milenin, E.A. Velikoivanenko, G.F. Rozynka, N.I. Pivtorak, *Collection of reports of the VII international conference.* (Kiev, 2014)
10. A.A. Smerdov, Herald of the Bauman Moscow State Technical University. Series Mechanical Engineering, **1**, 52-60 (2011) ISSN 0236-3941
11. A.A. Filatov, V.I. Kochetov, I.I. Veliyulin, R.R. Khasanov, Neftegaz Territory, **40**, 68–72 (2018) ISSN 2072-2745

Structural features of metal powders obtained by plasma spraying

Evgeny Gulikhandanov[1], *Sergey Ermakov*[2]

[1]Doctor of Engineering Science, Professor of the Institute of Mechanical Engineering, Materials and Transport SPbPU Saint Petersburg, Peter the Great St.Petersburh Politechnic University, RU-195251, St.Petersburg, Russia
[2]Engineer, Director of Scientific and Educational Center «Severstal-Politech» Saint Petersburg, Peter the Great St.Petersburh Politechnic University, RU-195251, St.Petersburg, Russia

Abstract. The development of additive technologies in the Russian Federation is constrained by the lack of indigenous base for the production of metal powders. SPbPU has developed, manufactured and introduced a plasma atomizer. It provides atomization of metal powders and alloys of various chemical compositions. The data on the main technological parameters of atomization of nickel alloy Inconel 718 and titanium alloy Ti-6A1-4V are presented in the paper. It is shown that by optimizing the spraying mode the possibility to obtain alloy powders with a high level of sphericity and a minimum number of particles with surface and internal defects. The chemical compositions of the obtained particles are almost identical to the composition of the initial atomized materials. Samples for the mechanical testing were made from the obtained powders by the SLM method. The results of mechanical properties tests show that the properties of samples made by 3D printing methods from powders sprayed by plasma atomization methods are identical to the properties of materials made and heat-treated by traditional methods.

1 Introduction

The beginning of the XXI century was marked by the development of new approaches and requirements for manufactured equipment. The most popular technologies of the past – technologies based on subtractive processing processes, such as turning, milling, and drilling – are becoming a thing of the past. Using subtractive technologies caused the metal utilization rate to reduce – more than half of the billet mass was consumed in the chips. Moreover, significant human and energy resources were spent. Subtractive technologies are being replaced by new – additive technologies based on prototyping parts by laser fusing of metal powders. These technologies can reduce the material consumption of products by two or more times, simplify the technological process of their production, and solve environmental safety issues.

Nowadays additive technologies are among the fastest growing technological processes – the annual growth rate of the use of 3D printing reaches, and in some years exceeds 20%. At the beginning of the century, the Russian Federation had the development rate of additive technologies was at the level of the highest world indicators. However, in the last decade, the intensity of development of additive technologies in the Russian Federation is gradually slowing down. One of the main reasons for this slowdown is the virtual absence of the indigenous production base of raw materials for additive machines i.e. metal powders. Currently, more than 70% of the required amount of metal powders are supplied from abroad i.e. the countries of the European Union, Canada, the United States and China. [1,2] However, the policy of sanctions imposed on our country by a number of Western countries, makes it difficult to obtain the necessary volume of powder product.

Despite the difficulties that have arisen in the Russian Federation, there is a steady demand for both additive machines and raw materials for the production using 3D printing methods. The main consumers of additive products in the Russian Federation are the most high-tech industries: aircraft and shipbuilding, space, medicine, and energy engineering. There has been the most common distribution of additive technologies in the experimental and experimental-industrial developments of these industries. These industries require production of one or more experimental products in the shortest possible time and conduct their accelerated tests. In these industries, additive prototyping based on 3D models can significantly speed up and reduce the cost of introducing new machines and mechanisms into production, including eliminating the production of complex tooling (molds, dies, etc.) necessary for the production of a prototype.

In our view, the further expansion of additive manufacturing at domestic enterprises should be linked to phasing out of external suppliers of the powder product. First of all, this applies to the pilot plant production. The creation period of a prototype is comparable to the purchase period of the powder

https://doi.org/10.1051/epjconf/202124804025

necessary for its production, and the creation of their own powder production facilities on the base of a plant. The powder production can rely on the plasma spraying of metal powders since the plasma spraying equipment is relatively cheap, does not require the allocation of specialized premises and highly qualified personnel. Currently, the Russian Federation has developed a number of plasma atomizers that fully meet the world level. These include atomizers designed and manufactured at Peter the Great St. Petersburg Polytechnic University. These atomizers are easily reconfigured for the powder production of various chemical and granulometric compositions (necessary for 3D printing), and have a high level of sphericity.

This paper presents the main technological parameters of the powder production of nickel - Inconel 718 and titanium - Ti-6Al-4V alloys and the research results of these powders, confirming the possibility of their use in additive machines of various types. The choice of materials was guided by the fact that these alloys are currently used as widely as possible in additive manufacturing. Due to the high stability of Inconel 718 mechanical properties at elevated temperatures this alloy is used extensively in gas turbines, jet engines, nuclear reactors and high-pressure vessels. Inconel 718 is used in oil and gas extraction and subsequent processing of raw hydrocarbon due to its high resistance in aggressive environments. However, in the application of this alloy, it is difficult to obtain products of complex shape. It happens due to its low-pressure workability because of extremely narrow temperature-speed range of plastic deformation and the complexity of mechanical processing due to the high tendency of this material to harden. One of the most effective manufacturing technologies for this alloy is the technology of layer-by-layer or direct laser fusing of powders in 3D printers [3].

The Ti-6Al-4V alloy is also widely used in products obtained by additive growing methods. It is used in products requiring the highest possible corrosion resistance and specific strength at a minimum density, so it is widely used in the aerospace industry, bioengineering, and medical technology.

2 Materials and methods

Plasma atomization is a kind of gas atomization in which the melting of a solid feedstock occurs under the influence of a high energy flow of ionized gas-plasma. Atomizers based on plasma atomization are the most compact and cheap systems for producing small-experimental batches of powders of various chemical compositions. The advantages of plasma atomization include a high sphericity level of the obtained powders, the absence or minimum number of defects typical for 3D printer powders, such as internal and external pores, satellites, and oxide films on the surface of the particles. The disadvantages of plasma atomization include the low productivity of the process (usually from 1 to 10 kg of powder per hour), due to the characteristics of the starting materials for plasma atomization, which are usually used as wires with a diameter of 1 to 3-4 mm [4-6].

The atomizer developed at SPbPU (Fig.1) [7] consists of several interconnected blocks. The main units of the installation are powder spraying units, the number of which can vary from one to three according to the type and purpose of the atomizer. In turn, the spray units consist of plasma generators – plasma torches, wire feed systems and plasma gas. The current supplied to the plasma torch, the wire feed rate and the volume of plasma gas supply can be independently adjusted for each of the spray units.

This makes it possible, if necessary, for example, to obtain initial powders for mechanical alloying [8], to simultaneously spray wires of several chemical compositions.

Depending on the volume and composition of the powder required at the moment, spraying on such atomizer can be carried out either by one or simultaneously by three independently operating spray units - (Fig.2).

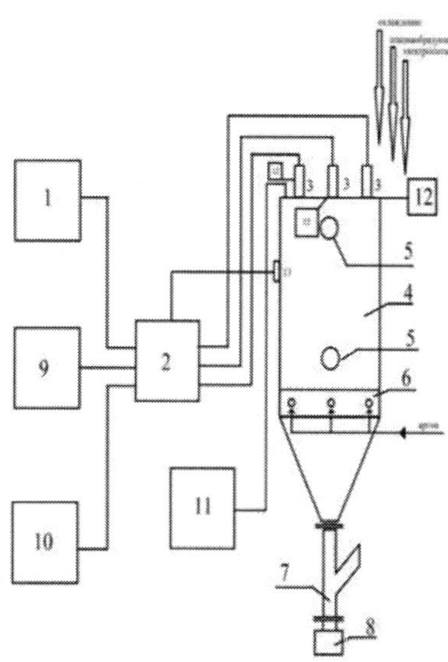

Fig. 1. Schematic circuit of a plasma atomizer: 1. Power supply, 2. Control and safety units, 3. Plasma torches (there is the layout of the atomizer with three spray units), 4. Spray column, 5. Peephole on the chamber side surface for visualizing the spray process, 6. "Fluidized bed" belt – a system of multidirectional injectors through which protective gas is supplied, 7. Plasma gas removal system during the spray process, 8. Powder collection system, 9. Gas system, 10. Plasma torch cooling system, 11. Column cooling system spraying, 12. Solid feedstock supply system, 13. Measuring equipment (control and safety device).

The atomizer also includes a spray column (a two-layer water-cooled sealed vessel where the process of spraying spheroidization and crystallization of powder particles takes place), a powder collection system that allows the collection, packaging and storage of the sprayed powder without access to aggressive environmental components, water, gas and energy supply systems, spraying control process, equipment for supplying the initial feedstock, plasma-forming and protective gas.

The indicators that determines the efficiency and profitability of the atomizers operation are the following: the yield of useful (marketable) products, the stability of the geometric and granulometric characteristics of the sprayed powders. The paper presents the research results of the technological parameters effect of spraying on the percentage of commercial fractions yield of two alloys powders with high affinity for oxygen - Inconel 718 and Ti-6Al-4V. The data on the influence of the technological parameters of spraying on the geometric shape and granulometric composition of the sprayed powders are also presented.

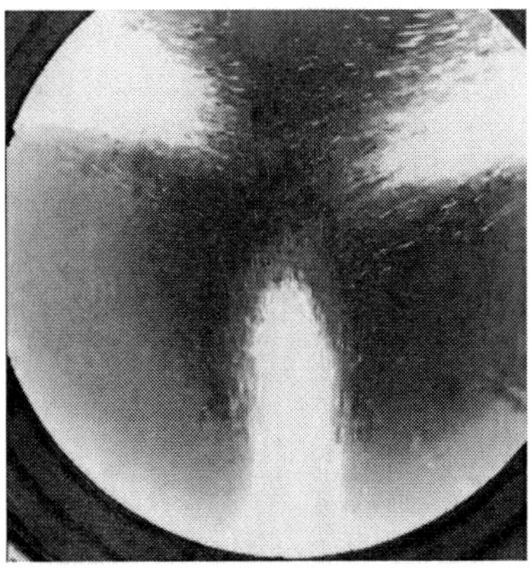

Fig. 2. Three independent powder spray torches – view through the peephole window in the upper bottom of the spray chamber.

During the experiments, the following sputtering parameters were varied: the current applied to the plasma torch in the range of 150-400 A, the speed (volume) of the plasma-forming gas – 30-60 l/ min, the wire feed speed – 1-5 m/min. As the constant factors of the experiments, the following factors were established: the flow rate of the protective gas 1.2 l/min per nozzle of the fluidized bed system and the gas temperature in the spray chamber determined by measuring the temperature on the inner wall of the spray chamber – 1050^{0}C. All experiments were carried out in the mode of simultaneous operation of three spray units operating in a single technological mode.

Wires of both alloys with a diameter of 1.8 mm were used for the research. The chemical compositions of the wires are given in table1.

Table 1. Chemical compositions of alloys (wt.%)

Ti-6Al-4V	C	Al	V	O	N	H
	0,63	6,1	4,2	0,10	0,03	0,012
Inconel 718	C	Ni	Cr	Nb	Mo	Ti
	0.07	53,2	20,4	5,03	3,12	1.10

High-purity argon was used as a plasma-forming and protective gas.

The spraying characteristics of the studied alloys are their high affinity for oxygen and nitrogen which requires total isolation of the spraying zone and the collection of hot powder from the external environment. The powder is increased, in relation to the sprayed iron alloys, quality control of the initial-sprayed wires and preparation of containers for storing finished products.

Given that plasma spraying is mainly used to produce so-called medium powders that is powders with sizes of 40-140 microns, the percentage of powders of these fractions in the total volume of the sprayed powder mass was used as the "yield of useful" parameter.

3 Results and discussion

The analysis of the Ti-6Al-4V alloy spraying modes showed the best combination of the geometric quality of the obtained powders and their granulometric composition. It is achieved by using the following mode: current strength of 275 A, plasma gas velocity of 45 l/min, wire feed speed of 2.5 m/min. When these conditions are met, the high sphericity of the powders is ensured, reaching about 93-96 %, the yield of the useful material is about 62% with an average size of powder particles d_{50} = 60-80 microns. The proportion of powder particles with satellites, pores and other defects did not exceed 1.5%.

If there is a deviation from the optimal mode, for example, with a decrease in the current supplied to the plasma torch, the percentage of useful output decreases sharply – about one and a half to two times, and the shape of the particles gradually shifts from spherical to rounded and even angular, the number of particles with satellites and other defects increases (GOST 25849-83) – Fig.3.

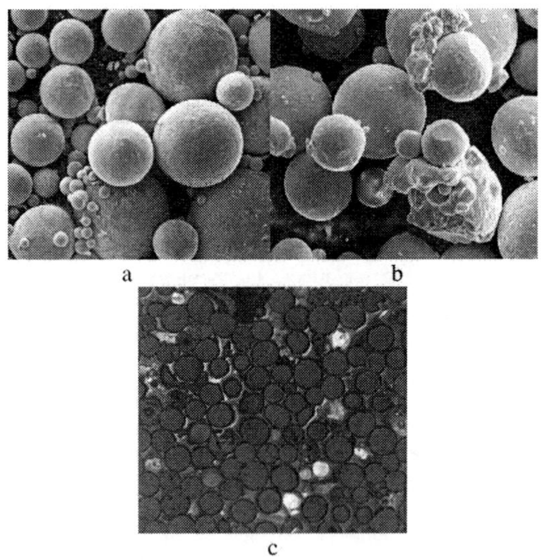

Fig.3. Particles of Ti-6Al-4V alloy powder sprayed according to the optimal mode (a) and with a deviation from it (current strength - 225 A) - an increase of x400 and (c) - a microslift of particles of the sprayed powder x200.

When spraying the Inconel 718 alloy wire, the optimal ratio of parameters was: current 200 A, plasma gas velocity 50 l/min, wire feed speed – 3.75 m/min. When spraying the Inconel 718 wire according to the specified parameters, the following results were achieved: sphericity of 95-97%, yield of at least 64% with an average size of powder particles d50 = 60-75 microns, the proportion of powder particles with various defects – no more than 1.0%. (Fig. 4). Deviations from the optimal mode, as well as in the case of a titanium alloy, lead to a decrease in the yield of the useful powder and a deterioration in the geometric quality of the resulting powder.

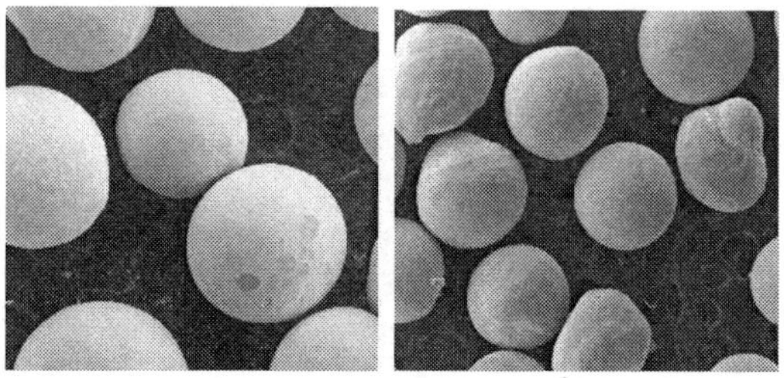

a b

Fig.4. Ti-6Al-4V alloy powder particles sprayed according to the optimal mode (a) and with a deviation from it (current - 170 A), magnification x400.

Thus, it was shown that plasma spraying provides a satisfactory yield percentage of commercial powder fractions – at the level of 60-65% of the mass of the sprayed metal. It allows to get powders with a minimum number of surface and internal defects, geometric shapes, granulometric composition of which allows them to be used in additive manufacturing. However, in addition to the geometric parameters of the powder quality it is necessary to confirm their chemical compliance with the sprayed materials. For this purpose, comparative studies of the chemical compositions of the initial wires and the resulting powders were carried out.

As an example, Fig.5 shows the results of chemical-spectral control of the wire and the powder particles of the Ti-6Al-4V alloy.

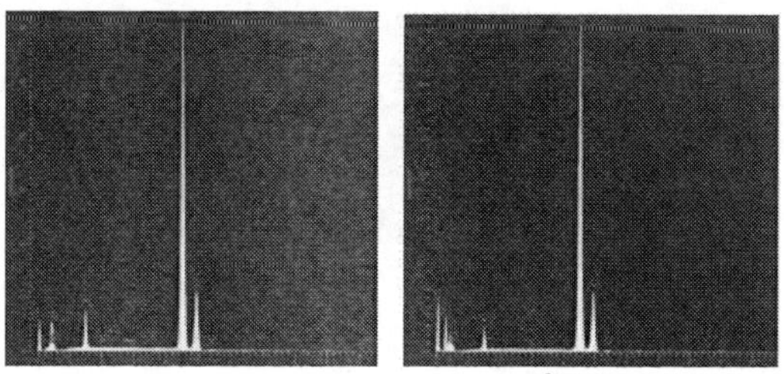

a b

Fig.5. Spectral analysis results of the initial wire (a) and the atomized powder (b) of the Ti-6Al-4V alloy.

Given the high affinity of the studied alloys to oxygen and other light impurities, special attention was paid to the change in the concentration of elements such as carbon, oxygen, nitrogen, and hydrogen during the chemical analysis.

It was found that the content of the main alloying elements and carbon in both alloys practically did not change during the wire spraying process. The results of the analysis of the gas content in the sprayed powder particles allow us to conclude that while ensuring the proper quality of the protective gas atmosphere in the spray column and in the powder collection system, their concentration does not exceed the level of requirements for the starting materials.

So, for example, the limiting concentration of oxygen in the Ti-6Al-4V alloy (ASTM F136-02a) is 0.13 wt.%, in the initial wire – 0.102 and in the powder – 0.109 wt.%; nitrogen – 0.05, 0.033 and 0.038 wt.%, respectively, of hydrogen – 0.012, 0.009 and 0.011 wt.%.

Thus, it was confirmed that plasma spraying does not lead to any significant changes in the chemical composition of the sprayed powders.

The most important indicator of the quality of powders, by which it is possible to confirm the legality of their use in the production of products using additive technologies, is the mechanical properties of samples made from them. For this purpose, using Selective Laser Melting (SLM) methods, proportional flat samples of type II GOST 1497-84 were made from sprayed powders.

Table 2 shows the mechanical properties results of samples made from sprayed powders. For comparison, the same table provides data on the regulatory requirements for the mechanical properties of compact alloys. For the Ti-6Al-4V alloy, these are the requirements of ASTM F136 and GOST R ISO 5832-3-2020, for the Inconel 718 alloy API Standard 6A718.

The heat treatment of the obtained samples was performed in accordance with the requirements of the regulatory documentation for these materials.

Table 2. Mechanical properties of samples at 20^0C

Alloy		σ_B	$\sigma_{0.2}$	δ_5
		MPa		%
Ti-6Al-4V	1*	885	820	12
	2	860	795	10
Inconel 718	1	165	124	18
	2	145	122	17

* 1 – actual properties of samples; 2 - regulatory documentation requirements (not less than)

Table 2 analysis confirms that the mechanical properties of samples made from sprayed powders by the SLM method meet the requirements for these alloys.

Based on the conducted experiments, it was shown that the method of plasma spraying of alloys with a high affinity for oxygen allows us to obtain high-quality spherical powders suitable for use in additive machines of various types.

4 Conclusions

1. Plasma atomization is found to obtain high-quality powders of nickel and titanium alloys, suitable for use in

additive machines of various types. The powders have a high degree of sphericity, the chemical composition of the obtained powders practically did not differ from the compositions of the original feedstock, the proportion of powders with surface and internal defects does not exceed 1.0–1.5%.

2. The basic energy-power parameters of atomization of titanium and nickel alloy wires, which provide the maximum percentage of the yield of commodity fractions of powders of the studied alloys, are determined.

3. The mechanical properties of products obtained by SLM methods from pulverized powders were shown to meet the requirements for these alloys manufactured and heat-treated according to traditional technologies.

References

1. E.N. Kablov, Metals of Eurasia, **3**, 3-21 (2017)
2. M.A. Zlenko, A.A. Popovich, I.N. Mutylina, St. Petersburg. Polytechnic University Press, 221 (2013)
3. M.Yu. Gryaznov, S.V. Shotin, V.N. Chuvildeev, Solid state physics. Vestnik Nizhegorodskogo universiteta im. n. I. Lobachevsky, **4**(1), 46-51 (2014)
4. T. Wohlers, *Wohlers Report 2014: Additive Manufacturing and 3D Printing State of the Industry* (Annual Worldwide Progress Report. Wohlers Associates Inc., 2014) ISBN 978-0-991333-2-6-4
5. A. Uriondo, M. Esperon-Miguez, S. Perinpanayagam, Proceedings of the Institution of Mechanical Engineers. Part G: Journal of Aerospace Engineering, **229**(11), 2132–2147 (2015) DOI: 10.18721
6. B.S. Ermakov, S.B. Ermakov, A. Pavlenko, S.A. Vologzhanina, IOP Conference Series: Materials Science and Engineering, **826**, 012007 (2020) DOI 10.1088
7. B.S. Ermakov, S.B. Ermakov, M.A. Odnoblyudov, V.G. Teplukhin, *Development of equipment for obtaining metal powders for additive technologies. New technologies for equipment and structures of the North and the Arctic* (Proceedings of the International Conference. Yakutsk, NEFU Publishing House, 2016)
8. A.A. Popovich, *Mechanochemical synthesis of refractory compounds. Ministry of Education of the Russian Federation, Far Eastern State Technical University (DVPI named after V. V. Kuibyshev)* (Vladivostok: Publishing House of DVSTU, 2003)

AUTHOR INDEX

Arkhangelskii Vladimir E. 342
Aung Pyae Phyo 334
Ávalos Fausto 191
Balashova L. M. 25
Balova Daria 293
Bayburova Minsariya 376
Bespalov Sergey 226
Bobrovskij Igor 301
Borisov Anatoly 297
Bublikov Jury 313, 319
Budko Pavel 171
Bychkova N. A. 222
Carrera Fausto Freire 200, 206
Chadrina Olga 200, 206
Chalovskaya Ekaterina 180, 212
Chechina Antonina 94, 159
Chekanin Alexander 153
Chekanin Vladislav 153
Chigarev Anatoli 313, 319
Chori Vladislav V. 50
Churbanova Natalia 94, 159
Dmitrieva Viktoria A. 240
Dolgov Nikita V. 338
Dolgov Vitalii A. 338, 342
Dumler Elena 381, 387
Egorov Aleksandr 247
Egorova Maia 247
Eliseeva Natalia 67
Elyseeva Julia 7
Erlhagen Wolfram 124
Ermakov Sergey 396
Ershov Artem 281, 324
Filippov Ilya 180
Garcia-Orellana Yelitza 206
Gavrilov Andrey 175
Georgiev Milen 67
Goydenko Vladimir 171
Grechishnikov Vladimir 324, 334
Gridnev M. A. 1
Grigoriev Sergey 352, 357
Gulikhandanov Evgeny 396
Gulin Denis A. 391
Gusarov A. V. 1
Hamdy Khaled 270, 275
Isaev Alexander 281, 334
Ismagilova Zulfiya 376
Ismagilova Zulfiya F. 391
Ivanov Igor 13
Ivanova Oksana 13, 195
Ivashin Sergey S. 338
Jiang Xiaohui 324
Kadymov Vagid 103, 164
Kamenov Renat 306
Kantemirov Igor F. 391
Kapustina Nataliya 313
Karabutov Nikolay 18
Karlova Tatiana V. 265
Karlova Tatyana Vladimirovna 187
Karpova E. E. 42
Kaverzneva Tatiana 212
Kekalo Anton 319
Khamzin Ramil 281
Kharlamov Sergei 200
Khmyrov R. S. 1
Klochihin Ilya 180, 212
Kolesnichenko I. I. 25
Kozochkin Mikhail 334, 347
Krit Boris 297
Kulagina Alevtina 236
Kulikov Gennady Yu. 124
Kulikova Maria V. 124
Kuptsov Vladimir 324, 347
Kuznetsov A. P. 368
Kuznetsov Vladimir 347
Kuznetsova Natalia Mikhailovna 187
Leonesio Marco 281
Leonesio Marko 347
Leonov Aleksandr A. 338
Levitskih Olesja 301
Lima Pedro M. 124
Logvin Vladimir A. 31
Lomas Washington 191
Loskutov A. I. 42
Loskutov S. A. 42
Luk'Yanov Aleksey 301
Makarova Tatyana 381, 387
Makhmudova Indira F. 391
Maksakova Svetlana 76
Mandel A. M. 42
Maracha Viacheslav 226
Maslov Andrey 347
Mazurov Mikhail 36
Melnik Yury 270, 275
Metel Alexander 270, 275
Migranov Artur 352, 363
Migranov Mars 352, 357
Mikhaylova Marianna V. 240
Minin Ilya 329, 347
Mitrofanov Evgenii 236
Morozkin Marian 347
Mosyanov Mikhail 319
Mustafaev Enver 270, 275
Nadykto A. B. 117
Nadykto B. A. 117
Nadykto O. B. 117
Nikishechkin Petr A. 338
Nikishechkin Petr. A. 342
Novoselova Olga 175, 255

AUTHOR INDEX

Okun'Kova Anna 313
Oshurko V. B. 42
Osinnii Aleksei 313, 319
Ovchinnikov Mikhail 297
Panferova Liya 236
Pavlenko Vitaliy D. 50
Petoukhov Sergey 56, 61
Petukhov Alexandr 148
Petukhova Elena 56
Pimushkin Ya. I. 288
Pivkin Petr 324, 329, 347
Pletnev Leonid 136, 141
Podkidyshev Alexey A. 342
Poltoranov Denis V. 265
Porvatov Artur 347
Quintana Iliana 191
Rai Maksim 363
Ramazanov Robert À. 391
Rechenko Denis 293, 306
Romanova Ekaterina 260
Rozhnov Aleksei 171
Ruzheynikov Anton 255
Sanchez Harry Saltos 206
Saracheva Diana 381, 387
Seitkulov Abdumalik 352, 357
Sekin A. A. 222
Senchenko Vladimir 212
Sevastianov Leonid 99
Shamanina Tetiana V. 50
Sharts A. A. 42
Shchetinin Eugene Yu 99
Skripnik Igor 212
Sosenushkin Evgeny 164, 195
Starodub Anton 67
Stebulyanin M. M. 288
Stepanyan Ivan 73
Stikhova Olga 217
Strelkov Sergey 76
Suminov Igor 297
Sushkevich Tamara 76
Svirin Vitaly 56
Tamayo Pablo 200
Terán Guillermo 191
Tikhonov Sergey 297
Tolokonnikov Georgy 88
Toporensky Alexey 73
Trapeznikova Marina 94, 159
Umnov Pavel I. 342
Uvarova Ludmila 260
Vakhitova Roza 381, 387
Verevkin Vladimir 61
Veselko S. G. 42
Vinogradenko Aleksei 171
Vinogradova Yuliya 195
Volkova Galina 175
Volosova Marina 363
Yanovskaya Elena 103, 195
Yaremko Natalia 109
Yaremko Oleg 109
Yusupdzhanov Vladimir 180
Yvarova Ludmila A. 31
Zelensky Alexander 347
Zheltov Sergey 136
Zibrov Petr 301
Zybina Olga 180